The Chronological Encyclopedia of Discoveries in Space

by
Robert Zimmerman

Oryx Press
2000

The rare Arabian Oryx is believed to have inspired the myth of the unicorn. This desert antelope became virtually extinct in the early 1960s. At that time, several groups of international conservationists arranged to have nine animals sent to the Phoenix Zoo to be the nucleus of a captive breeding herd. Today, the Oryx population is over 1,000, and over 500 have been returned to the Middle East.

© 2000 by Robert Zimmerman
Published by The Oryx Press
Oryx Press, 88 Post Road West, Westport, CT 06881
An imprint of Greenwood Publishing Group, Inc.
www.oryxpress.com

All rights reserved. No part of this publication may be reproduced or transmitted in any form or by any means, electronic or mechanical, including photocopying, recording, or by any information storage and retrieval system, without permission in writing from The Oryx Press.

Published simultaneously in Canada
Printed and bound in the United States of America

∞ The paper used in this publication meets the minimum requirements of American National Standard for Information Science—Permanence of Paper for Printed Library Materials, ANSI Z39.48, 1984.

Front cover photograph from Apollo 12 (November 14, 1969), courtesy of NASA

The lunar module *Intrepid*, photographed by Richard Gordon from the command module *Yankee Clipper*. Pete Conrad and Alan Bean in *Intrepid* are beginning their descent to the lunar surface. The landing site in the Ocean of Storms is to the west, just beyond the horizon. The large crater to the right is Herschel. The bright crater slightly right of center near the horizon is Lalande.

Back cover photographs from the Hubble Space Telescope, courtesy of NASA; see STS 61 (December 2, 1993)

Butterfly Nebula. Different colors indicate different elements: nitrogen is green, neutral oxygen is red, and doubly-ionized oxygen is blue.

Eta Carinae. One of the most gigantic stars in the Milky Way galaxy, Eta Carinae has a mass somewhere between 100 and 150 times that of the Sun. It exploded in 1841, becoming for a few years the second-brightest star in the sky. This photograph shows the present state of that eruption, with two hourglass-shaped bipolar jets streaming outward from the star's poles. Each small cloud within these jets is about the size of our solar system.

Eagle Nebula. Intense ultraviolet light, shining from several hot, massive, and young stars just beyond the top of the picture, has been boiling away a gigantic interstellar cloud of hydrogen gas. The dark pillars are the cloud's remains. Compare this close-up view, about one light year across, with the photograph on page xvii.

Orion Nebula. Near the image's center and among the blue wisps several proplyds are visible—disk-like clouds of material where stars and solar systems are now forming. This photograph covers an area about one light year across. Compare this view with ground-based photograph on page xviii.

Hourglass Nebula. Present theory postulates that the inner ring forms when a fast solar wind collides with an older, denser, and slower wind surrounding the star at its equator. The fast wind is then forced toward the star's poles, forming the hourglass-shaped bipolar jets. In this case, however, the central star is not in the nebula's center, confounding these theories.

Library of Congress Cataloging-in-Publication Data

Zimmerman, Robert, 1953–
    The chronological encyclopedia of discoveries in space / by Robert Zimmerman.
        p. cm.
    Includes bibliographical references and index.
    ISBN 1-57356-196-7 (alk. paper)
      1. Outer space—Exploration—History—Chronology.   I. Title.

QB500.262.Z56   2000
500.5′09—dc21

99-056080

To my parents,
who have never really understood my interest in wild and crazy things,
but encouraged it anyway.

Contents

Preface   vii

Introduction: Before 1957   xi

The Chronological Encyclopedia of Discoveries in Space   1
    1957–59   3
    1960–69   8
    1970–79   88
    1980–89   171
    1990–99   241

Appendix 1. Satellites and Missions Listed Alphabetically   345

Appendix 2. Satellites and Missions Listed by Subject   352

Appendix 3. Satellites and Missions Listed by Nation or Group of Nations   367

Glossary   376

Bibliography   381

Index   385

Preface

Science moves, but slowly, slowly, creeping on from point to point;

Slowly comes a hungry people, as a lion, creeping nigher,
Glares at one that nods and winks behind a slowly-dying fire.

Yet I doubt not thro' the ages one increasing purpose runs,
And the thoughts of men are widen'd with the process of the suns.

—Alfred, Lord Tennyson, *Locksley Hall*, 1842

The launch of Sputnik on October 4, 1957, began a revolution in the human race's understanding of Earth, the planets, the solar system, and the universe itself—a revolution that is still underway today. In less than half a century, every assumption about these phenomena has changed, solely because we have developed the capability to travel beyond Earth's atmosphere.

The Chronological Encyclopedia of Discoveries in Space catalogs the astonishing growth of scientific knowledge that has resulted from space exploration. Organized by launch, in chronological order, the *Encyclopedia* describes the goals—as well as the information gained—from every mission into space from every nation. Since the entries are organized by date, this reference work is useful for more than simply looking up a particular space mission. When the entries are read sequentially, a reader can experience the unfolding of knowledge as our understanding of the solar system and universe grew.

Sometimes, discoveries were unexpected and profound (such as the atmospheric drag on **Vanguard 1–March 17, 1958*** and the maneuvering difficulties experienced on **Gemini 4–June 3, 1965**). At other times, evidence was misleading, and scientists came to conclusions that later turned out to be fundamentally incorrect (see **Mariner 7–August 5, 1969**). Most often, the search for knowledge was slow and difficult, requiring repeated experiments and many missions before the incremental growth of data unveiled the answers to many mysteries (see, for example, the plant growth research begun with **Soyuz 35–April 9, 1980** and continuing through **STS 79–September 16, 1999**; see also the research into the origin of planetary craters, beginning with **Ranger 7–July 28, 1964** and continuing through **Mariner 10–March 16, 1975**).

Whether easy or hard, misleading or a revelation, this 40-year period of unprecedented progress has been a testament to the noble spirit of science and modern Western thought. As certain as some scientists were of their astronomical and geophysical theories on Monday, if the facts on Tuesday told them they were wrong, they abandoned their old ideas for better ones.

This book was written to meet a pressing need. In researching and writing about science, astronomy, and the history of space exploration, I have found it complicated—sometimes even impossible—to locate crucial information about the experiments and other scientific investigations carried out during past space missions. Indeed, many details that were common knowledge at the time a spacecraft was launched had, by 1999, become forgotten and obscure.

For example, in order to fully understand the goals and plans behind today's International Space Station, I sought accurate and detailed information about research conducted during past space station projects—a quest that typically required hours of searching, often in special libraries. *The Chronological Encyclopedia of Discoveries in Space* now consolidates such material into one information-packed reference source. It provides details about every space mission from every nation or consortium of nations, and it supplies information about a broad range of achievements in space—from manned communications to the creation of worldwide satellite communications; from planetary probes to the study of Earth's climate; and from military reconnaissance to space telescopes eyeing faraway black holes and galax-

---

* Boldface cross-references throughout the book indicate entries with additional or related information. Use the date to locate the entry in the chronology section.

Preface

ies—all topics of compelling interest both to researchers and to the general public.

In my research efforts, I also found it difficult to link specific scientific discoveries with the mission or missions responsible for them—even for breakthroughs regarded as important both historically and scientifically when they were achieved. For example, although NASA press releases always describe what the scientists on each space mission hope to learn, findings may be published months and even years afterward, often in obscure academic journals. For Soviet, Asian, and European projects, the data can be even harder to locate. Thus, I found that while existing reference sources discussed project goals extensively, they usually said nothing about project results. This *Encyclopedia* corrects that shortcoming by linking space discoveries with the missions on which those discoveries were actually made.

By bringing together this fascinating information and making it easy to find and organized within the context of history, I hope this book will be of value to high school and college students, teachers, historians, journalists, and all others who wish to trace "who did what, and when" in the first four decades of space exploration. In particular, the *Encyclopedia* will be of use to today's students, who often want to learn about the early history of space exploration but have not had available a single and easy-to-use reference source. When someone wants to know what spacecraft was the first to fly into space twice (see **Gemini 2–November 3, 1966***)*, or to learn what Alexei Leonov experienced during the first space walk (see **Voskhod 2–March 18, 1965***)*, or to discover when Arthur C. Clarke's 1949 concept of a network of geosynchronous communications satellites was finally completed (see **Intelsat 2B–January 11, 1967***)*, the *Encyclopedia* makes it easy to find the answers.

Above all, I hope that *The Chronological Encyclopedia of Discoveries in Space* will illustrate to readers how the challenge of space exploration has enriched humanity. The universe is a huge and complex place, and it will take every ounce of human determination to extend our reach and understanding beyond the middle latitudes and surface of our tiny planet Earth. If this book helps advance that effort in any way, I will be satisfied.

How to Use This Book

Entries in the Chronology

The Chronological Encyclopedia of Discoveries in Space covers the exploration of space in the twentieth century, from the launch of Sputnik in October 1957 through December 1999. Over 1,000 entries discuss the projects of every nation that has participated in the exploration of space: every manned mission, every scientific probe, every communications satellite project, every military mission, and every private commercial project during that period.

Each entry begins with a date (usually the date of the launch), followed by the name of the spacecraft, then the nation or nations involved. When several spacecraft were launched together on one rocket, their names are separated by a slash. When a spacecraft has more than one name, additional or alternative names are given in parentheses. Satellites deployed from an orbiting spacecraft are listed separately on an indented line.

October 6, 1981
SME / UoSat 1 (Oscar 9)
U.S.A. / U.K.

The above-cited entry discusses two satellites launched jointly on October 6, 1981: SME and UoSat 1 (which was also called Oscar 9). This launch was a cooperative effort between the United States and the United Kingdom.

August 2, 1991
STS 43, Atlantis Flight #9
TDRS 5
U.S.A.

The above-cited entry discusses a NASA space shuttle flight, STS 43, in which the Atlantis shuttle craft made its ninth trip into space. During the flight, a satellite called TDRS 5 was deployed.

When two or more spacecraft were launched on the same date but on separate rockets, each craft's launch time is given in Greenwich Mean Time (GMT) and the entries are listed chronologically. When a launch occurred on the same date that an already-launched probe reached another planet, the entries are listed alphabetically.

A few unmanned satellites have not been given individual entries. Exclusions include the many launches that were failures and achieved nothing (see, for example, the descriptions of Rangers 1 and 2 in **Ranger 3–January 26, 1962**). Others that were not given separate entries are the individual launches in a series of identical satellites, for which a group description made more sense (see, for example, **Intelsat 3B–December 19, 1968** and **Navstar 1–February 22, 1978**). Every successful mission during the first five years of space exploration has its own entry (see, for example, **Tiros 7–June 19, 1963**), but after 1962, only the first satellite launched in a new-generation design was given an entry (see **Tiros 8–December 21, 1963**).

Planetary probes have been given slightly different treatment. Some missions were complete failures (see the description of Mariner 8 in **Mariner 9–November 14, 1971**). Others succeeded in reaching orbit but failed before accomplishing their planetary-exploration objectives (see **Venera 1–February 2, 1961**). Furthermore, listing these missions by their launch date seemed misleading, as their place in space exploration history usually occurred not at launch but when they arrived at their goal. Therefore, planetary probes that reached orbit and were then successful in producing some scientific results, or that were significant in some historical manner, have entries—even if they failed prior to reaching their mission objective (see, for example, **Venera 2–November 12, 1965**). Successful planetary probes are also listed by the date when they *arrived*, not by when they were *launched*.

October 18, 1967
Venera 4, Venus Landing
U.S.S.R.

The above-cited entry discusses a Soviet craft that was launched on June 12, 1967. However, its significance centers on the fact that on October 18, 1967, it attempted to land on Venus, and during descent took readings of the Venusian atmosphere's composition, temperature, and pressure.

The various quotations in the text came from personal contacts, flight transcripts, press releases, newspaper articles, and the scientific publications listed in the bibliography. My research also included conversations and correspondence with many of the scientists directly involved with these space missions. Although I sought to describe the scientific findings and accomplishments of each project or mission, in some cases, the results simply could not be found—either because they were never published, because I could not discover them, or because they have not yet been made available. In such cases, I have described the mission's purpose or goal and have provided a cross-reference to related missions for which results are available.

Because of the broad scope of this work, it is possible that my characterization of a particular scientific discovery or description of a research effort is flawed. I request that readers who note mistakes contact The Oryx Press so that corrections can be made in later editions.

Special Features

Throughout the *Encyclopedia*, boldface cross-references indicate both prior and subsequent missions of relevance. These helpful cross-references will enable a reader to follow the satellites in a particular series, track an astronaut's or cosmonaut's career in space, and see how knowledge grew through research efforts that cut across nations and even decades.

More than 250 photographs and drawings illustrate the text. Color images in a special insert and on the back cover show particularly dramatic views recorded during various missions from 1966 through 1997.

Because the *Encyclopedia* is arranged chronologically, an alphabetical list of satellites is provided in Appendix 1, cross-referenced to the dates of their entries. A reader who knows the name of a mission—but not its date—can look up the name in Appendix 1 and see where to find its entry in the text.

In Appendix 2, missions are listed—both chronologically and alphabetically—by research subject. A reader interested in studying the developments in a specific research field, such as space astronomy, is provided with lists of all relevant missions, regardless of date or nation.

Appendix 3 lists the nations that have participated in space exploration, including a chronological list of the satellites and launches each nation has sponsored. These lists reveal not only which nations have participated in space exploration, but when their participation began, and how extensive their involvement has been.

A glossary is also included, defining special terms used in the text that may be unfamiliar to readers.

A bibliography provides the reader with most of the sources used in the preparation of this book.

A comprehensive index enables a reader to locate all information on a particular topic.

Acknowledgments

Specific thanks must be extended to everyone who works in the NASA History Office for making this book possible. Without their cheerful willingness to answer my endless questions about the most obscure of space satellites, I could never have completed this writing.

I also wish to thank Robert Tice of the Goddard Space Flight Center for his help in obtaining many of the photos in this book, as well as Janet Ormes and the entire staff of the Goddard Space Flight Center Library. I must also acknowledge Donna Felsenheld of the Photo Archives at NASA headquarters in Washington, D.C., Kathy Strawn of the Media Resource Center at the Johnson Space Flight Center in Houston, Texas, and Art Poland, the deputy project scientist of the SOHO project.

Nikolai Vlasov and the staff of the Russian Information Agency-Novosti in Washington, D.C., also deserve my gratitude for helping me obtain photographs of Soviet/Russian space achievements.

Thanks also to the California Institute of Technology and the Palomar Observatory; the European Space Agency; Eric Priest of the University of St. Andrews, Scotland; and principal investigators Kent Wood, Walter Smith, Horst Uwe Keller, and Martin Kessler.

Above all I must thank my editor Elizabeth Welsh, production manager Barbara Flaxman, designer Tom Brennan, and the entire staff at Oryx Press for helping me make this book better than I dreamed. Thanks to all.

Introduction: Before 1957

To understand the revolution in knowledge that space exploration made possible, it is useful to recall how little was known about the planets and the universe prior to Sputnik's launch. What follows is a brief summary of the astronomical and geophysical knowledge in 1957, giving the reader a baseline from which to appreciate the discoveries that were later made.

Because Earth's atmosphere blocked all infrared, ultraviolet, x-ray, and gamma radiation (see The Electromagnetic Spectrum, page xii), humanity's view of the universe was limited to what could be "seen" in visible and radio wavelengths. Furthermore, the view in visible light was severely hampered by Earth's atmosphere, which clouded and distorted the view of space.

In studying the solar system, astronomers depended almost entirely on a planet's albedo (the amount of light reflected off its surface) to determine details of its features and composition. Of the stars and other galaxies, even less was known. We stared out into the vastness of space, blind and unaware of the wonders hidden there.

The Sun
In 1957, the Sun was known only to emit radiation in visible light, although studies of comet tails in the mid-1950s suggested that a solar wind of unknown material might flow from the Sun. Although scientists knew the Sun had a magnetic field that switched polarity every 11 years, there was no understanding of how that field interacted with the Sun's surface, or whether it had any direct influence on the earth. Scientists were aware that during active periods of the Sun's 11-year cycle—when its surface was covered with a large number of sunspots and many solar flares occurred—that the earth's aurora would become enhanced and visible at lower latitudes, and that communications in the ionosphere would be hindered by radio interference. The exact cause of these phenomena, however, was unknown (see **Solrad 1–June 22, 1960** for the beginning of space solar research).

Furthermore, an accurate measure of the solar constant, the Sun's total radiance output, was unknown, and hence it was not known whether this might be a factor influencing the Earth's climate. See **Nimbus 7 (October 13, 1978)** for the first measurements of the solar constant.

Mercury
Known as the smallest of the planets, with a diameter of 3,000 miles, Mercury's day was thought to be 88 Earth days long, the same length as its year. This meant that—

Mercury photographed with the ground-based Catalina Observatory 61-inch telescope, ca. 1970. *NASA*

Introduction: Before 1957

The Electromagnetic Spectrum

	By Energy		By Wavelength		By Frequency	
Cosmic Rays	>1 TeV to 0.001 TeV	>1,000 GeV to 1 GeV				
Gamma Rays hard	1 GeV to 0.025 GeV	1,000 MeV to 25 MeV	0.0005 angstroms to 0.01 angstroms			
Gamma Rays soft		25 MeV to 1.24 MeV	0.01 angstroms to 0.1 angstroms			
X-Rays hard	1.24 MeV to 0.124 MeV	1,240 KeV to 124 KeV	0.1 angstroms to 100 angstroms	0.01 nm to 10 nm		
X-Rays soft		124 KeV to 0.124 KeV	100 angstroms to 1000 angstroms	10 nm to 100 nm		
Extreme Ultraviolet		124 eV to 12.4 eV	1000 angstroms to 4,000 angstroms	100 nm to 400 nm		
Ultraviolet			4,000 angstroms to 7,000 angstroms	400 nm to 700 nm		
Visible						
Near Infrared				0.7 microns to 1.3 microns	700 nm to 1,300 nm	0.0007 mm to 0.0013 mm
Infrared				1.3 microns to 7 microns	1,300 nm to 7,000 nm	0.0013 mm to 0.007 mm
Far Infrared				7 microns to 1,000 microns		0.007 mm to 1 mm	0.0007 cm to 0.1 cm
Microwave				1,000 microns to 25,000 microns	1 mm to 25 mm	0.1 cm to 2.5 cm	300 GHz to 12 GHz
Radar					25 mm to 100 mm	2.5 cm to 10 cm	10 cm	12 GHz to 3 GHz	12,000 MHz to 3,000 MHz
UHF-tv					100 mm to 1,000 mm	10 cm to 100 cm	0.1 m to 1 m	3 GHz to 0.03 GHz	3,000 MHz to 300 MHz
Radio VHF-tv, FM radio						100 cm to 1000 cm	1 m to 10 m		300 MHz to 30 MHz	300,000 KHz to 30,000 KHz
shortwave							10 m to 187 m		30 MHz to 1.6 MHz	30,000 KHz to 1,600 KHz
AM radio							187 m to 10,000 m	0.187 km to 10 km		1,600 KHz to 30 KHz
LF								10 km to 100 km		30 KHz to 3 KHz	30,000 Hz to 3,000 Hz
VLF											

↙ Atmospheric opacity, indicated by dots, lines, or black

Abreviations:
eV = electron Volts
KeV = One thousand electron Volts
MeV = One million electron Volts
GeV = One billion electron Volts
TeV = One trillion electron Volts

Hz = Hertz (or oscillations per second)
KHz = One thousand Hertz
MHz = One million Hertz
GHz = One billion Hertz

UHF = Ultra high frequency
VHF = Very high frequency
LF = Low frequency
VLF = Very low frequency

km = kilometers
m = meters
cm = centimeters
mm = millimeters
nm = nanometers

like the Moon—one side of the planet perpetually faced the Sun, with the other forever in darkness. Mercury's mass was not known for certain, but it was estimated to be about 0.04 of Earth's. This would make its density somewhere between that of Mars and the Moon.

Maps drawn by Giovanni Schiaparelli and others in the nineteenth century showed a number of large, dark patches scattered across Mercury's surface. The planet's very dark albedo, and its closeness to the Sun (which meant that astronomical observations had to take place in daytime or in the early evening, when Mercury was low on the horizon), made reliable study difficult, at best. A number of astronomers noted that the surface features seemed to vary in brightness while remaining fixed in position, and based on this, some scientists theorized that the planet had a very thin atmosphere.

See **Mariner 10 (March 16, 1975)** for the only probe to have made a close visit to Mercury.

Venus
Venus was thought to have a diameter of 7,700 miles, with a mass 0.81 of Earth's. Unlike Mercury, Venus's thick and cloudy atmosphere had been clearly documented. During inferior conjunction—when the planet's night side faced Earth and Venus resembled a thin crescent—the tips of that crescent often extended beyond 180 degrees, encircling the planet as the Sun's light was bent through Venus's atmosphere. Spectra of the upper layers of the atmosphere indicated very little oxygen (less than 0.001 the amount on Earth's surface). The makeup of the lower layers was unknown. Many speculated that Venus could harbor life, especially because it so resembled Earth in size and mass and had such a thick atmosphere.

Venus photographed with the ground-based 200-inch telescope at Palomar. *Caltech*

Because of its dense atmosphere, Venus's surface could not be observed, and hence the length of its day was unknown. Since spectroscopic measurements of the planet's opposite edges revealed little radial motion, most astronomers believed the planet rotated slowly. Radiometric observations of its night side, however, showed the release of a great deal of heat, leading some astronomers to infer that the planet rotated quickly.

For the first satellite to reach Venus successfully, see **Mariner 2 (August 27, 1962)**.

The Moon
At an average distance of 240,000 miles from Earth, the Moon is the only planetary body whose disk is visible to the naked eye, and so for centuries it had been studied extensively by astronomers. This Earth satellite has a diameter of 2,160 miles, with a mass 0.013 of Earth's.

Because the Moon's orbit around Earth matches the length of its day, 27.32 Earth days, the satellite always presents the same face earthward, with minor orbital variations making about 60 percent of its entire surface visible at various times. The remaining 40 percent remained concealed from earthbound observation.

The near side, however, had been extensively mapped, with features as small as one mile in diameter visible from the best Earth-based telescopes. Because the dark areas were at first thought to be either oceans or dried seabeds, they were called *maria*, from the Latin for sea (the singular is *mare*, pronounced MAR-ray). By 1957, scientists had concluded that these dark areas were solidified seas of lava.

The most prominent features on the lunar surface were the innumerable craters, some as large as 100 miles across with large central peaks. Some craters had bright and spectacular rays radiating outward, sometimes extending across most of the Moon's visible hemisphere. In the mid-1950s, most astronomers thought these lunar craters were volcanic in origin.

See **Luna 2 (September 12, 1959)** for the first satellite to reach the Moon. See **Ranger 7 (July 28, 1964)** for the beginning of comprehensive crater and lunar research.

Mars
Mars's mass was calculated to be 0.108 of Earth's. Its diameter was thought to be between 4,190 and 4,240 miles, the uncertainty due to observational interference caused by the atmospheres of both Earth and Mars.

The properties of Mars's atmosphere were unknown, though the little data available indicated that it was tenuous. Thin clouds were sometimes seen, and observations suggested that the atmosphere reached altitudes as high as 60 miles above the planet's surface. Spectra had indicated the presence of water vapor and possibly oxy-

Introduction: Before 1957

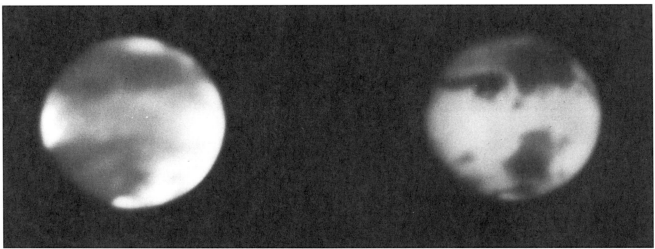
Mars; two images photographed through different filters using the ground-based 200-inch telescope at Palomar. *Caltech*

gen, with an upper limit no more than 0.01 of Earth's atmosphere. Measurements of the planet's radiation indicated surface temperatures ranging from –94°F to 86°F, depending on latitude and season.

To both scientists and the public, the surface was the red planet's most tantalizing feature. As its seasons changed, astronomers could see the waxing and waning of ice caps and bluish-green dark regions, suggesting the presence of water and vegetation. Several astronomers had carefully charted these dark regions, calling the long straight lines that connected large patches "canals." Although the name implied the existence of intelligent Martian life (beings who had created the canals for some purpose), scientists simply did not know. The canals themselves had an elusive nature, appearing differently to different astronomers, and changing over time.

Of Mars's two satellites—Deimos and Phobos, discovered in 1877—almost nothing was known. Deimos was thought to be about 10 miles across, with Phobos slightly larger.

For the first spacecraft to reach Mars, see **Mariner 4 (November 28, 1964)**.

Jupiter

The largest planet, Jupiter was thought to have an equatorial diameter of about 88,770 miles, more than 10 times that of Earth, and a mass equal to 320 Earth masses. Scientists believed that hydrogen and helium made up most of Jupiter's bulk. Spectroscopy of the planet's atmosphere also provided evidence of ammonia and methane. Radiometric observations indicated an average temperature of –216°F.

Since the 1600s, astronomers had observed parallel belts dividing Jupiter's so-called surface. Observations of these belts indicated that the planet rotated in just under 10 hours, but that different belts rotated at different rates, some completing their day five minutes faster.

The belts also varied in color and size. These facts, combined with the knowledge that Jupiter's mean density was only 1.35 greater than water, indicated that Jupiter was not a solid body like Earth, but a gaseous body whose small solid core of metallic hydrogen is buried many thousands of miles beneath the thick atmosphere.

Jupiter's most prominent feature was known was the Red Spot, located in the planet's southern hemisphere at approximately 20 degrees latitude. This reddish region measures about 30,000 miles east to west and about 7,000 miles north to south, though since the late nineteenth century, it has faded and shrunk somewhat. Although early theories postulated that the spot was a high-elevation plateau peeking up through the deep Jovian atmosphere, astronomers soon realized that if the spot were solid, it had to be floating in that atmosphere, as it

Jupiter and its Red Spot photographed with the ground-based 200-inch telescope at Palomar. Note Ganymede, one of Jupiter's moons, to the lower left. *Caltech*

apparently drifted relative to the belts in which it was embedded. Some scientists then proposed that the spot was an island of ice floating within the highly compressed gaseous atmosphere. Others theorized that the spot was a cloud formation hovering over a floating island, and that the changes in size and color took place because of changes in elevation.

In 1955 and 1956, bursts of radio emissions were detected coming from Jupiter. These were thought to be atmospheric disturbances similar to lightning storms on Earth, though of a vastly greater scale.

In 1957 Jupiter was known to have 12 satellites. The four largest—Io, Europa, Ganymede, and Callisto—are known as the Galilean satellites because they were the first objects discovered by telescope when Galileo identified them in 1610. By the twentieth century, a handful of surface markings had been identified on each. On Io, dark poles and dark north-south bands were seen. Europa and Ganymede had white polar caps with dark spots at lower latitudes. Callisto also had a white southern polar cap, with white spots surrounding its northern pole. These markings indicated that all four satellites rotated such that one side always faced Jupiter.

Except for their diameters, known to range from 14 to 150 miles, little was known about the remaining eight satellites.

See **Pioneer 10 (March 3, 1972)** for the first spacecraft to reach Jupiter.

Saturn
Famous for its spectacular rings, Saturn was also unusual because of its apparently flattened shape. While its equatorial diameter was 75,000 miles, its polar diameter was only 67,000 miles. The planet had a mass 95 times that of Earth. Despite this gigantic size, Saturn's mean density was measured at 0.13 of Earth's, making it less dense than water. If one could find an ocean large enough, Saturn would float.

While Saturn also had bands across its surface like Jupiter, they were less striking in appearance. The surface was generally yellowish in color, and the lack of surface detail made measuring the planet's rotation difficult. By the mid-twentieth century, Saturn's day was estimated to be slightly more than 10 hours long, and like Jupiter, different bands rotated at different rates.

Radiometric measurements revealed a surface temperature of about –238°F, and spectroscopic studies indicated the presence of ammonia and methane. Like Jupiter, Saturn was thought to be mostly hydrogen and helium, with small quantities of ice, ammonia, and methane floating in its atmosphere.

The famous Saturnian rings were thought to be unique. About 37,000 miles wide and only 10 miles thick, the rings were the thinnest object in the solar system. Three rings had definitely been identified, along

Saturn and its rings photographed with the ground-based 100-inch telescope at Palomar. *Caltech*

with Cassini's Division (the gap separating the two largest rings). Other features, including several additional rings and at least one other gap, had been reported periodically, but the discoveries were not considered conclusive. While careful study had revealed the rings to be partially transparent (stars could be seen through them), they were sufficiently dense to cast a shadow on Saturn's surface. From these observations, astronomers concluded that the rings were composed of many discrete objects of unknown size, which were orbiting Saturn independently.

Saturn was known to have nine satellites. Titan, the largest, was thought to have a diameter of approximately 3,000 miles, though this number was far from certain. It was also the only satellite known to have an atmosphere, containing methane and possibly ammonia. Iaputus's brightness varied significantly, indicating that one hemisphere was, for unknown reasons, very dark, while the other was very bright. Phoebe was also the only known object in the solar system orbiting in a retrograde direction.

See **Pioneer 11 (September 1, 1979)** for the first close approach to Saturn.

Uranus
Uranus's mass was estimated at 14.54 Earth masses, with a density 0.28 of Earth's. This density, along with its high albedo (second only to Venus), indicated a cloud-covered surface of high reflecting power. The planet's diameter had been measured tentatively as 29,400 miles. The surface temperature was estimated to be about –328°F.

Like Jupiter and Saturn, spectroscopy indicated the presence of methane and ammonia, while other measurements revealed hydrogen and helium. Through a telescope, Uranus appeared dark bluish-green, with very faint bands. By measuring the Doppler effect on the

Introduction: Before 1957

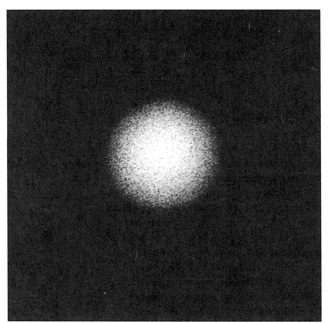
Uranus photographed with the ground-based Catalina Observatory 61-inch telescope, ca. 1970. *NASA*

planet's spectrum, astronomers in 1911 and 1912 estimated a rotation of 10.83 hours. This number was considered highly uncertain.

Only seven satellites were known, rotating Uranus at an inclination of 97.8 degrees to the plane of the solar system, thereby making this miniature system appear to revolve on its side. Little else was known of Uranus's moons.

See **Voyager 2 (January 24, 1986)** for the first close-up look at Uranus.

Neptune

Neptune's mass was known to be 17.2 Earth masses, with a diameter thought to be 27,600 miles. This diameter indicated a density 0.45 of Earth's. Like Uranus, Neptune had a high albedo, which indicated a cloud-covered surface. And like Uranus and the other gas giants of the solar system, scientists estimated that Neptune was composed mostly of hydrogen and helium; spectroscopy had also indicated the presence of methane. The surface temperature was estimated at about −337°F.

Only two satellites were known for Neptune. Triton, the larger of the two, was thought to have a diameter of about 3,000 miles and a mass 1.8 of the Moon's. The smaller, Nereid, had an estimated diameter of 180 miles, with a mass 0.00025 that of Triton.

Voyager 2 (August 25, 1989) took the only close-up images of Neptune.

Pluto

From measuring Pluto's albedo, astronomers thought that it was smaller than Earth, possibly even smaller than Mars. Its mass was estimated to be about 0.9 that of Earth.

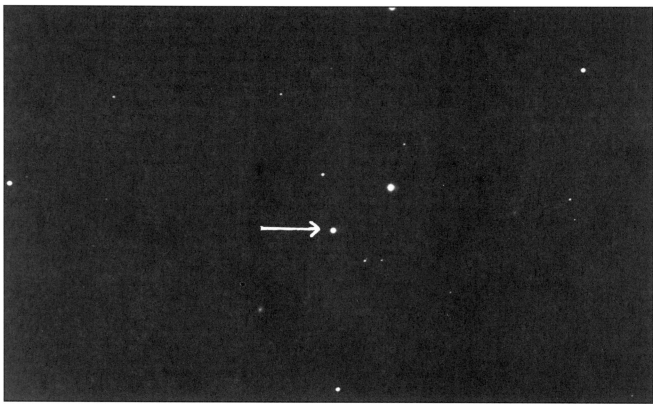
Pluto photographed with the ground-based 200-inch telescope at Palomar. *Caltech*

M31, the Andromeda Galaxy, photographed with the ground-based 200-inch telescope at Palomar. *Caltech*

Little else was known about this most-distant planet, which remains today the only planet not observed up close by spacecraft. However, photographs by the orbiting **Hubble Space Telescope** (see **STS 61–December 2, 1993**) have revealed details previously unknown.

Interplanetary Space
In 1957, no knowledge at all existed about the conditions between the planets. Some theories held that material was unevenly distributed through interplanetary space in the form of clouds of turbulent gas (see **Pioneer 5–March 11, 1960**). The number of micrometeorites floating in near-Earth space, as well as between the planets, was completely unknown. There were even concerns by some scientists that interplanetary space was so filled with these objects that space travel would require extensive shielding to be practical. See **Explorer 1 (February 1, 1958)**, **Mars 1 (December 1, 1962)**, **Explorer 16 (December 16, 1962)**, and **Pioneer 10 (March 3, 1972)** for micrometeorite research.

The Stars and Galaxies
Because Earth's atmosphere blocked all but visible light and radio wavelengths, humanity's understanding of the universe beyond the solar system was shaped by the gentle glow of the night sky. Astronomers studied starlight, observing how stars differed in color and brightness, some even pulsing over time. Just as for the Sun, no star was known to emit radiation other than visible light.

In 1957, only 20 years had passed since Edwin Hubble had proven that many of the faint nebulae in the sky were not objects inside the Milky Way but faint "island universes" of stars. The concept of galaxies was therefore very new, and many textbooks still confusingly referred to them as a type of planetary nebulae (clouds of gas that surrounded individual stars within the Milky Way galaxy).

There was great controversy concerning whether the universe had been formed in a single "big bang" or was forming continuously in a "steady state." Neither theory dominated, the facts being too sparse and unclear.

Galaxies were classed by shape, from the elliptical (round or football-shaped), to spirals (resembling whirlpools), to irregular. Knowledge about these different galaxy types was limited to what could be gleaned from visible-light spectroscopy.

Planetary nebulae were vague, unfocused objects. While scientists believed that they were a stage in the evolution of giant stars, photographs were too unclear to explain their origin or formation.

M16, the Eagle Nebula, photographed with the ground-based 200-inch telescope at Palomar. *Caltech*

Introduction: Before 1957

The Great Nebula in Orion photographed with the ground-based 200-inch telescope at Palomar. *Caltech*

This clouded perception of the universe changed as soon as instruments were placed above Earth's atmosphere. For the first gamma-ray space telescope, see **Explorer 11 (April 27, 1961)**. For the first ultraviolet space telescope, see **OAO 2 (December 7, 1968)**. For the first x-ray space telescope, see **Uhuru (December 12, 1970)**. For the first infrared space telescope, see **IRAS (January 26, 1983)**. And for the first telescope to operate in visible wavelengths above Earth's atmosphere, see the **Hubble Space Telescope** (see **STS 31–April 24, 1990**).

Earth
And Earth? If you looked in an astronomical textbook, you wouldn't have found it. While people understood intellectually that Earth was a planet, no one associated its study with the study of other planetary bodies—the concept of the Earth as a planet essentially did not exist. Earth science was broken into a number of smaller fields, such as geology, oceanography, and meteorology.

Geologists did not know how Earth's continents or oceans had formed. See **Landsat 1 (July 23, 1972)** and **Seasat (June 27, 1978)**. Even the planet's exact shape or the relative location of its land masses were not precisely known. See **Anna 1B (October 31, 1962)** and **Transit 1B (April 13, 1960)**.

The earth's topology was known only to a limited extent, and that knowledge was slow and difficult to obtain. Over land, mapping was either done by aerial photography or by surveyors on the ground, and except in technologically developed regions like Europe and the United States, many remote regions remained blank spots on the world's maps. Moreover, the ocean's topography, both on the surface and at the ocean floor, was completely unknown. See **Landsat 1 (July 23, 1972)**, **Seasat (June 27, 1978)**, and **Geosat (March 13, 1985)**.

Meteorology lacked a global view of the weather, its forecasts almost entirely dependent on data gathered locally on the continents of the Northern Hemisphere. Meteorologists were even unsure what Earth's weather and clouds would look like from space, or even if they would be visible. The concept of climate study was barely conceived. Conditions over the earth's vast oceans—including its currents, wind speeds, and temperatures at the surface—were simply unknown, except for the limited data gathered by ships. See **Tiros 1 (April 1, 1970)** for the beginning of this research.

Information about Earth's resources were also difficult to gather. In unsettled regions, it was extremely difficult for scientists to chart geological resources. Urban growth and land utilization could not be studied in depth and on a large scale. Ocean life and vegetation could not be tracked with any detail. See remote sensing satellites like **Seasat (June 27, 1978)** and **Nimbus 7 (October 13, 1978)** for the beginning of this research.

Details about Earth's magnetic field or magnetosphere were sketchy at best. Did it interact with the Sun in any way? What was its shape? Would it be a barrier for manned space exploration? See **Explorer 1 (February 1, 1958)** for the beginning of magnetospheric research, and **OGO 1 (September 5, 1964)** for a summary of the first wave of research.

There were many other questions, as well. What was the cause of the aurora seen in the high latitudes? (see **Injun 3–December 13, 1962** and **OGO 4–July 28, 1967**). What was the composition and size of the ionosphere, the highest layer of the earth's atmosphere? (see **Explorer 1–February 1, 1958** for the beginning of this research).

Cultural Knowledge
Space exploration during the last four decades has had other influences besides increasing our scientific knowledge. Prior to Sputnik, communications between continents required cable or the right ionospheric conditions, and connections were expensive, limited, and hard to maintain. The idea of picking up a telephone and instantly speaking to anyone else anywhere on the globe was considered the stuff of science fiction. See **SCORE (December 18, 1958)** for the beginnings of satellite communications.

Ship captains located themselves on the surface of the earth using sextants and mathematical calculations. Airplane pilots depended on ground-based tracking facilities. Hikers required compasses and a good sense of direction. Although all these techniques remain available, the advent of global positioning system (GPS) technology has transformed navigation on the ground and on the ocean. See **Transit 1B (April 13, 1960)** for the beginning of satellite-based navigation technology.

Politically, space exploration changed humanity's perspective in ways that no one in 1957 could have predicted. After the launch of Sputnik, the fear that space could be used for military domination fueled a feverish space race during the 1960s, whereby the United States and the Soviet Union aggressively competed for the bragging rights to outer space. As Neil Armstrong has stated, "It was a competition unmatched outside the state of war." Soviet Premier Nikita Khrushchev used Soviet space achievements to glorify communism. U.S. President John F. Kennedy committed the United States to placing a man on the Moon before the end of the 1960s. See **Mercury 2 (January 31, 1961)**, **Vostok 1 (April 12, 1961)**, and **Freedom 7 (May 5, 1961)** for the beginnings of this race.

The space race also involved the aggressive use of space for surveillance purposes. See **Discoverer 1 (February 28, 1959)**, **Samos 2 (January 31, 1961)**, **Midas 2 (May 24, 1960)**, and **Cosmos 4 (April 26, 1962)**. Prior to these satellites, surveillance could only be done by high-flying manned airplanes like the U2 (which could be shot down—as in the case of Gary Powers's flight over the Soviet Union—and were limited in their ability to obtain data). Such aerial reconnaissance could not cover very much area, and its technology was unable to penetrate cloud cover.

The fierceness of the U.S.-Soviet superpower competition had other ramifications as well. While the scientific data from almost all United States missions were made easily accessible almost immediately, many discoveries by Soviet scientists remained hidden behind a wall of secrecy for many decades. Even today it is difficult to track down data on many of the breakthroughs achieved by the Soviet space program.

Beginning with the publication of the first images of Earth from space (see **Lunar Orbiter 1–August 10, 1966**, **ATS 3–November 3, 1967**, **Apollo 8–December 21, 1968**), human beings' self-concept underwent a change. The photographs of Earth from an external perspective reinforced concepts of a single human culture and de-emphasized the notion of many competing and unique human societies.

Thus, beginning the late 1970s and following the United States triumph in winning the race to the Moon, Soviet secrecy and isolation behind its "iron curtain" began to ease; see **Soyuz 28 (March 2, 1978)**, **Soyuz T6 (June 24, 1982)**, and **Vega 1 (June 10, 1985)**. In the early 1990s, with the end of the Cold War and the collapse of the Soviet Union, that wall of secrecy finally fell, allowing scientists, engineers, and businesses from Russia, Ukraine, and many other former Soviet-bloc nations to participate openly and freely in research and commercial activities. See, for example, **Mars Polar Lander (January 3, 1999)**, **Cosmos 2349 (February 17, 1998)**, and **Bonum 1 (November 22, 1998)**.

The change in perspective that the view of Earth from space caused also changed the nature of almost all space missions. While in the 1960s, the two superpowers competed with each other to get into space, keeping their projects and efforts distinct and separate, by the 1990s almost no space mission is accomplished without some international cooperation. See, for example, **Geotail (July 24, 1992)** and the International Space Station (see **Zarya–November 20, 1998**).

This, then, was the state of knowledge and human culture on October 3, 1957. The human race was trapped inside a dark room, unable to look outside because the door to the greater universe was locked. The key was in the lock, however, and the door was about to open to reveal wonders undreamed of.

The Chronological Encyclopedia of Discoveries in Space

1957

October 4, 1957
Sputnik
U.S.S.R.

Weighing 185 pounds, Sputnik was the first artificially constructed satellite placed in Earth orbit. For three weeks, it transmitted a regular shortwave signal as it circled the globe once every 96.1 minutes in an elliptical orbit ranging from 141 miles to 587 miles in altitude.

Although little scientific data was obtained from Sputnik, this Soviet-built satellite was a historic engineering achievement, proving that it was possible to launch into orbit working machinery from which information could be gathered. It also galvanized interest in space research throughout the United States, which eventually led to the space race of the 1960s.

Sputnik's orbit decayed, and it burned up in the atmosphere on January 4, 1958.

November 3, 1957
Sputnik 2
U.S.S.R.

The second artificial satellite, Sputnik 2, weighed 1,109 pounds, 5.5 times more than **Sputnik (October 4, 1957)**. Its orbit was even more eccentric, ranging in altitude from 132 miles to 1,031 miles. Unlike the first Sputnik, Sputnik 2 was not merely an engineering achievement. The satellite also carried a small dog named Laika in a cabin designed to maintain sea-level atmospheric pressure, with sensors for measuring the dog's vital signs. For six days, scientists monitored Laika's heartbeat, blood pressure, and breathing rate. Then the spacecraft's oxygen supply ran out, and the dog died from asphyxiation. Laika's brief voyage in space demonstrated that animals do not need gravity to survive, at least for short periods.

Sputnik 2 also carried a geiger counter, which operated for seven days. As it passed over Earth-based receiving stations, the counter's signal was recorded. Although a 50 percent increase in detector counts occurred over Russia, the Soviet scientists concluded that this was an ordinary variation in energy flux coming from a burst of cosmic rays. The inclination of Sputnik 2's orbit prevented Soviet scientists from recognizing that the flux was actually evidence of the Van Allen radiation belts. When the spacecraft passed over Russia, it was cutting through the belts. See **Explorer 1 (February 1, 1958)**.

Sputnik 2 also showed that radio beacons from a spacecraft are detectable even when a satellite is below the horizon. Just as ground-based radio signals can bounce off the ionosphere and travel around the world, Sputnik 2's signal did the same from space. See **OV4-1T/OV4-1R (November 3, 1966)** for later research.

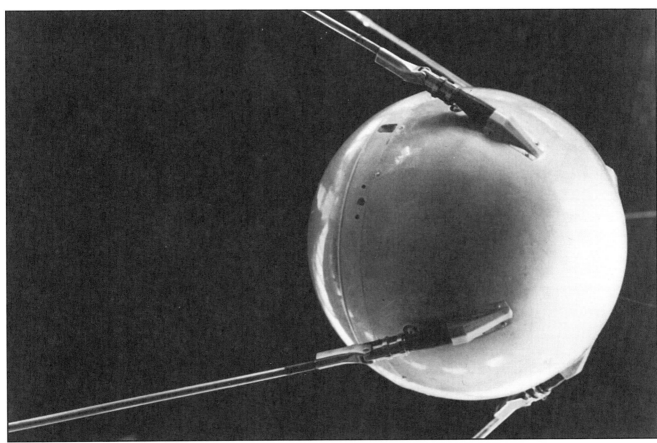

A mock-up of the Sputnik satellite. *NASA*

1958

February 1, 1958
Explorer 1
U.S.A.

Explorer 1, the first U.S. satellite, weighed only 11 pounds and operated for four months. Orbiting the earth every 107.2 minutes, with a perigee of 216 miles and an apogee of 1,155 miles, Explorer 1 carried a geiger counter for measuring radiation in space. As it circled the earth, data were recorded in two-minute spans as the satellite flew over 16 different receiving stations. These data, along with information obtained later by **Explorer 3 (March 26, 1958)**, **Explorer 4 (July 6, 1958)**, and **Pioneer 1 (October 11, 1958)**, confirmed the existence of what are now called the Van Allen radiation belts, named for the lead scientist on the Explorer 1 team, James Van Allen.

Explorer 1 also included an instrument for measuring the amount of micrometeorite material hitting the spacecraft. Daily counts indicated that for 8 out of every 24 hours, the impact rate increased significantly, a finding that was attributed to specific meteorite showers. Combining Explorer 1's micrometeorite data with information gathered later by a similar instrument on Explorer 3, the influx of cosmic dust into the earth's atmosphere was estimated to be over two million pounds per day.

March 17, 1958
Vanguard 1
U.S.A.

Vanguard 1, a tiny 6.4-inch probe weighing only 3.5 pounds, carried no instruments except a radio transmitter and the first solar cells for power in space. Studies of its orbit, which initially had a perigee of 404 miles and an apogee of 2,466 miles, led to a number of significant discoveries. Vanguard 1 was, at first, expected to remain in orbit about 2,000 years. However, the gravitational fields of the Moon and Sun, feeble as they are, modified the spacecraft's orbit, as did the radiation pressure of the Sun's light. Also slowing the spacecraft was the earth's atmosphere, which was found to expand and contract with the Sun's solar cycle. As the Sun became more active, the earth's atmosphere inflated, increasing the drag on the satellite and shortening its life expectancy. As of 1999, Vanguard 1 was expected to remain in orbit only another 200 years. This atmospheric effect was a significant discovery, and it has been the primary factor for predicting the life expectancy of every satellite since. The atmospheric data gathered by Vanguard 1 were later refined by **Echo 1 (August 12, 1960)**.

Two other discoveries were extrapolated from Vanguard 1's orbital motion. First, the earth is not, as had been predicted, a globe slightly flattened at the poles; it is pear-shaped. Second, the earth's magnetic field is strong enough to alter a satellite's spin over time.

March 26, 1958
Explorer 3
U.S.A.

Unlike **Explorer 1 (February 1, 1958)**, Explorer 3 (Explorer 2 had failed to reach orbit) carried a tape recorder, allowing scientists to collect data even when the satellite was out of contact with ground stations. Explorer 3 traveled extensively through the inner Van Allen belt, its geiger counter measuring the flux of charged particles held there by the earth's magnetic field. An instrument for measuring micrometeorite impacts confirmed the influx rate recorded earlier by Explorer 1. Both instruments operated until June 1958.

May 15, 1958
Sputnik 3
U.S.S.R.

Sputnik 3, which orbited the earth for seven months, carried detectors for measuring the charged particles of the ionosphere and the composition of the upper atmosphere. The data confirmed the existence of the Van Allen belts discovered by **Explorer 1 (February 1, 1958)** and **Explorer 3 (March 26, 1958)**.

The satellite also observed that the energy flux in the Van Allen belts appeared to have a downward motion, indicating that the magnetic lines of force were somehow guiding ionized particles into the earth's atmosphere. See **Injun 1 (June 29, 1961)**.

July 6, 1958
Explorer 4
U.S.A.

Based on data returned by **Explorer 1 (February 1, 1958)** and **Explorer 3 (March 26, 1958)**, the detectors on Explorer 4 were redesigned to measure a greater range of energy, although the instruments were still unable to identify the nature of particles causing the energy flux in the Van Allen belts. Explorer 4 operated for 10 weeks, helping to map the Van Allen belts more accurately and proving that the radiation detected in the belts followed the lines of force of the earth's magnetic field. The spacecraft repeatedly passed though the lower edges of an unexpected outer belt, which showed that the belts were more complex than first thought. Later data from **Pioneer 1 (October 11, 1958)** and **Pioneer 3 (December 6, 1958)** provided further information about the outer belt's shape.

Explorer 4 also monitored the nuclear explosion of Project Argus. Using an X-17 jet, three one-kiloton atomic bombs were detonated 300 miles above the South Atlantic in August and September. Immediately after each nuclear blast, auroral streamers were observed extending upward and downward along magnetic field lines. Explorer 4, along with 19 sounding rockets, observed energy fluxes that traced magnetic lines of force from the point of explosion. The charged ionized electrons quickly created an artificial radiation belt encircling the globe at about 4,000 miles elevation. That radiation belt remained stable for several weeks, proving that the earth's magnetic field can trap and direct the motion of electrons, as implied by the existence of the newly discovered Van Allen belts.

October 11, 1958
Pioneer 1
U.S.A.

Pioneer 1 was the first satellite launched by NASA (the National Aeronautics and Space Administration). Previous launches were built either by the U.S. Air Force, the U.S. Navy, or a hodge-podge of civilian agencies.

Though intended to enter lunar orbit, Pioneer 1 failed to reach the required velocity because of an incorrect firing angle, and it fell back to Earth. Despite this, the spacecraft climbed to an altitude of almost 80,000 miles, a record distance at the time. It carried two geiger counters, returning the first detailed (though brief) look at the outer Van Allen belt, indicating its presence at approximately 18,000 miles altitude. Pioneer 1's instrument for detecting micrometeorites registered only one impact in its first 15 hours of flight, leaving scientists doubtful that the device was operating properly.

December 6, 1958
Pioneer 3
U.S.A.

Though planned as a flyby of the Moon, the failure of Pioneer 3's first stage prevented this NASA spacecraft from reaching escape velocity. Much like **Pioneer 1 (October 11, 1958)**, the satellite fell back to Earth after reaching an altitude of 66,700 miles. (Pioneer 2 had failed to make it beyond Earth's atmosphere.)

Two geiger counters on this spacecraft measured radiation that had first been detected on **Explorer 1 (February 1, 1958)**. Combining Pioneer 3's data with previous information from Explorer 1, **Explorer 3 (March 26, 1958)**, **Explorer 4 (July 6, 1958)**, and Pioneer 1 made possible the first comprehensive maps of the Van Allen belts, showing that they consisted of an inner zone ranging from about 800 to 6,000 miles in altitude and an outer zone ranging from about 8,000 to 28,000 miles in altitude. Both zones are donut-shaped, are centered on the earth's equator, and follow the lines of force of the earth's magnetic field like iron filings around a bar magnet. The field's outer perimeter was estimated to be at about 36,000 miles altitude. However, later data from **Luna 1 (January 2, 1959)** and **Pioneer 4 (March 3, 1959)** contradicted this last figure.

December 18, 1958
SCORE
U.S.A.

SCORE (Signal Communications by Orbiting Relay Equipment), built by the U.S. Air Force, was the largest American satellite to date (154 pounds). SCORE also was the first in-orbit communications satellite, operating for 12 days. Radio messages were sent to the satellite, stored on a tape, and then retransmitted to Earth when commanded by a ground station. Such a design is called an active repeater communications satellite. The next communications satellite, **Echo 1 (August 12, 1960)**, was, instead, a passive reflector communications satellite. The next active repeater communications satellite was **Courier 1B (October 4, 1960)**.

1959

January 2, 1959
Luna 1
U.S.S.R.

Luna 1 was the first spacecraft to leave Earth orbit and to orbit the Sun. Its initial goal was to hit the Moon, but it missed its target by 3,700 miles, sending it into a heliocentric orbit where it was quickly beyond the range of ground communications. Nonetheless, Luna 1 discovered that the Moon lacks a magnetic field—a significant discovery because it implies that, unlike the earth's core, the lunar core is no longer active. Many scientists at the time believed that the Moon's many craters were volcanic in origin, so they had not expected it to have a frozen core.

Luna 1 also carried a sensor for measuring radiation in the Van Allen belts. Passing through the outer belt, it registered increased energy, as expected, but few high-energy particles—the first indication that the outer belt is variable over time. From the data, scientists also estimated that the outer perimeter of the magnetosphere—the structure of Earth's magnetic field—extended beyond the Van Allen radiation belts to about 40,000 miles in altitude, slightly higher than the altitude estimated from data gathered by **Pioneer 3 (December 6, 1958)**.

Approximately halfway to the Moon, the third-stage rocket that had launched Luna 1 was ordered to release a cloud of sodium gas that was visible on Earth to the unaided eye—equivalent in brightness to a six-magnitude star. This release aided in tracking the spacecraft and also permitted the study of gas in the earth's outer magnetosphere.

Luna 1 also measured the interplanetary cosmic ray flux. See **Venera 2 (November 12, 1965)** for results.

February 17, 1959
Vanguard 2
U.S.A.

Weighing 23.7 pounds, seven times more than **Vanguard 1 (March 17, 1958)**, Vanguard 2 operated for only 18 days. Excess solid rocket fuel in the separated third stage nudged the stage against the satellite, causing it to tumble. Launched by NASA, Vanguard 2 was intended to study the earth's cloud cover, but its tumbling made data-gathering difficult and very limited.

February 28, 1959
Discoverer 1
U.S.A.

Discoverer 1 was the first launch of an American military photosurveillance satellite. It was also the first spacecraft to orbit above the earth's poles. A test flight only, Discoverer 1 carried no camera or film; instead, it tested the rocket and satellite engineering. The project's design called for eventually putting a spacecraft in orbit for a short period, after which it would be returned to Earth and its film canister recovered. After Discoverer 1, the U.S. military flew 11 additional test missions over the next 17 months, all of which failed for one reason or another. The project's first complete success was **Discoverer 13 (August 10, 1960)**.

March 3, 1959
Pioneer 4
U.S.A.

Like **Pioneer 3 (December 6, 1958)**, Pioneer 4 was planned to be a flyby of the Moon. Unlike Pioneer 3, Pioneer 4 successfully reached escape velocity and become the first American spacecraft to enter solar orbit, passing within 37,000 miles of the Moon. Carrying the same detectors as Pioneer 3, this NASA spacecraft provided an even more complete picture of the Van Allen radiation belts—to a distance of over 400,000 miles. Pioneer 4's data indicated that the inner Van Allen belt was stable over time, its flux showing little difference from data returned by **Sputnik 2 (November 3, 1957)**, **Explorer 1 (February 1, 1958)**, **Explorer 3 (March 26, 1958)**, **Sputnik 3 (May 15, 1958)**, **Explorer 4 (July 6, 1958)**, **Pioneer 1 (October 11, 1958)**, and Pioneer 3. The outer belt, however, appeared to vary significantly (as inferred from **Luna 1–January 2, 1959**), the measured flux being significantly greater on Pioneer 4 than on Pioneer 3. Scientists posited that the trapped radiation in the inner Van Allen belt was ionized electrons and protons at several specific energies.

The size and shape of the earth's magnetic field were still unclear. Pioneer 4's data indicated that the field extended to at least 52,000 miles—about 16,000 miles more than indicated by Pioneer 3 and 12,000 miles more than indicated by Luna 1.

August 7, 1959
Explorer 6
U.S.A.

Launched by NASA into a highly elliptical orbit (perigee: 152 miles; apogee: 26,346 miles), Explorer 6's mission was to study the earth's magnetic field, cosmic and interplanetary radiation, and micrometeorite impact rates. Technical problems, including the failure of one of the spacecraft's four solar panels, limited the amount of data returned to Earth. Explorer 6 never operated at full power, and it had failed completely by October 1959.

Nonetheless, Explorer 6 confirmed that particles trapped in the earth's Van Allen belts were ionized electrons and protons. However, the spacecraft's apogee was too low to reach the interplanetary magnetic field, so the spacecraft only reconfirmed and further mapped the earth's magnetosphere.

Explorer 6 also carried a television camera to study the earth's cloud cover. Though of poor quality, this camera produced the first orbital photographs of the earth.

September 9, 1959
Big Joe
U.S.A.

This NASA mission used an Atlas rocket (dubbed Big Joe) to test the heat shield planned for use on the Mercury capsule, the first American spacecraft being designed to put humans in space. The stripped-down but functional "boilerplate" capsule was successfully recovered in the South Atlantic after a short suborbital flight. While the capsule's heat shield experienced temperatures of more than 10,000°F, the capsule's interior temperature never rose above 100°F. See **Little Joe 3 (December 4, 1959)** for later tests.

September 12, 1959
Luna 2
U.S.S.R.

Luna 2 was the first spacecraft to impact another world, hitting the Moon's surface on September 14, 1959, near the craters Aristides, Archimedes, and Autolycus. The spacecraft carried Soviet pennants. Observatories in Hungary and Sweden both reported seeing the impact's ejecta cloud. The Swedish observatory took photographs that showed a dark spot at the estimated impact site that had been visible for just under two minutes.

Luna 2 also carried three sensors for measuring the radiation of the Van Allen belts and the interplanetary medium. As had **Luna 1 (January 2, 1959)** and **Pioneer 4 (March 3, 1959)**, Luna 2's data suggested that there were variations in the energy spectrum and electron flux in the outer belt. It also confirmed that the Moon had no magnetic field or radiation belts like Earth, and that there was an energy flux in interplanetary space, implying but not proving the existence of a solar wind. See **Mariner 2 (August 27, 1962)**.

September 18, 1959
Vanguard 3
U.S.A.

Weighing 56 pounds, Vanguard 3, which was built by NASA, carried instruments for measuring solar radiation, the earth's magnetosphere, and the cosmic and interplanetary dust impacting the earth.

Because the spacecraft's apogee was only 2,326 miles, it never rose above the Van Allen belts and therefore could not gather any useful solar radiation data. Vanguard 3 did, however, provide further information for mapping the earth's magnetosphere and Van Allen belts.

Vanguard 3's cosmic dust sensors indicated a high variability in impact rate. During one brief period, the rate was as high as 1,900 impacts per hour, about 10 times more than had been inferred from data collected by **Explorer 1 (February 1, 1958)** and **Explorer 3 (March 26, 1958)**. **Explorer 16 (December 16, 1962)** continued this research.

A small section of the Vanguard 3 capsule was sealed to study how its contents would be affected by the orbital environment. Over the operational period of 70 days, the temperature in this compartment ranged from 28°F to 100°F, averaging around 68°F. The internal pressure remained constant, indicating that no meteorite penetration occurred.

October 4, 1959
Luna 3
U.S.S.R.

Luna 3 took the first photographs of the Moon's hidden far side. The spacecraft was launched into an eccentric figure-eight-shaped orbit that first swung it past the Moon on October 7, 1959, and then returned it to the vicinity of Earth on October 18th. As Luna 3 passed the Moon's far side, at its closest approach (about 3,800 miles) it took 29 photographs covering 70 percent of the far hemisphere. These photographs were processed onboard the spacecraft, then transmitted by facsimile as Luna 3 approached the earth on October 18th.

The images were indistinct, requiring computer enhancement to bring out details. They revealed an extremely rough surface, with many craters and only two dark areas resembling the mare (pronounced MAR-ray) regions on the near side. The Soviets named these two mares the Sea of Moscow

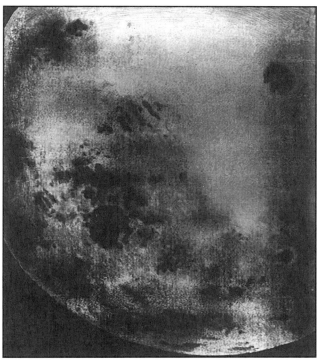

One of the first images sent back by Luna 3 of Moon's far side. North is at eleven o'clock. The dark areas to the west are the mare regions on the near side. The previously unseen far side is east of the large round mare, Smith's Sea, just southwest from the photograph's center. The dark spot in the northeast is the Moscow Sea. The dark spot in the southeast is Tsiolkovsky Crater. *RIA Novosti*

and the Sea of Tsiolkovsky (after Konstantin Tsiolkovsky, known as the father of Soviet space exploration).

Luna 3, like **Luna 2 (September 12, 1959)**, made brief measurements of the energy flux between the planets, inferring from this data the presence of a solar wind. See **Mariner 2 (August 27, 1962)**.

October 13, 1959
Explorer 7
U.S.A.

Explorer 7, like **Explorer 6 (August 7, 1959)**, was built by NASA and carried experiments for measuring the earth's magnetic field, micrometeorite impact, solar radiation, and cosmic radiation. As with **Vanguard 3 (September 18, 1959)**, the charged energy flux of the Van Allen belts prevented the measurement of solar radiation. The micrometeorite experiment also failed, for technical reasons.

Explorer 7 was, however, very successful in measuring the interaction of the earth's magnetic field, the charged particles in the Van Allen belts, and solar flares, working through August 1961. During several solar flare events, the energy released from the Sun was found to be strong enough to press down and squeeze the earth's magnetic field on the hemisphere nearer to the Sun, thereby pushing some of the charged particles from the Van Allen belts into the earth's atmosphere. This, in turn, caused an increase in auroral activity in latitudes lower than normal. This distortion of the earth's magnetic field because of solar activity was the first indication that the field is not symmetrical (see **Pioneer 5–March 11, 1960**).

Explorer 7 was also the first satellite to measure the solar radiation reflected off the earth's atmosphere. Infrared detectors indicated a strong correlation between heavy cloud cover and colder temperatures, suggesting that clouds absorbed more solar radiation, bouncing less back into space than clear skies. This also implied that a more sensitive infrared detector could map the earth's cloud cover, even at night (see **Nimbus 1– August 28, 1964**).

Additionally, Explorer 7's cosmic ray measurements showed that as altitude increased, so did the flux. These data were later confirmed by **Cosmos 17 (May 22, 1963)**.

December 4, 1959
Little Joe 3
U.S.A.

Little Joe 3 was the third suborbital test flight by NASA of the Mercury capsule; it used a small solid rocket booster, called Little Joe, rather than the much larger Atlas rocket used on **Big Joe (September 9, 1959)**. The suborbital flight was the first to carry a passenger, a rhesus monkey named Sam. The capsule rose to an altitude of 55 miles, splashing down in the Atlantic. While the flight lasted only 13 minutes, it took the U.S. Navy two hours to successfully recover the capsule and its passenger. Sam suffered no ill effects from his journey.

Sensors measured Sam's heart rate, breathing, dizziness, and disorientation. Biological specimens, including barley seeds, rat nerve cells, molds, and four beetle eggs, were attached to the outside of the capsule to measure how weightlessness and space radiation affected them.

See **Mercury 1A (December 19, 1960)** for further tests of the Mercury capsule.

1960

March 11, 1960
Pioneer 5
U.S.A.

Launched by NASA into solar orbit between Earth and Venus, Pioneer 5 studied the Sun's magnetic field, solar flares, and the nature of particles between the planets. It operated until June 1960, when it set a record for the longest communications link—22.5 million miles.

As Pioneer 5 traveled from 20,000 to 100,000 miles from the earth, it traversed a region of electromagnetic turbulence approximately 30,000 to 45,000 miles wide. Several scientists proposed that this was not merely the boundary of the earth's magnetic field, but a shock wave on the sunward side. They proposed that the earth's magnetic field was not symmetrical, but distorted by the Sun's solar wind pressing against it. This theory, also suggested by data obtained by **Explorer 7 (October 13, 1959)**, explained the conflicting data from **Pioneer 3 (December 6, 1958)**, **Luna 1 (January 2, 1959)**, and **Pioneer 4 (March 3, 1959)** about the size of the earth's magnetosphere.

The data from Pioneer 5 was not comprehensive enough, however, to prove this theory, nor was it complete enough to map the geomagnetic field's exact size and shape. A later launch, **Imp A (November 27, 1963)**, was specifically designed to answer these questions.

Pioneer 5 also studied the Sun's magnetosphere, showing that several theories about that environment were incorrect. Interplanetary space was not swept clear by a solar wind, and there were not irregular clouds of turbulent gas between the planets, as some theorists had posited. Instead, the data from Pioneer 5 indicated that the Sun's magnetic field was smooth and nearly uniform, at least during periods of low solar activity. This data was later confirmed by **Mariner 2 (August 27, 1962)**.

April 1, 1960
Tiros 1
U.S.A.

Tiros (Television and Infra-Red Observation Satellite) 1 was the world's first weather satellite. Built by NASA, it carried two black-and-white television cameras, one with a wide-angle lens and the other with a telephoto lens. When pointed straight down, the wide-angle lens covered a 750-square-mile area, while the telephoto lens covered one-tenth that area. The transmitted image had a resolution along each raster line of about 750 feet (for the telephoto lens) and 1.4 miles (for the wide-angle lens). Pictures were either sent directly to ground stations, or up to 32 pictures in sequence were stored on an onboard tape recorder for later retrieval. Tiros 1 operated for 2.5 months, until June 15, 1960, taking 22,952 photos of the earth's cloud cover. The satellite took pictures only during the daytime, and for only about one quarter of each orbit, covering from 55 degrees north latitude to 55 degrees south latitude.

Despite these limitations, the first pictures taken by Tiros 1 proved beyond doubt that clouds arrayed themselves in highly

A cloud mass over the northeastern United States, photographed during Tiros 1's first orbit. The dark area to the lower right is the Gulf of St. Lawrence. *NASA*

organized large-scale patterns, the most distinct of these being large cloud vortices, which corresponded to Earth-observed storm regions. Tiros 1 was able to track some of these storm patterns for as long as four days, thereby demonstrating the ability of weather satellites to forecast the weather. Meteorologists also found that the maps they drew of weather-front patterns were easily identified in the Tiros 1 pictures, demonstrating the ability of space-based observations to enhance and supplement Earth-based analysis.

More important, for the first time, meteorologists gained access to a global view of the earth's weather, observing weather patterns in regions previously out of range of ground-based weather stations, including the vast areas of the Atlantic and Pacific Oceans.

One entirely unexpected discovery from Tiros 1 was that the earth's magnetic field is strong enough to interact with the metal in a satellite and change the spacecraft's orientation in space. **Tiros 2 (November 23, 1960)**, as well as later Tiros satellites, carried equipment to make use of that effect.

April 13, 1960
Transit 1B
U.S.A.

Called a "space lighthouse" when launched, Transit 1B was the first navigational satellite, designed to provide ships, submarines, and airplanes precise data on their location, regardless of weather conditions or time of day. As the spacecraft orbited the earth, it transmitted a standard radio beacon, along with a code indicating its exact orbital position. By measuring the changing Doppler shift of the spacecraft's radio signal as it flew overhead and comparing that with the orbital data, a ship or airplane could then calculate its own position very precisely.

Launched by the U.S. Navy, Transit 1B (Transit 1A had failed at launch) was part of a planned four-satellite navigational network. Through 1973, the U.S. Navy launched approximately four Transit satellites a year. Because of higher power requirements, many used nuclear-powered batteries rather than solar cells. See **Transit 4A (June 29, 1961)**.

The technique used by the Transit satellites, dependent on the Doppler shift, required any moving ground station to know precisely its own speed and direction in order to get a good fix. With this information, position could be pinpointed to within 200 feet. If there was time to take multiple measurements, the error could be reduced to under 20 feet. This system, however, was insufficient for fast-moving airplanes, and it had large error margins. If a ship's speed was measured incorrectly by as little as one knot, the positional error grew to about 600 feet.

In the late 1960s, the U.S. Air Force and Navy began to experiment with a different navigation system. The newer technique, called the global positioning system (GPS), allowed fast-moving objects to calculate their positions, and it also was much more accurate. See **Timation 1** (**May 31, 1967**).

During its two months of operation, Transit 1B also performed the first engine restart in space. Its second-stage rocket burned for 4.3 minutes, then coasted for another 19 minutes before re-igniting, as planned, for an additional 13 seconds. Such restarts were essential if spacecraft were to be maneuverable in space for long periods of time.

May 15, 1960
Sputnik 4
U.S.S.R.

Sputnik 4 was the first test flight of the Vostok capsule, the spacecraft used 11 months later to put the first human, Yuri Gagarin, into space (**Vostok 1–April 12, 1961**). The entire orbital unit weighed over 10,000 pounds and consisted of two modules—a pressurized crew cabin and an equipment module. The two modules separated after one day in orbit, as planned. When Soviet scientists then attempted to fire the retro-rockets to deorbit the crew cabin, the engines misfired and raised the module's orbit instead. The crew cabin remained in orbit for five more years before its orbit decayed and it burned up in the atmosphere.

May 24, 1960
Midas 2
U.S.A.

Midas is an acronym for MIssile Detection And Surveillance. Midas 2 (Midas 1 had failed at launch) was the first early-warning satellite; built by the U.S. Air Force, it carried infrared sensors for detecting the heat released at launch by the rocket engines of an ICBM (inter-continental ballistic missile). If it worked as intended, the warning of a nuclear attack would be increased from 15 minutes to 30 minutes.

Midas 2's batteries failed after only two days of operation, after which the satellite ceased functioning. Following two additional test flights, the Midas satellites were made operational, and a dozen were launched through 1966. The design was soon replaced, however, because Midas's infrared detectors could not be trusted to distinguish between the heat of a missile's exhaust and high altitude clouds. See **Canyon 1** (**August 6, 1968**).

June 22, 1960
Transit 2A / Solrad 1
U.S.A.

The launch of Transit 2A and Solrad 1 marked the first time multiple satellites were placed in orbit by a single rocket.

• *Transit 2A* The second U.S. Navy navigational satellite, Transit 2A followed **Transit 1B** (**April 13, 1960**). The location data from both Transit 2A and Transit 1B allowed scientists to measure the earth's shape more precisely, apparently confirming **Vanguard 1's** finding (**March 17, 1958**) that the earth is pear-shaped. This research continued with **Anna 1B** (**October 31, 1962**), the first satellite dedicated solely to geodesy.

• *Solrad 1* The world's first orbiting solar observatory, Solrad 1 was designed to study the ionized radiation of the Sun, including its hydrogen Lyman-alpha radiation in the ultraviolet range of the spectrum (1,050–1,350 angstroms), as well as its soft x-ray radiation (2–8 angstroms). Solrad 1 was the first in a series of 11 U.S. Navy satellites launched from 1960 through 1976 that attempted to continuously monitor the Sun's ultraviolet and x-ray output using standardized sensors.

Solrad 1 gathered data for about nine months. Its results proved that solar x-ray emissions were caused by major solar flares and eruptions and that these events directly influenced the earth's ionosphere. The spacecraft determined the exact amount of x-ray output needed to trigger atmospheric disturbances and showed that these events were triggered quickly, sometimes in less than a minute. The satellite also proved that the active prominences seen periodically on the edge of the Sun produced the same x-ray radiation as sunspot flares seen in the middle of the Sun's disk, showing that the two types of events were actually the same.

The next satellite for solar research was **OSO 1** (**March 7, 1962**). The next successful Solrad satellite was **Solrad 6** (**June 15, 1963**).

August 10, 1960
Discoverer 13
U.S.A.

As part of the American military's surveillance program, Discoverer 13 was intended to be a short-term photosurveillance satellite whose film and camera would be recovered after their return to Earth. Since **Discoverer 1** (**February 28, 1959**), the U.S. military had attempted to launch and recover 11 other Discoverer satellites but had been unsuccessful each time. Some had failed during launch; most failed upon re-entry.

Discoverer 13 was the first successful recovery of a spacecraft from orbit. As it dropped through the atmosphere, a heat shield of ablative material protected the instrument package (which carried no film or camera, crammed instead with sensors for analyzing the spacecraft's flight). Discoverer 13's success prompted the military's first completely successful surveillance mission, **Discoverer 14** (**August 18, 1960**).

August 12, 1960
Echo 1
U.S.A.

The world's second communications satellite, Echo 1 was a 100-foot-diameter balloon built by NASA that opened and inflated upon reaching orbit. Unlike **SCORE (December 18, 1958)**, which was an active repeater satellite that required equipment to record and later transmit messages or data, Echo 1 was a passive reflector satellite. Television or radio signals could be bounced off it, enabling their transmission around the curvature of the earth regardless of the levels of interference from solar radiation. Less than two hours after launch, a short tape-recorded message by President Dwight Eisenhower was transmitted from the Jet Propulsion Laboratory in California to Bell Labs in New Jersey, bouncing off Echo 1 along the way. That same day saw intense solar activity, enough to interfere with regular radio communications and making such an air-to-air transmission impossible without the satellite's aid.

In its first few days of orbit, Echo 1 was used for a number of successful experiments in global communications. Two-way telephone conversations were relayed. Music (specifically, "America the Beautiful") was broadcast from the East Coast to the West Coast. Photographs were transmitted. The first transatlantic message via satellite, from Bell Labs in New Jersey to France, was completed. Short television programs were sent from Washington, D.C., to Europe.

Echo 1 also allowed scientists in the Soviet Union to make geodetic observations over a number of years. By making simultaneous sightings, Echo 1's motion and position were triangulated precisely, thereby pinpointing both the spacecraft's travels through the earth's gravitational field and the exact location of the ground-based observation points. Using this method, one ground station was located to within 200 feet.

These kinds of observations also allowed scientists to determine the density of the atmosphere to an altitude of 932 miles, more than twice as high as **Vanguard 1 (March 17, 1958)** had studied. Atmospheric drag on satellites had been predicted, but no one had anticipated that the atmosphere would expand or shrink with time. Yet, observations of Echo 1 showed that the atmosphere's density changed with time of day and solar activity. During daytime hours, the density increased in the upper atmosphere by as much as 30 times, while solar activity increased it by a factor of four or more. Seasonally, the density showed a 50 percent variation. The cause of these changes was believed to be the Sun and the solar cycle. In order to prolong or predict the life spans of future satellites, it was essential to better understand how and why this atmospheric variability occurred. See **Explorer 19 (December 19, 1963)** for further results.

Originally expected to last for, at most, a year, Echo 1 remained in orbit for almost eight years. Because of its large size and highly reflective surface, it was the first satellite that was

A mock-up of Echo 1 shows its size when fully deployed. *NASA*

easy to see with the unaided eye. In addition, small sample squares of the metal-coated Mylar material from which Echo was made were handed out nationwide to school children. Hence, for many years, it was the most popular space object ever launched. As the *Denver Post* noted when the satellite finally re-entered the atmosphere in May 1968, "So long as men continue to gaze at the stars at night, anywhere around the world, [Echo 1] will be remembered."

The next passive reflector satellite was **West Ford 2 (May 9, 1963)**. **Echo 2 (January 25, 1964)** was the next passive reflector balloon satellite. **Courier 1B (October 4, 1960)** was the next communications satellite.

August 18, 1960
Discoverer 14
U.S.A.

Discoverer 14 was the U.S. military's first successful photosurveillance satellite. Its film returned to Earth undamaged after circling the earth for 29 hours in a polar orbit that took it over the Soviet Union. Rather than allowing the instrument package to land in the ocean where its payload could be damaged, a C-119 Flying Boxcar airplane snatched it from the air at an altitude of 8,500 feet. Like **Discoverer 13 (August 10, 1960)**, Discoverer 14 was devoted entirely to military reconnaissance and carried no instruments for scientific research. Following two more test failures, **Discoverer 17 (November 12, 1960)** was the next successful U.S. surveillance satellite.

August 19, 1960
Sputnik 5
U.S.S.R.

Sputnik 5 was the second test flight of the Vostok space capsule that was intended to send humans into space (see **Sputnik 4–May 15, 1960**). It was also the first of four launches that included dogs in a pressurized crew capsule; other passengers aboard Sputnik 5 included rats, flies, plants, and fungi. The dogs, named Belka and Strelka, were observed throughout their flight by a television camera.

Sputnik 5 was the third capsule to be recovered from orbit, following **Discoverer 13 (August 10, 1960)** and **Discoverer 14 (August 18, 1960)**, and the first to bring living creatures back from orbit successfully (Sam's flight on **Little Joe 3–December 4, 1959** was suborbital). After a little over one day, its retro-rockets fired and the spacecraft parachuted safely back into the Soviet Union.

During the flight, sensors on the dogs measured heart rate, respiration, and blood pressure. Each dog was placed in its own compartment, a glass wall separating them so that they could see each other. During the initial moments of weightlessness, the dogs were seen to hang motionless, paws and heads drooping; they then adapted to the condition, moving about in their harnesses, one dog eventually even eating. Overall, Belka reacted more sharply to the experience, her breathing rate varying from 12 to 240 pants per minute, in contrast to the normal rate of 20 to 40. Two days after their return, the dogs were shown at a press conference. They displayed no ill effects from their short space flight and demonstrated that a human being could almost certainly survive a comparable journey. See **Sputnik 10 (March 25, 1961)** for a summary of biological results.

Sputnik 5 also carried a detector for measuring the Sun's x-ray radiation, which showed that for at least 24 hours, the x-ray output of the Sun was constant.

The next test flight of the Vostok space capsule was **Sputnik 6 (December 1, 1960)**.

October 4, 1960
Courier 1B
U.S.A.

Launched by the U.S. Army, Courier 1B was the second active repeater communications satellite, following **SCORE (December 18, 1958)**. Unlike passive reflector satellites (such as **Echo 1–August 12, 1960**)—which merely bounce transmissions off their surface for ground stations to receive—messages to Courier 1B were tape-recorded by the satellite for later transmission. Although this active repeater satellite could not perform instantaneous communications like Echo 1, it permitted more secure transmission of messages, a military concern. Furthermore, active repeater satellites are more economical, as they utilize much smaller ground transmission stations than passive reflectors.

After only three weeks of operation, Courier 1B's command system failed. The next active repeater communications satellite, **Telstar 1 (July 10, 1962)**, was also the first privately built communications satellite. The next military communications satellites were seven active repeater satellites launched as a group (**IDCSP 1 through IDCSP 7–June 16, 1966**).

November 3, 1960
Explorer 8
U.S.A.

Built by NASA and carrying seven experiments for studying the ionosphere from 250 to 1,000 miles elevation, Explorer 8 was powered by batteries instead of solar cells. It functioned for 54 days, returning data about the atmospheric drag on satellites, electron concentration and particle densities in the ionosphere, the earth's magnetosphere, and micrometeorite impact rates and impact energies. The spacecraft found that, as it flew through the ionosphere, a one-inch-thick plasma sheath of positive ionized particles developed around it, attracted by the satellite's slight negative potential. See **Gemini 4 (June 3, 1965)** for further research of this phenomenon.

November 12, 1960
Discoverer 17
U.S.A.

A military surveillance satellite (see **Discoverer 13–August 10, 1960**; **Discoverer 14–August 18, 1960**), Discoverer 17 was placed in a polar orbit to photograph the Soviet Union, then was recovered. It was the second Discoverer satellite to also carry scientific experiments (Discoverer 3 had carried four black mice, but it failed to reach orbit), including a sensor for studying solar flares and a package of biological specimens of artificially grown human cells from eye and joint tissues, gamma globulin blood protein, and bacterial spores and algae.

Serendipitously, the Sun emitted a tremendous solar flare during Discoverer 17's 50-hour flight, and the spacecraft was exposed to one of the most powerful bursts of radiation yet recorded. Despite this, the human cells seemed unaffected, though immediately after recovery they appeared to be in an advanced state of degeneration. One month later, the cultures had revived and were reproducing normally, indicating that humans could survive such bursts of energy, at least in the short run.

The next U.S. surveillance satellite was **Discoverer 18** (**December 8, 1960**).

November 23, 1960
Tiros 2
U.S.A.

Tiros 2 was the world's second weather satellite, following **Tiros 1** (**April 1, 1960**), with an identical television camera system. This NASA spacecraft operated until February 1, 1961, taking over 36,000 pictures of the earth's cloud cover.

Due to deposits on the lens, photographs taken with the wide-angle camera were of poor quality. Tiros 2's telephoto camera, however, functioned effectively. Pictures of the Gulf of St. Lawrence, as well as the bays, lakes, and rivers of Newfoundland, allowed scientists to study ice formation patterns, revealing ice sheets as they developed and changed. Later, New Zealand meteorologists correctly predicted a break in a six-week heat wave by using the satellite to observe a cold front's approach.

Having learned from Tiros 1 that the earth's magnetic field was powerful enough to affect a metal satellite's orientation in space, Tiros 2 carried a magnetic coil to take advantage of this effect. By controlling how this coil interacted with the earth's magnetic field, ground engineers could better adjust the satellite's attitude.

Tiros 2 also carried instruments to measure the radiation emitted and reflected by the earth, providing primitive measurements of the solar radiation reflected back into space and the temperature of the top atmospheric layers. This data was also used to map the cloud cover during night hours when the television camera was inoperable.

Ten months after launch, on September 28, 1961, engineers successfully restarted a pair of Tiros 2's attitude-control rockets so that its infrared sensors could continue to gather data.

The next weather satellite was **Tiros 3** (**July 12, 1961**).

December 1, 1960
Sputnik 6
U.S.S.R.

The third test flight of the Vostok space capsule (see **Sputnik 4–May 15, 1960**; **Sputnik 5–August 19, 1960**), Sputnik 6 was the second of four flights to place dogs in orbit. Like Sputnik 5, Sputnik 6 carried two dogs into space for one day. Unlike Sputnik 5, however, Sputnik 6's crew cabin burned up in the atmosphere when its retro-rockets did not shut off as intended, driving the spacecraft downward into the atmosphere at too steep an angle for its heat shield to work. See **Sputnik 10** (**March 25, 1961**) for a summary of results from Sputniks 5, 6, 9, and 10.

In addition to biological tests, Sputnik 6 made measurements of the Sun's x-ray output. Like Sputnik 5, it found that the radiation levels remained constant during the 24-hour flight, but at somewhat lower levels than had been recorded during the previous flight.

December 8, 1960
Discoverer 18
U.S.A.

Discoverer 18 was the first successful flight of the second-generation photosurveillance satellite of the U.S. Air Force. Along with its spy cameras, the spacecraft, like **Discoverer 17** (**November 12, 1960**), carried radiation dosimeters and artificially grown human cells and plant life. It also carried infrared radiometers and microwave band detectors for studying high-energy cosmic radiation.

Discoverer 18 flew a near-polar orbit for 3.1 days—one day longer than originally planned because everything was working so well. Then its retro-rockets fired and the re-entry capsule was recovered at 14,000 feet elevation near Hawaii, snatched out of the air by a C-119 Flying Boxcar aircraft.

December 19, 1960
Mercury 1A
U.S.A.

Following up earlier NASA tests (**Big Joe–September 9, 1959**; **Little Joe 3–December 4, 1959**), Mercury 1A was the first successful suborbital test of the complete Mercury package—the vehicle used by the United States to send the first American into space (see **Freedom 7–May 5, 1961**). The capsule reached a speed of almost 4,200 miles per hour, rising to an altitude of 135 miles and flying downrange 235 miles before splashing into the Atlantic Ocean, where it was recovered by U.S. Navy ships. The data collected were used to refine the design of the spacecraft.

The next Mercury capsule test was **Mercury 2** (**January 31, 1961**).

1961

January 31, 1961
Mercury 2
U.S.A.

Mercury 2 was a suborbital test flight of the NASA Mercury capsule design that was used to put Alan Shepard in space a few months later in **Freedom 7** (**May 5, 1961**). The Mercury 2 capsule carried a chimpanzee, Ham, and used the same environmental systems as the later manned capsule. Ham survived the flight with no ill effect, but because the capsule flew 40 miles higher than expected and its heat shield was lost after splashdown, NASA officials decided to fly an additional unmanned test flight (**Mercury 2A–March 24, 1961**) before undertaking a manned launch. This delay allowed the Soviets to put the first human in orbit (see **Vostok 1–April 12, 1961**). From that moment on, the United States and the Soviet Union found themselves participating in a nail-biting "space race." This race was escalated further in May when President John F. Kennedy committed the United States to "landing a man on the Moon and returning him safely to Earth" by the end of the decade.

January 31, 1961
Samos 2
U.S.A.

This Air Force surveillance satellite was launched into a polar orbit in an attempt to replace manned U-2 spy plane flyovers of the Soviet Union.

Unlike the Discoverer satellites (see **Discoverer 13–August 10, 1960; Discoverer 14–August 18, 1960**), which were designed to remain in orbit for about one week and then be recovered with their film intact, Samos 2 remained in orbit and transmitted television images back to Earth. During its one month of operation, the spacecraft took hundreds of photographs of Soviet military facilities with a resolution of about 20 feet. The images allowed the U.S. government to reduce its estimate of Soviet inter-continental ballistic missile (ICBM) capabilities by more than 50 percent.

Soon after the launch of Samos 2, the American military ceased releasing information about its reconnaissance satellites. Subsequent Samos, Midas, and Discoverer launches were not announced in advance, and little information was released about their purposes or designs. By November 1961, the very names "Samos" and "Midas" were no longer used in military publications. See also **Discoverer 36 (December 12, 1961)**, one of the last photosurveillance satellites launched under the Discoverer name.

By 1963, the Samos satellite had been redesigned and given the name KH-4. These new satellites were highly maneuverable, had more sophisticated cameras, and could stay in orbit longer. See **KH-4 9032 (April 18, 1962)**.

February 12, 1961
Venera 1
U.S.S.R.

Venera 1 was the first spacecraft launched toward the planet Venus. Although radio contact was lost before the probe reached its destination, Venera 1 did return information about the energy flux between the planets, first measured by **Luna 2 (September 12, 1959)**. Measurements showed that variations in the flux corresponded closely with variations in the earth's magnetic field, further strengthening the theory that the flux was evidence of a solar wind pushing against the earth's magnetosphere. However, because Venera 1's instruments could not measure the flux's direction or total energy, definitive proof of the existence of a solar wind would wait until **Mariner 2 (August 27, 1962)**.

February 16, 1961
Explorer 9
U.S.A.

Explorer 9 was a 12-foot-diameter balloon (similar to, though much smaller than, **Echo 1–August 12, 1960**), sent into orbit by NASA to study atmospheric density changes at 1,000 miles elevation. Shortly after launch, its transmitter failed, making direct telemetry readings impossible. Using optical and radar sightings, however, scientists were able to track the satellite during its three years in orbit, allowing them to measure the long-term effect of the atmosphere's density on the satellite's orbit. See **Explorer 19 (December 19, 1963)** for a summary of results.

Explorer 9 was also the first satellite placed into orbit from Wallops Island, Virginia, and the first satellite put into orbit using a solid-fuel rocket.

March 9, 1961
Sputnik 9
U.S.S.R.

Unlike the two previous Vostok test flights (**Sputnik 5–August 19, 1960; Sputnik 6 December 1, 1960**), Sputnik 9 carried only one dog, the place for the second dog occupied by a test dummy and a variety of biological specimens. The capsule was successfully recovered after one orbit, and the test dummy was safely ejected and parachuted to the ground, as was planned for a cosmonaut. See **Sputnik 10 (March 25, 1961)** for biological results.

March 24, 1961
Mercury 2A
U.S.A.

The Mercury 2A suborbital test of the Redstone rocket and Mercury capsule was carried out by NASA because of problems experienced during the previous Mercury test flight (**Mercury 2–January 31, 1961**), despite designer Wernher von Braun's protest that it was unnecessary and that NASA could safely proceed with the planned manned flight. The delay allowed Soviet cosmonaut Yuri Gagarin (**Vostok 1–April 12, 1961**) to beat American astronaut Alan Shepard (**Freedom 7–May 5, 1961**) as the first human in space.

March 25, 1961, 6:00 GMT
Sputnik 10
U.S.S.R.

Sputnik 10 was the fourth and last biological test of the Vostok capsule, following **Sputnik 5 (August 19, 1960)**, **Sputnik 6 (December 1, 1961)**, and **Sputnik 9 (March 9, 1961)**. Sputnik 10 carried a dog named Zvezdochka and a test dummy. Its flight lasted one orbit, after which the capsule was successfully recovered. As with Sputnik 9, the test dummy was ejected and parachuted to the ground.

In summary, all four of these flights found that short exposures to weightlessness and outer space had no long-term effect. The animals recovered quickly from the experience, though the dogs from the longer 24-hour flights (Sputnik 5 and Sputnik 6) exhibited biological changes in the cardiovascular and respiratory systems that took between 2 to 6 days to dissipate. There were also indications that the genetic make-up of the dogs and the reproductive systems of the biological specimens on the longer flights were affected, though the extent of these changes and their cause was unclear.

The next flight of the Vostok capsule was **Vostok 1 (April 12, 1961)**, the first manned flight into space.

March 25, 1961, 15:17 GMT
Explorer 10
U.S.A.

Explorer 10 flew an eccentric orbit (perigee: 137 miles; apogee: 112,530 miles) in an attempt by NASA to map the boundaries of the earth's magnetic field. Explorer 10's inclination, however, placed the spacecraft's orbit entirely within the field, and since it did not cross the boundary, it was impossible to gather data as planned. During its 60 hours of operation, it did pass close enough to the magnetopause—the barrier separating the earth's magnetosphere from interplanetary space—to make the first measurements of the magnetopause's properties and shape. It found that, within the earth's magnetic field, the magnetic lines of force are stable and parallel, curving away from the Sun, and implying a tail or wake. Periodically, however, Explorer 10 crossed regions where the magnetic field lines were highly disturbed. Scientists interpreted the area with disturbance as the magnetopause, the realm where the solar wind (still not proven to exist) collided with the earth's magnetic field. See **Explorer 12 (August 16, 1961)** for the first mapping of this boundary.

Though Explorer 10 never entered interplanetary space, it did confirm the energy fluxes measured on **Luna 2 (September 12, 1959)** and **Luna 3 (October 4, 1959)**. Explorer 10's data also indicated these energy fluxes seemed to come from the Sun at a speed of about 185 miles per second. This data further suggested the existence of a solar wind, which ultimately was confirmed by **Mariner 2 (August 27, 1962)**.

April 12, 1961
Vostok 1
U.S.S.R.

Vostok 1 was the first manned flight into outer space. Soviet cosmonaut Yuri Gagarin took off at 9:07 AM local time from the Baikonur cosmodrome in Kazakhstan (then within the Soviet Union) and circled the earth once for a total flight time of 1 hour and 48 minutes, landing southwest of the town of Engels near the village of Smelovka in the Saratov region of Russia.

The Vostok 1 capsule weighed over 12,000 pounds and was essentially the same spacecraft as had been used earlier in a series of unmanned test flights (see **Sputnik 10–March 25, 1961**). Because no one knew what reaction a human would have to weightlessness, Soviet engineers deactivated the capsule's controls. In an emergency, however, Gagarin could unseal an envelope and use a three-digit code within to reactivate them.

During his short flight, Gagarin ate some food and drank water, just to see what it would be like in zero gravity. Weightlessness made "everything easier to perform," he said later. "Handwriting did not change, though the hand was weightless. But it was necessary to hold the writing block, as otherwise it floated from the hands." He also noted that the capsule suddenly seemed much larger, as the upper corners normally ignored on Earth became available for storage and use.

One hour and 18 minutes into the flight, Vostok 1's retrorockets fired. These were part of a cone-shaped module attached to the base of the spherical pressurized crew cabin. As

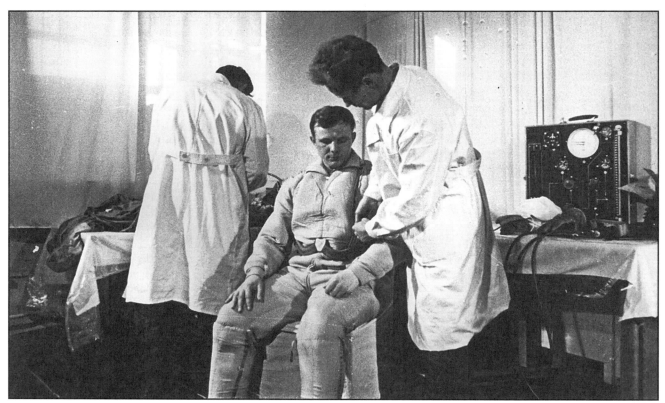

Yuri Gagarin during final preparations before his flight on Vostok 1. *RIA Novosti*

the spacecraft descended, the cables holding the two modules together were burned away by the atmosphere, as planned. The crew cabin, its center of gravity offset, then automatically righted itself with its heat shield facing downward. After the capsule's main parachutes unfurled and slowed its descent, Gagarin ejected, as planned, and parachuted to the ground.

The next man to fly in space was Alan Shepard on **Freedom 7 (May 5, 1961)**. The next Soviet manned flight was **Vostok 2 (August 6, 1961)**.

April 27, 1961
Explorer 11
U.S.A.

Built by NASA, Explorer 11 was the first space-based telescope for observing gamma rays, powerful radiation from the most energetic parts of the electromagnetic spectrum. Gamma rays cannot be studied on Earth because they are filtered out by the atmosphere.

The data gathered by Explorer 11 during slightly less than seven months of operation struck a damaging blow to the "steady state" theory, which postulated that, as the universe expanded, matter and anti-matter were being continuously created. If this theory were true, such newly-created matter and anti-matter must periodically collide, producing large amounts of gamma radiation from the resulting explosion. Explorer 11 should have detected this radiation coming from deep space. Instead, the satellite found that the levels of gamma radiation were much less than predicted by the "steady state" theory, thereby providing indirect support for the "big bang" theory. This evidence was further confirmed by **OSO 1 (March 7, 1962)**.

May 5, 1961
Freedom 7 (Mercury 3)
U.S.A.

On this space flight, astronaut Alan Shepard (**Apollo 14–January 31, 1971**) became the second human and the first American to travel in space.

Unlike Yuri Gagarin (**Vostok 1–April 12, 1961**), Shepard did not orbit the earth. Instead, he flew a short, 15-minute 22-second suborbital flight to an altitude of 115 miles, flying 302 miles downrange from Cape Canaveral at a maximum speed of 5,100 miles per hour. The Mercury 3 capsule, much smaller than Vostok 1 (2,844 pounds compared with 12,659), was steerable. During the flight and descent (when the g force increased to over 11), Shepard adjusted the capsule's pitch, roll, and yaw, proving that while weightless, a human could pilot a spacecraft manually.

Unlike Vostok 1 (which landed on the ground), Shepard's capsule splashed down in the Atlantic, where he was plucked from the water by sailors on the aircraft carrier *U.S.S. Lake Champlain*.

The next manned flight was the **Liberty Bell 7 (July 21, 1961)**.

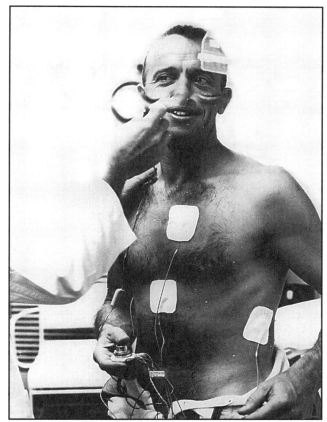
Alan Shepard during final preparations before flight. *NASA*

Shepard's view out the Mercury 3 capsule window, showing the curve of Earth's horizon, the Atlantic Ocean, and clouds. *NASA*

June 29, 1961
Transit 4A / Injun 1 / Solrad 3 (Greb)
U.S.A.

This triple-satellite launch by the U.S. Navy was only partially successful because the Injun 1 and Solrad 3 satellites failed to separate.

• *Transit 4A* Transit 4A was part of the U.S. Navy's first experimental navigational satellite system (see **Transit 1B–April 13, 1960**). It used the SNAP nuclear power source, the first use of atomic energy to provide long-term electrical power to a spacecraft. As a small amount of plutonium-238 decayed, it released heat that was used to produce electricity. This nuclear battery provided power to the satellite for about five years.

• *Injun 1* The Injun 1 satellite provided six months of data on the Van Allen belts, measuring the energy fluxes in both the inner and outer belts. It found that the inner belt seemed unchanged by significant solar activity, a finding later confirmed by data from **Relay 1 (December 13, 1962)**. The intensity within the outer Van Allen belts, however, was found to be highly variable, as much as 100 times greater during the day than at night. This variation in the outer belt implied that the earth's magnetic field was asymmetrical, compressed on the sunward side and extended in the anti-Sun direction, indicating the presence of a solar wind pressing against it. See **Alouette 1 (September 29, 1962)** for different conclusions. See **Imp A (November 27, 1963)** for more definitive proof.

Other data, when combined with results from **Sputnik 3 (May 15, 1958)**, showed that radiation particles were precipitating down into the atmosphere. Although this downward flux might be causing the aurora, the trapped radiation of the Van Allen belts was insufficient to produce the observed flux. See **Injun 3 (December 13, 1962)** for later research on the aurora.

Injun 1 also measured data for the **Starfish Test (July 9, 1962)**.

• *Solrad 3 (Greb)* Solrad 3 was intended to study x-radiation from the Sun and how it affected the earth's ionosphere. Because Solrad 3 failed to separate from Injun 1, it produced no data.

July 12, 1961
Tiros 3
U.S.A.

The world's third weather satellite following **Tiros 2 (November 23, 1960)**, Tiros 3's launch was timed by NASA to correspond to the hurricane season in North America. Unlike the previous weather satellites, Tiros 3 carried two wide-angle cameras instead of one wide-angle camera and one telephoto camera. Each camera could cover an area 62,000 miles square. Within days of launch, it took the first pictures of a developing storm in the Caribbean.

During its 4.5-month life span, Tiros 3 took over 36,000 pictures, photographing all six of the major hurricanes during the 1961 season. The satellite took the first pictures of a hurricane's birth, observing Hurricane Esther on September 10, 1961, two days before ground-based meteorologists iden-

Hurricane Betsy, 7 September 1961, 920 miles east of Cape Hatteras, North Carolina. Three days later, Tiros 3 identified the birth of Hurricane Esther to the southeast. *NASA*

tified it with reconnaissance aircraft. Tiros 3 then photographed the hurricane's development from a tropical depression southeast of the Carolinas to a full-blown hurricane.

Tiros 3 also demonstrated how satellites could be used to make weather observations over isolated islands such as New Zealand, which previously had few methods for obtaining detailed information of weather patterns in the surrounding Pacific Ocean.

The next weather satellite was **Tiros 4 (February 8, 1962)**.

July 21, 1961
Liberty Bell 7 (Mercury 4)
U.S.A.

Liberty Bell 7 was the second (and last) American suborbital manned flight launched by NASA. Astronaut Gus Grissom (**Gemini 3–March 23, 1965**; **Apollo 1–January 27, 1967**) duplicated Alan Shepard's earlier flight in **Freedom 7 (May 5, 1961)**, controlling the spacecraft's attitude and flight during his 15-minute 37-second journey. Upon splashdown, however, the spacecraft's hatch inadvertently blew free and—although Grissom escaped unharmed—the capsule sank in the ocean to a depth of over 10,000 feet. It was not recovered until July 1999.

The next manned mission was **Vostok 2 (August 6, 1961)**. The next American manned mission was **Friendship 7 (February 20, 1962)**.

August 6, 1961
Vostok 2
U.S.S.R.

The second Soviet manned space flight, Vostok 2 placed cosmonaut Gherman Titov in space for one full day, orbiting the earth 17 times.

During his flight, Titov was the first human to experience space sickness. He found that weightlessness made him feel as if he were upside down; that he was floating with his feet in the air and his head near the ground. By the fifth orbit, he began to feel nauseated and soon felt so sick that he was willing to end the flight. Because his spacecraft was at that moment out of touch with ground control, he could take little action, and by the time he regained communication, the nausea had subsided somewhat. He noted, however, that his vertigo was increased by sharp head motions, as well as by watching fast-moving objects. Titov continued to experience inner-ear problems following the flight. Based on his experience, the Soviets restructured their cosmonaut training program.

Unlike Gagarin (**Vostok 1–April 12, 1961**), Titov steered his spacecraft manually, adjusting Vostok 2's attitude several times during the flight.

After 10 hours of flight, Titov became the first human to sleep in space. Despite the nausea, Titov slept well. "I slept the sleep of the just," he later said of his eight-hour nap. "My sleep was good, without dreams. In contrast to Earth conditions, I didn't feel the necessity of turning from side to side." He did find it necessary, however, to fasten his arms in place to prevent them from floating out in front of his face.

After firing its retro-rockets, the equipment module took longer than expected to separate from the crew cabin. Nonetheless, re-entry went smoothly, and after 25 hours and 18 minutes in space, Titov ejected from the Vostok capsule at 4 miles elevation, safely parachuting to the ground.

The next manned mission was **Friendship 7 (February 20, 1962)**. The next Soviet manned flight was a dual mission lasting almost 4 days. See **Vostok 3 (August 11, 1962)** and **Vostok 4 (August 12, 1962)**.

August 16, 1961
Explorer 12
U.S.A.

Launched by NASA, a magnetometer on Explorer 12 gave the first clear picture of the magnetopause following its initial discovery by **Explorer 10 (March 25, 1961)**. After nine orbits, the boundary region between the earth's magnetic field and interplanetary space ranged from 32,000 miles to 48,000 miles from the earth's center, depending on the satellite's position, once again suggesting that the field was distorted by the solar wind.

By studying the energetic electrons outside the magnetopause, scientists were able to map the bow shock of the magnetic field, finding it to be about 8,000 to 12,000 miles across. As predicted, there was a sharp transition from the bow shock into interplanetary space. **Imp A (November 27, 1963)** mapped this feature more thoroughly.

Scientists had previously thought that the inner Van Allen belt consisted of a mixture of protons and electrons, and the outer belt was made up mostly of low energy electrons. Explorer 12 found, instead, that the composition of the outer belt was similar to that of the inner belt. In comparing data from **Explorer 14 (October 2, 1962)**, the proton population was found to be stable over time, indicating that the variability of the outer belt was due almost entirely to the electron population. The experiment also found that the outer belt was concentrated in the equatorial regions, was stable over time, and had large variations of energy flux.

Data relating to magnetic storms were also gathered by Explorer 12 during its four months of operation. During solar flares, the earth's magnetic field also fluctuated, for reasons that were not clear. Explorer 12 found large increases in charged electrons in the Van Allen belts during two magnetic storms, indicating that ionized electrons instead of protons were a dominant factor in causing the storms.

The archived data from this satellite was later used to see if the Sun's solar cycle had any effect on the earth's magnetic field. See **OGO 1 (September 5, 1964)**, **Elektron 3/Elektron 4 (July 11, 1964)**, and **Injun 4 (November 21, 1964)**.

October 21, 1961
Midas 4 / West Ford 1
U.S.A.

- *Midas 4* A U.S. Air Force early-warning satellite for the detection of nuclear attack (see **Midas 2–May 24, 1960**), Midas 4 was also the last American military satellite mission for which any public information was released. For security reasons, later military surveillance missions were announced publicly as nondescript numbers or acronyms, such as DSP 647 or IMEWS, with little information about their designs or purposes. See **Samos 2 (January 31, 1961)**.

- *West Ford 1* West Ford 1 was an experimental passive reflector communications satellite like **Echo 1 (August 12, 1960)**, built by the U.S. Air Force. Instead of being a large reflective balloon, however, West Ford 1 was a belt of 80 million tiny copper wires. After reaching orbit, the satellite was to release these filaments, which were then to form an orbiting cloud 9 by 18 miles across. West Ford 1 was a failure when the belt was not released as planned. This test was followed by the more successful **West Ford 2 (May 9, 1963)**.

November 15, 1961
TRAAC
U.S.A.

The Transit Research And Attitude Control (TRAAC) satellite was a U.S. Navy research satellite launched in coordination with the **Starfish Test (July 9, 1962)**. It carried experimental packages for testing the degradation of orbiting solar cells and transistors. These experiments found that for the first eight months of flight, the solar cells degraded about 2 percent per month. Immediately following the Starfish Test, however, the degradation rate increased to about 1 percent per day. See also **Anna 1B (October 31, 1962)**.

November 29, 1961
Mercury 5
U.S.A.

Mercury 5 was the first attempt by the United States to place a living creature (a chimpanzee named Enos) into Earth orbit, more than four years after Laika's orbital flight on **Sputnik 2 (November 3, 1957)**. The two-orbit NASA flight tested the Mercury capsule for John Glenn's orbital mission, **Friendship 7 (February 20, 1962)**.

Enos had been trained to pull levers that turned off randomly switched-on lights. He did this successfully during the flight, despite the failure of an electronic switch that was supposed to give him electric shocks if he failed in a task. Instead, the device shocked the puzzled "chimponaut" even when he did things right.

Originally planned to be a three-orbit flight, the failure of one attitude thruster forced NASA to bring the capsule home one orbit early. Both Enos and the capsule were safely recovered after splashdown in the Atlantic.

December 12, 1961
Discoverer 36 / Oscar 1
U.S.A.

• *Discoverer 36* This U.S. Air Force surveillance satellite, which was recovered after 4.1 days in orbit, was one of the last launched under the Discoverer name. Because the American military wished to restrict the release of information about its future military surveillance launches, this well-publicized program was officially ended in early 1962. See **Samos 2 (January 31, 1961)**.

In the 1960s, these military surveillance satellites, under the general designation KH (Keyhole), were also undergoing redesign. By 1963, the U.S. military replaced the wide-angle Samos satellites with the KH-4, KH-4A, KH-4B, and KH-5 satellites. The Discoverer satellites were, in turn, replaced with the telephoto KH-6, KH-7, and KH-8 satellites, which were more reliable, could stay in orbit longer, and produced better pictures. The first successful launch of this new design was **KH-7 1 (July 12, 1963)**.

By the 1990s, this early American surveillance satellite program had been given the name Corona, and its image archive from 1960 to 1972 was made available to the public in 1995.

• *Oscar 1* Financed and built entirely by private amateur shortwave radio users, Oscar (Orbiting Satellite Carrying Amateur Radio) 1 was a nonpaying piggyback passenger on Discoverer 36, which was allowed to drift away upon reaching orbit. Oscar 1 was also the first in a series, which through 1998 has placed 30 private satellites in orbit performing radio experiments and providing communications links for ham radio operators worldwide.

Oscar 1 was very simple, weighing only 10 pounds and carrying a 0.1-watt transmitter that broadcast a signal from which only the inner temperature of the satellite could be determined. More than 5,000 ham radio operators worldwide took measurements, recording the spacecraft's temperature daily during its 18-day life span.

The next Oscar satellite was **Oscar 2 (June 2, 1962)**.

1962

January 26, 1962
Ranger 3
U.S.A.

Built by NASA, Ranger 3 was the first U.S. attempt to hit another planet with a probe (Rangers 1 and 2 had been Earth-orbit test flights, both of which had failed). Ranger 3 missed the Moon by about 20,000 miles and went into orbit around the Sun.

Ranger 3 carried out one successful scientific experiment: an onboard gamma-ray spectrometer indicated that a diffuse flux of gamma radiation was coming from outside the solar system. The data was too imprecise, however, to draw any detailed conclusions. See **OSO 1 (March 7, 1962)** for later gamma-ray research.

The next Ranger flight, **Ranger 4 (April 23, 1962)**, was the first American space probe to impact the Moon.

February 8, 1962
Tiros 4
U.S.A.

The fourth NASA weather satellite, following **Tiros 3 (July 12, 1961)**, Tiros 4 had a life span of 4.5 months and returned 32,593 cloud-cover photos. The satellite studied the formation of hurricanes, atmospheric flow patterns, and the ice formation in rivers and oceans.

Like Tiros 3, Tiros 4 carried infrared sensors for measuring the heat radiated by the earth's cloud cover. Although the data was not detailed enough for weather forecasting, it showed that such measurements closely matched the photographed patterns of cloud cover and could be used to record the changing cloud cover at night (see **Nimbus 1–August 28, 1964**).

The next weather satellite was **Tiros 5 (June 19, 1962)**.

February 20, 1962
Friendship 7 (Mercury 6)
U.S.A.

Friendship 7 was the first American manned orbital flight, sending John Glenn (see also **STS 95–October 29, 1998**) around the earth three times. During his 4-hour 55-minute flight, Glenn performed a number of maneuvering tests on the Mercury capsule. For the last half of the flight, failure of the automatic control system required him to fly the spacecraft manually. This was similar to the problem that had forced Enos's **Mercury 5 (November 29, 1961)** flight to be cut short by one orbit. On Friendship 7, however, a human being could correct the problem, flying the ship by hand to conserve fuel and thereby allowing the flight to continue.

Sensors that measured Glenn's heart rate, breathing, blood pressure, and temperature throughout the flight indicated that his body responded normally to both launch and weightlessness. Unlike Gherman Titov on **Vostok 2 (August 6, 1961)**, Glenn experienced no space sickness. In fact, no American astronaut experienced that problem until **Apollo 8 (December 21, 1968)**. Glenn was also the first American to urinate in space. (While Shepard and Grissom had urinated during their missions, neither had done so while in flight.) His urine was collected and analyzed after splashdown, and when compared with preflight samples, it appeared completely normal.

During the flight, Glenn made astronomical and meteorological observations. As he passed over Africa for the first time, he reported seeing dust storms. Later, he identified thunderstorms and lightning over Zanzibar and the Indian Ocean. When he observed his first orbital sunset, he note the presence of "a brilliant blue band clear across the horizon," later named the airglow layer. During the night portion of each orbit, he could see stars with no problem and identified a num-

April 18, 1962

Sunset with airglow band along Earth's horizon, seen from Friendship 7. *NASA*

ber of constellations. He also could see the black horizon of the dark Earth against an even blacker sky.

Glenn also observed what he called "fireflies" swarming about his capsule. "They swirl around the capsule and then depart back the way I am looking," Glenn reported. "There are literally thousands of them!" While many doubted what Glenn reported, his experience was later confirmed and explained by astronaut Scott Carpenter on **Aurora 7 (May 24, 1962)**.

During the flight's second orbit, a sensor light indicated that the capsule's landing bag (for use during splashdown) had prematurely deployed. If the bag had indeed deployed, the capsule's heat shield also might have become detached and could break free during re-entry. If this were to happen, the capsule would burn up, killing Glenn. In order to prevent this, mission control decided to leave the spacecraft's retro-rocket package attached to the heat shield during re-entry. Their hope was that the straps that held the retro-rockets to the heat shield would hold the whole assembly in place.

In actuality, the sensor light had malfunctioned and the landing bag had not been released. Glenn splashed down safely in the Atlantic Ocean near Puerto Rico. During re-entry, he saw chunks of material fly past his windows as the retro-rocket package burned away.

The next manned space flight was **Aurora 7 (May 24, 1962)**.

March 7, 1962
OSO 1
U.S.A.

The first orbiting astronomical observatory, OSO (Orbiting Solar Observatory) 1 was dedicated to studying the Sun. It was also the first of NASA's series of eight OSO satellites whose purpose was to observe the Sun as it went through an entire solar cycle, supplementing the observations of the Solrad satellites (see **Solrad 1–June 22, 1960**). These OSO satellites were launched at regular intervals through 1975.

In its 77 days of operation, OSO 1 observed 140 solar flares, measuring the ultraviolet radiation and x-radiation re- leased in each. It found that during these active events, ultra-violet and x-ray emissions could be four times more intense than normal.

OSO 1 also mapped the sky in gamma rays, once again looking for proof of the "steady state" theory, as had **Explorer 11 (April 27, 1961)**. Like Explorer 11, OSO 1's negative results helped confirm the "big bang" theory of the universe's formation.

The next OSO satellite was **OSO 2 (February 3, 1965)**.

March 16, 1962
Cosmos 1
U.S.S.R.

Cosmos 1 was the first successful launch in the Cosmos series of Soviet space vehicles, a general name used for any space-craft injected into Earth orbit. Through the late 1990s, over 2,000 Cosmos satellites have been launched.

Cosmos 1 carried a radio beacon operating at several frequencies. Scientists on the ground could use the beacon to track the satellite precisely, thereby indirectly studying the ionosphere by how it affected the spacecraft's orbit. The satellite burned up in the atmosphere after two months.

April 6, 1962
Cosmos 2 (Sputnik 12)
U.S.S.R.

Similar to **Cosmos 1 (March 16, 1962)**, scientists used Cosmos 2's radio beacon to study the earth's ionosphere. The satellite remained in orbit until August 1963.

April 18, 1962
KH-4 9032
U.S.A.

KH-4 9032 was the first satellite in a new U.S. Air Force program to replace the Samos wide-angle photo-reconnaissance satellites (see **Samos 2–January 31, 1961**). Both KH-4 and KH-5 satellites were more maneuverable, able to stay in orbit longer (an average of 23 days), and had better camera systems. Over 100 of these satellites were launched through 1971 under four different designs, designated KH-4, KH-4A, KH-4B, and KH-5. Initially, the wide-angle-lens cameras had a resolution of about 450 feet and could cover about 200 square miles. Later, the cameras in the KH-4B had resolutions under 100 feet. The KH-5 camera, meanwhile, was identical to the camera used by the Lunar Orbiters (see **Lunar Orbiter 1– August 10, 1966**).

The KH-4 and KH-5 satellites took wide-angle images, working in conjunction with the telephoto images taken by the KH-6, KH-7, and KH-8 satellites, which had resolutions under five feet. If a KH-4 satellite saw something that required a more detailed observation, a KH-7 satellite was launched to get a better picture. See **KH-7 1 (July 12, 1963)** for details on the satellites with telephoto equipment.

After 1971, the KH-4 through KH-5 design was replaced by the KH-9, also called Big Bird. See **KH-9 1 (June 15, 1971)**.

April 23, 1962
Ranger 4
U.S.A.

Built by NASA, Ranger 4 was the second spacecraft after **Luna 2 (September 12, 1959)**—and the first American spacecraft—to impact another world, hitting the Moon's far side on April 26, 1962. Because of a computer failure, however, the spacecraft returned no data. After the failures of Ranger 5 (which missed the Moon entirely when its solar panels failed and it lost all power) and **Ranger 6 (June 30, 1964)**, the United States finally obtained the first close-up pictures of the Moon with **Ranger 7 (July 28, 1964)**.

April 24, 1962
Cosmos 3
U.S.S.R.

Data from Cosmos 3 (which remained in orbit until October 1962), as well from **Cosmos 5 (May 28, 1962)** and **Cosmos 17 (May 22, 1963)**, allowed scientists to map the lower altitudes of the Van Allen radiation belts. This work discovered what is now called the South Atlantic anomaly. See Cosmos 17 for details.

April 26, 1962, 10:04 GMT
Cosmos 4
U.S.S.R.

This military photo-reconnaissance satellite was the Soviet Union's first. It was similar in concept to the American Discoverer series of telephoto Keyhole satellites (see **Discoverer 36–December 12, 1961**). Built on assembly-line principles so that they would be relatively cheap, simple, and quick to manufacture and launch, the Soviets sent up an average of about 35 of these satellites per year to observe political hotspots or to eye specific American military facilities. Although the Cosmos cameras did not have the resolution of the American cameras, they were adequate for basic reconnaissance.

Over the next 30 years, the Soviets developed several generations of telephoto satellites (as well as wide-angle satellites), much like those of the United States. Each was launched on short notice, orbited for a short period (usually from one to two weeks), then recovered in Soviet territory. Often, these satellites were launched to take reconnaissance photos of specific events. In 1983, for example, Cosmos 1504 was used to conduct photo-reconnaissance of the American invasion of Grenada.

By the 1980s, the service life of many of these Cosmos satellites had increased from two weeks to two months. They had also become very maneuverable, with ground control able to adjust their orbit very precisely, permitting their positioning over specific political trouble spots. Some satellites carried multispectral cameras recording both surveillance images and Earth resource photography, surveying the agricultural regions of the Soviet Union during the spring and summer months. Later satellites also performed mapping (including topographic maps for the military), geodesy, cosmic-ray and gamma-ray experiments, and other scientific research. Because of the short service life and the military nature of the Cosmos satellites, however, little information is available that describes what was learned from these experiments.

Ironically, after the fall of the Soviet Union, the design of the later Cosmos surveillance satellites was enlisted as part of a commercial operation launched in partnership with an American company. See **Cosmos 2349 (February 17, 1998)**.

April 26, 1962, 18:00 GMT
Ariel 1
U.K. / U.S.A.

Ariel 1 was the first satellite built by more than one country, and the first built by a country other than the United States or the Soviet Union. Scientists from the United Kingdom built and designed Ariel 1's experiments, while the satellite's launch and structure were provided by NASA. Ariel 1's goal was to study the ionosphere, x-rays from solar flares, and cosmic ray flux in near-Earth space.

While Ariel 1 provided some data in its first few months of operation, its instruments were disabled when it flew through the radiation belt created by the **Starfish Test (July 9, 1962)**. After that date, it provided little useful data.

The next satellite to study the ionosphere was **Alouette 1 (September 29, 1962)**. The next joint U.S.A.-U.K. satellite was **Ariel 2 (March 27, 1964)**.

May 24, 1962
Aurora 7 (Mercury 7)
U.S.A.

Aurora 7 was the second American manned orbital flight, following **Friendship 7 (February 20, 1962)**. Like John Glenn, Scott Carpenter completed three orbits of the earth for a total flight time of 4 hours, 56 minutes, during which he performed several scientific experiments.

Carpenter took photographs of the earth's cloud cover using a number of different types of film. Meteorologists wished to compare these photographs with the television images being transmitted by **Tiros 4 (February 8, 1962)**.

Another experiment had Carpenter release a 30-inch multicolored balloon, attached to a 100-foot tether, to test judging distances and colors in space. He had trouble at first identifying the balloon (which did not inflate completely), and he later misjudged its distance.

One medical experiment had Carpenter shake his head from side to side with his eyes closed in an attempt to duplicate Gherman Titov's space sickness (**Vostok 2–August 6, 1961**). Like Glenn, however, Carpenter felt no space sickness.

Carpenter also saw the "fireflies" noted by John Glenn on Friendship 7. He concluded that they probably were "frost from a thruster"—frozen fuel particles drifting about the spacecraft.

An accumulation of minor problems and mistakes caused Carpenter to overshoot his landing point by about 250 miles. For these and other reasons, Carpenter never again flew as an astronaut for NASA.

The next manned mission was **Vostok 3 (August 11, 1962)**. The next American manned mission was **Sigma 7 (October 3, 1962)**.

May 28, 1962
Cosmos 5 (Sputnik 15)
U.S.S.R.

Cosmos 5 carried sensors to monitor the artificial radiation belt produced by the **Starfish Test (July 9, 1962)**, remaining in orbit until May 1963. Cosmos 5's data was also combined with observations from **Cosmos 3 (April 24, 1962)** and **Cosmos 17 (May 22, 1963)** to map the lower altitudes of the Van Allen radiation belts. See Cosmos 17 for details.

June 2, 1962
Ferret 1 / Oscar 2
U.S.A.

- *Ferret 1* This U.S. Air Force satellite was the first to do Ferret-type electronic surveillance, that is, "ferreting out" the electronic signals of radar equipment, voice communications, and missile telemetry. Sixteen such satellites, most launched piggyback with other satellites, were orbited through 1971. **Canyon 1 (August 6, 1968)** and **Jumpseat 1 (March 21, 1971)** eventually superseded the Ferrets.

- *Oscar 2* Oscar (Orbiting Satellite Carrying Amateur Radio) 2 was identical in operation to **Oscar 1 (December 12, 1961)**. For 18 days, more than 6,000 amateur radio operators worldwide measured the satellite's inner temperature. The next private ham radio satellite was **Oscar 3 (March 9, 1965)**.

June 19, 1962
Tiros 5
U.S.A.

The fifth NASA weather satellite, following **Tiros 4 (February 8, 1962)**, Tiros 5 returned over 58,000 cloud-cover images while tracking the 1962 hurricane season over North America. Its infrared sensors, however, failed to function.

Tiros 5 operated for 10.5 months and, working in conjunction with **Tiros 6 (September 18, 1962)**, was able to provide meteorologists with their first almost continuous television coverage of the world's cloud cover. The developing data archive from all the Tiros satellites allowed meteorologists to begin classifying the different cloud structures seen from space.

July 9, 1962
Starfish Test
U.S.A.

On July 9, 1962, the United States ignited a 1.4-megaton nuclear explosion at 250 miles elevation. Like Project Argus (see **Explorer 4–July 6, 1958**), the Starfish explosion was focused on studying the behavior of trapped particles in the magnetosphere as the energized debris from the atomic explosion formed temporary and artificial radiation belts not unlike the Van Allen belts.

Very quickly, four different satellites recorded the formation of a transient radiation belt. In its first orbit following the explosion, **Cosmos 5 (May 28, 1962)** recorded a hundredfold increase in energy flux. **Ariel 1 (April 26, 1962)** showed that this high-energy flux quickly appeared at unexpectedly high altitudes. Both **TRAAC (November 15, 1961)** and **Injun 1 (June 29, 1961)** provided good pre- and post-explosion data, measuring the change in energy flux. Both satellites also mapped the belt, with TRAAC finding what one scientist called "a puddle of fission debris sitting on top of the atmosphere in the Pacific, continuously emitting electrons into this transient belt." While Injun 1 mapped the belt to about 600 miles altitude, **Telstar 1 (July 10, 1962)**, launched one day after the explosion, mapped the belt up to 3,500 miles altitude.

The radiation belt from the Starfish Test decayed slowly over a number of years, its dissipation tracked by many satellites.

July 10, 1962
Telstar 1
U.S.A.

AT&T's Telstar 1 was the first communications satellite built and financed by a private company, rather than the government. Launched by NASA, this commercial active repeater satellite (like **Courier 1B–October 4, 1960**) operated in low-Earth orbit until November 1962, when its command systems failed due to degradation of the transistors from Van Allen belt radiation. Ground controllers were able to restart the satellite for about six weeks in early 1963 before it failed permanently.

Telstar 1 had one television channel, 60 two-way voice circuits (only 12 of which could be used at one time because of ground equipment limitations), and 600 one-way voice circuits. The satellite was a 34.5-inch sphere covered with solar cells, weighing 170 pounds. Over 400 transmission tests were performed from stations in the United States, the United Kingdom, and France, sending multichannel telephone, telegraph, facsimile, and television signals. On Telstar 1's first day in orbit, its signal was broadcast for 15 minutes on the three American television networks, showing conversations between U.S. Vice President Lyndon B. Johnson and Fred Kappel, AT&T's chairman. This signal was also picked up in France.

Telstar 1 also carried detectors for studying the trapped radiation in the Van Allen belts, and it immediately provided data on the **Starfish Test (July 9, 1962)**, which had been conducted the previous day. The data showed how the electrons released from the explosion and then trapped by the earth's magnetic field slowly decayed over time. At low and

July 28, 1962

Telstar 1 prior to launch. *NASA*

high altitudes, the decay times were short, while in the middle altitudes they were long. The detectors also confirmed what **Injun 1 (June 29, 1961)** had found: that the energy fluxes of the outer Van Allen belt were directly related to solar activity.

The next communications satellite, built by NASA, was **Relay 1 (December 13, 1962)**. AT&T later built and launched **Telstar 2 (May 7, 1963)**. These two satellites confirmed Telstar 1's findings concerning the Van Allen belts.

July 28, 1962
Cosmos 7
U.S.S.R.

Cosmos 7—like **Cosmos 4 (April 26, 1962)** and the American Discoverer satellites (see **Discoverer 1–February 28, 1959**)—was a short-term surveillance satellite, remaining in space for four days, after which its film canister was recovered.

The spacecraft also measured the radiation left over from the **Starfish Test (July 9, 1962)** to make sure the levels were safe for the **Vostok 3 (August 11, 1962)** and **Vostok 4 (August 12, 1962)** manned missions.

August 11, 1962
Vostok 3
U.S.S.R.

Vostok 3's mission, which lasted just under four days, was the longest manned space flight to date. Launched one day prior to **Vostok 4 (August 12, 1962)**, it also was part of the first dual space flight.

Andrian Nikolayev (**Soyuz 9–June 1, 1970**), along with Pavel Popovich on Vostok 4, performed a number of medical experiments developed in response to Gherman Titov's space sickness on **Vostok 2 (August 6, 1961)**. Sensors on Nikolayev's body measured his heart rate, respiration, blood pressure, and temperature. To gauge his reaction to weightlessness, he removed his harness and floated free for long periods, doing specific exercises. Nikolayev reported that when he turned his head quickly, the motion caused him to unintentionally extend his arms in the opposite direction. In another test, Nikolayev and Popovich (in Vostok 4) closed their eyes while floating freely in their capsules, attempting to judge the drift of their bodies. They both found it impossible to determine.

Nikolayev experienced no problems from weightlessness. He ate, drank, and slept normally during the flight, and experienced no ill effects afterward. He (as well as Popovich on Vostok 4) also performed the first public television broadcast from space, showing an audience in the Soviet Union and Europe how both he and other objects floated in zero gravity.

August 12, 1962
Vostok 4
U.S.S.R.

Vostok 4 was the second half of the world's first dual space mission. When Pavel Popovich (**Soyuz 14–July 3, 1974**) in Vostok 4 joined Andrian Nikolayev in **Vostok 3 (August 11, 1962)** in orbit, the two spacecraft were less than four miles apart, although neither had the ability to make the precise orbital changes required for rendezvous. The cosmonauts reported seeing each other's capsules, and they performed the first direct radio communications between orbiting spacecraft.

As Vostok 3 and 4 circled the earth, their orbits slowly changed, moving them away from each other. This drift was caused partly because of atmospheric drag and partly because of unexpected and tiny irregularities in the earth's gravitational field, subtly modifying each spacecraft's orbit as it passed over these density variations.

Popovich performed the same medical experiments as Nikolayev in Vostok 3, reporting no ill effects. After almost three days in orbit, he landed mere minutes after Nikolayev in Vostok 3. The Soviets later duplicated this dual mission with **Vostok 5 (June 14, 1963)** and **Vostok 6 (June 16, 1963)**.

The next manned mission was **Sigma 7 (October 3, 1962)**.

August 27, 1962
Mariner 2
U.S.A.

Built by NASA, Mariner 2 was the first planetary probe to reach another planet and successfully return data—Mariner 1 had failed on launch, and Venera 1 failed before it reached Venus.

As it traversed the space between Earth and Venus, Mariner 2 also provided the first direct observations of the Sun's magnetic field and solar wind, confirming what **Luna 2 (September 12, 1959)**, **Luna 3 (October 4, 1959)**, **Venera 1 (February 12, 1961)**, and **Explorer 10 (March 25, 1961)** had inferred. Mariner 2's measurements, combined with later data from **Imp A (November 27, 1963)** and **Vela 1A/Vela 1B (October 17, 1963)**, allowed scientists to develop a rough outline of the solar wind's properties, at least during the quiet part of the Sun's solar cycle. Mariner 2 measured solar wind speeds ranging from 200 to 500 miles per second, with an average velocity of 300 miles per second. These wind speeds fluctuated sharply for reasons that were not clear. The wind's average temperature was 270,000°F, with temperature variations matching the wind's velocity changes.

A cosmic dust experiment on Mariner 2 found that, in interplanetary space, the flux of particles was much less than seen in near-earth space (where the density was about 10,000 times greater), and that there was no comparable increase in dust particles as the spacecraft approached Venus. This confirmed observations by **Pioneer 5 (March 11, 1960)** and showed that no clouds of matter floated between Earth and Venus. The cause of the high density near the earth was unknown, however.

Mariner 2 flew past Venus in mid-December (see **Mariner 2, Venus Flyby–December 14, 1962**).

September 18, 1962
Tiros 6
U.S.A.

Tiros 6 continued NASA's first-generation study of the earth's cloud cover, begun with **Tiros 1 (April 1, 1960)** and completed with **Tiros 7 (June 19, 1963)**. Tiros 6 returned over 66,000 cloud-cover images during its 13-month life span and acted as a link between **Tiros 5 (June 19, 1962)** and Tiros 7. The pictures from these three satellites gave scientists their first almost continuous view of the world's cloud cover. See Tiros 7 for results.

September 29, 1962
Alouette 1
Canada / U.S.A.

Alouette 1, designed to study the ionosphere looking down from space, was the first satellite entirely designed and built by a country other than the United States or the Soviet Union. NASA provided only the rocket and launch facilities. It was also the first non-U.S. satellite launched from Vandenberg Air Force Base in California.

During its 10 years of data-gathering, Alouette 1 found that the outer Van Allen belt varied, shrinking as it went from day to night. Some scientists attributed this to the short lifetime of the outer belt's particles, not to the pressure of the solar wind, as had been theorized from data first seen on **Injun 1 (June 29, 1961)**.

Alouette 1 also observed the artificial radiation belts created by three high-altitude Soviet nuclear explosions set off in October and November 1962. See **Starad (October 26, 1962)**.

The next spacecraft involved in ionospheric research was **Explorer 20 (August 25, 1964)**.

October 2, 1962
Explorer 14
U.S.A.

Explorer 14 continued the magnetosphere studies of previous NASA Explorer satellites, following **Explorer 12 (August 16, 1961)**. (Explorer 13 had failed to reach its required orbit and returned no data.) It also supplemented data from **Ariel 1 (April 26, 1962)** and **Alouette 1 (September 29, 1962)**. It operated through October 1963.

One of Explorer 14's specific discoveries was "spikes" of energy in or near the transition zone between interplanetary space and the earth's magnetic field, revealing the irregular and complex nature of the magnetosphere.

In comparing Explorer 14's data with that from Explorer 12, it was found that the proton population of the outer Van Allen belt was stable over time. Explorer 14 discovered, however, that the electron population was highly variable, the fluxes changing by many orders of magnitude over hours. Variations could be attributed to several different causes, including variations because of solar rotation, the solar cycle, magnetic storm effects, and radiation field decay during quiet periods. See **38C (September 28, 1963)** for solar rotation results. See **OGO 1 (September 5, 1964)** for solar cycle results.

October 3, 1962
Sigma 7 (Mercury 8)
U.S.A.

Sigma 7, the fifth American manned space flight, doubled the length of the previous Mercury mission, **Aurora 7 (May 24, 1962)**, orbiting the earth six times and placing astronaut Wally Schirra (**Gemini 6–December 15, 1965**; **Apollo 7–October 11, 1968**) in space for a total of 9 hours and 13 minutes.

Sigma 7 was focused on perfecting the engineering of space flight rather than on performing scientific experiments. Schirra believed that these additional tasks had contributed to Scott Carpenter's re-entry problems on Aurora 7. He was, therefore, resolved to run a perfect mission. While in orbit, he ran numerous tests of the capsule's maneuvering rockets, using less fuel than either Carpenter on Aurora 7 or John Glenn on **Friendship 7 (February 20, 1962)**. He found that using visual references out his windows, in both day and night, worked best for steering the spacecraft.

The only scientific experiments performed involved medical research. In one test, Schirra closed his eyes and reached out to touch a sequence of selected points on his instrument panel. He found that even in zero gravity, his fingers could come within two inches of the intended spot.

During his re-entry, Schirra manually controlled the capsule's attitude, becoming the first astronaut to splashdown in the Pacific.

While Schirra showed no ill effects from his flight, postflight debriefing noted several short-term medical changes not observed with previous astronauts. For example, when Schirra went from a prone to a standing position, his heart rate suddenly rose from 70 to 100 beats per minute. Short-term changes in his blood pressure were also noted.

The next manned flight was **Faith 7 (May 15, 1963)**.

October 26, 1962
Starad
U.S.A.

Starad was launched by the U.S. Air Force specifically to study the artificial radiation belt created by the **Starfish Test (July 9, 1962)** several months after the explosion. Operating through January 1963, the spacecraft showed that the artificial radiation belt decayed very slowly over time.

October 27, 1962

Coincidentally, the Soviets detonated three high-altitude nuclear explosions on October 22nd, October 28th, and November 1st, enabling Starad to study three additional artificial belts as they formed. Less than one hour after the October 28th blast, Starad detected a sharp rise in background radiation. Over time, the satellite was able to make estimates of the blast's altitude, as well as to roughly describe the immediately expanding magnetic bubble of plasma after the explosion, followed by its eventual breakup and distribution into a belt of ionized particles trapped in the earth's magnetic field. See also **Explorer 15** (**October 27, 1962**).

October 27, 1962
Explorer 15
U.S.A.

Built by NASA, Explorer 15 measured the energy fluxes of the Van Allen belts, mapping their shape, size, and variability during its three months of operation. See **OGO 1** (**September 5, 1964**) for a summary of results through 1964.

Explorer 15 also mapped and tracked the slow decay of the artificial radiation belt formed by the **Starfish Test** (**July 9, 1962**), finding that four months after the explosion, about 1 percent of the electrons released were still trapped in the earth's magnetic field. The satellite also studied the three artificial belts created by Soviet high-altitude nuclear explosions set off on October 22, October 28, and November 1, 1962. Unlike the Starfish belts, the radiation belts created by the Soviet explosions decayed much more rapidly, for reasons that were not clear.

October 31, 1962
Anna 1B
U.S.A.

Anna 1B (which stands for Army, Navy, NASA, and Air Force) was the first dedicated geodetic satellite, built and launched by the United States military. Geodesy is the science of precisely measuring the earth's shape and size, and the accurate positioning of various surface landmarks relative to one another. It was hoped that Anna 1B would do this to a precision better than 25 feet, significantly more precise than previous earthbound measurements, which had an accuracy of between 50 to 1,000 feet. The satellite beamed a series of five very bright light flashes, each flash separated by 5.6 seconds. Researchers triangulated their position by tracking the flashes against the background of stars as the satellite orbited the earth. The position was confirmed by measuring the changing Doppler shift of the spacecraft's radio beacon as it moved.

Although the data from Anna 1B did allow some refinements to what was known about the overall shape of the earth, the difficulties in observing the light flashes made tracking difficult, and fewer measurements were made than had been expected.

Anna 1B also carried an experimental package for testing solar cells and transistors in orbit, similar to those carried on **TRAAC** (**November 15, 1961**). As with TRAAC, the experiment found that higher radiation levels caused a serious increase in solar cell degradation. See **38C** (**September 28, 1963**).

After the launch of Anna 1B, the geodetic research program was transferred from the Department of Defense to NASA, with **Explorer 22** (**October 10, 1964**) being the next civilian satellite. In the military, this research continued with **Secor 1** (**January 11, 1964**).

November 1, 1962
Mars 1
U.S.S.R.

This first attempt to send a spacecraft to Mars failed when, at 66 million miles, communications were lost. Along the way, however, Mars 1 provided data about the earth's magnetic field. The data indicated that the geomagnetic field extended out at least 8,000 miles from the earth's surface.

Mars 1 also measured the micrometeoroid flux in interplanetary space, finding that up to about 25,000 or 30,000 miles from the earth, the rate of impact was high, after which it decreased significantly. Mars 1 later passed through the Taurids meteor stream, registering about one hit every two minutes, and a second unidentified meteor stream with a similar impact rate.

Mars 1 also measured the solar wind, confirming results from **Mariner 2** (**August 27, 1962**). The next mission to Mars was **Mariner 4** (**November 28, 1964**). The next Soviet Mars mission was **Zond 2** (**November 30, 1964**).

December 13, 1962, 4:04 GMT
Injun 3 / Surcal 1A / Calsphere
U.S.A.

- *Injun 3*   Active for almost one year, this satellite, built by the U.S. Air Force and U.S. Navy, continued the detailed study of the Van Allen belts. Injun 3 also helped narrow the search for the unknown origin of the aurora, the curtain-like glowing lights seen in the Arctic and Antarctic skies, especially during periods of intense solar activity. Based on earlier satellite data from **Injun 1** (**June 29, 1961**)—Injun 2 had failed at launch—scientists knew that the Van Allen radiation belts produced insufficient energy to generate the aurora. Injun 3 found that electrons, not protons, were the dominant particle in all auroral events, and that these electrons were being accelerated downward along magnetic field lines in the higher latitudes above 600 miles elevation. Injun 3 also showed that, during large auroral events, the energy flux in the inner Van Allen belt increased rather than decreased, strong evidence that the energy source for the aurora came from beyond the earth's magnetic field. See **Elektron 3** (**July 11, 1964**) for further results.

- *Surcal 1A and Calsphere*   These two satellites were the first launches in a series of U.S. Navy–U.S. Army satellites used to calibrate their ground stations and satellite surveillance networks. Through 1971, almost two dozen of these small spherical satellites were placed in orbit, usually piggybacked on other spacecraft. The satellites, weighing from 2 to 100 pounds, were also used to study the drag of the atmosphere on each satellite's orbit.

December 13, 1962, 23:30 GMT
Relay 1
U.S.A.
Built by a commercial firm, the Radio Corporation of America (RCA), Relay 1 was NASA's first communications satellite. Its purpose was to test the technology needed to build practical and usable low-Earth-orbit communications satellites. Like **Telstar 1 (July 10, 1962)**, Relay 1 was an active repeater satellite; it was more complex, however, carrying two identical repeaters in case one failed. Each repeater's capacity included two narrowband channels and one wideband channel. Each wideband could handle 300 one-way voice conversations or one television signal. The narrowband's capacity was limited to 12 two-way telephone conversations because of limited ground equipment.

Because of power problems, Relay 1 transmitted its signals for only one day, relaying telephone and television signals from the Americas to Europe, and telephone signals from North America to South America. See **Telstar 2 (May 7, 1963)**, **Syncom 2 (July 26, 1963)**, and **Relay 2 (January 21, 1964)** for later communications satellites.

Relay 1 also carried sensors for studying the Van Allen belts. Its data indicated, as had **Injun 1 (June 29, 1961)**, that the energy flux of the inner belt was stable and unaffected by solar storms, while the outer belt fluctuated depending on solar magnetic activity.

December 14, 1962
Mariner 2, Venus Flyby
U.S.A.
Launched by NASA in August 1962, Mariner 2 flew past Venus on December 14, 1962, at a distance of about 21,580 miles, becoming the first probe to successfully send back data from another planet (see also **Mariner 2–August 27, 1962**).

The spacecraft carried three radiometers for measuring the temperature at the top and bottom of Venus's cloud cover and at the surface of the planet. These instruments proved that Venus is a hot planet, with surface temperatures exceeding 800°F, on both the day and night sides of the planet. At the top of the cloud cover, the temperature was –60°F, rising to –30°F at the cloud cover's bottom. This data suggested that the cold upper cloud cover hovered from 45 to 60 miles above the planet's surface.

The instruments also discovered a large area in the planet's southern hemisphere where the temperature was approximately 20°F colder. Earth-based radar also indicated a strong radar reflection in this same area. The data suggested a specific surface feature, possibly a very tall mountain or range of mountains.

At about 21,000 miles from the surface of Venus, the spacecraft measured no evidence of a magnetic field, indicating that the planet rotates slowly. Furthermore, Mariner 2 detected no significant change in the solar wind flux as it approached Venus, which would have happened if the planet had a magnetic field similar to Earth's.

The overall data indicated that Venus's day was very long and rotated in a retrograde direction, so that the Sun rose in the west and set in the east. Taking 243 days to complete one rotation, the retrograde direction, combined with the planet's orbit around the Sun, made each Venus day actually last about 117 days.

The next successful mission to Venus was Venera 4 (**Venera 4, Venus Landing–October 18, 1967**), which made the first direct probe of the Venusian atmosphere.

December 16, 1962
Explorer 16
U.S.A.
This flight was the third attempt by NASA to study the meteorite threat to manned and unmanned missions in low-Earth orbit (the previous two efforts had failed at launch). **Mariner 2 (August 27, 1962)** data had indicated that the micrometeorite flux in near-Earth space was 10,000 times greater than that in interplanetary space. NASA needed to know how dangerous this flux was in order to design safe spacecraft.

After seven months, Explorer 16's sensors recorded over 15,000 micrometeoroid impacts. An array of patches on the spacecraft's exterior, each patch of a different material and thickness, was punctured several dozen times. Although the thinnest patches were pierced, the micrometeorite particles could not get through the thickest test patches, stainless steel .005-inch thick. Based on this data, engineers could design spacecraft to withstand the degree of impact experienced by Explorer 16.

The next satellite for measuring micrometeoroid impact rates was **Explorer 23 (November 6, 1964)**.

1963

April 3, 1963
Explorer 17 (Atmosphere Explorer)
U.S.A.
Built by NASA, Explorer 17 studied the earth's atmosphere and the density of the ionosphere. Over a four-month period, scientists used the satellite to precisely map density changes in the upper atmosphere, finding that they corresponded closely to the Sun's 27-day rotation period and confirmed results from **Echo 1 (August 12, 1960)**. See **Explorer 19 (December 19, 1963)** for a summary of results.

The next American atmospheric research satellite was **Explorer 32 (May 25, 1966)**.

May 7, 1963
Telstar 2
U.S.A.
The world's second private satellite, Telstar 2, an active repeater satellite, was built and launched by AT&T, with NASA providing only the launch facilities. It was nearly identical to **Telstar 1 (July 10, 1962)**, except that its transistors were more resistant to radiation damage, and the orbit was higher so that the satellite was less exposed to Van Allen belt radiation. Telstar 2 operated for two years, relaying signals between North America and Europe, including portions of both the Democratic and Republican parties' presidential nominating conventions in 1964, as well as television programs featuring

politicians in France, the United Kingdom, and the United States.

Telstar 2 also carried instruments for studying the Van Allen radiation belts. It confirmed results from previous satellites (**Injun 1–June 29, 1961**; **Relay 1–December 13, 1962**) that the outer belt was directly affected by magnetic storms.

For complex political reasons, more than a decade would pass before private American commercial companies reentered the satellite communications business. Other than four tiny Oscar satellites built by amateur radio enthusiasts (**Oscar 3–March 9, 1965**; **Oscar 4–December 21, 1965**; **Oscar 5–January 23, 1970**; **Oscar 6–October 15, 1972**), Telstar 2 was the last wholly private communications satellite launched until Western Union launched **Westar 1 (April 13, 1974)**.

May 9, 1963
West Ford 2 / Dash 1
U.S.A.

- *West Ford 2* Following **West Ford 1 (August 21, 1961)**, West Ford 2 was the U.S. Air Force's second attempt to test a passive reflective satellite made from a belt of copper wires rather than a reflective balloon such as **Echo 1 (August 12, 1960)**. West Ford 2 consisted of 350 million copper needles that, when released, formed a cloud 9 miles by 18 miles across, off of which ground stations could bounce signals. During its first four months of operation, voice and other data were transmitted from California to Westford, Massachusetts (from which came the project's name). When the belt became fully extended, its density dropped, thereby reducing the amount of data that could be transmitted. These results, combined with the increasing success of active repeater satellites like **Telstar 2 (May 7, 1963)**, brought an end to the use of passive reflective satellites as communications tools in space.

- *Dash 1* This U.S. Air Force satellite was a 1-foot-diameter balloon designed to study atmospheric drag in low-Earth orbit. Dash (Density And Scale Height) 1 also carried an experiment to investigate the damage to solar cells from radiation in space. This satellite relayed data to Earth for three months. See **Dash 2 (October 5, 1966)**.

May 15, 1963
Faith 7 (Mercury 9)
U.S.A.

Faith 7 was the final Mercury mission, putting Gordon Cooper (**Gemini 5–August 21, 1965**) into space for 34 hours and 19 minutes, orbiting the earth 22 times. During his flight, he conducted several scientific experiments. In one, a xenon-gas discharge lamp was released from the capsule. Cooper was supposed to track the flashing light visually, testing a human's ability to identify such blinking lights against a space background. After releasing the lamp, Cooper failed to spot the light for one full orbit, but was later able to identify it easily as it drifted more than a dozen miles away.

To measure how the spacecraft shielded Cooper from radiation, a number of emulsion packs were distributed throughout the capsule, each capable of recording the radiation level in its area.

Cooper also took numerous photographs, including infrared cloud photos and colored pictures of the horizon. The astronaut was astonished at the identifiable detail on the earth's surface: "I could detect individual houses and streets in the low humidity and cloudless areas such as the Himalayan mountain area, the Tibetan plain, and the southwestern desert of the U.S. ... I saw what I took to be a vehicle along a road in the Himalayan area and in the Arizona–West Texas area. I could see the dust blowing off the road, then could see the road clearly, and when the light was right, an object that was probably a vehicle." That such details could be recognized carried great significance for later remote-sensing Earth satellites, beginning with **Landsat 1 (July 23, 1972)**. Cooper's sight was tested again on Gemini 5.

Cooper was the first American to sleep in space, sleeping fitfully for over eight hours. "I woke up one time from about an hour's nap with no idea where I was," he remarked later. "It took me several seconds to orient myself to where I was and what I was doing."

By his nineteenth orbit, Faith 7's automatic pilot had failed, and Cooper manually fired the retro-rockets and piloted the spacecraft to a pinpoint splashdown, only four miles from the *U.S.S. Kearsarge*. Though he recovered quickly, Cooper felt dizzy when he first exited the capsule. The medical data indicated that his cardiovascular system had adapted itself to weightlessness, therefore requiring an adjustment period once back on Earth.

The next American manned mission was **Gemini 3 (March 23, 1965)**. The next manned mission was **Vostok 5 (June 14, 1963)**.

May 22, 1963
Cosmos 17
U.S.S.R.

Using information acquired by Cosmos 17 combined with data from **Cosmos 3 (April 24, 1962)** and **Cosmos 5 (May 28, 1962)**, scientists were able to map the lower altitudes of the Van Allen radiation belts, discovering what is now called the South Atlantic anomaly. The Van Allen radiation belt makes its closest approach to the earth's surface over the south Atlantic Ocean near South America. Significantly higher radiation levels were measured as low as 200 miles elevation. This discovery has influenced the planning of all subsequent manned and unmanned missions, since the radiation in the Van Allen belts could be harmful to humans, as well as damaging to sensitive electronic equipment.

June 14, 1963
Vostok 5
U.S.S.R.

Flown in conjunction with **Vostok 6 (June 16, 1963)**, Vostok 5 was the first half of the Soviet Union's second dual space mission, following the paired flight of **Vostok 3 (August 11, 1962)** and **Vostok 4 (August 12, 1962)**. Valeri Bykovsky (**Soyuz 22–September 15, 1976**; **Soyuz 31–August 26, 1978**) orbited the earth for a record-setting 4 days, 23 hours, and 6 minutes.

Bykovsky's flight was originally planned to last eight days. However, liftoff was delayed repeatedly because of intense solar activity. The Soviets did not want to send the cosmonaut into space when the threat from radiation was high. Then, once launched, a premature separation of the Vostok capsule from its third stage placed the crew cabin in too low an orbit. The lower altitude plus the increased solar activity combined to increase the atmospheric drag on the capsule, causing its orbit to decay and shortening the mission. With additional problems in the capsule's waste management system, the decision was made to bring Bykovsky home early.

As with previous Soviet flights, Bykovsky's heartbeat, respiration, blood pressure, brain waves, and temperature were continually monitored.

June 15, 1963
Solrad 6 / Lofti 2A / Radose
U.S.A.

- *Solrad 6*   Solrad 6 was part of a long-term U.S. Navy project attempting continuous monitoring of the ultraviolet and x-ray output of the Sun, using standardized sensors (see **Solrad 1–June 22, 1960**). Solrad 6 gathered data for only about one month. The next Solrad satellite was **Solrad 7A (January 11, 1964)**.

- *Lofti 2A*   This U.S. military communications research satellite performed very low frequency (VLF) radio experiments. It carried two VLF receivers working in the 18-KHz waveband, testing how radio signals at this frequency propagated through the atmosphere.

- *Radose*   A joint U.S. Navy–U.S. Army satellite, Radose carried a radiation dosimeter into low-Earth orbit to make radiation measurements.

June 16, 1963
Vostok 6
U.S.S.R.

Vostok 6 carried Valentina Tereshkova, the first woman to fly in space. Like the previous Soviet dual mission, Vostok 6 and **Vostok 5 (June 14, 1963)** did not have the capability to rendezvous in space, although immediately after Vostok 6 reached orbit, the two spacecraft were only three miles apart (see also **Vostok 3–August 11, 1962** and **Vostok 4–August 12, 1962**).

For just under three days, Tereshkova orbited the earth in Vostok 6. Exactly what happened on her flight is unclear. According to some reports, her mission was originally planned to last one day but was extended to three because things were going so well. Other reports, however, indicate that Tereshkova experienced severe nausea and vertigo, although television broadcasts showed her looking well. Another Soviet description says that even though she always appeared to be in good spirits, her electroencephalogram readings on the second and third days indicated a slowing of her brain activity. Some descriptions of her flight say that she manually adjusted the capsule's orientation. Others say that she never did this—that ground control was dissatisfied with her performance.

Whatever happened on Vostok 6, Tereshkova never flew in space again. Five months later, in November 1963, she and cosmonaut Andrian Nikolayev (**Vostok 3–August 11, 1962; Soyuz 9–June 1, 1970**) married. The couple had their first child, a girl, in June 1964, the first human born of two cosmonauts. Their daughter grew up showing no side effects from her parents' experiences, demonstrating that, at least in the short run, space flight did no obvious harm to the human body.

Nineteen years would pass before another female cosmonaut would fly in space. See **Soyuz T7 (August 19, 1982)**.

Left to right, Andrian Nikolayev, Valentina Tereshkova, Pavel Popovich, and Sergei Korolev. Nikolayev and Popovich had flown on the dual flight of Vostok 3 (August 11, 1962) and Vostok 4 (August 12, 1962). Korolev was the chief designer of the Soviet space program. *RIA Novosti*

The next manned mission was **Voskhod 1** (October 12, 1964).

June 19, 1963
Tiros 7
U.S.A.

Tiros 7 was the last of the first-generation weather satellites launched by NASA. Over its almost 2.5-year life span, Tiros 7 returned over 150,000 cloud-cover images, the most pictures taken by any weather satellite to date. Its data, overlapping that of **Tiros 6 (September 18, 1962)**, allowed an almost continuous view of the world's cloud cover.

By 1963, scientists had used satellite data to track the birth of over 100 hurricanes and typhoons, as well as to forecast their movement, issuing more than 1,000 storm warnings. Meteorological theories predicting the origin of hurricanes were revised based on these new data.

The Tiros satellites (see **Tiros 1–April 1, 1960; Tiros 2–November 23, 1960; Tiros 3–July 12, 1961; Tiros 4–February 8, 1962; Tiros 5–June 19, 1962; Tiros 6**) also gave scientists their first view of the earth's overall weather patterns, allowing the development of more accurate models for predicting the weather. Tropical regions—where hurricanes are born, but for which weather data had been difficult if not impossible to gather—had become readily observable.

The photographs also allowed the mapping of the jet stream and other large-scale wind patterns invisible from the ground. For the first time, meteorologists saw how land features could change cloud patterns. Wave clouds and vortex disturbances were easily discernible in the lee of islands and mountain ranges.

Large cloud features that had been unknown from ground observations were also discovered by the Tiros satellites. Cell patterns, believed in 1963 to be related to convection cells, had simply been too large for ground-based observers to view.

This first generation of weather satellites also showed that satellite observations could unveil details about geological features. Land features in Africa, snow cover in the Alps and Rockies, and ice sheets in the St. Lawrence River were all observed in Tiros photographs.

Tiros 8 (December 21, 1963) tested the equipment for the next generation of weather satellites; see **Nimbus 1 (August 28, 1964)**.

June 26, 1963
Syncom 1
U.S.A.

Built by NASA, Syncom 1 was intended to be the world's first geosynchronous communications satellite. Although the initial communications tests were successful immediately after the satellite reached low-Earth orbit, the spacecraft was lost 20 seconds after firing its engines to lift it into geosynchronous orbit. See **Syncom 2 (July 26, 1963)**.

June 27, 1963
Hitch Hiker 1
U.S.A.

Placed in a polar orbit by the U.S. Air Force, Hitch Hiker 1 collected data about the Van Allen belts, including energy fluxes, spectra, and pitch angles (the direction the energized particles are traveling), confirming and refining data collected from **Injun 3 (December 13, 1962)** and **Relay 1 (December 13, 1962)**. During its three months of operation, the satellite noted that variations in the outer belt ranged across one to two orders of magnitude, correlating with both solar and geomagnetic activity. During solar flares, the intensity levels in the outer belt increased, quickly dropping to pre-storm levels during quiet periods of solar activity. Hitch Hiker 1 also recorded evidence that the **Starfish Test (July 9, 1962)** released significant energy into the Van Allen belt.

June 28, 1963
GRS
U.S.A.

The Geophysical Research Satellite (GRS) was designed and built by the Cambridge Research Laboratories for the U.S. Air Force to study the composition of the ionosphere between 260 and 745 miles altitude. Because of the failure of the spacecraft's power supply, only 13 orbits (less than one day) of data were accumulated.

July 12, 1963
KH-7 1
U.S.A.

KH-7 1 was the first successful launch in a series of U.S. Air Force photo-reconnaissance military satellites that replaced the Discoverer series (see **Discoverer 36–December 12, 1961**). The telephoto cameras on the KH-7s had a resolution of under five feet, and the satellites flew as low as 76 miles—about 25 miles lower than normal orbital altitudes—to obtain pinpoint photographs of foreign military installations. Over 90 KH-7s were launched through 1984; these satellites stayed in orbit for an average of 5.3 days. A later version, KH-8, carried a greater supply of fuel and was able to stay in orbit for a much longer period, averaging 50 days. The KH-8 cameras had a resolution as small as six inches. Both satellites also carried infrared and multispectral recording capabilities and four recoverable capsules for returning exposed film.

The KH-7 and KH-8 telephoto satellites worked in conjunction with the wide-angle KH-4, KH-4A, KH-4B, and KH-5 photo-reconnaissance satellites. If the panoramic pictures taken by a wide-angle camera showed anything that needed to be seen in greater detail, a KH-7 was launched to take higher resolution pictures of that area. See **KH-4 9032 (April 18, 1962)** for details on the wide-angle satellites.

The American telephoto KH-7 and KH-8 satellites were eventually replaced by the KH-11 series. See **KH-11 1 (December 19, 1976)**.

July 18, 1963
TRS 4
U.S.A.
This U.S. Air Force satellite carried technology experiments for testing how the trapped radiation of the Van Allen belts damaged a variety of different materials. This classified information was used in the design of later military satellites.

July 26, 1963
Syncom 2
U.S.A.
Syncom 2 was the world's first geosynchronous communications satellite. (**Syncom 1–June 26, 1963** had failed to reach its proper orbit.) Built and launched by NASA, Syncom 2 orbited the earth at an altitude of 22,302 miles and at a 32.7-degree inclination. At this altitude, a satellite will take one day to orbit the earth, thereby remaining visible 24 hours a day from the part of the earth over which it is positioned. Because Syncom 2's orbit was inclined 32.7 degrees, however, it did not remain stationary over the same point, but drifted north and south as the day passed. This inclined orbit required less weight and fuel to reach. To further save on fuel, the satellite was shaped like a drum and spin-stabilized—designed to operate as it spun on its side relative to the earth.

Syncom 2's broadcast capacity was similar to that of **Relay 1 (December 12, 1962)**. The satellite operated for over three years and was still functional as late as April 1969. During its life span, NASA used it to perform over 100 public demonstrations, as well as a large number of engineering tests. For example, the satellite telecast the 1963 International Telecommunications Union conference where the frequencies for satellite communications were first being allocated.

As a result of Syncom 2's success, **Syncom 3 (August 19, 1964)** was launched into stationary geosynchronous orbit.

September 28, 1963
38C
U.S.A.
A military surveillance satellite for measuring radiation levels over Soviet territory, 38C was also used to study the outer Van Allen belt. As had other satellites, 38C confirmed that solar activity had a significant effect on the energy flux and particle densities in the Van Allen belts. The satellite found that there was a measurable 27-day cycle in energy levels in the outer belt, closely matching solar wind data later returned from **Vela 2A/Vela 2B (July 17, 1964)**. These data further confirmed the effect of the solar wind on the shape of the earth's magnetic field. See **Imp A (November 27, 1963)**.

38C also carried a package similar to that on **Anna 1B (October 31, 1962)** for testing solar cells and transistors in orbit. See also **83C (December 13, 1964)**.

October 17, 1963
Vela 1A / Vela 1B
U.S.A.
Vela 1A and Vela 1B were the first two in a series of United States military satellites designed to detect atmospheric nuclear explosions that were in violation of the 1963 Atomic Test Ban Treaty. They carried x-ray, gamma-ray, and neutron particle detectors for measuring the radiation emitted from an atmospheric nuclear explosion.

The elliptical orbit of Vela 1A and Vela 1B, and their onboard equipment, also allowed these satellites to make the first measurements of the earth's magnetosphere at high latitudes, cutting through both the bow shock and transition zone (the shock wave and turbulence where the solar wind hits the magnetosphere, first seen by **Pioneer 5–March 11, 1960**; see also **Imp A–November 27, 1963**), as well as the interplanetary space beyond. See **Vela 2A/Vela 2B (July 17, 1964)** for the results from the first Vela cluster.

November 1, 1963
Polyot 1
U.S.S.R.
Polyot 1 was a test flight of anti-satellite (ASAT) interceptor control and propulsion systems. The spacecraft made over 350 firings of its engines, changing both its orbital inclination and apogee many times to test the ability of ground controllers to precisely maneuver a robot satellite. See **Polyot 2 (April 12, 1964)** for the next test flight.

November 27, 1963
Imp A (Explorer 18)
U.S.A.
In order to better understand the shape of the earth's magnetosphere, NASA's Imp (Interplanetary Monitoring Platform) satellites were given orbits that had them travel repeatedly in and out of the magnetic field.

Imp A's orbit, for example, was very eccentric, taking it just under four days to travel from a perigee of 119 miles to an apogee of 122,792 miles. This wide-ranging orbit was chosen based on data returned from previous satellites (see **Explorer 12–August 16, 1961**). It allowed the spacecraft to map the topography of the earth's magnetosphere and prove that the magnetic field was not symmetrical. During Imp A's first 47 orbits, data showed that as the solar wind collided with the magnetosphere, it pressed against the field and created a region of turbulence, called a bow shock. The magnetosphere was distorted, not unlike water impacted by the bow of a moving ship. In the sunward direction, the field was compressed, the bow shock region developing at about 40,000 miles from the earth's center. Away from the Sun, a "magnetic tail," or wake, trailed away for at least 120,000 miles, and probably millions of miles more (the tail's end was not measured). See **Mariner 4 (November 28, 1964)** for further tail measurements.

Imp A's travels through the earth's magnetosphere also recorded how the field is divided in half by a neutral sheet at the field's equator, with the magnetic lines of force of its northern and southern hemispheres charged in opposite directions. During its 1.5 years of operation, Imp A also demonstrated that as the Moon traveled through the earth's magnetic tail, it produced its own electromagnetic wake.

Imp A (which was also called Imp 1) was the first of 10 Imp satellites launched over 10 years, all focused on studying the earth's magnetosphere at high altitudes. All but one of these satellites were placed in high or eccentric Earth orbit extending at least halfway to the Moon, with **Imp E (July 19, 1967)** placed in high lunar orbit. These orbits were often complementary, covering different latitudes to enhance previously returned data. Imp 8 was unique in that it operated for almost 25 years, far exceeding its expected service life. All worked with the OGO satellites (see **OGO 1–September 5, 1964**).

Later Imp satellites provided scientists with detailed measurements of the solar wind's temperature and velocity as it sped past the earth. They found that as the solar wind approached the magnetosphere's bow shock, the wind formed waves. These waves were not unlike the waves that form on a beach—as the solar wind impacted the magnetosphere, its material piled up and then deflected sideways around the bow shock.

These satellites also found evidence of atomic oxygen in the magnetotail. This oxygen was thought to be accelerated outward from near the earth's poles along magnetic field lines. What caused this outflow, however, was unclear.

For a summary of additional results, see **OGO 5 (March 4, 1968)** and **OGO 6 (May 5, 1969)**.

In 1967, solar radiation sensors on the Imp satellites were also used to check the accuracy of data being collected by **Surveyor 3 (April 17, 1967)** on the lunar surface, as well as to monitor the Sun's activity prior to the **Apollo 8 (December 21, 1968)**, **Apollo 9 (March 3, 1969)**, and **Apollo 11 (July 16, 1969)** missions, in order to insure the safety of the astronauts.

December 19, 1963
Explorer 19
U.S.A.

This NASA spacecraft was identical to **Explorer 9 (February 16, 1961)**: a 12-foot-diameter balloon for studying atmospheric density variations at orbital elevations. Because the first satellite was still active, Explorer 19 was placed in a different orbit to obtain comparison measurements. It remained in orbit until May 1981.

By 1964, scientists were able to combine the data from these two satellites, as well as from **Echo 1 (August 12, 1960)**, and reach some general conclusions about the changing density of the earth's atmosphere over time. At altitudes of 125 to 500 miles, the atmospheric density and temperature were found to be strongly influenced by either Earth's day-night cycle or by the Sun's varying surface activity.

The day-night effect was measured as these satellites circled the earth. On the sunlight side, the density reached a sharp peak around 2:00 PM, then dropped to a flat minimum between midnight and dawn. The highest density increase was seen at the equator and fell off steadily in higher latitudes. The density variation also increased with elevation: the higher the elevation, the greater the variation.

The influence of solar activity on atmospheric density was found to be potentially greater than the day-night cycle (a revision of conclusions drawn from Echo 1's data). At 375 miles elevation, the atmospheric density dropped by 32 times as the Sun approached its solar minimum. The Sun's 27-day rotational period also caused the atmosphere to bulge and shrink, as did specific solar flares and magnetic storms. The largest magnetic storms caused density to fluctuate by a factor of six.

The overall temperature of the upper atmosphere also fluctuated as the Sun's cycle went from maximum to minimum, dropping from a daytime average of 3,100°F to about 1,100°F. Nighttime temperatures were 25 percent less. Individual magnetic storms also caused temperature increases of as much as 900°F.

Scientists theorized that these variations were directly caused by the heating effect of extreme ultraviolet radiation at wavelengths between 200 and 1,000 angstroms. This radiation, which appeared to vary significantly with solar activity, was absorbed by the atmosphere at about 60 to 120 miles elevation, and hence influenced the atmosphere's temperature and density directly.

One other small seasonal variation in density seemed directly connected with the earth's orbit around the Sun.

The next atmospheric density balloon satellite, **Explorer 24 (November 21, 1964)**, operated as the solar cycle went from minimum to maximum.

December 21, 1963
Tiros 8
U.S.A.

Though very similar to NASA's seven previous Tiros satellites (see **Tiros 7–June 19, 1963**), Tiros 8's second camera system was a more advanced design developed for use on the second-generation weather satellite **Nimbus 1 (August 28, 1964)**. While the images from previous Tiros satellites could be processed only by very sophisticated centralized facilities, photographs taken on this camera system could be downloaded at low cost by local weather bureaus. Though the satellite's orbit prevented it from viewing much of the earth's surface and therefore limited the system's usefulness, Tiros 8 demonstrated its feasibility.

Tiros 8 was operational for more than 3.5 years and returned more than 100,000 cloud-cover images. It was followed in 1965 by two similar Tiros satellites, after which the program was replaced by the Nimbus, ESSA, and NOAA satellites. See Nimbus 1, **ESSA 1 (February 3, 1966)**, and **NOAA 1 (December 11, 1970)**.

1964

January 11, 1964
GGSE 1 / Secor 1 / Solrad 7A
U.S.A.

• *GGSE 1* This small U.S. Navy satellite was the first test of the use of long extendable rods for orienting spacecraft. GGSE (Gravity Gradient Stabilization Experiment) 1 had one pole, which when lengthened or retracted changed the spacecraft's center of gravity, causing its spin rate and orientation relative to the earth to change as well. With one pole, this satellite could orient itself in only two dimensions. **GGSE 3 (March 9, 1965)** tested a three-axis system.

• *Secor 1* This satellite was the first of 11 geodetic satellites launched by the U.S. Army over five years to fix precisely the locations of a number of points on the earth (such as the location of several Pacific Islands) and tie these points to the topographic survey of the United States.

- *Solrad 7A*   As part of a U.S. Navy project to study the Sun (see **Solrad 1–June 22, 1960**), Solrad 7A's sensors observed solar radiation from 2 to 60 angstroms, gathering data until September 1964 in the extreme ultraviolet and soft x-ray wavelengths. Because the satellite had no onboard attitude control system, it could only collect data for about 200 minutes per day. It showed that the waveband from 44 to 60 angstroms was the most sensitive for measuring solar x-ray events.

The spacecraft found that during this quiet period of the Sun's solar cycle, the x-ray flux was very low, matching measurements from earlier sounding-rocket flights during a previous quiet period in 1953. Yet the data also showed that following solar flares, the x-ray flux increased, suggesting that the flares were the source of the x-ray emissions.

The next Solrad satellite was **Explorer 30 (November 19, 1965)**.

January 21, 1964
Relay 2
U.S.A.

A follow-up to NASA's **Relay 1 (December 13, 1962)**, Relay 2 was similar in design, although its equipment was modified to be more reliable and resistant to radiation. The satellite operated until May 1965 and was used by seven countries on four continents for tests of television and telephone transmissions.

Relay 2 also continued the study of the slow decay of the artificial radiation belt created by the **Starfish Test (July 9, 1962)**.

January 25, 1964
Echo 2
U.S.A.

Built by NASA, Echo 2 was a larger version of **Echo 1 (August 12, 1960)**. Once in orbit, this passive reflector satellite expanded to a 135-foot balloon weighing 500 pounds. Like Echo 1, Echo 2 merely reflected any radio transmissions beamed at it.

Echo 2 was the first satellite used jointly by the United States and the Soviet Union. As the satellite expanded into a balloon, Soviet observers photographed it and relayed this information to the United States. Later joint studies of the satellite's brightness and motion through space contributed to geodetic and atmospheric density research. Soviet scientists also measured the satellite's radio transmissions during the spring of 1964 to make studies of the earth's shape.

Echo 2 re-entered the atmosphere on June 7, 1969. All subsequent communications satellites, except for a few military test satellites (see **OV1-8–July 14, 1966**), have been active repeaters rather than passive reflectors.

January 30, 1964
Elektron 1 / Elektron 2, 9:36 GMT
U.S.S.R.

These two Soviet research satellites were launched to study the Van Allen radiation belts and the earth's magnetosphere. They were the first dual Soviet launch, taking place four years after **Transit 2A/Solrad 1 (June 22, 1960)**. When released, the two satellites were placed into different but complementary orbits, Elektron 1 ranging from 250 to 4,000 miles altitude and Elektron 2 ranging from 3,500 to 39,000 miles. Elektron 1 studied the inner radiation belt and Elektron 2 the outer. Data from the two satellites were later supplemented by information gathered by **Elektron 3/Elektron 4 (July 11, 1964)**; see that entry for a summary of results.

Elektron 2 made detailed measurements of the Sun's x-ray output for two months and showed that the output varied considerably over short daily time periods. These variations correlated well with other variations of solar activity, including the 27-day solar rotation. The overall flux, however, was considerably less than that seen on **Sputnik 5 (August 19, 1960)** and **Sputnik 6 (December 1, 1960)**, indicating that the Sun's 11-year cycle also affected the amount of x-ray radiation it emitted. This data also confirmed observations made by the **OSO 1 (March 7, 1962)** and **Ariel 1 (April 26, 1962)** satellites.

January 30, 1964, 15:49 GMT
Ranger 6
U.S.A.

Built by NASA, Ranger 6 was the third spacecraft, following **Luna 2 (September 12, 1959)** and **Ranger 4 (April 23, 1962)**, to impact another world, hitting the Moon on February 2, 1964, in the Sea of Tranquility. However, because its television camera failed, Ranger 6 returned no pictures.

The fifth American attempt to impact and photograph the lunar surface was made by **Ranger 7 (July 28, 1964)**.

March 18, 1964
Cosmos 26
U.S.S.R.

This satellite helped map the earth's magnetic field, remaining in orbit until September 1964. See **Cosmos 49 (October 24, 1964)** for results.

March 27, 1964
Ariel 2
U.K. / U.S.A.

Ariel 2 was the second joint United States–United Kingdom satellite, following **Ariel 1 (April 26, 1962)**. NASA supplied the satellite body and launch facilities, while British scientists built the experiments. The satellite operated until November 1964.

The satellite carried three sensors. One measured galactic radio noise known to radiate from faint celestial objects. A second recorded micrometeorite hits in low-Earth orbit. The third charted the atmosphere's ozone profile. This last sensor showed that the atmospheric ozone distribution through the middle and equatorial latitudes was almost constant from 25 to 40 miles elevation. Practically no variations were measured.

The next British satellite, **Ariel 3 (May 5, 1967)**, was built entirely in the United Kingdom.

April 2, 1964
Zond 1
U.S.S.R.

Intended to carry out a flyby of Venus, communications with Zond 1 were lost on May 14, 1964, preventing the return of any significant data. The spacecraft flew past Venus at a distance of about 60,000 miles and entered solar orbit.

April 12, 1964
Polyot 2
U.S.S.R.

This second Soviet test flight of an anti-satellite (ASAT) interceptor, following **Polyot 1 (November 1, 1963)**, made extensive maneuvers during its first day in orbit, including large changes to its orbital inclination and apogee. These changes were tests to see how precisely a robot satellite could be controlled from the ground—a question directly related to the development of the rendezvous and docking systems to be used by the **Soyuz 1 (April 23, 1967)** manned spacecraft.

July 11, 1964
Elektron 3 / Elektron 4
U.S.S.R.

Like **Elektron 1/Elektron 2 (January 30, 1964)**, these two Soviet satellites were launched together and then placed into different orbits in order to study the Van Allen radiation belts and the earth's magnetosphere. Elektron 3's orbit ranged from 250 to 4,000 miles, while Elektron 4 traveled from 275 to 41,000 miles.

Elektron 3 joined Elektron 1 in studying the inner Van Allen belt, accumulating data that agreed with results from **Explorer 12 (August 16, 1961)**. Both satellites also measured large energy fluxes and the acceleration of particles above the southern auroral zone at about 4,000 miles elevation. Since the energy source of the aurora was unknown, and since the radiation in Van Allen belts alone could not have created it, these data, combined with other data from **Injun 3 (December 13, 1962)**, **Mariner 2 (August 27, 1962)**, and a number of other satellites, indicated that some complex interaction of the solar wind and the earth's magnetic field was involved. During solar flares and high solar activity, auroral events increased. Ionized particles flowing outward from the Sun hit the earth's magnetosphere at 40,000 miles elevation, somehow causing the aurora at 60 miles elevation.

Elektron 4, with Elektron 2, studied the outer Van Allen belt at high latitudes. Data suggested that the outer belt electrons found by Explorer 12 were not permanently trapped radiation as previously thought, but transient particles drifting in from the edge of the magnetosphere and geomagnetic tail. These two satellites also recorded evidence of both low- and high-energy electrons in the outer belt.

The two satellites confirmed that several different solar events cause variations in the outer belt. During magnetic storms, the belt was pushed inward. Long-term changes in the size of the Van Allen belts also took place in conjunction with the Sun's 11-year cycle. Since the late 1950s and earlier 1960s, the belts had expanded as the Sun's activity declined. See Explorer 12.

Elektron 2 and Elektron 4 also found, as did the Vela satellites (see **Vela 2A/Vela 2B–July 17, 1964**), that the outer perimeter of the magnetosphere at high latitudes moved in during the day and out at night, indicating that the solar wind was pressing against it and deforming it.

The next satellite to study the earth's magnetosphere, **OGO 1 (September 5, 1964)**, began a U.S. six-satellite campaign to systematically map its environment.

July 17, 1964
Vela 2A / Vela 2B
U.S.A.

Designed by the U.S. Air Force to uncover secret nuclear weapons tests by detecting their energy release, Vela 2A and Vela 2B followed **Vela 1A/Vela 1B (October 17, 1963)**, creating a four-satellite cluster orbiting at high altitude (over 62,000 miles above the earth) to cover as much of the earth's surface as possible. Two more Vela satellites were added to the cluster in 1965.

Each satellite's nearly circular orbit took it across previously unexplored parts of the earth's magnetosphere, as well as through the transition zone and bow shock separating the magnetosphere from interplanetary space. When in interplanetary space, the Vela satellites could also provide detailed information about the solar wind.

Measurements from the first four Vela spacecraft noted, as had **Mariner 2 (August 27, 1962)**, that the solar wind speed fluctuated between a base velocity of about 200 miles per second to over 400 miles per second. These peaks now also appeared to take place at recurring 27-day intervals, closely corresponding to data from satellite **38C (September 28, 1963)**. **Mariner 4 (November 28, 1964)** confirmed these fluctuations. Although the explanation was unknown, the cycle did correspond closely with the Sun's rotation.

In crossing the bow shock and transition zone between interplanetary space and the earth's magnetosphere, the Vela spacecraft also supplemented data from **Imp A (November 27, 1963)**, proving unequivocally the existence of the bow shock by measuring sudden and significant changes in the detected solar wind particles, some of these changes taking place in less than two seconds. As the solar wind hit the bow shock, the particles moved in a more random pattern, creating turbulence and temperature increases. The particles then aligned themselves with magnetic lines of force to stream around the magnetosphere.

In October, the Vela spacecraft, along with **OGO 1 (September 5, 1964)** and the second Imp satellite (see **Imp A–November 27, 1963**), measured the magnetopause from widely different points in space and found its properties to be homogeneous wherever observed—similar in intensity, energy flux, and magnetic structure across large distances.

Vela instruments also found additional evidence of a dawn-to-dusk asymmetry in the shape of the earth's magnetic field, further proof of the solar wind's effect upon it. This asymmetry was also found to be skewed about 3 to 5 degrees from the ecliptic plane, caused by the earth's orbit around the Sun.

Through 1970, the U.S. military launched eight additional Vela satellites. The last four, beginning with **Vela 5A/Vela 5B (May 23, 1969)**, carried improved data storage and command capabilities, meaning that each satellite was able to gather more data.

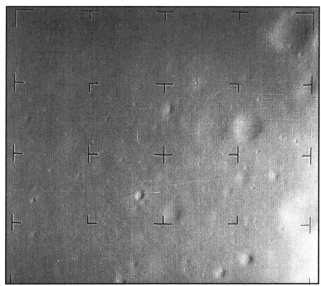
The last image taken by Ranger 7 prior to impact. *NASA*

July 28, 1964
Ranger 7
U.S.A.

The fourth NASA spacecraft to hit the Moon (the previous one being **Ranger 6–January 30, 1964**), Ranger 7 took the first close-up pictures of the lunar surface, returning 4,316 photos before crashing near the Sea of Clouds in one of the rays emanating from the crater Tycho.

Scientists immediately noted how the features on the Moon remained constant as the scale decreased—the number and shape of the craters did not change as Ranger 7 approached the surface. They also noted the eroded nature of the photographed features. As project scientist Gerard Kuiper noted, "The rounded features of the large number of secondary craters are new. Also their large number." The photographs also indicated that the surface where Ranger 7 hit was not covered with a deep layer of dust, which suggested that the design of the Apollo program's lunar lander was practical (see **Apollo 9–March 3, 1969**). This conclusion, however, remained disputed until both **Luna 9 (January 31, 1966)** and **Surveyor 1 (May 30, 1966)** soft-landed on the Moon.

Ranger 7's pictures did not solve the mystery of whether the Moon's craters were formed by volcanoes or by impact, although some scientists argued that the photographs showed the lunar mares to be lava flows. Astrogeologist Eugene Shoemaker noted that the ejecta from the Tycho Crater possibly made Ranger 7's impact area atypical for the Moon, and also that the photos suggested the lunar surface was constantly being "sand blasted" by high-speed particles from space.

The next Ranger flight was **Ranger 8 (February 17, 1965)**.

August 14, 1964
KH-7-10
U.S.A.

This Keyhole surveillance satellite, in orbit for only nine days, was one of the few American military surveillance satellites to carry scientific experiments after 1961. KH-7-10 carried instruments for studying the Van Allen belts, and while in orbit, it measured both this and the dissipating radiation belt that had been created by the **Starfish Test (July 9, 1962)**.

August 19, 1964
Syncom 3
U.S.A.

Syncom 3 was the world's first "stationary" geosynchronous satellite. Orbiting the earth at an altitude of 22,238 miles and an inclination of 0 degrees, from the perspective of an Earth-based observer, Syncom 3 remained stationary in the sky. It was positioned over the Pacific.

Like its predecessor **Syncom 2 (July 26, 1963)**, Syncom 3's transmission capacity was similar to that of **Relay 1 (December 13, 1962)** and **Relay 2 (January 21, 1964)**. The satellite was used by NASA to test geosynchronous communications satellite technology, as well as to perform a number of public communications demonstrations, the most famous of which was the live broadcast of portions of the 1964 Summer Olympics from Japan to the United States and Europe.

The satellite functioned until May 1965. It was followed by **Early Bird (April 6, 1965)**, the first operational geosynchronous communications satellite.

August 22, 1964
Cosmos 41
U.S.S.R.

Cosmos 41 was the first launch of a prototype Soviet communications satellite, testing the design of the Molniya-1 satellite (see **Molniya 1-01–April 23, 1965**). Its highly eccentric orbit (perigee: 636 miles; apogee: 24,338 miles), with an inclination of 68.4 degrees, became the standard orbit for most Soviet communications satellites, since it provided better coverage of the Soviet Union than a geosynchronous orbit.

The satellite also carried instruments for studying the Van Allen belts, finding a correlation between the belts' intensity and magnetic storms.

August 25, 1964
Explorer 20
U.S.A.

Built by NASA, Explorer 20 was the first American satellite dedicated to ionospheric research since **Explorer 8 (November 3, 1960)**. It continued the work done by **Ariel 1 (April 26, 1962)** and **Alouette 1 (September 29, 1962)**, operating until July 1966. Explorer 20 transmitted radio signals across six frequencies from 1.5 to 7.22 MHz, deducing ionospheric conditions by how each signal was modified as it was propagated through the atmosphere. While Alouette 1 had gathered vertical readings, Explorer 20 collected horizontal data, studying the ionosphere's small irregularities and fine structure. The propagation data was also used to develop communications satellite technology, determining what frequencies penetrated the ionosphere with the least interference. Later communications satellites, such as **Early Bird (April 6, 1965)**, were designed based upon this research.

The next ionospheric research satellites were **Explorer 31/Alouette 2 (November 29, 1965)**.

August 28, 1964, 8:36 GMT
Nimbus 1
U.S.A.

Nimbus 1 was the first second-generation NASA weather satellite, following the Tiros satellites (see **Tiros 1–April 1, 1960** and **Tiros 8–December 21, 1963**). While Tiros satellites could take usable photographs for only about one-third of each orbit and could not observe higher than 65 degrees latitude, Nimbus 1's orbit was intended to enable it to observe the earth's entire surface each day. Such an orbit, called a sun-synchronous polar orbit, was inclined to the equator 98 degrees with a period lasting about 98 minutes. From this perch, a different strip of surface would come in view with each orbit as the earth rotated beneath the spacecraft. The orbit's length also matched the earth's rotation rate, so that "local time" on the earth surface below the satellite was always the same. Thus, the Sun's angle and surface shadows remained consistent for all images.

Because the rocket's second stage cut off early during launch, however, Nimbus 1's orbit was more eccentric than intended, preventing it from taking full advantage of this orbit to record the first global picture of the earth's cloud cover within a 24-hour period. **Tiros 9 (January 22, 1965)**, the next American civilian weather satellite, would accomplish this task.

Nonetheless, Nimbus 1 returned over 27,000 cloud-cover images through September 1964, using two camera systems significantly more advanced than those on the first seven Tiros satellites. Like Tiros 8, Nimbus 1 carried one camera whose pictures could be quickly downloaded by local weather bureaus. These pictures covered an area approximately 1,000 miles on a side.

Nimbus 1's second camera system—a three-camera array with a center camera and two side cameras angled outward at 35 degrees, creating a total field of 107 degrees—was designed to optimize use of the sun-synchronous orbit so that the entire Earth surface could be photographed each day.

The satellite also carried a more advanced array of infrared sensors for mapping the earth's cloud cover at night. The cloud-cover infrared pictures were almost as detailed as the daytime television cameras, demonstrating that round-the-clock study of the earth's weather was possible. For example, these sensors discovered Typhoon Ruby one day before it was found by Earth-based reconnaissance.

Using this basic design, an additional six Nimbus satellites and two Tiros satellites were launched over the next decade. Also using this design were the ESSA satellites (see **ESSA 1–February 3, 1966**), operated by the Environmental Science and Services Administration (the forerunner of the National Oceanic and Atmospheric Administration), and the DMSP satellites (see **DMSP Block-4A F1–January 19, 1965**). All of these satellites were placed in sun-synchronous polar orbits, and all provided the first regular and complete cloud-cover imagery of the earth.

While the Tiros and ESSA satellites provided routine data for local weather bureaus, NASA's Nimbus satellites tested new instruments for the long-term study of the earth's global climate. From April 1969 through 1993, these Nimbus satellites maintained a continuous series of Earth meteorological observations. By the end of the Nimbus program in the early 1990s, the most useful Nimbus instruments were also being used on other satellites. See **Nimbus 7 (October 13, 1978)** for a program summary.

August 28, 1964, 16:19 GMT
Cosmos 44
U.S.S.R.

Cosmos 44 is believed to have been the first Soviet weather satellite. The Soviets launched two more test missions in 1965 and 1966 before releasing any information about their weather satellite program. See **Cosmos 122 (June 25, 1966)**.

September 5, 1964
OGO 1
U.S.A.

By the launch of OGO 1, the first satellite in NASA's Orbiting Geophysical Observatory (OGO) series, scientists had more than a half decade of data about the interaction of the solar wind and the earth's magnetosphere, gathered from a plethora of satellites (see **Explorer 4–July 6, 1958; Explorer 7–December 13, 1959; Injun 1–June 29, 1961; Explorer 12–August 16, 1961; Ariel 1–April 26, 1962; Alouette–September 29, 1962; Explorer 14–October 2, 1962; Explorer 15–October 27, 1962; Injun 3–December 13, 1962; Elektron 1/Elektron 2–January 30, 1964; Ariel 2–March 27, 1964; Explorer 25–November 25, 1964**) and covering periods of both high (1961) and low (1965) solar activity. They had found significant long-term changes in energy levels for the outer Van Allen belt during this period, with an decrease in higher energy protons and an increase in lower energy protons as the Sun became less active.

By 1964, the general structure of the magnetosphere and how the solar wind affected it were well determined, though many specific questions remained. The source of the aurora was still unclear. Several theoretical models of the magnetosphere existed to explain the data, though none worked perfectly.

Thus, magnetosphere research was now being planned with this existing knowledge in mind. OGO 1 was the first in a six-spacecraft program dedicated to a systematic study of the

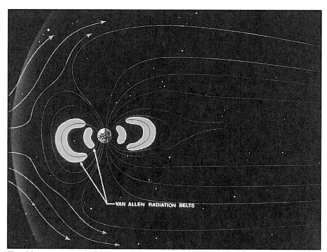

Earth's magnetic field as understood in 1964. *NASA*

earth's magnetosphere and the Van Allen belts. These six satellites, OGO 1 through OGO 6, were launched on a yearly basis from 1964 through 1969. OGO 1, 3, and 5 traveled in high-apogee eccentric orbits so as to observe the outer perimeters of the magnetosphere, while OGO 2, 4, and 6 traveled in low polar orbits to observe the inner magnetosphere at high latitudes. The goal of all six was to maintain continuous observations of the magnetosphere as the Sun went from its quiescent stage in 1964 to its most active stage in 1969.

With the exception of the Vela satellites (see **Vela 2A/Vela 2B–July 17, 1964**) and a handful of others, the bulk of American magnetosphere research for the next six years took place on the OGO and Imp satellites (see **Imp A–November 27, 1963**), their orbits carefully coordinated to cover as much of the magnetosphere as possible.

Because OGO 1's attitude control system failed, about half of its instruments could not get usable data. Nonetheless, OGO 1 was able to complete a number of its experiments, refining what was known about the magnetosphere and Van Allen belts. In conjunction with Vela 2A/Vela 2B and the second Imp satellite, OGO 1 measured the intensity and energy flux of the magnetopause over large distances, proving that the magnetopause was homogeneous; similar wherever observed.

See **OGO 5 (March 4, 1968)** and **OGO 6 (May 5, 1969)** for a summary of results.

October 10, 1964
Explorer 22 (Beacon Explorer B)
U.S.A.

Built by NASA, Explorer 22 had two scientific objectives: (1) to study how radio signals propagate through the ionosphere at high latitudes (between 550 to 675 miles elevation); and (2) to better determine the size and shape of the earth and to locate precisely the position of various surface points relative to each other. This second task was a follow-up of the geodetic work done on **Anna 1B (October 31, 1962)**.

In both experiments, ground researchers listened to the spacecraft's radio beacon. By measuring how that signal was changed, or propagated, as it traveled through the ionosphere, they obtained a vertical profile of the ionosphere's total electron content, for that time and place, while also mapping the size, shape, and location of the earth's surface by calculating changes in the satellite's orbit.

An identical satellite, **Explorer 27 (April 29, 1965)**, provided NASA with a second set of orbital data.

October 12, 1964
Voskhod 1
U.S.S.R.

Voskhod 1 was the first spacecraft to carry more than one person. To outdo the American two-man Gemini program (see **Gemini 3–March 23, 1965**), Premier Nikita Khrushchev ordered the Soviet space program to quickly achieve two space firsts: put a multi-person capsule in space and complete the first space walk (**Voskhod 2–March 18, 1965**). To build a three-man capsule quickly, Soviet engineers took the Vostok capsule (**Vostok 1–April 12, 1961**) and modified it. The seats were placed sideways so that the crew sat perpendicular to the instrument panel during launch, the crew members wore no spacesuits, the ejector seats were removed, and the rocket's escape tower was eliminated.

During Voskhod 1's 24-hour, 17-minute flight, Konstantin Feokistov (who was not a cosmonaut but one of the spacecraft's designers), test pilot Vladimir Komarov (**Soyuz 1–April 23, 1967**), and physician Boris Yegorov orbited the earth just under 17 times. Although no detailed reports were published, Yegorov was said to have taken the first in-space blood samples from his crewmates. The cosmonauts were also reported to have seen John Glenn's "fireflies" (see **Friendship 7–February 20, 1962**).

For re-entry, Voskhod 1 differed from Vostok in two ways. First, an additional solid retro-rocket package was mounted on the crew cabin in case the primary retro-rockets failed. Second, another set of rocket engines was added to slow the capsule as it made its final descent to Earth. Earlier Vostok cosmonauts had ejected from the capsule at about four miles altitude and parachuted to the ground; with three men onboard, however, this was impossible. Just prior to Voskhod 1's landing, the additional rockets fired, slowing the capsule's descent and making it possible for the cosmonauts to remain safely inside as the spacecraft hit the ground.

In accordance with Khrushchev's orders, the next and last Voshkod flight (Voskhod 2) included the world's first space walk. Khrushchev, however, was removed from office even before the three cosmonauts on Voskhod 1 returned to Earth. His removal was, in part, a result of his insistence that the Soviet space program fly such risky missions as Voskhod 1. After the flight of Voskhod 2, the Voskhod program was abandoned, and the Soviet space program was restructured around the Soyuz spacecraft.

October 24, 1964
Cosmos 49
U.S.S.R.

This satellite, along with **Cosmos 26 (March 18, 1964)**, made the first continuous maps of the earth's magnetic field, covering 75 percent of the earth's surface at an altitude of about 200 miles. See **OGO 2 (October 14, 1965)** and **OGO 4 (July 28, 1967)**.

November 6, 1964
Explorer 23
U.S.A.

A follow-up to **Explorer 16 (December 16, 1962)**, Explorer 23—also built by NASA—gathered micrometeoroid data for a period of one year. Its data confirmed the micrometeorite impact rate found on the earlier spacecraft and indicated that the protective aluminum skin of Earth-orbiting spacecraft needed to be about 0.1 inch thick to resist almost all impacts.

The next American satellite to measure the micrometeorite impact rate in low-Earth orbit was **Pegasus 1 (February 16, 1965)**.

November 21, 1964
Explorer 24 / Explorer 25 (Injun 4)
U.S.A.

• *Explorer 24*   Built by NASA and identical to **Explorer 9 (February 16, 1961)** and **Explorer 19 (December 19, 1963)**, Explorer 24 was a 12-foot-diameter balloon whose orbital changes were studied to map the density changes in the upper atmosphere over time. Based on satellite studies from 1957 to 1964, the effect of the solar cycle from maximum to minimum had been examined. Explorer 24 was intended to follow the atmosphere's changes as the Sun returned to its maximum in 1968–69. The satellite, which re-entered the atmosphere on October 18, 1968, confirmed that as the Sun's activity increased, the upper atmosphere expanded, increasing its density and thereby increasing the drag on satellites and causing their orbits to lose altitude. The next and last balloon satellite in this program was **Explorer 39 (August 8, 1968)**. The next balloon satellite was **Pageos 1 (June 24, 1966)**.

• *Explorer 25 (Injun 4)*   This NASA satellite was launched to conduct a continuous study of the Van Allen radiation belts (see **Explorer 10–March 25, 1961**; **Explorer 15–October 27, 1962**) as the Sun went through its 11-year solar cycle. See **OGO 1 (September 5, 1964)** for a summary of what had been discovered through 1964. The satellite also made continuing measurements of the artificial radiation belt created by the **Starfish Test (July 9, 1962)**, tracking its steady dissipation during its more than two years of operation.

November 28, 1964
Mariner 4
U.S.A.

Built by NASA, Mariner 4 was the third attempt to send a probe to Mars and the first to be successful; Mariner 3 had failed at launch, while **Mars 1 (November 1, 1962)** failed before reaching Mars.

On its journey to Mars, Mariner 4 used several instruments to study the solar wind. It found that wind speed, direction, and density values were comparable to those measured on **Mariner 2 (August 27, 1962)**, **Imp A (November 27, 1963)**, and the Vela spacecraft (see **Vela 2A/Vela 2B–July 17, 1964**), with the solar wind speed showing abrupt, repetitive, and unexplained fluctuations. A correlation between wind bursts and solar flares explained only some of these oscillations. See **OGO 5 (March 4, 1968)** for additional findings.

Measurements of high-energy particles indicated that during the quiet period of the Sun's solar cycle (which corresponded to the Mariner 4 mission), cosmic ray activity from outside the solar system increased. Scientists theorized that when the Sun was more active, it tended to block interstellar cosmic rays from entering the solar system.

Data from Mariner 4 also identified a limit on the length of the earth's magnetic tail, first mapped by Imp A. When Mariner 4 was about 13 million miles from the earth, it passed through the predicted location of this tail but recorded no evidence of its existence. Mariner 4 also helped confirm the size and shape of the earth's Van Allen belts as measured by Imp A.

Mariner 4 carried a cosmic dust experiment, similar to that carried by Mariner 2. Mariner 2 had found that the space between Earth and Venus was practically devoid of dust particles—0.0001 of what was measured near Earth. Because Mariner 4 was approaching the asteroid belt, and because it would also pass close to two known meteoroid streams, scientists had expected the cosmic dust flux toward Mars to become greater. The results, however, were almost identical to what had been seen by Mariner 2 until about 1.2 astronomical units (AU). Beyond this distance, the flux steadily increased, reaching its maximum as the spacecraft passed 1.38 AU (equal to Mars's perihelion). Scientists attributed the subsequent drop to Mars sweeping clear its orbital path. Moreover, the flux did not increase as the spacecraft swung past Mars, a result unlike Earth's, but similar to Venus's.

Mariner 4 reached Mars seven and a half months later. See **Mariner 4, Mars Flyby (July 15, 1965)**.

November 30, 1964
Zond 2
U.S.S.R.

Zond 2 was the second Soviet probe to Mars, following **Mars 1 (November 1, 1962)**. Like Mars 1, Zond 2 failed when communications were lost five months into the flight. The spacecraft flew past Mars on August 6, 1965, at a distance of about 1,000 miles.

Before contact was lost, the spacecraft sent back solar wind data, confirming results from **Mariner 2 (August 27, 1962)**. During one solar magnetic storm, the spacecraft registered significant increases in wind fluxes.

The next probe to Mars was **Mariner 6, Mars Flyby (July 31, 1969)**. The next Soviet mission to Mars was **Mars 2 (November 27, 1971)**.

December 10, 1964
Cosmos 51
U.S.S.R.

During almost a year in orbit, Cosmos 51 studied the luminosity of the stars in both ultraviolet and visible light.

December 13, 1964
83C
U.S.A.

This military-sponsored research satellite carried sensors to map the earth's magnetic field, much like the Soviet Union's **Cosmos 26 (March 18, 1964)** and **Cosmos 49 (October 24, 1964)** satellites. It confirmed and more precisely measured the South Atlantic anomaly (see **Cosmos 17–May 22, 1963**) over Brazil, and it mapped the sky in ultraviolet wavelengths. Because 83C had no data storage capability, however, the information it gathered could only be recorded when the satellite was in contact with ground stations.

83C also carried an experimental package for testing the degradation of solar cells and transistors while in orbit. The results from 83C and earlier satellites led to modifications to this equipment in which degradation would be minimized. See **38C (September 28, 1963)** for earlier experiments.

December 15, 1964
San Marco 1
Italy / U.S.A.

Using an American four-stage Scout rocket launched from Wallops Island, Virginia, San Marco 1 was Italy's first satellite, making Italy the fifth country to orbit its own satellite, after the Soviet Union (**Sputnik–October 4, 1957**), the United States (**Explorer 1–February 1, 1958**), the United Kingdom (**Ariel 1–April 26, 1962**), and Canada (**Alouette 1–September 29, 1962**). San Marco 1 carried two experiments to measure the changing density of the atmosphere between 125 to 200 miles elevation. Both experiments operated for about two weeks and showed that, during its lifetime, the spacecraft traveled through a bulge in the atmosphere, located in the Southern Hemisphere. Such data helped scientists forecast the life expectancy of satellites as the atmosphere dragged them back to Earth.

San Marco 1 was the first of four Italian satellites in the 1960s and 1970s that studied atmospheric density. See **San Marco 2 (April 26, 1967)**.

December 21, 1964
Explorer 26
U.S.A.

This NASA satellite continued the study of the Van Allen radiation belts, accumulating data through May 1967 on their energy, size, and shape to supplement what had been learned from previous satellites (see **OGO 1–September 5, 1964**). Explorer 26's data suggested that the "slot"—the name given to the gap between the two belts—appeared to have moved outward between 1958 and 1965. Scientists attributed this to the solar cycle—as the Sun became less active, the earth's magnetic field had expanded. Explorer 26 also found, as had a number of other satellites (see **Cosmos 41–August 22, 1964**), that during magnetic storms, electron and proton detection in the belt increased.

Though a number of additional Explorer satellites were launched to study the Van Allen belts, the OGO satellites and Imp satellites (see **Imp A–November 27, 1963**) handled the bulk of American magnetosphere research from 1964 through 1974.

See **OGO 5 (March 4, 1968)** and **OGO 6 (May 5, 1969)** for a result summary.

1965

January 19, 1965, 5:03 GMT
DMSP Block-4A F1
U.S.A.

This Defense Meteorological Satellite Program (DMSP) satellite was the first in a continuing series of United States military weather satellites, launched in pairs and placed in complementary sun-synchronous polar orbits (see **Nimbus 1–August 28, 1964**), with the first in the pair crossing the equator in the early morning and the second in the afternoon.

DMSP satellites are similar in concept to Nimbus 1. They carry both visible and infrared imaging systems, with ground resolution as small as 1.5 miles, photographing the entire Earth surface twice each day. Through December 1999, the U.S. government has launched 39 DMSP satellites, with early satellites having a service life of two years and later satellites functioning as long as five years.

Though intended for military use, the data from these satellites are also available to the public. Overall, four different generations of satellite designs have been launched, each carrying increasingly more sophisticated and sensitive instrumentation for measuring temperature, precipitation, and cloud cover, along with ultraviolet, x-ray, and gamma-ray emissions from the earth's atmosphere. Also monitored are changes in the earth's ionosphere, its aurora, and its magnetosphere, as well as the changing levels of atmospheric drag on the satellite.

The DMSP satellites were supplemented by the ESSA (see **ESSA 1–February 3, 1966**) and NOAA (see **NOAA 1–December 11, 1970**) satellites, all of which (after 1964) were sun-synchronous polar orbiting satellites.

January 19, 1965, 9:03 GMT
Gemini 2
U.S.A.

Gemini 2 was NASA's second and final unmanned suborbital test flight of the Gemini capsule prior to its first manned flight, **Gemini 3 (March 23, 1965)**. Originally scheduled to fly in August 1964, Gemini 2 experienced a long series of delays, including three hurricanes, being struck by lightning, equipment problems, and an engine shutdown during its first launch attempt in December. When finally launched and recovered on this date, it successfully tested the Gemini capsule's engines and re-entry capabilities. This same capsule later became the first object to enter space twice. See **Gemini 2 (November 3, 1966)**.

January 22, 1965
Tiros 9
U.S.A.

The Tiros 9 weather satellite continued NASA's program to develop a system for continuous meteorological observation from space. See **Tiros 8 (December 21, 1963)** and **Nimbus 1 (August 28, 1964)** for more information.

As the first Tiros satellite successfully placed in a sun-synchronous orbit (see Nimbus 1), Tiros 9 took the first global weather picture within a 24-hour period on February 13, 1965 (see page 38).

Tiros 9 operated through February 1967. See **ESSA 1 (February 3, 1966)** and **NOAA 1 (December 11, 1970)** for later U.S. weather satellites.

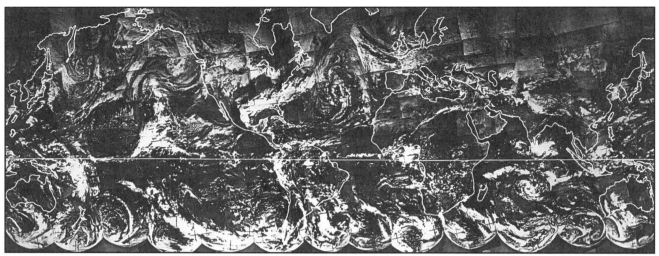
The first global view of Earth's weather, assembled from 450 images taken by Tiros 9 during a single 24-hour period on 13 February 1965. *NASA*

February 3, 1965
OSO 2
U.S.A.
OSO 2 was NASA's second Orbiting Solar Observatory, following **OSO 1** (**March 7, 1962**). The satellite studied solar flares and the ultraviolet and x-ray radiation released by them. Other sensors photographed the Sun's corona in white light, searched the sky for gamma-ray sources, and observed both terrestrial lightning and the zodiacal light.

OSO 2 functioned for nine months. It found that the brightness of the zodiacal light, a faint streak of glow seen rising up from the horizon shortly after sunset, was much less at the poles than previously thought. OSO 2 also found that lightning strikes were more frequent over land than over the sea, a discovery that was confirmed three decades later by **TRMM** (**November 28, 1997**).

The next solar satellite was **Explorer 30** (**November 19, 1965**). The next OSO satellite was **OSO 3** (**March 8, 1967**).

February 11, 1965
LES 1
U.S.A.
LES (Lincoln Experimental Satellite) 1 was the first in a series of experimental military satellites designed and built by the Massachusetts Institute of Technology's Lincoln Laboratory. All but **LES 3** (**December 21, 1965**) were designed to test tactical communications for military operations.

Because of a launch rocket failure, LES 1 was placed in an incorrect orbit. Although telemetry indicated that its equipment was operating properly, the incorrect orbit made further experimentation impossible. See **LES 2** (**May 6, 1965**).

February 16, 1965
Pegasus 1
U.S.A.
As had its predecessor, **Explorer 23** (**November 6, 1964**), Pegasus 1 tested the micrometeorite impact rate in low-Earth orbit. Launched into space by NASA on a Saturn 1 rocket, Pegasus 1 was the third largest object put into space, following the large-diameter balloons of **Echo 1** (**August 12, 1960**) and **Echo 2** (**January 25, 1964**), 100 and 135 feet in diameter, respectively. Like the two Echo satellites, Pegasus's large size made it visible to the unaided eye. To maximize the spacecraft's ability to measure the micrometeorite flux, Pegasus 1 deployed two 14-foot-wide panels creating a wingspan of 96 feet. These large panels provided over 2,000 square feet of impact surface—almost 100 times more than the area provided by Explorer 23.

Pegasus 1 was the first of a three-satellite Pegasus program to study the micrometeorite flux. See **Pegasus 3** (**July 30, 1965**) for a summary of results.

February 17, 1965
Ranger 8
U.S.A.
Ranger 8, following **Ranger 7** (**July 28, 1964**), was NASA's second probe to impact the Moon and successfully return photographs. The spacecraft beamed back 7,137 photos before hitting the lunar surface in the Sea of Tranquility. The pictures showed a cratered surface practically identical to that seen from Ranger 7.

One new feature, dubbed "dimple craters," was identified. Some argued these craters were a kind of collapsed depression, as are seen sometimes in lava fields, while others claimed they were merely degraded secondary-impact craters. The mystery of the origin of the lunar craters—whether volcanic or impact—remained unsolved.

However, like Ranger 7, Ranger 8's photographs did demonstrate that there were areas of the Moon's surface smooth enough to land on. The project scientists proposed that the surface was made up of a light, frothy material, not unlike "crunchy snow," and they felt confident that it was strong enough to support a manned spacecraft, at least at the two Ranger impact sites.

The next and last Ranger mission was **Ranger 9** (**March 21, 1965**).

**March 9, 1965
Oscar 3 / GGSE 2 / GGSE 3 / Dodecapole 1
U.S.A.**

• *Oscar 3*   Oscar 3 was the third privately financed amateur radio satellite (see **Oscar 1–December 12, 1961; Oscar 2–June 2, 1962**). More sophisticated than the previous two Oscars, Oscar 3 was an active repeater satellite, able to relay radio communications. Over its 16-day life span, it enabled 176 radio contacts between the two coasts of the United States, as well as between the United States and Europe. The next amateur radio satellite was **Oscar 4 (December 21, 1965)**.

• *GGSE 2 and GGSE 3*   These two satellites were part of a joint NASA–U.S. Navy program to test the use of extendable booms to reorient satellites in space. By lengthening or retracting the booms, the satellite's center of gravity could be changed, causing its spin and orientation relative to the earth to change as well. While **GGSE 1 (January 11, 1964)** and GGSE 2 tested a two-axis system using one boom, GGSE 3 was the first satellite to use three booms to achieve three-axis control. Further tests continued with **Triad 1 (September 2, 1972)**.

• *Dodecapole 1*   This U.S. military satellite was a 12-sided polyhedron. Like **Surcal 1A/Calsphere (December 13, 1962)**, it was used to calibrate the military's ground stations and surveillance network.

**March 18, 1965
Voskhod 2
U.S.S.R.**

The second and final flight of the Voskhod spacecraft lasted just over one day, during which the first space walk was accomplished. Like **Voskhod 1 (October 12, 1964)**, Voskhod 2 was a modified Vostok capsule. For the space walk, an inflatable airlock (which was sold for $90,500 at a Sotheby's auction in December 1993) was attached to the crew cabin, fitting over the capsule's main escape hatch. By the end of the first orbit, this 6-foot by 3-foot chamber was extended, allowing space-suited Alexei Leonov (**Soyuz 19–July 15, 1975**) to climb inside, close the hatch, and slowly drift out into space, tethered to Voskhod 2 by a thin 20-foot cord. Voskhod 2's commander, Pavel Belyaev, photographed the space walk from inside the crew cabin.

Alexei Leonov spent almost 10 minutes floating about 15 feet from the capsule. He later remarked, "I didn't experience fear. There was only a sense of the infinite expanse and depth of the universe." As he glided along, he easily identified the Black Sea, the Ural Mountains, and the Ob and Yenisey Rivers.

Leonov discovered that in order to orient his body properly, he needed specific visual cues, such as the television camera attached to the airlock hatch (which indicated to him its "top"). Just as the cosmonauts of **Vostok 3 (August 11, 1962)** and **Vostok 4 (August 12, 1962)** had found, without gravity, Leonov had difficulty judging the location and drift of his body.

When Leonov tried to return to the capsule, there were problems. The camera kept getting in his way, and, more important, his spacesuit had expanded in the vacuum of space

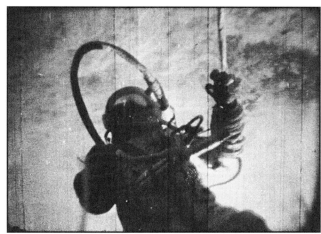
Alexei Leonov during the first space walk. *RIA Novosti*

and no longer fit through the hatch. In desperation, after struggling for about eight minutes to squeeze through, he partly depressurized his suit, which allowed him to get back inside.

Leonov's space walk prompted NASA to quickly reschedule its own first space walk to the second manned Gemini mission, **Gemini 4 (June 3, 1965)**.

On re-entry, Voskhod 2's automatic system for aligning the spacecraft's retro-rockets failed, and for the first time, a Soviet spacecraft's re-entry was piloted manually, using the backup retro-rockets attached to the top of the crew cabin (see Voskhod 1). The spacecraft landed safely but missed its planned target by many miles, landing 750 miles northeast of Moscow in the dense forests of the Ural Mountains. For two days, Leonov and Belyaev waited in the snow-covered forest while helicopters dropped them supplies and skiers chopped open a field 12 miles away so that rescue helicopters could land.

This flight was the last Soviet manned mission until the unveiling of the Soyuz spacecraft, **Soyuz 1 (April 23, 1967)**. The next manned mission was **Gemini 3 (March 23, 1965)**.

**March 21, 1965
Ranger 9
U.S.A.**

Ranger 9 was NASA's last Ranger mission, hitting the Moon on the floor of Crater Alphonsus, returning 5,814 images, and repeating the success of **Ranger 7 (July 28, 1964)** and **Ranger 8 (February 17, 1965)**.

The pictures resolved little about the nature of the lunar surface. Scientists were still unsure exactly how much weight the Moon's surface could bear (though they were generally confident that it could bear a spacecraft's weight), and they still disputed the depth of the dust layer. The question of the origin of the craters also remained unsolved. Ranger 9's impact site, just north of Crater Alphonsus's central peak, had been chosen because in the 1950s several astronomers had sighted gaseous emissions there, evidence of volcanic activity. Though Ranger 9's images identified one new feature, dark "halo craters" on the crater floor that appeared volcanic, most of the features closely resembled those seen on Ranger 7 and Ranger 8.

March 23, 1965
Gemini 3
U.S.A.

Gemini 3 was the first of 10 American Gemini missions that were launched every two months through 1966. While the sole purpose of the Mercury missions (see **Faith 7–May 15, 1963**) had been to send a human into space, the two-person Gemini missions were designed to flight-test the specific techniques necessary for completing a mission to the Moon. These tasks included in-orbit maneuvers, the rendezvous and docking of two spacecraft, survival in space for two weeks, and space walks during which a number of basic construction chores were accomplished. All of the techniques tested in the Gemini program have been used extensively, not only during the Apollo Moon flights (see **Apollo 11–July 16, 1969**), but also by later missions such as **Skylab (May 14, 1973)** and the International Space Station (see **STS 88–December 4, 1998**). The Soviets, too, refined these techniques during their Salyut (see **Salyut 1–April 19, 1971**) and **Mir (February 20, 1986)** space station programs.

The astronauts for Gemini 3 were Gus Grissom (**Liberty Bell 7–July 21, 1961**; **Apollo 1–January 27, 1967**) and John Young (**Gemini 10–July 18, 1966**; **Apollo 10–May 18, 1969**; **Apollo 16–April 16, 1972**; **STS 1–April 12, 1981**; **STS 9–November 28, 1983**). Grissom, in memory of the sinking of the Liberty Bell 7 capsule, wished to name Gemini 3 "Molly Brown" (taken from the just-released Hollywood musical, *The Unsinkable Molly Brown*). Although the American public liked the idea, officials at NASA did not, and they outlawed the naming of spacecraft by astronauts (until it became necessary again during the Apollo program in order to distinguish between the command module and the lunar module).

Gemini 3 was essentially a shakedown flight of the new Gemini capsule. The flight also performed the first orbital changes ever accomplished by a manned vehicle. During the three-orbit flight, the astronauts used the spacecraft's thrusters to change their orbit three different times. First they lowered its orbit from 100 by 140 miles to 98 by 105 miles. Then they changed its inclination by one-fiftieth of a degree. Finally, they lowered their perigee to 52 miles, an altitude low enough that if their retro-rockets failed, atmospheric drag would still bring them back to Earth.

During the mission, human blood samples were irradiated, and it was found that a combination of zero gravity and radiation acted together to increase damage to living substances. The exact relationship and full extent of the risk was unclear, however, and the test was repeated on **Gemini 11 (September 12, 1966)**.

John Young also carried an unofficial experiment, bringing with him a corned beef sandwich to show that a person could eat ordinary food in space. Halfway through the flight, he suddenly took this out and handed it to Grissom. The commander grinned, took a bite, and then put the sandwich away for fear of having thousands of crumbs floating through the capsule. Later, NASA officials criticized Young's action and banned "corned beef sandwiches" from future flights.

The next Gemini mission, **Gemini 4 (June 3, 1965)**, was the first American space flight to last more than one day.

April 6, 1965
Early Bird (Intelsat 1)
International / U.S.A.

Early Bird, or Intelsat 1, was the first geosynchronous communications satellite owned by the International Telecommunications Satellite (Intelsat) Consortium. This organization was founded by 23 Western countries in an effort to promote and regulate the communications satellite industry. Over the years, Intelsat would grow to 140 members, launching eight generations of satellites covering the entire free world.

Early Bird used the same design as **Syncom 3 (August 19, 1964)**, though its power and bandwidth were increased. The satellite had capacity for 240 two-way voice conversations, as well as one two-way television circuit. It was launched into a stationary geosynchronous orbit over the Atlantic and provided commercial service until 1969, becoming part of the first constellation of Intelsat satellites (see **Intelsat 2A–October 26, 1966**; **Intelsat 2B–January 17, 1967**; **Intelsat 2C–March 23, 1967**; **Intelsat 2D–September 28, 1967**). This first cluster of satellites provided service over the Atlantic and Pacific Oceans, essentially covering most of the Western world.

Early Bird's first significant worldwide use was to broadcast to Europe the launch of **Gemini 4 (June 3, 1965)**. The next generation of Intelsat satellites began with **Intelsat 3A (December 19, 1968)**.

April 23, 1965
Molniya 1-01
U.S.S.R.

Molniya 1-01 was the first officially announced Soviet communications satellite, following the unannounced test flight of **Cosmos 41 (August 22, 1964)**. Molniya means "lightning" in Russian, or "news flash" in the vernacular.

Because of the high latitude of Soviet territory, an equatorial geosynchronous satellite would have been poorly positioned for communications. The Molniya satellites were instead launched into eccentric orbits at high inclination. Molniya 1-01's orbit had a perigee 334 miles over the Western Hemisphere and an apogee 24,420 miles over the Soviet Union. This placed the satellite within reach of almost all ground stations in Soviet territory for a period of about nine hours each day. By adding additional satellites to the constellation (see **Molniya 1-02–October 14, 1965**; **Molniya 1-03–April 25, 1966**), communications could be maintained 24 hours per day over the entire Soviet Union.

Molniya 1-01 had the capacity to either broadcast one television signal or transmit an undisclosed number of telephone transmissions. On its first day of operation, it broadcast television programs from Moscow to Vladivostok.

Over the next 32 years, 90 Molniya 1 satellites were launched, an average of about four per year. At first, a three-satellite constellation was maintained. Later, this was expanded to four, then to eight in the 1970s. Although the technology of these satellites was certainly updated over time, few specifics have been released. Beginning in 1994, a more advanced version of the Molniya 1 satellite was developed, of which five have been launched as of December 1999.

In 1974, a slightly different communications satellite, **Molniya 2 (November 24, 1971)**, was launched. This second satellite constellation, which was later replaced by the Molniya 3 constellation (see **Molniya 3-01–November 21, 1974**), took over the civilian communications needs of the Soviet Union, leaving the Molniya 1 satellites to handle military and government communications.

April 29, 1965
Explorer 27 (Beacon Explorer C)
U.S.A.

This NASA geodetic satellite was a twin of **Explorer 22 (October 10, 1964)**. Its radio beacon, which functioned for three years, was used to study the ionosphere and changes in the satellite's orbit, thereby mapping the earth's shape and size.

Lasers were also used for the first time with Explorer 27 to measure the earth's shape. A ground station would beam a laser at the satellite, and the laser's light would bounce off the satellite's surface of silica reflectors. By calculating the light's travel time to and from the Explorer 27, scientists triangulated precisely both their position and the satellite's.

The next geodetic satellite was **Explorer 29 (November 6, 1965)**. Further laser tracking tests were also done on **Gemini 7 (December 4, 1965)**.

May 6, 1965
LES 2
U.S.A.

LES 2 was the second experimental satellite built by the Massachusetts Institute of Technology's Lincoln Laboratory for the U.S. Air Force; it followed the failure of **LES 1 (February 11, 1965)**. LES (Lincoln Experimental Satellite) 2 was identical to LES 1—designed to test solid-state transmitters and small ground-based mobile terminals for tactical military communications. Though initially intended for use in military operations, these experiments eventually led to today's modern telephone satellite technology (see **Iridium 4 through Iridium 8–May 5, 1997**). LES 2 functioned for two years, then was followed by **LES 3/LES 4 (December 21, 1965)**.

May 25, 1965
Pegasus 2
U.S.A.

Pegasus 2 was the second of three large satellites launched by NASA to measure precisely the micrometeorite flux in low-Earth orbit. Like its predecessor **Pegasus 1 (February 16, 1965)**, Pegasus 2 deployed two huge panels, giving the spacecraft a wingspan of 96 feet. These panels were designed to detect impacts and measure their depth of penetration. See **Pegasus 3 (July 30, 1965)** for a summary of results.

June 3, 1965
Gemini 4
U.S.A.

Gemini 4 was the second manned Gemini mission, following **Gemini 3 (March 23, 1965)**. The astronauts were Jim McDivitt (**Apollo 9–March 3, 1969**) and Ed White (**Apollo 1–Janu-

Ed White. *NASA*

ary 27, 1967). The Gemini 4 flight lasted just over four days and included the first American space walk, hurriedly added to the schedule following cosmonaut Alexei Leonov's space walk on **Voskhod 2 (March 18, 1965)**. The original plans had called for Ed White to open his hatch and merely poke his head out of the capsule. Instead, White not only floated free, attached to the spacecraft by a 25-foot tether, but he also maneuvered using a hand-held jetpack powered by small puffs of oxygen. While Leonov's spacesuit had been self-contained, holding enough oxygen for about 45 minutes, White's air supply was provided through the tether line. If he somehow were

A Gemini 4 photograph of the mouth of the Colorado River. The white streaks are saltpan beds. Sand dunes of the Sonoran Desert can be seen along the shore. *NASA*

to become disconnected, White had a nine-minute emergency air supply in his spacesuit.

During the space walk, White found the maneuvering device useful, but he ran out of fuel almost immediately. McDivitt discovered that White's movements quickly changed the capsule's attitude, requiring him to do a lot more steering. Once White climbed back inside, he found that when weightless, the simple task of locking a hatch was not so simple. Every time he tried to pull the locking latch down, he drifted upward in the zero gravity. McDivitt finally had to hold White's legs so that he could get enough leverage.

During their first orbit, the astronauts attempted to steer their capsule closer to the spent second stage of their rocket, which was floating in orbit about 650 feet away. They found this maneuver unexpectedly difficult. Whenever McDivitt tried to close the distance by applying thrust in the direction implied by his earthbound instincts, he found that the distance between the two craft actually increased. These difficulties gave the first indication that rendezvous and docking in space were not simple tasks—indeed, it would take almost the entire Gemini program to perfect these maneuvers (see **Gemini 5–August 21, 1965; Gemini 6–December 15, 1965; Gemini 8–March 16, 1966; Gemini 9–June 3, 1966; Gemini 10–July 18, 1966**).

The Gemini 4 astronauts took over 100 weather and terrain photographs of the earth. The terrain pictures proved the usefulness of geological study from orbit, easily showing erosion details of sand dunes in the Sahara, ancient lava flows along the Niger River, and the geological fault structure of the Rift Valley in Ethiopia.

The weather photographs provided high quality reference data for comparison with the television pictures from the Tiros and Nimbus weather satellites. Meteorologists were still developing the science of interpreting satellite weather images, and the photos from Gemini 4 helped them clarify their interpretations. See **Nimbus 1 (August 28, 1964)** for details about these American weather satellites.

In order to design safe docking in space, scientists needed more information about the electrostatic charge on a spacecraft's exterior, first discovered by **Explorer 8 (November 8, 1960)**. On Gemini 4, a sensor found this plasma charge to be much higher than expected. Whether the electrical field was real or was caused by instrument effects remained unclear.

Extensive and detailed measurements were made of radiation levels within the spacecraft. The tests used small portable sensors placed in specific positions throughout the capsule. During certain orbits, the astronauts moved these sensors into new positions to see if the dosages were affected by the capsule's structure. The results indicated that in certain situations, the radiation level was less than predicted, and that the radiation came almost exclusively from cosmic ray sources, not from the Van Allen belts. The data also showed methods for improving spacecraft shielding to protect future astronauts.

Gemini 4 also tested a number of new spacecraft systems. Dehydrated food packs were developed for feeding the astronauts four meals a day, and a first-generation defecation system had been developed in addition to the urination system.

The most important experiments on Gemini 4, however, were the medical tests. Based on a number of Soviet scientific papers outlining both bone loss and depleted cardiac response following the longer Vostok flights (see **Vostok 3–August 11, 1962; Vostok 4–August 12, 1962; Vostok 5–June 14, 1963; Vostok 6–June 16, 1963**), as well as the results from the two prior Mercury flights (**Sigma 7–October 3, 1962; Faith 7–May 15, 1963**), a small exercise device had been included on Gemini 4. Attached to a bungee cord was a handle and foot strap. Each astronaut pulled this cord during several scheduled exercise periods daily, and on the day prior to re-entry, they used it hourly. Whether this exercise helped them re-adapt to Earth gravity, however, was unclear.

Upon re-entry, medical tests revealed that after four days in space, both McDivitt and White had sustained bone loss of between 8 and 10 percent. Their blood plasma volume had also declined. These results suggested serious risks during longer space flights.

The next Gemini mission, **Gemini 5 (August 21, 1965)**, was an eight-day mission, designed to prove that humans could survive in space long enough to travel back and forth from the Moon.

July 15, 1965
Mariner 4, Mars Flyby
U.S.A.

Launched by NASA on November 28, 1964, Mariner 4 was the first spacecraft to reach Mars and successfully return data (see **Mariner 4–November 28, 1964**).

The most significant discoveries from Mariner 4's flyby of Mars were in the 22 photographs beamed back to Earth showing about 1 percent of the Martian surface. The first picture, showing the planet's limb (the horizon of a planet as seen from space), captured a smudge above the horizon which was thought to be a thin cloud of frozen carbon dioxide crystals hovering at

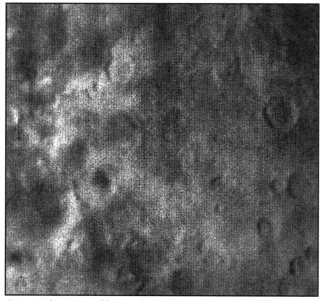

A cratered region on Mars at about 10 degrees south latitude, south of Amazonis Planitia (which is about 40 degrees west of Olympus Mons). See also the Mars map with Viking 1 (June 19, 1976). *NASA*

about 18 miles elevation. Images 2 through 6 showed what appeared to be, as noted in a later NASA report, "a desert landscape with a few roundish spots that might be dried up lakes." Picture 7, however, revealed a surface as dead and as cratered as the Moon, with 10 craters scattered randomly throughout the image. All the remaining pictures revealed similar features, differing from a lunar landscape only in that white patches could be seen on crater rims, which scientists thought might possibly be frost. Scientists concluded, therefore, that the roundish spots in the first few photographs were craters, not dried lakebeds.

Mariner 4's pictures showed nothing of the famous canals of Mars. The craters appeared ancient and well preserved, suggesting that little erosion had taken place in several billion years. This, in turn, suggested that the planet's atmosphere had never been thicker than seen at present, and that liquid water had never flowed on its surface. Without a thick atmosphere and liquid water, it seemed difficult—if not impossible—for Martian life to have developed.

Before Mariner 4, theorists had proposed that Mars would have a magnetic field about one-tenth that of Earth's. However, the spacecraft's magnetometer observed practically no Martian magnetic field, nor did it find evidence of trapped radiation belts encircling the planet like the earth's Van Allen belts. The absence of a magnetic field indicated that Mars probably did not have a molten core, and therefore did not have an active and changing topology like the earth. Mariner 4's photographs further confirmed this conclusion.

Mariner 4 found the Martian atmosphere to be very thin, less than 1 percent Earth's at the planet's surface, with a weak ionosphere comparable to Earth's at night. Furthermore, the temperature readings at 75 to 125 miles elevation were much colder than expected, approximately −100°F. Based on this data, the Martian atmosphere was thought to consist primarily of carbon dioxide with little or no nitrogen.

While scientists were reluctant to state definitively that life did not exist on Mars, the photographic and experimental evidence from this flight indicated a bleak and dead world, closely resembling that of the Moon. See **Mariner 6, Mars Flyby (July 31, 1969)** and **Mariner 7, Mars Flyby (August 5, 1969)** for the next successful missions to Mars, and how those missions changed these conclusions.

July 16, 1965
Proton 1
U.S.S.R.

Proton 1 was the first test launch of the powerful Soviet Proton rocket, which was capable of putting heavier payloads into orbit than earlier Soviet rockets. At just under 27,000 pounds, the satellite and last-stage booster were the heaviest objects placed in orbit by the Soviets to date—exceeded only by the American test launch of Saturn 1 the previous year (see **Pegasus 1–February 16, 1965**).

The Proton 1 satellite carried the first Soviet gamma-ray telescope, the second such orbiting telescope since **Explorer 11 (April 27, 1961)**. It could detect gamma-ray emissions up to 50 MeV and cosmic rays up to 100 TeV.

The satellite operated for 45 days. Its orbit then decayed on October 11, 1965. No data from the gamma-ray telescope was ever published, though the Soviets announced that it made the first detection of cosmic rays with energies in the range of 100 TeV.

The next Soviet gamma-ray telescope was orbited during a second Proton rocket test. See **Proton 2 (November 2, 1965)**.

July 18, 1965
Zond 3
U.S.S.R.

This lunar flyby mission took additional and higher resolution pictures of the Moon's far side, photographing about 30 percent of the far side not seen by **Luna 3 (October 4, 1959)** and leaving only about 5 percent of the Moon still unseen. Twenty-five visible-light photographs and three ultraviolet-light photographs were taken, as well as ultraviolet spectrograms of the far side.

Zond 3 also made measurements of the cosmic ray flux as the spacecraft traveled outward into interplanetary space. See **Venera 2 (November 12, 1965)** for results.

July 20, 1965
ORS 3
U.S.A.

This U.S. Air Force Orbiting Research Satellite (ORS) was the first (despite being numbered 3) satellite in a program of military research satellites that did technology and engineering tests in orbit. ORS 3 carried five sensors for measuring x-rays, gamma rays, and cosmic rays in the earth's magnetosphere from 350 to 69,000 miles elevation. Data was collected regularly during its first month in orbit, then intermittently for the next few months. The results were used to modify the design of later satellites. See **ERS 16 (June 9, 1966)** for later research.

July 30, 1965
Pegasus 3
U.S.A.

Pegasus 3 was the third in NASA's three-satellite program (begun with **Pegasus 1–February 16, 1965**) whose purpose was to obtain complete measurements of the micrometeorite flux in low-Earth orbit. Like its predecessors, Pegasus 3 unfolded two 14-foot-wide panels creating a wingspan of 96 feet and exposing over 2,000 square feet of surface area to meteorite impact. All three satellites operated for more than three years, compiling data that proved that the threat of impact in low orbit is minimal. The accumulated data from the three Pegasus spacecraft revealed that aluminum panels 0.016 inch thick resisted puncture from almost all impacts.

August 21, 1965
Gemini 5
U.S.A.

Gemini 5 was the longest space flight to date, breaking the record set by Valeri Bykovsky on **Vostok 5 (June 14, 1963)**. The astronauts were Gordon Cooper (**Faith 7–May 15, 1963**) and Pete Conrad (**Gemini 11–September 12, 1966; Apollo 12–November 14, 1969; Skylab 2–May 25, 1973**). For just under eight days, these two men orbited the earth.

October 5, 1965

This flight was also the first to use fuel cells to provide power instead of batteries. Because of a failure in maintaining proper pressure in the oxygen tank that supplied these fuel cells, however, the flight was almost terminated after six orbits. The problem forced the astronauts to power down for most of the eight-day flight, preventing them from completing many of their planned experiments. For example, one experiment involved releasing a mini-satellite, the Radar Evaluation Pod, on the second orbit in order to practice rendezvous techniques using newly designed radar equipment. The radar in the capsule's nose cone detected the beacon in the pod and successfully determined its distance and speed. The plan was to then use this data to try to rendezvous with the pod. While the radar worked, the fuel cell problem prevented the rendezvous.

Like the previous American manned mission, **Gemini 4 (June 3, 1965)**, Gemini 5 focused mostly on medical tests. One medical experiment had Cooper wear a cuff around each thigh, with Conrad acting as the control. During the first four days of the mission, this cuff inflated two minutes out of every six, with the hope that it would help Cooper readapt to Earth gravity and avoid the high pulse rate experienced by earlier astronauts after re-entry (see **Sigma 7–October 3, 1962**). After landing, however, the response of both men was about the same, with increased heart rates and lower blood pressure that took several days to return to normal. The cuffs had had no effect.

Another experiment tested how weightlessness altered the inner ear's ability to judge orientation. Wearing goggles, each astronaut tried to align an adjustable white line to what they believed was the pitch axis of the capsule. The test found that the astronauts were able to orient themselves, even without gravity, though each man responded differently to weightlessness. After the flight, their sense of balance also quickly returned to normal.

A visual acuity test involved testing the ability of the astronauts to see small details on the surface of the earth. After Gordon Cooper's experience on **Faith 7 (May 15, 1963)**, in which he claimed he could see cars and even people, scientists and military experts wished to test Cooper again to see exactly how much detail could be identified from space. A pattern of marks had been laid out both in Texas and Australia, both of which the astronauts were easily able to recognize. They also spotted the launch of a Minuteman missile, noting its smoke trail, as well as the smoke of a rocket sled at Holoman Air Force Base in New Mexico.

Another experiment measured the radiation released by the exhaust nozzles of the launch rocket. This was a military test performed in conjunction with the Midas early warning system (see **Midas 2–May 24, 1960**).

The next two Gemini flights, **Gemini 6 (December 15, 1965)** and **Gemini 7 (December 4, 1965)**, flew in conjunction, testing both long-term survival in space as well as rendezvous techniques.

October 5, 1965
OV1-2
U.S.A.

This satellite, designed and built by the U.S. Air Force, was the first launch in a family of five designs dubbed Orbiting Vehicles (OVs). These spacecraft were standardized to minimize cost. Over the next six years, the Air Force launched over 20 OVs to test their design and to conduct research into the space environment. Though OV research often collected data on the earth's magnetic field, its magnetosphere, the Van Allen radiation belts, and the earth's atmosphere, the data was gathered primarily for the purpose of improving future military satellite designs. Hence, results were rarely published in scientific journals.

OV1-2 collected radiation data at high polar latitudes. The satellite carried an x-ray detector, a magnetometer, and other sensors for measuring proton-electron densities. The satellite operated until April 1967, when it was shut off due to equipment failures.

October 14, 1965, 12:58 GMT
OGO 2
U.S.A.

OGO (Orbiting Geophysical Observatory) 2 was the second of six NASA satellites dedicated to a systematic study of the earth's magnetic field (see **OGO 1–September 5, 1964**). Carrying 20 experiments, OGO 2 was placed in a low polar orbit to observe the magnetosphere at high latitudes. Because of a failure in the spacecraft's attitude control system, however, only five experiments generated useful data.

One of the successful experiments, a detailed survey of the earth's magnetic field, found an excellent correlation with data obtained by the **Cosmos 49 (October 24, 1964)** spacecraft, especially in the satellites' observations of the South Atlantic anomaly over Brazil (see **Cosmos 17–May 22, 1966**).

For additional results, see **OGO 6 (June 5, 1969)**.

October 14, 1965, 19:41 GMT
Molniya 1-02
U.S.S.R.

Molniya 1-02 was the second Soviet communications satellite, joining **Molniya 1-01 (April 23, 1965)** in a similar orbit but positioned further west in order to link the U.S.S.R. with Europe. See Molniya 1-01 for program details.

November 2, 1965
Proton 2
U.S.S.R.

The launch of Proton 2 was the second test of the Soviet heavy-lift Proton rocket, following **Proton 1 (July 16, 1965)**. It also carried the second Soviet gamma-ray telescope into orbit. Like its predecessor, it operated for only a few months before its orbit decayed. As with Proton 1, Proton 2's research results have not been published.

November 6, 1965
Explorer 29 (GEOS 1)
U.S.A.

Explorer 29 was also called the Geodetic Earth Orbiting Satellite (GEOS A or GEOS 1). It was designed by NASA specifically for geodetic research; that is, to map precisely the earth's gravitational field and shape to within 30 feet accuracy and to locate its irregularities and anomalies. It was a more advanced version of previous satellites like **Anna 1B (October 31, 1962)** and **Explorer 27 (April 29, 1965)**. Throughout 1966, synchronized satellite observations were made from the United States, the Soviet Union, Iran, and throughout Europe as Explorer 29 passed overhead. The spacecraft carried several different types of beacons to increase the ease of observation, including optical beacons and laser reflectors. These observations also confirmed the connection between atmospheric density changes and solar activity (see **Explorer 19–December 19, 1963**). As the Sun moved from its active state in 1958 to its quiet state in 1965, the size of the earth's atmosphere also shrank, reducing the orbital drag on satellites, enabling them to stay in orbit longer.

The next geodetic satellite was **Pageos 1 (June 24, 1966)**. See **Geos 2 (January 11, 1968)** for this program's continuation.

November 12, 1965
Venera 2
U.S.S.R.

This Soviet attempt to fly past Venus failed when communications were lost on January 24, 1966. The spacecraft passed Venus on February 27, 1966, at a distance of about 15,000 miles.

On its way to Venus, however, Venera 2 compiled data on the interstellar cosmic ray flux. It found that since **Luna 1 (January 2, 1959)**, this flux had increased significantly, more than 160 percent outside the earth's magnetosphere. This increase suggested that as the Sun entered its quiescent period in the solar cycle, the solar wind was less effective in sweeping the cosmic radiation from the solar system.

Venera 2's measurements were also taken in conjunction with **Pioneer 6 (December 16, 1965)**. Both spacecraft registered the same brief fluctuations in cosmic ray fluxes, despite being separated by a very great distance. There was, however, a time difference in the data, with the spacecraft farther from the Sun recording the changes later in time. The results suggested that the solar wind was somewhat homogeneous, and that its influence required time to travel outward from the Sun.

November 16, 1965
Venera 3
U.S.S.R.

Although Venera 3 successfully hit Venus on March 1, 1966, becoming the first spacecraft to impact another planet, the mission returned no data because communications had failed on January 7, 1966.

Prior to this communications failure, Venera 3 did transmit solar wind observations, confirming the particle densities and wind speeds seen on **Zond 3 (July 18, 1965)**. Furthermore, because Venera 3 and **Pioneer 6 (December 16, 1965)** were in orbit at the same time, their data were comparable, showing the widespread similarity of solar wind events across great distances.

The next mission to Venus, Venera 4 (**Venera 4, Venus Landing–October 18, 1967**), was the first spacecraft to return data from inside that planet's atmosphere.

November 19, 1965
Explorer 30 (Solrad 8)
U.S.A.

Explorer 30, also called Solrad 8, was part of a U.S. Navy project (see **Solrad 1–June 22, 1960**) making the first effort to continuously monitor the Sun's output in ultraviolet and x-ray wavelengths. Explorer 30 was also launched as the United States contribution to the International Quiet Sun Year project, and it was the first in a three-satellite joint Navy-NASA solar research project.

Unlike the previous Solrad satellites, Explorer 30 carried an active attitude control system as well as a data storage system. The spacecraft not only could continuously maintain its orientation relative to the Sun, but it could also record data for later transmission to Earth. Unfortunately, the data storage system failed after one month, so subsequent measurements of the Sun's output could be obtained only while the spacecraft was in contact with ground stations.

Explorer 30 gathered solar data for almost two years, until August 1967, covering almost the entire quiet period of the Sun's solar cycle. It found that an increase in the Sun's background x-ray output could be used to predict solar flare activity. It also proved that sunspots produced powerful x-ray emissions.

The next Solrad satellite was **Solrad 9 (March 5, 1968)**. The next solar satellite was **Pioneer 6 (December 16, 1965)**.

November 26, 1965
Asterix
France

Asterix, launched by a French rocket from a facility in Algeria, was the first satellite placed in orbit by a nation other than the United States or the Soviet Union. Its primary purpose was to test the rocket's operation, and it performed no scientific research while in orbit. Asterix was followed by a three-satellite test program, the second and third launches of which placed geodetic science satellites in orbit. See **Diademe 1 (February 8, 1967)**.

November 29, 1965
Explorer 31 / Alouette 2
U.S.A. / Canada

Explorer 31 and Alouette 2 were launched together by NASA to perform complementary research into the composition of the ionosphere. These two satellites followed **Alouette 1 (September 29, 1962)** and **Explorer 20 (August 25, 1964)** and were specifically designed to extend the data from the earlier missions, making additional measurements at higher elevations.

Both satellites were placed in polar orbits ranging from 300 to 1,850 miles altitude. For the first month, their orbits were close enough together that scientists could compare their data and correlate their instruments. Later, scientists used the satellites' growing separation to measure different but similar regions of the ionosphere. The satellites found a low-density region of electrons above the winter pole at about 1,850 miles altitude.

The next ionospheric research satellite was **France 1 (December 6, 1965)**.

December 4, 1965
Gemini 7
U.S.A.

Gemini 7's nearly 14-day-long mission set an endurance record for space flight that lasted until **Soyuz 9 (June 1, 1970)**. The astronauts were Frank Borman (**Apollo 8–December 21, 1968**) and Jim Lovell (**Gemini 12–November 11, 1966; Apollo 8–December 21, 1968; Apollo 13–April 11, 1970**). The flight was also part of the first rendezvous between two manned vehicles; see **Gemini 6 (December 15, 1965)**.

Gemini 7 carried 20 experiments, with medical research taking priority. Detailed records were kept of the astronauts' food intake for the entire flight, as well as for nine days afterward. Heart rate, respiration, blood pressure, and temperature were tracked for each man, as was Borman's brain activity during his first two sleep sessions.

In general, the results confirmed what had been learned from **Gemini 5 (August 21, 1965)**: humans could easily adapt to weightlessness for several weeks, though it took several days after landing for them to return to normal. In fact, Borman and Lovell experienced fewer problems than previous astronauts. They lost less bone mass, and no loss was detected in their red cell and blood plasma mass. Their heart rates upon landing, while increased, were also not as high as had been seen previously.

Earth observation tests were also performed during the flight. Borman and Lovell were the first to observe from space both the launch and re-entry of a missile.

In an optical communications test, the astronauts fired a laser beam at several selected sites to see if this could be used to transmit information. Though cloud cover prevented the first test, and the signal was too weak on the second, the test did prove that laser beacons could be used at orbital distances.

After Gemini 6 completed its rendezvous tests during Gemini 7, **Gemini 8 (March 16, 1966)** followed with the first docking in space.

December 6, 1965
France 1
France / U.S.A.

Much like **Ariel 1 (April 26, 1962)** and **Alouette 1 (September 29, 1962)**, France 1 was a joint project of the United States and a second country, this time, France. The satellite studied very low frequency transmissions through the ionosphere, measuring their propagation as they traveled through different regions. France 1 also measured electron densities. Both experiments continued the research of **Alouette 2 (November 29, 1965)**.

The next ionospheric-propagation research satellite was **Ariel 3 (May 5, 1967)**. **Explorer 32 (May 25, 1966)** did additional research into the low atmosphere.

December 15, 1965
Gemini 6
U.S.A.

Gemini 6, whose sole purpose was to test in-space rendezvous techniques, performed the first such maneuver by coming within 10 feet of **Gemini 7 (December 4, 1965)**. The astro-

Rendezvous: Gemini 7 as seen from Gemini 6. *NASA*

nauts were Wally Schirra (**Sigma 7–October 3, 1962; Apollo 7–October 11, 1968**) and Tom Stafford (**Gemini 9–June 3, 1966; Apollo 10–May 18, 1969; Apollo 18–July 15, 1975**).

Achieving this goal, however, involved overcoming repeated failures and problems. The original plans had called for an October 25 launch of both the Gemini 6 capsule and an Atlas-Agena target vehicle. When the target vehicle failed to reach orbit, the new plans called for a rendezvous during the two-week Gemini 7 mission. However, when Gemini 6 next tried to launch on December 12, 1965, the Titan rocket ignited as planned, then shut down while still on the launchpad. Engineers later discovered that a loose electrical plug, along with the failure to remove a temporary dust cover, had caused the stall.

Gemini 6 finally reached orbit on its third try and within six hours achieved the first rendezvous between two manned vehicles. Along the way, Gemini 6 made eight engine burns, changing the spacecraft's altitude, orbit shape, inclination, and speed. During final rendezvous maneuvers, both visual and radar data were used to guide the spacecraft.

Gemini 6 and Gemini 7 flew in formation for approximately four hours. Then Gemini 6 returned to Earth, having completed one day in orbit. The next Gemini flight, **Gemini 8 (March 16, 1966)**, achieved the first hard docking in space.

December 16, 1965
Pioneer 6
U.S.A.

Pioneer 6 was the first of four NASA spacecraft designed to study the solar wind from solar orbit (see **Pioneer 7–August 17, 1966; Pioneer 8–December 13, 1967; Pioneer 9–November 8, 1968**). These four spacecraft formed a cluster that orbited the Sun and measured the solar wind from widely separated points in space. See Pioneer 9 for a summary of results.

Pioneer 6, working in conjunction with the third Imp satellite (see **Imp A–November 27, 1963**), proved that the Sun's magnetic field rotated in unison with the 27-day-long solar day. Fifty-seven minutes after Pioneer 6 saw the Sun's magnetic field reverse, the Imp satellite observed the same reversal. The time difference matched the Sun's rotation speed. The reversal was caused by the wavy equatorial sheet that separated the oppositely charged northern and southern hemispheres of the Sun's magnetic field. As the magnetic field rotated, one of the sheet's waves rolled past the two spacecraft, thereby causing the field's charge to reverse.

Other observations made jointly with **Venera 2 (November 12, 1965)** showed that short-term variations in the cosmic ray flux from interplanetary space were measurable across great distances. Similarly, solar wind measurements from **Venera 3 (November 16, 1965)** and Pioneer 6 were virtually identical, showing the same variations at corresponding times.

Pioneer 6 was originally designed for a six-month life span; however, it continued to provide data through 1990, supplementing such later satellites as **ISEE 1/ISEE 2 (October 22, 1977)** and **Solar Max (February 14, 1980)**.

December 21, 1965
LES 3 / LES 4 / Oscar 4
U.S.A.

Because the Titan 3C launch rocket failed to separate properly, these three satellites were placed in improper orbits.

• *LES 3* Built by MIT's Lincoln Laboratory for the U.S. Air Force, LES (Lincoln Experimental Satellite) 3 was not an experimental communications satellite like **LES 2 (May 6, 1965)**. Instead, its payload was a beacon for testing how specific radio frequencies were transmitted and received from space satellites.

• *LES 4* Similar to LES 2, LES (Lincoln Experimental Satellite) 4 was an experimental communications satellite launched by the U.S. Air Force to test advanced technology for tactical military communications that made use of small mobile antennas on the ground to communicate via orbiting satellites. This technology was an early test of today's mobile telephone satellite systems (see **Iridium 4 through 8–May 5, 1997**). Though the initial incorrect orbit prevented LES 4's solar cells from recharging, after five days in orbit the spacecraft's orientation changed enough to allow the satellite to power up and function. **LES 5 (July 1, 1967)** continued these experiments.

• *Oscar 4* An active repeater communications satellite, Oscar 4 was the fourth privately-built amateur radio satellite (see **Oscar 1–December 12, 1961; Oscar 2–June 2, 1962; Oscar 3–March 9, 1965**). Oscar 4 had numerous problems, not only because of its incorrect orbit. Its solar cells and a timing switch in its transmitter also failed. Nonetheless, Oscar 4 functioned for three months and managed a few two-way communications, including the first direct satellite link between the United States and the Soviet Union. The next amateur radio satellite was **Oscar 5 (January 23, 1970)**.

1966

January 31, 1966
Luna 9
U.S.S.R.

After a number of failed attempts to soft-land a spacecraft on the Moon (Lunas 5, 7, and 8 hit the Moon but failed to land safely, while Lunas 4 and 6 missed the Moon), Luna 9 became the first spacecraft to land on another world and successfully transmit data back to Earth. Luna 9 landed in the Ocean of Storms on February 3, 1966. To accomplish its landing, the spacecraft carried no rockets, but instead was protected by giant airbags as it bounced against the lunar surface.

Luna 9 operated for three days on the Moon's surface, taking several panoramic photographs of the surrounding terrain. These images revealed that the spacecraft had probably landed on the western slope of a crater about 50 feet across and about 2 feet deep. Furthermore, between its first and second transmissions, the spacecraft's position shifted, turning about 3 degrees, with its tilt increasing by about 6.5 degrees. This change indicated that the spacecraft was settling into the ground. It also allowed scientists to make stereoscopic analysis of the area captured by the panoramas.

Luna 9's photographs, and the fact that the spacecraft did not sink below the surface, showed that the lunar surface—at least in this one location—was not coated with a deep layer of dust, as had been theorized by some scientists (see **Ranger 6–July 28, 1964; Ranger 7–February 16, 1965; Ranger 8–March 21, 1965**).

Luna 9 also carried sensors for measuring cosmic ray radiation, and it found that the in-space flux decreased by almost 40 percent upon landing.

The next lunar soft landing was **Surveyor 1 (May 30, 1966)**. The next Soviet soft landing was **Luna 13 (December 21, 1966)**.

February 3, 1966
ESSA 1
U.S.A.

Launched by ESSA, the Environmental Science and Services Administration (the forerunner of NOAA, the National Oceanic and Atmospheric Administration), ESSA 1 continued the Tiros satellite program (see **Tiros 7–June 19, 1963; Tiros 8–December 21, 1963**), its management moving from NASA to ESSA. These satellites provided close-up images of the earth's cloud cover to facilitate forecasting by local weather bureaus.

Through 1969, ESSA launched nine ESSA satellites. They were orbited as pairs, each placed in sun-synchronous orbits (see **Nimbus 1–August 28, 1964**), where together they imaged the earth's cloud cover twice a day under identical lighting conditions. These satellites also tested a variety of new imaging systems, taking pictures in several spectral bands in both visible and infrared light. Images monitored cloud cover, atmospheric and sea surface temperatures, vegetation, and atmospheric aerosols.

The satellites also tested engineering designs for increasing their orbital life spans. Generally, ESSA satellites operated for an average of two years before a system failure made

February 22, 1966
Cosmos 110
U.S.S.R.

Cosmos 110 was a biological test mission that used the third Voskhod capsule, which had originally been built for a long-duration manned flight (see **Voskhod 1–October 12, 1964**). Two dogs were placed in orbit for almost 21 days to study their reaction to space and high radiation. The apogee of the spacecraft's orbit was intentionally high, 550 miles, so that the animals repeatedly passed through the lower Van Allen radiation belt.

The dogs, named Veterok and Ugolek, were held in separate compartments. Sensors constantly monitored each dog's pulse, respiration, and blood pressure. Continuous electrocardiograms and electroencephalograms recorded their heart and brain activity. Television cameras recorded their every move. As a control, two other dogs were similarly confined on the ground and subjected to the same battery of tests. Moreover, Ugolek acted as an in-space control, with Veterok subjected to a number of medications designed to counter the effects of radiation.

Both dogs returned to Earth unharmed, though they experienced considerable physical deterioration from their flight. Of particular concern was the significant loss of bone calcium, as well as loss of muscle strength. In addition, though the dogs were fed intravenously, they both experienced weight loss. It took about 8 to 10 days for the dogs to return to their preflight condition. This was comparable to what had been found after **Gemini 4 (June 3, 1965)**, but showed more serious consequences than had been seen after **Gemini 5 (August 21, 1965)** and **Gemini 7 (December 4, 1965)**.

March 16, 1966, 15:07 GMT
Gemini 8 Agena Target
U.S.A.

This spacecraft was the final stage of the Atlas-Agena rocket, designed as the target vehicle for the Gemini docking missions. It lifted off 1 hour and 40 minutes ahead of **Gemini 8 (March 16, 1966)**, which used it to achieve the first in-space hard docking.

When Gemini 8 was forced to undock and abort its mission, ground controllers fired the Agena's rockets and placed it in a high "parking" orbit where **Gemini 10 (July 18, 1966)** later rendezvoused with it to recover a micrometeorite experiment.

March 16, 1966, 16:48 GMT
Gemini 8
U.S.A.

Following the successful rendezvous of **Gemini 6 (December 15, 1965)** and **Gemini 7 (December 4, 1965)**, Gemini 8 accomplished the first in-space hard docking of two different spacecraft, followed less than half an hour later by the first in-space crisis requiring an emergency landing. The astronauts were Neil Armstrong (**Apollo 11–July 16, 1969**) and Dave Scott (**Apollo 9–March 3, 1969; Apollo 15–July 26, 1971**).

Six orbital maneuvers over six hours brought Gemini 8 within 150 feet of its target (see **Gemini 8 Agena Target–March 16, 1966**). Armstrong aligned the two spacecraft, slowly inched the Gemini capsule to within two feet of the docking port, then eased the two craft together.

Twenty-seven minutes later, a short-circuit caused one of Gemini 8's attitude control thrusters to fire uncontrollably, causing the two linked spacecraft to gyrate wildly. At first, the astronauts thought the problem was coming from the Agena vehicle, so they undocked from it. This only worsened the gyrations. With the capsule spinning 60 times a minute, its actual structure was threatened. The astronauts meanwhile were experiencing dizziness and blurred vision.

Gemini 8 approaching the Agena target. *NASA*

Gemini 8 and the Agena target about to dock. *NASA*

The problem was finally solved when the astronauts shut down the capsule's attitude control system and used the re-entry control system to regain control of the spacecraft. This choice, however, required them to abort the rest of their planned three-day mission. On their seventh revolution, they fired their retro-rockets and splashed down in the Pacific Ocean, where they were safely recovered.

The next manned docking attempt was **Gemini 9 (June 3, 1966)**.

March 30, 1966
OV1-4 / OV1-5
U.S.A.

These two satellites, part of the U.S. Air Force's Orbiting Vehicle (OV) program (see **OV1-2–October 5, 1965**), performed space research for the military.

- *OV1-4* This satellite carried three experiments. One tested the ability of four different types of thermal coatings to endure the harsh environment of space. The other two experiments studied how chlorella algae and duckweed plants grew in weightlessness. OV1-4 gathered data for 30 days.

- *OV1-5* This satellite tested five different infrared sensors for remote observations of the earth. OV1-5's instruments were early experiments of the kinds of optical equipment used on later Earth resource satellites such as **Landsat 1 (July 23, 1972)**.

March 31, 1966
Luna 10
U.S.S.R.

Luna 10 was the first spacecraft to go into orbit around another world, operating in lunar orbit for about two months. Although Luna 10 carried no cameras, it had several instruments for studying the lunar environment. It detected evidence of a very faint and weak magnetic field around the Moon, caused not by a lunar magnetic field but by the Moon as it plowed its way through the Sun's magnetic field. Other Luna 10 data, however, left some doubt about this conclusion. See **Imp E (July 19, 1967)**.

Measurements of changes in the spacecraft's orbit also indicated that the Moon's gravitational field was asymmetrical on both the near and far sides. See **Lunar Orbiter 1 (August 10, 1966)**.

Luna 10 also detected evidence of the earth's magnetic tail to lunar distances, about 240,000 miles, the farthest distance yet measured. See **Imp D (July 1, 1966)**.

See **Luna 12 (October 22, 1966)** for a summary of results from the Soviet lunar orbiters.

April 22, 1966
OV3-1
U.S.A.

Part of the U.S. Air Force research series of Orbiting Vehicles (OVs) (see **OV1-2–October 5, 1965**), OV3-1 carried six experiments to measure the radiation within the Van Allen belts.

April 25, 1966
Molniya 1-03
U.S.S.R.

Molniya 1-03 was the third Soviet communications satellite, joining **Molniya 1-01 (April 23, 1965)** and **Molniya 1-02 (October 14, 1965)** to complete the first Molniya satellite cluster. The spacecraft also carried television cameras and took cloud-cover images from just under 25,000 miles elevation. See Molniya 1-01 for program details.

May 25, 1966
Explorer 32 (Atmosphere Explorer B)
U.S.A.

Also called Atmosphere Explorer B, Explorer 32 was the second in NASA's five-satellite project to study the upper atmosphere, following **Explorer 17 (April 3, 1963)**. Explorer 32 operated through December 1966.

Data from these two Explorer spacecraft revealed that although the atmosphere's density fluctuated in synch with the Sun's 27-day rotation, the change was caused by events taking place at lower altitudes than explored by either spacecraft. Most of the Sun's ultraviolet radiation was absorbed at these lower altitudes, somewhere below 100 miles, and the added energy was what apparently caused the changing density of the atmosphere.

The next three satellites in this atmospheric research project, beginning with **Explorer 51 (December 16, 1973)**, attempted to explore those lower regions.

May 30, 1966
Surveyor 1
U.S.A.

Surveyor 1 made the second soft landing on another planet, following **Luna 9 (January 31, 1966)**. Like Luna 9, Surveyor 1 landed in the Ocean of Storms, coming to rest on June 2, 1966, just south of the lunar equator.

Because the spacecraft was NASA's first attempt to soft-land on another planet, it was designed mostly as an engineering test. Other than its television cameras, Surveyor 1 carried only three small experiments for measuring the ground's bearing strength, temperature, and radar reflectivity. These data were

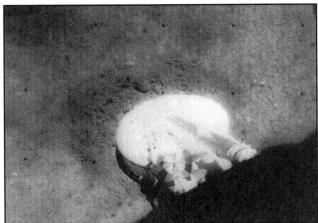

Note the disturbed soil around Surveyor 1's landing pad. The dark area in the lower right is the spacecraft's own shadow. *NASA*

June 1, 1966

crucial for any manned landing attempt, answering questions that the Ranger spacecraft had not (see **Ranger 6–July 28, 1964; Ranger 7–February 15, 1965; Ranger 8–March 21, 1965**).

During its month of operation, Surveyor 1 transmitted over 11,000 pictures. These images revealed a flat, level, almost featureless plain, studded with small craters ranging from several inches across to several hundred feet in diameter. These craters all resembled those seen from the Ranger spacecraft.

The images also revealed a surface littered with coarse blocks and rock fragments. Surrounding one crater was a field strewn with blocks. The blocks were generally made of a material that was brighter than the ground on which they rested. A few exhibited melt features either from impact or volcanic action. Some showed slightly rounded edges. Many more were fractured with clearly visible cracks. All implied a blanket of debris ejected from the various visible craters.

Careful study of the photographs of the spacecraft's landing pads indicated that the lunar surface below them was composed of a granular soil-like material.

The nature of the material on the lunar surface was unknown, but its behavior upon impact resembled that of damp, fine-grained soil made up of a weaker layer about one inch deep coating a hard underlying layer. All underlying material exposed by the spacecraft's landing was darker than the surface. Since the appearance beneath all three landing pads appeared similar, scientists concluded that the composition of the "soil" was homogeneous, at least in the area across the spacecraft's width. And while its nature was unknown, its bearing strength was clearly enough to sustain the weight of a manned spacecraft.

Temperature readings indicated a surface temperature averaging around 180°F, about 50°F higher than predicted from Earth-based measurements.

The next successful lunar landing was **Luna 13 (December 21, 1966)**, followed by **Surveyor 3 (April 17, 1967)**.

June 1, 1966
Gemini 9 ATDA
U.S.A.

Because of the failure of the Agena target vehicle during the mission of **Gemini 6 (December 15, 1965)**, NASA had developed this simple backup, which was called the Augmented Target Docking Adapter (ATDA). Made from the Agena docking cone and a simple package of small attitude thrusters, it had no rocket engines and was lifted into orbit entirely by the Atlas rocket.

When the Atlas-Agena target vehicle for **Gemini 9 (June 3, 1966)** failed to reach orbit on May 17, 1966, this ATDA backup was placed in orbit. However, the protective nosecone shroud did not release after launch. Partially open and resembling what Gemini 9 astronaut Tom Stafford called "an angry alligator," the shroud blocked the docking port and prevented **Gemini 9** from docking with the ATDA.

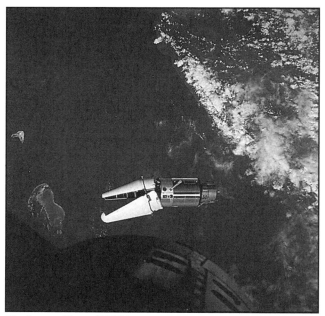

The "angry alligator." *NASA*

June 3, 1966
Gemini 9
U.S.A.

Gemini 9 was NASA's third attempt to achieve a completely successful in-space hard docking, following **Gemini 6 (December 15, 1965)** and **Gemini 8 (March 16, 1966)**. The astronauts were Tom Stafford (**Gemini 6–December 15, 1965; Apollo 10–May 18, 1969; Apollo 18–July 15, 1975**) and Eugene Cernan (**Apollo 10–May 18, 1969; Apollo 17–December 7, 1972**).

Despite three successful rendezvous maneuvers, Gemini 9 was unable to dock because the nosecone shroud on the Agena target vehicle (**Gemini 9 ATDA–June 1, 1966**) failed to release, thereby blocking the docking port. Therefore, instead of docking, the astronauts practiced precise maneuvers in space. In one maneuver, Stafford delicately held the Gemini 9 capsule mere inches away from the slowly rolling ATDA as he sustained the two spacecraft's precise orientation.

During the three-day Gemini 9 mission, Eugene Cernan performed the third space walk in history, which lasted two hours and seven minutes. Cernan discovered that even simple movements became exceedingly complex in microgravity. His every gesture changed the spacecraft's attitude. He also found the handholds attached to the capsule's exterior inadequate for holding him in place while he worked. As he struggled to mount a redesigned jetpack for maneuvering in space (see **Gemini 4–June 3, 1965**), he began to sweat, and this increased moisture completely fogged the inside of his visor. Because Cernan was essentially blinded by the condensation, he and Stafford decided to forego testing the jetpack, which was then not tested in space again until **Skylab 3 (July 28, 1973)**. At the end of the space walk, Cernan groped his way back to the capsule's hatch. There he attempted to take some photographs before returning to the cabin but found it difficult to frame his pictures with a fogged visor.

The next attempt to dock in space was **Gemini 10 (July 18, 1966)**.

June 7, 1966
OGO 3
U.S.A.

OGO 3 was part of NASA's six-satellite American program dedicated to systematically studying the magnetosphere at widely separate locations and latitudes. Like **OGO 1 (September 5, 1964)**, OGO 3 was placed in an eccentric orbit (perigee: 12,126 miles; apogee: 63,881 miles) to study high altitudes. The spacecraft carried 23 different experiments.

OGO 3's data, when combined with that from OGO 1, **Elektron 2 (January 30, 1964)**, and **Elektron 4 (July 11, 1964)**, measured the long-term changes to the Van Allen belts, finding that since **Explorer 12 (August 16, 1961)**, the radiation in the belts had intensified as the Sun began the transition from its quiet to its active state.

The spacecraft crossed the bow shock and magnetopause over 500 times, measuring the energy flux of the charged particles and their density at the bow shock, data that indicated the nature of the particle interaction between the solar wind and the magnetosphere. Often the bow shock swept back and forth past the spacecraft several times during a single orbit, indicating the sudden influence of the solar wind on the magnetosphere's shape and size.

See **OGO 5 (March 4, 1968)** for a summary of results.

June 9, 1966
ERS 16 (ORS 2)
U.S.A.

The Environmental Research Satellite (ERS) 16—also called ORS 2 (see **ORS 3–July 20, 1966**)—was one of two U.S. Air Force satellites launched in 1966 to test metal-to-metal bonding experiments while in orbit. Using four solenoid valves and a solenoid actuator powered by batteries and solar cells, a variety of metal surfaces were brought together and subjected to an electrical current. The hard vacuum of space combined with weightlessness facilitated the adhesion, producing welds in space without melting.

June 10, 1966
OV3-4
U.S.A.

Part of the U.S. Air Force's research series of Orbiting Vehicles (OVs) (see **OV1-2–October 5, 1965**), OV3-4 tested the hazards to living tissues posed by radiation in space. The results were used by the Air Force in its manned military program, the Manned Orbital Laboratory, which was eventually cancelled.

June 16, 1966
IDCSP 1 through IDCSP 7 / GGTS 1
U.S.A.

- *IDCSP 1 through IDCSP 7*   These seven satellites, part of the Initial Defense Communications Satellite Program (IDCSP) of the U.S. military, were the first American military communications satellites since **Courier 1B (October 4, 1960)**. Because of the previous military satellite failures, the IDCSP active repeater satellites were specifically designed for simplicity. They worked automatically, requiring no ground commands. Spin stabilized and launched in large numbers, they required no engines to maintain their positions or orientation in space. Used for strategic worldwide communications, generally with large ground-based antennas, each satellite could transmit up to 11 two-way voice circuits or 1,550 teletype messages at the same time.

Through June 1968, the military successfully orbited 26 IDCSP satellites, launched in groups of seven, eight, three, and eight. By 1967, the cluster was used to transmit high-speed digital data between Vietnam and Washington, D.C. By 1968 it was declared officially operational. The average life span of each satellite continued well beyond the predicted six years, with one still operating as late as 1977.

To supplement the IDCSP satellites, the U.S. built and launched **Skynet 1 (November 22, 1969)** for the United Kingdom, the first successful geosynchronous military satellite.

The next American military satellite system for providing worldwide strategic communications were the DSCS II satellites (see **DSCS II 1/DSCS II 2–November 3, 1971**). The next U.S. military satellite was **TACSAT (February 9, 1969)**, the first operational satellite for relaying tactical mobile communications.

- *GGTS 1*   Placed in geosynchronous orbit, GGTS (Gravity Gradient Test Satellite) 1 continued the military's program to test the use of extendible booms for orienting spacecraft, begun with **GGSE 1 (January 11, 1964)**. By retracting or extending booms, the satellite's center of gravity was changed, thereby causing its spin and orientation to change relative to the earth. Because of GGTS 1's high orbit (and the resulting weakness of the earth's gravity), changes to its orientation were slow. It took 40 days for its booms to bring the satellite to its desired spin and orientation, and engineers discovered that the magnetic fields of the earth and Sun, as well as the solar wind, prevented the spacecraft from pointing directly down at the center of the earth. Instead, its angle was off center by 16 degrees. The next test satellite, **Dodge 1 (July 1, 1967)**, was redesigned accordingly.

June 24, 1966
Pageos 1
U.S.A.

Similar to **Echo 1 (August 12, 1960)** and **Echo 2 (January 25, 1964)**, Pageos 1 was a 100-foot-diameter balloon that inflated once in orbit. During its 10 years in orbit, scientists tracked the decay of its initial orbit (perigee: 1,835 miles; apogee: 3,235 miles) in order to calculate very precisely the earth's shape, as well as changes to the earth's atmospheric density.

Just as they had with the previous balloon satellites, the Soviets put together a vast network of ground stations across Europe, the Soviet Union, and even Mongolia to make synchronized observations of Pageos 1. These observations allowed them to triangulate precisely the position of both the satellite and the ground stations for measuring the earth's shape.

The next atmospheric density balloon satellite was **Dash 2** (**October 5, 1966**).

June 25, 1966
Cosmos 122
U.S.S.R.

Cosmos 122 was what the Soviets called a "second stage" test flight of a Soviet weather satellite, the first Soviet weather satellite about which any detail is known. See **Cosmos 44** (**August 28, 1964**) for earlier tests. Like the American **Tiros 4** (**February 8, 1962**), Cosmos 122 carried two television cameras for photographing the cloud cover during the day, and an infrared detector for measuring the cloud cover at night. Also like the Tiros satellites, Cosmos 122 used the earth's magnetic field to help orient the spacecraft (see **Tiros 1–April 1, 1960**).

The next Soviet weather satellite test was **Cosmos 144** (**February 28, 1967**).

July 1, 1966
Imp D (Explorer 33)
U.S.A.

Imp D was designed by NASA to gather information on the earth's magnetosphere in lunar orbit. Although an incorrect engine burn prevented it from reaching lunar orbit, Imp D was still able to obtain much of its planned data on the earth's magnetosphere. It was turned on and off during optimum data-collecting periods through September 1971. From this data, scientists measured the tail of the earth's magnetic field out to 320,000 miles, finding that it was still over 160,000 miles wide at that point.

See **OGO 5** (**March 4, 1968**) and **OGO 6** (**May 5, 1969**) for additional results.

July 6, 1966
Proton 3
U.S.S.R.

This flight was the third test of the Soviet heavy-lift Proton rocket, and like its predecessors (see **Proton 1–July 16, 1965**), Proton 3 was one of the heaviest objects ever placed in orbit, weighing almost 27,000 pounds. It carried a gamma-ray telescope, which operated for about two months.

Proton 4 (**November 16, 1968**) tested an even more powerful rocket.

July 14, 1966
OV1-8
U.S.A.

Part of the U.S. Air Force research series of Orbiting Vehicles (OVs) (see **OV1-2–October 5, 1965**), OV1-8 performed radio communications tests for the military, placing in orbit a 30-foot-diameter balloon covered with fine aluminum wires to act as a passive reflective communications satellite. This test was a variation on the work done with **West Ford 2** (**May 9, 1963**).

July 18, 1966, 20:38 GMT
Gemini 10 Agena Target
U.S.A.

Launched 1 hour and 41 minutes before **Gemini 10** (**July 18, 1966**), this Agena target vehicle functioned exactly as designed. Once the two spacecraft were docked, its engines were used to raise Gemini 10 to 474 miles altitude, a human altitude record.

July 18, 1966, 22:22 GMT
Gemini 10
U.S.A.

After three attempts (**Gemini 6–December 15, 1965**; **Gemini 8–March 16, 1966**; **Gemini 9–June 3, 1966**), Gemini 10 finally accomplished a completely successful docking in space. The astronauts were John Young (**Gemini 3–March 23, 1965**; **Apollo 10–May 18, 1969**; **Apollo 16–April 16, 1972**; **STS 1–April 12, 1981**; **STS 9–November 28, 1983**) and Mike Collins (**Apollo 11–July 16, 1969**).

Less than six hours into the flight, the spacecraft made a clean and successful docking with the **Gemini 10 Agena Target** (**July 18, 1966**). The astronauts then used the Agena's rocket engine to lift themselves into a 183-mile by 474-mile orbit, which carried them higher than any human had ever flown. There they abandoned their Agena and rendezvoused with the **Gemini 8 Agena Target** (**March 16, 1966**) that had been left in a parking orbit after Gemini 8's abrupt return to Earth. Mike Collins then made an exciting and successful space walk from Gemini 10 to the Gemini 8 Agena to recover a micrometeorite experiment that had been exposed to space for four months. Close inspection of the package revealed four impacts larger than 4 microns, with one 200 microns wide and 35 microns deep.

Collins completed one other space walk, or EVA (extra-vehicular activity), during this three-day flight. Based on what had been learned from Eugene Cernan's experience on **Gemini 9** (**June 3, 1966**), NASA had decided to divide EVAs into smaller chunks. Furthermore, Collins carefully coated the inside of his visor with an anti-fogging solution just prior to opening the hatch. He then stood up in the capsule's hatch, where he took 20 ultraviolet pictures of the Milky Way, as well as test photos of color patches to see if the film registered the same hues in the vacuum of space. The photographs, while unusable for astronomical purposes, illustrated the technical requirements for such research.

The next manned mission was **Gemini 11** (**September 12, 1966**).

August 4, 1966
OV3-3
U.S.A.

Part of the U.S. Air Force research series of Orbiting Vehicles (OVs) (see **OV1-2–October 5, 1965**), OV3-3 carried seven experiments for measuring charged particles. Its purpose was to gauge the threat posed by these ions to equipment.

Earth above a lunar horizon, photographed from Lunar Orbiter 1. *NASA*

August 10, 1966
Lunar Orbiter 1
U.S.A.

Lunar Orbiter 1 was the first American spacecraft to orbit another world, and the second overall following **Luna 10 (March 31, 1966)**. The spacecraft's mission was primarily to photograph the southern part of what NASA labeled "the Apollo zone of interest," a region of potential landing sites and approach landmarks for the planned Apollo manned landing (see **Apollo 11–July 16, 1969**). This zone of interest covered 5 degrees latitude north and south and 45 degrees longitude east and west of the Moon's near side center. Lunar Orbiter 1 was to concentrate on 10 specific sites within that zone. To do this, the spacecraft was placed in a low-altitude equatorial orbit, preventing the observation of higher latitudes. Nonetheless, Luna Orbiter 1 photographed almost 50 percent of the far side of the Moon, returning the first pictures of that surface since **Zond 3 (July 18, 1965)** and the first high-resolution photographs ever. In total, 413 pictures were taken.

The images revealed a lunar crust significantly fractured and faulted, with evidence of a highly active past that included much volcanic activity—the mare regions on the near side were relatively smooth, suggesting that these were the youngest features on the Moon with little recent impact activity. The far-side images, however, showed a much rougher and more heavily cratered surface, with notably fewer mare regions. Although these photographs did not settle the argument about whether the lunar craters were formed from volcanoes or from impact, the crater evidence from the far side of the Moon seemed to favor the impact theory.

Lunar Orbiter 1 also took the first photograph of the earth above a lunar horizon. In this image, both the east coast of the United States and Antarctica are visible.

One instrument on board Lunar Orbiter 1 measured the frequency of meteorite impacts in the vicinity of the Moon. Registering no impacts during the mission's 80 days in lunar orbit, the detector indicated that the impact rate was probably less than that seen near Earth.

Another instrument measured radiation levels, finding that these corresponded to those of interplanetary space. During several solar flares, however, the rates jumped appreciably. This data was correlated with information returned from **Pioneer 7 (August 17, 1966)**, allowing scientists to measure the relative size and strength of at least one solar flare event.

During its service life, scientists used the spacecraft's changing orbit around the Moon to try to calculate precisely the Moon's shape and gravitational field. They found that the Moon was shaped like a pear, with a bulge in its north pole and a depression at its south pole. These variations made predicting the spacecraft's exact orbit difficult and thus increased the difficulty of planning manned missions. More data was required to reduce the size of the errors.

After 2.5 months of operation, Luna Orbiter 1 was deliberately sent crashing onto the Moon to avoid interfering with **Lunar Orbiter 2 (November 6, 1966)**.

August 17, 1966
Pioneer 7
U.S.A.
Pioneer 7 was the second in NASA's cluster of four solar-orbiting Pioneer satellites designed to study the solar wind, following **Pioneer 6 (December 16, 1965)**. See **Pioneer 9 (November 8, 1968)** for a summary of results.

Pioneer 7 also provided data about the tail of the earth's magnetic field, finding evidence for its existence more than four million miles downwind from the earth. Some scientists interpreted the data as a "magnetospheric wake," not unlike the wake following a boat as it plows its way through the water. Other scientists, however, disagreed, finding no evidence from Pioneer 7 of the earth's magnetic tail at these distances.

August 24, 1966
Luna 11
U.S.S.R.
Luna 11 was the second Soviet spacecraft to be placed in lunar orbit following **Luna 10 (March 31, 1966)**, and the third following the American **Lunar Orbiter 1 (August 10, 1966)**. Like Luna 10, Luna 11 carried no cameras. Instead, it studied the composition of the Moon and space around it using spectrometers and magnetometers. See **Luna 12 (October 22, 1966)** for a summary of results.

September 12, 1966, 13:12 GMT
Gemini 11 Agena Target
U.S.A.
This Agena target vehicle was launched 1 hour and 26 minutes before **Gemini 11 (September 12, 1966)** and was its docking target.

September 12, 1966, 14:38 GMT
Gemini 11
U.S.A.
Gemini 11 broke the altitude record set by **Gemini 10 (July 18, 1966)** while further refining the engineering challenges of orbital rendezvous. The astronauts were Pete Conrad (**Gemini 5–August 21, 1965; Apollo 12–November 14, 1969; Skylab 2–May 25, 1973**) and Richard Gordon (**Apollo 12–November 14, 1969**).

The astronauts successfully docked with the **Gemini 11 Agena Target (September 12, 1966)** during their first orbit. On the mission's second day, the astronauts used the Agena to lift them to a record altitude of 850 miles. From this height, the earth's curve was clearly visible, and the two astronauts took over 300 pictures of the cloud cover, geology, and airglow layer.

Gemini 11 also repeated the blood sample–radiation experiment conducted on **Gemini 3 (March 23, 1965)**. The results, however, did not match Gemini 3's, finding instead that there was no increase in radiation damage during weightlessness.

Richard Gordon performed two space walks, once again finding zero gravity a difficult place to work. When he opened the hatch, both he and everything unfastened in the capsule was sucked toward space. Pete Conrad had to grab a leg strap on Gordon's spacesuit to prevent him from drifting away. Later, Conrad had to pull him back using his umbilical cord. The arduous nature of the work caused both Gordon and his spacesuit to overheat, leading him to terminate the first space walk after only 33 minutes.

During this first EVA, however, he managed to attach a tether to the Agena *(see color section, Figure 21)*. Later in the flight, when the two spacecraft separated, this tether allowed the vehicles to rotate around a common center of gravity. For about 90 minutes, they swung about like a dumbbell, using centrifugal force to create artificial gravity. This experiment was repeated on **Gemini 12 (November 11, 1966)**.

During Gordon's second space walk, which was conducted standing up in the capsule's hatch, he recorded the ultraviolet spectrum of the star Canopus, as well as other ultraviolet star images, the first ever made. See **Gemini 12 (November 11, 1966)** for more ultraviolet astronomy. Gordon also recovered a nuclear emulsion experiment placed on the outside of the capsule. During the daytime portion of the EVA, when star photography was impossible, he floated in the open hatch, watching the earth slowly drift by—and drifting to sleep himself.

On the next (and final) Gemini mission, Gemini 12, Buzz Aldrin became the first human to work successfully during an EVA.

October 5, 1966
Dash 2
U.S.A.
Density and Scale Height (DASH) 2 was an eight-foot-diameter balloon launched by the U.S. military to measure long-term atmospheric density in low-Earth orbit, complementing the data from **Pageos 1 (June 24, 1966)**. During this period of increasing solar activity, atmospheric drag quickly changed the satellite's circular orbit to an eccentric one, and in 1971, its perigee had dropped so low that the spacecraft re-entered the atmosphere.

The next atmospheric density satellite was **San Marco 2 (April 26, 1967)**.

October 22, 1966
Luna 12
U.S.S.R.
The third Soviet lunar orbiter, following **Luna 10 (March 31, 1966)** and **Luna 11 (August 24, 1966)**, was different in that it carried cameras for photographing the lunar surface. For unknown reasons, however, few pictures from this spacecraft have ever been released.

The flux of dust particles from all three Soviet lunar orbiters was found to be greater than that seen in interplanetary space by **Mariner 2 (August 27, 1962)** and **Mariner 4 (November 28, 1964)**. Using spectrometers, the three Soviet orbiters studied the composition of the lunar surface to a depth of about one foot, finding that the lunar mare regions were similar to Earth basalt—solidified molten rock. The lunar surface was generally dominated by the elements oxygen, magnesium, aluminum, and silicon.

The spacecraft also measured the gamma radiation at the lunar surface, finding that the measured radiation exceeded that of Earth's, with 90 percent of what was recorded caused by the bombardment of cosmic rays.

The next Soviet lunar orbiter was **Luna 14** (April 7, 1968). The next lunar orbiter was **Lunar Orbiter 2** (November 6, 1966).

October 26, 1966
Intelsat 2A
International / U.S.A.

Intelsat 2A was the second communications satellite owned by the consortium known as Intelsat (see **Early Bird–April 6, 1965**); it was intended to complement Early Bird's position over the Atlantic by being placed above the Pacific. When Intelsat 2A's final rocket stage malfunctioned, however, the satellite failed to reach geosynchronous orbit. Despite its low orbit, Intelsat 2A was still usable about 4 to 8 hours per day, and it operated until it was replaced by **Intelsat 2B** (**January 11, 1967**).

October 28, 1966
OV3-2
U.S.A.

Part of the U.S. Air Force research series of Orbiting Vehicles (OVs) (see **OV1-2–October 5, 1965**), OV3-2 carried five experiments for studying particles in the Van Allen belts. The satellite's launch was timed so that it was operating during the November 12, 1966, solar eclipse, in order to determine the effect to its equipment of a sudden loss of solar energy. The findings were used for improving the designs of military surveillance satellites.

November 3, 1966
OV4-1T / OV4-1R / Gemini 2
U.S.A.

- *OV4-1T and OV4-1R*   These two satellites were part of the U.S. Air Force research series of Orbiting Vehicles (OVs). See **OV1-2** (**October 5, 1965**).

Called the Whispering Gallery experiments, these satellites tested the limits of satellite communications, even when two satellites were on opposite sides of the earth. (An earlier attempt to do this failed when the launch rocket failed to properly separate the two satellites; see **LES 3/LES 4–December 21, 1965**.) As early as **Sputnik 2** (**November 3, 1957**), it had been found that, just as ground-based radio signals can be bounced off the ionosphere to extend communication links beyond the horizon, signals from orbiting satellites can do the same.

For this experiment, OV4-1T was the transmitting satellite while OV4-1R was the receiving satellite. OV4-1T transmitted continuously, and ground controllers successfully used OV4-1R over 30 times through January 1967 to test their ability to pick up its transmission at distances ranging from a few hundred miles to orbital positions on opposite sides of the earth. These tests were also the first direct unmanned-satellite-to-unmanned-satellite crosslink, though manned flights had achieved direct spacecraft-to-spacecraft communications beginning with **Vostok 3** (**August 11, 1962**) and **Vostok 4** (**August 12, 1962**).

- *Gemini 2*   This launch reused the capsule from the **Gemini 2** (**January 19, 1965**) flight to test equipment for the Air Force's Manned Orbital Laboratory (MOL) program—which made it the first object to fly into space twice. According to MOL designs, a manned Gemini capsule and laboratory eventually would be sent together into orbit, with access to the laboratory through a hatch cut into the capsule's heat shield and covered with the heat shield's ablative material. This flight tested the hatch and heat shield's combined ability to protect the capsule during re-entry. When the capsule was recovered, the engineers found that the hatch had been melted shut by the heat of re-entry, demonstrating that the design worked.

The next time a spacecraft made a second trip into space was the second flight of the space shuttle Columbia. See **STS 2** (**November 12, 1981**).

November 6, 1966
Lunar Orbiter 2
U.S.A.

Lunar Orbiter 2 was the second American spacecraft to orbit the Moon. Like **Lunar Orbiter 1** (**August 10, 1966**), it was placed in a low equatorial orbit so that it could photograph landing sites and landmarks in what NASA called "the Apollo zone of interest," an area bounded by 5 degrees north and south latitude and 45 degrees east and west longitude on the Moon's near side. Within this zone, the United States planned to land the first manned spacecraft on the Moon (see **Apollo 11–July 16, 1969**).

Lunar Orbiter 2's objective was to photograph 13 potential landing sites within that zone; during its 11 months in lunar orbit, the spacecraft took 422 images, both of the 13 sites of greatest interest and of 17 secondary sites. Several of Lunar Orbiter 2's photographs were used by **Surveyor 6** (**November 7, 1967**) to study the geology of that spacecraft's landing site. One photograph identified the impact site of **Ranger 8** (**February 17, 1965**).

The photographs revealed evidence of volcanic activity in the Ocean of Storms. The images also uncovered a number of tall pinnacle-like objects, ranging from 20 feet to 75 feet in height, casting long shadows across the western edge of the Sea of Tranquility.

Three micrometeoroid impacts were registered during Lunar Orbiter 2's life span. Though causing no serious damage, they indicated an increased particle flux in lunar space since Lunar Orbiter 1, possibly from the annual Leonid meteor shower in mid-November.

The Moon's interior and shape was further mapped, indicating that although it is pear shaped, it also has additional bulges around its equator, one located on the near side closest to the earth. These bulges, in later years dubbed "mascons" (for mass concentrations), implied that the interior is irregular, not homogeneous as had previously been believed. These irregularities also explained the aberrations to the orbital paths of both Lunar Orbiter 1 and **Luna 10** (**March 31, 1966**).

The next lunar orbital mission was **Lunar Orbiter 3** (**February 5, 1967**).

**November 11, 1966, 19:12 GMT
Gemini 12 Agena Target
U.S.A.**
This spacecraft, the last Agena Target Vehicle launched, was used by **Gemini 12 (November 11, 1966)** for rendezvous and docking practice.

**November 11, 1966, 20:53 GMT
Gemini 12
U.S.A.**
Gemini 12, focused specifically on the problem of extravehicular activities, completed the American Gemini Program. The astronauts were Jim Lovell (**Gemini 7–December 4, 1965; Apollo 8–December 21, 1968; Apollo 13–April 11, 1970**) and Buzz Aldrin (**Apollo 11–July 16, 1969**).

Unlike previous space walks, Gemini 12's EVAs were designed less to accomplish particular tasks than to study the difficulties of working in space. On the Gemini 12 flight, Aldrin became the first human to work during a space walk. He performed a number of carefully designed tasks, such as turning a bolt, connecting and disconnecting plugs, and cutting wires, so that engineers could learn how the human body reacted to the experience. Based on previous space walk difficulties, the spacecraft was provided with extensive handrails, and a waist tether held Aldrin to his work stations (not unlike the safety line used by a high rise window washer). Furthermore, two-minute rest periods were scheduled to allow Aldrin to cool down after each task.

Aldrin performed each task several times, progressively reducing his hold on the spacecraft so that ground controllers could measure his body's reaction to the increased strain. He also completed two work sessions, once during darkness and once in daylight. At the end of the space walk, he also managed to wipe the outside surface of Jim Lovell's window, removing a thin film that obscured the view. This film had developed on every previous flight, and Aldrin's cleaning rag was used to find out how to eliminate this problem.

Aldrin also performed two standup EVAs, during which he continued the ultraviolet astronomy begun on **Gemini 11 (September 12, 1966)**. He recorded the ultraviolet spectrum of the star Sirius, as well as ultraviolet images of many dim stars, the last such images until **OAO 2 (December 7, 1968)**. This data was used to calibrate solar spectra taken by suborbital rockets. Aldrin also recovered a micrometeorite package attached to the outside of the spacecraft and took cloud-cover photographs of the earth.

The rendezvous with the Agena target vehicle, though almost routine by this flight, was done manually when the rendezvous radar equipment failed. Lovell twice docked the capsule to the Agena. Later he carefully adjusted Gemini 12's orbit so that the astronauts could make the first observations of a solar eclipse from space.

Lovell and Aldrin also repeated the tether experiment from Gemini 11, and their spacecraft rotated while attached to the Agena for almost 4.5 hours. Though the two spacecraft experienced a bounce between them, the astronauts had no trouble damping this oscillation using Gemini 12's attitude control system.

Other than its use by the military NOSS satellites (see **NOSS 1–April 30, 1976**), this flight was the last attempt until **Seds (March 30, 1993)** to fly two spacecraft connected by a tether. A more sophisticated but unsuccessful tether experiment was attempted twice by the space shuttle (**STS 46–July 31, 1992; STS 75–February 22, 1996**).

The next planned American manned flight, **Apollo 1 (January 27, 1967)**, caused the death of three astronauts. The next manned mission, **Soyuz 1 (April 23, 1967)**, caused the death of one cosmonaut.

**November 28, 1966
Cosmos 133
U.S.S.R.**
This unmanned mission was the first test flight of the Soyuz spacecraft. This craft, which became the foundation of the Soviet and Russian manned space program for the next several decades, comprised three separate modules. The rear equipment module was similar to the service modules on the American Gemini and Apollo capsules, carrying the spacecraft's retro-rockets and fuel tanks. The middle descent module, a bell-shaped crew compartment with a heat shield at its base, was designed to return cosmonauts to Earth. The front module, a spherical capsule called the orbital module, was used only in space and had a docking port for linking up with other Soyuz spacecraft.

This first Soyuz design used solar energy rather than the fuel cells and batteries used by previous manned capsules. Two wing-like solar panels unfolded on each side of the equipment section once Soyuz reached orbit.

Cosmos 133 remained in orbit for two days. During its flight, the spacecraft's attitude control system malfunctioned, wasting fuel and making it impossible to orient the spacecraft precisely for re-entry. Because Cosmos 133 would have landed in China instead of the Soviet Union, the self-destruct command was given, exploding the unmanned spacecraft before impact.

The next Soyuz test flight was **Cosmos 140 (February 7, 1967)**.

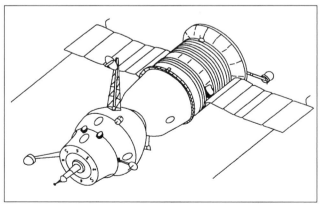

The original Soyuz spacecraft. *NASA*

December 7, 1966
ATS 1
U.S.A.
ATS (Applications Technology Satellite) 1 was designed and built by NASA to test geosynchronous communication and meteorological technology, following **Syncom 2 (July 26, 1963)** and **Syncom 3 (August 19, 1964)**. ATS 1 was capable of relaying 1,200 one-way voice circuits or one color television channel. In tests, the satellite achieved the first air-to-air and air-to-ground communications relay using an orbital satellite. The spacecraft also tested system noise, delay, frequency response, and other communications satellite engineering problems.

Though designed for a three-year life span, ATS 1 was in use by a number of customers for almost 20 years. For example, in later years, ATS 1—in combination with **ATS 3 (November 5, 1967)**—was used by Alaska and many Pacific islands for communications purposes. In 1970, the University of Alaska established a program using ATS 1 to link the state's many remote villages with its major hospitals using local home-made ground stations. Patients getting medical advice through these links tripled within a year. In 1971, the University of Hawaii used ATS 1 to establish a communications network throughout the South Pacific. A number of additional ground stations were built in Hawaii, Australia, New Zealand, Fiji, and various remote islands, using ATS 1 about 20 hours per week for communications. When linked with the Alaskan system, doctors in the South Pacific were able to confer with doctors in Alaska.

ATS 1's meteorological experiments laid the foundation for the later GOES meteorological satellites (see **SMS 1–May 17, 1974**). Its television camera took a global picture of the earth's daylight cloud cover every 30 minutes. Images had a resolution of two miles.

December 21, 1966
Luna 13
U.S.S.R.
Like both **Luna 9 (January 31, 1966)** and **Surveyor 1 (May 30, 1966)**, Luna 13 soft-landed in the Moon's Ocean of Storms. And like both previous landers, Luna 13 found itself surrounded by a smooth terrain with few distinct features. Based on photographic analysis, the spacecraft had landed, like Luna 9, inside a crater about 50 feet across and 2 feet deep.

The terrain was dominated by many small craters ranging in diameter from one inch to nine feet. Like Surveyor 1, Luna 13 found the surface littered with rocks, most of which appeared to be ejecta from other nearby impacts.

Also visible were mysterious thin linear features running across the surface in all directions. Sometimes these lines could be seen cutting through craters. Though the make-up of these veins was unknown, Soviet scientists suggested that they were made of a material more durable than the surrounding rock.

Luna 13 carried an instrument for impacting the lunar surface three feet from the spacecraft and measuring the instrument's depth of penetration. From this experiment, it was inferred that the ground's density at Luna 13's landing site was considerably less than Earth's and less than the mean density of the Moon. As had Surveyor 1's scientists, the Soviet scientists concluded that the ground was made of porous or lightly bonded granular rocks.

Another instrument measured the temperature at the landing site. At lunar noon, the temperature read 242°F.

The next successful lunar soft-landing was **Surveyor 3 (April 17, 1967)**.

1967

January 11, 1967
Intelsat 2B
International / U.S.A.
Intelsat 2B replaced **Intelsat 2A (October 26, 1966)**, relaying communication signals from its geosynchronous orbit above the Pacific Ocean. It was capable of handling 240 simultaneous two-way voice circuits.

Combined with **Early Bird (April 6, 1965)** over the Atlantic and the Soviet Union's Molniya satellites (see **April 23, 1965**), reliable and instantaneous worldwide communications became possible for the first time across the entire globe.

Intelsat 2B was the second satellite in Intelsat's first four-satellite cluster (see Early Bird for more details). The next Intelsat satellite was **Intelsat 2C (March 23, 1967)**.

January 27, 1967
Apollo 1
U.S.A.
Apollo 1 was planned to be the first manned flight of the third-generation American space capsule; it was to have been flown by Gus Grissom (**Liberty Bell 7–July 21, 1961**; **Gemini 3–March 23, 1965**), Ed White (**Gemini 4–June 3, 1965**), and Roger Chaffee. Problems delayed the launch from late 1966 until February 1967. Then, during a simulated countdown on January 27, 1967, a fire broke out inside the capsule, killing all three men. The subsequent NASA investigation attributed the tragedy to the difficulty of opening the hatch; the use of flammable materials in the capsule, combined with an extremely combustible atmosphere of pure oxygen; and a general over-confidence and sloppiness in quality control.

As a result, the capsule's hatch was redesigned so that astronauts could flee quickly in an emergency. On the launchpad, a mixed atmosphere of 40 percent nitrogen and 60 percent oxygen was introduced to lessen the threat of fire during a launch—with the nitrogen then purged once the capsule was in space. Over 2,500 items were removed from the capsule and replaced with nonflammable materials. And NASA's quality-control systems were overhauled to prevent sloppy workmanship.

The next Apollo flight, unmanned **Apollo 4 (November 9, 1967)**, was the first full test of the Saturn 5 rocket. The next manned Apollo mission was **Apollo 7 (October 11, 1968)**. The next manned mission was **Soyuz 1 (April 23, 1967)**.

February 5, 1967
Lunar Orbiter 3
U.S.A.

This NASA spacecraft, following **Lunar Orbiter 2 (November 6, 1966)**, continued the survey of landing sites for the planned American manned mission to the Moon (see **Apollo 11–July 16, 1969**). Unlike the first two orbiters, which had made preliminary study of about two dozen possible landing sites, Lunar Orbiter 3 instead made a closer re-inspection of the 12 most promising sites. In total, 211 photographs were taken during its nine months in lunar orbit. From these photos, eight locations were picked as suitable for a future Apollo manned landing. The pictures also located **Surveyor 1 (May 30, 1966)**, already on the lunar surface, and **Surveyor 3 (April 17, 1967)**, which landed while Lunar Orbiter 3 was still in operation. These images were used by **Apollo 12 (November 14, 1969)** when it made its pinpoint landing near Surveyor 3.

Like **Lunar Orbiter 1 (August 10, 1966)**, Lunar Orbiter 3 recorded no micrometeoroid hits, which further strengthened the conclusion that the three impacts registered by Lunar Orbiter 2 were due to the Leonid meteor shower, not the normal flux of particles in lunar space.

The next American unmanned orbiter, **Lunar Orbiter 4 (May 4, 1967)**, was placed in a polar orbit in order to make a more complete photographic survey of the Moon.

February 7, 1967
Cosmos 140
U.S.S.R.

Cosmos 140 was the second unmanned orbital test flight of the Soyuz spacecraft, following **Cosmos 133 (November 28, 1966)**. Once again, the spacecraft's attitude control system malfunctioned. This time, however, enough control was maintained that when the retro-rockets were fired near the end of the second day in space, the spacecraft was able to land within the U.S.S.R. It still missed its landing target by several hundred miles, and the re-entry, steeper than planned, burned a foot-wide hole in the heat shield.

Despite these problems, the next flight of the Soyuz spacecraft, **Soyuz 1 (April 23, 1967)**, was manned, resulting in the first fatality in space.

February 8, 1967
Diademe 1
France

This third test flight of France's rocket system was also the first to carry an operational scientific satellite. Diademe 1 was used to pinpoint the precise position of a number of locations in Europe for the European topological survey. The positional fixes were accurate to within 300 feet.

February 15, 1967
Diademe 2
France

Identical to **Diademe 1 (February 8, 1967)**, Diademe 2 operated for two months, providing a second orbital reference for the geodetic mapping of Europe.

This launch was the last from France's launch facility in Hamaguir, Algeria. According to the terms of Algeria's independence, France was required to evacuate the site. Future French space launches took place in French Guiana. See **DIAL-MIKA/DIAL-WIKA (March 10, 1970)** for the first South American launch.

February 28, 1967
Cosmos 144
U.S.S.R.

Cosmos 144, a third-generation Soviet weather satellite, was launched to operate in conjunction with **Cosmos 156 (April 27, 1967)**. See that entry for details.

March 8, 1967
OSO 3
U.S.A.

OSO 3 was NASA's third Orbiting Solar Observatory, following **OSO 2 (February 3, 1965)**. OSO 3 was soon joined in orbit by **OSO 4 (October 15, 1967)**, with both satellites studying the Sun. OSO 3 functioned until July 1968.

Like OSO 2, OSO 3 attempted to observe the entire Sun's face in extreme ultraviolet light and soft x-rays, from 1 to 400 angstroms, as well as hard x-rays in the 7.7 to 210 KeV energy range. It confirmed in more detail results from **Solrad 1 (June 22, 1960)**, showing how surface eruptions on the Sun quickly produced high-energy x-rays. In addition, it showed that not all solar x-ray bursts were caused by flares.

The satellite also studied high-energy gamma rays (produced by cosmic ray interactions with interstellar matter) having energies greater than 50 MeV. The data showed the earth to be bright in gamma radiation, as was the galactic plane.

March 16, 1967
Cosmos 148
U.S.S.R.

Cosmos 148 was the first Soviet electronic surveillance satellite, similar to satellites in the American **Ferret (June 2, 1962)** series. It was designed to detect electronic signals, including radar, missile telemetry, and voice communications. Over 100 such satellites were launched over the next decade, and—as in the American programs—several generations of increasingly complex and heavier designs would follow.

See **Cosmos 198 (December 27, 1967)** for a related Soviet electronic surveillance satellite.

March 23, 1967
Intelsat 2C
International / U.S.A.

This geosynchronous satellite was the third in Intelsat's first four-satellite cluster designed to provide communications services over the Atlantic and Pacific Oceans (see **Early Bird–April 6, 1965**). It was placed over the Atlantic Ocean, where it augmented Early Bird.

The cluster was completed with the launch of **Intelsat 2D (September 28, 1967)**.

April 17, 1967
Surveyor 3
U.S.A.

Built by NASA, Surveyor 3 achieved the second successful American soft-landing on the Moon following **Surveyor 1 (May 30, 1966)** (Surveyor 2 had failed), and it was the fourth lunar soft-landing overall (see **Luna 9–January 31, 1966** and **Luna 13–December 21, 1966**).

Surveyor 3 landed in the Ocean of Storms, as had the three previous lunar lander spacecraft. Like Luna 9 and Luna 13, it landed inside a crater, in this case, one measuring about 650 feet across. Like Surveyor 1, Surveyor 3 carried sensors to track its landing precisely. These instruments indicated that the spacecraft actually bounced twice while landing, moving laterally about 65 feet the first time and about 36 feet the second. Then the spacecraft settled into place, sliding about a foot before coming to rest. Photographs recorded the second impact marks, confirming the detector's data. Photographs also showed that the footpads did not penetrate the ground more than two inches.

The spacecraft took 6,315 photographs over two weeks. Though its view was limited (being inside a crater), the sloping walls that surrounded it allowed the camera to get a better view of nearby features. The terrain seemed quite similar to that seen from Surveyor 1 and from the two Soviet landers, though Surveyor 3 did see more large boulders 30 to 50 feet across. Fractures and sharp, jointed rocks were clearly visible. Photographs of the spacecraft's color filter wheel (a disk showing the rainbow of colors in the visible spectrum so that scientists on earth could calibrate the spacecraft's color images) indicated that the Moon was essentially gray, even in areas of impact and disruption. As seen by previous landers, the rocks were lighter-colored than the ground, and the ground was brighter than the underlying materials exposed by the spacecraft's landing pads. The darker surface layer was estimated to be very thin.

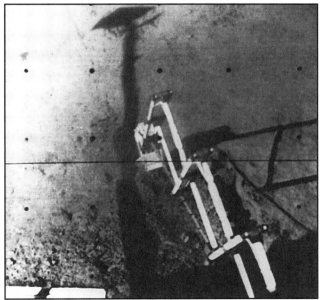

Surveyor 3's extendable arm and scoop testing lunar soil. *NASA*

On the spacecraft's fourth day of operation, a solar eclipse took place in which the earth moved in front of the Sun. Pictures showed the earth's atmosphere bending the Sun's light, encircling the earth with a faint halo of varying brightness, including several bright "beads" of light. When compared with stereographic mosaics of the earth's cloud cover taken by one of the ESSA satellites (see **ESSA 1–February 3, 1966**), these beads correlated with clear regions between clouds.

Surveyor 3 carried a short extendible arm for scooping up handfuls of lunar soil and for digging a trench visible to the cameras. This instrument allowed scientists to study the soil's behavior and the scoop's ability to penetrate the ground. The scoop dug four trenches (widening two several additional times), impacted the ground 13 times, and dumped soil on the spacecraft's footpad once. As seen previously, the undersoil was darker than the surface and appeared homogeneous wherever tested. The impact test found the scoop penetrating the surface anywhere from 0.5 to 2.5 inches, depending on drop height and point of contact. Some sections of soil appeared more resistant to penetration, indicating a harder consistency. Below a depth of 3 inches, the lunar soil appeared to get considerably stronger or denser.

When the scoop was used to dump soil on the spacecraft's footpad, it got the soil by biting into a nearby lump. The lump showed no resistance, demonstrating that it was merely a clump of loose material. The dumped soil on the footpad also broke apart into a loose pile, not unlike densely packed sand. No pebbles could be seen to the limit of the camera's resolution. As the scoop worked, the lunar soil appeared to adhere to the inside of the scoop, sometimes remaining in place even after the scoop was shaken. Overall, the surface soil appeared composed of a very fine particulate material.

The daily surface temperature at Surveyor 3's landing site was very similar to that found by Surveyor 1 and Luna 13. Surveyor 3 also made measurements during the solar eclipse, recording a drop in temperature of about 300°F during that three-hour event, from over 220°F to –80°F.

Surveyor 3 was later visited by the astronauts on **Apollo 12 (November 14, 1969)**, who photographed the spacecraft and retrieved its scoop and other pieces of equipment for study back on Earth.

The next lunar soft-landing was **Surveyor 5 (September 8, 1967)**.

April 23, 1967
Soyuz 1
U.S.S.R.

This first manned flight of the Soyuz spacecraft was originally planned as a dual mission. One day after Soyuz 1—piloted by Vladimir Komarov (**Voskhod 1–October 12, 1964**)—reached orbit, a second Soyuz craft would lift off with three cosmonauts aboard. The two spacecraft would then dock, and two cosmonauts from the second Soyuz would space-walk to the first. After this crew transfer, the two spacecraft would undock and return to Earth. The only two orbital test flights of the Soyuz spacecraft, **Cosmos 133 (November 28, 1966)** and **Cosmos 140 (February 7, 1967)**, both had had serious malfunctions of their attitude control systems. Political pressure, however, forced this third orbital flight to be manned.

Vladimir Komarov. *RIA Novosti*

Problems occurred almost as soon as Komarov reached orbit. One of the Soyuz solar panels failed to unfold, cutting the spacecraft's electrical supply in half. Then the automatic attitude control system malfunctioned. The spacecraft had been designed for ground controllers to do most of the in-space maneuvers, and Komarov found the manual system, as designed, difficult to use.

By this point, the second Soyuz flight had been canceled as ground controllers recognized that they might not be able to get Komarov home safely. For 17 orbits, the cosmonaut struggled to keep his spacecraft from tumbling uncontrollably. Then he manually fired the retro-rockets, and as the descent module tore its way through the searing heat of re-entry, he steered the craft by hand in order to keep the heat shield in position to protect him.

Though he miraculously survived re-entry, Komarov was killed when his main parachute failed to deploy. He released his reserve chute, but its lines became tangled with the still-attached drogue chute, and the capsule hit the ground at more than 400 miles per hour.

Before the Soviets again committed a cosmonaut to space, they flew five unmanned test flights of Soyuz, twice performing automatic docking in space in order to perfect the spacecraft's maneuvering abilities. See **Cosmos 186 (October 27, 1967)** and **Cosmos 188 (October 30, 1967)**. The next Soviet manned mission was **Soyuz 3 (October 26, 1968)**. The next manned mission was **Apollo 7 (October 11, 1968)**.

April 26, 1967
San Marco 2
Italy / U.S.A.

This satellite was similar in design and purpose to Italy's first satellite, **San Marco 1 (December 15, 1964)**, performing atmospheric density measurements in the lower ionosphere. It was also the first satellite launched from a floating ocean platform, located off the coast of Kenya in the Indian Ocean. Both satellite and platform (a modified oil-drilling pier) were built by Italy. An American Scout rocket was used to put the satellite into orbit.

San Marco 2 operated for 10 days, measuring atmospheric densities at the equator between 125 and 200 miles altitude. This data contributed to a growing archive showing the drag of the atmosphere on satellites, thereby improving scientists' ability to predict their orbital life span.

This successful launch resulted in a U.S.-Italian joint agreement whereby three small NASA satellites would be launched by Italy from their San Marco platform. See **Uhuru (December 12, 1970)** for the first. The next Italian satellite to study atmospheric densities was **San Marco 3 (April 24, 1971)**. The next satellite for the study of atmospheric density was **Spades (July 11, 1968)**.

April 27, 1967
Cosmos 156
U.S.S.R.

Cosmos 156 was a third-generation Soviet weather satellite, launched to operate in conjunction with **Cosmos 144 (February 28, 1967)**. Because of the location of the Soviet Union's launch sites, a sun-synchronous orbit (see **Nimbus 1–August 28, 1964**) required Soviet rockets to launch over populated regions. To avoid this, both satellites were instead placed in a non-sun-synchronous polar orbit, oriented so that, working together, their images covered half the earth's cloud cover in a 24-hour period. The test flights of Cosmos 144 and Cosmos 156 proved this strategy's feasibility.

The two satellites also tested television cameras and infrared sensors for observing the earth's cloud cover. Each photographed about 8 percent of the earth's surface per orbit, with their infrared detectors observing about 20 percent.

The Soviets launched four more test weather satellites in 1967 and 1968, and then followed them with **Meteor 1-01 (March 26, 1969)**, the first official Soviet weather satellite.

April 28, 1967
OV5-1 / OV5-3 / ERS 18
U.S.A.

• *OV5-1 and OV5-3* These two satellites were part of the U.S. Air Force research series of Orbiting Vehicles (OVs). See **OV1-2 (October 5, 1965)**. OV5-1 studied the x-ray radiation released by the Sun, both during quiet periods and solar flares, while OV5-3 performed friction tests in space on 16 different metals and Teflon samples. To create the friction, a device not unlike a car windshield wiper swiped these samples repeatedly.

• *ERS 18* ERS (Environmental Research Satellite) 18, a U.S. Air Force satellite, measured high-energy radiation throughout the earth's magnetosphere, from 5,500 to 69,000 miles altitude. Three detectors monitored x-ray, gamma-ray, and cosmic radiation, data that were used to improve future military satellite design.

May 4, 1967
Lunar Orbiter 4
U.S.A.

Unlike **Lunar Orbiter 1 (August 10, 1966)**, **Lunar Orbiter 2 (November 6, 1966)**, and **Lunar Orbiter 3 (February 5, 1967)**, which had been placed by NASA in low equatorial lunar orbits in order to study potential landing sites for future manned Apollo missions (see **Apollo 11–July 16, 1969**), Lunar Orbiter 4 was placed in a high-altitude polar orbit in order to provide a wider coverage of the entire lunar landscape at resolutions under 300 feet. During its five months of operation, it took 381 photographs covering 99 percent of the Moon's near side, including both polar regions—the first photographic survey of another planetary body from an orbiting spacecraft.

The images had 10 times the resolution of any Earth-based photographs. They also included the first vertical overhead views of both polar regions and the Moon's limbs, a perspective impossible from Earth. The photographs revealed a crater-pocked surface, showing large primary craters as well as nearby secondary craters. Close-up pictures of the Orientale Basin showed a cracked and deformed surface, indicative of a gigantic impact so powerful it caused multi-ringed ripples in the Moon's crust. From these images, scientists began developing the physics of impact cratering—how different-sized objects moving at different speeds cause different types of craters. Such theories are necessary to not only understand how the Moon's surface formed as it did, but also to understand the surface of the earth and every other solid-body planet in the solar system.

Every later Apollo mission used these photographs for making its flight plan. The next and last Lunar Orbiter mission, **Lunar Orbiter 5 (August 1, 1967)**, used these photographs to focus its mission on specific sites of scientific interest.

Two micrometeorites hit Lunar Orbiter 4 during its life span. Neither did any damage. The count was consistent with the previous three missions, further reinforcing the belief that a manned spacecraft would face little danger from meteor impact while in lunar orbit.

May 5, 1967
Ariel 3
U.K. / U.S.A.

Unlike the first two British satellites, **Ariel 1 (April 26, 1962)** and **Ariel 2 (March 27, 1964)**, Ariel 3 was built entirely by the United Kingdom. The earlier satellites had carried British-built experiments on American-built satellites. For Ariel 3, the United States provided only the rocket and launch facilities.

Ariel 3 operated until September 1969. It studied the atmosphere's distribution of oxygen and ozone at 60 miles altitude and higher, the density and temperature of the ionosphere, the intensity of very-low-frequency radiation, the levels of galactic radio noise emitted at frequencies too low to measure from the ground, and the terrestrial radio noise levels measurable at the altitude of the satellite's orbit. Its data was coordinated with other ionospheric research satellites, such as **Aurorae (October 3, 1968)** and **Explorer 32 (May 25, 1966)**.

The next British research satellite was **Ariel 4 (December 11, 1971)**. See also **Explorer 51 (December 16, 1973)**.

May 31, 1967
Timation 1
U.S.A.

This U.S. Navy satellite performed the first test of a global positioning system (GPS), designed to replace the earlier system established by **Transit 1B (April 13, 1960)**. To obtain a fix using the global positioning system, a ship or ground station needed to know precisely the spacecraft's orbit, and both the ground and satellite had to use accurate clocks that were synchronized. The ground receiver measured precisely the travel time of a satellite signal from the satellite to the ground, then multiplied this travel time by the speed of light to calculate the distance. With multiple satellites providing multiple distances in different directions, a ground receiver's position could be triangulated in three dimensions almost instantly. With a single satellite, several measurements were necessary over a longer period of time.

Timation 1 used a quartz clock accurate to one microsecond. Several ships tested their ability to determine their ground position using this single satellite, getting fixes with an accuracy within 1,200 feet. Both the Doppler shift technique (used by the Transit satellites, in which a moving ship had to know its own velocity and travel direction to get an accurate fix) and the GPS range-triangulation technique were tested. The tests also found that solar radiation and ionospheric radiation interfered with the satellite.

The second GPS satellite was **Timation 2 (September 30, 1969)**.

June 16, 1967
Cosmos 166
U.S.S.R.

Cosmos 166 was a solar observatory. The spacecraft carried detectors for observing solar x-rays and flares. Its goal was to localize the source of the Sun's x-ray radiation. A spectrometer also tried to pinpoint the composition of solar flares. The research was performed in order to learn how to predict and gauge solar activity in order to plan manned space missions.

Cosmos 166 recorded evidence of x-ray solar flares not accompanied by the visible flares seen on Earth, and it confirmed data gathered by **Elektron 2 (January 30, 1964)**.

June 29, 1967
Aurora 1
U.S.A.

This military research satellite was placed in a polar orbit to study the aurora and the electromagnetic radiation levels in the polar regions. The data was used to improve the design of military satellites.

July 1, 1967
DATS / LES 5 / Dodge 1
U.S.A.

• *DATS* The Despun Antenna Test Satellite (DATS) performed antenna tests in geosynchronous orbit for future U.S. military communications. Since spin-stabilized satellites were simpler to make and less prone to failure—requiring no attitude control system to keep the spacecraft oriented properly—the military wished to develop a communications satellite that could be spin stabilized yet still keep its antennas pointed toward the earth at all times. DATS tested technology to do this, using a dual spin configuration. While the satellite's outer frame spun like a gyroscope, keeping DATS's orientation stable, its antennas were mounted on an inner frame that remained oriented to the earth.

After this test flight, the technology was modified and used in the DSCS II satellites (see **DSCS II 1/DSCS II 2–November 3, 1971**) that replaced the IDCSP satellites (see **IDCSP 1 through IDCSP 7–June 16, 1966**). A large number of future communications satellites adopted variations of this design, including the **Anik C** (see **STS 5–November 11, 1982**) and **Telstar 3** (**July 28, 1983**) satellites and their descendants.

• *LES 5* A solid-state experimental communications satellite developed by the Massachusetts Institute of Technology's Lincoln Laboratory, LES (Lincoln Experimental Satellite) 5 continued the engineering studies begun with **LES 1** (**February 11, 1965**), developing a tactical communications system using satellites to link small mobile ground radios. This technology was an early forerunner of today's mobile telephone satellite systems (see **Iridium 4 through Iridium 8–May 5, 1997**).

LES 5 tested designs for increasing a satellite's transmission power and explored the use of solid-state technology in space. During its four-year life span, Lincoln Laboratory, as well as the U.S. military, tested both mobile and fixed communications facilities on land, sea, and air. **LES 6** (**September 26, 1968**) continued these tests.

• *Dodge 1* This military satellite was a follow-up to **GGTS 1** (**June 16, 1966**), testing the use of extendible booms for maintaining a satellite's orientation in space. Dodge 1 used 10 beryllium-copper booms, a four-inch-diameter flywheel, and a magnetic damping system to compensate for the Sun's solar wind. The satellite also carried two television cameras, with wide-angle and close-up lenses, and on July 25, 1967, it took the first color photograph of the earth as an almost-full disk.

July 19, 1967
Imp E (Explorer 35)
U.S.A.
Built by NASA, Interplanetary Monitoring Platform (Imp) E was placed into a high lunar orbit (perigree: 1,500 miles; apogee 6,000 miles) in order to study the interaction of the Moon with the magnetic fields of the Sun and the earth. This continued work begun with **Luna 10** (**March 31, 1966**). Operating through June 1973, Imp E found no evidence of an independent lunar magnetic field: when the spacecraft was outside the earth's magnetic tail, the measured magnetic field was indistinguishable from that previously seen in interplanetary space. The spacecraft also saw no evidence of a lunar bow shock. When it passed behind the Moon, the solar flux dropped to near zero, indicating that the Moon had no ability to direct the Sun's magnetic flux around its horizon. The data strongly suggested that the Moon's interior was cold and solid, with little ability to conduct electricity.

July 28, 1967
OGO 4
U.S.A.
The Orbiting Geophysical Observatories (OGO) was a NASA program of six satellites for systematically studying the earth's magnetosphere. See **OGO 1** (**September 5, 1964**).

Like **OGO 2** (**October 14, 1965**), OGO 4 was placed in a low-altitude polar orbit. It carried 20 experiments and was the first OGO satellite to maintain its attitude in space precisely for more than two months, functioning perfectly for 18 months. When combined with data from **OGO 5** (**March 4, 1968**) and **OGO 6** (**June 5, 1969**), OGO 4's data made possible the continuous study of the magnetosphere from 1964 until the end of the OGO program in late 1971. This period covered the solar cycle during its rise from a quiescent to an active stage. See OGO 5 and OGO 6 for a summary of results.

OGO 4 discovered that the orientation of the earth's magnetic field lines actually helped funnel electrons from the solar wind through the bow shock and down to the earth's surface at the polar caps, funnels that were termed the polar cusps. See the illustration with OGO 5. OGO 4 also discovered that the aurora was a continuous phenomenon, easily observable in space during the day, with little distinction between the northern and southern lights. The aurora was also found to be similar in both visible and ultraviolet light, and it seemed linked to the solar wind's passage through the polar cusps down onto the poles. See **Aurorae** (**October 3, 1968**) for further details.

OGO 4 also made the first worldwide measurements of atmospheric nitric oxide, showing that at about 50 miles altitude between 40 degrees latitude north and south, the atmosphere's gas content was stable.

August 1, 1967
Lunar Orbiter 5
U.S.A.
The last of NASA's Lunar Orbiter series (see **Lunar Orbiter 1–August 10, 1966; Lunar Orbiter 2–November 6, 1966; Lunar Orbiter 3–February 5, 1967; Lunar Orbiter 4–May 4, 1967**), this American probe took detailed photographs of 50 specific locations on the Moon's near side that were of interest to scientists, operating until January 1968. Five of these locations were prime landing sites for the first manned mission (see **Apollo 11–July 16, 1969**). Images showed boulders, some as small as 15 feet across, and the trails they left as they rolled down the sides of craters.

Close-up images of the crater Tycho showed geological features (flows, fractures, and domes) that scientists generally associate with impact.

The spacecraft also took pictures of the lunar far side, increasing to almost 90 percent the far-side coverage.

September 8, 1967

Tycho Crater. *NASA*

Close-up of melt features on Tycho Crater's floor. *NASA*

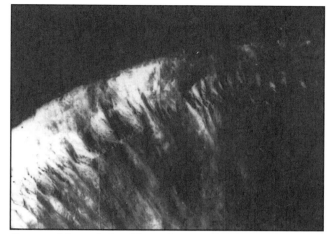

Close-up of the north rim of Dawes Crater, on the northwest edge of the Sea of Tranquility. Large boulders up to 500 feet across can be seen across the escarpment, as well as collapse and avalanche features. *NASA*

The photographs from all five Lunar Orbiters revealed that the lunar highlands and its mare lowland regions were chemically distinct. The highlands were a mix of impact craters and volcanic flows. The lowlands were generally filled with what appeared to be volcanic basalt. The interior of the Moon seemed less differentiated than the interior of the earth, with its gravitational center slightly displaced toward the earth.

At the completion of this mission, 99 percent of the lunar surface had been photographed. **Surveyor 5 (September 8, 1967) Surveyor 6 (November 7, 1967)**, and **Surveyor 7 (January 7, 1968)** all used Lunar Orbiter 5 photographs for studying the geology of their landing sites.

The next lunar orbiter was **Luna 14 (April 14, 1968)**.

September 7, 1967
Biosatellite 2
U.S.A.

Biosatellite 2 was the second attempt by the United States and NASA to launch and recover a satellite dedicated to biological research (the first satellite's retro-rockets failed, and the capsule was stranded in orbit). The Biosatellite 2 capsule carried a wide variety of specimens, including bacteria samples, insects such as fruit flies and flour beetles, and plant life such as wheat seedlings. It remained in space for two days, after which it returned to Earth for examination and testing.

The mission had 13 experiments, all designed around two questions: First, what is the effect of weightlessness on living organisms, and second, does exposure to high levels of gamma radiation combined with weightlessness have an increased detrimental effect on living things? In addition to several sets of control specimens on Earth, two identical sets were launched into space, one in a compartment shielded from radiation and exposed only to weightlessness, the second subjected to gamma radiation from an 85-strontium nitrate source.

The results showed that although the weightless environment induced changes to some of the biological specimens, radiation and weightlessness did not combine to increase the risk to life. Furthermore, some of the negative changes in some specimens were possibly attributable to the violent nature of launch and recovery, not to weightlessness.

Nonetheless, the tests did show that certain life forms were adversely affected by microgravity. The wheat seedlings exhibited mutations, as did the fruit flies. However, other specimens, such as frog eggs, developed in weightlessness without any problem or deformity.

The next unmanned biological satellite was **Biosatellite 3 (June 29, 1969)**.

September 8, 1967
Surveyor 5
U.S.A.

Built by NASA, Surveyor 5 was the third American soft-landing on the Moon, following **Surveyor 3 (April 17, 1967)**. (Surveyor 4 had failed just prior to its touchdown on the Moon.) It was similar to the previous spacecraft except that the soil-sampling scoop was replaced with an alpha-backscatter instrument for determining the relative abundance of elements in the lunar soil. Also added to the experiment package was a small bar magnet to measure the presence of magnetic material. Surveyor 5 was the first spacecraft to soft-land in a region other than Ocean of Storms, landing in the southwest part of the Sea of Tranquility.

Surveyor 5 operated successfully through two lunar days, lasting until its second lunar sunset on October 24, 1967. After that date, transmissions continued sporadically through two more of the harsh lunar days, until the spacecraft completely failed in December 1967. In that time, it took over 19,000 pictures, revealing a crater-pocked surface strewn with many fragmented boulders and small rocks, not dissimilar from previous Surveyor landing sites, or those of **Luna 9 (January 31, 1966)** and **Luna 13 (December 21, 1966)**. The soil material was finely grained and slightly cohesive, though the difference in brightness between the surface and underlying layers was not as great as it had been at the previous two Surveyor sites.

The spacecraft landed on the slope of a small rimless depression, a "dimple" crater photographed by **Ranger 8 (August 17, 1965)**. Photo analysis and soil behavior suggested that this depression formed when the fine-grained surface material flowed downward into a cavity or fissure, creating a sink. A chain of smaller nearby dimple craters supported this theory, indicating that the underground fissure was probably trending to the northwest.

The alpha-backscatter sensor found that the lunar soil was 58 percent oxygen, 18 percent silicon, 6 percent aluminum, and 13 percent a surprising number of elements heavier than silicon. The closest terrestrial match to this mix of elements was found in basalts. This finding led scientists to infer that the mare region of the Sea of Tranquility was a lava field, filled with basaltic volcanic flows and littered with ejecta materials from meteoroid impacts.

The bar magnet experiment confirmed that about 1 percent of the surface material was iron, less than expected if the lunar surface had been formed entirely from meteorite impact; this was further indication that the mare region was volcanic in origin.

Because the evidence at the Sea of Tranquility was so similar to that found at the Ocean of Storms, scientists felt confident they could apply their growing knowledge to all the mare regions. Furthermore, the fact that the geological processes at the lunar mare seemed to so resemble terrestrial geology showed for the first time that Earth-based geological knowledge could be extrapolated to extraterrestrial data. "Apparently, the geochemical processes on Earth do not differ greatly from their lunar counterparts despite environmental differences between the two bodies," noted the project scientists in their preliminary report.

Temperature readings were in close agreement with previous findings, indicating that the lunar weather was pretty much the same everywhere—ranging from 0°F to 220°F during the daylight hours.

During its first sunset, Surveyor 5 took photographs of the Sun's corona, making the first measurements to a distance of 10 to 30 solar radii from the Sun's center and identifying streamers extending out for at least 6 solar radii.

The next lunar soft-lander: **Surveyor 6 (November 7, 1967)**.

September 28, 1967
Intelsat 2D
International / U.S.A.

This geosynchronous satellite supplemented **Intelsat 2B (January 11, 1967)** in providing communications services over the Pacific Ocean. In combination with the Atlantic Ocean satellites of **Early Bird (April 6, 1965)** and **Intelsat 2C (March 23, 1967)**, this four-satellite constellation provided 480 two-way voice circuits and 2 two-way television channels over each ocean. See Early Bird for more details.

Intelsat, the international consortium responsible for building and maintaining this system, supplemented these satellites with the launch of **Intelsat 3B (December 18, 1968)**, the first successful launch in its third generation of communications satellites.

October 15, 1967
OSO 4
U.S.A.

OSO 4 was part of NASA's Orbiting Solar Observatory program, joining **OSO 3 (March 8, 1967)** in orbit. It functioned until May 1968, though it was reactivated for a short period on March 7, 1970, to observe a solar eclipse.

The satellite studied the extreme ultraviolet radiation from the Sun, from 300 to 1,400 angstroms, as well as the electrons and protons in the earth's magnetosphere. Spectroheliograms of the Sun were obtained in x-ray and ultraviolet light, which were then digitally converted into the first coarse "pictures" of the Sun. Such images and the technology used to produce them eventually led to the kinds of images produced by satellites like **Soho (December 2, 1995)**.

The next OSO satellite was **OSO 5 (January 22, 1969)**.

October 18, 1967
Venera 4, Venus Landing
U.S.S.R.

Launched on June 12, 1967, Venera 4 made the first direct observations of the Venusian atmosphere, descending into its night side near the equator. As the spacecraft entered Venus's atmosphere, parachutes unfurled and several instruments took readings. Data was transmitted for 93 minutes as the spacecraft dropped about 20 miles through the atmosphere.

Prior to this flight, scientists estimated that the planet's surface temperature was about 800°F, that its atmospheric pressure could be anywhere from a few atmospheres to several hundred, and that carbon dioxide constituted anywhere from a few percent to 50 percent of the atmosphere, with the assumption that nitrogen made up the remainder.

As Venera 4 descended through the atmosphere, it registered temperatures rising from 77°F to 520°F at about 30 degrees per mile as the spacecraft fell. The maximum atmospheric pressure equaled just under 18 times Earth's surface pressure, or 265 pounds per square inch. The atmosphere was also found to be 90–95 percent carbon dioxide, less than 7 percent nitrogen, 1 percent oxygen, and between 0.1 to 0.7 percent water.

These results, however, were in conflict with those of **Mariner 5, Venus Flyby (October 19, 1967)**. Both the surface temperature and atmospheric pressure values were significantly less than indicated by Mariner 5 (520°F versus 800°F and 18 atmospheres versus 100 atmospheres). Scientists disputed whether Venera 4's instruments ceased operating because the satellite had impacted the surface, or because something destroyed the spacecraft at about 12 miles elevation. One theory even proposed that the spacecraft hit the top of a very high mountain.

Venera 4 confirmed results from **Mariner 2, Venus Flyby (December 14, 1962)** that Venus had no measurable magnetic field. The planet, however, greatly disturbed the Sun's magnetic field, creating a shock front as the planet plowed through it.

See **Mariner 5, Venus Flyby (October 19, 1967)** for a summary of results.

October 19, 1967
Mariner 5, Venus Flyby
U.S.A.

Built by NASA and launched on June 14, 1967, Mariner 5 reached Venus one day after Venera 4 entered the Venusian atmosphere **(Venera 4, Venus Landing–October 18, 1967)**. Mariner 5 flew past the planet at a distance of about 2,585 miles. As Mariner 5 swept behind Venus, scientists measured the propagation of the satellite's signal as it traveled through the planet's atmosphere on its way to Earth.

Mariner 5's data indicated that Venus's surface temperature was about 800°F and that its surface atmospheric pressure was approximately 100 times greater than Earth's, or about 1,450 pounds per square inch. These results conflicted with Venera 4's data, suggesting to many scientists that Venera 4's instruments had ceased operating not when it hit the planet's average surface level (what we would call "sea level" on Earth), but when it was still 12 miles above it.

Venera 4's signal, however, had cut off sharply and in a manner that indicated it had hit the surface. Some scientists theorized that Venera 4 may have landed on a high mountain. Radar measurements from Earth, however, indicated that Venus was topographically much smoother than the earth, with no elevation differences greater than three miles. The explanation for this mystery required the arrival of later Venus probes.

Other Mariner 5 results generally confirmed Venera 4's data. The atmosphere was 90 percent carbon dioxide with traces of water vapor and a very low amount of oxygen (0.01 percent). Knowing Venus's elongated day (117 Earth days long) and the composition of its atmosphere, scientists then attempted to infer the atmosphere's circulation pattern. They theorized that over the planet's daylight side, the air was slowly rising, its wind speed increasing as the air flowed to the planet's dark side. When it reached a small 100-mile area centered at midnight, the air quickly sank, starting the cycle over again. Additional satellite probes were necessary to test this theory.

Mariner 5's data indicated that, excluding the top ionospheric layer for both planets, Venus's atmosphere was much shallower than Earth's, only about 100 miles high in contrast to Earth's 200 miles. Both Mariner 5 and Venera 4 found that Venus had virtually no magnetic field. As the solar wind flowed past, the planet's lack of a magnetic field prevented the formation of an earthlike magnetosphere with its bow shock at about 40,000 miles altitude on the earth's sunward side and its Van Allen belts of trapped ionized particles (see **OGO 1– September 5, 1964**; **OGO 5–March 4, 1968**). Instead, the solar wind plows into the very top layers of Venus's ionosphere at an elevation of about 300 miles, where a bow shock forms. The solar wind is then guided around the planet, creating a small protective bubble.

The composition of Venus's white clouds remained unclear, though the thickness of the cloud layer was estimated to be a little over one mile. While many scientists still believed that the cloud layer consisted of water droplets, the properties of water did not easily fit all the data. Nor was there very much water around to supply the clouds. Venera 4 had discovered that a little less than one percent of the Venusian atmosphere was water vapor, making Venus very dry, much drier than the scientists had expected, with only about 0.0001 of the water on Earth.

This absence of water also caused theoretical problems. Scientists had assumed that because Earth and Venus had formed near each other, had the same size, and were both covered with clouds, that they both had approximately the same amount of water. Yet they seemingly did not. Although the Sun's ultraviolet radiation could have destroyed much of Venus's water, splitting the gaseous water molecules and allowing the hydrogen to escape into space, the oxygen released from such a process should still have remained on Venus. Yet, the planet apparently lacked this oxygen.

Additionally, prior to the launch of Venera 4 and Mariner 5, scientists had assumed that nitrogen, because it was plentiful in Earth's atmosphere, would be plentiful in Venus's, making up as much as 50 percent of its composition. This assumption was wrong. Nitrogen, though the second largest constituent of Venus's atmosphere, only constituted about 7 percent of the total, and later Venus probes found even this number high. These contradictions indicated that there were fundamental differences between the make-up and formation of Earth and Venus.

Based on the knowledge beamed back from these two spacecraft, however, scientists could safely conclude that the planet's searing surface temperature was caused directly by its thick atmosphere of carbon dioxide. The carbon dioxide absorbed the Sun's radiation and trapped it there—the so-called "greenhouse" effect. How the carbon dioxide got there, however, was still unclear.

While much had been learned about Venus's atmosphere, practically nothing was known about the surface and make-up of the planet.

The next spacecraft to visit Venus were Venera 5 and 6 **(Venera 5, Venus Landing–May 16, 1969** and **Venera 6, Venus Landing–May 17, 1969)**.

October 27, 1967
Cosmos 186
U.S.S.R.

Cosmos 186 was the first flight of the refurbished Soyuz spacecraft following the death of Vladimir Komarov in **Soyuz 1 (April 23, 1967)**. The four-day test flight achieved the first unmanned automatic docking in space, linking Cosmos 186 with **Cosmos 188 (October 30, 1967)**.

During its first three days of flight, Cosmos 186 made a number of orbital changes. Then Cosmos 188 joined it in orbit. This second spacecraft became the target Soyuz, with Cosmos 186 tracking, rendezvousing, and docking with it. After remaining linked for several hours, the spacecraft separated. Cosmos 186 successfully returned to Earth on October 31, 1967.

October 30, 1967
Cosmos 188
U.S.S.R.

Cosmos 188 was the target Soyuz craft for **Cosmos 186 (October 27, 1967)**. Sixty-eight minutes after liftoff, Cosmos 188 was found by Cosmos 186, which then docked with it. For 3.5 hours, the two spacecraft remained linked. They then separated, with Cosmos 188 continuing in orbit for another three days, after which it successfully returned to Earth without problems.

Three more Soyuz unmanned test flights, including a second dual mission with automatic docking, were completed in 1968 before the next Soviet manned mission: **Soyuz 3 (October 26, 1968)**.

November 5, 1967
ATS 3
U.S.A.

Like **ATS 1 (December 7, 1966)**, ATS (Applications Technology Satellite) 3 was built by NASA to perform engineering tests on stationary geosynchronous satellites, and like ATS 1, ATS 3 exceeded its three-year design life many times, functioning for more than 20 years (ATS 2 had failed at launch).

The spacecraft carried two different communications experiments, testing communications at 4 MHz and 135 MHz frequencies.

ATS 3 also carried two television cameras, one black-and-white and one color, for taking meteorological photographs of the earth's cloud cover. The color camera, which took a picture every 30 minutes, produced the first color images of the earth as a globe *(see color section, Figure 20)*.

This color camera system predated the camera system used on GOES satellites (see **SMS 1–May 17, 1974**), while the black-and-white real-time camera was the prototype for similar cameras used on later NOAA satellites (see **NOAA 1–December 11, 1970**).

The next successful Applications Technology Satellite was ATS 5 (August 12, 1969).

November 7, 1967
Surveyor 6
U.S.A.

Built by NASA, Surveyor 6 was the fourth American spacecraft to successfully soft-land on the Moon, following **Surveyor 5 (September 8, 1967)**. Surveyor 6 landed in a nearly flat, heavily cratered, mare region in the Sinus Medii. Pictures from **Lunar Orbiter 2 (November 6, 1966)** and **Lunar Orbiter 5 (August 1, 1967)** were used to locate the landing site and to supplement the study of its geology. When combined with data from the Ocean of Storms (see **Luna 9–January 31, 1966; Surveyor 1–May 30, 1966; Luna 13–December 21, 1966; Surveyor 3–April 17, 1967**) and the Sea of Tranquility (see Surveyor 5), scientists could compare results from three widely separated mares on the lunar surface.

Over a two-week period, the spacecraft transmitted over 30,000 pictures to Earth, once again revealing a flat pebble- and rock-strewn terrain pockmarked with many craters. Once again, the lunar material was very fine grained and the surface was brighter than the underlying layers. The daytime surface temperatures ranged between 0°F and 220°F.

Only slight differences from previous Surveyors were found. Unlike Surveyor 5's landing site, no dimple craters were seen at Surveyor 6's. The bar magnet experiment found much less magnetic material, indicating less iron in the soil. The alpha-backscatter experiment, when compared with Surveyor 5's results, found slightly less oxygen in the soil (57 compared with 58 percent), and more silicon (22 compared with 18 percent). The overall soil composition matched quite closely, however.

In general, Surveyor 6's results confirmed the discoveries of previous landers. As noted by the project scientists, "From all these similarities, it is reasonable to think that the maria [mares] form a reasonably homogeneous class of objects, and hence they are all basaltic in composition." Beyond this, however, there was disagreement among the scientists about the exact type of basalt and its origin.

Furthermore, no one yet knew exactly what caused the difference in brightness between the surface and underlying layers, though many theories existed. Nor did any scientist have an explanation for the origin of the fine-grained lunar soil. Some proposed it was caused by endless micrometeorite impact. Others posited that it was volcanic in origin and bleached on the surface by radiation.

On November 17, one week after landing, the spacecraft's engines were fired up and Surveyor 6 lifted off the surface, moving laterally 8 feet to a new landing spot, the first lunar "sightseeing" trip. This short hop allowed scientists to photograph and study the initial landing site and to test the behavior of rocket engines in lunar gravity.

The last American robot soft-lander to arrive on the Moon (as of December 1999) was **Surveyor 7 (January 7, 1968)**.

November 9, 1967
Apollo 4
U.S.A.

Apollo 4 was the first "all-up" test flight of the Saturn 5 rocket with an Apollo capsule attached. Because of a change in the numbering system, Apollo 4 was the first Apollo launch following the Apollo 1 fire. The Saturn 5 was the largest and most powerful rocket ever built (as of December 1999). Weighing 6 million pounds on the launchpad, each of its five F-1 first-stage engines produced 1.5 million pounds of thrust, for a total thrust of 7.5 million pounds. When the 1 million pounds of thrust from the five second-stage J-2 engines and the 200,000 pounds of thrust from the single third-stage J-2 engine were added, the whole rocket generated 8.7 million pounds of thrust, enough energy to send more than 50 tons to any point in the solar system.

Following the **Apollo 1 (January 27, 1967)** launchpad fire, the Apollo capsule had been significantly overhauled to reduce the threat of fire. Its basic design was similar to Gemini—made of two components, a command module and service module—though Apollo was larger and more sophisticated.

The command module was where the astronauts lived and worked and was the only part of the rocket that returned to Earth. Shaped like a wide-mouthed ice-cream cone, its base was covered with an ablative heat shield. The service module carried the spacecraft's fuel tanks, fuel cells, and other miscellaneous support equipment. It also carried the Service Propulsion System, or SPS engine, capable of putting the spacecraft into lunar orbit and blasting it homeward to Earth.

Apollo 4 liftoff at Kennedy Space Center, Florida. *NASA*

During Apollo 4's one-day flight, the command and service modules were lifted to an altitude of over 11,000 miles, then aimed to re-enter the earth's atmosphere using what was called the double skip maneuver. A return from the Moon had the capsule traveling over 25,000 miles per hour as it approached the earth. To slow this speed enough for passengers to survive re-entry, the double skip maneuver used the earth's atmosphere, the spacecraft cutting through the atmosphere like a stone skipping across the water. Travelling at just under 25,000 miles per hour, Apollo 4 successfully bounced out of the atmosphere after its initial contact, then fell earthward again and landed safely by parachute in the Pacific Ocean. This re-entry technique was later used by every mission to the Moon, as well as by the **Mars Pathfinder (July 4, 1997)** in its landing on Mars.

The next full test flight of the Saturn 5 was **Apollo 6 (April 4, 1968)**.

November 29, 1967
Wresat 1
Australia / U.S.A.

Built by the Weapons Research Establishment of Australia, this small satellite was launched from Woomera, South Australia, on a modified American Redstone rocket in order to study solar radiation. It functioned for five days, returning limited data on the Sun's ultraviolet and x-ray emissions.

December 5, 1967
OV3-6
U.S.A.

Also called ATCOS 2 for ATmospheric COmposition Satellite, this satellite was part of the U.S. Air Force research series of Orbiting Vehicles (OVs). See **OV1-2 (October 5, 1965)**. The satellite operated for four months, gathering data on the atmosphere's composition in low-Earth orbit, measuring both charged and uncharged particles.

December 13, 1967
Test and Training Satellite 1 / Pioneer 8
U.S.A.

• *Test and Training Satellite 1* This NASA satellite was the first of three satellites used to simulate downlinks from an Apollo spacecraft. With this satellite, NASA tested its ground tracking network for downloading voice and telemetry prior to the launch of the first manned Apollo spacecraft, **Apollo 7 (October 11, 1968)**.

• *Pioneer 8* Pioneer 8 was the third in a cluster of four Pioneer spacecraft sent into solar orbit to study the solar wind. **Pioneer 6 (December 16, 1965)** was the first. See **Pioneer 9 (November 8, 1968)** for summary of results.

December 27, 1967
Cosmos 198
U.S.S.R.

Cosmos 198 was the Soviet Union's first successful Radar Ocean Reconnaissance Satellite (RORSAT). Its surveillance purpose was to detect the electronic signals, including radio and radar, of oceangoing vessels. RORSATs track these signals by using powerful radar scanners to intercept and track their targets. They hence had high power requirements and used a nuclear battery of radioactive uranium-235 to produce their electrical power.

To prevent this radioactive fuel from contaminating the earth's atmosphere when the satellite's orbit decayed, the satellite's battery compartment separated prior to re-entry and was lifted into a higher parking orbit where batteries were expected to remain for approximately 500 years. Twice, however, this procedure failed. See **Cosmos 954 (September 18, 1977)** and **Cosmos 1402 (August 27, 1982)**.

Approximately three dozen RORSATs were placed in orbit through March 1988. They were often launched in conjunction with EORSATs. See **Cosmos 699 (December 24, 1974)** for details on the EORSAT program.

1968

January 7, 1968
Surveyor 7
U.S.A.

Built by NASA, Surveyor 7 was the last American robot ship, following **Surveyor 6 (November 7, 1967)**, to soft-land on the Moon prior to the manned landing of **Apollo 11 (July 16, 1969)**. Unlike the previous Surveyors, Surveyor 7 was focused less on studying the geology of potential manned landing sites and more on learning about the lunar geology. The spacecraft landed a little less than 20 miles north of the crater Tycho on the outside flank of its rim. This site in the highlands of the Moon had been chosen specifically because it appeared overlain with ejecta from the crater—hence material that originated in the lunar interior.

This debris field was far more complex than any previous landing site. Surveyor 7's photographs, combined with enlargements of a **Lunar Orbiter 5 (August 1, 1967)** high-resolution picture of the landing site, showed a number of different geological features. A patterned terrain of thin lines, reminiscent of what had been seen by **Luna 13 (December 21, 1966)**, indicated that an underlying fault pattern had been covered by debris blasted from the Tycho impact. Over time, the debris had settled into these faults, creating the pattern of lines. Many craters of all sizes were visible, both with and without rims. Flows were also visible. Also identifiable were at least eight closed depressions with flat floors, dubbed playas. These dark, irregularly shaped playas had few craters and resembled miniature mares. They were also geologically reminiscent of basaltic lava pools seen in Hawaii.

Before Surveyor 7 and the Lunar Orbiter images, scientists were unsure of the cause of these flows and lunar craters. Some theorized the Tycho crater to be a volcano caldera. Others were convinced it was an impact remnant. The flows surrounding Surveyor 7 could either be lava flows from a volcano or a melt caused during or after an impact. After Surveyor 7, most scientists believed that Tycho and the features at the landing site had been caused by an impact.

As seen with all the previous Surveyor craft, the lunar topsoil was a fine-grained, slightly cohesive material. As with **Surveyor 3 (April 17, 1967)**, the soil adhered to spacecraft components, almost as if it were wet snow sticking to a shovel. Scientists also noted that the adhesion increased during the lunar day. Based on data from all the Surveyor spacecraft, scientists proposed that this fine-grained topsoil, or regolith as they named it, had formed from "the repetitive bombardment of the lunar surface by meteoroids and by secondary lunar fragments." The difference between the brighter surface and the darker underlying layers was attributed to "scrubbing of the particles on the surface by impact and by radiation." The older underlying material was darker because it had been sitting on the surface longer, and was therefore "varnished" more thoroughly. Brighter grains were younger, and had been in place a shorter period of time. This theory also explained the bright rays emanating out from young craters like Tycho: with age and further bombardment, the rays faded.

The alpha-backscatter experiment found the soil to have a composition similar to that found at the **Surveyor 5 (September 8, 1967)** and Surveyor 6 sites. Oxygen, silicon, and aluminum dominated. The bar magnet experiment found less iron content than the previous missions, though a small dark rock actually attached itself to the magnet. Scientists posited that this was an iron-nickel meteorite fragment that had not been vaporized upon impact or that it was a buried fragment that had been thrown out from the Tycho impact.

The basaltic composition of the lunar rock seen on all five Surveyor missions argued that the meteorites found on Earth did not come from the Moon. Meteorites on Earth were either largely iron and nickel ore or undifferentiated chondritic stone, neither of which had been found at any landing site on the Moon. Furthermore, the nature of the observed lunar material showed that it had been melted and differentiated, indicating an active volcanic past (though how active and when was still unclear). The Moon hence more resembled the earth than meteorites. This meant that most meteorites came from outside the Earth-Moon system, and that their origin was still unknown.

Above all, the most significant conclusion coming from the Surveyor, Lunar Orbiter, and Luna missions was that the Moon was not the changeless, airless, dead planet imagined previously but a planet whose surface had undergone significant erosion across eons. And although the precise cause of the lunar craters was still unknown, the details were becoming less cloudy. The lunar mares were regions of vast lava flows, flooding over older features. Their make-up resembled Earth basalts, though some significant differences were apparent. Meteoroid impacts played an important part in shaping and eroding the lunar surface, both on a micro and macro scale. Many major crater features on the lunar surface could now be assigned clearly to impact events or their subsequent consequences, not volcanic eruption.

The next soft-landing on the Moon, Apollo 11, was also the first manned mission to land on another world.

January 11, 1968
Geos 2
U.S.A.

Geos (Geodetic Earth Orbiting Satellite) 2, also called Explorer 36, was the second Geos satellite (see **Explorer 29–**

November 6, 1965). It was launched by NASA to help map and pinpoint the topography of the earth. Geos 2's data allowed scientists to precisely predict the orbital paths of satellites, which dip and rise as they fly over the large-scale, unseen distortions in the earth's sphere. While now known to be very small, less than .001 percent, these distortions also needed to be identified so that navigational global positioning systems (GPSs) could work (see **Timation 1–May 31, 1961**). This data was also necessary so that geologists could use GPS satellites to map the overall movement of the plates that form the earth's crust. For example, Geos 2 proved the existence of an ocean depression above the Puerto Rico Trench. The Trench, one of the deepest points on the earth, caused the water surface above it to sag, actually creating a general drop of about 80 feet in sea level.

January 22, 1968
Apollo 5
U.S.A.

Apollo 5 was the first in-space test of the Lunar Module (LM, pronounced "lem"), the spacecraft used in the American space program to land humans on the Moon. During the flight, several successful test firings of the LM's descent- and ascent-stage engines were made. In one case, the abort of a lunar landing was simulated: the descent engine fired for several seconds, then shut down as the ascent engine simultaneously ignited and burned until its tanks were dry.

The next Lunar Module orbital test was manned: **Apollo 9 (March 3, 1969)**.

March 2, 1968
Zond 4
U.S.S.R.

This five-day unmanned flight, using a modified Soyuz descent module, tested techniques for bringing a capsule back safely from lunar distances. The spacecraft was lifted to an elevation of 200,000 miles and then aimed to re-enter the earth's atmosphere using the double skip maneuver (see **Apollo 4–November 9, 1967**). However, the spacecraft's guidance system failed, sending Zond 4 into an uncontrolled ballistic re-entry. Though the craft would have splashed down in the ocean and would have been recoverable, the decision was made for Zond 4 to self-destruct, and the capsule exploded at 7.5 miles altitude above the Gulf of Guinea.

The next Soviet test of this modified lunar capsule was **Zond 5 (September 15, 1968)**.

March 4, 1968
OGO 5
U.S.A.

OGO 5 was the fifth of NASA's six American Orbiting Geophysical Observatory satellites dedicated to studying the earth's magnetosphere (see **OGO 1–September 5, 1964**). Like OGO 1 and **OGO 3 (June 7, 1966)**, OGO 6 was placed in a highly eccentric orbit (perigee: 168 miles; apogee: 92,000 miles) so that the spacecraft cut across a wide range of altitudes. It also gathered solar wind data when beyond the magnetosphere.

By the completion of OGO 5's mission in late 1971, scientists had more than five years of accumulated and continuous

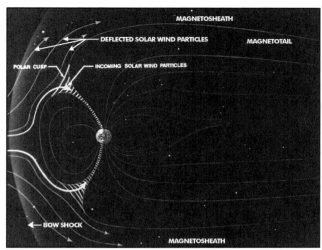

Earth's magnetosphere, with bow shock and polar cusps. *NASA*

data on the magnetosphere. The magnetosphere's complex interaction with the solar wind was now understood in great detail, as research had traced that interaction from a quiet period in the solar cycle to an active period. A model of the magnetosphere could include the precise energies and densities of particles at many altitudes, how those particles moved through the field, and how solar flares and the solar cycle influenced the shape and configuration of the field. The two Van Allen belts did appear to exist, though the gap grew and dwindled depending on time of day (the gap became most empty at midnight and noon).

Detailed studies on how solar magnetic storms affect the magnetosphere had found that the energy from the storm accumulated in the tail of the magnetosphere, where it was then released to cause substorms in the earth's field.

The bow shock and magnetopause were again shown to fluctuate in position frequently and over significant distances, depending on solar events and location. As had happened on **OGO 3 (June 7, 1966)**, the bow shock sometimes swept past OGO 5 several times during a single orbit. Scientists were thus able to develop a firm understanding of this region, including the energy, speed, and density of solar wind particles as they collided and interacted with the magnetosphere.

Solar wind data continued to show abrupt changes in speed and density, as shown by data from **Mariner 4 (November 28, 1964)**. Scientists disputed whether these solar wind changes were the result of flares on the Sun's surface or the existence of two separate solar winds, one fast and the other slow. See **Pioneer 9 (November 8, 1968)** for later results.

OGO 5 also carried a sensor for observing high-energy gamma rays. When the Vela satellites discovered the existence of gamma-ray bursts (see **Vela 5A/Vela 5B–May 23, 1969**), scientists reviewed the data from OGO 5 and found almost 200 of these mysterious unexplained bursts. The events appeared to concentrate along the galactic plane, implying that they were local Milky Way events. See **Solrad 11A/Solrad 11B (March 15, 1976)** for later results.

See **OGO 6 (June 5, 1969)** for a summary of the low-altitude polar orbiting OGO satellites.

March 5, 1968
Solrad 9
U.S.A.
This satellite was the ninth in a long-term U.S. Navy project to obtain continuous measurements of the x-ray output of the Sun using standardized sensors. See **Solrad 1 (June 22, 1960)**. It was also the second of a three-satellite joint project with NASA.

Solrad 9 operated until 1974. Its solar flare data, combined with that received from **OSO 5 (January 22, 1969)**, provided continuous coverage from March 1968 until July 1970. These data were used by NASA in planning its manned lunar missions, to anticipate solar activity with radiation that could injure Moon-walking astronauts.

The next Solrad satellite was **Solrad 10 (July 8, 1971)**.

April 4, 1968
Apollo 6
U.S.A.
Apollo 6 was the second unmanned test flight of the complete Saturn 5 rocket, following **Apollo 4 (November 9, 1967)**. See Apollo 4 for a description of the Saturn 5.

Unlike NASA's previous test, Apollo 6 had a number of problems, none of which were mission-destroying but all of which were serious. During launch, the rocket experienced violent pogo oscillations along its entire length, vibrations that tore a panel free and would have injured any astronaut onboard. Two engines of the second stage shut down prematurely, and a third stage did not re-ignite on command.

These problems prevented NASA from executing a full double skip trajectory as performed by **Apollo 4 (November 9, 1967)** and attempted by the Soviets with **Zond 4 (March 2, 1968)**. The command module was lifted to an altitude of 13,800 miles and driven back toward the earth at a speed of about 22,400 miles per hour—not as fast as desired, but the best possible under the circumstances. Apollo 6 splashed down and was safely recovered.

Despite these difficulties, the Apollo 6 flight was the last unmanned test of the Saturn 5. After considerable research, NASA engineers felt confident they had solved each problem. The pogo effect occurred because the natural resonance of the rocket matched the natural resonance of the engines. Shock absorbers were added to dampen this vibration. The three engine failures were caused when fuel lines shook apart during launch. These lines were given additional cushioning.

The next use of the Saturn 5 was to fly astronauts to the Moon; see **Apollo 8 (December 21, 1968)**. This manned lunar mission was preceded by a single manned test flight in Earth orbit: **Apollo 7 (October 11, 1968)**.

April 6, 1968
OV1-13
U.S.A.
OV1-13 was part of the U.S. Air Force research series of Orbiting Vehicles (OV)s. See **OV1-2 (October 5, 1965)**. OV1-13 carried nine experiments, seven of which measured high-energy radiation below 5,000 miles elevation. The other two experiments performed engineering research. One tested the feasibility of using flexible solar panels that unrolled once reaching orbit. This experiment led to additional tests of this technology on **STP 1 (October 17, 1971)**. OV1-13's second engineering test was a variation of the materials friction experiment carried on **OV5-3 (April 28, 1967)**. In this case, the samples tested how quickly bearings made of different materials wore out in space.

April 7, 1968
Luna 14
U.S.S.R.
Luna 14 was the fourth Soviet lunar orbiter, following **Luna 12 (October 22, 1966)**. The spacecraft's primary research goal was to study the uneven lunar gravitational field, mapping the anomalies called mascons. It carried no cameras, though it did have instruments for measuring the magnetic field of the earth and the solar wind at lunar distances.

Although no termination date was ever announced, the spacecraft apparently functioned at least through the end of April. Little else has ever been published about Luna 14's research findings.

The next Soviet lunar mission was **Luna 15 (July 13, 1969)**.

May 17, 1968
IRIS
Europe / U.S.A.
This satellite was the first built by the European Space Research Organization (ESRO), which at that time was a 10-nation association that was the forerunner of today's European Space Agency. The satellite was built by ESRO and launched by NASA as part of a joint program.

IRIS (International Radiation Investigation Satellite) carried seven experiments for studying cosmic radiation and the solar wind and the Sun's x-ray output. For its first six months in orbit, the satellite was able to gather data almost continuously, storing information when not in contact with ground stations for later downloads. It confirmed results from **OGO 4 (July 28, 1967)** that the aurora was caused by the increased downward flux of charged particles from the solar wind as these particles followed the earth's magnetic field lines down onto the poles.

The next ESRO satellite was **Aurorae (October 3, 1968)**.

July 4, 1968
RAE 1
U.S.A.
Built by NASA, Radio Astronomy Explorer (RAE) 1 was the first radio telescope launched into space. The satellite used extendible antenna booms, four 750 feet long and two 60 feet long, to pick up radio emissions from deep space. These booms used engineering tested by the GGSE (see **GGSE 1–January 11, 1964**), **GGTS 1 (June 16, 1966)**, and **Dodge 1 (July 1, 1967)** satellites. This flight, however, was the first to use such booms as antennas for detecting radio signals. Once in orbit, the longest booms were extended in stages to their maximum length of 750 feet (1,500 feet tip-to-tip), testing their ability to stabilize the spacecraft's attitude. This wingspan made RAE 1 the largest single object ever placed in space, through Deember 1999.

The spacecraft's sensors observed radio signals in the 0.2–20 MHz range, a frequency range for which ground-based radio astronomy was difficult due to atmospheric interference.

During its three years of operation, the antennas showed that most radio emissions coming out of the sky came from within the galactic plane, rather than from the Milky Way's halo or beyond. The spacecraft found that while many bright stars were silent in radio frequencies, other fainter objects such as quasars, galaxies, and pulsars generated loud radio emissions.

RAE 1 also detected radio emissions thought to be generated by the earth's magnetosphere, resembling those generated by Jupiter. These unexpectedly powerful emissions often saturated the spacecraft's sensors, rendering them inoperable between 25 and 40 percent of the time.

Finally, the satellite measured numerous low-frequency solar bursts, indicating that the Sun was a much more active source of radio emissions than predicted.

The next radio telescope was **RAE 2 (June 10, 1973)**.

July 11, 1968
Spades (OV1-15) / Cannonball 1 (OV1-16)
U.S.A.

• *Spades (OV1-15)* Part of the U.S. Air Force research series of Orbiting Vehicles (OVs) (see **OV1-2–October 5, 1965**), Spades, like Cannonball 1, studied the relationship between atmospheric density and solar radiation, continuing work done by **Explorer 24 (November 21, 1964)** and other satellites.

Spades made the first measurements of changes in temperature and density at the low altitudes (between 90 and 125 miles), confirming once again that solar activity had a direct effect on the upper atmosphere. This data was crucial in predicting the re-entry of **Skylab (May 14, 1973)** in 1979. It also measured the average levels of oxygen at elevations from 75 to 600 miles.

• *Cannonball 1 (OV1-16)* Part of the U.S. Air Force research series of Orbiting Vehicles (OVs) (see **OV1-2–October 5, 1965**), Cannonball 1 performed tests on the drag of the atmosphere in low-Earth orbit. The satellite—a shell of solid brass 2 feet in diameter and weighing 600 pounds—was the densest object launched to that date. Because a satellite of this size at 100 miles altitude was normally pulled down to Earth very quickly, Cannonball 1's high density allowed it to resist the atmosphere's drag, keeping it in orbit long enough (36 days) for ground researchers to gather data on the atmosphere's density at those elevations.

The next atmospheric density monitoring satellite was **Explorer 39 (August 8, 1968)**. The Cannonball experiment was repeated with **Cannonball 2 (August 7, 1971)**.

August 6, 1968
Canyon 1
U.S.A.

This second-generation surveillance and early warning satellite was designed by the U.S. Air Force to replace the satellites of the unsuccessful Midas program (see **Midas 2–May 24, 1960**), as well as those of the Ferret program (see **Ferret 1–June 2, 1962**). A total of six Canyon satellites were successfully launched, the last reaching orbit in May 1977. Unlike the Midas satellites, Canyon 1 did not use infrared sensors. Instead, it was a more sophisticated version of the Ferret series, designed to monitor the electronic signals and telemetry broadcast by Soviet and Chinese military. Not only could missile launches be detected, but the radio, walkie talkie, and long distance telephone conversations of the Soviet and Chinese military could be recorded.

The next-generation surveillance satellite for monitoring electronic signals was **Rhyolite 1 (March 6, 1973)**. Also replacing the Canyon satellites were the Chalet satellites (see **Chalet 1–June 10, 1978**).

August 8, 1968
Explorer 39 / Injun 5
U.S.A.

• *Explorer 39* This was the last inflatable balloon satellite launched by NASA, similar to **Echo 1 (August 12, 1960)**, **Explorer 9 (February 16, 1961)**, **Explorer 19 (December 19, 1963)**, and **Explorer 24 (November 21, 1964)**. Twelve feet across, the balloon studied the changes in the atmosphere's density and temperature at polar latitudes during a period of increasing solar activity, giving scientists baseline data for predicting future changes to the density of the upper atmosphere. The data continued the atmospheric monitoring of **Spades/Cannonball 1 (July 11, 1968)**.

The next atmospheric density satellite was **San Marco 3 (April 24, 1971)**.

• *Injun 5* Built by NASA to continue the long-term study of the interaction of the earth's magnetosphere and solar wind, Injun 5 was placed in a polar orbit to monitor the magnetosphere's changing flux of charged particles during a period of increasing solar activity, finding evidence of ionized particles of carbon, nitrogen, and oxygen trapped in the magnetic field lines of the outer Van Allen belt.

See **OGO 5 (March 4, 1968)** and **OGO 6 (May 5, 1969)** for a summary of results.

September 15, 1968
Zond 5
U.S.S.R.

Zond 5 was the second unmanned test of the spacecraft the Soviets hoped would send two cosmonauts around the Moon before the Americans. Like the prior test flight of **Zond 4 (March 2, 1968)**, Zond 5 used a modified descent module from a Soyuz spacecraft (see **Cosmos 133–November 28, 1966**).

Zond 5 became the first spacecraft to fly past the Moon, return to Earth, and be successfully recovered. During re-entry, however, the guidance system failed (as it had on Zond 4), causing the spacecraft to make an uncontrolled ballistic re-entry through the earth's atmosphere. Unlike Zond 4, however, no self-destruct command was given, and the capsule splashed down in the Indian Ocean where it was successfully recovered. Nonetheless, the re-entry would have seriously injured—if not killed—any human crew.

The next Soviet attempt to send a capsule around the Moon and return it safely to Earth was **Zond 6 (November 10, 1968)**.

September 26, 1968
OV2-5 / OV5-2 / LES 6
U.S.A.

• *OV2-5 and OV5-2*   Both satellites were part of the U.S. Air Force research series of Orbiting Vehicles (OVs). See **OV1-2 (October 5, 1965)**.

OV2-5 was the first non-communications satellite placed in geosynchronous orbit. The spacecraft collected data on particle and radiation levels at 22,000 miles altitude, including measuring solar wind fluxes.

OV5-2 worked in conjunction with OV2-5, simultaneously studying particles and radiation of the Van Allen radiation belts. The data of both spacecraft allowed scientists to gauge the immediate effect of solar activity on the radiation in the Van Allen belts.

• *LES 6*   Built by MIT's Lincoln Laboratory for the U.S. Air Force, LES (Lincoln Experimental Satellite) 6 was essentially identical to **LES 5 (July 1, 1967)** and was used in a similar series of tactical communications experiments, testing the use of satellites as communications links for small mobile antennas and transmitters, some as small as a soldier's backpack. The satellite operated until 1976, providing tactical communications after **Tacsat 1 (February 9, 1969)** failed in 1971. It was turned off in order to avoid interfering with **Marisat 1 (February 19, 1976)**, its replacement. LES 6 remained operational, however, and was turned on for additional tests in 1978, 1983, and 1988.

Based on the engineering tests performed, the military launched Tacsat 1, intended as the first test of a tactical satellite for use with small mobile ground antennas in the field.

The last two Lincoln Experimental satellites were **LES 8/LES 9–March 15, 1976**. (LES 7 was cancelled before launch.)

October 3, 1968
Aurorae
Europe / U.S.A.

This satellite, launched by NASA, was the second built by the European Space Research Organization (ESRO), a 10-nation association and a forerunner of today's European Space Agency, following **IRIS (May 17, 1968)**. Aurorae's research goal was to study the aurora and other ionospheric phenomena in the polar regions, continuing work of **Ariel 3 (May 5, 1967)**. Operating for 18 months, its data further confirmed what **OGO 4 (July 28, 1967)** had seen, that the aurora was closely linked to the funneling of solar wind material along magnetic field lines down onto the pole. The data also suggested that the aurora formed a halo over the pole. See **OGO 5 (March 4, 1968)** and **OGO 6 (May 5, 1969)** for a summary of results.

ESRO's next successful satellite to study the aurora was **ESRO 4 (November 22, 1974)**. ESRO's next satellite was **HEOS 1 (December 5, 1968)**.

October 11, 1968
Apollo 7
U.S.A.

Apollo 7 was the first American manned space flight since **Gemini 12 (November 11, 1965)**; it followed the **Apollo 1 (January 27, 1967)** launchpad fire that killed Grissom, White, and Chaffee. The 10-day mission was an Earth-orbit checkout of the Apollo command and service modules. The astronauts were Wally Schirra (**Sigma 7–October 3, 1962**; **Gemini 6–December 15, 1965**), Walt Cunningham, and Don Eisle. Apollo 7 had literally no technical problems, though all three crew members caught colds. The mission's success, which NASA labeled "101 percent successful," made possible a flight to the Moon by **Apollo 8 (December 21, 1968)**.

October 19, 1968
Cosmos 248
U.S.S.R.

This spacecraft was the target for the first successful test of an anti-satellite (ASAT) military system in which one satellite rendezvoused with a second, then destroyed it. See **Cosmos 249 (October 20, 1968)** and **Cosmos 252 (November 1, 1968)** for a description of these tests.

The only previous ASAT test flights were the Soviet **Polyot 1 (November 1, 1963)** and **Polyot 2 (April 12, 1964)** missions.

October 20, 1968
Cosmos 249
U.S.S.R.

Cosmos 249 was the first anti-satellite (ASAT) interceptor ever flown. It was designed to enter Earth orbit, rendezvous with its target, and then destroy it. Because both target and interceptor are in independent orbit, this type of ASAT is called co-orbital.

In this test, Cosmos 249 successfully rendezvoused several times with its test target, **Cosmos 248 (October 19, 1968)**. These maneuvers tested the spacecraft's homing and radar equipment. Cosmos 249 was then destroyed in a test of its self-destruct mechanism. Cosmos 249 was followed by **Cosmos 252 (November 1, 1968)**, which homed in on Cosmos 248 and destroyed it.

October 25, 1968
Soyuz 2
U.S.S.R.

Soyuz 2 was the unmanned target vehicle for the manned **Soyuz 3 (October 26, 1968)**. However, no docking took place between the two spacecraft. See Soyuz 3 for details.

During its three-day mission, Soyuz 2 made several orbital maneuvers and was then recovered on October 28, 1968, after making an uncontrolled ballistic re-entry through the earth's atmosphere.

October 26, 1968
Soyuz 3
U.S.S.R.

Soyuz 3 was the first Soviet manned space flight after the death of Vladimir Komarov on **Soyuz 1 (April 23, 1967)**. Georgi Beregovoi spent almost four days in orbit, testing the refurbished Soyuz spacecraft. An attempt to dock with the unmanned **Soyuz 2 (October 25, 1968)** target vehicle failed, however. Although the automatic rendezvous system worked, bringing the two spacecraft within 700 feet of each other, when Beregovoi activated his manual controls, the automatic systems that kept the two spacecraft aligned became deactivated. He could bring his spacecraft close to Soyuz 2, but he could not dock with it.

During the remaining three days of Beregovoi's flight, he made numerous Earth observations, sighting cyclones, typhoons, and three different forest fires. He identified the development of a thunderstorm near the equator. He also gave an extended televised public tour of the inside of the Soyuz spacecraft, a first for the normally secretive Soviet space program.

The next Soviet manned mission, the dual flight of **Soyuz 4** and **5 (January 14–15, 1969)**, finally achieved the manned docking and crew transfer mission planned for Soyuz 1. Before this, however, a two-man flight around the Moon using a Zond capsule, which had been tentatively scheduled for early December 1968, was canceled after the problematic flight of **Zond 6 (November 10, 1968)**. **Apollo 8 (December 21, 1968)** was, therefore, the next manned mission—and the first manned flight to circle the Moon.

November 1, 1968
Cosmos 252
U.S.S.R.

Cosmos 252 successfully accomplished the first anti-satellite (ASAT) interceptor mission, following tests by **Cosmos 249 (October 20, 1968)**. In this Soviet ASAT design, the interceptor, Cosmos 252, reached orbit as an independent satellite. It then rendezvoused with its target, **Cosmos 248 (October 19, 1968)**, and destroyed it. This is called a co-orbital ASAT, because both target and interceptor are orbital satellites.

The Soviets successfully performed four ASAT intercepts over the next three years, after which they suspended further tests due to the signing of the SALT treaty in May 1972. They then resumed ASAT tests in 1976, launching 13 interceptors through 1982.

The only comparable American ASAT program is its direct-ascent system; see **Solwind P78-1 (February 24, 1979)** and **MHV (September 13, 1985)**.

November 8, 1968
Pioneer 9
U.S.A.

Pioneer 9 was the last of NASA's Pioneer solar satellites, joining **Pioneer 6 (December 16, 1965)**, **Pioneer 7 (August 17, 1966)**, and **Pioneer 8 (December 13, 1967)** in orbit around the Sun. Originally planned for a half-year life span, this network of satellites spent almost two decades studying the solar wind, as well as solar weather and sunspot activity. They provided the first detailed look at the solar wind, first discovered by **Mariner 2 (August 27, 1962)**. Working in conjunction, these satellites found that the solar wind was a turbulent flow made up of two jet streams—a slow wind and fast wind—though the source on the Sun of these winds was unknown. See **Helios 2 (January 15, 1976)**.

The spacecraft also found that the tail of the earth's magnetic field not only points away from the Sun due to solar wind pressure, but extends out for millions of miles. Moreover, they discovered that during the Sun's 11-year solar cycle, its increasing and decreasing activity correlated directly with a decrease and increase of interstellar cosmic radiation. During peak activity periods, the Sun acted to shield the inner solar system from cosmic rays. This confirmed data from **Venera 2 (November 12, 1965)**.

The four satellites together formed a solar wind forecasting network. Because the constellation continuously observed both the near and far sides of the Sun, they provided NASA with regular and reliable reports of the Sun's activity. This information was then used to plan all the manned lunar missions.

In August 1972, during the most intense solar storm yet recorded, Pioneer 9 was fortuitously lined up with **Pioneer 10 (March 3, 1972)**, though the second spacecraft was 132 million miles away on its way to Jupiter. By correlating the changes to the solar wind as it roared past these two spacecraft, scientists learned that the wind lost about half its velocity but increased its temperature as it flew outward. See Pioneer 10.

Pioneer 9 operated for 18 years, the shortest period of the four Pioneer satellites designed to study the solar wind.

November 10, 1968
Zond 6
U.S.S.R.

Using a modified unmanned Soyuz descent module, this mission was the second lunar test flight of the capsule the Soviets hoped to use to beat the Americans to the Moon, after **Zond 5 (September 15, 1968)**. The ship was sent on a circumlunar orbit, looping around the Moon on November 14th to return to Earth on November 17th—the first spacecraft to successfully return from the Moon using the earth's atmosphere to slow its speed, skipping once through the upper atmosphere above Antarctica before landing in Russia. This same technique was used by all Apollo missions to the Moon (see **Apollo 8–December 21, 1968**), as well as by **Mars Pathfinder (July 4, 1997)** in its landing on Mars. Unlike **Zond 4 (March 2, 1968)** and Zond 5, Zond 6's re-entry guidance system did not fail, and the spacecraft was able to maintain its course.

Earthrise above a lunar horizon, from Zond 6. The numbers refer to text in the original Soviet press release. *RIA Novosti*

Despite this successful recovery, several technical problems during the flight would have killed a human crew. On the return voyage, a gasket blew, which caused the cabin to depressurize, and upon re-entry, the main parachute canopy was ripped when it released prematurely. These failures forced the Soviets to cancel a planned manned circumlunar flight that had been tentatively scheduled for early December 1968.

Zond 6 nonetheless did bring back the first high-quality black-and-white photographs of the earth from lunar distances. Previous photographs (see **Lunar Orbiter 1–August 10, 1966**; **Lunar Orbiter 5–August 1, 1967**; **ATS 3–November 5, 1967**) had been beamed back as television scans, reducing image resolution. Zond 6's film was instead returned to Earth for development. The pictures included the second image of the earth rising above a lunar horizon (see **Lunar Orbiter 1–August 10, 1966**).

The next Soviet unmanned flyby mission of the Moon was **Zond 7** (**August 8, 1969**).

November 16, 1968
Proton 4
U.S.S.R.

This flight was the first test of the Soviet Union's second-generation heavy-lift Proton rocket, capable of putting over 40,000 pounds (20 tons) in orbit. On this test flight, the rocket placed 17 tons in orbit, slightly more than launched on **Apollo 7** (**October 11, 1968**) and the heaviest unmanned satellite to date. Its payload included a cosmic ray detector. Like its predecessor (see **Proton 1–July 16, 1965**), this launch was mostly a test of the rocket systems.

Proton became the heavy-lift mainstay of the Soviet space program, placing in orbit all of its Salyut (see **Salyut 1–April 19, 1971**) space stations, as well as **Mir** (**February 20, 1986**) and all its modules.

December 5, 1968
HEOS 1
Europe / U.S.A.

Launched by NASA, Highly Eccentric Orbit Satellite (HEOS) 1 was the third satellite of the European Space Research Organization (ESRO), a 10-nation association and a forerunner of today's European Space Agency. Intended to study the Sun's magnetic field, its solar wind, the cosmic radiation percolating into the solar system from galactic space, and how all these phenomena changed the earth's magnetosphere, the satellite was placed in an eccentric orbit, with a perigee of 12,440 miles and an apogee of 126,000 miles. From this orbit, HEOS 1 repeatedly traveled through and beyond the bow shock of the earth's magnetosphere.

In its first few weeks of operation, HEOS 1 was one of a number of satellites providing radiation data used in planning the **Apollo 8** (**December 21, 1968**) mission to the Moon.

The satellite's six sensors all functioned for its first 16 months in orbit, after which they were turned off one by one over the next five years. See **HEOS 2** (**January 31, 1972**) for results.

December 7, 1968
OAO 2
U.S.A.

Built by NASA, Orbiting Astronomical Observatory (OAO) 2 was the first successful orbiting ultraviolet telescope (OAO 1 had failed almost immediately after reaching orbit). Designed to photograph stars in the ultraviolet range of the spectrum (1,000–4,250 angstroms), OAO 2's camera system had a two-degree-square field of view. During the spacecraft's 16 months of operation, it took over 8,500 images (covering about 10 percent of the sky), taking the ultraviolet measurements of more than 5,000 stars—the first ultraviolet astronomical images since **Gemini 12** (**November 11, 1966**).

The brightness of the zodiacal light was accurately measured, showing a distinct symmetry along the elliptic plane and a brightness equal to about 30 10-magnitude stars per square degree.

The first ultraviolet observations of a comet were made. Both Comet Tago-Sato-Kosaka and Comet Bennett were photographed, confirming the existence of a large halo of hydrogen atoms surrounding each comet's nucleus, which resulted from the evaporation of water-ice on the comet's surface. This discovery proved the theories that described comets as composed mostly of ice and snow.

Ultraviolet spectrograms of Mars and Jupiter were made. The Martian measurements indicated an atmosphere consisting almost entirely of carbon dioxide, with an atmospheric pressure of 5.5 millibars, confirming data returned from **Mariner 4, Mars Flyby** (**July 15, 1965**). The measurements of Jupiter indicated the presence of ammonia and the possibility of organic molecules (benzene, purines, and pyrimidines) in the Jovian atmosphere.

The first observations of the Crab Nebula in the ultraviolet were made, finding that the slope of the ultraviolet flux did not seem to correspond to the slope predicted by previous x-ray measurements. Other ultraviolet observations of a number of different star types revealed information about their make-up. The eclipsing binary star CW Cephei, for example, was found to have light variations far more complex and asymmetrical than expected, indicating the presence of a thin shell of material surrounding one of the stars. Observations of other stars indicated the possibility of shock waves or tidal distortions across their surfaces.

Many stars with similar magnitudes in visible light exhibited widely differing magnitudes in the ultraviolet range. These variations indicated that the interstellar medium between the earth and each star varied widely in density and content, causing the ultraviolet light from each star to reach the earth at a different magnitude.

The first ultraviolet observations of a nova were also made. Nova Serpentis erupted on February 13, 1970, and was studied by OAO 2 during its first 60 days of activity.

The next ultraviolet telescope was **Thor Delta 1A** (**March 12, 1972**). The next American ultraviolet telescope was **Copernicus** (**August 21, 1972**).

December 19, 1968
Intelsat 3B
International / U.S.A.

Intelsat 3B was the first successful launch in Intelsat's second cluster of geosynchronous communications satellites. It was also Intelsat's third-generation design, following **Early Bird (April 6, 1965)** and the Intelsat 2 satellites (see **Intelsat 2A–January 11, 1967**). See Early Bird for more details about Intelsat.

Though similar in design to the Intelsat 2 satellites, Intelsat 3 satellites had almost four times the capacity, increased from 240 to 1,200 simultaneous two-way voice circuits. The Intelsat 3 satellites could also handle four television circuits each.

Through July 1970, Intelsat successfully launched five Intelsat 3 satellites, placing them over the Atlantic, Pacific, and Indian Oceans, thereby becoming the first single organization to provide global communications service. At this cluster's peak operation, it could handle 3,600 conversations across the Atlantic Ocean at the same time.

Beginning in 1972, this second cluster was replaced by Intelsat's fourth generation of satellites. See **Intelsat 4 F2 (January 26, 1971)**.

December 21, 1968
Apollo 8
U.S.A.

Apollo 8 was the first manned mission to reach and orbit another world. Astronauts Frank Borman (**Gemini 7–December 5, 1965**), Jim Lovell (**Gemini 7–December 4, 1965; Gemini 12–November 11, 1966; Apollo 13–April 11, 1970**), and Bill Anders orbited the Moon 10 times over a 20-hour period on Christmas Eve. During two television broadcasts from lunar orbit, an audience estimated at over one billion people saw views of a stark, lifeless Moon. During the evening telecast from lunar orbit, the astronauts read the first 10 verses of Genesis.

Because this flight involved great risks, few scientific experiments were planned. The mission was the first manned use of the Saturn 5 on only its third launch. Furthermore, this flight followed the problematic **Apollo 6 (April 4, 1968)** test flight. Apollo 8 was also the first manned mission designed to use the double skip technique for re-entry (see **Apollo 4–November 9, 1967**). Because of these risks, Apollo 8 focused on proving the ability of the Apollo spacecraft to transport three humans to lunar orbit and return them safely to Earth.

Nonetheless, the astronauts' photographs and their eyewitness reports on the appearance of the lunar surface helped confirm that impacts had caused almost all lunar craters. Lovell: "These new craters look like pick-axes striking concrete, leaving a lot of fine haze dust." Anders: "The back side looks like a sand pile my kids have been playing in for a long time. It's all beat up, no definition. Just a lot of bumps and holes." These observations also added support to the theory proposed after **Surveyor 7 (January 7, 1968)** that the lunar surface had been significantly shaped by the erosional effect of micrometeoroid impacts over eons.

Moreover, the astronauts' photographs and their contemplation of Earth from lunar distances were culturally significant, changing the human perspective of the earth, at least in the United States. Bill Anders's photograph of the first earthrise ever witnessed by humans has probably been one of the most reproduced space pictures ever taken. It was placed on a six-cent stamp, and later adorned every issue of the *Whole Earth Catalog*. Nature photographer Galen Rowell said in 1995 that it was "the most influential environmental photograph ever taken."

During the flight, Borman experienced nausea and diarrhea, making him the first American to become ill in space. Whether the cause was space sickness or a reaction to a sleeping pill remains unclear.

The next American space flight, **Apollo 9 (March 3, 1969)**, tested in Earth orbit the lunar module—the vehicle designed to land humans on the Moon and then to leave it safely.

Anders: "Just a lot of bumps and holes." *NASA*

Earthrise on Apollo 8's fourth orbit. *NASA*

1969

January 14, 1969
Soyuz 4
U.S.S.R.

Soyuz 4, in conjunction with **Soyuz 5 (January 15, 1967)**, achieved the manned docking and crew transfer originally planned for **Soyuz 1 (April 23, 1967)**. Piloted by cosmonaut Vladimir Shatalov (**Soyuz 8–October 13, 1969; Soyuz 10–April 23, 1971**), Soyuz 4 tracked, rendezvoused, and docked with the target spacecraft, Soyuz 5. Then two of the three cosmonauts on Soyuz 5, Yevgeni Khrunov and Alexei Yeliseyev, space-walked to Soyuz 4, thereby completing the first in-space crew transfer in history. The two ships then undocked, and Soyuz 4 returned to Earth the next day with three cosmonauts.

January 15, 1969
Soyuz 5
U.S.S.R.

Soyuz 5 was the target vehicle for **Soyuz 4 (January 14, 1969)**. It was launched with three cosmonauts on board, Boris Volynov (**Soyuz 21–July 6, 1976**), Yevgeni Khrunov, and Alexei Yeliseyev (**Soyuz 8–October 13, 1969; Soyuz 10–April 23, 1971**). After Soyuz 4 had successfully docked with Soyuz 5 on January 16th, Khrunov and Yeliseyev put on spacesuits and transferred by space walk to Soyuz 4. This space walk, the first for the Soviets since Alexei Leonov's on **Voskhod 2 (March 18, 1965)**, included the first orbital test of the spacesuit the Soviets had designed for lunar surface operations. The two spacecraft then undocked, with Volynov returning to Earth in Soyuz 5 on January 18, 1969. (According to some reports, after Volynov fired his retro-rockets, the Soyuz service module failed to separate. He was spared death when its attaching struts either broke or burned off during re-entry, after which he was able to quickly reorient the descent module to position its heat shield in front.)

The next Soviet space walk did not take place until nine years later; see **Soyuz 26 (December 10, 1977)**. The next Soviet space flight, the first and only three-spacecraft mission yet flown, was **Soyuz 6, 7, and 8 (see October 11, 1969)**.

January 22, 1969
OSO 5
U.S.A.

The fifth Orbiting Solar Observatory, following **OSO 4 (October 15, 1967)**, OSO 5 continued NASA's systematic study of the Sun through a complete solar cycle. The satellite was able to monitor the Sun's surface continuously in visible, ultraviolet, x-ray, and gamma-ray wavelengths. X-ray spectroheliograms were obtained.

One instrument measured the x-ray output during solar flares. Combined with information from **Solrad 9 (March 5, 1968)**, these two satellites provided coverage of every solar event from March 1968 to July 1970. Scientists used the data to learn the details surrounding the Sun's x-ray flares—these events seemed to begin in small confined regions and then quickly erupt to cover large areas, both on the Sun's surface as well as into its corona.

The next OSO satellite was **OSO 6 (August 9, 1969)**.

January 30, 1969
ISIS 1
Canada / U.S.A.

The International Satellite for Ionospheric Studies (ISIS) 1 was the third satellite built by Canada and launched by NASA. It continued the research of **Alouette 2/Explorer 31 (November 29, 1965)** into the composition of the ionosphere in polar latitudes. ISIS 1 also provided data for improving radio communications during ionospheric disturbances.

ISIS 1 was joined by a second ISIS satellite (ISIS 2) in April 1971, the last Canadian-built research satellite through December 1999. The two satellites, designed to cover complementary regions, were placed in polar orbits; ISIS 1's altitude ranged from approximately 375 miles to 2,200 miles, while the second satellite hovered at about 900 miles.

Data were collected from both satellites through 1990: by NASA and Canadian scientists from launch to 1979, by Canadian scientists through 1984, and by Japanese researchers through 1990.

See **Dynamics Explorer 1/Dynamics Explorer 2 (August 3, 1981)** for a results summary.

February 9, 1969
Tacsat 1
U.S.A.

Tacsat 1 used technology developed by the Lincoln Experimental Satellites, **LES 2 (May 6, 1965)**, **LES 4 (December 21, 1965)**, **LES 5 (July 1, 1967)**, and **LES 6 (September 26, 1968)**. These U.S. Air Force experimental communications satellites, forerunners of today's telephone satellite systems (see **Iridium 4 through Iridium 8–May 5, 1997**), had tested the use of satellites to link small mobile receivers/transmitters, some the size of a soldier's backpack. Tacsat 1, though still experimental, provided tactical communications for troops in the field. It worked in conjunction with the military's global communications satellites that used large fixed antennas—first the IDCSP satellites (**June 16, 1966**) and then the DSCS satellites (**November 3, 1971**).

Tacsat 1 was placed in geosynchronous orbit over the Pacific. Its capacity was about 40 voice messages or several hundred teletype messages. Because it required a higher power output so that small ground receivers could pick up its relayed signal, Tacsat 1 needed a greater area of solar panels, hence it was the largest communications satellite launched to date, 11 feet high and 9 feet across, with a total surface area of 25 square feet.

In initial tests with some 20 stations on ships and islands in the Pacific, portable transmitters weighing 22 pounds and receivers weighing only 6 pounds (including batteries) communicated with each other. Once operational, the satellite was used extensively during the war in Vietnam, as well as during Apollo capsule-recovery efforts.

Tacsat 1's service ended when its attitude control system failed in December 1971. LES 6 was then used until the launch of **Marisat 1 (February 19, 1976)**, one of three interim tactical communications satellites launched pending the establishment of the FLTSATCOM cluster (see **FLTSATCOM 1–February 9, 1978**), the U.S. military's first operational tactical communications system.

March 3, 1969
Apollo 9
U.S.A.

This flight was the first manned mission to include the lunar module (or LM, pronounced "lem"), and it was the first complete test of the entire package designed to take humans to and from the Moon safely. For 10 days, Jim McDivitt (**Gemini 4–June 3, 1965**), Dave Scott (**Gemini 8–March 16, 1966**; **Apollo 15–July 26, 1971**), and Rusty Schweickart performed Earth orbital flight tests of the command module, lunar module, and service module.

Schweickart and Scott also performed a 40-minute space walk, with Schweickart standing in the LM's hatch and Scott in the command module's. During this EVA (extra-vehicular activity), Schweickart used, for the first time, the self-contained spacesuit intended for lunar exploration. Scott, meanwhile, recovered several samples that had been attached to the exterior of the command module to test their reaction to the vacuum of space. Both men took multispectral photographs of the earth, including images of both North and South America.

The LM was given a thorough check-out. Its life-support system was activated and tested, as was its navigational system. Its ascent and descent engines were fired several times each. The astronauts performed the first tests of the Apollo docking system, and on the fifth day in space, the two spacecraft undocked, with Schweickart and Scott piloting the lunar module and McDivitt in the command module. For two hours, the two spacecraft performed maneuvers, and then redocked.

Early in the flight, both McDivitt and Schweickart experienced nausea and motion sickness. Schweickart vomited twice. Their symptoms, resembling those of Frank Borman on **Apollo 8 (December 21, 1968)** and Gherman Titov on **Vostok 2 (August 6, 1961)**, indicated that the initial adaptation to the weightless environment requires a period of time, at least several days. Their reaction also suggested that different indi-

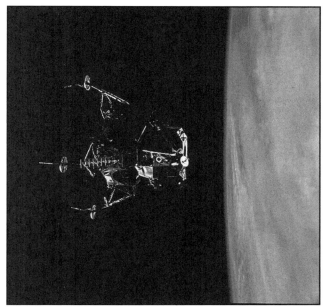

The Lunar Module as seen from the Apollo 9 command module. *NASA*

viduals adapted differently to weightless—Scott, for example, experienced little motion sickness.

The next Apollo flight, **Apollo 10 (May 18, 1969)**, took this entire package to within 10 miles of the lunar surface.

March 18, 1969
OV1-17 / OV1-18 / OV1-19 / OV1-17A (Orbiscal 2)
U.S.A.

These four satellites were part of the U.S. Air Force's Orbiting Vehicle (OV) research series. See **OV1-2 (October 5, 1965)**.

• *OV1-17* On OV1-17, eight experiments studied the horizon's airglow, the magnetic field, and ionized particles in low-Earth orbit. The satellite also tested a sampling of thermal coatings and an experimental CdS solar cell.

• *OV1-18* On OV1-18, 12 experiments tested the propagation of radio communication signals through the ionosphere.

• *OV1-19* While the OV1-17 and OV1-18 were placed in complementary low-Earth orbits about 300 miles in elevation, OV1-19 was lifted into the Van Allen belts at 3,000 miles altitude to study the effect that the trapped radiation there had on equipment.

• *OV1-17A (Orbiscal 2)* This satellite was actually the launch rocket's last stage. A radio beacon transmitting at 9 MHz and 13.25 MHz was attached to its outside skin, and once it reached orbit, researchers used this beacon to test the propagation of the signal through the earth's ionosphere at between 100 and 200 miles altitude.

March 26, 1969
Meteor 1-01
U.S.S.R.

Meteor 1-01 was the first official Soviet weather satellite, following nine known test flights (see **Cosmos 156–April 27,**

Dave Scott standing in the Apollo 9 command module hatch, in a photo taken by Rusty Schweickart from the hatch of the Lunar Module. *NASA*

May 16, 1969

1967). Meteor 1-01 was comparable in design to the American Nimbus, ESSA, and NOAA satellites (see **Nimbus 1–August 28, 1964**; **ESSA 1–February 3, 1966**; **NOAA 1–December 11, 1970**).

Through 1981, the Soviet Union launched 31 Meteor 1 satellites, maintaining a cluster of two or three operating satellites over the Soviet Union at all times. Because of the vast size of the U.S.S.R., a single satellite in geosynchronous orbit could not see the nation in its entirety. Moreover, the location of the country's launch site made a sun-synchronous polar orbit (see **Nimbus 1–August 28, 1964**) inadvisable—at launch, a rocket was required to travel over heavily populated areas. To overcome these obstacles, Meteor 1 satellites were placed in a low-Earth orbit, about 300 to 400 miles elevation, with near polar orbits inclined to the equator by 81 degrees. While not sun-synchronous, the satellites could cover the entire Soviet Union by working in either pairs or trios and staggering their orbits.

Twice a day, these satellites photographed the cloud, ice, and snow cover in the Soviet Union in visible light. Images covered a 300-square-mile area, with a resolution of 0.75 miles. The spacecraft also carried a scanning infrared radiometer for measuring cloud, snow, and ice distribution in the 8–12 micron waveband. The data was transmitted to over 50 ground stations, which then beamed it to the U.S.S.R. Hydrometeorological Center in Moscow. There the data was processed and released within 90 minutes.

In 1975, the Soviets launched a next generation satellite: **Meteor 2-01 (July 11, 1975)**.

May 16, 1969
Venera 5, Venus Landing
U.S.S.R.

Launched on January 5, 1969, Venera 5 was very similar in design to Venera 4 (**Venera 4, Venus Landing–October 18, 1967**). As it approached Venus, it released a three-foot-diameter descent capsule, which had been strengthened based on data returned from Venera 4 so that it could resist the high temperature and pressure of Venus's atmosphere.

As the capsule entered the Venusian atmosphere, it transmitted data for 53 minutes beginning at an altitude of 37 miles and continuing until the high atmospheric pressure crushed it at 15 miles. See **Venera 6, Venus Landing (May 17, 1969)** for results.

May 17, 1969
Venera 6, Venus Landing
U.S.S.R.

Venera 6, launched on January 10, 1969, was identical to Venera 5 (**Venera 5, Venus Landing–May 16, 1969**). Its descent capsule hit the Venusian atmosphere on May 17th, transmitting data for 51 minutes as it dropped from 31 to 7 miles elevation.

Based on the data received from both Venera spacecraft, the atmospheric pressure and temperature on Venus's surface were estimated. Venera 5 indicated a surface pressure 140 times greater than Earth's, while Venera 6 indicated one 60 times greater. Surface temperature estimates ranged from 750°F to 1,000°F. The atmosphere itself was 93 to 97 percent carbon dioxide, with traces of nitrogen and other inert gases. Oxygen comprised less than 0.4 percent. These results confirmed data beamed back from **Mariner 5, Venus Flyby (October 19, 1967)** and **Venera 4, Venus Landing (October 18, 1967)**. The data also proved that Venera 4 had stopped transmitting while still a dozen miles above the Venusian "sea" level, not when it impacted the surface.

With the completion of these two missions, the basics of the Venusian atmosphere were known. The next mission to Venus, **Venera 7, Venus Landing (December 15, 1970)**, was the first to return data from the surface.

May 18, 1969
Apollo 10
U.S.A.

Apollo 10 was the second manned mission to leave Earth orbit and travel to the Moon, following **Apollo 8 (December 21, 1968)**. Crewed by Tom Stafford (**Gemini 6–December 15, 1965**; **Gemini 9–June 3, 1966**; **Apollo 18–July 15, 1975**), Eugene Cernan (**Gemini 9–June 3, 1966**; **Apollo 17–December 7, 1972**), and John Young (**Gemini 3–March 23, 1965**; **Gemini 10–July 18, 1966**; **Apollo 16–April 16, 1972**; **STS 1–April 12, 1981**; **STS 9–November 28, 1983**), Apollo 10 was the final dress rehearsal for the planned lunar landing.

During this eight-day mission, the astronauts spent two days circling the Moon, giving the lunar module (LM) its first flight tests in lunar orbit. With Young manning the command module, Stafford and Cernan flew the LM to within 10 miles of the lunar surface. Both the descent and ascent stages of the module were tested, as well as docking maneuvers in lunar gravity.

The astronauts' observations of the lunar surface matched those of the Apollo 8 crew. The lunar surface was heavily impacted and largely colorless, though its tint changed depending on the Sun's angle. Based on lunar photographs and observations made by the crews of Apollo 8 and Apollo 10, the landing site for **Apollo 11 (July 16, 1969)** was chosen—a flat area in the Sea of Tranquility.

May 23, 1969
OV5-5 / OV5-6 / OV5-9 / Vela 5A / Vela 5B
U.S.A.

• *OV5-5, OV5-6, and OV5-9*   These three satellites were part of the U.S. Air Force's Orbiting Vehicle (OV) research program (see **OV1-2–October 5, 1965**) and were placed in high Earth orbits, between 37,000 and 43,000 miles altitude, to study the region of space where the solar wind meets the earth's magnetosphere. Each carried different sensors for studying magnetic fields, solar flares, the solar wind, and x-ray radiation.

• *Vela 5A and Vela 5B*   These two satellites were the ninth and tenth in the American military's first cluster designed to detect nuclear explosions that violated the 1963 Atomic Test Ban treaty. See **Vela 2A/Vela 2B (July 17, 1964)** for program information.

Vela 5A and Vela 5B, upgraded versions of the earlier satellites, were joined by two more satellites in 1970 to complete the constellation. All four satellites carried increased data-storage capability, a better command system, and more electrical power. These improvements allowed them to gather more data, which resulted in the serendipitous discovery of the mysterious gamma-ray bursts.

In 1972, three scientists from the University of California—Ray Klebesadel, Ian Strong, and Roy Olson—reviewed the data gathered over a period of three years beginning in July 1969, looking for any unusual gamma-ray emissions. While the spacecraft had been designed to look for such energy releases from atomic explosions on the ground, they could just as easily detect such bursts coming from deep space.

The scientists discovered 16 bursts of gamma-ray energy that had come from 16 different directions in the sky. The bursts had lasted from less than a second to about 30 seconds. The bursts had no connection to any other celestial events such as supernovae or solar flares. Very quickly, scientists realized that these energy bursts had not occurred close to Earth, and therefore had to have been caused by natural events of great energy.

No one knew exactly how far away these events were. The Vela sensors were not accurate enough or fast enough for astronomers to use other kinds of telescopes to study the gamma-ray bursts. By the time evidence of a burst was documented, the burst was long over. Thus, no data corresponding to any other wavelength, from radio to visible to x-ray frequencies, could be found to match the bursts. Without this information, scientists could not determine the distance or the make-up of the gamma-ray bursts, or anything about their cause.

For the next 22 years, gamma-ray detectors on a number of spacecraft would detect these bursts, finding an average of about one per month, scattered over the entire sky. See **OGO 5 (March 4, 1968)**, **Helios 2 (January 15, 1976)**, and **Solrad 11A and 11B (March 15, 1976)**. Over 2,000 papers were written trying to unlock the secret of their cause. Theories ranged from emissions in the cometary Oort cloud surrounding the solar system to starquakes on the surface of neutron stars to the merger of neutron stars. Not until the launch of the **Compton Gamma Ray Observatory** (see **STS 37–April 5, 1991**) was progress made in solving the mystery of gamma-ray bursts.

June 5, 1969
OGO 6
U.S.A.

OGO 6 was the last of NASA's American Orbiting Geophysical Observatories (see **OGO 1–September 5, 1964**). Along with **OGO 2 (October 14, 1965)** and **OGO 4 (July 28, 1967)**, OGO 6 was placed in a polar low-Earth orbit to study high-latitude phenomena. The spacecraft carried 26 experiments and operated successfully for two years, at which time the entire OGO program was terminated. For results, see also **OGO 5 (March 4, 1968)**.

Continuing the work begun by **Cosmos 26 (March 18, 1964)** and **Cosmos 49 (October 24, 1964)**, the polar orbiting OGOs mapped the earth's magnetic field at low orbit. The anomalies in that field were charted with extreme precision, and it was found that they were directly related to geological features in the earth's crust, suggesting that an even more detailed map of the magnetic field could be used to discover oil and mineral deposits. **Magsat (October 30, 1979)** continued this research.

Based on accumulated data from an almost 10-year time span (see **Explorer 19–December 19, 1963**; **Spades–July 11, 1968**), models for forecasting the atmosphere's expansion and contraction due to the solar cycle had become reasonably accurate, with the atmosphere expanding during active solar periods and contracting during quiet solar periods. With this information, scientists were able to precisely predict the orbital decay of satellites at various altitudes.

OGO 6 data allowed the mapping of many specific features of the thermosphere (60 to 300 miles altitude) and the geomagnetic field. Densities of neutral nitrogen, oxygen, and helium in the solar polar regions were measured, revealing a surprising range of concentrations, depending upon latitude and time. This fact indicated that the high atmosphere went through extensive mixing and changing, involving convection currents.

In the exosphere (above 300 miles altitude), the atmosphere's temperature was found to be no longer dependent upon altitude. Instead, the temperature varied depending on latitude, time of day, season, solar activity, and geomagnetic fluxes.

OGO 6's data confirmed what OGO 4 and **Aurorae (October 3, 1968)** had found—that the aurora's formation was directly caused by solar wind particles being directly downward to the earth's poles by its magnetic field. See **OGO 5 (March 4, 1968)** for map.

With the completion of the OGO program in late 1971, American research into the earth's magnetic field continued with the launch of two more Imp satellites (see **Imp A–November 27, 1963**), followed by **Explorer 45 (November 11, 1971)**. The next major systematic program to map the ionosphere, the magnetosphere, and their interaction with the Sun was begun with **Intercosmos 1 (October 14, 1969)**. **Azur (November 8, 1969)** was the next satellite to study the aurora.

June 29, 1969
Biosatellite 3
U.S.A.

Built by NASA, Biosatellite 3 was a follow-up to **Biosatellite 2 (September 7, 1967)**. It carried onboard one primate, a pigtail monkey named Bonnie.

The mission was initially scheduled to last 30 days but was cut short after nine days when Bonnie developed dehydration and related circulation problems. These problems occurred because the monkey's small body size and her immobilization in the small capsule (with the many sensors attached to her body), combined with a significant loss of body weight and the pooling of liquids normally experienced by mammals when weightless. Eight hours after Biosatellite 3's recovery, Bonnie died, probably from a massive heart attack.

Because of the flight's premature end, its data relating to long-term bone marrow loss and the brain's reaction to weightlessness were less useful.

The next unmanned satellite dedicated to biological research in space was the **Orbiting Frog Otolith (November 9, 1970)** mission.

July 13, 1969
Luna 15
U.S.S.R.

Luna 15 was launched as an unmanned lunar sample-return mission; it was the last Soviet attempt to beat the United States and **Apollo 11 (July 16, 1969)** to the Moon. The spacecraft reached lunar orbit, but after a number of unsuccessful maneuvers, it crashed. The next Soviet attempt to land and return lunar samples to Earth, **Luna 16 (September 12, 1970)**, was the first unmanned probe to accomplish the task, following two American manned missions.

July 16, 1969
Apollo 11
U.S.A.

Apollo 11 achieved the first human landing on another world. Neil Armstrong (**Gemini 8–March 16, 1966**) and Buzz Aldrin (**Gemini 12–November 11, 1966**) spent 2.5 hours on the lunar surface, while Mike Collins (**Gemini 10–July 18, 1966**) remained in lunar orbit in the command module.

Left to right, Neil Armstrong, Mike Collins, and Buzz Aldrin. *NASA*

The lunar module landed on July 20, 1969, in the flat open mare of the Sea of Tranquility, meeting President John F. Kennedy's call to land a man on the Moon before the end of the 1960s *(see color insert, Figure 22)*.

Armstrong's first words as he stepped onto the lunar surface were "That's one small step for man, one giant leap for mankind." The astronauts set up an American flag and left a plaque behind that read, "Here men from the planet Earth first set foot upon the Moon, July 1969 AD. We came in peace for all mankind."

Both astronauts performed a series of motion tests to see how the one-sixth lunar gravity and their bulky spacesuits affected their mobility. While on Earth, they and their suits would have weighed about 350 pounds, while on the Moon, their weight was less than 60 pounds. They found movement in this gravity environment easier than expected. Aldrin discovered that he unconsciously leaned forward at an impossible angle to compensate for the mass of his suit. He also found that when he moved quickly across the surface, he literally floated from

Buzz Aldrin standing on lunar surface. *NASA*

step to step, and that he could actually rest while gliding through the air. With each landing, the powder on the ground splashed away in all directions in perfect ballistic arcs, imitating crater ejecta.

The airless environment on such a small world made the horizon appear very close and crystal clear, which emphasized the reality that they were standing on a spherical planet, not a flat surface.

During their lunar EVA (extra-vehicular activity), Armstrong and Aldrin gathered 46 pounds of lunar material, getting a good sampling of the landing site. Armstrong found that the density of the mare soil was inconsistent. Soft powder covered everything, but its depth varied. In some places, it

Buzz Andrew's foot on the lunar surface. *NASA*

Setting up experiments; the post to the right of the flag is the television camera. *NASA*

was very thin and the soil beneath was hard and resistant to his shovel. He noticed that some of the small craters had shiny clumps, like polished metal or blobs of molten solder. He also identified numerous microcraters on the surface of larger rocks.

Both astronauts described the lunar surface as a colorless gray terrain, with a wide array of scattered boulders. The soil was "fine and powdery," and Armstrong noted that it would "adhere in fine layers like powdered charcoal to the sole and sides of my boots." This cohesive phenomenon, which made the lunar soil almost appear "wet" to the astronauts, confirmed what had been seen on **Surveyor 3 (April 17, 1967)** and **Surveyor 7 (January 7, 1968)**. One of Aldrin's tasks was to photograph the surface before, during, and after a footprint, to document how the soil interacted with his boot. He noted how the underlying surface was consistently darker, confirming results from the previous unmanned probes.

Later, by assaying radioactive isotopes in the rock samples, their age was established to be around 3.65 billion years, which in turn dated the age of the lava field of the Sea of Tranquility.

Another experiment had Aldrin try to obtain core samples by hammering two metal tubes into the ground. As the two men had found when they set up the American flag, the first four to six inches were soft and easy, after which the ground became hard and very resistant. The most Aldrin could pound either tube into the ground was nine inches.

Aldrin also extended and then retrieved a solar wind detector, a flat sheet of aluminum that remained exposed to the Sun during the Moon-walk. This detector was subsequently extended on each of the next four lunar landings. See **Apollo 16 (April 16, 1972)**.

Together, Aldrin and Armstrong set up a solar-powered seismometer for measuring moonquakes. This instrument functioned for three weeks. See **Apollo 15 (July 16, 1971)** for results. They also set up a mirrored reflector, at which Earth-based lasers were beamed to precisely measure the distance between the earth and the Moon. Because of uncertainties about the location of the landing site, followed by instrument and weather problems, the first accurate reflections were not received for several days. These data measured the Moon's distance to within 13 feet. Later reflectors were set up during **Apollo 14 (January 31, 1972)** and Apollo 15, allowing ground scientists to combine the data from three widely separated positions on the lunar surface to precisely track the Moon's orbital motion over time.

All told, the astronauts took about 100 pictures during their Moon-walk, some of which are among the most famous photographs ever taken. During their Moon-walk, they had a short conversation with President Richard Nixon. Because Apollo 11 was the first to the Moon, its scientific research had been kept simple and uncomplicated. Later missions had more ambitious goals. The next manned lunar landing, **Apollo 12 (November 14, 1969)**, began this process by attempting a pinpoint landing adjacent to **Surveyor 3 (April 17, 1967)**.

July 31, 1969
Mariner 6, Mars Flyby
U.S.A.

Built by NASA and launched on February 25, 1969, Mariner 6 was the second spacecraft to successfully beam back pictures of Mars, following **Mariner 4, Mars Flyby (July 15, 1965)**. The spacecraft was part of a dual flyby mission with Mariner 7. As Mariner 6 flew past the red planet, it sent back 75 images of the Martian surface. See **Mariner 7, Mars Flyby (August 5, 1969)** for results.

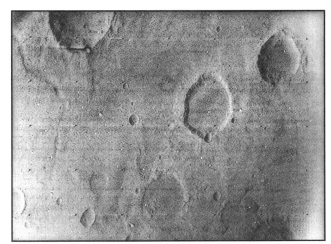

Mariner 6 image of the surface of Mars (detail). See also page 82. *NASA*

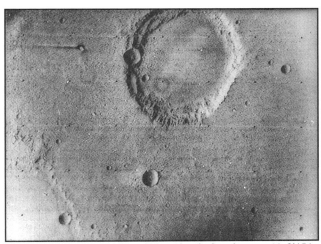

Mariner 6 image of the surface of Mars (detail). See also page 82. *NASA*

August 5, 1969

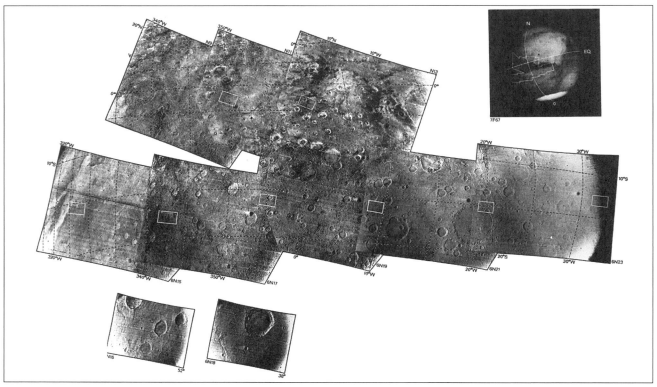

Mosaic of Mariner 6 images of the surface of Mars. See also page 81. *NASA*

August 5, 1969
Mariner 7, Mars Flyby
U.S.A.

Built by NASA and launched on March 27, 1969, Mariner 7 was the third spacecraft to fly past Mars and beam back pictures, returning 126 images of the Martian surface. Flown as part of a dual flyby mission with **Mariner 6, Mars Flyby (July 31, 1969)**, these two spacecraft allowed scientists to see about 10 percent of the Martian surface, a tenfold increase from the coverage of **Mariner 4, Mars Flyby (July 15, 1965)**. The pictures also improved the resolution to less than 1,000 feet.

For Earth astronomers, the most interesting Martian features had always been the dark areas, since their changing shape and color indicated the possibility of vegetation. Hence, the cameras on both Mariner 6 and Mariner 7 were aimed at these regions. The 201 total pictures revealed a cratered surface not unlike the Moon, confirming the results from Mariner 4. Except for one area of chaotic terrain and the almost featureless floor of the Helas region, Mars showed little that was radically different from the Moon's highland regions. "Craters are the dominant landform on Mars and . . . their occurrence is not correlated uniquely with latitude, elevation, or albedo markings," noted one report.

The atmospheric pressure measured by both spacecraft was low, from 3.5 to 6.5 millibars. Except for thin haze, few clouds were seen in the Martian atmosphere. The southern polar cap had a temperature of about –190°F, matching that of carbon dioxide ice and indicating that the polar caps were not water. The planet also had virtually no magnetic field.

Based on the data sent back by the three successful Mariner flyby missions, scientists concluded that Mars was a dead planet that offered no evidence of having sustained life. The project scientists concluded that "it would seem unlikely that Mars ever experienced a terrestrial phase in which primitive oceans could have been present."

The next successful mission to Mars, **Mariner 9, Mars Orbit (November 14, 1971)**, changed these conclusions dramatically.

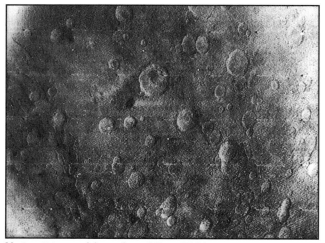

Mariner 7 image of the surface of Mars. *NASA*

Mariner 7 image of the surface of Mars. *NASA*

August 8, 1969
Zond 7
U.S.S.R.
The third Soviet unmanned circumlunar test flight, following **Zond 6 (November 10, 1968)**, Zond 7 tested equipment for a planned (but never launched) manned flight around the Moon. The capsule, a modified Soyuz descent module, was safely recovered on August 14, 1969. Unlike the previous Zond missions, Zond 7 was completely successful. Had it carried men, it would have flown them around the Moon and back to Earth without injury.

The spacecraft flew about 1,232 miles above the lunar surface, taking both color and black-and-white photographs of the Moon and the earth. The photos from this flight, as well as those of Zond 6 and **Zond 8 (October 20, 1970)**, were used to produce an atlas of the far side of the Moon.

August 9, 1969
OSO 6 / PAC 1
U.S.A.
• *OSO 6*   Built by NASA, OSO 6 was the sixth Orbiting Solar Observatory, joining **OSO 5 (January 22, 1969)** in orbit. OSO 6 functioned until December 1972. Like its predecessors, OSO 6 studied the Sun continuously in ultraviolet and x-ray radiation, allowing scientists to refine their understanding of solar flare events and the different elements involved. The next OSO satellite was **OSO 7 (September 29, 1971)**.

• *PAC 1*   The Package Attitude Control (PAC) 1 satellite, built by NASA, was attached to the last stage of the Delta launch rocket. Once in orbit, it performed engineering tests on the use of a flywheel or gyroscope combined with a magnetic boom to orient itself to the earth's magnetic field for the purpose of providing spacecraft attitude control without fuel. Variations on this flywheel concept became standard satellite technology, and was later used, for example, on **IUE (January 26, 1978)**, **Skylab (May 14, 1973)**, and the **Mir (February 20, 1986)** space station (see **Kvant 1–March 31, 1987**).

August 12, 1969
ATS 5
U.S.A.
ATS 5 was the fifth satellite in the Applications Technology Satellite program, the third to reach orbit successfully (following **ATS 3–November 5, 1967**), and the last of the program's first generation. It was only slightly different from the previous ATS satellites. Built and launched by NASA, ATS 5's goal was to perform communications tests while in geosynchronous orbit. One experiment tested the atmospheric effects of satellite-to-ground radio signals. Another tested the use of satellites in aircraft communication.

During launch, the satellite was spin-stabilized, with plans to de-spin it once it was placed in orbit. In orbit, however, the satellite began to tumble in a secondary direction that its attitude engines could not eliminate. The satellite therefore pointed toward the earth for only a short period each orbit, limiting the success of its experiments.

The next and last ATS was **ATS 6 (May 30, 1974)**.

September 30, 1969
Timation 2 / SOICAL Cone / SOICAL Cylinder
U.S.A.
• *Timation 2*   Launched by the U.S. Navy, this satellite was the second test flight of global positioning system (GPS) technology, following **Timation 1 (May 31, 1967)**. Timation 2 was given an eighth-inch additional lead shielding to protect its equipment, its broadcast wattage was tripled, and its attitude control system was improved.

Additional navigational and location tests were done by several Navy ships, using several different techniques. Fixes were accurate to under 200 feet. It was found that Doppler techniques did not work well with moving targets—users needed to know their precise direction and velocity to get a good fix. The tests also showed that the ionosphere had significant irregularities and changed tremendously over days. Any positioning system needed to account for these variations to improve its accuracy. The satellite's broadcasting frequency also went through a large shift because a solar flare swept protons through its instrumentation. The next GPS test satellite was **NTS 1 (July 14, 1974)**.

• *SOICAL Cone and SOICAL Cylinder*   These two small military satellites, one shaped like a cone and the other a cylinder, were used to calibrate and test radar techniques for identifying the shape and size—and hence the purpose—of unidentified satellites in orbit.

October 1, 1969
Boreas
Europe / U.S.A.
Identical in design to **Aurorae (October 3, 1968)**, this European Space Research Organization (ESRO) satellite was placed in orbit by NASA to study the aurora and the ionosphere in the polar regions. Because the satellite wound up in a lower than expected orbit, it operated for only 52 days before its orbit decayed, which limited its results.

October 11, 1969
Soyuz 6
U.S.S.R.

Along with **Soyuz 7 (October 12, 1969)** and **Soyuz 8 (October 13, 1969)**, Soyuz 6 was part of the only three-spacecraft mission flown to date. Manned by Georgi Shonin and Valery Kubasov (**Soyuz 19–July 15, 1975; Soyuz 36–May 26, 1980**), Soyuz 6 was a five-day flight to observe and film the planned docking operations between Soyuz 7 and Soyuz 8. Though the docking failed, the three spacecraft flew in tandem for over a day. On Soyuz 6's last day in orbit, Kubasov performed a series of vacuum welding experiments, the second welds done in space following **ERS 16 (June 9, 1966)**. When the welded samples were later analyzed on Earth, they showed that welding in space was more difficult than expected. None of the welds were acceptable, though the electron-beam welds had worked the best. See **Soyuz T12 (July 17, 1984)** for later electron beam welding experiments.

October 12, 1969
Soyuz 7
U.S.S.R.

Soyuz 7 was originally planned to dock with **Soyuz 8 (October 12, 1969)** while **Soyuz 6 (October 11, 1969)** observed. Though the three spacecraft flew in tandem for almost a day, the docking failed. A helium pressurization integrity test on all three spacecraft prior to liftoff had damaged their rendezvous electronics.

During their five days in orbit, the Soyuz 7 cosmonauts, Anatoli Filipchenko (**Soyuz 16–December 2, 1974**), Vladislav Volkov (**Soyuz 11–June 6, 1971**), and Viktor Gorbatko (**Soyuz 24–February 7, 1977; Soyuz 37–July 23, 1980**), took photographs of the earth and sky in both visible, ultraviolet, and infrared wavelengths.

October 13, 1969
Soyuz 8
U.S.S.R.

Soyuz 8 was the third part of the three-spacecraft mission with **Soyuz 6 (October 11, 1969)** and **Soyuz 7 (October 12, 1969)**. Manned by Vladimir Shatalov (**Soyuz 4–January 14, 1969; Soyuz 10–April 23, 1971**) and Alexei Yeliseyev (**Soyuz 5–January 15, 1969; Soyuz 10–April 23, 1971**), Soyuz 8 flew in tandem with the other two spacecraft for about a day, but did not dock because of the failure of its rendezvous electronics prior to launch.

This triple mission put seven humans in space concurrently, a record that was not matched until **July 15, 1975**, when the Apollo 18–Soyuz 19 joint mission flew at the same time the crew of **Soyuz 18B (May 24, 1975)** occupied **Salyut 4 (December 26, 1974)**. The record was not broken until the launch of **STS 41C (April 6, 1984)**.

The next Soviet manned mission was **Soyuz 9 (June 1, 1970)**, a long-duration flight.

October 14, 1969
Intercosmos 1
U.S.S.R. / East Europe

This satellite was the first in a joint program between the Soviet Union and its communist allies, with scientists from Czechoslovakia, Bulgaria, Hungary, East Germany, Poland, Romania, and Cuba participating (though the bulk of the cost and work was carried by the Soviet Union).

Over the next 25 years, 26 Intercosmos satellites were launched. Most studied the earth's magnetosphere, the ionosphere, and how both interacted with solar flares and the solar wind, augmenting data from the Solrad satellites (see **Solrad 1–June 22, 1960**), the Imp satellites (see **Imp A–November 27, 1963**), the OGO satellites (see **OGO 1–September 5, 1964**), the Aureol satellites (see **Aureol 1–December 27, 1971**), and the Prognoz satellites (see **Prognoz 1–April 14, 1972**). Intercosmos satellites also worked with European satellites such as **HEOS 2 (January 31, 1972)**. In studying the ionosphere, many also did propagation research on extreme low frequency (ELF) and very low frequency (VLF) signals in low-Earth orbit, as well as the ionospheric airglow seen at orbital elevations. Several also carried packages to measure the micrometeorite flux near the earth.

Intercosmos 1 was typical of most Intercosmos satellites: its sensors studied the ionosphere's content along with the Sun's output of x-ray and ultraviolet radiation during solar flares. Its data suggested that the upper atmosphere contained fewer oxygen molecules than expected, and that the amount changed depending on time of day. Later satellites made detailed measurements of hundreds of x-ray solar flares not visible from the earth. Others measured the electrical field structure of the magnetosphere's polar cusps that caused the auroral halos above the poles.

Through 1976, the Soviet Union launched about two Intercosmos satellites a year to maintain continuous observations of the Sun and ionosphere, placing them in a variety of equatorial and polar orbits. After 1976, the launch rate slowed, the satellites having become more sophisticated, with more sensitive equipment and longer life spans.

Beginning in 1978, a second subsatellite built by Czechoslovakia was often included to provide additional ionospheric data. See **Magion 1 (October 28, 1978)**.

Germany's **Azur (November 8, 1969)** next augmented the international cluster of magnetospheric research satellites. For a summary of what these satellites learned, see **Dynamics Explorer 1/Dynamics Explorer 2 (August 3, 1981)**.

A handful of Intercosmos satellites did other research. One made cosmic ray observations using a recoverable capsule (**Intercosmos 6–April 7, 1972**), another did engineering tests of new satellite designs (**Intercosmos 15–June 19, 1976**), and two performed weather and Earth resource observations (**Intercosmos 20–November 1, 1979; Intercosmos 21–August 7, 1981**).

November 8, 1969
Azur
West Germany / U.S.A.
Azur was the first German-built satellite, launched by NASA under a cooperative agreement similar to the joint projects with Canada (see **Alouette 1–September 29, 1962**), the United Kingdom (see **Ariel 1–April 26, 1962**), and Italy (see **San Marco 1–December 15, 1964**). The satellite researched the inner zones of Van Allen belts at about 2,000 miles altitude and at polar latitudes. From this height, it also studied how the Sun's solar wind caused the earth's aurora as it flowed down along the polar cusps (see **OGO 4–July 28, 1967**). Operating until July 1970, its data augmented that gathered by the Imp (see **Imp A–November 27, 1963**), OGO (see **OGO 1–September 5, 1964**), and Intercosmos (see **Intercosmos 1–October 14, 1969**) satellites.

Japan's **Shinsei (September 28, 1971)** was the next magnetospheric research satellite. **Aeros 1 (December 16, 1972)** was the next German satellite. See **Dynamics Explorer 1/ Dynamics Explorer 2 (August 3, 1981)** for a summary of results.

November 14, 1969
Apollo 12
U.S.A.
Apollo 12 was the second manned flight to the Moon, following **Apollo 11 (July 16, 1969)**. During launch, the Saturn 5 rocket was hit by lightning, temporarily shorting out the guidance and navigation system, the fuel cells, the electrical system, and other equipment. Despite this, by the end of the second orbit and just prior to the rocket burn that sent the spacecraft to the Moon, all systems were back in operation. During the journey to the Moon, the astronauts also checked out the LM (lunar module) to make sure it had not been damaged. It had not.

While Richard Gordon (**Gemini 11–September 12, 1966**) remained in lunar orbit, Pete Conrad (**Gemini 5–August 21, 1965; Gemini 11–September 12, 1966; Skylab 2–May 25, 1973**) and Alan Bean (**Skylab 3–July 28, 1973**) spent two days on the Moon. During their stay, they took two separate Moon-walks, each about four hours long. Conrad, the third man to walk on the Moon and one of the shortest astronauts, said this as he stepped off the LM: "Whoopee! Man, that may have been a small one for Neil, but it's a long one for me!"

During the first Moon-walk, the two men set up a package of experiments designed to function after their departure. Similar to those installed on **Apollo 11 (July 16, 1969)**, the instruments included a seismometer, a magnetometer, a solar wind detector, and two ion detectors to measure the almost nonexistent lunar atmosphere. The seismometer functioned for eight years, after which it was shut down intentionally. See **Apollo 15 (July 26, 1971)** for results.

On their second Moon-walk, the astronauts took a 1.5-mile hike across the lunar surface. One of Apollo 12's main scientific achievements was the astronauts' inspection of **Surveyor 3 (April 17, 1967)** and their recovery of some of its

Pete Conrad removes parts of Surveyor 3. The Apollo 12 Lunar Module can be seen on the horizon on the crater's opposite rim. *NASA*

November 14, 1969

A series of trenches made by the scoop can be seen just above the footpad. The multiple footpad impressions were caused when Surveyor 3 settled into position upon landing. *NASA*

parts. Apollo 12's landing site, right next to Surveyor 3, had been chosen specifically to prove that a pinpoint lunar landing could be accomplished—demonstrating that later lunar missions could be sent almost anywhere.

Conrad had landed the LM only 600 feet away from Surveyor 3, alighting on the opposite side of the same crater rim. After hiking past four different craters and obtaining numerous rock samples, Conrad and Bean finished their surface exploration at the robot ship. While Bean took over 50 photographs of Surveyor 3, the trenches made by the probe's scoop, and the nearby terrain that the probe had photographed two and a half years earlier, Conrad used shears to remove Surveyor 3's camera, scoop, television cable, and assorted aluminum tubes. Later analysis of these parts revealed no serious environmental damage from their long-term exposure to the lunar environment. The parts were bleached by the Sun during their 31 months of exposure, and coated everywhere with lunar dust sprayed on them during the probe's initial landing, as well as during the nearby landing of Apollo 12's LM.

Scientists also looked for evidence of micrometeoroid impacts on these parts but failed to find any. This lack placed an upper limit on the rate of micrometeoroid impacts, which agreed with rates indicated by **Pioneer 8 (December 13, 1967)**, **Pioneer 9 (November 8, 1968)**, and a number of other satellites. The photographs of the surface surrounding Surveyor 3 also showed no evidence of additional craters or impacts since it took its own pictures. The new photographs showed no changes of any kind, indicating the extremely slow rate of erosion on the lunar surface.

Radiation damage to the recovered equipment matched predicted rates for interstellar cosmic radiation. Solar wind deposits matched closely that of the solar wind detectors of both Apollo 11 and Apollo 12, though a higher level of helium was detected. Scientists attributed the difference to the longer observation period of Surveyor 3.

A single bacterium, *Streptococcus mitis*, was discovered still surviving inside the probe's camera. Scientists concluded that it had been there prior to launch, and that the bacterium survived because prelaunch vacuum tests freeze-dried it, a condition then maintained in the airless lunar environment.

During this hike, Conrad twice rolled rocks down into a crater and photographed the trail they left in the dust while earthbound scientists measured the reaction of the seismometer. Bean observed that every object around them was heavily impacted and hence severely eroded. Both men collected samples continually.

Returning to the LM, the astronauts dismantled the solar wind collector and finished their EVA. Their second Moonwalk had covered a mile and a half and had taken them as far as a quarter of a mile from the LM. They had collected 75 pounds of lunar samples. These samples, once dated, indicated an age about 500 million years younger than the samples brought back from Apollo 11. Hence, the volcanic activity that had formed the Ocean of Storms had occurred long after that of the Sea of Tranquility. The rocks, while basalt, also differed in type from the Sea of Tranquility samples, exhibiting a wider range of texture and composition. While half the rocks from Apollo 11 had been breccia, only a few Apollo 12 were that type.

Based on the data brought back from the first two lunar landings, it appeared that these two maria regions were volcanic lava fields, but that the melting had taken place at different times and involved somewhat different materials. The data

Collecting a sample. *NASA*

The Apollo 12 Lunar Module during its initial approach to the Moon. See also caption on copyright page. *NASA*

also showed that while the Moon had once been active, its interior had been dead for most of its history.

Once Conrad and Bean had returned to lunar orbit, docked with the command module, and rejoined Gordon, the LM was sent on a trajectory to crash on the lunar surface. This impact allowed scientists to test and calibrate the seismometer on the surface.

The next Apollo mission, **Apollo 13 (April 11, 1970)**, failed to land on the Moon because of an oxygen tank explosion. The next successful sample return mission was the unmanned **Luna 16 (September 12, 1970)**, the first robot ship to visit another world and return samples. The next manned lunar landing was **Apollo 14 (January 31, 1971)**, which made the first landing in the lunar highlands.

November 22, 1969
Skynet 1
U.K. / U.S.A.

Built and designed by the United States for the British military, Skynet 1 worked first in conjunction with the American IDCSP satellites (see **IDCSP 1 through IDCSP 7–June 16, 1966**) and later with their successors, the DSCS II satellites (see **DSCS II–November 3, 1971**). It was the first operational military geosynchronous satellite, and it functioned for three years.

While the lower orbiting IDCSP satellites provided links for large ground antennas, the United Kingdom needed a satellite for the smaller antennas on its oceangoing vessels. Positioned over the Indian Ocean, Skynet 1 filled this gap. Two NATO-built satellites, dubbed NATO 1 and NATO 2, augmented Skynet 1, placed respectively in geosynchronous orbit over Europe and over the Atlantic, in 1970 and 1971. These satellites were later supplemented by the American DSCS II military communications satellites (**November 3, 1971**). The second NATO satellite operated until August 1976 when it was replaced by the **NATO 3A (April 22, 1976)** cluster of satellites.

Skynet 1 was superseded by **Skynet 2 (November 22, 1974)**.

1970

January 23, 1970
ITOS 1 / Oscar 5
U.S.A. / Australia

• *ITOS 1*   This Improved TIROS Operational Satellite was a prototype for NASA's second-generation sun-synchronous weather satellites, following the early Nimbus (see **Nimbus 1–July 28, 1964**) and ESSA (see **ESSA 1–February 3, 1966**) satellites. ITOS 1 carried two different camera systems, each with two cameras. One system allowed individual local weather bureaus to download images directly. Each image covered an area 1,950 by 1,500 miles, with resolution as good as 1.8 miles. The second system took pictures automatically, each image covering a 1,900- by 1,900-mile-area, with a resolution of 1.8 miles. These images were stored on an onboard tape recorder (whose memory could hold as many as 38 photographs) and downloaded when the satellite passed over a ground station.

Both systems had been used before but had flown alternatively on earlier ESSA satellites. By placing both on ITOS 1, the weather satellite was given greater redundancy, as well as better data-gathering capability. The spacecraft also carried two radiometers (working in both visible and infrared wavelengths) for measuring surface and cloud-top temperatures. Temperature changes could be detected in areas as small as 2.5 miles, covering temperatures from −126°F to 134°F.

Finally, a sensor measured the proton flux of particles reaching the poles from the Sun. This experiment confirmed that the earth's magnetic field lines helped to direct solar particles downward onto the poles. See **OGO 4 (July 28, 1967)**.

ITOS 1 operated until June 1971. Its success cleared the way for **NOAA 1 (December 11, 1970)**, the first operational American second-generation sun-synchronous weather satellite.

• *Oscar 5*   Financed and developed entirely by the radio amateurs of Melbourne University in Australia, this was the fifth privately built amateur radio satellite, following **Oscar 4 (December 21, 1965)**. For 23 days, this satellite transmitted data that allowed ham radio operators to measure the status of the satellite's batteries, as well as its position in space relative to the Sun. Oscar 5 was followed by **Oscar 6 (October 15, 1972)**.

February 4, 1970
SERT 2
U.S.A.

Space Electric Rocket Test (SERT) 2 was NASA's second test of an ion engine, which uses small amounts of electricity to generate charged particles that are then beamed away from the spacecraft to produce a thrust. Although the acceleration is small, it requires a very small amount of fuel and is continuous over a long period. The result is that tremendous speeds can eventually be achieved. The first test, during a suborbital launch on July 20, 1964, had shown that a fast-moving stream of ionized particles could be produced.

SERT 2 tested the long-term use of an ion engine in space, using mercury as a fuel and firing its small thruster (generating less than 0.006 pounds of thrust) for a period of at least six months. The spacecraft also tested whether the ion engine interfered with radio communications and the rest of the satellite's electronics.

In 1970, the thruster operated for five months, failing because of an electrical short-circuit. During its operation, the spacecraft's orbit was raised approximately 50 miles.

In 1974, ground controllers were able to reactivate the satellite's ion engine, and they proceeded to perform six years of additional experiments, running the engine for more than 24 months while making 540 engine restarts.

Following these later successful tests, an ion engine was installed on **Telstar 401 (December 16, 1993)**, the engine's first operational use in a commercial satellite. Before this, however, the Soviet Union installed ion engines in its EORSAT surveillance satellites (see **Cosmos 699–December 24, 1974**). In the late 1990s, Pan American Satellites used a xenon ion engine in its communications satellites (see **Pan American Satellite 1–June 15, 1988**).

Later, **Deep Space 1 (October 24, 1998)** was the first spacecraft to use a xenon ion engine to leave Earth orbit to rendezvous with an asteroid.

February 11, 1970
Ohsumi
Japan

With this launch, Japan became the fourth country to launch its own satellite into orbit, following the Soviet Union (**Sputnik–October 4, 1957**), the United States (**Explorer 1–February 1, 1958**), and France (**Asterix–November 26, 1965**). Ohsumi (meaning "rising Sun" in Japanese) was designed to transmit telemetry covering the test launch of the Japanese rocket, and therefore it operated for only 36 hours before its batteries died.

Japan's first operational satellite, **Shinsei (September 28, 1971)**, studied the solar wind and the earth's magnetosphere.

March 10, 1970
DIAL-MIKA / DIAL-WIKA
France / West Germany

This first joint space flight involving France and West Germany was the first test of France's second-generation rocket, the Daimant B; the first dual launch by a French rocket (making France the third country to accomplish this after the United States and the Soviet Union); and the first orbital launch from France's launch facility in French Guiana.

The French DIAL-MIKA satellite was an engineering test for monitoring the new rocket's performance, while the German DIAL-WIKA satellite carried four instruments for measuring the earth's magnetosphere in low-Earth orbit. The scientific package operated for a little over two months.

April 11, 1970
Apollo 13
U.S.A.

Because of an explosion of one of the service module's oxygen tanks two days into the flight, Apollo 13 never landed on the Moon. Astronauts Jim Lovell (**Gemini 7–December 4, 1965;**

Gemini 12–November 11, 1966; Apollo 8–December 21, 1968), Fred Haise (Enterprise, Flight 4–August 12, 1977), and Jack Swigert survived using the LM (lunar module) as a lifeboat. It took four days of nail-biting tension for the crippled command/service module and LM to swing around the Moon and return safely to Earth.

An investigation afterward revealed that the oxygen tank's internal thermometer had been damaged during preflight tests and had ceased functioning. In later preflight tests, the unmonitored temperature inside the tank rose high enough to burn away wire insulation. During the flight, the exposed wires caused the spark and explosion.

The next manned mission to the Moon was **Apollo 14** (**January 31, 1971**). The next successful mission to the Moon was **Luna 16** (**September 12, 1970**), the first unmanned mission to return lunar material to Earth.

April 24, 1970
Mao 1
China

Mao 1, China's first satellite, tested the launch capabilities of the first-generation Long March rocket. A dummy satellite, Mao 1 was artificially weighted to simulate the mass of a full-scale satellite. Its equipment allowed it only to broadcast telemetry and music back to Earth, for just a few days.

This flight made China the fifth country, following the Soviet Union (**Sputnik–October 4, 1957**), the United States (**Explorer 1–February 1, 1958**), France (**Asterix–November 26, 1965**), and Japan (**Ohsumi–February 11, 1970**) to launch its own satellite.

China placed a second dummy satellite in orbit in 1971 and then launched a handful of test satellites for military surveillance purposes through 1981. Its first nonmilitary satellite was **STW 1** (**April 8, 1984**).

April 25, 1970
Strela 1 (Cosmos 343)
U.S.S.R.

Strela military communications satellites resembled the American IDCSP satellites in concept and design (see **IDCSP 1 through IDCSP 7–June 16, 1966**), comprising a cluster of 24 low-Earth-orbiting satellites with full Earth coverage. Eight satellites were generally launched by a single carrier rocket, and through 1997, an average of 16 to 24 satellites were launched per year. From 1970 to 1997, three generations of Strela satellites were developed, the last appearing in 1985.

June 1, 1970
Soyuz 9
U.S.S.R.

Soyuz 9, manned by Andrian Nikolayev (**Vostok 3–August 11, 1962**) and Vitali Sevastyanov (**Soyuz 18B–May 24, 1975**), set a new duration record of almost 18 days in space, exceeding the record set by **Gemini 7** (**December 4, 1965**).

Like Gemini 7, Soyuz 9 was devoted to medical experiments to see how the human body was affected by the environment of space. The most significant discovery was the apparent negative consequences of spin stabilization. The orientation of Gemini 7 in space had generally been controlled using the spacecraft's attitude jets. Soyuz 9, however, used spin stabilization, a technique whereby a spacecraft is spun, not unlike a top, to keep its orientation consistent and its solar panels always facing the Sun. Such a technique had been used on unmanned communications satellites such as **Syncom 2** (**July 26, 1963**) to save on fuel.

This spin also created centrifugal force inside the spacecraft, a kind of artificial gravity. The force, however, existed in different amounts, depending on position and location. This caused the cosmonauts to experience some motion sickness, and when they returned to Earth, they were surprisingly debilitated. Unlike the Gemini 7 astronauts, Nikolayev and Sevastyanov could not stand after landing, despite the much larger size of the Soyuz spacecraft and scheduled daily exercise periods during the entire flight. Both commented that upon returning to Earth's gravity, it felt like they were under several g's of force—and it took both men 10 days before their bodies returned to normal.

During the flight, the cosmonauts performed a number of other experiments. They tested navigation techniques, and on the eighth day, they very precisely calculated their position using Earth landmarks. On the thirteenth day of the flight, they made joint weather observations with a ship in the Indian Ocean and with four Soviet Meteor weather satellites (see **Meteor 1-01–March 26, 1969**).

The next Soviet manned flight, **Soyuz 10** (**April 23, 1971**), flew in conjunction with the launch of the first laboratory in space, **Salyut 1** (**April 19, 1971**). The endurance record of Soyuz 9 lasted until **Soyuz 11** (**June 6, 1971**).

September 12, 1970
Luna 16
U.S.S.R.

Luna 16 was the first successful robot mission to reach another world and return soil samples to Earth. It was also the third spacecraft, following **Apollo 11** (**July 16, 1969**) and **Apollo 12** (**November 14, 1969**), to return lunar samples. The spacecraft soft-landed in the Sea of Fertility.

In total, Luna 16 brought back a little more than a quarter of a pound of material. The spacecraft landed on September 20, 1970, then drilled out a hermetically sealed ground sample core a little over a foot long. Luna 16's upper stage then lifted off and returned the core to Earth, where it was recovered in the Soviet Union.

Comparison of Luna 16's sample with the mare material from Apollo 11 and Apollo 12 revealed an almost identical make-up, indicating that the Sea of Fertility was composed of a surface layer that resembled Earth basalt.

In December 1993, three small pebbles from this sample were sold at a Sotheby's auction for $442,500.

The next lunar landing was the unmanned **Luna 17** (**November 10, 1970**), the first planetary probe to use a separate robot rover with television cameras.

October 20, 1970, 5:44 GMT
Cosmos 373
U.S.S.R.
Cosmos 373 was the target for a test of an ASAT (anti-satellite) system to intercept and destroy orbiting satellites. See **Cosmos 375 (October 30, 1970)**.

October 20, 1970, 19:55 GMT
Zond 8
U.S.S.R.
Zond 8 was the last Soviet circumlunar flight, following **Zond 7 (August 8, 1969)**. The spacecraft was a modified Soyuz descent module, and like the previous flights, it carried a payload of small biological specimens to test the spacecraft's life-support systems.

After flying around the Moon, Zond 8 re-entered the earth's atmosphere. Because the spacecraft's guidance system failed, however, Zond 8 was unable to achieve the necessary re-entry angle to perform a double-skip re-entry off the atmosphere. Instead, it experienced 20 g's of force as it made a ballistic re-entry. Nonetheless, the spacecraft was successfully recovered in the Indian Ocean.

Like Zond 7, Zond 8 took color photographs of the earth and the Moon. These were combined with photographs taken on the previous two Soviet circumlunar flights, **Zond 6 (November 10, 1968)** and Zond 7, to produce an atlas of the Moon's far side.

This spacecraft is the last probe to fly past the Moon and return directly to Earth, as of December 1999. The next lunar flyby mission would not take place for another 20 years: Japan's **Hiten (January 25, 1990)**.

October 30, 1970
Cosmos 375
U.S.S.R.
This radar-guided satellite was the interceptor on a test of an ASAT system for intercepting and destroying orbiting satellites. It successfully targeted and destroyed **Cosmos 373 (October 20, 1970)**.

November 9, 1970
OFO / RM 1
U.S.A.
• *OFO*   OFO (Orbiting Frog Otolith) was the last American unmanned mission dedicated to biological research. Built by NASA, OFO sent two bullfrogs into orbit, testing their otolith sensor cells, which are used by their inner ear for balance. After six days in orbit, both frogs died. However, data revealed that their inner ears, considered similar to humans' inner ears, had within the first three days of flight easily adapted to weightlessness, suggesting that the motion sickness experienced by some astronauts (see **Vostok 2–August 6, 1961; Apollo 8–December 21, 1968; Apollo 9–March 3, 1969**) was a temporary phenomenon.

The next unmanned biological satellite was the Soviet **Bion 1 (October 31, 1973)**.

• *RM 1*   This small NASA satellite carried two experiments. RM (Radiation/Meteoroid) 1's radiation instrument tested new sensors for measuring radiation dosages. Though its results did not correlate well with measurements made using standard equipment, its data was used to more fully map the radiation levels of the South Atlantic anomaly (see **Cosmos 17–May 22, 1963**).

The meteoroid test measured the micrometeoroid impact rates for objects with masses about 10^{-14} grams in weight. During its 90 days of operation, it recorded two hits, one less than predicted for a particle of this size but within the expected impact rate.

November 10, 1970
Luna 17
U.S.S.R.
Luna 17 soft-landed on the Moon in the Sea of Rains, where it conducted observations for almost 11 months. The heart of the spacecraft was a small lunar rover, called Lunokhod 1, which carried four television cameras and a probe for doing impact and soil density tests. This eight-wheeled rover could travel up to 5.5 feet per minute in both forward and reverse.

During its 11 months of operation, the rover traveled more than 6.5 miles, taking over 20,000 pictures and 200 panoramas. The rover traveled across numerous craters and depressions while passing scores of boulder piles. Five hundred soil density tests were done, as well as 25 chemical tests of the

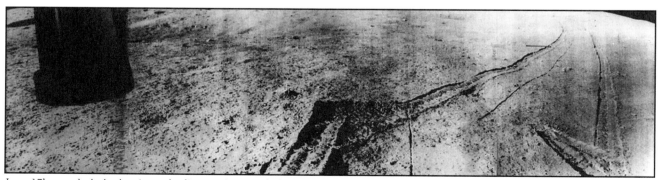

Luna 17's rover looks back at its tracks disappearing over the horizon. *RIA Novosti.*

lunar soil. This exploration found that the Sea of Rains was a typical mare terrain, a relatively flat basalt lava field strewn with countless craters and impact ejecta.

The spacecraft also carried an x-ray telescope, scanning the sky in a three-degree-wide band as the Moon rotated. This instrument mostly tested engineering designs for building larger lunar x-ray telescopes.

In December 1993, Luna 17 (still sitting on the Moon) was auctioned at Sotheby's for $68,500.

The next lunar landing was **Apollo 14 (January 31, 1971)**, achieving the first landing and sample return mission in the lunar highlands.

December 11, 1970
NOAA 1
U.S.A.

Built by NASA, NOAA 1 was operated by NOAA (National Oceanic and Atmospheric Administration). This agency, formerly called the Environmental Science and Services Administration (ESSA), includes the National Weather Service. It is assigned the task of maintaining weather satellite coverage for the United States.

NOAA 1 was the first of the second-generation U.S. weather satellites, following the ESSA satellites (see **ESSA 1–February 3, 1966**) and using designs first tested on **ITOS 1 (July 23, 1970)**.

Over the next three decades, NOAA launched 15 NOAA satellites, each carrying increasingly advanced instrumentation for monitoring the earth's daily weather patterns. All these satellites were placed in sun-synchronous orbits (see **Nimbus 1–July 28, 1964**). They operated in pairs, with the "morning" satellite crossing the equator at local times of 7:30 AM and 7:30 PM, and the "afternoon" satellite at 2:30 AM and 2:30 PM. Together the two satellites would cover the entire Earth surface four times daily.

The multispectral imaging system on later satellites scanned a swath 1,700 miles wide with a resolution of two-thirds of a mile. The system includes four different spectral bands. Two bands, one visible (0.58–0.68 microns) and one infrared (0.72–1.10 microns), mapped surface vegetation. Two other wavebands provide thermal infrared measurements for monitoring volcanoes, fires, surface and atmospheric temperatures, and cloud cover. Other instruments measured a vertical temperature profile of the atmosphere as high as 30 miles elevation.

Later NOAA satellites also carried instruments to measure the earth's total radiation gain and loss over time; monitor the ozone concentration within the atmosphere to an accuracy of 1 percent; observe the interaction of the Sun, its solar wind, and the earth's magnetosphere; and survey the x-ray radiation produced in the earth's upper atmosphere.

Other instruments on later NOAA satellites included a data system for transmitting weather information collected by many automatic ground sensors (such as balloons and buoys) and relaying this information to manned ground stations. This technology, developed by the French, was a direct descendent of the **Peole (December 12, 1970)** and **Eole (August 12, 1971)** experiments and was first tested on **Tiros N (October 13, 1978)**.

Starting in 1974, NOAA sun-synchronous low orbit satellites began working in conjunction with geosynchronous weather satellites to provide both close-up and wide-angle views of the earth's global weather patterns. See **SMS 1 (May 17, 1974)** for the start of this program.

In addition to providing daily weather reports, scientists have used these satellites for climate research. For example, using over two decades of NOAA infrared data, scientists in 1999 discovered that the Sahara Desert in Africa was *not* growing southward because of human land use, as had been believed. In fact, the desert's overall size had not changed at all from 1980 through 1999. The yearly variations of the desert's southern border were caused, instead, by variations in rainfall: in wet years the desert shrank, and in dry years it grew.

Beginning with the eighth NOAA satellite, these spacecraft also carried a SARSAT-COSPAS search-and-rescue receiver/transmitter for quickly locating the distress beacons of ships and aircraft in trouble. See **Cosmos 1383 (June 30, 1982)** for more details.

December 12, 1970, 10:54 GMT
Uhuru (Explorer 42 or SAS 1)
U.S.A. / Italy

Also known as Small Astronomical Satellite (SAS) 1, Uhuru ("freedom" in Swahili) was the first U.S. satellite launched by another country. It was placed in orbit using an American Scout rocket, launched by Italy from its San Marco launch platform off the coast of Kenya.

Uhuru was also the first orbiting astronomical x-ray telescope that produced significant results. During its two-year life span, the satellite generated the first reasonably complete map of the x-ray sky, discovering almost 200 x-ray sources with a surprisingly wide range of behavior, from the first likely black hole candidates to radio galaxies emitting energy in vast amounts. Many of these objects had no visible counterpart and were complete mysteries as to their make-up or origin. Others, though corresponding to visible stars, behaved in a manner that was totally unexpected. One object named Scorpio X-1 generated 100 times more energy in x-rays than in visible light. Another object, Cygnus X-1, was believed the invisible black hole companion to a dim visible star. Many of these galactic x-ray sources varied in output, including the famous supernova remnant of the Crab Nebula. The Crab's x-ray output was found to pulse 30 times per second, in sync with its already measured radio pulses.

These x-ray pulsars were a new class of pulsars, indicating more complex processes were involved in the creation of black holes and neutron stars than the star's simple collapse after a supernovae. That many were like Cygnus X-1, part of a binary system, implied that the interaction of two stars was somehow involved. The central region of our galaxy was revealed to be crowded with these x-ray sources, including its very center.

Beyond the galaxy were many other x-ray sources, associated with quasars, galaxies, and even clusters of galaxies. Some radio galaxies (known to emit radio radiation) were also found to emit energy in x-rays, indicating that powerful and violent events were taking place within them. Several quasars were also identified as major x-ray emitters, as well as a special

December 12, 1970

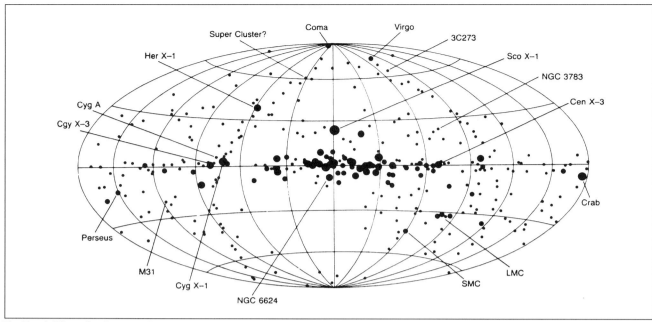

The Fourth Uhuru Catalog map of the x-ray sky. The map's equator matches the plane of the Milky Way galaxy. *NASA*

class of galaxies dubbed Seyfert galaxies, known for their tight, compact shapes and large nuclei. Uhuru also identified the first superclusters of galaxies, bound together by thin, hot, x-ray-emitting gas of hydrogen and helium.

The next x-ray telescope was **Thor Delta 1A (March 12, 1972)**.

December 12, 1970, 12:57 GMT
Peole
France / U.S.A.

The Preparation for Eole (Peole) satellite was a prototype satellite in a French–U.S. research project testing methods for gathering meteorological data from weather balloons. See **Eole (August 16, 1971)**.

December 15, 1970
Venera 7, Venus Landing
U.S.S.R.

Venera 7 was the first spacecraft to transmit data from the surface of Venus, functioning for 23 minutes before the harsh Venusian environment destroyed it. Launched on August 17th, it took four months to reach Venus.

The spacecraft was similar in design to previous Soviet Venus landers (see **Venera 4, Venus Landing–October 18, 1967; Venera 5, Venus Landing–May 16, 1969; Venera 6, Venus Landing–May 17, 1969**). Its descent capsule, however, had been strengthened, and it was now reinforced to withstand for 90 minutes atmospheric pressures exceeding 180 Earth atmospheres (over 2,600 pounds of pressure per square inch) and temperatures as great as 1,000°F. A shock absorption system had also been added to further protect the instruments upon landing.

Once on the surface, Venera 7 recorded temperatures of approximately 900°F and an air pressure about 90 times greater than Earth's, equal to about 1,300 pounds per square inch.

This data confirmed and refined the estimates that had been extrapolated from data returned on Venera 5, Venera 6, and **Mariner 5, Venus Flyby (October 19, 1967)**.

The next Venus probe, **Venera 8, Venus Landing (July 22, 1972)**, made the first readings of soil content, giving humanity its first hint at what lay below the Venusian atmosphere.

1971

January 26, 1971
Intelsat 4 F2
International / U.S.A.

Intelsat 4 F2 was the first successful launch of Intelsat's fourth-generation geosynchronous communication satellite cluster, following the Intelsat 3 satellites (see **Intelsat 3B–December 19, 1968**). See **Early Bird (April 6, 1965)** for a description of Intelsat.

Unlike the Intelsat 2 (see **Intelsat 2B–January 11, 1967**) and Intelsat 3 satellites, Intelsat 4 satellites used a narrow beam design, based on what had been learned on **Tacsat 1 (February 9, 1969)**. Capacity for each satellite was increased significantly to between 3,000 and 9,000 simultaneous two-way voice circuits, depending upon the satellite's configuration. The satellites' television channel capacity was increased from 4 to 12.

Through May 1975, seven Intelsat 4 satellites were placed in orbit above the Atlantic, Pacific, and Indian Oceans. Each satellite had a service life of between 9 and 12 years.

The increasing demand for telephone and television service across the Atlantic Ocean forced Intelsat to supplement the Intelsat 4 cluster in 1975. See **Intelsat 4A F1–September 26, 1975**.

**January 31, 1971
Apollo 14
U.S.A.**

A nine-day flight, Apollo 14 was the third manned flight to the Moon, following **Apollo 11 (July 16, 1969)** and **Apollo 12 (November 14, 1971)**. The crew included Alan Shepard (**Freedom 7–May 5, 1961**), Edgar Mitchell, and Stuart Roosa.

Prior to the astronauts' arrival, the third stage of their discarded Saturn 5 rocket was intentionally crashed into the Moon so that the seismometer from Apollo 12, still working, could measure the seismic vibrations. After their departure, the same was done with the abandoned lunar module (LM). The resulting vibrations, which lasted almost an hour and a half, were measured by both the Apollo 12 and Apollo 14 seismometers to calibrate their instruments. See **Apollo 15 (July 26, 1971)** for overall seismic results.

Shepard and Mitchell landed on the Moon on February 5th. During their 33-hour stay, they completed two Moon-walks, each almost 10 hours long, during which they collected 95 pounds of lunar samples.

On their first EVA (extra-vehicular activity), the two men set up a package of experiments similar to that left by the Apollo 12 astronauts, including a solar wind collector, laser reflector, two seismometers, and several sensors for measuring the particles hitting the Moon—from either the Sun, interplanetary space, or interstellar space. For one of the seismometers, the men had to lay out a string of sound detectors, called geophones, across a distance of 300 feet. Following the astronauts' liftoff, five separate mortar shells were fired at distances of 500, 1,000, 3,000, and 5,000 feet along the string so that these instruments could be calibrated. In another test, Mitchell walked along the string, and every 15 feet, he set off a small explosive charge into the ground. Of the 21 charges, the ground vibrations of nine were measured by the geophones.

During the second Moon-walk, Shepard and Mitchell took a 1.7-mile walk in an attempt to reach the rim of 1,100-foot-wide Cone Crater, about 3,600 feet from the LM. There, scientists hoped the astronauts would find ejecta blasted out from the deep inside the Moon when the crater was formed.

During their hike, Shepard and Mitchell gathered samples, took photographs, and used a magnetometer to measure the local magnetic field at various points. However, the strange, alien terrain made route-finding difficult, and they failed to reach the rim, stopping a mere 65 feet away. Despite their failure to actually look directly down into the crater, they did collect samples from the crater's ejector blanket, including several core samples.

At the end of this Moon-walk, Shepard took out a golf club to see how far he could hit a golf ball. The first swing missed, the second tipped the ball a few feet away, but the third connected and sent the ball "miles and miles and miles," according to Shepard. In the low lunar gravity, a golf ball travels six to seven times farther than on Earth.

Roosa remained in orbit. He took 140 high-resolution photographs of the lunar surface, showing details as small as seven feet. He also photographed the earth-lit night side of the Moon with high-speed film. Much of Roosa's research focus was on the Descartes Crater highland region, where **Apollo 16 (April 16, 1972)** would land.

Alan Shepard inspects the first large boulder seen on the way to Cone Crater. *NASA*

While traveling to and from the Moon, the astronauts performed the first in-space experiments in materials processing. One experiment showed the effect of adding baffles to a tank used for liquids to limit sloshing in weightlessness. Another measured heat flow through different materials—containers of water, sugar solution, and carbon dioxide gas were heated, and the temperature change over time was measured.

Other experiments proved that weightlessness could be a valuable tool in producing alloys that were impossible to make on Earth. Eighteen different materials that normally did not form evenly mixed alloys in gravity were melted to see how the materials mixed in weightlessness. The tests showed that in weightlessness, such immiscible samples did mix more evenly. This materials research was continued on **Skylab 4 (November 16, 1973)**.

The first electrophoresis experiments were attempted to see if weightlessness could be used to separate biological samples, such as the dissimilar components of human blood. Under this process, an electric field was applied to the substance, and the differently shaped objects within it reacted, moving at different speeds depending on their shape, thereby becoming separated. On Earth, gravity overwhelmed the force of the electric field, making the process unworkable. If perfected in weightlessness, such techniques could be invaluable for medical research. On this flight, the test results were inconclusive, due mostly to the experimental nature of the equipment. Later tests on **Apollo 18 (July 15, 1975)** were more productive.

During the flight, Mitchell performed his own private experiment in extrasensory perception. Mitchell randomly chose a number of different symbols, and then without words attempted to communicate his choices to four friends on the ground. Two of the four got 51 out of 200 cards correct, while the other two were less successful. Overall, the results were about the same as chance.

The next manned lunar landing was **Apollo 15 (July 26, 1971)**.

February 26, 1971
Cosmos 398
U.S.S.R.

Cosmos 398 is believed to have been an orbital test flight of a Soviet lunar lander module, similar to **Apollo 5 (January 22, 1968)**. Little is known about this lander, however, and some are even unsure that it actually existed. What is known is that Cosmos 398 made several maneuvers, ejected a module of some type, and ended up in an eccentric Earth orbit ranging from 115 to 1258 miles.

A last test of the lunar lander thought to have been part of Cosmos 398's mission took place with **Cosmos 434 (August 12, 1971)**.

March 21, 1971
Jumpseat 1
U.S.A.

The Jumpseat series of U.S. Air Force surveillance satellites replaced the Ferret series (see **Ferret 1–June 2, 1962** for details). Seven Jumpseat satellites were launched through 1983. Each was placed in an eccentric orbit similar to the Soviet Molniya domestic communication satellites (see **Molniya 1-01–April 23, 1965**). With a perigee of about 200 miles over the Western Hemisphere and an apogee of about 24,500 miles over the Eastern Hemisphere, a Jumpseat satellite spent about 12 hours a day over Soviet territory. During this time, it "ferreted" out radars and antiballistic missile electronics while also listening to military voice communications and other electronic signals.

April 19, 1971
Salyut 1
U.S.S.R.

Salyut 1 was the first laboratory in space, and the first of six Salyut orbiting stations. Weighing over 20 tons, Salyut 1 was the most massive payload launched by the Soviet Union to this date. Its interior space was over 3,200 cubic feet.

Two telescopes conducted astronomy, though optical observations could only be done while Salyut 1 was in Earth's shadow, about 30–35 minutes per orbit. One cosmonaut oriented Salyut while a second aimed the telescope. Spectrographic studies were made with the optical telescope, fitted with two mirrors, 2 and 11 inches in diameter. A gamma-ray telescope observed radiation over 100 MeV in energy, locating sources within an error box 1 degree wide.

Other instruments observed cosmic rays, micrometeoroid fluxes, and the earth's ionosphere.

The laboratory contained one docking port, through which the cosmonauts entered the station (a space walk was no longer required—see **Soyuz 4–January 14, 1970** and **Soyuz 5–January 15, 1970**. During its life span, Salyut 1 was visited twice:

Soyuz 10–April 23, 1971 to April 25, 1971 (2 days)
Soyuz 11–June 6, 1971 to June 29, 1971 (24 days)

Salyut 1 remained in orbit for 180 days; it was deorbited on October 16, 1971. The next successful space laboratory was **Skylab 1 (May 14, 1973)**. The next Soviet space laboratory was **Salyut 3 (June 25, 1974)**.

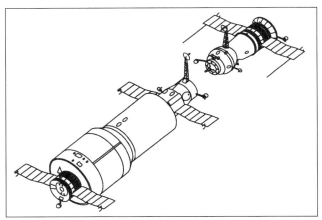

Diagram of a Soyuz craft *(right)* approaching Salyut 1 *(left)*. NASA

April 23, 1971
Soyuz 10
U.S.S.R.

Soyuz 10, the first mission to **Salyut 1 (April 19, 1971)**, was a failure. Its crew was Vladimir Shatalov (**Soyuz 4–January 14, 1969; Soyuz 8–October 13, 1969**), Alexei Yeliseyev (**Soyuz 5–January 15, 1969; Soyuz 8–October 13, 1969**), and Nikolai Rukavishnikov (**Soyuz 16–December 2, 1974; Soyuz 33–April 10, 1979**).

Soyuz 10 was able to soft dock with Salyut 1; that is, the docking probe on Soyuz could be linked with the docking drogue on Salyut. The spacecraft and laboratory could not, however, achieve an airtight seal with electrical link-up. Because of this failure to hard dock, the newly designed Soyuz docking system could not be swung aside to let the cosmonauts enter the laboratory.

After two docking attempts and two days in orbit, the cosmonauts were forced to return to Earth because of fuel and power limitations. The next and last flight to Salyut 1 was **Soyuz 11 (June 6, 1971)**.

April 24, 1971
San Marco 3
Italy / U.S.A.

The third Italian-built satellite, following **San Marco 2 (April 26, 1967)**, San Marco 3 was similar in purpose though more sophisticated, studying the density, composition, and temperature of the atmosphere between 140 and 550 miles elevation. It also continued the research of **Explorer 39 (August 8, 1968)**.

San Marco 3 was launched from the Italian-built San Marco floating platform off the coast of Kenya, using an American Scout rocket. The satellite operated for seven months, its data on changing atmospheric densities used to better forecast the long-term drag on satellites.

The next Italian satellite to study atmospheric densities in low-Earth orbit was **San Marco 4 (February 18, 1974)**. The next satellite to study atmospheric density was **Cannonball 2 (August 7, 1971)**.

May 5, 1971
IMEWS 1 (DSP-647 1)
U.S.A.

IMEWS (Integrated Missile Early Warning Satellite) 1 was the first launch of a second-generation U.S. Air Force infrared early warning satellite that replaced the unsuccessful Midas series (see **Midas 2–May 24, 1960**).

Placed in a high orbit just below geosynchronous, the satellite used cameras and an infrared telescope to quickly detect the launch of ICBMs. Because the satellite's infrared sensors could accidentally confuse the reflected heat from high altitude clouds with that of a missile's rocket plume, television cameras were added to permit ground controllers to verify any triggered alarm. The system also included ultraviolet sensors and gamma-ray detectors for monitoring underground nuclear tests.

In order to provide global coverage, three IMEWS satellites were necessary. A total of four IMEWS were launched, about one per year, the last reaching orbit in December 1975. This early-warning system was then replaced by a third-generation system. See **IMEWS 6 (June 26, 1976)**.

June 6, 1971
Soyuz 11
U.S.S.R.

Ending with the death of its three cosmonauts, Soyuz 11 nevertheless completed the first occupancy of **Salyut 1 (April 19, 1971)**, the first orbiting laboratory. Unlike **Soyuz 10 (April 23, 1971)**, which failed to hard dock with Salyut 1, Soyuz 11 successfully linked up, its crew beginning a record 24-day mission, exceeding the record set on **Soyuz 9 (June 1, 1970)**. The crew, Georgi Dobrovolsky, Viktor Patsayev, and Vladislav Volkov (**Soyuz 7–October 12, 1969**), had originally been the back-up crew supporting the prime crew of Alexei Leonov, Valery Kubasov, and Pyotr Kolodin. When Kubasov came down with a lung problem a week before launch, the Soviets replaced the entire crew, as was their policy.

The first few days of the flight were devoted to activating and checking out Salyut 1's systems. The cosmonauts then devoted the remaining three weeks to a medical, biological, astronomical, and meteorological studies.

Because of the health difficulties experienced by the Soyuz 9 cosmonauts upon returning to Earth, the Soyuz 11 cosmonauts had a more carefully designed exercise program. Salyut 1 provided them a treadmill, and samples of their blood and exhaled air were taken before, during, and after exercise. All three men periodically wore special suits that attempted to simulate Earth gravity. The "penguin" suit forced anyone wearing it to use their muscles—the suit naturally tried to force the body into a fetal position. The "chibis" suit, a more sophisticated version of the cuffs tried on **Gemini 5 (August 21, 1965)**, forced blood circulation into the lower limbs—in weightlessness, the body's fluids were redistributed to the upper body, causing the face to puff up.

Both the gamma-ray and optical telescopes were used several times. The optical telescope took spectrograms of the stars Beta Centauri and Alpha Lyrae, revealing several absorption

The Soyuz 11 crew. *Left to right*, Vladislav Volkov, Viktor Patsayev, and Georgi Dobrovolsky. *RIA Novosti*

lines not seen previously. The gamma-ray telescope measured the general gamma-ray flux near the earth, while other instruments measured the charged particle flux at Salyut 1's low-Earth orbit.

The cosmonauts made numerous Earth observations, taking photographs of weather patterns and geological features.

On June 29, 1971, the crew packed up and undocked from Salyut 1, firing their retro-rockets. As they ejected Soyuz 11's orbital module, however, a vent for equalizing the capsule's atmosphere with the earth's after the release of the main parachutes accidentally opened, allowing the air in the descent module to escape. Although the cosmonauts attempted to close the vent, the manual lever took too long to work. After landing, the rescue party found the three men dead in their couches.

This disaster prompted a major redesign of the Soyuz spacecraft, ending three-man missions until the flight of **Soyuz T2 (June 5, 1980)** almost a decade later. It also led to the cancellation of an additional manned mission to Salyut 1 before the space laboratory was deorbited on October 11, 1971.

The Soviets flew two short unmanned test flights of the redesigned Soyuz before their next manned mission, **Soyuz 12 (September 27, 1973)**, which was a two-day check-out flight prior to its use as a ferry to the next Soviet space lab, **Salyut 3 (June 25, 1974)**. Before Salyut 3's launch, however, the American space laboratory **Skylab 1 (May 14, 1973)** had been launched and occupied by three different crews.

June 15, 1971
KH-9 1
U.S.A.

This U.S. Air Force photo-reconnaissance satellite replaced the wide-angle KH-4/KH-4A/KH-4B/KH-5 series (see **KH-4 9032–April 18, 1962**). Nicknamed "Big Bird," the KH-9 series included a telescope mirror more than six feet in diameter. Once in orbit, this mirror unfolded, its shape held precisely in place by computer. This telescope produced pictures with resolution better than one foot. From four to six film capsules were periodically ejected and recovered on Earth.

The KH-9 satellites had longer life spans than earlier photo-reconnaissance satellites, reaching 275 days by the program's end. KH-9s also had a second imaging system, with a resolution under three feet. Instead of returning film in recoverable capsules, this system transmitted its data digitally.

Twenty KH-9 satellites were launched through April 1986. A separate satellite was usually launched piggyback with them, designed for eavesdropping on radio and microwave transmissions or detecting foreign radar facilities, much like **Canyon 1 (August 6, 1968)**.

The KH-9 series was later augmented by the more advanced KH-11 series. See **KH-11 1 (December 19, 1976)**.

July 8, 1971
Solrad 10
U.S.A.

Solrad 10 was the tenth in a series of U.S. Navy satellites designed to continuously monitor the x-ray and ultraviolet output of the Sun using standardized sensors. See **Solrad 1 (June 22, 1960)** for program information. It was also the third in a three-satellite joint project with NASA.

While carrying sensors similar to those on previous Solrad satellites, Solrad 10 also covered a wider range of wavelengths (0.08–1,600 angstroms) including the extreme ultraviolet and hard x-ray regions. The spacecraft operated until 1974. Like **Solrad 9 (March 5, 1968)**, Solrad 10 was also used to monitor solar activity in conjunction with manned missions, including **Apollo 15 (July 26, 1971)** and **Skylab (May 14, 1973)**.

The next Solrad satellites were **Solrad 11A/Solrad 11B (March 15, 1976)**.

July 26, 1971
Apollo 15
U.S.A.

This space flight was the fourth manned mission to the Moon, following **Apollo 14 (January 31, 1971)**, and the fifth overall to return samples to Earth. It was also the second to land in the lunar highlands, this time exploring the foothills of the Apennine Mountains near Hadley Rille. Apollo 15 lasted just over 12 days and was crewed by Dave Scott (**Gemini 8–March 16, 1966; Apollo 9–March 3, 1969**), Al Worden, and Jim Irwin. Scott and Irwin used a lunar rover with camera attached to travel almost 20 miles during three Moon-walks, the second use of a wheeled vehicle after **Luna 17 (November 10, 1970)**.

Before the first Moon-walk, however, Scott did a half-hour site survey by opening the LM (lunar module) hatch and standing up outside the spacecraft. To the north was the 11,500-foot Mount Hadley. To the south was 11,000-foot-high Hadley Delta. During this review of the landing site, he noted that the nearby mountain slope of Hadley Delta appeared to be covered with a pattern of straight lines, similar to that seen by **Luna 13 (December 21, 1966)** and **Surveyor 7 (January 7, 1968)**, caused either by a series of overlapping layers or an underground structure of faults. He also noted how smooth and eroded all the mountains and features appeared. "I see no jagged peaks of

The sheer, 11,500-foot face of Mount Hadley, eight miles away. Note the faint, diagonally oriented strata or lines on the mountain side. *NASA*

any sort," he said, despite being surrounded by high mountains on almost all sides.

On their first EVA (extra-vehicular activity), which lasted six hours, Scott and Irwin traveled toward the edge of Hadley Rille. Along the way, they obtained dust and rock samples and drilled core tubes. At one point, they sampled the top of a boulder, then rolled it over to sample the underside. They finished this first Moon-walk by assembling a package of instruments near the LM, similar to those on **Apollo 11 (July 16, 1969)**, **Apollo 12 (November 14, 1969)**, and Apollo 14. They set up a seismometer, magnetometer, and two sensors to measure any gases released by the interaction of the solar wind with the lunar surface. They also built a solar wind experiment, as well as a laser reflector for measuring precisely the distance between the earth and the Moon. One new instrument that measured the heat flowing from the interior of the Moon found that the flow was about half that of the earth's. Other sensors proved that the highlands were chemically different than the mare basins, richer in aluminum, poorer in magnesium, and composed mostly of the mineral pyroxene.

On their second EVA, covering six miles of lunar terrain, the goal was a mountain named Hadley Delta. Using the rover, they drove three miles from the LM onto the mountain's slope, climbing about 300 feet above the valley floor. Twice they stopped beside craters to take photographs, then collect samples. They also obtained core samples at each stop. At one point, the slope was so steep that while Scott hammered a rock sample off a boulder, Irwin had to hold the rover to keep it from sliding downhill.

Other samples were collected when Irwin used a rake to gather walnut-sized fragments out of the ground. A number of green-colored samples were found to be made up of tiny spherules of iron and magnesium-rich glass formed early in the Moon's history in fountains of volcanic fire, then with time

Irwin prevents the rover from sliding (note the wheel off the ground) while Scott collects a sample. *NASA*

buried deep in the lunar interior. Impacts later threw this material free to resettle on the lunar surface.

Of the samples collected on this second Moon-walk, the most significant was #15415, a white crystalline chunk of anorthosite that had floated to the surface of the Moon when its crust was molten. While the two astronauts found three other pieces of anorthosite during this EVA, as well as pieces of basalt and breccia flung up from the mare valley below, this sample was later dubbed the "Genesis Rock." Earth-based laboratories dated its age to be 4.5 billion years, proving that it had congealed almost immediately after the Moon's formation. Scientists also found that, during its long history, the rock had been shocked twice by hard impacts.

At the LM, the two astronauts finished their Moon-walk by completing the setup of their surface experiments. The heat flow experiment was installed. Scott drilled down about eight feet, the deepest core sample yet. Because of drill problems, he had trouble extricating the core and left this job for the next EVA. Irwin, meanwhile, dug a trench and photographed the collapse behavior of the lunar dirt. Based on his data, the regolith of the Apollo 15 landing site was estimated to be the hardest tested to date.

The third Moon-walk began with the extraction of the eight-foot core drilled by Scott the day before. This core was found to contain over 50 separate layers, with its bottommost layer 500 million years old. The composition, density, and make-up of these layers helped scientists determine how lunar soil developed, while also giving mining engineers a better overview of the soil for future quarrying efforts.

The EVA then headed to the edge of Hadley Rille. Along the way, the astronauts picked up the youngest sample yet found on the Moon, a million-year-old rock ejected from a small young crater.

Based on seismic readings, scientists had estimated that at the Apollo 15 landing site, the lunar regolith was generally about 15 feet deep. At the rim of Hadley Rille, however, this topsoil thinned out, exposing a rock-strewn slope of bedrock tilting gently and with increasing steepness into the rille. Scott and Irwin parked clear of this boulder field and spent almost an hour gathering samples. They obtained two core samples. They described the rille's far wall, which they could see was subdivided by layers, indicating repeated volcanic flows.

This last EVA ended with final packing at the base of the LM. Scott also performed one last experiment. He held up a feather and his geology hammer, and released both simultaneously. Both hit the ground simultaneously, proving—as had Galileo centuries before—that gravity acts equally on all objects, regardless of size or weight.

Scott's last task was to drive the rover about 300 feet from the LM so that its television camera could broadcast their lift-off. At this spot, he also left a Bible and a small plaque commemorating the 14 individuals, both American and Soviet, who had died to that date participating in the exploration of space.

As Scott and Irwin were exploring the lunar surface, Al Worden remained in orbit, performing his own research. Unlike previous missions, one whole compartment of the Apollo service module was dedicated to scientific experiments. Shortly after entering lunar orbit, the outside panel was jettisoned, exposing this equipment to space. One camera could take pictures with a resolution as small as three feet. A second took wide-angle panoramas. A mass spectrometer studied lunar space at between 67 and 200 miles elevation. Three other spectrometers mapped the geological make-up of the surface in gamma, x-ray, and alpha-particles, finding that most of the radioactive materials on the Moon were concentrated in the Imbrium and Procellarum (Ocean of Storms) mare regions.

During Worden's observations, he noticed what appeared to be lava cinder cones in the region of Littrow near the Sea of Serenity. This region was explored later by **Apollo 17 (December 7, 1972)**.

Just before leaving lunar orbit, the astronauts released a small satellite, Particles and Fields Subsatellite (PFS) 1, to relay communications between the earth and the experiments on the lunar surface. PFS 1 also measured the uneven lunar gravity field, more precisely mapping its mascons. It found that though the Moon did not have an active magnetic field, there were small remnant fields located over a number of craters.

Also jettisoned was the LM, which was then crashed onto the lunar surface so that the seismometers from Apollo 12, Apollo 14, and Apollo 15 could calibrate their sensors. The first two seismometers had been accumulating data for almost 20 months, measuring the quakes produced by many minor meteor impacts, as well as the impacts of Saturn 5 rocket stages and lunar modules. With the addition of Apollo 15's seismometer, quakes could be better triangulated, and the data showed that about 80 percent of all lunar earthquake activity was centered about 500 miles below the surface at a point 370 miles south-southwest of the Apollo 12 and Apollo 14 landing sites, under the Sea of Moisture. Based on the low level of earthquake activity and the lack of a planetary magnetic field, scientists inferred that the Moon was essentially solid—dead internally for many eons.

The 170 pounds of rocks returned to Earth by this mission also helped complete the lunar historical timeline. From material brought back from Apollo 11, scientists could date the formation of the Sea of Tranquility to approximately 3.65 billion years. Apollo 12 had dated the Ocean of Storms at about 3.60 billion years. Apollo 14 had dated the highlands region near Hadley Rille to be about 3.85 billion years old. And Apollo 15 had found both the youngest and oldest rocks, 1 million and 4.5 billion years old, respectively, stretching the Moon's age back to the very beginnings of the solar system. From this data, scientists now estimated that the cratered lunar highlands was the Moon's oldest known terrain, formed between 4.5 to 3.6 billion years ago and revealing a period early in the solar system's history when violent impacts dominated. Pictures of Mars from **Mariner 9, Mars Orbit (November 14, 1971)** confirmed this conclusion. This accretion period was later followed by the volcanic events that flooded the lunar lowlands about 3.6 billion years ago, forming the mare regions.

During the return flight to Earth, Worden made a space walk to recover the film magazines from the outside cameras, as well as a micrometeoroid detector. The impact rate from this last experiment correlated well with previous experiments, with no impacts larger than 200 microns in diameter.

After splashdown, Irwin had trouble readjusting to Earth gravity. The hard preflight training had created a potassium deficiency for him and for Dave Scott, which on the Moon had caused heart irregularities. It took both men almost two weeks for their cardiovascular systems to return to normal.

The next successful lunar landing was the unmanned **Luna 20 (February 14, 1972)**, returning about a quarter of a pound of soil from the Apollonius highlands near the Sea of Fertility. The next manned mission was **Apollo 16 (April 16, 1972)**, which landed in the Descartes region.

August 7, 1971
OV1-20 / Cannonball 2 / OV1-21 / Musketball 1 / Mylar Balloon / Rigidsphere / Gridsphere 1 / Gridsphere 2 / LCS 4
U.S.A.

These eight satellites were the last launch in the U.S. Air Force's Orbiting Vehicle (OV) research series (see **OV1-2–October 5, 1965**). The OV program gathered engineering data to improve the design of military satellites.

• *OV1-20* This satellite functioned as the in-orbit launchpad for Cannonball 2, the second Cannonball experiment (see **Cannonball 1–July 11, 1968**). It also carried two experiments to measure high-energy particles during its three-week orbital life.

• *Cannonball 2* Similar in design to Cannonball 1, this two-foot-wide, high-density sphere weighed 800 pounds. It was injected into an orbit with a perigee of only 82 miles, so that ground researchers could measure the atmospheric friction dragging it out of orbit.

• *OV1-21* This satellite provided an in-orbit launchpad for five satellites, Musketball 1, Gridsphere 1, Gridsphere 2, Mylar Balloon, and Rigidsphere. It also carried sensors to measure oxygen density and changes in the atmosphere's density.

• *Musketball 1* A one-foot-diameter sphere, this satellite carried a radar tracking beacon. Placed in an orbit with a perigee of 85 miles, Musketball 1 supplemented the atmospheric drag tests of Cannonball 2.

• *Mylar Balloon* This inflatable balloon, which expanded to a seven-foot diameter, was placed in orbit at about 530 miles to measure atmospheric drag.

• *Rigidsphere* This two-foot-wide aluminum sphere acted as a control for the atmospheric drag measurements obtained from Cannonball 2, Musketball 1, Mylar Balloon, Gridsphere 1, and Gridsphere 2. Researchers could measure the effect of atmospheric drag and compare the differences. The research from these satellites allowed military satellite designers to predict exactly how much fuel was needed to maintain surveillance satellites at low altitudes.

The next atmospheric density satellite was **Explorer 51 (December 16, 1973)**.

• *Gridsphere 1 and Gridsphere 2* These two inflatable spherical balloons, similar in design to **OV1-8 (July 14, 1966)**, were covered with fine aluminum wires to act as passive reflector communications satellites. Placed in orbit at about 500 miles elevation, both spheres were also used to measure atmospheric drag.

- *LCS 4*   Built by Lincoln Laboratory at the Massachusetts Institute of Technology, LCS (Lincoln Calibration Sphere) 4, a 3.5-foot-diameter magnesium sphere, was part of a long-term program to calibrate ground tracking radar systems. Because the sphere's size and shape were known precisely, ground controllers could gauge the accuracy of their radar systems and adjust them accordingly.

August 12, 1971
Cosmos 434
U.S.S.R.
Cosmos 434 was the second and last orbital test flight of a Soviet lunar lander, following **Cosmos 398 (February 26, 1971)**. Very little is known about this spacecraft. It made several maneuvers, ejected a module, and ended up in a high orbit that ranged from 6,800 to 8,700 miles in altitude.

August 16, 1971
Eole
France / U.S.A.
Eole, designed to gather data from weather balloons released on Earth, was built by the French space agency and launched by NASA. It was a follow-up to **Peole (December 12, 1970)**. During its first five months in orbit, over 500 weather balloons were released above Argentina, with Eole relaying to a station in France the temperature and air pressure recorded by each balloon over time. The satellite also relayed each balloon's changing location, which provided data on atmospheric wind speeds.

Though built to have a life span of only six months, the satellite functioned for two years. After completion of the balloon experiment Eole was used to relay data from a number of other satellites. Its data collection system was later redesigned and flown on the NOAA weather satellites. See **NOAA 1 (December 11, 1970)** and **Tiros N (October 13, 1978)**.

September 2, 1971
Luna 18
U.S.S.R.
Intended as the Soviet Union's second unmanned lunar sample return mission, following **Luna 16 (September 12, 1970)**, Luna 18 ceased functioning upon impact with the lunar surface.

The next lunar sample return mission was **Luna 20 (February 14, 1972)**.

September 28, 1971, 4:00 GMT
Shinsei
Japan
Shinsei ("new star" in Japanese) was Japan's first operational satellite. It collected data for about four months on solar flares, the earth's magnetosphere, and the ionosphere, augmenting data from the Imp (see **Imp A–November 27, 1963**), OGO (see **OGO 1–September 5, 1964**), and Intercosmos (see **Intercosmos 1–October 14, 1969**) satellites.

See **Dynamics Explorer 1/Dynamics Explorer 2 (August 3, 1981)** for results. The next magnetospheric research satellite was **Explorer 45 (November 15, 1971)**.

September 28, 1971, 10:00 GMT
Luna 19
U.S.S.R.
Luna 19 was a lunar orbiter, designed for photographing the Moon's surface while mapping its gravitational field. The spacecraft also carried instruments for measuring the solar wind, lunar surface radiation, and Earth's magnetosphere at lunar distances.

The spacecraft operated for more than a year, orbiting the Moon more than 4,000 times and taking photographs of many specific surface features. By studying the changes to Luna 19's orbit, scientists more precisely located the mascons that caused variations in the Moon's gravitational field. Luna 19 also recorded a direct correlation between solar flares and sudden increases in solar wind protons impacting the spacecraft.

The next lunar orbiter was **Luna 22 (May 29, 1974)**, which also was the last lunar orbiter intended solely for lunar research for almost 20 years.

September 29, 1971
OSO 7
U.S.A.
OSO 7 was the seventh Orbiting Solar Observatory built by NASA, following **OSO 6 (August 9, 1969)**. Operating until May 1973, OSO 7 obtained spectroheliograms of visible and ultraviolet radiation while monitoring the Sun in x-ray and gamma-ray wavelengths (from 1 KeV to 10 MeV). It also made detailed daily observations of the corona from 3 to 10 solar radii out from the Sun.

OSO 7 found that the streams of material emanating from the Sun changed drastically from day to day. These "helmet streamers," as scientists dubbed them because of their shape, were linked with surface regions of high activity and extreme ultraviolet radiation.

OSO 7 also monitored the celestial sky for hard x-ray emissions in the 1 to 60 KeV range, studying objects like the Crab Nebula and the black hole candidate Hercules X-1 discovered by **Uhuru (December 12, 1970)**. The periodic variations of emissions indicated that Hercules X-1 was a binary system with some sort of interaction taking place between its two objects.

The next and last OSO satellite was **OSO 8 (June 21, 1975)**.

October 17, 1971
STP 1 (ASTEX)
U.S.A.
Space Technology Project (STP) 1, also called ASTEX, was part of the U.S. Air Force's Space Test Program. It performed engineering tests on the use of flexible solar panels, a continuation of research from **OV1-13 (April 6, 1968)**. It also carried instruments for measuring the charged particles in the earth's ionosphere as well as the Sun's radio emissions. The information was used to gauge the interference to radio communications caused by solar activity and to develop technology for overcoming such interference. Because the solar panels worked so successfully, the satellite continued to radio back data for more than a year.

October 28, 1971
Prospero
U.K.

With this launch from the Woomera rocket range in Australia, the United Kingdom became the sixth nation to launch its own satellite into orbit, following the Soviet Union (**Sputnik–October 4, 1957**), the United States (**Explorer 1–February 1, 1958**), France (**Asterix–November 26, 1965**), Japan (**Ohsumi–February 11, 1970**), and China (**Mao 1–April 24, 1970**). It was also the last satellite launched by the U.K. Following this flight, the British government decided that it would be cheaper to launch future British satellites on either U.S. or European Space Agency (ESA) rockets.

Prospero itself was mostly an engineering satellite that tested designs for future satellite construction. It also carried one scientific experiment to measure the micrometeoroid flux in low-Earth orbit.

The next new rocket system was launched by the European Space Agency (see **Cat 1–December 24, 1979**). The next independent nation to join the space community was India (see **Rohini 1B–July 19, 1980**).

November 3, 1971
DSCS II 1 / DSCS II 2
U.S.A.

The Defense Satellite Communications System (DSCS) satellites were the second generation of American military communications satellites, replacing the IDCSP satellites (see **IDCSP 1 through IDCSP 7–June 16, 1966**). They also supplemented the geosynchronous military communications satellite cluster begun with **Skynet 1 (November 22, 1969)**.

The DSCS II satellites had a significantly greater capacity than the IDCSP satellites: 1,300 two-way voice circuits versus 11. They were usually launched in pairs and operated anywhere from eight months to over 10 years. In total, 12 DSCS II satellites were launched through 1989, positioned over the Indian, Atlantic, and Pacific Oceans. The last eight, launched from 1977 through 1989, kept a cluster of four operating satellites with two orbiting spares in orbit at all times.

In 1982, the U.S. military began launching its third-generation communications system. See **DSCS III (October 30, 1982)**.

November 14, 1971
Mariner 9, Mars Orbit
U.S.A.

Built by NASA and launched on May 30, 1971, this unmanned probe was the first spacecraft to orbit the red planet. Its mission was to make a planetwide photographic survey of Mars. It was originally part of a two-spacecraft mission, but Mariner 8 failed at launch.

In September, while Mariner 9 and two Soviet probes, **Mars 2 (November 27, 1971)** and **Mars 3 (December 2, 1971)**, were approaching Mars, Earth-based telescopes noted the appearance on Mars of what were then called "yellow" clouds (so called because they are most visible when photographed through a yellow filter). Soon these clouds spread through the mid-southern latitudes, and by October, they covered the entire planet.

Three days before Mariner 9 reached Mars, it took its first pictures, revealing a dust storm–shrouded planet, with only a bright southern polar cap and four mysterious black smudges slightly north of the equator.

On November 14th, the spacecraft fired its rockets and slid into orbit around Mars. Unlike the two Soviet probes that arrived shortly thereafter, Mariner 9's design allowed ground controllers to shut the spacecraft down, save fuel, and wait for the planetwide dust storm to settle. During this waiting period, the spacecraft did take a few pictures, including the first close-up image of Deimos, one of Mars's moons. It showed an irregular-shaped object pocked with craters. Later pictures of another moon, Phobos, showed an even more irregular shape with even more craters.

By the middle of December, the dust storm had eased enough that Mariner 9 could begin photographing the surface. The dark smudges at first appeared to be large craters, 40 to 50 miles wide, but strangely located at the tops of four separate mountains. After a short hesitation, NASA scientists announced that these were not craters, but the calderas of four gigantic volcanoes, the first non-terrestrial volcanoes ever discovered and far larger than any volcanoes on Earth. The largest, Olympus Mons (formerly called Nix Olympica), was nearly 17 miles high (three times higher than Mount Everest) and almost 400 miles across. Both **Mariner 6, Mars Flyby (July 31, 1969)** and **Mariner 7, Mars Flyby (August 5, 1969)** had photographed it as they had approached the planet in 1969, but at the time no one realized what it was. Near it sits the three other volcanoes, also larger than anything seen on Earth. All together, this volcano cluster forms the Tharsis bulge, rising about five to six miles above the average altitude of the rest of the Martian surface.

More surprises followed. Just east of the four volcanoes was an enormous canyon that made the Grand Canyon appear trivial in comparison. Valles Marineris (named in honor of Mariner 9) was 2,500 miles long, 400 miles wide, and 6 miles

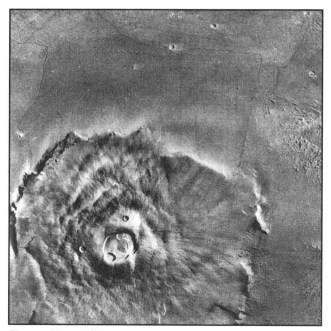

Olympus Mons. *NASA*

November 14, 1971

deep. If placed across the United States, it would cover the length of the country.

Mars quickly turned out to be quite different from the dead, Moon-like planet indicated by Mariner 4 (**Mariner 4, Mars Flyby–July 15, 1965**), Mariner 6, and Mariner 7. Mariner 9 photographed riverlike channels with tributaries and erosion patterns remarkably reminiscent of Earth's waterways. Some craters looked as if the meteors had splattered in mud. Others showed wind patterns and evidence of sand dunes. The north polar cap was made of carbon dioxide and ice, was surrounded by a huge field of sand dunes, and showed evidence of a constantly changing shape and size.

The periodic wave of darkening that earlier astronomers (and science fiction writers) had hoped was the ebb and flow of seasonal vegetation was instead deposits from the annual dust storms. The planet's northern hemisphere generally had few craters, sat at a lower elevation, and had volcanoes and huge canyons. The southern hemisphere was crater covered and generally higher in elevation. Everywhere scientists looked, they could see waterlike channels and erosional features, evidence of catastrophic flooding and hidden subsurface water.

By counting and noting how the craters on Mars's southern highlands overlay one another, scientists were able to make a rough outline of the planet's early history. This data indicated that, much like the Moon, Mars had experienced a period of intense bombardment early in its history. Despite the red planet's giant volcanoes, the origin of the solar system's craters was now undeniably assigned to impact, and that origin spoke volumes about the nature of the solar system's beginnings. These conclusions were further confirmed by Mari-

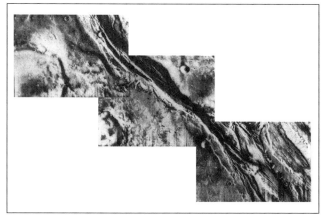

River channels? *NASA*

ner 10's observations of Mercury (see **Mariner 10, Mercury Flyby #3–March 16, 1975**).

In total, Mariner 9 operated for almost a year, beaming back a total of 7,349 pictures. The images left scientists humbled, for they showed that the experts had been wrong about Mars—not once, but twice. There were no canals, no beautiful alien cities. But Mars was also not a dead planet like the Moon. Its geology was strangely earthlike—in an alien and majestic way. And it exhibited features that scientists could only explain by giving the planet a past with running water, a thicker atmosphere, and an environment almost as active as Earth's.

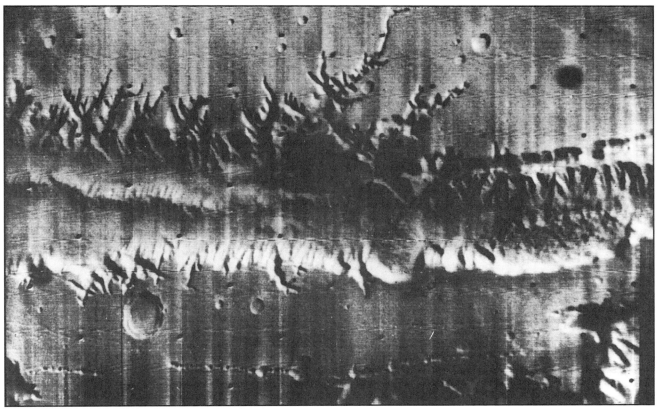

Valles Marineris. *NASA*

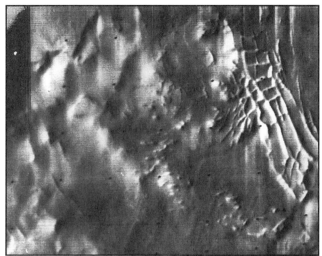

"Inca City." Some features were not analogous to any Earth formation. *NASA*

The two Soviet Mars missions, Mars 2 and Mars 3, arrived at Mars on November 27, 1971 and December 2, 1971 respectively.

November 15, 1971
Explorer 45
U.S.A. / Italy

Built by NASA and launched by Italy from its floating San Marco launchpad off the coast of Kenya, Explorer 45 (also called the Small Scientific Satellite 1) spent three years studying how solar flares and the Sun's magnetic storms disturb the earth's magnetosphere. Its detailed measurements supplemented data from the Imp (see **Imp A–November 27, 1963**), OGO (see **OGO 1–September 5, 1964**) and Intercosmos (see **Intercosmos 1–October 14, 1969**) satellites.

During a giant solar flare in the summer of 1972, for instance, the magnetosphere's bow shock was pushed to within 15,000 miles of the earth's surface, or about one-third the normal distance. This same flare was also seen by the Pioneer solar orbiting satellites (see **Pioneer 9–November 8, 1968**, as well as **Pioneer 10–March 3, 1972**).

Explorer 45 also measured the magnetosphere's ring current, a westward-traveling electric current circling the earth's equator at several thousand miles elevation. The satellite also made the first detailed analysis of how the charged particles in the magnetosphere interact with the waves of energy emanating from the Sun.

See **Dynamics Explorer 1/Dynamics Explorer 2 (August 3, 1981)** for results. **Ariel 4 (December 11, 1971)** continued this ionospheric and magnetospheric research.

November 24, 1971
Molniya 2-01
U.S.S.R.

Molniya 2-01 was the first launch in a Soviet second-generation communications satellite cluster. While the Molniya 1 satellites (see **Molniya 1-01–April 23, 1965**) were designed to handle military and government communications, Molniya 2 satellites handled civilian telephone and television needs within the Soviet Union. Seventeen Molniya 2 satellites were launched through 1977, eventually to be replaced by the slightly improved the Molniya 3 satellites. See **Molniya 3-01 (November 21, 1974)**.

November 27, 1971
Mars 2
U.S.S.R.

Launched on May 19, 1971, Mars 2 was similar to the Soviet Venera probes (see **Venera 5, Venus Landing–May 16, 1969**), including both an orbiter and lander. The Mars 2 descent module, however, was destroyed upon impact with the Martian surface, and though the Mars 2 orbiter module did work, automatically taking pictures of the Martian surface for several months, its pictures were useless because Mars during that time was engulfed in a planetwide dust storm (see **Mariner 9, Mars Orbit–November 14, 1971**). See **Mars 3 (December 2, 1971)** for a result summary.

December 2, 1971
Mars 3
U.S.S.R.

Identical to **Mars 2 (November 27, 1971)** and also similar to the Soviet Venera probes (see **Venera 5, Venus Landing–May 16, 1969**), Mars 3 included both a lander and an orbiter. It was launched from the Soviet Union on May 28, 1971.

While the Mars 3 lander successfully soft-landed, becoming the first spacecraft to land on Mars and transmit its signal back to Earth, the transmission ceased after only 20 seconds. No data of any use was received. Scientists theorized that the spacecraft's instrumentation had been damaged by the planetwide dust storm that at that moment covered Mars.

Like the Mars 2 orbiter, the Mars 3 orbiter functioned perfectly, automatically taking pictures of the Martian surface over the next few months. Because of the dust storm, however, its pictures showed no details, and were useless. See **Mariner 9, Mars Orbit (November 14, 1971)**.

Both Mars 2 and Mars 3 did return some measurements of the Martian atmosphere. They recorded surface temperatures ranging from 55°F at midday to –166°F at night. The atmospheric pressure on the surface averaged between 5.5 and 6 millibars, in contrast to Earth's 1,000 millibar surface air pressure.

The Soviet Union once again sent a fleet of unmanned probes to Mars in 1973. Once again, all failed. See **Mars 4, Mars Flyby (February 10, 1974)**.

The next successful missions to Mars, **Viking 1 (June 19, 1976)** and **Viking 2 (August 7, 1976)**, included two orbiters and two landers and made the first direct measurements of its surface material.

December 11, 1971
Ariel 4
U.K. / U.S.A.

Ariel 4 was similar to **Ariel 3 (May 5, 1967)**, except that it also carried one American experiment for measuring charged particles. The spacecraft studied the earth's ionosphere, magnetosphere, and the aurora. Its data supplemented information from the Imp (see **Imp A–November 27, 1963**), OGO (see **OGO 1–September 5, 1964**), and Intercosmos (see **Intercosmos 1–October 14, 1969**) satellites.

See **Dynamics Explorer 1/Dynamics Explorer 2 (August 3, 1981)** for results. The next ionospheric-magnetospheric research satellite was **Aureol 1 (December 11, 1971)**.

December 27, 1971
Aureol 1
France / U.S.S.R.

Aureol 1 was the first in a series of joint Soviet and French research satellites. The rocket and launch facilities were provided by the Soviet Union, while France contributed the satellite experiments. Besides the Aureol satellites, this joint program launched two astronomical satellites, **Signe (June 17, 1977)** and **Gamma (July 11, 1990)**.

Through 1981, three Aureol satellites were launched to study the charged particles traveling along the magnetosphere's polar cusp and causing the aurora. These satellites worked in conjunction with the Solrad satellites (see **Solrad 1–June 22, 1960**), the Imp satellites (see **Imp A–November 27, 1963**), the OGO satellites (see **OGO 1–September 5, 1964**), and the Intercosmos satellites (see **Intercosmos 1–October 14, 1969**).

See **Dynamics Explorer 1/Dynamics Explorer 2 (August 3, 1981)** for results. The next magnetospheric research satellite was **HEOS 2 (January 31, 1972)**.

1972

January 31, 1972
HEOS 2
Europe / U.S.A.

Launched by NASA, the Highly Eccentric Orbit Satellite (HEOS) 2 was the fifth satellite of the European Space Research Organization (ESRO), a 10-nation association that was the forerunner of today's European Space Agency. Similar to its predecessor, **HEOS 1 (December 5, 1968)**, HEOS 2 studied the earth's magnetosphere and its interaction with the solar wind, working in conjunction with the Solrad (see **Solrad 1–June 22, 1960**), Imp (see **Imp A–November 27, 1963**), OGO (see **OGO 1–September 5, 1964**), and Intercosmos (see **Intercosmos 1–October 14, 1969**) satellites.

HEOS 2 was placed in an eccentric polar orbit (perigee: 252 miles; apogee: 149,231 miles) so that it repeatedly passed in and out of the magnetosphere. Furthermore, its apogee was specifically placed over the North Pole. The first satellite to fly through such regions, it explored an area that was like a continental divide of magnetic lines of force. On one side the solar wind pushed the earth's magnetic lines downward and toward the earth. On the other side the wind swept them backward and out into the magnetosphere's long tail. In between was the polar cusp, a neutral area that allowed solar wind particles to penetrate directly onto the earth's surface at the poles,

as discovered by OGO 4 (July 28, 1967). HEOS 2 studied this region at high altitude near the magnetosphere's bow shock, mapping in detail its structure and formation during the spacecraft's year and a half of operation.

See **Dynamics Explorer 1/Dynamics Explorer 2 (August 3, 1981)** for results. **Prognoz 1 (April 14, 1972)** continued solar and magnetospheric research.

February 14, 1972
Luna 20
U.S.S.R.

Luna 20 was the second unmanned probe, following **Luna 16 (September 12, 1970)**, to successfully soft-land on the Moon and return soil samples to Earth. It was the third to land in the lunar highlands, following **Apollo 14 (January 31, 1971)** and **Apollo 15 (July 26, 1971)**, and the sixth overall to return samples.

Luna 20 landed in the Apollonius highlands near the Sea of Fertility, where it used a drill to dig about 10 inches into the ground and obtain a core sample weighing somewhere between an eighth and a quarter of a pound (Soviet publications never specified its exact weight). A television camera broadcast this entire operation to Earth, then made panoramic images of the spacecraft's landing site.

Compared with Luna 16, Luna 20 drilled into much harder ground. The sample itself was mostly anorthosite, not unlike the rock found in the highlands during the Apollo 15 mission—ancient crystals from the earliest period of the Moon's formation. Unexpectedly, the samples also contained pulverized metallic iron with a molecular structure that prevented it from rusting like terrestrial iron, which would make a superb construction material if found in large quantities.

The next lunar sample return mission was **Apollo 16 (April 16, 1972)**. The next successful unmanned lunar sample return mission was **Luna 24 (August 9, 1976)**. Luna 24 was also the last lunar soft landing, as of December 1999.

March 3, 1972
Pioneer 10
U.S.A.

Built by NASA, this American spacecraft was the first probe beyond Mars, and the first to reach the regions of space beyond Pluto. Eventually it will be the first object of human origin to escape the solar system.

Pioneer 10, though aimed at Jupiter, also had more mundane goals. Its purpose was to learn if spacecraft could safely travel through the asteroid belt between Mars and Jupiter, and then survive passage through near-Jupiter space. This second goal applied directly to future missions, as no rocket was available that could propel an unmanned spacecraft to the outer planets of Saturn and beyond or to the high polar regions above the Sun. (The Saturn 5 rocket, which had the power to send a spacecraft to these places, was never intended for unmanned exploration, and its manufacture was also being discontinued at the end of the Apollo program.) To reach these distances, NASA engineers hoped to use the deep gravitational well of Jupiter to slingshot spacecraft throughout the solar system. Before they could risk expensive projects on such a trajectory, however, they needed to know how dangerous near-Jupiter space was to spacecraft instrumentation. Pioneer 10 would assess this for them.

While journeying to Jupiter, Pioneer 10 made repeated measurements of the solar wind. Two solar storms, on August 2, 1972, and August 7, 1972, emitted enough energy in one hour to have powered the United States for 100 million years. The wave of energy from these flares was recorded by Pioneer 10, as well as by the four solar orbiting Pioneer spacecraft (see **Pioneer 9–November 8, 1968**). What was found was that the wind speed of this blast wave had dropped from over 2 million miles per hour at about 1 AU (astronomical unit) from the Sun to half that by the time it reached Pioneer 10 at about 2 AU. Correspondingly, the wind's temperature had increased, indicating that the wind's speed was being converted to thermal energy. This data gave the scientists a benchmark for mapping the solar wind's dissipation as it traveled outward.

Pioneer 10 emerged from the asteroid belt in February 1973. During its journey, the spacecraft found that the belt contained fewer small particles than expected and that the impact count was only slightly higher than that seen in the inner solar system, indicating that future spacecraft could travel through the asteroid belt with little fear. The route of **Pioneer 11 (April 6, 1973)** to Jupiter was then set, based on this data. Though Pioneer 10's sensors detected a few nearby objects between 4 to 8 inches in diameter, no impacts damaged the spacecraft. The data did show that near both the earth and Mars, space had been swept clear of small particles. Close to Jupiter, however, the particle count increased, with the gas giant's vast gravity gathering a swarm of small particles around it.

After a 22-month journey, Pioneer 10 finally reached Jupiter on December 3, 1973, zipping past at a speed of 82,000 miles per hour. See **Pioneer 10, Jupiter Flyby–December 3, 1973** for further details.

March 12, 1972
Thor Delta 1A
Europe / U.S.A.

Thor Delta 1A was the sixth satellite of the European Space Research Organization (ESRO), a 10-nation association that was the forerunner of today's European Space Agency. It was placed in orbit by a U.S. Thor Delta rocket, hence its name.

Thor Delta 1A's seven instruments continued the astronomical observations of **OAO 2 (December 7, 1968)** and **Uhuru (December 12, 1970)**. Two instruments mapped the ultraviolet sky; two measured x-ray radiation, targeting the galactic center, the Crab Nebula, and the Sun; two measured gamma-ray emission from the Sun; and one measured the flux of high energy interstellar cosmic rays.

The satellite operated for a little more than a year, making 2.5 scans covering 95 percent of the ultraviolet sky. The output of numerous stars were recorded repeatedly to establish a baseline of data. Spectra provided information about the abundance of various metals within many of these stars.

Thor Delta 1A's data were augmented by the next ultraviolet–x-ray space telescope, **Copernicus (August 21, 1972)**.

April 7, 1972
Intercosmos 6
U.S.S.R. / Poland

The sixth research satellite launched jointly by the Soviet Union and its communist allies (see **Intercosmos 1–October 14, 1969**), Intercosmos 6 performed astronomical research for Soviet and Polish scientists, studying high-energy cosmic rays as they hit the earth's atmosphere at approximately 60 miles altitude. Unlike other astronomical research satellites, Intercosmos 6's experimental equipment, which used a nuclear spark chamber to record high energy particles, was placed in a recoverable capsule, and after four days, it returned to Earth for study.

The satellite's photographic plates detected the highest-energy cosmic ray particle yet recorded, exceeding one million GeV. Such high-energy emissions were not measured again until the launch of the Compton Gamma Ray Observatory (see **STS 37–April 5, 1991**).

April 14, 1972
Prognoz 1
U.S.S.R.

Like both HEOS satellites (see **HEOS 1–December 5, 1968**) and most of the Imp satellites (see **Imp A–November 27, 1973**), the Prognoz ("forecast" in Russian) satellites were placed in eccentric orbits with their apogees halfway between the earth and the Moon. There they monitored low-energy particles in the magnetosphere, the solar wind, and the gamma and x-ray output of the Sun, looking for solar flares. Because of limitations in their sensors, however, they could detect only the most energetic x-ray solar flares.

Through 1980, the Soviet Union launched eight Prognoz satellites, with at least one and usually two functioning at any one time. Some carried experiments built by scientists from France, Czechoslovakia, Hungary, and Sweden. Their data further confirmed results from the Imp satellites, as well as augmenting the Intercosmos (see **Intercosmos 1–October 14, 1969**), Solrad (see **Solrad 1–June 22, 1960**), and OGO (see **OGO 1–September 5, 1964**) satellites.

A comparison of 20 different solar flare events observed by later Prognoz satellites linked these x-ray burst events with radio emissions detected by ground based radio telescopes, indicating that such outbursts also produced radio emissions.

The last two Prognoz satellites did additional research. **Prognoz 9 (July 1, 1983)** was also a radio telescope, and **Prognoz 10 (April 26, 1985)** incorporated instruments and experiments from the Intercosmos program.

See **Dynamics Explorer 1/Dynamics Explorer 2 (August 3, 1981)** for results. **ESRO 4 (December 11, 1972)** continued research into the aurora and ionosphere.

April 16, 1972
Apollo 16
U.S.A.

Apollo 16 was the fifth manned mission to the Moon, following **Apollo 15 (July 26, 1971)**, and the seventh sample-return mission, following **Luna 20 (February 14, 1972)**. It lasted 11 days, including 3 days of exploration on the lunar surface. The crew consisted of John Young (**Gemini 3–March 23,**

1965; Gemini 10–July 18, 1966; Apollo 10–May 18, 1969; STS 1–April 12, 1981; STS 9–November 28, 1983), Charles Duke, and Ken Mattingly (STS 4–June 27, 1982; STS 51C–January 24, 1985).

Young and Duke landed the lunar module in the region of hills surrounding the crater Descartes, near the Sea of Nectar. There, scientists hoped the astronauts would recover rock samples of volcanic activity *other* than the events that formed the large maria regions. Stuart Roosa on **Apollo 14 (January 31, 1971)** had identified what appeared to be cinder cones in this region, while other scientists had spotted other signs of both recent and ancient volcanic activity. This evidence indicated that it was still reasonable to believe that many of the Moon's craters had formed from volcanism, not asteroid impact.

On Apollo 16's first EVA (extra-vehicular activity), the usual ground experiments were deployed, including a seismometer, a magnetometer, a solar wind detector, and assorted other instruments. These sensors confirmed most of the results from previous missions. A heat-flow experiment similar to one deployed from Apollo 15 failed when a cable was accidentally pulled free. The cumulative data from the solar wind detector, installed on all four landings, provided scientists an estimate of the solar wind's helium and neon content.

Young and Duke then drove their lunar rover to nearby Plum Crater, and for about two hours, gathered samples from two locations. They found no evidence of recent volcanic activity. The rocks collected from this locale were all breccia (impact rocks) ejected from highland crater sites, comparable to material found on the previous two highland missions, Apollo 14 and Apollo 15.

During the second EVA, Young and Duke traveled 2.5 miles south to a cluster of craters on the slope of Stone Mountain. Their journey took them to 500 feet elevation, the highest any human has yet climbed on the lunar surface. They made three stops to collect samples, none of which showed evidence of volcanic activity. In fact, everything appeared to be breccia from meteorite impacts, or anorthosite crystals dating from the Moon's earliest formation.

During Apollo 16's third EVA, the two astronauts made a traverse to North Ray Crater, where scientists once again hoped they would find evidence of volcanic debris. While there, the astronauts obtained samples from "House Rock," the largest boulder yet found on the Moon. As before, their samples were all impact related or early rock from the Moon's beginning.

With its 160 pounds of lunar samples, Apollo 16 put an end to the debate over whether the Moon's craters had been formed by impact or volcanoes. The mission had investigated a region of craters that appeared from a distance to be volcanic in origin. Yet the returned data showed nothing but impact debris. From this finding grew a whole new understanding of the history of the solar system, as well as the complex process surrounding cratering. Scientists now knew for certain that the early history of the solar system involved a period of violent impact as the untold billions of objects surrounding the young Sun aggregated to form planets. See **Mariner 10, Mercury Flyby #3 (March 16, 1975)**.

While Young and Duke were on the lunar surface, Mattingly remained in orbit, doing multispectral photography of the Sun's corona and the Moon using cameras attached to the service

John Young prepares to hammer a sample from rock on the rim of Plum Crater. *NASA*

Charles Duke watches while John Young photographs a panorama of Plum Crater. *NASA*

module. During the return flight home, Mattingly performed a space walk to retrieve the film magazines from these cameras.

The command module also carried a package of biological specimens, including bacterial spores, protozoa cysts, plant seeds, shrimp eggs, and insect eggs, to measure the effect of weightlessness and cosmic radiation and determine the amount of shielding necessary to protect humans. The experiment found that a single hit of a cosmic heavy ion or cosmic ray seriously damaged the shrimp eggs and inhibited the growth of the spores. Whether weightlessness had any additional effect was unclear.

Just prior to leaving lunar orbit, Apollo 16 released Particles and Fields Subsatellite (PFS) 2. This small probe re-

July 22, 1972

Shadow Rock, where the astronauts collected samples from its permanently shadowed base. *NASA*

mained in orbit for about six months, relaying information about the Moon's magnetic field and confirming what a similar subsatellite released by Apollo 15 had found: that the Moon had no active magnetic field, only small remnant fields associated with craters and large basins.

This subsatellite and the lunar module eventually spiraled down to the lunar surface, providing further seismic data for the four seismometers functioning there. Additionally, three weeks after the astronauts' departure, the largest seismic quake was measured on the Apollo 16 seismometer. Scientists concluded from this data that the lunar crust under the Apollo 16 highlands landing site was several miles thicker than the crust under the mare regions.

The next manned mission to the Moon was **Apollo 17 (December 7, 1972)**. Through December 1999, this remains the last manned mission to another world.

July 22, 1972
Venera 8, Venus Landing
U.S.S.R.

Venera 8 was the second spacecraft to successfully land on Venus and return data, and it was the first to return information about the soil of Venus, transmitting data for 50 minutes. Similar in construction to Venera 7 (**Venera 7, Venus Landing–December 15, 1970**), Venera 8 was launched on March 27, 1972.

During the spacecraft's descent through Venus's atmosphere, a number of readings were taken. Wind speeds ranged from hurricane-force gusts of 225 miles per hour above 30 miles altitude to 2 miles per hour below 6 miles. Traces of ammonia were measured at altitudes between 20 and 30 miles.

Once on the surface, both temperature and surface air pressure readings matched closely with data returned by Venera 7—900°F and 90 atmospheres. The spacecraft had landed in the early Venusian morning, soon after sunrise, and sensors indicated that, at that moment, only 1 percent of the Sun's light managed to penetrate the planet's cloud cover and reach the surface. The visibility was hence limited to less than two-thirds of a mile.

Venera 8 carried a gamma spectrometer, which measured the natural radioactive emissions of the Venusian rocks directly below the spacecraft, the first measurement of the planet's make-up. The data implied that the rocks were granite, resembling the most ancient material found on the earth's continents. Like granite, the materials under Venera 8 had been formed at great depth and were later changed by further geological processes.

The next mission to Venus was Mariner 10's flyby (**Mariner 10, Venus Flyby–March 16, 1973**). The next Venusian landing was **Venera 9, Venus Landing (October 22, 1975)**.

July 23, 1972
Landsat 1
U.S.A.

Built by NASA, Landsat 1 (until 1974, called ERTS or Earth Resources Technology Satellite) was the first satellite entirely dedicated to Earth resources study. Its inspiration came from the infrared cameras developed for weather satellites like **Tiros 1 (April 1, 1960)** and **Nimbus 1 (August 28, 1964)**, and from multispectral photography taken on manned missions such as **Apollo 9 (March 3, 1969)**. Also influencing its development was the photo-reconnaissance satellites, **Midas 2 (May 24, 1960)**, KH-4/KH-4A/KH-4B/KH-5 (see **Discoverer 36–December 12, 1961** and **KH-4 9032–April 18, 1962**), and KH-9 (see **KH-9 1–June 15, 1971**).

Landsat 1 operated for 5.5 years, until January 1978. During its service life, it produced over 300,000 Earth images. Two identical spacecraft were launched in 1975 and 1978 in order to maintain continuous coverage through 1984.

These first-generation Landsat satellites orbited the earth 14 times a day at about 570 miles altitude in a sun-synchronous orbit (see **Nimbus 1–August 28, 1964**). After 18 days, each satellite had observed the entire surface of the earth from 81 degrees north to 81 degrees south latitude. With two working satellites, this coverage was achieved in half the time. Repeated passes allowed the spacecraft to photograph surface areas often obscured by clouds. Because the orbit was sun-synchronous—the Sun's angle always remained the same—photographs of the same areas could be easily compared for changes in surface features.

Because of the failure of tape recorders on the first two Landsats, data was generally relayed earthward only when the satellites were over ground stations. Thus, most images from the early Landsats were limited to the land masses over North and South America, northern Europe, Japan, and Australia, where numerous ground stations existed.

The satellites carried two-camera systems, a multispectral scanner and a return-beam vidicom. Because of technical failures of the latter on all three Landsats, only the multispectral scanner returned useful images, of which many hundreds of thousands are still of great value. The multispectral scanner covered four spectral bands, red and green in the visible wavelengths and two near-infrared wavelengths. It scanned an area about 115 miles by 115 miles with each image, resolving objects as small as 260 feet across. The large area covered, about 13,225 square miles, allowed the study of regional

August 21, 1972

Oil slick in Gulf of Suez, as detected by Landsat 1. *NASA*

geology and resources. To cover a comparable area with aerial photography would have required several thousand individual pictures.

Because this mission was the first dedicated to remote sensing, much of the early research involved learning how to interpret the reflection of different features in different spectral bands. Depending on their composition, surface features could appear either bright or dark, depending on wavelength. Identifying how various vegetation, urbanized areas, minerals, and geological features appeared in each spectral band required careful analysis. To aid in this effort, many multispectral photographs taken on **Skylab (May 14, 1973)** were compared with Landsat 1 pictures. Aerial high-altitude images were also used for comparison.

Though these three satellites are no longer in operation, their images are still available as a valuable archive of surface features in the 1970s and 1980s. The images revealed many details previously unseen. Different crop types could be identified; geological features, such as faults and joints, could be quickly identified and mapped; water resources could be identified, including pollution and river drainage patterns; and snow, ice, and glacier growth and shrinkage were observable *(see color section, Figure 23)*.

In one case, Landsat 1 images showed that the simple presence of fences in a semiarid region of Africa acted to reduce overgrazing and prevent the encroachment of the desert. In another, Landsat 1 photographed the underwater sedimentary deposits in Lake Superior, revealing that a freshwater intake line, built by a local municipality at a cost of $8 million, had been placed within these deposits, making it useless. Future projects could avoid such failures with Landsat images.

Above all, Landsat pictures provided, for the first time, detailed imagery of some of the most remote regions in the world. Previously blank areas on the world's geological maps could finally be filled in, and overall patterns of geology and topography could be studied in detail.

Landsat 4 (July 16, 1982) was the second-generation American Earth resource satellite. The next Earth resources satellites were two Indian satellites (see **Bhaskara 1–June 7,** 1979), filling the coverage gaps of the second-generation Landsats.

August 13, 1972
Explorer 46
U.S.A.

Explorer 46 was a follow-up to **Explorer 16 (November 16, 1962)**, **Explorer 23 (November 6, 1964)**, and Pegasus missions (see **Pegasus 1–February 16, 1965**) that studied the micrometeoroid impact rates in low-Earth orbit. This NASA spacecraft tested the impact of very small particles, weighing less than 10^{-8} gram, on the more delicate finishes being put on newer satellites. It also tested a technique for providing protection to satellite surfaces in which two thin layers are separated by a one-half-inch space.

After a year in orbit, it was found that the double-wall system worked twice as well as predicted, which meant it provided six times the shielding of a single wall. The rate of impacts was also found to be much less than predicted from previous experiments, though it also seemed to vary with time.

August 21, 1972
Copernicus (OAO 3)
U.S.A.

Built by NASA, Copernicus, or Orbiting Astronomical Observatory (OAO) 3, followed **OAO 2 (December 7, 1968)** while also augmenting **Thor Delta 1A (March 12, 1972)**. Copernicus carried one ultraviolet telescope and three x-ray telescopes and functioned for over nine years, until February 1981. The mirror for the ultraviolet telescope was 32 inches across, the largest yet launched. The instruments covered the electromagnetic spectrum from 0.05 to about 1,450 angstroms and from about 1,650 to 3,000 angstroms.

Over its long life, Copernicus made thousands of observations of many of the same x-ray objects first identified by **Uhuru (December 12, 1970)**. For example, observations of the x-ray object Cygnus X-1 confirmed what scientists had suspected since Uhuru had first seen it—that it was actually a black hole, the first such object confirmed to exist. Moreover, within a month of launch, Copernicus found that the rotation rate of another x-ray object, Cygnus X-3, was actually increasing at an easily measurable rate, providing even stronger evidence that the evolution of such objects was linked somehow to their rotation and to the accretion of material onto its surface from a binary companion.

The spacecraft discovered that the immense dust clouds of the galaxy were also filled with vast amounts of molecular hydrogen, and that there were surprisingly large quantities of deuterium among those hydrogen molecules. The large amount of deuterium, formed during the "big bang," was seen by cosmologists as evidence that the universe was "open," that it would continue to expand forever rather than eventually condensing into a new singularity leading to another "big bang."

Copernicus's discovery of these gigantic clouds also indicated that the interstellar material within the Milky Way was much more unevenly distributed than scientists had expected. This discovery implied that a more sensitive telescope observing in the extreme ultraviolet from 100 to 1,000 angstroms would be able to see through those clouds and map nearby galactic space. See **EUVE (June 7, 1992)**.

Though Copernicus detected phosphorus, chlorine, manganese, carbon, nitrogen, oxygen, magnesium, silicon, sulfur, and iron, it also found evidence that, compared with the Sun, the galaxy contained less aluminum, oxygen, and carbon than expected, and that the interstellar clouds were also more rarefied (despite their uneven distribution).

The next space-based astronomical telescope, **Explorer 48 (November 15, 1972)**, studied the gamma-ray sky. The next satellite with ultraviolet and x-ray telescopes was **ANS 1 (August 30, 1974)**. See also **Fuse (June 24, 1999)**.

September 2, 1972
Triad 1
U.S.A.

Triad 1, a U.S. Navy satellite performed technology tests for improving the Navy's Transit satellite navigation system (see **Transit 1B–April 13, 1960**). The satellite, made of three sections connected by booms, also tested gravity-gradient attitude control techniques previously tried on the GGSE satellites (see **GGSE 1–January 11, 1964**). A second Triad test satellite was launched in 1975, and the engineering results were used by U.S. Navy NOSS reconnaissance satellites (see **NOSS 1–April 30, 1976**).

October 2, 1972
Radsat / Radcat
U.S.A.

• *Radsat* This U.S. military satellite performed technology tests on a variety of thermal coatings to determine how the radiation in the Van Allen belts affected them. It also made a map of the background gamma radiation over the whole Earth, and the ultraviolet airglow at the equator. All these data were used in improving satellite engineering.

• *Radcat* This U.S. military radar calibration satellite provided a target so that ground-based radar facilities could calibrate their equipment. Unlike the **Calsphere (December 13, 1962)** satellites, Radcat was not a sphere but a flat, five-square-yard target.

October 15, 1972
Oscar 6
U.S.A. / International

The sixth in a series of satellites built by amateur radio enthusiasts, following **Oscar 5 (January 23, 1970)**, Oscar 6 was the first built by AMSAT (the Radio Amateur Satellite Corporation), an organization of ham radio enthusiasts from the United States, Australia, West Germany, Canada, and Japan. Oscar 6 used solar cells in order to prolong its life expectancy, and hence it continued to operate until 1977. The satellite could both transmit and store data, such as weather warnings and emergency calls, in its small 896-bit memory bank. Oscar 6 was followed by AMSAT's next satellite, **Oscar 7 (November 15, 1974)**, with which it operated in conjunction during the period from 1974 to 1977.

November 10, 1972
Anik A1 (Telesat 1)
Canada / U.S.A.

The Anik A satellites (Anik means "brother" in the Inuit language) were owned by Telesat Canada, a semiprivate corporation formed by the Canadian government expressly to launch and operate communications satellites covering Canada. These satellites made it possible for isolated villages in the northern reaches of Canada to obtain telephone, television, and radio service.

Through May 1975, Telesat Canada launched three Anik A satellites, built by the American Hughes Aircraft company and launched by NASA. Each satellite had a capacity of up to 960 one-way voice circuits or up to 12 color television channels.

Because of the delay in getting permission in the United States to launch communications satellites owned by private American companies (see **Westar 1–April 13, 1974**), several American companies subleased space on Anik A1 for U.S. domestic operations, pending permission to launch their own satellites (see **RCA Satcom 1–December 12, 1975**).

Telesat Canada followed the Anik A satellites with two experimental communications satellites, **Hermes (January 17, 1976)** and **Anik B (December 15, 1978)**. It then replaced the Anik A constellation with the **Anik D (August 26, 1982)** and **Anik C** (see **STS 5–November 11, 1982**) clusters, the second cluster designed expressly for launch on the U.S. space shuttle.

November 15, 1972
Explorer 48 (SAS 2)
U.S.A. / Italy

The second U.S. small astronomical satellite (SAS 2) built by NASA and launched by Italy from its San Marco launch platform off the coast of Kenya (see **Uhuru–December 12, 1970**), Explorer 48 studied celestial gamma-ray sources emitting energy from 30 to 300 MeV, observing both specific astronomical objects, as well as the general gamma-ray background of the sky. Though many previous satellites had carried gamma-ray sensors (**Explorer 11–April 27, 1961**; **OGO 5–March 4, 1968**; **Vela 5A/Vela 5B–May 23, 1969**), Explorer 48 had much greater resolution. Its data also augmented observations from the gamma-ray telescope on the **Salyut 1 (April 19, 1971)** station.

During its half year of operation, the spacecraft mapped this radiation along the galactic plane, both toward and away from the galactic center. The denser regions of the galaxy, including its central region and spiral arms, were found to be rich in gamma-ray emissions. Overall, about 25 to 30 gamma-ray sources were identified within the Milky Way, including the first detection of high energy gamma radiation from the Crab Nebula. The radiation appeared unevenly distributed throughout the universe and closely tied to large galactic features such as spiral arms and galactic nuclei, though astronomers also thought that some radiation was coming from outside galaxies.

Based on these data, scientists theorized that the high-energy cosmic rays entering the solar system from interstellar space almost certainly came from past supernova explosions, as well as the resulting pulsars, neutron stars, and black hole binaries both inside and outside the galaxy. Because the gamma-ray bursts discovered by Vela 5A/Vela 5B had not yet been recognized when Explorer 48 operated, however, this theory did not include these events and was therefore not entirely accurate. To separate gamma-ray bursts from emissions from other events required more advanced space-based high-energy telescopes than carried by Explorer 48—telescopes able to resolve the directional source of gamma rays with greater precision. See **COS B (August 9, 1975)**, the next gamma-ray telescope.

November 22, 1972
ESRO 4
Europe / U.S.A.

ESRO 4 was the seventh satellite of the European Space Research Organization (ESRO), a 10-nation association that was a forerunner of today's European Space Agency. Launched by NASA, ESRO 4 continued the research of **Aurorae (October 3, 1968)**, studying the earth's ionosphere and aurora from polar orbit, as well as monitoring the solar wind influx over the poles. Operating for more than a year, its data also augmented observations by the Solrad (see **Solrad 1–June 22, 1960**), Imp (see **Imp A–November 27, 1963**), OGO (see **OGO 1–September 5, 1964**), Intercosmos (see **Intercosmos 1–October 14, 1969**), Aureol (see **Aureol 1–December 27, 1971**), and Prognoz (see **Prognoz 1–April 14, 1972**) satellites.

See **Dynamics Explorer 1/Dynamics Explorer 2 (August 3, 1981)** for results. **AEROS 1 (December 16, 1972)** continued research into the aurora and ionosphere.

December 7, 1972
Apollo 17
U.S.A.

Apollo 17 was the sixth manned flight to land on the Moon, following **Apollo 16 (April 16, 1972)**, and the eighth mission overall to return samples. Lasting over 12 days, its crew was Eugene Cernan (**Gemini 9–June 3, 1966; Apollo 10–May 18, 1969**), Harrison Schmidt, and Ron Evans. Schmidt was the first professionally trained geologist to fly in space.

The landing site was a 4.5-mile-wide valley nestled between two mile-high mountains in what was called the Taurus-Littrow region. This area, studied in detail by Al Worden on **Apollo 15 (July 26, 1971)**, showed evidence of volcanic cinder cones. The scientists wanted one last try at finding evidence of recent volcanism on the Moon. One valley wall also showed evidence of a landslide, which scientists hoped had exposed very ancient rocks.

During Apollo 17's first EVA (extra-vehicular activity), a heat flow experiment was deployed to confirm lunar interior temperatures measured by Apollo 15. Also assembled was a cosmic ray detector, a sensor for measuring the lunar atmosphere, a micrometeoroid detector, and a gravity wave detector. This last instrument failed to function, however, due to a design flaw.

Cernan and Schmidt then drove a lunar rover to a nearby crater to obtain rock samples and a deep core sample. The samples were either breccias or coarse-grained basalt—magma that had taken a long time to solidify, indicative of material

Harrison Schmidt sets up surface experiments during the first EVA. *NASA*

that had formed deep underground. No surface lava, which would have hardened quickly and therefore would have been fine-grained, was found.

During the second EVA, Cernan and Schmidt traveled farther from their lunar module than any previous lunar explorer, driving 3.67 miles to the west and up a slope, then 0.6 miles beyond to areas where boulders had rolled down from the south wall of the Taurus-Littrow valley. There, they found mostly breccias, as well as one sample later determined to be nearly 4.5 billion years old. Along the way, they stopped several times to grab samples while still sitting in the rover.

At their second stop, Ballet Crater, Cernan obtained a deep core, while Schmidt took photographs and collected a variety of small samples. Some of these samples were later dated as 109 million years old and were thought by some to be ejecta from Tycho Crater, far to the south and halfway around the lunar globe.

At their third stop, Shorty Crater, the men discovered orange soil along the crater's rim. Core and rock samples revealed this to be ancient volcanic debris. A three-foot core sample showed the deepest sections to be black, the middle sections gray, and the surface sections orange. All were made of tiny spherules of volcanic glass, produced not by recent volcanic activity, but very ancient, primordial events. These spherules had been deposited by a volcanic fire fountain 3.5 billion years ago, then covered by later lava and breccias. When an impact had created Shorty Crater 19 million years ago, it had blasted this deeply buried material up onto the surface. And because the physics of impact cratering puts the deepest ejecta onto a crater's rim, the orange soil was regarded as debris from the deepest and therefore most ancient lunar events.

At Camelot Crater, Cernan and Schmidt found coarse-grained basalts. Many boulders showed a volcanic history, but all were indicative of magma slowly hardening deep underground.

December 7, 1972

The goal of Apollo 17's third Moon-walk was an avalanche slope on the northern wall of Taurus-Littrow, first noticed in Apollo 15 photographs, with the main target one large boulder at the end of a 1,500-foot trail of depressions and scrape marks caused when the rock had crashed downhill. The boulder itself had split into five pieces before coming to a rest.

Dubbed "Tracy's Rock" in honor of Cernan's daughter, the astronauts spent about an hour collecting samples from the top, sides, and even underneath rock sections. The boulder was breccia with evidence that it had once melted and also that it had once been deep underground. As the mountains around it formed, it had been partly injected with lava

The landing site. In the distance, the south wall of the Taurus-Littrow valley rises 8,200 feet. *NASA*

Climbing the north wall of the Taurus-Littrow valley on the third EVA. The white-streaked floor now visible at the base of the large mountain is the avalanche rubble that the astronauts had visited during their second EVA. The LM is out of frame, to the left. *NASA*

December 7, 1972

Schmidt collects a sample from Tracy's Rock. The LM is located just to the right of the boulder's peak—a tiny speck in the dark floor of the valley.
NASA

material. Then, approximately 3.8 billion years ago, it was ejected from below by an impact. Later, it broke free from its perch and tumbled downward, bouncing across the mountainside, breaking into five pieces, and coming to a rest.

Later breccia samples in this locale showed the same evidence of ejected molten rock. Cernan also sampled a boulder covered with shocked glass from ancient lunar crust, ejected to the surface by impact. This sample was so similar to the "Genesis Rock" found on the Apollo 15 mission that scientists believe it proves the heterogeneous nature of the lunar interior.

At the Van Serg Crater, rocks were highly shocked, very crumbly, and not basalt. They appeared to be a kind of breccia, but unfamiliar to any of the geologists. Core samples revealed that these rocks were soil that had been instantly compacted by impact. The impactor had landed in an unusually thick layer of regolith, and rather than ejecting the deep underlying layers of basalt, it had made "instant rock."

Cernan and Schmidt had spent 22 hours wandering across the lunar surface, traveling 19 miles. They took over 2,200 pictures and collected almost 250 pounds of lunar material, out of a total of 835 pounds returned by all six Apollo landings. Their results added detail and emphasis to earlier work: scientists were now writing the lunar geological history with impact as the major shaping phenomenon. Volcanism had been a process only during the Moon's earliest history. Later melting events had been caused by the heat of asteroid collisions. For more information, see **Mariner 10, Mercury Flyby #3 (March 16, 1975)**.

Ron Evans in the command module, meanwhile, was doing observations with his own array of scientific instruments, including an infrared radiometer, an ultraviolet spectrometer, a lunar sounder for making an electromagnetic map of lunar space, and a panoramic and a high resolution camera. He took over 1,600 pictures with the panoramic camera and over 3,000 with the mapping camera. During the flight home, he performed a space walk to recover these film canisters from the service module research compartment. He had used the laser altimeter over 3,700 times to map the exact height of the lunar surface. The infrared radiometer and ultraviolet spectrometer also swept the lunar surface, mapping its make-up and content.

Apollo 17 was the last manned mission to the Moon, as of December 1999. The next American manned mission was **Skylab 2 (May 25, 1973)**, a 28-day mission to the orbiting space laboratory **Skylab (May 14, 1973)**. The next American mission to the Moon would not take place for 22 years: **Clementine (January 25, 1994)**.

The next mission to the Moon was the unmanned **Luna 21 (January 8, 1973)**, the second Soviet long-term robot lunar

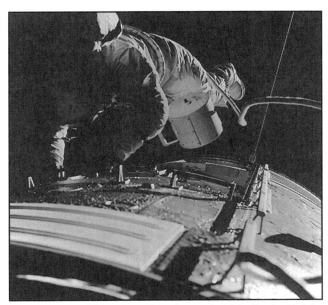

Evans recovering film canisters from the exterior of the Service Module.
NASA

rover mission. The next successful sample return mission was **Luna 24 (August 9, 1976)**.

December 16, 1972
Aeros 1
West Germany / U.S.A.

This satellite was the second German research satellite, following **Azur (November 8, 1969)**. It was placed in a sun-synchronous orbit (see **Nimbus 1–August 28, 1964**) by a NASA rocket. From this position, the satellite studied the upper ionosphere, measuring its temperature, density, and charged particles, as well as how the solar flux changed it. Aeros 1 also measured the Sun's ultraviolet output. Its observations were done in conjunction with the next German satellite, **Aeros 2 (July 16, 1974)**; see that satellite for more details.

Hawkeye (**June 3, 1974**) continued research into the ionosphere and magnetosphere.

1973

January 8, 1973
Luna 21
U.S.S.R.

Luna 21 was the second Soviet long-term robot rover to explore the lunar surface, following **Luna 17 (November 10, 1970)**. It soft-landed in Le Monnier Crater on the eastern edge of the Sea of Serenity not far from the **Apollo 17 (December 7, 1972)** landing site, beginning a four-month, 23-mile exploration of the crater, in which fissures and small craters were studied along the way. During that period, the rover took over 80,000 individual television images and 86 panoramic images.

The rover carried a probe to test the soil's density, mechanics, and structure at various sites. It also carried a solar x-ray detector, a sensor to measure visible and ultraviolet light, a magnetometer, a radiometer, and a detector and reflector for Earth-Moon laser reflection experiments.

The ultraviolet light detector discovered that the Moon was not an ideal place for astronomical research during its daytime hours. Measurements of the lunar sky revealed a glow from a layer of dust particles that interfered with sky observations.

During its first lunar night (January 29th), the spacecraft measured a temperature of –297°F, the coldest temperature yet detected on the lunar surface.

Luna 21 was the last robot rover mission to the Moon through December 1999. The next similar robot rover mission was a Martian probe, **Mars Pathfinder (December 4, 1996)**.

The next lunar mission was an orbiter, **Luna 22 (May 29, 1974)**. The next successful sample return mission was **Luna 24 (August 9, 1976)**. Luna 24 was also the last lunar landing mission as of December 1999.

March 6, 1973
Rhyolite 1
U.S.A.

Rhyolite 1 was a U.S. Air Force satellite designed to eavesdrop on microwave transmissions—a more sophisticated version of the Canyon satellites (see **Canyon 1–August 6, 1968**).

Rhyolite 1 carried several antennae for monitoring different wavebands, the largest of which was a 70-foot-diameter dish that unfolded once in orbit.

Stationed in geosynchronous orbit above Burma (now known as Myanmar), it monitored about 50 channels, including the UHF and microwave frequencies used by Soviet and Chinese military for radio, walkie-talkie, long distance telephone, and spacecraft telemetry transmission. The telemetry data provided early warning of any Soviet and Chinese missile launches, a more reliable warning system than the infrared sensors of the Midas satellites (see **Midas 2–May 24, 1960**).

To complete the system, three more Rhyolite satellites were launched, the last reaching orbit in April 1978. Of the four, two were spares while two constantly monitored the ground.

April 6, 1973
Pioneer 11
U.S.A.

Built by NASA and launched as a follow-up to **Pioneer 10 (March 3, 1972)**, Pioneer 11 not only made the second flyby of Jupiter, its course took it past Saturn, providing the first close-up views of the ringed planet. The spacecraft was identical to Pioneer 10, and during its 22 years of operation, it also provided detailed information about the Sun's magnetosphere and solar wind at distances as great as four billion miles from Earth.

On its journey out from Earth, Pioneer 11 confirmed Pioneer 10's data that the asteroid belt was not as dense as once feared. It also provided a second long distance detector for measuring solar wind speeds and solar flares.

Pioneer 11 reached Jupiter on December 4, 1974. Five years later, on September 1, 1979, it reached Saturn. Finally, on September 30, 1995, NASA ceased communications with the spacecraft, its power supply having dropped too low to operate its instruments. For details, see **Pioneer 11, Jupiter Flyby–December 4, 1974**; **Pioneer 11, Saturn Flyby–September 1, 1979**; and **Pioneer 11, Shutdown–September 30, 1995**.

May 14, 1973
Skylab
U.S.A.

Skylab was the second orbiting science laboratory, following **Salyut 1 (April 19, 1970)**. It was also the first American space station.

The Skylab program used Apollo program capsules and rockets that had originally been built to fly to the Moon. For instance, Skylab itself was built from the structure of the third stage of a Saturn 5 rocket, which was not needed if the rocket's payload was not leaving Earth orbit. Until May 20, 1995, when the Spektr module was added to Mir (see **Spektr–May 20, 1995** and **Mir–February 20, 1986**), Skylab was the heaviest single object to orbit the earth. It weighed over 82 tons and had about 12,700 cubic feet of habitable space—an interior volume that was not exceeded until Mir's last module was added (see **Priroda–April 23, 1996**). It was over four times heavier than Salyut 1 with more than three times the interior room.

The laboratory comprised five sections: the orbital workshop, where the astronauts lived and worked; the telescope

May 14, 1973

mount, which was the station's solar observatory; the multiple docking adapter, where Apollo capsules docked; the airlock module, which connected the docking adapter to the workshop; and the instrument unit, which only functioned during and immediately after launch to control the deployment of the station in space.

For orientation, Skylab used three gyroscopes, a variation of the gyroscope system tested on **PAC 1 (August 9, 1969)**, with thrusters for making specific and precise changes during experimental work.

Skylab's solar observatory included an ultraviolet spectrometer and spectrograph, an x-ray telescope, and a visible light chronograph. The station's Earth observation camera package included three different multispectral cameras. One camera took photographs simultaneously in six different wavelengths, each image covering about 4,000 square miles, with a resolution ranging from 78 to 223 feet, depending on wavelength. Another camera took high-resolution color photographs, resolving objects on the ground with dimensions as small as 50 feet. A third instrument scanned the ground as Skylab flew over, observing a 42.5-mile-wide swathe across 13 different wavelengths, including ultraviolet, visible, and infrared light.

Almost 90 different experiments were performed during Skylab's usable life span by the three manned missions. This research focused on six main areas of science: solar physics, Earth resources, medical research, material science, space engineering, and astronomy. See these flights for further details:

Skylab 2–May 25, 1973 to June 22, 1973 (28 days)
Skylab 3–July 28, 1973 to September 25, 1973 (59 days)
Skylab 4–November 16, 1973 to February 8, 1974 (84 days)

During the station's initial hours, however, it appeared to many that the entire project would be a failure. During launch, a shield—intended to protect the laboratory from meteoroids and shade it from the Sun—was ripped from the station. Furthermore, the two main solar panels attached to the orbital workshop failed to deploy—one panel was apparently inoperative, while the other worked but could not generate electricity. Engineers guessed that it had not unfolded when planned.

The loss of the shield caused temperatures in the laboratory to rise to intolerable levels, over 120°F. The failure of the two solar panels meant that the laboratory only had about 40 percent of its power supply.

With the station's interior temperature rising, ground controllers needed to orient the laboratory so that as little direct sunlight as possible hit its surface. With the loss of two solar panels, they also needed to maximize the sunlight hitting its four remaining solar panels to generate enough electricity to keep the station functioning.

While controllers juggled with these contradictory requirements, shifting Skylab's orientation again and again, NASA mission control delayed by 10 days the launch of the first manned flight to the station, Skylab 2, in order to determine how to repair the damage and save the project.

To shade the station from the hot sunlight, NASA engineers came up with a simple idea: an umbrella. A canopy of Mylar and nylon was attached to an expandable framework of telescoping rods, not unlike a giant beach umbrella. The as-

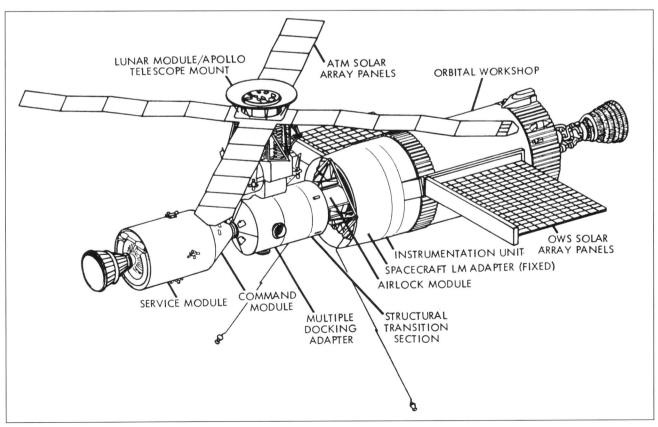

Skylab. *NASA*

tronauts would enter the station and insert this sunshade through an access tube airlock normally reserved for scientific experiments. As they pushed it out, they would attach additional rods until the umbrella was far enough from the station to be opened. Then they would pull it back, positioning it so that it hung just above the workshop.

Repairing the solar panels would first require a visual inspection to find out what had happened. Then the astronauts would attempt to fix the problems, performing several space walks if necessary. See Skylab 2 for details about this repair effort.

Though NASA had intended to incorporate use of Skylab with the space shuttle program, delays and a lack of funds prevented any shuttle missions to the station before atmospheric drag caused Skylab's orbit to deteriorate. On July 11, 1979, still weighing 77 tons, Skylab burned up in the atmosphere over the Indian Ocean, with pieces crashing into the outback of Australia—the largest being a one-ton piece recovered near the town of Kalgoorlie.

The next orbiting space laboratory was **Salyut 3** (**June 25, 1974**).

May 25, 1973
Skylab 2
U.S.A.

Skylab 2 was the first manned mission to the **Skylab** (**May 14, 1973**) orbiting laboratory, crewed by Pete Conrad (**Gemini 5– August 21, 1965; Gemini 11–September 12, 1966; Apollo 12–November 14, 1969**), Joe Kerwin, and Paul Weitz (**STS 6–April 4, 1983**). The flight's goal was to repair and activate the space laboratory, and then to complete four full weeks in space, setting a new manned endurance record, exceeding that of **Soyuz 11** (**June 6, 1971**).

As the Apollo capsule approached the station, the three astronauts provided both a televised and verbal description of its condition. One main solar panel was completely gone—it had been torn off when the meteoroid shield had come free during the launch. The second solar panel was only partly deployed—a single metal strap from the lost meteoroid shield had tangled with it, preventing it from unfolding more than 15 degrees.

The crew soft-docked with the station so they could eat a meal prior to the start of repairs. They then undocked, and Conrad maneuvered the capsule close to the jammed solar panel. Paul Weitz opened the Apollo capsule's hatch and, holding a telephone repairman's 10-foot-long tool for pulling and prying at wires, he attempted to free the stuck panel. For 37 minutes, Weitz worked on the strap without success. His effort caused the entire 82-ton Skylab laboratory to rock back and forth, requiring Conrad to ease the Apollo capsule away out of fear of a collision. Finally, the crew gave up on the solar panel and closed the hatch so that Conrad could bring the capsule around to redock with the laboratory.

Their spacecraft, however, refused to redock. Seven times Conrad tried to dock, using five different techniques. In each case, the docking latches that would give him a hard-dock refused to close. In a last-ditch attempt, the astronauts donned their spacesuits, depressurized the capsule, and opened the command module hatch to get access to the electronic circuits that controlled the latches. By resetting these circuits, the latches could finally be made to work, and a successful hard-dock was achieved. The astronauts then began their first sleep period in space, still confined to the small Apollo capsule.

The next day, their second in space, the astronauts entered Skylab. In both the docking adapter module and the airlock module, the temperature was a comfortable 50°F. In the workshop, however, the air temperature was a blistering 123°F. Weitz, the first to go in, noted calmly that the laboratory climate reminded him of being "in the desert." During the next few hours, with Joe Kerwin standing by in the Apollo command module, Weitz and Conrad deployed the umbrella sunshade through the scientific airlock (see **Skylab–May 14, 1973** for a description). With the parasol in place, the laboratory's temperature immediately began to decrease, and over the next 12 hours, the temperature dropped by about two degrees per hour. By the end of next day, the laboratory was a balmy 90°F, with the temperature still dropping.

For the next two weeks, the three astronauts unpacked equipment, set up experiments, and discussed plans for releasing the stuck solar panel. Without that panel, the laboratory had only about 40 percent of its electrical power, curtailing research.

The medical experiments, however, could go on. This research included use of the first zero gravity "scale," or body mass measuring device (as it was officially called). To weigh themselves, the astronauts sat on its seat and latched themselves in. The seat then began oscillating, at a speed dependent upon the mass placed on it. Measuring the pulses determined the astronaut's weight.

Other medical tests included use of the lower body negative-pressure device. This instrument, similar in concept to the "chibis" suit used by the Soviets on Soyuz 11, forced blood to circulate in the astronaut's lower limbs, approximating the body's circulation on Earth. It also measured the state of each man's cardiovascular system, finding that the astronauts' circulatory systems underwent subtle changes adapting to weightlessness. These changes all quickly disappeared once the men returned to Earth.

Photographs taken during the flight confirmed that in zero gravity there was a redistribution of fluids in the body, with fluids shifting to the body's upper half. Previous astronauts had reported that their faces felt "full" or red, as if they were hanging upside down and blood was rushing to their heads. The photographs of the astronauts' faces clearly showed this bloated state.

Also tested was an exercise bicycle, which the astronauts found difficult to use because the harness for holding them on the bike was inadequate. The three men found a more effective exercise routine was to run along the curve of the laboratory's inside wall. As their speed increased, centrifugal force drove their feet against its surface.

Interestingly, none of the astronauts on Skylab 2 experienced space sickness. This result contrasted sharply with what the Soviets had experienced on their long-term flights (see **Soyuz 9–June 1, 1970** and Soyuz 11), as well what the American doctors had expected. The assumption prior to Skylab 2's launch had been that the very large interior space of Skylab would cause vertigo and space sickness. See **Skylab 3** (**July 28, 1973**) for further motion sickness results.

Other medical tests revealed that all three astronauts lost bone and mineral mass during the flight. "In general, [this loss] followed the loss patterns observed in . . . bed-rested subjects," one researcher explained. The later and longer Skylab flights would determine whether such bone losses stabilized with time, or continued unabated as long as the human body was exposed to weightlessness.

On June 7th, the fourteenth day of the mission, Pete Conrad and Joe Kerwin made an audacious space walk to free the jammed solar panel. Because the stuck panel was located on a part of Skylab where no space walk had been planned, no handholds had been installed there. In order to work in this exposed area, the two men assembled a 25-foot-long pole, fastening the near end to the station and attaching an ordinary heavy-duty wire cutter to its far end. A cord ran from this tool back to Kerwin, who could make the cutter's jaws shut by pulling on it.

For almost an hour, the two astronauts struggled with this clumsy tool, trying to slip the wire cutter's jaws around the metal strap that was preventing the solar panel from unfolding. Finally, they got the cutter in position. Next, Conrad climbed out along the pole, his only attachment to the station his 55-foot-long umbilical cord. He carried with him a second rope, which he tied to a small bracket. This bracket also held the solar panel in a closed position. Once the astronauts had cut the strap, they would use the rope to pull the bracket free so that the panel could deploy.

With Conrad holding the pole in place, Kerwin pulled on the wire cutter rope, trying to get its jaws to snip the strap. After several minutes of unsuccessful struggle, Conrad climbed back along the pole to the wire cutter to see if he could help increase Kerwin's leverage. With both men pulling, the wire cutter worked and the strap snapped. As the panel popped open about five degrees, however, it flung Conrad off the station and into space. The astronaut tumbled away, grasping at his umbilical cord to slowly pull himself hand over hand back to Kerwin.

Then came a second task. The two astronauts grabbed the rope tied to the small bracket and pulled, trying to snap it open so that the panel could fully deploy. Once again, the length of the rope reduced their leverage, and once again Conrad climbed out along the pole. By placing the rope over his shoulder and standing upright against the laboratory's wall, he could increase Kerwin's leverage enough that the bracket finally snapped. The panel unfolded to its planned 90-degree open position, increasing Skylab's power output from four kilowatts to seven.

The final task on this space walk required Joe Kerwin to climb onto the Apollo telescope mount and retrieve and reload several magazines of film. The mount was a separate structure that, after launch, had been deployed at a 90-degree angle from Skylab's long axis (see diagram under **Skylab–May 14, 1973**). As Kerwin climbed up the mount's superstructure, he was overwhelmed with a feeling of exposure, as if he were climbing on the girders of an unfinished skyscraper. All previous space walks, although outside in space, had taken place beside a small capsule only slightly larger than the astronauts themselves. Thus, the men had no easy way of judging size and distance, and their view therefore had not seemed much different from what they were used to seeing through the window of an airplane. Kerwin, however, hung about 25 feet above a gigantic 48.5-foot-long structure, with the earth 270 miles below him. For the first time, an astronaut has some perspective on how high above the ground he was. Kerwin found it somewhat unnerving.

With the solar panel deployed, the orbiting laboratory had sufficient power to complete most of its experiments. Metallurgy research, in particular, required this increased power, because the experiments used several small electrical furnaces to melt a variety of alloys and crystals to see how they resolidified in weightlessness, continuing research begun on **Apollo 14 (January 31, 1971)**. In one experiment, a crystal of indium and antimony doped with tellurium resolidified with surface and internal features never before seen. Different experiments tested how other metals refroze in zero gravity, finding that their crystal structure changed when solidified. Welding experiments found that nickel dissolved much more rapidly in liquid silver-copper alloys than it did on Earth. This increased dissolution speed indicated that saturated alloys could be produced better in space.

Over their entire 28-day flight, the astronauts took extensive multispectral photographs of the earth, including many areas in the United States and South America. Because multispectral photography was a new research tool, much of this photography was to find out how different kinds of Earth terrains and vegetation looked when photographed under different wavelengths. For example, scientists found that open fields with growing crops could be distinguished from plowed fields using two specific wavelengths. Similarly, evergreen and deciduous forests could be recognized because they also photographed differently under different wavelengths. This data helped redesign the number and type of wavelengths used by subsequent multispectral cameras, especially those used on later Landsat satellites (see **Landsat 4–July 16, 1982**).

By precisely measuring the color and brightness variations produced by different land and water surfaces, computer programs could also be written that quickly mapped an area's surface features. These same programs also enhanced each image's contrast and details, bringing out information not immediately observable. Such data processing was used with later Earth observation satellites, such as Landsat 4 and **Spot 1 (February 22, 1986)**.

All this Earth resources photography was done in conjunction with the multispectral photographs taken by **Landsat 1 (July 23, 1972)**. Because Skylab's cameras had a higher resolution, the two facilities complemented each other.

The astronauts also made many solar observations, discovering coronal mass ejections, vast eruptions of matter flung outward from the Sun at great speed that emitted energy in ultraviolet, visible, and x-ray wavelengths. Their photographs of these events in profile were spectacular—some ejections were longer than the entire diameter of the Sun *(see color section, Figure 1)*. The images indicated that these coronal mass ejections were linked with solar flares, active regions on the Sun's surface, and loops of material that leaped outward from these regions.

Before returning to Earth, Conrad and Weitz made one last space walk on June 19th. Conrad climbed onto the tele-

June 10, 1973

scope mount to try to get a small battery-operated unit to work. He succeeded in getting it to turn on by taking a hammer and literally whacking it several times. The two astronauts also deployed a small test sheet of material intended to replace the sunshade. NASA already knew that the parasol would need replacement, as its material was unsuited for long exposures to the vacuum and radiation of space. The next Skylab crew, due in about a month, would pick up this sheet to see how conditions in space affected it.

Finally, the astronauts removed several test samples of different surface coatings for protecting the outside of the Apollo capsule that had been placed there prior to launch to see how they fared after a month's exposure in space. Analysis of the samples was used in improving later satellite and spacecraft design.

The Skylab 2 astronauts returned to Earth on June 22nd. It took almost three weeks for the three men to completely recover from their month in space. Each experienced lower back pain and muscle soreness in the lower part of the body. Moreover, all three suffered physically during the long period (over 24 hours) during which they were at sea following their recovery. Future recoveries were timed to reduce the period onboard ship to less than a day. Later astronaut crews also significantly increased their daily exercise schedule in order to shorten this recovery period.

The next Skylab mission, **Skylab 3 (July 28, 1973)**, more than doubled the human in-space flight record.

June 10, 1973
RAE 2
U.S.A.

Built by NASA, Radio Astronomy Explorer (RAE) 2 was the second space-based radio telescope, following **RAE 1 (July 4, 1968)**. Like its predecessor, RAE 2 used four long antenna booms able to extend to a tip-to-tip wingspan of 1,500 feet.

Skylab during a final inspection by the Skylab 2 crew. Note parasol and missing solar panel (see Skylab 2–May 25, 1973). *NASA*

Unlike its predecessor, RAE 2 was placed in lunar orbit, where the Moon's body could shield it from radio interference from the earth. Such interference had saturated RAE 1's receivers from 25 to 40 percent of the time.

During its operational life of more than one year, RAE 2 confirmed that radio impulses were being emitted by the magnetospheres of both Jupiter and Saturn. The cause of these bursts, however, remained unknown until Pioneer 10 (see **Pioneer 10, Jupiter Flyby–December 3, 1973**).

This mission, while not designed to study the Moon, was the last American flight to that satellite for more than two decades, until **Clementine (January 25, 1994)**.

The next orbiting radio telescope was **Prognoz 9 (July 1, 1983)**. See also **Soyuz 32 (February 25, 1979)**.

July 28, 1973
Skylab 3
U.S.A.

This 59-day mission, the second occupancy of **Skylab (May 14, 1973)**, was crewed by Alan Bean (**Apollo 12–November 14, 1969**), Jack Lousma (**STS 3–March 22, 1982**), and Owen Garriott (**STS 9–November 28, 1983**). The mission doubled the manned endurance record in space set by **Skylab 2 (May 25, 1973)**. Because this eight-week flight lasted so long, after its fourth week, ground control extended the mission on a weekly basis, depending on the medical status of each man.

Unlike the astronauts on Skylab 2, all three men on Skylab 3 experienced several days of severe space sickness soon after launch. On the first day in the laboratory, Lousma had vomiting fits, while Bean and Garriott became dizzy and nauseated. After a night's sleep, all three men felt better, but by noon of the second day, the nausea and dizziness had returned. For the next three days, these symptoms forced the astronauts to reduce their workload, including the postponement of a space walk to replace the deteriorating sunshade parasol installed during the Skylab 2 mission. Doctors on the ground advised the astronauts to restrict their head motions to help their bodies' balance systems to adapt to weightlessness, suggestions based on what was known of Gherman Titov's experience on **Vostok 2 (August 6, 1961)**.

The medical researchers later theorized that the space sickness in the Skylab 3 crew occurred because the men had entered the large interior space of Skylab almost immediately after reaching orbit. The crew of Skylab 2, because of their repair effort, had been confined to their Apollo capsule for over 22 hours, and this extra time in a small space had helped them adapt.

By the sixth day in orbit, the symptoms had subsided and the three men resumed a full workload. On this same day, the service module of their Apollo spacecraft sprung its second fuel tank leak (a small leak immediately after launch had disabled one set of the capsule's attitude thrusters). Because of the possibility that the capsule was no longer maneuverable, the ground began to prepare **Skylab 4 (November 16, 1973)** for a possible rescue flight. As it turned out, the leaks were isolated and the astronauts were able to return safely in their own capsule (though to save fuel, they made no flyover inspection of Skylab as they left).

Hurricane Ellen over the Atlantic Ocean on September 20, 1973. *NASA*

On the tenth day, Garriott and Lousma did their first space walk, replacing the parasol with an A-frame shade. Once outside the laboratory, they assembled two 55-foot-long poles which they attached to a bracket to form a V. This frame was then placed against the side of the station and, using two cords, the astronauts drew the new shade up along the poles, covering both the deteriorating parasol and the side of the laboratory.

On this same space walk, the two men also reloaded the film magazines on the telescope mount and deployed and retrieved a number of micrometeoroid and materials experiments. Original plans were for these experiments to be deployed by inserting them through the scientific airlock that was now blocked by the parasol and the sunshade, forcing the astronauts to place them manually on the station's exterior. During their 59-day mission, two additional space walks took place in which the astronauts reloaded and recovered film from the telescope mount.

One of the most fascinating discoveries from Skylab 3 came from a high school student. Seventeen-year-old Judith Miles from Lexington, Massachusetts, had sent two spiders, Arabella and Anita, up to the station to see if weightlessness affected their ability to weave webs. Both spiders adapted well to zero gravity, building ordinary webs and demonstrating that they did not use gravity as a reference system for web-spinning. The spiders' ability to consistently size the thickness of each thread, however, seemed hampered by the lack of gravity.

The interior space of Skylab was so large that Bean, Lousma, and Garriott were able to test-fly a backpack maneuvering unit while still inside the laboratory. This unit had been built by the Air Force for use during Eugene Cernan's **Gemini 9 (June 3, 1966)** space walk. Not only did the unit work, but Garriott was able to fly it without any prior training.

In Earth resource research, the astronauts continued the multispectral photography begun on Skylab 2, taking 39 scans. Besides extensive coverage in the United States, these scans included projects to map the resources of Paraguay, central Africa, and three dozen other countries. In Peru, they photographed the unexplained gigantic lines etched into the Plains of Nazca. In Africa, where a drought was causing starvation and death, the photographs taken by the astronauts were immediately used to find underground water supplies. Thousands of lives are believed to have been saved because of this information.

The astronauts tracked several hurricanes, including the formation of Hurricane Brenda as it moved through the Caribbean. They took multispectral images of the Great Barrier Reef off Australia and measured rainfall on that continent.

Solar observations continued, including more detailed photographs of coronal mass ejections. In one instance, Bean was doing solar observations at the moment an event took place.

In an effort to reduce the recovery period after their return to Earth, the astronauts' daily workout time had been increased from 30 minutes to one hour. As a result, the physical condition of all three men returned to normal less than a week after recovery. Both their heart and muscular systems showed no ill effects from weightlessness, though exercise seemed necessary to reduce its short-term impact.

All three men, however, showed a steady loss of bone mass during their 59-day flight. Despite the increased exercise, the human body's production of bone material seemed directly tied to the presence of gravity. The next mission, Skylab 4, tested this hypothesis for a period of almost three months.

After 59 record-breaking days in space, the astronauts returned to Earth on September 25th. Their in-space endurance record, however, lasted only until the flight of **Skylab 4 (November 16, 1973)**, launched less than two months later.

September 27, 1973
Soyuz 12
U.S.S.R.

Soyuz 12 was the first manned test flight of the redesigned Soyuz spacecraft, following the death of Dobrovolsky, Patsayev, and Volkov on **Soyuz 11 (June 6, 1971)**. In order to allow cosmonauts to wear spacesuits during re-entry as well as make room for additional life support equipment, the Soyuz capsule no longer supported three-man flights. Other changes included the removal of the Soyuz solar panels, which were replaced with batteries carrying power for about two days. Soyuz, now dedicated as a ferry between the ground and Soviet space laboratories, would get its batteries recharged by the station's solar panels.

The crew of this two-day flight was Vasili Lazarev (**Soyuz 18A–April 5, 1975**) and Oleg Makarov (**Soyuz 18A–April 5, 1975; Soyuz 27–January 10, 1978; Soyuz T3–November 27, 1980**). Though their flight was dedicated to checking out Soyuz, the two men also made Earth observations, taking simultaneous photographs of the earth using optical and multispectral cameras. The photographs were used to study forest and crop conditions as well as lake pollution.

The next Soviet manned mission, **Soyuz 13 (December 18, 1973)**, was the first in a series of four scientific Soyuz missions leading up to and following **Soyuz 19–Apollo 18 (July 15, 1975)**, a Soviet-American joint mission. The next test of the new Soyuz capsule was **Cosmos 613 (November 30, 1973)**.

October 31, 1973
Bion 1 (Cosmos 605)
U.S.S.R.

This three-week flight initiated a Soviet program of biological space research that included 11 Bion satellites launched over a 23-year period. These satellites placed a variety of biological specimens into orbit for periods of several days to several weeks, then returned them to Earth for study. The first few missions utilized unused Voskhod capsules first built in the early sixties for manned flight, then abandoned after the overthrow of Nikita Khrushchev (see **Voskhod 1–October 12, 1964**). Later missions used newly built and modified Voskhod capsules.

Bion 1 carried several dozen rats, a sampling of tortoises, a mushroom bed, four beetles, and living bacterial samples into space for three weeks. Upon return to Earth, scientists found that the rats had experienced a deterioration of tissue and of respiration, loss of body temperature, decrease in muscle and bone mass, and a reduction in appetite. Moreover, changes to their spleens, kidneys, and endocrine glands were noted. None of these changes, however, were permanent, and all the rats soon returned to normal. Neither the tortoises nor the fruit flies showed any significant changes from the flight. The mushrooms, however, grew in a deformed manner, with long twisted stems and large root systems.

The next Bion satellite was **Bion 2 (October 22, 1974)**.

November 3, 1973
Mariner 10
U.S.A.

Built by NASA, Mariner 10's primary mission goal was to take the first close look at Mercury. To do this at less cost, the spacecraft took advantage of a "gravity assist" from Venus, flying past that planet and using its gravity to redirect the spacecraft to Mercury—the first time this maneuver was ever attempted. It has since become a standard technique for either reaching other worlds or adjusting a spacecraft's orbit. See for example **Pioneer 11 (April 6, 1973)**, **Voyager 1 (March 5, 1979)**, **Voyager 2 (July 9, 1979)**, **Vega 1, Venus Flyby (June 10, 1985)**, **Vega 2, Venus Flyby (June 14, 1985)**, Galileo (see **STS 34–October 18, 1989**), Ulysses (see **STS 41–October 6, 1990**), and **Nozomi (July 4, 1998)**.

Mariner 10 flew past Venus on February 5, 1974, then continued on to Mercury, its solar orbit looping past the solar system's innermost planet three times, on March 29, 1974, September 21, 1974, and March 16, 1975. See **Mariner 10, Venus Flyby–February 5, 1974** and **Mariner 10, Mercury Flyby #3–March 16, 1975**.

November 16, 1973
Skylab 4
U.S.A.

Specifically scheduled to catch Comet Kohoutek as it made its closest approach to the Sun, Skylab 4 lasted 84 days and was the longest human space flight until **Soyuz 26 (December 10, 1977)**. This endurance record remained the American record for more than 20 years, until Norman Thagard spent 116 days on **Mir (February 20, 1986)**. See **Soyuz TM21 (March 14, 1995)** and STS 71 **(June 27, 1995)**.

The Skylab 4 mission was manned by Ed Gibson, Bill Pogue, and Jerry Carr, the first American all-rookie crew since **Gemini 8 (March 16, 1966)**. It was also the last manned mission to the **Skylab (May 14, 1973)** orbiting laboratory.

Unlike the **Skylab 3 (July 28, 1973)** crew, Gibson, Pogue, and Carr did not enter the orbital laboratory on their first day in space. They docked, but then slept in the Apollo capsule. This schedule was an attempt to prevent or minimize space sickness. Nonetheless, Bill Pogue and Jerry Carr both experienced nausea during the first three days of the flight, with Pogue vomiting.

Once again, the daily exercise workout was increased, from 60 minutes on Skylab 3 to 90 minutes on Skylab 4. Once again, the recovery period after re-entry was less than a week. Once again the astronauts experienced a continuous loss of bone mass throughout their flight, similar to that experienced by patients confined to bed rest.

Materials research confirmed more thoroughly results from **Apollo 14 (January 31, 1971)** showing that certain materials which on Earth did not form alloys could be made to mix in weightlessness. Other tests found that small single spherical crystals with few defects could be grown in space. The astronauts also repeated the crystal melt experiments performed on earlier Skylab missions, confirming the previous results: some crystals resolidified differently in zero gravity.

Earth observations included 39 multispectral scans and 2,000 photographs, focusing mostly on the United States, Mexico, and South America. Also studied, however, were regions in Europe, Asia, Africa, and Australia.

Observations of the Sun included both a solar eclipse and the largest coronal mass ejection ever seen. In general, astronomers found that during the eight months of Skylab observation, coronal mass ejections took place much more often than expected, at a rate of one every few days, and varied significantly in power and behavior. Other images included detailed pictures of gigantic loops of ionized particles streaming away from the Sun, apparently along the Sun's magnetic field lines. These images also revealed several new features, such as coronal holes and the spicules or streamers near the Sun's poles. Moreover, the first spectrum of a sunspot was obtained. Daily observations across as wide a range of spectrum with as much detail did not occur again until the launch of **Soho (December 2, 1995)**.

To maximize the observations of Comet Kohoutek, a special far-ultraviolet camera was built and carried into orbit by the astronauts. Over 200 hours were spent photographing the comet, the first ever observed in detail from space. Two space walks were also performed so that the astronauts could photograph it. These observations revealed a spike of material being ejected in the direction of the Sun. The comet's tail was also found to contain water vapor. Two other space walks were to reload film magazines and retrieve other experiments for return to Earth. Almost all of the particles collected on the micrometeoroid packages were from the solar wind. No material was detected from interstellar space.

The astronauts returned to Earth on February 8, 1974, after completing 84 days in space. In total, the laboratory had been occupied for 171 days, during which over 128,000 pictures of the Sun and 46,000 pictures of the earth were taken.

The next long-term mission to an orbiting laboratory was **Soyuz 14 (July 3, 1974)**. The next American space flight was part of the Apollo-Soyuz joint flight, **Apollo 18 (July 15, 1975)**.

November 30, 1973
Cosmos 613
U.S.S.R.

This 60-day unmanned flight tested the ability of the Soyuz space capsule to survive in space over a long period. The Soviet space program required such testing before allowing any manned crew to use the capsule for a comparable time period.

The Soyuz capsule was recovered successfully on January 29, 1974. The Soviets performed one more unmanned test of its Soyuz ferry spacecraft in May 1974 before launching **Salyut 3 (June 25, 1973)**, their second successful space laboratory.

December 3, 1973
Pioneer 10, Jupiter Flyby
U.S.A.

Built by NASA and launched on March 3, 1972, Pioneer 10 took 22 months to finally reach Jupiter, dipping to within 81,000 miles of the giant planet's cloud tops and rushing past at a speed of 82,000 miles per hour (see also **Pioneer 10–March 3, 1972**).

As it dropped into Jupiter's gravity well, Pioneer 10 mapped its magnetosphere, which was much larger than expected: Pioneer 10 crossed Jupiter's bow shock at 4 million miles from the planet's center. At that moment, the speed of the solar wind dropped from 1 million to 500,000 miles per hour, accompanied by a tenfold increase in temperature, from 90,000°F to 900,000°F. One day later, Pioneer 10 passed out of this turbulent region and through the magnetopause, entering Jupiter's magnetic field where the solar wind could not penetrate. The bow shock was thus much more compressed and much closer to the magnetosphere than scientists had expected. Upon leaving Jovian space, Pioneer 10 passed through this bow shock 17 different times as the changing pressure from the solar wind caused the planet's magnetosphere to bulge and shrink.

The spacecraft found that Jupiter's magnetosphere was tilted 11 degrees from the planet's axis and that its polarity was the opposite of Earth's—a compass would point south on Jupiter instead of north. Closer to the planet, the field became extremely complex, with indications of four and possibly eight poles instead of two. Scientists believed this complexity was a reflection of the tangled flow and circulation patterns of Jupiter's liquid metallic hydrogen interior.

Scientists were able to infer Jupiter's interior based on subtle changes in Pioneer 10's flight path; they concluded that 85 to 90 percent of Jupiter was liquid hydrogen, with most of the remainder helium. Whether this helium mixed with the liquid hydrogen, or formed another layer below the hydrogen, was unknown. Scientists theorized that although the planet might have a small iron-silicate core, its interior was probably made of two layers of hydrogen liquid, a 28,000-mile-deep inner core of liquid metallic hydrogen at a temperature of 54,000°F, and an upper layer of liquid hydrogen about 19,000 miles deep at a temperature of 3,600°F. The transition zone between these two layers floated at an elevation of about 28,000 miles above the planet's center, at 20,000°F, and at pressures equal to three million Earth atmospheres, or 22,000 tons per square inch.

The properties of metallic hydrogen were such that it could conduct electricity and heat. This innermost layer, churned by the planet's rapid 10-hour rotation period, therefore acted as a dynamo for Jupiter's powerful magnetic field.

Above these hydrogen layers was Jupiter's atmosphere, made of a series of thin layers 600 miles deep. From the planet's surface up, the layers were water droplets, ice crystals, ammonium hydrosulfide crystals, ammonia crystals, and gaseous hydrogen. These layers were what formed the horizontal bands visible from Earth and photographed by Pioneer 10. Scientists labeled the dark bands belts and the light bands zones. Pioneer 10 revealed that the lighter colored zones were higher in altitude than the darker belts—the dark color was actually the shadow cast on the belt by the adjacent zone. In each zone, gases rose at its center and sunk at its edges, causing the swirling hook-shaped storms visible at the zone-belt intersection (see color section, Figure 5).

The first close-up photograph was taken of Jupiter's famous Red Spot, located in the southern hemisphere (see color section, Figure 4). Some theories for the Red Spot's origin had suggested that it was caused by the atmosphere's flow over the top of a high mountain. Since Pioneer 10's data indicated that Jupiter was almost all liquid, this theory now seemed unlikely. Pioneer 10 also photographed a smaller red spot in the northern hemisphere at a similar latitude. In both photographs, the spots resembled gigantic rotating storm systems, not flow patterns over surface features.

Pioneer 10 took pictures of two of Jupiter's moons, Europa and Ganymede, as it passed through the Jovian system. Neither picture showed much detail, though the Ganymede image revealed several dark circular regions.

As the spacecraft cut through Jupiter's powerful magnetic field, it found that the field was distorted by the movement through it of the planet's innermost four satellites—Amalthea, Io, Europa, and Ganymede. Jupiter's regular radio bursts, detected from Earth, were caused by Io's interaction with the planet's magnetic field—a powerful electric current, equal to 10 trillion watts, flowed to and from Jupiter and Io along the planet's magnetic lines of force.

Io also appeared to have an ionosphere. This yellow glow that surrounded the planet was thought to be sodium, and the planet was believed to be covered with sodium-rich salt flats. Scientists postulated that the radiation from Jupiter's magnetosphere interacted with these salt flats, releasing sodium molecules into space to surround the planet.

The masses of the four Galilean moons (Io, Europa, Ganymede, and Callisto, discovered by Galileo in 1610) was calculated. When compared to the Moon, these masses were Io, 1.21; Europa, 0.65; Ganymede, 2.02; and Callisto, 1.46. Io was much heavier than expected, and the density of the satellites apparently declined with distance from Jupiter. Thus,

Io and Europa were believed to be rocky bodies, while Ganymede and Callisto were thought to be mostly water ice.

Having survived its passage through the Jovian system, Pioneer 10 headed out of the solar system at a speed of about 25,000 miles per hour. Its path took it into the tail of the Sun's heliosphere, aimed at the star Aldebaran in the constellation Taurus. It continued to function for another 24 years, until March 31, 1997, when its nuclear batteries were practically depleted. See **Pioneer 10, Shutdown (March 31, 1997)** for what was learned as the spacecraft became the first object made by humans to travel beyond the orbit of Pluto.

The next spacecraft to reach Jupiter was Pioneer 11 (**Pioneer 11, Jupiter Flyby–December 4, 1974**).

December 16, 1973
Explorer 51 (Atmosphere Explorer C)
U.S.A.

This satellite was the third in NASA's five-satellite project to study the atmosphere between 75 and 200 miles elevation. See **Explorer 17 (April 3, 1963)** and **Explorer 32 (May 25, 1966)**, for the previous two satellites.

Placed in a high latitude orbit inclined to the equator 68 degrees, Explorer 51 was designed to measure how the upper atmosphere's density, composition, and temperature were changed by solar activity. The satellite worked in conjunction with **San Marco 4 (February 18, 1974)**, which orbited at the equator and took similar measurements.

During its life span, Explorer 51's perigee was periodically lowered to 75 miles, the lowest orbital altitude of any NASA spacecraft, excluding tests by **Gemini 3 (March 23, 1965)**. At this height, the spacecraft could obtain measurements of the atmosphere that previously had been unattainable. After several days of data gathering, the satellite's perigee would be raised so that its orbit would not decay.

The satellite found that at altitudes between 81 and 186 miles, the atmosphere fluctuated, with highly variable winds blowing as much as 175 miles an hour. The air pressure at this altitude was only one one-hundred-thousandth of sea level; therefore, the gusts had little effect on the spacecraft.

The fourth satellite in this atmospheric research project was **Explorer 54 (October 6, 1975)**.

December 18, 1973
Soyuz 13
U.S.S.R.

Soyuz 13 was an eight-day manned mission dedicated to astronomical and Earth observations. The spacecraft, once again using solar panels, was a variant of the Soyuz craft planned for the joint mission of **Soyuz 19-Apollo 18 (July 15, 1975)**.

The cosmonauts, Pyotr Klimuk (**Soyuz 18B–May 24, 1975; Soyuz 30–June 27, 1978**) and Valentin Lebedev (**Soyuz T5–May 13, 1982**), took multispectral and optical photographs of the earth, similar to those taken on **Soyuz 12 (September 27, 1973)**.

Because Soyuz 13 performed no dockings, it carried an ultraviolet telescope in its docking port. This telescope took spectrograms of thousands of stars, including ultraviolet spectrums of several planetary nebulae. The space flight was also timed so that it could make observations of the recently discovered Comet Kohoutek.

In Soyuz 13's orbital module, a self-contained system was tested for producing water, food, and oxygen using bacteria. Another experiment measured the redistribution of body fluids during weightlessness, trying to find out if the disorientation experienced by some cosmonauts early in space flight was caused by the increased blood flow to the brain.

The next manned space flight, **Soyuz 14 (July 4, 1974)**, resumed operations in a space laboratory, **Salyut 3 (June 25, 1974)**.

1974

February 5, 1974
Mariner 10, Venus Flyby
U.S.A.

Built by NASA and launched on November 3, 1973, Mariner 10's primary mission was to take the first close-up photographs of the surface of Mercury (see **Mariner 10–November 3, 1973**). To reach the innermost planet, however, the spacecraft first flew past Venus, using that planet's gravity to redirect its journey through the solar system.

Passing only 3,600 miles above Venus's clouds, Mariner 10 took 4,165 pictures of the cloud cover. Taken using an ultraviolet filter, the images gave scientists their first view of the planet's atmospheric weather patterns, revealing an unexpected circulation pattern that made it necessary to discard all previous theories.

Wind speeds were inferred by comparing how features changed position from picture to picture. Near the equator, the wind speed was about 225 miles per hour. Despite these hurricane winds, large cloud patterns, including a dark Y-shaped cloud turned sideways that covered an entire hemisphere, held their shape as they circled the planet. The atmosphere therefore appeared to rotate as a unit every four days, 60 times faster than the planet below it and in the opposite direction the planet was turning. The planetwide atmospheric features, such as the Y-shaped cloud, seemed to be waves rippling through that atmosphere.

At the subsolar point (the position of the planet closest to the Sun), the cloud pattern was chaotic, filled with many large convection cells of rising air, surrounded by spreading waves of cloudlike ripples covering hundreds of miles. Images of other parts of the atmosphere also indicated the possible existence of a circulation pattern, similar to Earth's, where the heat at the equator made the air rise, flow north and south toward the poles, where it became colder and sank toward the ground, then flowed back toward the equator. This pattern, though indicated by Mariner 10's pictures, was not unequivocally proven.

Propagation measurements were done as Mariner 10 passed behind Venus and its radio signal cut through the planet's atmosphere. The measurements confirmed that Venus had less topographic relief than Earth. This experiment also revealed that the planet's cloud layer was actually at least two layers. The top layer, at about 37 miles altitude, was thin and broken into many clouds by fast-moving winds. The lower layer, from

Ultraviolet image showing Venus's global cloud structure. *NASA*

22 to 32 miles altitude, was thick and opaque. Data also indicated that other layers might exist between these two.

In the six years since **Venera 4 (October 18, 1967)** and **Mariner 5, Venus Flyby (October 19, 1967)** had first sent back data about Venus, many theories had been proposed about the composition of the planet's white clouds. Scientists had finally abandoned the notion that these were water vapor clouds, especially since water simply could not explain all their data. In 1972, Godfrey Sill proposed that the clouds were about 75 percent sulfuric acid. The dark features visible in Mariner 10's ultraviolet pictures indicated the presence of sulfur, which helped confirm this theory.

Venus's atmosphere was found to contain hydrogen, oxygen, and helium. The hydrogen came from the solar wind. The quantity of oxygen was much greater than had been seen by previous probes, indicating that it might be fossil evidence of water. The exact amount of oxygen in Venus's atmosphere, however, was still unclear.

During Mariner 10's approach to Venus, it traveled through the wake created as the planet plowed through the solar wind. The information from Venera 4 and Mariner 5 was correct—although Venus had no magnetic field, its atmosphere still interacted with the solar wind to create a bow shock and a long tail (or wake) trailing out from behind the planet, although this structure was found to be much smaller than Earth's.

With the accumulating data from Mariner 10 and its predecessors, scientists were able to make a rough outline of the atmosphere that surrounded Venus. However, a detailed understanding of the actual planet still eluded them. Only **Venera 8, Venus Landing (July 22, 1972)** had studied the ground at its landing site, but this had only indicated that the rocks were ancient crust material resembling granite. Little else was known about the planet's solid body. No pictures of the Venusian surface had yet been taken, and no Venusian rocks had ever been seen.

The next probes to Venus, **Venera 9, Venus Landing (October 22, 1975)** and **Venera 10, Venus Landing (October 25, 1975)**, changed this, beaming back the first pictures of Venus's surface, along with more data about its composition.

Mariner 10, its path through space changed by Venus's gravity, continued on to Mercury, looping past the solar system's innermost planet three times, on March 29, 1974, September 21, 1974, and March 16, 1975. See **Mariner 10, Mercury Flyby #3 (March 16, 1975)** for more information about what the spacecraft learned concerning Mercury.

February 10, 1974
Mars 4, Mars Flyby
U.S.S.R.

Mars 4, launched on July 21, 1973, was the first of four spacecraft sent to Mars by the Soviet Union in 1973 (see **Mars 5, Mars Orbit–February 12, 1974; Mars 6, Mars Landing–March 12, 1974; Mars 7, Mars Flyby–March 9, 1974**). Because of the failure of a computer chip when exposed to the environment of space, however, all four spacecraft miscarried, with only Mars 4 and Mars 5 returning any data.

The computer chip failure prevented Mars 4's rockets from placing the spacecraft in orbit around Mars as planned. As it flew past the planet, the probe did manage to take a single strip of images.

February 12, 1974
Mars 5, Mars Orbit
U.S.S.R.

Mars 5, launched on July 25, 1973, was the second of four Soviet spacecraft sent to Mars in 1973 (see **Mars 4, Mars Flyby–February 10, 1974; Mars 6, Mars Landing–March 12, 1974; Mars 7, Mars Flyby–March 9, 1974**). Though it was successfully inserted into Mars orbit, it only functioned for a few days, returning images of a small section of the planet's southern hemisphere. During its short period of operation, it also measured an average temperature on the Martian surface of –45°F and an atmospheric surface pressure of 6 millibars (less than .001 of the atmospheric pressure on Earth)—findings similar to previous measurements.

February 16, 1974
Tansei 2
Japan

This flight was Japan's first test of a rocket using a guidance system able to place its satellite into a precise orbit. The dummy satellite, whose only purpose was to transmit the spacecraft's telemetry and orbit, was successfully orbited with an apogee of 1,925 miles, only slightly lower than the planned 2,160 miles.

February 18, 1974
San Marco 4
Italy / U.S.A.

This fourth Italian satellite, following **San Marco 3 (April 24, 1971)**, completed Italy's first space research project, dedicated to studying atmospheric densities at altitudes between 130 and 500 miles. The satellite was launched from Italy's floating San Marco launch platform off the coast of Kenya. An American Scout rocket was used to place it in orbit.

More sophisticated than the previous three satellites, San Marco 4 not only measured the changing air density, but it also carried two American-built mass spectrometers to monitor air temperature and composition. The satellite, placed in an equatorial orbit, worked in conjunction with **Explorer 51 (December 16, 1973)**, which made similar measurements at a similar altitude but at higher latitudes. The two satellites together showed how much time was necessary for solar events, such as magnetic storms and solar flares, to change the atmosphere's density at different latitudes.

The next Italian satellite, **Sirio 1 (August 25, 1977)**, did communications research. The next atmospheric density satellite was **Explorer 54 (October 6, 1975)**.

March 9, 1974
Mars 7, Mars Flyby
U.S.S.R.

This Soviet lander was one of four spacecraft launched toward Mars by the Soviet Union in 1973 (see **Mars 4, Mars Flyby–February 10, 1974; Mars 5, Mars Orbit–February 12, 1974; Mars 6, Mars Landing–March 12, 1974**). Launched on August 9, 1973, Mars 7 missed Mars entirely, due to a computer chip failure. The spacecraft failed to return any data.

March 9, 1974
Miranda
U.K. / U.S.A.

Built by British engineers and launched by NASA, Miranda was an engineering flight to test gyros for maintaining a satellite's orientation in three dimensions (see **PAC 1–August 9, 1969**). Such a system, which saves both fuel and weight, was used on **IUE (January 26, 1978)**, the Hubble Space Telescope (see **STS 31–April 24, 1990**), and many other later satellites. A similar design was also developed independently by the Soviets for **Mir (February 20, 1986)**.

March 12, 1974
Mars 6, Mars Landing
U.S.S.R.

Mars 6 was the last of four Soviet spacecraft launched to Mars in 1973 (see **Mars 4, Mars Flyby–February 10, 1974; Mars 5, Mars Orbit–February 12, 1974; Mars 7, Mars Flyby–March 9, 1974**). Launched on August 5, 1973, it was the third human craft to impact the Martian surface, following **Mars 2 (November 27, 1971)** and **Mars 3 (December 2, 1971)**.

Mars 6 failed, however, to return any useful data. During its descent through the Martian atmosphere, the spacecraft transmitted for 150 seconds, but the data was unreadable because of a failed computer chip. When the lander's retro-rockets fired, all contact was permanently lost.

After seven failed attempts, the Soviet Union ceased sending missions to Mars for 15 years. Their next effort was **Phobos 1 (July 7, 1988)** and **Phobos 2 (July 12, 1988)**.

The next successful missions to Mars, **Viking 1 (June 19, 1976)** and **Viking 2 (August 7, 1976)**, included two orbiters and two landers.

March 26, 1974
Cosmos 637
U.S.S.R.

Cosmos 637 was the first Soviet communications satellite placed in geosynchronous orbit. It performed tests leading to the eventual establishment of several different Soviet geosynchronous communications satellite systems. The next test flight of a Soviet geosynchronous communications satellite was **Molniya 1S (July 30, 1974)**.

April 3, 1974
Cosmos 638
U.S.S.R.

This unmanned Soyuz test flight checked out the capsule configuration to be used during the **Soyuz 19–Apollo 18 (July 15, 1975)** joint flight. The spacecraft remained in orbit for 10 days, duplicating the orbital maneuvers planned for the joint flight, and was successfully recovered on April 13, 1974.

Because the Americans could place no limits on media coverage, the Soviets decided to run this test entirely in secret, which was their normal procedure. After one more unmanned six-day test flight in August, they flew **Soyuz 16 (December 2, 1974)**, the first manned use of this Soyuz configuration.

April 13, 1974
Westar 1
U.S.A.

Westar 1, owned and operated by Western Union, was the first wholly private American communications satellite launched in more than a decade, since **Telstar 2 (May 7, 1963)**. This pause in the development of commercial satellites was because of political disagreements in the United States about how private satellite business should be regulated, as well as perceived conflicts with the international consortium Intelsat (see **Early Bird 1–April 6, 1965**), of which the United States was a member.

Following almost seven years of debate in both the Federal government and private industry, the Federal Communications Commission (FCC) finally decided in 1972 to introduce what it called an "Open Skies" policy, "authorizing any qualified entity, subject only to specified technical and antitrust criteria, to launch and operate a domestic satellite system." Westar 1 was the first private American satellite launched under that policy and also the first private satellite placed in geosynchronous orbit.

Westar 1 was the first of a three-satellite cluster, the last of which was launched in 1988. Each was similar to the Canadian Anik A satellites (see **Anik A1–November 10, 1972**), with capacity for 1,200 one-way voice circuits or 12 television channels, and all three remained in use for about 10 years.

In 1982 Western Union began launching a second series of slightly modified Westar satellites; see **Westar 4 (February 26, 1982)**.

May 17, 1974
SMS 1
U.S.A.
Built by NASA, Synchronous Meteorological Satellite (SMS) 1 was a geosynchronous satellite for photographing an entire hemisphere of the Earth's surface and atmosphere in both visible and infrared light, an experimental forerunner of the GOES satellites (see **GOES 1–October 16, 1975**). SMS 1's wide-angle images worked in conjunction with the close-up pictures taken by the Landsat (see **Landsat 1–July 23, 1972**) and the NOAA (see **NOAA 1–December 11, 1970**) satellites.

SMS 1's images had a resolution of two-thirds of a mile in visible light and five miles in infrared. Every 30 minutes, it accumulated enough image data to produce a hemispheric picture of the earth's cloud patterns and temperature profile. The images were generally resampled, reducing their resolution to about five miles in order to minimize the data transmitted. At a global scale, better resolution was unnecessary. These images allowed the spacecraft to monitor air pressure, temperature, rain, snow, river levels, and ocean currents.

The orbiting plan called for two such satellites in orbit at all times, positioned over the western Atlantic and eastern Pacific in order to cover the entire United States and its offshore areas.

A second SMS satellite was launched in February 1975, with both SMS satellites later being replaced by the more advanced GOES satellites.

May 29, 1974
Luna 22
U.S.S.R.
Luna 22 was the seventh Soviet unmanned lunar orbiter, following **Luna 19 (September 28, 1971)**. It orbited the Moon for 18 months.

The spacecraft carried cameras for mapping the lunar surface, sensors for measuring the surface's gamma-ray emissions and composition, and micrometeorite and cosmic ray detectors.

Luna 22's micrometeoroid detector found that impact rates varied significantly over time, with evidence suggesting that the impacts were related to specific meteor showers and the kick-up of dust from the Moon's surface because of these showers.

Luna 22 was the last lunar orbiter intended solely for lunar research for almost 20 years, until **Clementine (January 25, 1994)**.

The next lunar lander was **Luna 24 (August 9, 1976)**, a sample return mission. Luna 24 was also the last lunar soft-landing mission, as of December 1999.

May 30, 1974
ATS 6
U.S.A.
ATS (Applications Technology Satellite) 6 was part of a NASA program testing geosynchronous communication satellite technology, following **ATS 5 (August 12, 1969)**. ATS 6 carried eight different experiments, investigating signal propagation while also demonstrating the feasibility of several new technologies.

Previous communications satellites had used frequencies below 10 GHz, since the atmosphere has little effect on signals at these frequencies. Above 10 GHz, rain and various gases can interfere with satellite transmissions. In the 1970s, however, the steadily growing demand for satellite communications and a shortage of bandwidth was making these higher frequencies more attractive to satellite companies, thereby requiring a number of tests to determine the limits of interference and methods for minimizing it. ATS 6 tested transmissions at a number of frequencies, including NASA tests at 20 GHz and 30 GHz and Comsat Corporation tests at 13 GHz and 18 GHz. The Comsat tests continued for a full year, with 24 ground stations scattered throughout the eastern part of the United States gathering data on the range of weather changes that produced interference.

Later satellites did further propagation tests in a number of other frequency bands, with **Hermes (January 17, 1976)** doing transmission tests at 11 and 12 GHz and receiver tests at 14 GHz; **Kiku 2 (February 3, 1977)** doing transmission tests at 1.7, 12, and 34.5 GHz; and **Sirio (August 25, 1977)** doing both transmission and receiver tests at 11 and 17 GHz, respectively.

ATS 6's design was unusual. Its antenna was a 30-foot parabolic reflector that unfolded like an umbrella once in orbit, and its solar cells coated the outside of two half drums rather than being the more typical flat panels.

For its first year, ATS 6 orbited above the United States so that ground stations could measure the effects of weather on these various frequencies. During this period, ATS 6 performed joint tests with **GEOS 3 (April 9, 1975)**. From its perch in geosynchronous orbit, ATS 6 maintained contact with GEOS 3 as it passed in low-Earth orbit, thereby becoming the first satellite to track another orbiting satellite while it verified communication techniques planned for the joint **Apollo 18–Soyuz 19 (July 15, 1975)** mission.

In its second year, ATS 6 was placed over the Middle East, where India used it to broadcast experimental instructional television and NASA did additional tests using European ground terminals, as well as several NASA weather satellites. After July 1976, it was slowly moved eastward until it reached a position over the eastern Pacific Ocean.

During that slow transfer and until it was turned off in 1979, the satellite was used by the U.S. Department of Health, Education, and Welfare; NASA; and the Corporation for Public Broadcasting. Doctors in the remote areas of Alaska, the Rocky Mountains, and Appalachia used the satellite to communicate medical advice and treatments.

Appalachian students and teachers also used the satellite under a program developed by the Appalachian Community Services Network (ACSN). Started in 1975, ACSN used ATS 6 to distribute video educational courses throughout Appalachia as well as allow students and teachers to participate in teleconferences. ACSN (which changed its name to The Learning Channel) continued these services using Hermes as well as **RCA Satcom 1 (December 13, 1975)**, on which Home Box Office (HBO) was also distributed widely.

ATS 6 was the last solely NASA experimental communications satellite for almost 20 years, until **ACTS (September 12, 1993)**. The next experimental communications satellite in which NASA had any participation was Hermes.

June 3, 1974
Hawkeye (Explorer 52)
U.S.A.

Built by the University of Iowa and launched by NASA, Hawkeye was a continuation of the older Injun series of satellites, the last of which had been **Injun 5 (August 8, 1968)**. Like the eccentric orbiting Imp (see **Imp A–November 27, 1963**), HEOS (see **HEOS 1–December 5, 1968**), and Prognoz (see **Prognoz 1–April 14, 1972**) satellites, Hawkeye was placed in an eccentric polar orbit ranging from about 300 to 78,000 miles elevation. From this position, the satellite continued the monitoring of the interaction of the solar wind with the earth's magnetosphere, operating through 1977 and augmenting the Solrad (see **Solrad 1–June 22, 1960**), OGO (see **OGO 1–September 5, 1964**), Intercosmos (see **Intercosmos 1–October 14, 1969**), Aureol (see **Aureol 1–December 27, 1971**), and Prognoz (see **Prognoz 1–April 14, 1972**) satellites.

See **Dynamics Explorer 1/Dynamics Explorer 2 (August 3, 1981)** for results. The next ionospheric satellite was **Aeros 2 (July 16, 1974)**.

June 25, 1974
Salyut 3
U.S.S.R.

Salyut 3 was the second successful Soviet space laboratory (Salyut 2 broke apart shortly after lift-off). Like **Salyut 1 (April 19, 1971)**, the station weighed over 20 tons, and it contained about the same interior space for research. Salyut 3, however, had significantly larger solar panels to provide more power, and these could be rotated to face the Sun, thereby allowing the laboratory to be pointed away from the Sun for long periods whenever the research required it.

Salyut 3 was dedicated to military research and included a separate re-entry capsule resembling the American Mercury capsule, which returned safely to Earth on September 22, 1974. Because of Salyut 3's military nature, little was made public about its design or contents. Rather than carry astronomical telescopes (as Salyut 1 had), Salyut 3 was equipped with high-resolution surveillance cameras capable of seeing objects as small as one foot.

Two spacecraft visited Salyut 3:

Soyuz 14–July 3, 1974 to July 19, 1974 (16 days)
Soyuz 15–August 25, 1974 to August 28, 1974 (2 days)

Salyut 3 remained in orbit for 90 days. It was then deorbited, burning up over the ocean on September 27, 1974. The next Soviet space laboratory was **Salyut 4 (December 26, 1974)**.

July 3, 1974
Soyuz 14
U.S.S.R.

This 16-day flight was the first manned mission to **Salyut 3 (June 25, 1974)** and the first Soviet manned mission to an orbiting laboratory since the deaths of cosmonauts Dobrovolsky, Patsayev, and Volkov on **Soyuz 11 (June 6, 1971)**. The cosmonauts were Pavel Popovich (**Vostok 4–August 12, 1962**) and Yuri Artyukhin.

Because Salyut 3 was dedicated to military reconnaissance, much of the work done during Soyuz 14 was not revealed, though the cosmonauts did continue some research begun on **Soyuz 9 (June 1, 1970)** and **Soyuz 11 (June 6, 1971)**.

Medical research involved electrocardiograms and blood and respiratory samples. The cosmonauts' lung and heart capacity was tested, and the blood circulation to their brains was studied. And both men used the "penguin" suits first tested during Soyuz 11.

Multispectral photographs were taken of Soviet ore deposits, water pollution, crop resources, and changes in ocean ice sheets. Comparative observations of weather patterns were performed jointly with a Meteor weather satellite (see **Meteor 1-01–March 26, 1969**). Spectrograms were made of the earth's horizon, including its twilight airglow and aurora.

Popovich and Artyukhin experienced some of the same discomforts as had the Soyuz 9 cosmonauts and the Skylab astronauts (see **Skylab 2–May 25, 1973; Skylab 3–July 28, 1973; Skylab 4–November 16, 1973**) upon return to Earth. Their blood pressures and pulses were low, and though immediately after landing they were able to walk without assistance, it took them four to six hours before they could do this in a normal manner, and several days for their bodies to return entirely to normal.

The second flight and last flight to Salyut 3 was **Soyuz 15 (August 26, 1974)**.

July 14, 1974
NTS 1 (Timation 3)
U.S.A.

NTS 1 (Navigation Technology Satellite) was a combined U.S. Navy–U.S. Air Force test flight of global positioning system (GPS) technology, the third such test following **Timation 2 (September 30, 1969)**.

Based on the test results from the previous satellites, NTS 1 was given more power and an even better attitude control system, and it was placed at a higher orbit to increase precision. Its internal clock was upgraded to two rubidium atomic clocks, accurate to within one part per 100 billion per day.

The fourth test flight of this technology was **NTS 2 (June 23, 1977)**. This launch also inaugurated the establishment of the first fully operational global positioning system.

July 16, 1974
Aeros 2
West Germany / U.S.A.

The third German-built satellite, Aeros 2 was placed in orbit by NASA. It was similar in design and purpose to its predecessor, **Aeros 1 (December 16, 1972)**, carrying five experiments to study the upper atmosphere and monitor the Sun's ultraviolet output.

The combined data from the two satellites showed that there was a connection between the aurora and specific temperature variations during periods of increased solar magnetic activity.

The combined data from the two satellites also found significant discrepancies between the measurements of different satellites, showing the importance of precise instrument calibration, without which reliable results were impossible.

Taiyo (February 24, 1975) continued research into the ionosphere and magnetosphere.

July 30, 1974
Molniya 1S
U.S.S.R.

Placed in geosynchronous orbit, Molniya 1S was the second Soviet test geosynchronous communications satellite, following **Cosmos 637 (March 26, 1974)**. Results from both experimental flights led to the launch of three different Soviet geosynchronous satellite clusters. See **Raduga 1 (December 22, 1975)**, **Ekran 1 (October 26, 1976)**, and **Gorizont 1 (December 19, 1978)**. All three clusters also supplemented the Molniya satellites (see **Molniya 1-01–April 23, 1965**; **Molniya 2-01–November 24, 1971**; **Molniya 3-01–November 21, 1974**).

August 26, 1974
Soyuz 15
U.S.S.R.

Soyuz 15, intended as a three-week-plus manned mission to **Salyut 3 (June 25, 1974)**, ended up returning to Earth after only two days. The cosmonauts, Gennady Sarafanov and Lev Demin, were unable to dock with the space laboratory, and since the Soyuz ferry depended on the station for power (no longer using solar panels and carrying batteries with only two days' power reserve—see **Soyuz 12–September 27, 1973**), the mission was aborted.

Because Salyut 3 had a life expectancy of only six months, there was no time for a third flight to this station. On September 23, 1975, the recoverable module returned to Earth, and the laboratory was deorbited on January 24, 1975. The next manned flight, **Soyuz 16 (December 2, 1974)**, tested components for the joint Soviet-American flight, **Soyuz 19–Apollo 18 (July 15, 1975)**. The next Soviet mission to a space laboratory was **Soyuz 17 (January 11, 1975)**, docking with **Salyut 4 (December 26, 1974)**.

August 30, 1974
ANS 1
Netherlands / U.S.A.

The Astronomical Netherlands Satellite (ANS) 1 was a joint project between the Netherlands and the United States. The Dutch built the satellite and experiments, and NASA provided the rocket and launch services.

ANS 1 was an astronomical ultraviolet and x-ray telescope, following up the discoveries of **Uhuru (December 12, 1970)** and supplementing **Copernicus (August 21, 1972)**. Though intended for a circular orbit about 325 miles elevation, the satellite ended up in an eccentric orbit ranging from 165 to 731 miles altitude. The higher apogee placed the spacecraft in the Van Allen radiation belts during part of each orbit, where background radiation blinded its ultraviolet and x-ray telescopes.

Despite this limitation, ANS 1 did make some discoveries during its 20 months of operation. It found that the Large Magellanic Cloud contained a large amount of dust and appeared to be a young galaxy, about 3 billion years old. The spacecraft also discovered that the powerful x-ray emissions coming from the galaxy Centaurus A came from its nucleus, not its outer lobes.

ANS 1 also studied Cygnus X-1, thought to be a black hole, finding that the object's x-ray steady output actually consisted of two states of low intensity and high intensity. Scientists theorized that Cygnus X-1 was surrounded by a rotating disk of debris (not dissimilar to the solar system) left over from objects previously torn apart and pulled into the black hole. The rotation of this material combined with the further accretion of this material onto Cygnus X-1 was thought to cause the x-ray emissions and their periodic change in output.

The spacecraft also observed x-rays being emitted by the first-magnitude star Sirius, thought to come from the corona surrounding the star. Other flare stars, known to increase in brightness for short periods, were also found to emit x-ray flares.

Also discovered were x-ray bursters. Unlike the x-ray sources first mapped by Uhuru, x-ray bursters released x-rays randomly in short staccato bursts. These objects were thought to be binary star systems in which one star was a neutron or pulsar, and the bursts somehow caused when this star's gravitational field pulled material from its binary companion. One object, however, called the rapid burster, remained inexplicable. Scientists were not sure what caused its rapid but random bursts of energy, lasting about a month and then fading into a period of inactivity.

The next orbiting x-ray telescope was **Ariel 5 (October 5, 1974)**.

October 5, 1974
Ariel 5
U.K. / U.S.A.

Launched by NASA, Ariel 5 was the fifth British-built research satellite; it joined both **Copernicus (August 21, 1972)** and **ANS 1 (August 30, 1974)** to form a network of astronomical x-ray telescopes, following **Uhuru (December 12, 1970)**. Its purpose was to observe and map the sky in x-ray radiation. Its all-sky pinhole camera allowed it to continuously monitor the entire sky, thereby providing an "early warning" capability for the other satellites. Thus, Ariel 5 would identify an x-ray burst, and the other satellites would then join in, making concurrent observations.

Ariel 5 found that of the 200 or so x-ray objects mapped by Uhuru, 16 had disappeared in the ensuring two years while and 19 new objects had appeared. Some of these new objects were very transient, lasting only a few weeks. One of these transient objects was at the galactic center and had been quiescent since February 1974.

Ariel 5's more sensitive instruments also detected x-ray emissions from two well known supernova remnants, one called Casseopia A and the other the remains from the supernova discovered by Danish astronomer Tycho Brahe in 1572.

Ariel 5 and its sister x-ray telescopes clearly showed that the sky was filled with many violent objects, some steady sources of x-ray emissions; others bursting only rarely, doing so at random and unpredictable times. The bursters, however, were 10 times more numerous than scientists had predicted, and when they burst they sometimes became the brightest x-ray sources in the sky, brighter even than the Sun.

The surprisingly changing nature of the x-ray sky demonstrated the necessity of monitoring the x-ray sky continuously. The next two spacecraft to carry x-ray instruments, **Explorer 53 (May 7, 1975)** and **OSO 8 (June 21, 1975)**, maintained this sky monitoring, with the x-ray telescope **HEAO 1 (August 12, 1977)** taking over from them.

October 22, 1974
Bion 2 (Cosmos 690)
U.S.S.R. / International

Like **Bion 1 (October 31, 1973)**, Bion 2 carried biological specimens into space in a Voskhod capsule that had originally been intended for the Voskhod manned program, which was abandoned after the overthrow of Khrushchev (see **Voskhod 1–October 12, 1964**).

The capsule, which carried several albino rats, remained in orbit for three weeks. During the flight, researchers from the Soviet Union, Czechoslovakia, and Romania subjected rats both on the ground and on Bion 2 to an identical daily dose of gamma radiation. After landing, more of the space rats had developed lung problems, and their bone marrow and blood content had changed more than the earthbound control specimens. These results indicated that the changes seen in blood circulation and bone mineralization in animals and humans after space flight apparently weakened the body, leaving it more vulnerable to damage from radiation.

The next Bion satellite, **Bion 3 (November 25, 1975)**, was the first joint Soviet-American unmanned mission.

October 28, 1974
Luna 23
U.S.S.R.

Luna 23 attempted to be the third robot sample return mission to the Moon, following **Luna 20 (February 14, 1972)**. It landed in the Sea of Crises, and though the spacecraft transmitted for three days from the lunar surface, its drill was damaged upon landing and was unable to obtain any samples. For this reason, the Soviets decided to leave the return capsule on the Moon.

The next Soviet lunar sample return mission, **Luna 24 (August 9, 1976)**, was the third unmanned probe to successfully return samples from the Moon. It was also the last probe, as of December 1999, to land softly on the Moon.

November 15, 1974
Oscar 7 / Intasat 1
U.S.A. / International / Spain

- *Oscar 7*   Oscar 7 was the second satellite, following **Oscar 6 (October 15, 1972)**, financed and built by AMSAT (the Radio Amateur Satellite Corporation), an organization of amateur radio operators from the United States, Australia, Canada, and West Germany. Oscar 7 was a small active repeater communications satellite, the seventh in a series of such ham radio satellites and very similar to its predecessor. Oscar 7 was able to send and receive messages between the Western Hemisphere and Europe using two transponders and four telemetry transmitters. It transmitted radio messages and relayed television signals and weather pictures, sometimes working in conjunction with Oscar 6. Solar cells provided long-term power, and the satellite operated until it was turned off in 1981. It was followed by AMSAT's third satellite, **Oscar 8 (March 5, 1978)**.

- *Intasat 1*   Spain's first satellite, Intasat 1 was part of a joint Spanish-American project whereby the United States provided the rocket and launch services and Spain provided the satellite. Intasat 1 carried a beacon transmitting at 40 MHz for propagation and ionospheric research. See **ATS 6 (May 30, 1974)** for more details.

November 21, 1974
Molniya 3-01
U.S.S.R.

Essentially the same as the Molniya 2 satellites (see **Molniya 2-01–November 24, 1971**), Molniya 3 satellites—first launched with Molniya 3-01—are still in use in the late 1990s, handling civilian telephone and television within the Soviet Union and leaving the government and military communications to the Molniya 1 satellites (see **Molniya 1-01–April 23, 1965**). The Molniya 3 constellation was first maintained as a four-satellite cluster, then grew to eight satellites in the 1980s. Through December 1999, almost 50 Molniya 3 satellites have been launched.

November 23, 1974
Skynet 2
U.K. / U.S.A.

Skynet 2 replaced **Skynet 1 (November 22, 1969)**, which had provided ground and ocean global communications for the British military, working in conjunction with **DSCS II 1/DSCS II 2 (November 3, 1971)** and two NATO satellites. Developed in Britain with U.S. assistance, it was placed in a geosynchronous orbit over the Indian Ocean. In 1977 its onboard command system failed, making it impossible to maintain the satellite's position above a specific longitude. It was still used through 1987, however, as it drifted in geosynchronous orbit above the Eastern Hemisphere. It was replaced by the Skynet 4 satellite cluster (see **Skynet 4B–December 11, 1988**).

December 2, 1974
Soyuz 16
U.S.S.R.

Soyuz 16 was a six-day dress rehearsal of the Soviet-American joint mission, **Soyuz 19–Apollo 18 (July 15, 1975)**, lacking only the American half of the flight. The crew was Anatoli Filipchenko (**Soyuz 7–October 12, 1969**) and Nikolai Rukavishnikov (**Soyuz 10–April 23, 1971; Soyuz 33–April 10, 1979**).

Soyuz 16 simulated, as much as possible, the conditions of the Soviet-American joint flight. Solar panels were reinstalled to provide power. The Soviet half of the androgynous docking system, built by the United States, was attached to the docking port, and to this was docked an additional ring to simulate a docked Apollo spacecraft. After extensive soft and hard docking tests, on the flight's fifth day, this ring was successfully blown off with explosive bolts, simulating the emergency procedure that would be used if separation failed during the joint flight.

The atmosphere of the Soyuz capsule was also changed to more closely match that of an Apollo spacecraft. For most of the flight, the oxygen-nitrogen mix was changed from 25-75 percent to 40-60 percent, while the atmospheric pressure was reduced from normal atmospheric pressure of 14.6 pounds per square inch to 10.4. Then, during a simulated docking, the cosmonauts donned their spacesuits and sealed the orbital module, as would be done during the joint flight as the Apollo spacecraft rendezvoused and docked with Soyuz.

The cosmonauts also conducted plant research. They observed the growth of branching fungi to see if the 90-minute day-night cycle in orbit affected their growth. They attempted to grow several plant shoots, with mixed results. They tested the use of two different bacteria to regenerate oxygen and produce biomass. The waste products of one bacterium were used by the other for food, which increased its mass by 35 times during the flight.

Continuous polarized photographs of the earth's horizon were also taken for three orbits to study the atmosphere's composition.

Following **Soyuz 19–Apollo 18 (July 15, 1975)**, the Soviets flew their back-up Soyuz capsule on its own solo research flight; see **Soyuz 22 (September 15, 1976)**.

December 4, 1974
Pioneer 11, Jupiter Flyby
U.S.A.

Built by NASA, Pioneer 11 was the second human spacecraft to fly out beyond the orbit of Mars, and the second to reach Jupiter, following **Pioneer 10, Jupiter Flyby (December 3, 1973)**. On December 2nd it passed within 26,725 miles of Jupiter's cloud tops.

Most of Pioneer 11's data confirmed what Pioneer 10 had found. Pioneer 11's photographs, however, revealed the northern polar regions of Jupiter for the first time, showing that the horizontal zones and belts of the gas giant's equatorial regions disappeared at higher latitudes. Instead, the weather patterns became chaotic, a speckled atmosphere of numerous swirls and cells *(see color section, Figure 6)*.

Pioneer 11 also confirmed that the Red Spot of Jupiter was a three-century-old storm, not a flow pattern over a surface feature. Not only was the smaller second red spot seen by Pioneer 10 in the northern hemisphere now gone, but the famous Red Spot in the southern hemisphere showed that it had rotated counterclockwise in the ensuing year *(see color section, Figure 7)*.

Pioneer 11 also took more pictures of the Galilean moons of Io, Europa, Ganymede, and Callisto. As with Pioneer 10, the images were too distant to show much detail, though they did reveal that Io was orange with dark reddish markings near one pole.

Like Pioneer 10, during Pioneer 11's traverse of the Jovian system it crossed the bow shock to Jupiter's magnetosphere three times. These multiple crossings took place because of the push and pull between the solar wind and the magnetosphere.

Having sped past Jupiter, Pioneer 11 headed out toward Saturn. It reached the ringed planet five years later (see **Pioneer 11, Saturn Flyby–September 1, 1979**). Beyond Saturn it would continue out ahead of the Sun, sending back solar wind data for another 16 years, until its battery power dropped so low that it could no longer operate its instruments (see **Pioneer 11, Shutdown–September 30, 1995**).

The next spacecraft to reach Jupiter was **Voyager 1 (March 5, 1979)**.

December 10, 1974
Helios 1
U.S.A. / West Germany

This joint American-German project placed two solar research satellites (see **Helios 2–January 15, 1976**) in orbit around the Sun at distances closer than previously achieved, inside the orbit of Mercury. German and American scientists built the satellite and the United States provided the rocket and launch facilities.

Helios 1 operated for 12 years, getting as close to the Sun as 28.58 million miles during its 22 orbits. To withstand the heat and radiation at this distance, the spacecraft was built to tolerate temperatures exceeding 930°F.

Helios 2 (January 15, 1976) approached several million miles closer to the Sun. See that satellite for a summary of results.

December 18, 1974
Symphonie 1
France / West Germany / U.S.A.

Built as a joint project of France and West Germany and launched by NASA, Symphonie 1 was the first half of a two-satellite communications cluster placed in geosynchronous orbit above the Atlantic, the second satellite reaching orbit in August 1975. Because the two satellites broadcast in interleaved frequencies, they could be placed very close together and essentially function as a single satellite. One satellite relayed signals from Europe to North America while the other relayed from North America to Europe. Each satellite had a capacity for 600 one-way voice circuits or one color television signal.

Both satellites remained in operation until the mid-1980s, when they were replaced by the Telecom 1 satellites (see **Telecom 1A–August 4, 1984**). During their service life, both India and China used them to perform communications experiments.

December 24, 1974
Cosmos 699
U.S.S.R.

Cosmos 699 was the first test of the Soviets' electronic ocean reconnaissance satellite (EORSAT) series. These satellites were designed to detect communications and radar from ocean-going vessels. EORSATs were passive, merely listening for other signals, and they worked in conjunction with the active RORSAT electronic reconnaissance satellites (see **Cosmos 198–December 27, 1967**), which used powerful radar signals to track and identify unknown oceangoing vessels. If a ship tried to jam the RORSAT radar signal, the EORSAT would then hear it.

The EORSAT series was also the first to use an ion engine to maintain the spacecraft's position and orbit, similar to the engine first tested on **SERT 2 (February 4, 1970)**. This engine made possible extremely accurate positioning, which was necessary to precisely pinpoint the location of moving ships.

Over 44 EORSATs were launched through December 1997, with two operational satellites generally in orbit at the same time. Often they were launched jointly with the RORSAT satellites.

December 26, 1974
Salyut 4
U.S.S.R.

Like the previous Soviet space laboratories, Salyut 4 weighed over 20 tons. It was similar in design to **Salyut 1 (April 19, 1971)**, and unlike the previous military station, **Salyut 3 (June 25, 1974)**, it was dedicated entirely to civilian research. It carried an increased number of more sophisticated instruments for doing space research, including a teletype machine so that ground controllers could relay information and instructions without the need for verbal communication.

The astronomical instruments included a solar telescope for taking simultaneous photographs and spectrograms of the Sun, two x-ray telescopes (covering 1–60 angstroms), an infrared spectrometer, an interferometer, an ion-mass spectrometer, an ionic angular position indicator, a photometer, a luminescence meter, and multispectral and optical cameras.

Medical instruments included a blood analyzer, a bone-tissue density monitor, a pulmonary ventilation recorder, brain blood-level monitor, and an electrical muscle stimulator.

Biological experiments included a small facility for studying cellular, microbe, and plant growth under weightless conditions.

Also included was an experiment to investigate the behavior of liquids in zero gravity.

During its first year in orbit, two Soyuz spacecraft carrying two-man crews docked with the station:

Soyuz 17–January 11, 1975 to February 9, 1975 (30 days)
Soyuz 18B–May 24, 1975 to July 26, 1975 (63 days)

In addition, **Soyuz 20 (November 17, 1975)** docked unmanned for 90 days with the space station, carrying a payload of biological specimens.

Salyut 4 was deorbited on February 2, 1977, after spending a year in a parking orbit. The next space laboratory was **Salyut 5 (June 22, 1976)**.

1975

January 11, 1975
Soyuz 17
U.S.S.R.

This flight was the first to the **Salyut 4 (December 26, 1974)** space laboratory. It was manned by Alexei Gubarev (**Soyuz 28–March 2, 1978**) and Georgi Grechko (**Soyuz 26–December 10, 1977; Soyuz T14–September 18, 1985**). Soyuz 17's mission lasted 30 days, making it the third longest flight in history—after **Skylab 3–July 28, 1973** (59 days) and **Skylab 4–November 16, 1973** (84 days)—and breaking the previous Soviet record of 24 days, set by **Soyuz 11 (June 6, 1971)**.

On two occasions, the cosmonauts used Salyut 4's solar telescope to observe the Sun, obtaining spectrograms and optical photographs of sunspots and faculae. They also repaired the telescope's orientation system and recoated the telescope's mirror with aluminum, finding that it had become contaminated during the station's first few weeks in orbit.

The x-ray telescope was used to observe the Crab Nebula as well as the black hole candidate Vela X-1, measuring the x-ray flux for each on three different days.

On February 5th, the cosmonauts used the infrared spectrometer to measure water vapor and ozone in the earth's upper atmosphere. They did this by taking spectrograms of the Sun's light at sunrise and sunset as it cut through the atmosphere. The absorption lines indicated the atmosphere's content. Because their data covered such a short period, no long-term conclusions about any atmospheric changes were possible. However, the experiment proved the validity of the equipment's design for future missions.

The cosmonauts also tested a temporary water regeneration unit, designed to capture, treat, and reuse the water vapor inside the laboratory. This recycling was so successful that the unit was permanently installed in future Soviet space stations.

Among the biological experiments, the cosmonauts grew plants, bacteria, fruit flies, and frog embryos. Of the plant experiments, the cosmonauts were successful in getting peas to sprout within three weeks.

Medical research continued. On this flight, the cosmonauts dedicated 2.5 hours per day to physical exercise. Salyut 4 boasted both a bicycle machine and treadmill. The men also periodically wore the "chibis" suit first used on Soyuz 11, which forced blood into the body's lower half in order to simulate the earth environment. Two other suit designs for forcing the men to use their muscles were also tested. Samples of blood and exhale air were repeatedly collected, and the cosmonauts' bone density was measured regularly. The data confirmed earlier results from **Skylab 2 (May 25, 1973)**, Skylab 3, and Skylab 4.

Both Gubarev and Grechko took about a week to adapt to the weightless environment. Upon their return, they fared much better than previous Soviet cosmonauts after long space flights; their experience was more in line with that of the American astronauts on the Skylab missions. Both men had lost a few pounds, but they recovered quickly and had no difficulty walking immediately after landing.

The next Soviet manned flight to Salyut 4 was the aborted **Soyuz 18A (April 5, 1975)**. The next successful mission was **Soyuz 18B (May 24, 1975)**.

February 6, 1975
Starlette
France

This geodetic satellite was spherical and covered with a skin of 60 laser reflectors. Its core of uranium 238 gave it a high mass so that it had a long orbital life. Using ground lasers, scientists used the satellite to study how its orbit was changed by subtle variations in the earth's gravitational field. This research continued the work of **Anna 1B (October 31, 1962)**, the GEOS satellites (see **GEOS 1–November 6, 1965**); **Diademe 1 (February 8, 1967)**; **Diademe 2 (February 15, 1967)**.

Starlette showed that the earth's crust actually changed shape in response to tides caused by both the Sun and the Moon.

The next geodetic satellite was **Lageos 1 (May 4, 1976)**.

February 24, 1975
Taiyo (SRATS)
Japan

Taiyo (also called SRATS for Solar Radiation And Thermospheric Satellite) was Japan's fourth successful satellite launch, the previous being **Tansei 2 (February 16, 1974)**. The satellite studied the structure and composition of the earth's upper atmosphere and the Sun's x-ray output. Its data augmented the Solrad (see **Solrad 1–June 22, 1960**), Imp (see **Imp A–November 27, 1963**), OGO (see **OGO 1–September 5, 1964**), Intercosmos (see **Intercosmos 1–October 14, 1969**), Aureol (see **Aureol 1–December 27, 1971**), and Prognoz (see **Prognoz 1–April 14, 1972**) satellites.

See **Dynamics Explorer 1/Dynamics Explorer 2 (August 3, 1981)** for results. **Aura (September 27, 1975)** continued ionospheric research.

March 16, 1975
Mariner 10, Mercury Flyby #3
U.S.A.

Built by NASA and launched on November 3, 1973, Mariner 10's primary mission was to take the first close-up photographs of the surface of Mercury (see **Mariner 10–November 3, 1973**). To reach the solar system's innermost planet, however, the spacecraft first flew past Venus on February 5, 1974, in order to obtain a gravity assist (see **Mariner 10, Venus Flyby–February 5, 1974**).

Mariner 10's route through the inner solar system then took it looping past Mercury three separate times, on March 29, 1974; September 21, 1974; and March 16, 1975. During these three flybys, Mariner 10 was able to photograph a little less than half of Mercury's surface.

The most astonishing discovery made by Mariner 10 about Mercury, however, was that the planet had a magnetic field. Scientists had assumed that Mercury would have no such field because, like Venus, Mercury rotated very slowly. Theory had said that a magnetic field could only be generated by the internal dynamo of the molten core of a rotating planet.

Mercury's magnetic field was found to be one-sixtieth of Earth's, far stronger than expected. Somehow generated internally by the planet, it produced a magnetosphere, bow shock, and magnetotail much like Earth's. Because scientists still believe that Mercury's core is no longer molten, they remain puzzled by the cause of this field.

Mercury's surface temperature was measured and was found to range from 370°F in the late afternoon to –300°F at dawn. The measured variations in temperature between light and dark areas also indicated that the soil of Mercury was very

A mosiac of Mercury as photographed by Mariner 10 as it approached the planet on all three flybys. The detail (see below) shows a close-up of the weird terrain. *NASA*

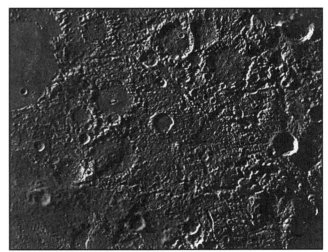

The weird terrain of Mercury. *NASA*

similar to the Moon's, a porous dust regolith battered by eons of impact.

The half of Mercury that Mariner 10 photographed appeared very Moon-like, its surface coated with innumerable

March 16, 1975

craters and a few small regions similar to the maria areas on the Moon. Like Venus, however, Mercury was found to be very spherical, with less surface irregularity than the earth.

Mariner 10 also photographed several new and unique geological features, including a jumbled region scientists dubbed "weird terrain." Located directly opposite from the largest impact site on Mercury (Caloris Basin), this weird terrain looked like a complex pattern of broken mounds and cracks. Scientists theorized that the seismic shock from the Caloris Basin impact had traveled through the planet, shaking apart the surface on the planet's opposite hemisphere.

Also photographed by Mariner 10 were many cliff faces or scarps, some as high as 2 miles and as long as 300 miles. Based upon their appearance, geologists theorized that these scarps were caused by the shrinking of Mercury's outer crust as its interior core solidified and cooled.

By counting craters on Mercury's photographed surface, scientists found that the impact rate appeared very similar to that seen on both the Moon (see **Surveyor 7–January 7, 1968**; **Apollo 17–December 7, 1972**) and Mars (see **Mariner 9, Mars Orbit–November 14, 1971**). These established impact rates on several widely scattered planets enabled scientists to outline roughly the early history of the inner solar system. From 4.5 to 4 billion years ago, the planets came together, during which time their interiors were heated and differentiated into

A mosiac of Mercury as photographed by Mariner 10 as it departed the planet on all three flybys. The detail (next column) shows a close-up of the Caloris Basin. *NASA*

The concentric rings of the Caloris Basin can be seen radiating out from the terminator on the left. *NASA*

heavy cores and lighter mantles. From 4 to 3.3 billion years ago was a period of frequent and continuous bombardment, creating the cratered surfaces of Mercury, the Moon, and Mars. During this period, the inner solar system was cleared of many minor planetary objects, leaving the four major planets and a scattered collection of asteroids. During the latter part of this epoch, the Moon experienced a period of volcanism, in which large areas were flooded with lava, obscuring older craters. Both Earth and Mars later experienced tectonic processes, which on Earth almost completely obscured its ancient craters. On Mars this period produced the planet's giant volcanoes and deep rift valleys.

As of December 1999, Mariner 10 is the only spacecraft to have visited Mercury. The next probe, dubbed Messenger, is a NASA mission scheduled for launch in 2004 that—after two flybys of Venus and two flybys of Mercury—will enter Mercury orbit in 2009.

April 5, 1975
Soyuz 18A
U.S.S.R.

Soyuz 18A, intended for a 60-day mission on **Salyut 4 (December 26, 1974)**, carried Oleg Makarov (**Soyuz 12–September 27, 1973**; **Soyuz 27–January 10, 1978**; **Soyuz T3–**

November 27, 1980) and Vasili Lazarev (**Soyuz 12–September 27, 1973**). It was also the first manned launch failure in which the abort escape systems activated and the passengers survived.

During liftoff, the third stage completed firing, cut off, but then failed to separate completely. Then the second stage fired automatically, sending the spacecraft out of control. The ground control activated the abort sequence, which first separated Soyuz 18A from the rocket, then separated the descent module and released its parachutes. The cosmonauts landed 119 miles downrange after only 21 minutes of flight, unintentionally completing the world's longest manned suborbital flight, exceeding **Liberty Bell 7 (July 21, 1961)** by 6 minutes. Though both men survived, Lazarev received internal injuries and never flew again.

The next flight to Salyut 4 was **Soyuz 18B (May 24, 1975)**. The next use of a rocket's abort escape systems took place on **Soyuz T10A (September 27, 1983)**

April 9, 1975
Geos 3
U.S.A.

The Geodynamics Experimental Ocean Satellite (Geos) 3 was a continuation of NASA's geodetic program, following **Explorer 29 (November 6, 1965)** and **Geos 2 (January 11, 1968)**. Geos 3's goal was to map the surface of the ocean more accurately, a direct follow-up to Geos 2. Geos 3 carried more precise instrumentation and was able to detect changes in elevation to between 3 and 15 feet. Not only did it detect depressions and rises in the ocean's surface, it was used to map wave heights and tidal patterns. For example, the spacecraft was able to precisely track the changing location of the Gulf Stream, noting that the stream's eastern edge was actually three feet higher than the rest of the ocean. Such sea surface data, including wave height and wind speed, were later made available to researchers on an almost real-time basis, thereby providing meteorologists more information for making weather predictions.

Geos 3 also mapped the features and elevations of the ice sheet of Greenland, finding evidence of large-scale waves moving across its surface, apparently correlated with underlying bedrock features. This discovery indicated that better radar data from later satellites like **Seasat (June 27, 1978)** and **Radarsat (November 4, 1995)** would be able to map the hidden topography of these icebound land areas.

Moreover, Geos 3 performed communication tests with **ATS 6 (May 30, 1974)**, confirming the feasibility of the ATS 6-to-low-orbit-satellite communications necessary for the joint flight of **Apollo 18-Soyuz 19 (July 15, 1975)**.

April 19, 1975
Aryabhata
India / U.S.S.R.

Designed and built in India, Aryabhata (named after a sixth-century Indian astronomer) was the first Indian satellite. It was launched by a Soviet rocket in the Soviet Union and carried three experiments, an x-ray detector, a gamma-ray sensor, and electron and ultraviolet sensor for measuring radiation in the ionosphere. The gamma-ray detector, intended to measure solar emissions, functioned for a little less than one day. The satellite's other data indicated that intergalactic cosmic rays also contributed to the Van Allen belt radiation.

The next Indian satellite, **Bhaskara 1 (June 7, 1979)**, studied India's Earth resources.

May 7, 1975
Explorer 53 (SAS 3)
U.S.A. / Italy

The third and last in NASA's small astronomical satellite (SAS) program, Explorer 53 was launched by Italy from its San Marco launch platform off the coast of Kenya (see **Uhuru–December 12, 1970**). Explorer 53 carried x-ray and ultraviolet telescopes for doing astronomical observations, joining **Copernicus (August 21, 1972)**, **ANS 1 (August 30, 1974)**, and **Ariel 5 (August 30, 1974)** in the sky's first network of x-ray telescope satellites.

Explorer 53 could detect variations in x-ray output as short as 0.1 millisecond, a thousand times better than Uhuru. This capability allowed it to measure the fast changes and sudden bursts of the newly discovered x-ray bursters and x-ray transients (objects that emit a steady output of x-ray emissions for a period of time, and then fade away). The spacecraft's x-ray instruments could detect radiation across a greater spectral range, from 0.1 KeV to 50 KeV, allowing it to measure structural details within individual x-ray sources.

During its four years of operation, Explorer 53 provided further confirmation of the existence of x-ray bursters, objects that periodically release strong bursts of x-ray energy, usually reaching a peak in one second and decaying to zero over about 10 seconds. It showed that the bursters were located along the galactic plane, proving that they resided within the Milky Way, with many centered close to the center of the galaxy.

The spacecraft's pointing accuracy also made possible the identification of many optical counterparts to the x-ray sources, proving that this radiation was being produced by small stellar-sized objects.

OSO 8 (June 21, 1975) joined this network of orbiting x-ray telescopes, followed by the launch of the second-generation x-ray telescope, **HEAO 1 (August 12, 1977)**.

May 17, 1975
Pollux / Castor
France

Pollux tested the use of small attitude hydrazine thrusters, while Castor was designed to measure changes to the density of the atmosphere over a five-year period during a time of increasing solar activity.

The next satellite to study atmospheric density was **Explorer 55 (November 20, 1975)**.

May 24, 1975
Soyuz 18B
U.S.S.R.

Manned by Pyotr Klimuk (**Soyuz 13–December 18, 1973**; **Soyuz 30–June 27, 1978**) and Vitali Sevastyanov (**Soyuz 9–June 1, 1970**), this 63-day flight set a new Soviet space endurance record, second only to that of the American **Skylab 4 (November 16, 1973)**. It reactivated **Salyut 4 (December**

May 24, 1975

26, 1974), which had been powered down after the **Soyuz 17 (January 11, 1975)** cosmonauts left it on February 9, 1975.

Communications experiments similar to those conducted on **Gemini 7 (December 4, 1965)** were performed on Soyuz 18B: a ground-based laser beam was bounced off a reflector on the outside of Salyut 4 to track precisely the laboratory's position.

The cosmonauts also attempted to grow a garden of onions and peas, and in July, Sevastyanov celebrated his 40th birthday by eating a meal that included spring onions grown in this greenhouse. Several self-regulated experiments that had been started by **Soyuz 17 (January 11, 1975)** were reactivated, including beetle and fruit fly studies. The fruit flies bred through several generations, each generation showing increasing evidence of developmental changes.

At one point during the flight, the cosmonauts performed 10 straight days of Earth observations, taking 2,000 photographs with the laboratory's optical and multispectral cameras. Salyut 4's spectrometers measured pollution levels in the atmosphere over the Soviet Union and detected ore deposits and crop levels. These observations covered about 3.3 million square miles of the U.S.S.R.

Aurora images were made using the spectrograph and cameras. The inner Van Allen belt was studied. The atmosphere's density at Salyut 4's altitude was measured to see how it changed during the then-quiet part of the Sun's solar cycle. Oxygen was measured at altitudes between 150 to 170 miles. Temperatures between 800°F and 3,100°F were recorded in the upper layers of atmosphere. Spectral studies of the earth's atmosphere repeated the measurements of the ozone, water vapor, and nitrogen-oxide levels first made during Soyuz 17.

Salyut 4's solar telescope made over 600 spectrograms and photographs of the then-quiet Sun, including observations of a solar disturbance detected on June 18th. The data showed a direct link between loops of solar material seen at the Sun's limb, and the Sun's magnetic field lines. These loops were also seen to be directly connected to sunspots and faculae, indicating that the surface features were caused as the field lines cut through the Sun's surface. See **OSO 8 (June 21, 1975)**.

The x-ray telescope was used to study a number of known x-ray sources, including black hole candidates in Scorpio, Virgo, Cygnus, and Lyra. The x-ray flux of Scorpius X-1 was found to be several thousand times greater than the total output of the Sun.

The cosmonauts also continued the medical studies performed on Soyuz 17. They were subjected to regular blood and respiratory tests, as well as electrocardiograms and kinetocardiograms. They exercised a minimum of two hours a day, using the treadmill and bicycle machine. In the last 10 days of the flight this workout was increased significantly. Each used the "penguin" suit for exercising their muscles and the "chibis" suits for increasing the blood pressure in the lower parts of their bodies. Upon return to Earth on July 26th, both men walked from their capsule, refusing to be carried. They had only lost a few pounds, and within two days, both had regained this weight and were apparently back to normal, playing tennis, running, and swimming.

Soyuz 18B was the last manned mission to Salyut 4, though **Soyuz 20 (November 17, 1975)**, containing a number of biological specimens, would remain docked with the space laboratory for 90 days. The next Soviet space laboratory was **Salyut 5 (June 22, 1976)**. The next manned mission was **Soyuz 19 (July 15, 1975)**.

June 21, 1975
OSO 8
U.S.A.
Built by NASA, OSO 8 was the eighth and last Orbiting Solar Observatory, following **OSO 7 (September 29, 1971)**. These first-generation solar observatories operated in low-Earth orbit, observing solar ultraviolet radiation from 150 to 4,000 angstroms and soft x-rays from 0.13 to 35 KeV. It also carried several astronomical experiments studying the sky from 2 to 5,000 KeV to join an already orbiting network of x-ray telescopes (see **Explorer 53–May 7, 1975**).

OSO 8 studied the Sun during the quiet period of its solar cycle. The birth and early development of one flare was studied in detail. The increase of the x-radiation during this event showed a steady growth superimposed over many large bursts and detectable quickly after the appearance of the flare. The data suggested that the emissions were associated with the emergence of flux tubes of material, following the loops of the Sun's magnetic field lines up and then down from its surface.

X-ray astronomy indicated that some supernova remnants could be the source of some of the sky's x-ray emissions. A half dozen of the most likely black hole candidates were studied, including Cygnus X-1, Hercules X-1, and Scorpius X-1. The increased database for these objects further confirmed that each were almost certainly binary systems with an accretion disk surrounding an invisible but massive object.

The next satellite to study the Sun was **Aura (September 27, 1975)**. The next x-ray telescope was **HEAO 1 (August 12, 1977)**.

July 11, 1975
Meteor 2-01
U.S.S.R.
After two years of testing, this second-generation weather satellite design replaced the Meteor 1 satellite cluster (see **Meteor 1-01–March 26, 1969**), beginning in 1977. Meteor 2 satellites were placed in a slightly higher orbit than Meteor 1 satellites, 550 miles versus 350 miles. They carried scanning cameras for photographing the earth's surface twice a day in both visible and infrared light, documenting the daily cloud, snow, and ice cover. Visible light scans covered swathes about 1,500 miles wide with a resolution of less than two-thirds of a mile. The spacecraft also measured both cloud-top and water temperatures twice a day, as well as the radiation hitting the earth from space.

Like their predecessors, two or three Meteor 2 satellites were necessary to maintain an operational cluster covering the entire Soviet Union. Through 1993, the Soviet Union/Russia launched 21 Meteor 2 satellites.

Beginning in 1988, the Soviet Union began replacing its Meteor 2 satellites with a third-generation weather satellite. See **Meteor 3-01 (October 24, 1985)**.

July 15, 1975, 10:30 GMT
Soyuz 19 (ASTP)
U.S.S.R.
Soyuz 19 was the Soviet half of the Apollo-Soyuz Test Project (ASTP), the first joint space flight of the United States and the Soviet Union, intended to demonstrate a new desire for cooperation between the two countries. This six-day flight included the first docking between spacecraft from two different countries. The American half of this joint flight was **Apollo 18 (July 15, 1975)**. Soyuz 19 was crewed by Alexei Leonov (**Voskhod 2–March 18, 1965**) and Valery Kubasov (**Soyuz 6–October 11, 1969**; **Soyuz 36–May 26, 1980**).

To make the docking possible, the Soviet Union agreed to make a number of engineering changes to its Soyuz spacecraft, while the U.S. agreed to build a docking module with a compatible docking system. The Soyuz spacecraft's interior atmosphere was changed from an oxygen-nitrogen mix of 25-75 percent to 40-60 percent, with a reduction in air pressure from 14.6 to 10.4 pounds per square inch. Furthermore, to accommodate the American space program's safety rules following the fire on **Apollo 1 (January 27, 1967)**, the Soviets fireproofed the interior of Soyuz 19. Finally, the American-built androgynous docking port replaced the standard Soyuz docking port.

Because the United States required that everything about this flight be made public, the launch of Soyuz 19 was the first Soviet liftoff broadcast live to the world. The first docking of the two spacecraft took place on July 17th, followed by a second docking on July 19th. In between, there were several crew exchanges with gift giving, speeches, and public broadcasts to the world. See Apollo 18 for details.

The next manned space flight was **Soyuz 21 (July 6, 1976)** to **Salyut 5 (June 22, 1976)**.

July 15, 1975, 18:00 GMT
Apollo 18 (ASTP)
U.S.A.
Manned by Tom Stafford (**Gemini 6–December 15, 1965**; **Gemini 9–June 3, 1966**; **Apollo 10–May 18, 1969**), Vance Brand (**STS 5–November 11, 1982**; **STS 41B–February 3, 1984**; **STS 35–December 2, 1990**), and Deke Slayton, this nine-day mission was a joint flight with **Soyuz 19 (July 15, 1975)**. The two craft performed the first docking between American and Soviet spacecraft.

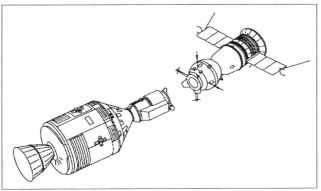

Apollo 18 with Docking Module (*left*) docking with Soyuz 19 (*right*). *NASA*

To make the docking possible, the Soviet Union had agreed to make several engineering changes to the Soyuz spacecraft (see Soyuz 19 for details). In return, the United States agreed to build a special airlock and docking module. While the Apollo spacecraft used a pure oxygen atmosphere pressurized at 5 pounds per square inch, the Soyuz spacecraft used a mixed atmosphere of 25-75 percent oxygen-nitrogen, pressurized at 14.6 pounds per square inch, or normal sea level air pressure. Though Soyuz 19's air pressure was reduced to 10.4 pounds per square inch with a 40-60 oxygen-nitrogen mix, an airlock was still needed to separate these two different atmospheres.

The American built docking module provided this airlock, being a pressure vessel with an internal cabin about 10 feet long and 5 feet wide. It had a standard Apollo docking port on one end, and a specially built androgynous docking port on its other end. Both the U.S. and Soviet standard docking systems used incompatible designs of the same concept: a male probe on the active spacecraft searched out and locked with a female drogue on the passive craft. The androgynous design allowed either spacecraft to be the active port, so that any craft with this unit could dock with any other craft. While the agreement for developing this unit said it would be incorporated as standard equipment on future missions, this flight was its last use. A modified version was later used on the **Kristall (May 31, 1990)** module of **Mir (February 20, 1986)**.

The docking module was carried into orbit much as the lunar module had been, attached to the top of the Saturn 1B's last rocket stage and just below the Apollo spacecraft. After reaching orbit, the Apollo capsule separated from this configuration, turned around, docked with the module, and extracted it from the rocket stage.

The two spacecraft docked twice, testing the androgynous docking system once with Apollo as the active unit and once with Soyuz. Because the Apollo spacecraft's fuel capacity was larger and its control systems more sophisticated, it made all major orbital maneuvers for both dockings. In total, the two craft remained docked for almost two days, exchanging visits and gifts and airing several broadcasts worldwide. Though science was performed on this flight, its central purpose was political. The joint flight demonstrated the policies of *détente* and peaceful coexistence advocated at this time by the leaders of both countries.

One experiment took place between the two dockings. The Apollo spacecraft placed itself between the Sun and Soyuz 19, creating an artificial eclipse and allowing the cosmonauts to photograph the Sun's corona to within 25 solar radii of the Sun's surface. This test proved that by blocking out the Sun artificially, coronal research was possible.

Because the two-man crew of **Soyuz 18B (May 24, 1975)** was also occupying **Salyut 4 (December 26, 1974)** at this time, the joint flight placed seven humans in space, matching the record set during the triple mission of Soyuz 6, 7, and 8 (see **Soyuz 8–October 13, 1969**). This total was not matched again until **STS 4 (June 27, 1982)** joined **Soyuz T5 (May 13, 1982)** and **Soyuz T6 (June 24, 1982)** in orbit.

Apollo 18 was the last use of the Apollo equipment originally designed to send Americans to and from the Moon. It was also the last American manned mission for six years, until the first test flight of the space shuttle Columbia on **STS 1**

(April 12, 1981). Because of this anticipated gap in American manned missions, Apollo 18 conducted a number of additional experiments while in orbit.

The flight's most significant discovery involved research into the process of electrophoresis (see **Apollo 14–January 31, 1971**). Using electricity, the astronauts attempted to isolate different kidney cells, each with a unique ability to produce useful chemicals—such as urokinase (which dissolves blood clots and can also treat breast cancer) and erythropoietin (which is used to treat anemia). The results suggested that in-space electrophoresis could produce these difficult-to-harvest drugs. Further electrophoresis experiments were next performed on **STS 4 (June 27, 1982)**, followed by more sophisticated attempts to produce erythropoietin on **STS 61B (November 26, 1985)** and urokinase on **STS 56 (April 8, 1993)**.

August 9, 1975
COS B
Europe / U.S.A.

Cosmic B (COS B) was the first operational satellite built by the European Space Agency, formed from the European Space Research Organization (ESRO). It was launched by NASA into an eccentric orbit in order to map the gamma radiation of the sky. Its sensors could detect both x-ray and gamma rays, from 20 MeV to 5 GeV. This was a significant improvement over ESRO's previous satellite, **Thor Delta 1A (March 12, 1972)**, as well as the American **Explorer 48 (November 15, 1972)**, both of which observed x-rays from 30 to 300 MeV.

During its six-year life span, COS B discovered a number of new gamma-ray sources, distinct from the gamma-ray bursts first found by **Vela 5A/Vela 5B (May 23, 1969)**, though just as mysterious in their origin. Because they were clustered along the galactic plane, they were at least known to reside within the Milky Way, and approximately half were linked to x-ray source objects.

The center of the Milky Way itself was in general the greatest source of gamma rays. COS B also found that two other extragalactic objects, the quasar 3C-273 and the Seyfert galaxy NGC4151, both emitted gamma rays. Whether this radiation was caused by starlike point sources (such as black holes or pulsars) or was merely the general background radiation of these galactic nuclei was as yet unknown.

The next satellite with a gamma-ray burst detector was **Helios 2 (January 15, 1976)**. The next satellite dedicated to gamma-ray astronomy was **HEAO 3 (September 20, 1979)**.

September 9, 1975
Kiku 1 (JETS 1)
Japan

Kiku 1 was also called JETS (Japanese Engineering Test Satellite) 1. It tested the Japanese N1 launch rocket and Japanese satellite ground tracking equipment. Also tested were the satellite's extendible antennas.

This flight was the first of six Japanese Engineering Test Satellites. See **Kiku 2 (February 23, 1977)**.

September 26, 1975
Intelsat 4A F1
International / U.S.A.

The increased demand for international telephone and television service across the Atlantic Ocean forced Intelsat (see **Early Bird–April 6, 1965**) to develop the Intelsat 4A satellites, an upgrade of its Intelsat 4 satellites (see **Intelsat 4 F2–January 26, 1971**) launched to increase the capacity of the Intelsat 4 cluster.

Through March 1978, Intelsat successfully launched five Intelsat 4A satellites, positioning the first three over the Atlantic Ocean and the last two over the Indian Ocean. All had a service life of almost a decade.

Similar in design to the Intelsat 4 satellites, Intelsat 4A satellites nonetheless had almost double the capacity; each was able to relay between 6,000 and 15,000 two-way voice circuits, depending on the satellite's particular configuration at any given time. With these five satellites added to the Intelsat 4 cluster through the mid 1980s, the telephone capacity across the Atlantic during this period exceeded 30,000 two-way voice circuits at any time. This was a 125-fold increase in the 10 years since Early Bird.

Service demands, however, forced Intelsat to increase its capacity even further. See **Intelsat 5 F2 (December 6, 1980)**.

September 27, 1975
Aura
France

This satellite carried three sensors for monitoring the solar ultraviolet radiation in the spectral range from 174 to 1,315 angstroms. It also measured the absorption of this radiation by the atmosphere and magnetosphere.

Aura operated until December 1976. Its data supplemented observations of the Solrad (see **Solrad 1–June 22, 1960**), OGO (see **OGO 1–September 5, 1964**), Intercosmos (see **Intercosmos 1–October 14, 1969**), Aureol (see **Aureol 1–December 27, 1971**), and Prognoz (see **Prognoz 1–April 14, 1972**) satellites, as well as **OSO 8 (June 21, 1975)**.

See **Dynamics Explorer 1/Dynamics Explorer 2 (August 3, 1981)** for results. **ESA–Geos 1 (April 20, 1977)** continued Aura's magnetospheric research.

October 6, 1975
Explorer 54 (Atmosphere Explorer D)
U.S.A.

The fourth in a five-satellite research project by NASA to study the atmosphere at elevations between 75 and 200 miles, Explorer 54 was similar to the program's previous satellite, **Explorer 51 (December 16, 1973)**. Placed in a near-polar orbit (inclination 90 degrees) with perigee of 100 miles, the satellite's orbit was periodically dipped to 75 miles altitude to make measurements. The polar orbit also allowed the spacecraft to reach regions unattainable by Explorer 51.

Explorer 54 functioned for 3.5 months, failing prematurely when its power supply malfunctioned. Nonetheless, its data confirmed that solar wind particles are able to enter the polar regions of the earth's upper atmosphere directly, as seen by **OGO 4 (July 28, 1967)**. The polar cusps, the region where the earth's protective magnetic field lines transitioned from lying parallel to the earth's surface to pointing in toward the poles, was found to be complex, turbulent, and closely linked to the generation of aurora. See **Dynamics Explorer 1/Dynamics Explorer 2 (August 3, 1981)**.

The next and last satellite in this NASA atmosphere research program was **Explorer 55 (November 20, 1975)**.

October 16, 1975
GOES 1
U.S.A.

Geostationary Operational Environmental Satellite (GOES) 1 followed the SMS satellites (see **SMS 1–May 17, 1974**) that had been built and managed by NASA. GOES 1, though built by NASA, was operated by the National Oceanographic and Atmospheric Administration (NOAA) to provide data for weather forecasts in the United States. Located in geosynchronous orbit, GOES 1 photographed the earth's surface and atmosphere in both visible and infrared light, taking hemispheric images every 30 minutes, 24 hours a day, with a resolution of about two-thirds of a mile in visible light and about five miles in infrared. Because these large-scale photos usually did not require such high resolution, the images were reprocessed to reduce their resolution to about five miles in order to lessen the amount of data transmitted.

GOES 1 was launched as part of an international agreement between the United States, the European Space Agency (ESA), Japan, and the Soviet Union to place in orbit a string of geosynchronous weather satellites providing images of the globe's cloud cover every 30 minutes in the middle and equatorial latitudes. In order to cover the United States and its offshore areas, the U.S. agreed to place two GOES satellites in orbit, one positioned over the western Atlantic and the other over the eastern Pacific. Both the Japanese and European satellites were launched in 1977 (see **Himawari 1–July 14, 1977; Meteosat 1–November 23, 1977**). Because the Soviet satellite for this coordinated international effort, **Elektro 1 (October 31, 1994)**, was not launched until 19 years later, GOES 1 was moved above the Indian Ocean in 1978 to complete the worldwide coverage. In 1982, the Indian communications-weather satellite **Insat 1A (April 10, 1982)** further supplemented the coverage over the Indian Ocean.

Through 1997, a total of 10 GOES satellites were launched to maintain coverage of the United States. The fourth through seventh carried additional equipment for measuring atmospheric temperatures in 12 spectral bands. The eighth, ninth, and tenth are second-generation spacecraft, carrying equipment able to make simultaneous images and temperature soundings in visible and infrared wavebands. The "imager" photographs the earth in visible wavelengths and four infrared wavebands (ranging from 3.80 to 12.5 microns). These different bands show cloud cover both day and night, the heat from volcanoes and fires, water vapor, and the surface temperature of clouds. The "sounder" measures temperatures, water vapor, and ozone levels across 19 different wavelengths in the visible and infrared.

The GOES satellites also carry instruments for monitoring the Sun and the Sun-Earth interaction of the earth's magnetosphere. Solar flares, the earth's magnetic field, and the Sun's x-radiation are routinely measured to anticipate changes in the earth's atmosphere caused by these phenomena.

GOES hemispheric pictures are taken in conjunction with the close-up weather images snapped by the NOAA satellites (see **NOAA 1–December 11, 1970**). While GOES satellites

High-resolution infrared image (before data reduction) showing storm patterns over the southeastern United States and Cuba on 20 March 1976. *NASA*

provide a wide-angle view of Earth weather patterns, NOAA satellites provide the detailed high-resolution images of specific local regions.

GOES images are also used in conjunction with the close-up Earth resources images taken by the Landsat satellites (see **Landsat 1–July 23, 1972; Landsat 4–July 16, 1982**). While GOES images provide the wide-angle view, and NOAA images provide close-ups, Landsat images provide extreme close-ups of the earth. All three together make possible a detailed survey of the earth's geological and atmospheric environment.

As a result of the unified multiple design of the GOES and NOAA satellites, useful weather predictions have been extended from one or two days to almost a week. Furthermore, meteorologists can now accurately track and predict the course of hurricanes, monitor changes in the path of the jet streams, locate forest fires, and map changes in snow cover that predict snow melt.

Beginning with the seventh GOES satellite launched in 1987, each satellite also carries a SARSAT-COSPAS search and rescue receiver-transmitter for quickly locating the distress beacons of ships and aircraft in trouble. See **Cosmos 1383 (June 30, 1982)**.

October 22, 1975
Venera 9, Venus Landing
U.S.S.R.

Four times heavier than previous Soviet Venusian landers (see **Venera 8, Venus Landing–July 22, 1972**), Venera 9 succeeded in transmitting to Earth the first picture of the harsh Venusian surface. Unlike the previous Soviet probes to Venus, Venera 9 (along with its companion, **Venera 10, Venus Landing–October 25, 1975**) included both an orbiter and a lander. The orbiter was used to relay the lander's signal back to Earth.

The spacecraft was launched on June 8, 1975, soft-landing on Venus on October 22nd during daylight hours. For 53 minutes, it beamed back data, including a panoramic photograph showing a rocky surface of sharp flat stones. To the surprise of everyone, the spacecraft's floodlights were not needed to illuminate these pictures. The sunlight at the surface equaled that of a cloudy day on Earth.

The panorama used a wide-angle fish-eye lens that showed a close-up of the ground in the center of the frame, with a long view to the horizon at each upper corner. Analysis of this panorama indicated that Venera 9 had landed on a 15- to 20-degree slope scattered with small rocks. While a few stones appeared to be soft and irregularly shaped breccia, most were more sharply edged, as if they had been broken apart. Several resembled a series of flat slabs cemented together. Most of the stones appeared to lie on top of the ground, though a few appeared partially embedded in the soil. The soil itself seemed to be fine-grained and porous, not unlike that seen on the Moon.

Venera 9 had apparently landed on a talus slope whose surface was being naturally reworked over time as rocks either slid downward or were repositioned by wind.

Three days later, Venera 10 (**Venera 10, Venus Landing–October 25, 1975**) landed and repeated the feat. See that entry for a summary of results from both spacecraft.

October 25, 1975
Venera 10, Venus Landing
U.S.S.R.

Launched on June 14th, Venera 10 soft-landed on Venus about 1,200 miles south of Venera 9 (**Venera 9, Venus Landing–October 22, 1975**). The spacecraft transmitted data for 65 minutes, including the second picture ever taken of the Venusian surface.

The photograph revealed a terrain of flat boulders, generally larger than seen at the Venera 9 landing site. The spacecraft itself apparently landed on a rock less than 10 feet across, with three similar boulders nearby. The horizon line, only visible in the right upper corner of the frame, was estimated to be no more than 1,000 yards away. It was vague and indistinct, with two horizontal lines indicating some form of atmospheric haze.

Venera 10's landing site appeared to be a gently rolling plain, covered with large flat blocks, two of which appeared to fit together, separated only by a crack. A test of the hardness of the boulder on which the spacecraft sat indicated that it was dense and hard.

Both Venera spacecraft carried gamma-ray spectrometers for detecting the make-up of a small piece of Venusian soil below the spacecraft. Both instruments indicated that the surface under them resembled basalt, not granite as had been inferred from Venera 8 measurements. From this data, it appeared that volcanic activity had flooded the Venera 9 and Venera 10 landing sites with a lava not dissimilar from that found on both the earth and Moon.

The two spacecraft also confirmed the atmospheric measurements of the previous Soviet landers (see **Venera 7, Venus Landing–December 15, 1970**; **Venera 8, Venus Landing–July 22, 1972**) as well as the American flyby probes (see

The surface of Venus, as photographed by Venera 9. *NASA*

The surface of Venus, as photographed by Venera 10. *NASA*

Mariner 5, Venus Flyby–October 19, 1967; **Mariner 10, Venus Flyby**–**February 5, 1974**). Temperatures ranged from 850°F to 870°F. Wind speeds were about nine miles per hour. Atmospheric pressure ranged between 82 to 89 atmospheres, equal to between 1,200 and 1,300 pounds per square inch.

In 1978, four spacecraft visited Venus—the long-term radar orbiter **Pioneer Venus Orbiter (December 4, 1978)**; two landers, **Venera 11 (December 25, 1978)** and **Venera 12 (December 21, 1978)**; and **Pioneer Venus Probe (December 9, 1978)**, which dropped five probes into the planet's atmosphere. The radar charts produced by Pioneer Venus Orbiter gave humankind its first planetary map of Earth's so-called sister planet.

November 17, 1975
Soyuz 20
U.S.S.R.

Soyuz 20, an unmanned mission, tested the endurance capabilities of the Soyuz spacecraft in anticipation of Soviet manned missions that would exceed 60 days. See **Salyut 6 (September 29, 1977)** and **Soyuz 29 (June 15, 1978)**. Soyuz 20 contained a biological package of flies, tortoises, vegetable seeds, maize, cacti, and plants. For 90 days, this spacecraft remained docked to **Salyut 4 (December 26, 1974)**. The survival and recovery of its specimens on February 16, 1976, proved the Soyuz spacecraft's durability while also measuring the effect of cosmic radiation on such specimens.

November 20, 1975
Explorer 55 (Atmosphere Explorer E)
U.S.A.

Explorer 55 was the last of a five-satellite NASA project to study the atmosphere between 75 and 200 miles altitude, following **Explorer 54 (October 6, 1975)**. Like the previous satellites in this program, Explorer 55 carried a sophisticated propulsion system which it used to periodically and temporarily lower its orbit to 75 miles altitude. While **Explorer 51 (December 16, 1973)** had studied the middle latitudes, and Explorer 54 had studied polar latitudes, Explorer 55 was placed in an equatorial orbit, its inclination 20 degrees.

Explorer 55 functioned for 5.5 years, gathering data about the changing density of the atmosphere and the ultraviolet radiation of the Sun. It also made the first attempt to find out whether the earth's ozone layer was being damaged by the chlorofluorocarbons (CFCs) used in aerosol spray cans and air conditioning equipment.

The satellite found that the lower ionosphere was much more active than expected. Both this satellite and **Explorer 51 (December 16, 1973)** showed that ultraviolet radiation from the Sun varied significantly with the solar cycle and that its unexpected rapid increase as the Sun became more active in the late 1970s caused the atmosphere to expand more rapidly than expected. As the ultraviolet radiation was absorbed by the atmosphere, it caused an expansion and thus an increase in density at orbital altitudes above 100 miles. This greater density explained why the orbit of **Skylab (May 14, 1973)** experienced an unpredicted rapid decay.

Explorer 55 also found that at the equator, the ozone layer generally remained unchanged throughout the year. There was no indication of a decrease in total ozone. See **Nimbus 7 (October 13, 1978)** for follow-up data.

The next satellite to study atmospheric density was **San Marco D/L (March 25, 1988)**.

November 25, 1975
Bion 3 (Cosmos 782)
U.S.S.R. / U.S.A.

Bion 3 continued the Soviet biological space research program begun with **Bion 1 (October 31, 1973)**. This particular mission was the first joint Soviet-U.S. unmanned space flight, launched under the same cooperative program as the **Apollo 18/Soyuz 19 (July 15, 1975)** manned mission.

The Bion 3 capsule carried 14 biological experiments into space (four of which were American), with specimens including white rats, tortoises, guppies, and plant and fruit fly tissues. The specimens were separated into two identical groups, one kept in a fixed compartment and the other kept in a revolving compartment to create centrifugal force. The mission remained in orbit for 20 days and was then recovered safely.

Bion 3's most interesting experiment attempted to see if guppies could be fertilized in space. Though it was unclear whether weightlessness was the cause, the scientists found that fertilization was hindered.

The next Bion mission was **Bion 4 (August 3, 1977)**.

December 13, 1975
RCA Satcom 1
U.S.A.

Built and operated by Radio Corporation of America (RCA), the Satcom geosynchronous satellites were the second wholly private constellation of American satellites, following Western Union's Westar satellites (see **Westar 1–April 13, 1974**). Satcom satellites provided telephone and television broadcasting capacity to the U.S. domestic market.

RCA's biggest customers, however, were cable television channels like Home Box Office (HBO) and the Learning Channel. Using **ATS 6 (May 30, 1974)**, the Learning Channel had been providing satellite broadcast educational services in the Appalachian region under its previous name, the Appalachian Community Service Network (ACSN). Satcom 1 allowed this cable channel to distribute its programming nationwide—hence the name change. Similarly, pay-television providers like HBO quickly realized that they could use geosynchronous satellites to reach every cable system through the United States. Free television stations like WTBS in Atlanta and WOR in New York also realized that they could earn additional revenue by using satellites to distribute their product outside their local area. The sudden growth of cable in the 1980s is directly attributable to the launch of satellites like Satcom 1. Similarly, radio programs such as National Public Radio and Rush Limbaugh found they, too, could distribute their product nationwide with the use of geosynchronous communications satellites. Eventually, the antennas of over 10,000 radio stations would get product from satellites.

Through January 1982, RCA successfully orbited four Satcoms, each with a capacity of 1,000 one-way voice circuits or 48 television broadcasts. In 1982 it began supplementing

December 22, 1975
Raduga 1
U.S.S.R.

Raduga 1 was the first operational geosynchronous communication satellite launched by the Soviet Union following geosynchronous tests by **Cosmos 637 (March 26, 1974)** and **Molniya 1S (July 30, 1974)**. Raduga means "rainbow" in Russian. Through 1996, 32 Raduga satellites were launched and positioned at a variety of longitudes above the Soviet Union and to its east and west. Three additional satellites using a modified design were launched beginning in 1989.

Though little has been published about Raduga, each satellite's capacity is thought to have included one television circuit and up to 1,000 two-way voice circuits. It is also thought that these satellites were dedicated mostly to military communications.

In 1985, the Soviet Union supplemented this constellation with **Altair 1 (October 25, 1985)**, the first of two geosynchronous satellites for use by its space program. Then, in 1994, the Russian Federation began replacing both the Raduga and Altair satellites with the Luch satellites (see **Luch 1–December 16, 1994**).

1976

January 15, 1976
Helios 2
U.S.A. / West Germany

Helios was part of a two-satellite joint American-German project (see **Helios 1–December 10, 1974**) in which German and American scientists built the experiments, Germany built the satellites, and NASA provided the rocket and launch facilities. Both satellites were placed in solar orbit, approaching the Sun closer than any satellite through December 1999, with Helios 2 getting as close as 26 million miles during its four-year life span.

During their close approaches, both Helios satellites found that the Sun's magnetic field had more structure than seen at Earth distances. The solar wind was less turbulent, more streamlined, and more local in nature. Events detected by one spacecraft were often not seen by the other, even when they were separated by only 30 degrees of solar longitude.

The fast- and slow-moving streams of solar wind, first identified by the solar Pioneer satellites (see **Pioneer 9–November 8, 1968**), were seen to be clearly distinct. The fast wind appeared to come from solar surface features called coronal holes seen in higher latitudes, while the origin of the slow wind was less clear. These observations were later confirmed by **STS 56 (April 8, 1993)**, as well as by the interstellar travelers Pioneer 10, Pioneer 11, Voyager 1, and Voyager 2 (see **Pioneer 11, Shutdown–September 30, 1995**).

In the early years of both spacecraft's journeys, they observed the Sun during the quiet period of its 11-year cycle, finding that the Sun's magnetic field lines flowed from the Sun

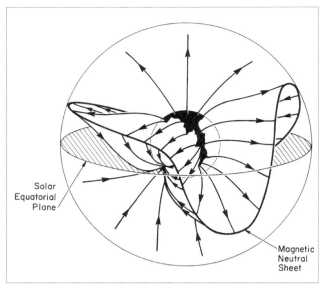

The Sun's magnetic field lines. The black area is a coronal hole. *NASA*

like water from a garden hose, vast spirals unwinding as the solar wind dragged them outward. The magnetic field's oppositely charged poles were separated by a neutral sheet at the equator, pleated like a ballerina's skirt.

As the Sun grew more active, the pleats in the ballerina's skirt became more and more pronounced until the entire field became almost chaotic. Solar shock waves emanated out from the Sun with increasing frequency, disturbing the shape of the entire solar magnetic field. These shock waves explained why, during solar active periods, interstellar cosmic radiation dropped—the shock waves acted to shield the inner solar system (see **Venera 2–November 12, 1965** and **Pioneer 9–November 8, 1968**).

Helios 2 also carried a sensor for detecting gamma-ray bursts, first discovered by scientists processing data from **Vela 5A/Vela 5B (May 23, 1969)**. The sensor's pointing accuracy was the best yet flown, able—when combined with data compiled by Solrad 11A and Solrad 11B—to pinpoint bursts to within an arc second. See **Solrad 11A/Solrad 11B (March 15, 1976)** for results.

January 17, 1976
Hermes (CTS 1)
U.S.A. / Canada / Europe

Hermes, also called Communications Technology Satellite (CTS) 1, was a joint experimental communications project of Canada and NASA. Canada developed the satellite, NASA launched it, and ground stations in Canada performed the experiments. The European Space Agency also built several satellite instruments.

Until it was turned off in 1979, Hermes tested the use of experimental high frequencies for satellite communications from geosynchronous orbit, performing transmission tests at 11 and 12 GHz, and receiver tests at 14 GHz. See **ATS 6 (May 30, 1974)** for more details about these propagation experiments.

Hermes also tested new solar panel designs, high powered transmission equipment, and an advanced three-axis attitude control system.

The next Telesat Canada satellite, **Anik B (December 15, 1978)**, also performed propagation tests. It was also used to augment the Anik A satellite cluster (see **Anik A1–November 10, 1972**).

February 19, 1976
Marisat 1
U.S.A.

Marisat 1 was the first in a three-satellite cluster dedicated to providing communications for U.S. Navy activities. Marisat 1 was also used for tactical military communications, replacing **LES 6 (September 26, 1968)** and **Tacsat 1 (February 9, 1969)**. Placed in geosynchronous orbit above the Atlantic Ocean, Marisat 1 was joined by two other Marisat satellites later in 1976, the second placed over the Pacific and the third above the Indian Ocean.

Though intended for military use, the satellites were also available for commercial use. With time, and the launch of the military FLTSATCOM satellites (see **FLTSATCOM 1–February 9, 1978**), the military use was reduced and commercial activity increased.

In 1982, the three Marisat satellites were transferred to the International Maritime Satellite Organization (Inmarsat). This organization, formed in 1979, is similar to Intelsat (see **Early Bird–April 6, 1965**), comprising a consortium of over 80 nations for establishing a communications satellite network for ships and offshore industries. In order to establish its constellation as quickly as possible, Inmarsat leased space on the three Marisat satellites.

Because the Marisat satellites had a small capacity (less than a few dozen channels for two-way voice communications, or several hundred teletyped messages), Inmarsat also leased space on the three MARECS satellites (see **MARECS A–December 20, 1981**), using the Marisat satellites as in-orbit spares. Inmarsat later launched its own satellite constellation (see **Inmarsat 2–October 30, 1990**).

February 29, 1976
ISS 1
Japan

Ionospheric Sounding Satellite (ISS) 1 studied the propagation of radio noise across four frequencies in the ionosphere to better understand how radio shortwave transmissions were influenced by atmospheric changes. The satellite only functioned for about a month; overheating in its solar panels caused it to fail.

March 15, 1976
LES 8 / LES 9 / Solrad 11A / Solrad 11B
U.S.A.

- *LES 8 and LES 9* These satellites in the Lincoln Experimental Satellite (LES) program were experimental military communications satellites, built and operated by the Massachusetts Institute of Technology's Lincoln Laboratory. They were the last in this program and followed **LES 6–September 26, 1968**. (LES 7 was cancelled before launch.) LES 8 and LES 9, which were practically identical, tested a variety of engineering and narrow-beam frequency tests from geosynchronous orbit, including ground-to-satellite as well as satellite-to-satellite transmissions. They also were the first communications satellites to use nuclear power sources to generate electricity. Initially, one was placed over the Atlantic and the other over California. Later, both satellites were positioned over California. They were operational through 1989.

- *Solrad 11A and Solrad 11B* These twin satellites were the last in a series of satellites built by the U.S. Navy for continuously monitoring the x-ray and ultraviolet output of the Sun. See **Solrad 1 (June 22, 1960)**. Unlike earlier satellites in this series, Solrad 11A and 11B were orbited at the same time in complementary orbits at 61,000 miles altitude. From this perch above the earth's magnetosphere, the two satellites together provided 24-hour full-time coverage of the Sun. They also carried a larger number of more sensitive sensors for measuring ultraviolet, x-ray, and gamma-ray emissions.

Both satellites monitored the Sun's flare and x-ray output as the solar cycle moved from its quiet to its active stage, the second solar cycle observed from space since the launch of Solrad 1.

The satellites' gamma-ray burst detectors, working with **Helios 2 (January 15, 1976)**, were able to pinpoint these bursts to within an arc second. Despite this accuracy, no optical counterpart for any gamma-ray burst was found, and the cause of the bursts, first discovered by **Vela 5A/Vela 5B (May 23, 1969)**, remained a complete mystery. The bursts themselves seemed to occur at a rate of about one per month. No other information about their origin or make-up was obtained, and the accumulating evidence left it unclear whether the events were concentrated along the plane of the Milky Way or scattered across the entire sky. See **HEAO 1 (August 12, 1976)** for more gamma-ray burst research.

Future American solar research continued on **Solar Max (February 14, 1980)**. Solar observations were also conducted by the ISEE satellites (see **ISEE 1/ISEE 2–October 22, 1977**) and several Soviet satellites (see **Intercosmos 1–October 14, 1969**; **Prognoz 1–April 4, 1972**).

April 22, 1976
NATO 3A
NATO / U.S.A.

This satellite was the first of four NATO military communications satellites that replaced NATO 2 (see **Skynet 1–November 22, 1969**). These satellites provided both narrow- and wide-beam communications channels for both ground and ship military communications over Europe and the eastern United States. Of the four satellites, the first two performed hardware tests and were retired. The second two were used either for communications or were reserved as spares. Through 1978, the U.S. military used one of these satellites for global strategic communications, pending the completion of the first DSCS II satellite cluster (see **DSCS II 1/DSCS II 2–November 3, 1971**). In 1991 **NATO 4 (January 8, 1991)** was launched, and the NATO 3 satellites were placed in parking orbits as spares.

April 30, 1976
NOSS 1 (Whitecloud 1)
U.S.A.
Navy Ocean Surveillance Satellite (NOSS) 1 was also called Whitecloud 1. It provided the U.S. Navy surveillance information on the location of both U.S. and Soviet ships. The satellite was actually three subsatellites floating several hundred feet apart and connected together by fine wires, technology tested by **Triad 1 (September 2, 1972)**.

Through 1987, the Navy launched nine NOSS satellites, usually configured in clusters of one main satellite and three subsatellites. In 1990, this system, now called NROSS (for Navy Remote Ocean Sensing System), was upgraded. These satellites, of which four clusters have been launched, carry microwave sensing equipment for obtaining data on ocean weather and wave patterns, similar to the instrumentation on the NOAA weather satellites (see **NOAA 1–December 11, 1970**). In 1990, a second-generation cluster of ocean surveillance satellites was begun (see **USA 59–June 8, 1990**).

May 4, 1976
Lageos 1
U.S.A.
Lageos (Laser Geodetic Satellite) 1, placed by NASA in a high circular orbit about 3,660 miles above the poles, was a dense, two-foot-wide sphere weighing 900 pounds and covered with laser reflectors. Continuing work of **Starlette (February 6, 1975)**, Lageos's higher density and elevation made it a stable platform for measuring motions and positions on the earth's surface. By beaming lasers at Lageos from ground stations and measuring the time it took for the laser signals to return, scientists could very accurately determine their position.

Using this satellite, scientists made the first measurements of the motion of the earth's continental plates, proving the theory of plate tectonics. Each plate's velocity and direction were measured, not only allowing scientists to map precisely the earth's earthquake and volcanic hot zones, but also helping scientists understand how the continents have shifted over eons.

The satellite also showed how these plate movements subtly deformed the earth's shape, ranging from one to two inches during any one year. Moreover, the data allowed scientists to calculate the viscosity of the earth's deep mantle.

The next geodetic satellite was Lageos 2 (see **STS 52– October 22, 1992**).

May 13, 1976
Comstar 1
U.S.A.
This geosynchronous communications satellite was developed and operated by Comsat General Corporation under a lease from AT&T. It was the first of four Comsat satellites launched through 1984, and the first AT&T satellite launched since **Telstar 1 (July 10, 1962)** and **Telstar 2 (May 7, 1963)**. Comstar satellites formed the third private cluster serving the American domestic telephone market, following the Westar satellites (see **Westar 1–April 13, 1974**) and the RCA Satcom satellites (see **RCA Satcom 1–December 13, 1975**). Each satellite handled about 1,500 one-way voice circuits. Two satellites operated through the late 1980s while two were held as spares.

All four satellites also had beacon transmitters at 19 and 28 GHz for performing propagation tests at these frequencies. See **ATS 6 (May 30, 1974)** for the background behind these experiments.

In 1983, AT&T developed its second-generation communications satellite, **Telstar 3 (July 28, 1983)**, to replace the Comstar satellites.

May 22, 1976
P76-5
U.S.A.
Built by the U.S. Air Force, this experimental satellite performed propagation research, measuring the effects of atmospheric conditions and solar activity on radio signals. The satellite carried a beacon that broadcast at 10 different frequencies, ranging from 137 MHz to 2.89 GHz.

June 2, 1976
SDS 1
U.S.A.
Satellite Data Systems (SDS) 1 was a U.S. Air Force test communications satellite, placed in an orbit similar to the Soviet Molniya communications satellites (see **Molniya 1-01–April 23, 1965**), with a perigee of 285 miles and an apogee of 24,417 miles over the North Pole. It tested communications techniques from ground stations and from satellite to satellite.

After one more test flight, the Air Force made this system operational, launching 7 satellites through February 1987. Its configuration included a two-satellite complex for providing low-orbit communications between the ground and its polar region Strategic Air Command. The SDS satellites were also used to relay data from the military's KH-11 satellites (see **KH-11 1–December 11, 1976**).

This first-generation system was replaced with the SDS 2 satellites. See USA 40 on **STS 28 (August 8, 1989)**.

June 19, 1976
Intercosmos 15
U.S.S.R. / East Europe
Part of a continuing series of satellites launched by the Soviet Union and containing experiments built by scientists from communist nations (see **Intercosmos 1–October 14, 1969**), Intercosmos 15 was unlike the previous satellites in that instead of conducting research of the earth's atmosphere and magnetic field, the spacecraft performed engineering tests on several new satellite designs. Soviet, Czechoslovakian, East German, Hungarian, and Polish scientists participated.

The primary test was to check out a telemetry system for beaming data back to Earth. This new system allowed individual research facilities to directly monitor the spacecraft's systems, rather than having one ground station download the data and then distribute it to other researchers.

June 19, 1976
Viking 1, Mars Orbit/Landing
U.S.A.
Viking 1 was one of two identical American spacecraft designed to search for life on Mars, the other being **Viking 2 (August 7, 1976)**. Both spacecraft consisted of a combined

June 19, 1976

orbiter and lander. The two orbiters carried cameras to photograph and map the red planet's surface at a higher resolution than **Mariner 9 (November 14, 1971)**. Each lander carried cameras; sensors to measure temperature, wind speed, and air pressure; and a robot arm to dig out a soil sample, place it in a hopper, and then conduct three different experiments to test for evidence of life.

Built by NASA and launched on August 20, 1975, Viking 1 arrived in Mars orbit on June 19, 1976. From this perch, it took detailed photographs of the planet's surface and discovered that the planned landing site in the Chryse Planitia was too rugged. This plains region was north of what appeared to be the outflow of the vast Valles Marineris canyon, and scientists hoped it would show evidence of the catastrophic floods they believed had shaped Mars's riverlike channels. After several weeks of photographic search, a smoother landing site was chosen for Viking 1, still in Chryse Planitia but further west.

The lander touched down on Mars on July 20, 1976, the seventh anniversary of the **Apollo 11 (July 16, 1969)** lunar landing. During its descent, its sensors found the atmosphere to be mostly carbon dioxide with about 3 percent nitrogen—about three times the amount of nitrogen estimated by Mariner 9.

After landing, the lander immediately took a picture of the ground beside one footpad, showing a pebble-strewn surface of soil. Its second picture, a panorama, revealed a field of sharply broken rocks resembling basalt and lava deposits. After some fine-tuning of the picture's color, the Martian sky was found to be a gloriously alien reddish pink, not blue. A second panorama showed dunelike drifts of sand.

On the spacecraft's eighth day on the surface, its robot arm dug out a trench and dumped a soil sample into the experiment hopper. Then, three robot laboratories went to work. The first to obtain results was the gas-exchange experiment. In this test, the soil sample was placed in a sealed compartment and drenched with what the scientists playfully labeled "chicken soup," a broth of earthlike nutrients and water. The assumption was that any Martian life would lap-up this food and begin reproducing, causing a change in the compartment's contents. These changes would be monitored by a mass spectrometer, which recorded the spectrum emitted by the gas molecules in the compartment. To the mystification of the scientists, within 2.5 hours, the instrument displayed a burst of oxygen, which then leveled off.

Next was the labeled-release experiment. In this test, the sample was also exposed to a nutrient solution, but in a smaller amount and of slightly different composition, resembling materials found in meteorites and interstellar clouds. The test also used radioactive carbon-14 to measure the soil's reaction to the nutrient, a different and more sensitive method than used by the gas-exchange experiment. Like the gas-exchange experiment, the labeled-release test almost immediately registered a sudden increase in carbon-14, something that could only happen if something in the soil was interacting with the nutrient.

At this moment, the scientists seemed to have evidence of something "very much like a biological signal," but they were hesitant to declare that alien life had been discovered until all three experiments had returned results. Furthermore, neither

Viking 1's scoop arm. *NASA*

the gas-exchange nor the labeled-release experiment had reacted in the manner expected for a population of microbes. Scientists had assumed that any life would feed on the nutrients, thereby causing a slow population rise that would increase with time. This increase would, in turn, cause each experiment to show an increasing change in the compartment's content. Instead, both showed a quick change that soon leveled off.

The third test, called the pyrolytic release experiment, merely tested the soil for any evidence of organic (or carbon-based) material. The assumption here was that Martian life would almost certainly be founded on the carbon atom, like Earth life. Carbon's atomic structure is uniquely able to form a literally uncountable array of molecules. If life had existed—or did exist—on Mars, scientists believed that its soil would carry some evidence of carbon. The experiment could detect carbon atoms in quantities as small as a few parts per billion.

Apparently, the robot arm had failed to drop soil into this experiment's hopper during its first digging operation. Scientists, and the general public, had to wait several additional days until the experiment performed a second run. The results, however, showed absolutely no evidence of organic material, something most scientists believed would not be possible if there was, or had been, microbiotic life in the Martian soil.

If one test showed no evidence of life but the two others did, what was happening? Moreover, the identical experiments on Viking 2's lander came up with the same results, compounding the mystery. After much disagreement and chemical experimentation on Earth, most scientists concluded that the outcomes produced in the gas-exchange and labeled-release experiments were the result of chemical reactions, not biological life processes. It appeared that the Martian soil contained substances that were highly reactive with water, such as hydrogen peroxide (normally used as a bleach on Earth). On Earth, these materials, called oxidants, are not found naturally in the soil because so much water is present. On Mars however, the ultraviolet radiation from the Sun produced these oxidants, and the environment was so dry that they survived.

Not all scientists on the Viking team agreed, so the published results for both missions stated that "no conclusions were reached concerning the existence of life on Mars." The

June 19, 1976

dissenters, however, were in the minority. Following Viking, most researchers believed that no life existed on Mars.

Both landers had been designed for a 90-day service life. Both exceeded this greatly, with Viking 1's lander lasting until November 1982 and Viking 2's lander surviving until April 1980. During their life span, both spacecraft acted as Martian meteorological stations, giving scientists a rough idea of the annual weather cycle on the northern lowlands of Mars.

During the summer months, the temperature at Viking 1's Chryse plains site ranged from –28°F in the late afternoon to –118°F just before dawn. Viking 2's site, farther north in the Utopia plains, generally was about 10°F colder. Air pressure averaged between 6 and 8 millibars (compared to Earth's 1,000 millibars). Wind speeds were light, ranging from 4 to 15 miles per hour.

As winter approached, both landing sites experienced weather fluctuations, including increasing wind speeds and air pressure. In the winter, the air pressure at both sites increased to 9 and 10 millibars respectively, caused by the evaporating carbon dioxide drifting south from the north pole.

Major global dust storms occurred twice in early 1977. During these storms, the wind speeds at Viking 1's landing site increased to 38 miles per hour, with gusts as high as 58 miles per hour. The lander cameras, however, showed no evidence of any material being lifted and carried through the air, mostly because the atmosphere's thinness made such transport rare in human time scales. After the first storm, a thin layer of ice was seen coating the rocks at the Viking 2 landing site. A third dust storm appeared to be developing in 1982, just as the Viking project was ending.

The composition of the Martian atmosphere at both landing sites averaged 95 percent carbon dioxide, 2.7 percent nitrogen, and 1.6 percent argon, with traces of oxygen, carbon monoxide, and water vapor. Further tests of the Martian soil at both landing sites showed its composition to resemble iron-rich clay.

Even as both landers continued to beam their data back to Earth, the two orbiters began their planetwide photographic survey of the red planet. Both exceeded their planned service life. Viking 2's orbiter operated until August 1980, while Viking 1's worked until July 1978. During their life span, they transmitted to Earth more than 46,000 photographs of the Martian surface, covering the entire planet at a resolution ranging from 500 to 1,000 feet. In selected areas, the images saw objects as small as 25 feet across.

The pictures showed that Mars did not seem to have plate tectonics like Earth. Hence, its geological features tended to grow larger and were older. The famous giant volcanoes and deep canyons seen by Mariner 9 were the result.

Everywhere the spacecraft's camera looked, it recorded evidence of water and flooding. On the walls of some canyons, the images showed thick-layered deposits, indicating that some form of sedimentary process had occurred. The many Martian channels often resembled Earth river channels, indicating large flood run-offs or the outflow from large water basins. And many craters appeared to have landed in mud flats, splattering material away from them rather than shattering the ground.

Fractured terrain and complex grooved areas, while caused by some form of deformational process, resembled nothing seen on Earth. Though some association with cratered regions could be seen, the origin of these regions was unknown.

While the images showed nothing of the seasonal wave of darkening that had been previously observed from Earth, they did reveal small changes caused by wind deposits during the summer season. Many images revealed sand dunes, visible within craters, canyons, and on plains.

The pictures also showed in detail the two Martian polar caps, revealing a region of layered deposits surrounding each pole. The caps themselves were made of a permanent layer of water ice covered by a temporary winter layer of frozen carbon dioxide.

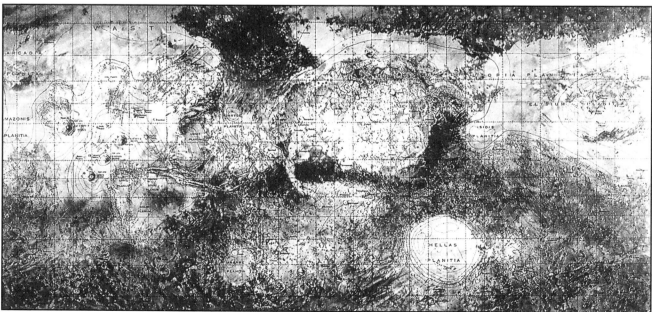

Global map of Mars, 1976. *NASA*

June 19, 1976

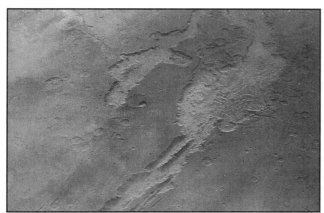
The eastern part of Valles Marineris. *NASA*

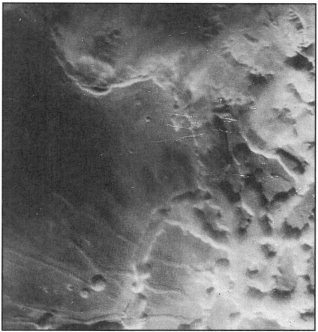
The outwash from Valles Marineris on its eastern end. *NASA*

The discovery that both caps were composed not only of carbon dioxide but also of water ice was a surprise. Scientists had assumed that most of Mars's surface water had disappeared long ago, or was frozen underground. Instead, the planet had enough water in its ice caps that, if distributed evenly as liquid throughout the planet, it would cover its surface to a depth of 30 to 130 feet.

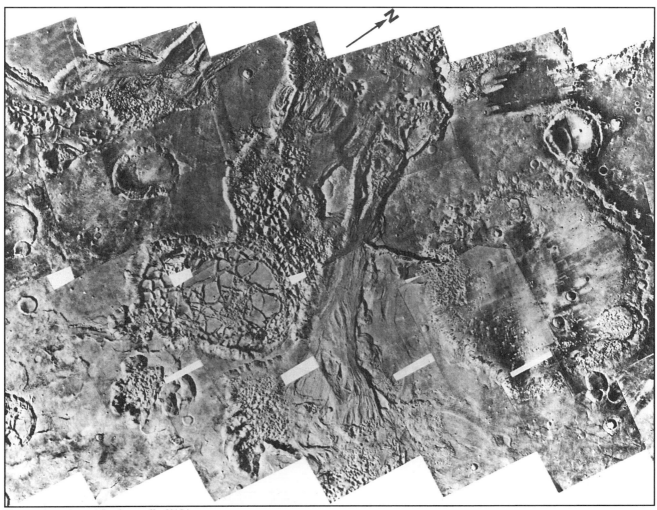
Fractured terrain and flood run-offs. *NASA*

The images also indicated that the two ice caps were not identical, due to the eccentricity of Mars's orbit. Because the northern hemisphere was closer to the Sun in the winter and farther away in the summer, its seasons were more moderate. Conversely, the southern hemisphere's seasons were more extreme. The images indicated that while the carbon dioxide ice in the northern hemisphere evaporates each summer, exposing the permanent water ice below, the carbon dioxide ice in the southern hemisphere never completely evaporates.

Scientists estimated that, each year, about 30 percent of the Martian carbon dioxide atmosphere was processed through the planet's polar caps, with this seasonal freezing and evaporation producing the layered terrain and the vast fields of sand dunes surrounding each cap.

The orbiters also photographed the global dust storms of 1977, as well as the processes that seemed to trigger them. The first storm began in a long narrow valley, Claritas Fossae, south of the Tharsis volcano bulge. Within days, it had spread to cover the entire planet.

The orbiters found that certain regions experienced regular dust storms, a few of which would grow to the global proportions seen in the 1956, 1971, 1973, and 1977 storms. The global storms always occurred during the southern spring and summer when the planet was at perihelion. The additional heat from the Sun, combined with the greater atmospheric pressure caused by the evaporated polar caps, generated convection cells of rising air in certain locations. These cells, the source of more energetic winds, also lifted dust into the atmosphere, which acted to increase the atmosphere's ability to absorb the Sun's radiation, thereby speeding up the cycle and causing several small storms to quickly merge into one huge one. After a short period, the storm would die out, for unknown reasons. Also unknown is why these storms only happened in certain years and are not an annual occurrence.

In general, the Viking orbiters confirmed what Mariner 9 had seen: the surface of Mars had experienced repeated catastrophic flooding in its ancient past. For water to have flowed like a liquid, however, the atmosphere of Mars had to have once been much thicker. How the Martian atmosphere changed, and why, remains the greatest mystery surrounding the red planet.

Viking 1 and Viking 2 were last the American probes to Mars for 17 years, until **Mars Observer (September 25, 1992)**. Twenty-one years passed before another lander, **Mars Pathfinder (July 4, 1997)**, arrived on the Martian surface.

June 22, 1976
Salyut 5
U.S.S.R.

Salyut 5 was the second and last military Salyut mission, following **Salyut 3 (June 25, 1974)**. Like **Salyut 3**, Salyut 5 weighed about 20 tons and included a separate Mercury-like capsule that could return cargo to Earth. Like Salyut 3, the military nature of the station meant that many details of its design and much of the work performed within it were kept secret, even through December 1999.

During its 411 days in orbit, Salyut 5 was visited three times by manned Soyuz spacecraft:

Soyuz 21–July 6, 1976 to August 25, 1976 (50 days)
Soyuz 23–October 14, 1976 to October 16, 1976 (2 days)
Soyuz 24–February 7, 1977 to February 25, 1977 (18 days)

Salyut 5 was deorbited on August 8, 1977. The next space laboratory was **Salyut 6 (September 29, 1977)**.

June 26, 1976
IMEWS 6
U.S.A.

IMEWS 6 was the first satellite in a third-generation infrared early-warning cluster launched by the U.S. military. See **IMEWS 1 (May 5, 1971)** for the previous generation. The newer satellites produced about one-third more electrical power and had a longer life expectancy. Nine were launched through February 1997.

July 6, 1976
Soyuz 21
U.S.S.R.

Manned by Boris Volynov (**Soyuz 5–January 15, 1976**) and Vitaly Zholobov, Soyuz 21 was a 50-day mission to **Salyut 5 (June 22, 1976)**.

Soyuz 21 was dedicated mostly to military reconnaissance. The cosmonauts made high resolution surveillance photographs of the earth, as had been done with **Salyut 3 (June 25, 1974)**. They also observed a massive Soviet military maneuver staged in the eastern part of the Soviet Union during their flight.

The cosmonauts performed scientific work, as well. They studied the formation of alloys in zero gravity. An ingot of bismuth, lead, tin, and cadmium was melted and allowed to solidify in zero gravity, forming a ball about the size of a match head. A homogeneous compound of dibenzyl and tolane was attempted. Later the cosmonauts did smelting and soldering experiments using high-grade nickel and manganese solder to weld stainless steel tubes.

The growth of crystals was also studied. Twice during the flight, three crystal seeds were placed in three different containers and their growth monitored. The samples were then returned to Earth for study.

The swimming techniques of two guppies in weightlessness (one of whom was pregnant) were studied, and it was found that after several days, they adapted to weightlessness. Fish eggs were also incubated, and they developed normally.

As had been done on previous Soyuz flights (see **Soyuz 18B–May 24, 1975**), the cosmonauts took comprehensive optical and multispectral photographs of Earth resources, studying geological deposits and weather patterns.

Tests were conducted of fuel pump designs, using capillary and surface tension rather than electricity and fuel pumps. This work was directly related to Soviet plans to refuel future space laboratories in order to extend their orbital life (see **Progress 1–January 20, 1978**). At another point, the cosmonauts donned suits and depressurized the station, replacing its atmosphere entirely.

Medical research continued. In addition to the standard exercises and monitoring of blood, heart, brain, and lung, a new mass meter or "scale" allowed the cosmonauts to monitor their weight. This flight also included an efficiency study, examining how weightlessness affected the cosmonauts' use of time.

Originally planned to last for two months, the flight was cut short because of problems in the station's environmental control system. In the last few weeks, an "acrid odor" became increasingly worse and finally forced the cosmonauts to evacuate the station, making an emergency landing on August 25, 1976, at least 10 days ahead of schedule.

The next manned space flight was **Soyuz 22 (September 15, 1976)**, which used the back-up spacecraft from the **Soyuz 19–Apollo 18 (July 15, 1975)** joint flight. The next flight to Salyut 5 was **Soyuz 23 (October 14, 1976)**.

July 8, 1976
Palapa 1
Indonesia / U.S.A.

Developed by Hughes Communications for Indonesia and launched in the United States, Palapa 1 was Indonesia's first satellite, providing communication services across its far-flung territories. It was also the first geosynchronous communications satellite launched for a country outside of North America, the Soviet Union, or Europe. Palapa 1 provided service to Indonesia as well as its neighbors, Singapore, Malaysia, Thailand, Burma, and the Philippines. Similar to **Anik A (November 10, 1972)** and **Westar 1 (April 13, 1974)**, Palapa 1 had capacity for 600 2-way voice circuits or 12 television channels.

In 1977, Indonesia launched a second and identical Palapa satellite (Palapa 2) to augment Palapa 1. Both remained in operation until the launch of Indonesia's second-generation Palapa B satellites (see **STS 7–June 18, 1983**).

August 7, 1976
Viking 2
U.S.A.

Identical in design to **Viking 1 (June 19, 1976)**, Viking 2 was launched by NASA on September 9, 1975. After entering Mars orbit on August 7, 1976, it dropped its lander to the surface on September 3, 1976, placing it in Utopia Planitia, a plains region in the middle northern latitudes about 4,500 miles to the northeast of the Viking 1 landing site. Once on the ground, its photographs revealed a very flat plain covered with innumerable rocks generally larger than those seen at the Viking 1 site *(see color section, Figure 29)*. No sand dunes were visible.

Close-up of pitted rocks; the Viking 2 lander footpad is at the lower right. *NASA*

These rocks at Utopia, covered abundantly with many small pits, resembled volcanic debris in which small gas bubbles in the molten lava caused cavities to form. See Viking 1 for more details.

August 9, 1976
Luna 24
U.S.S.R.

Luna was the last spacecraft, through December 1999, to soft-land on the Moon and return samples. It was the third successful Soviet probe to accomplish this deed, following **Luna 16 (September 12, 1970)** and **Luna 20 (February 14, 1972)**, and the ninth overall, including the American manned missions.

The spacecraft touched down in the Sea of Crisis, close to the **Luna 23 (October 28, 1974)** landing site. This location was close to one of the large mascons that perturbed the Moon's gravitational field. Luna 24's drill then proceeded to obtain a core sample six feet long, which was stored in the return capsule in one-foot sections. The total sample weight equaled a little more than one-third of a pound.

Luna 24 returned to Earth on August 22, 1976. Once analyzed, its lunar samples showed evidence that a series of layering events had deposited the basalt in the Sea of Crisis. The material itself was different than other samples in that its basalts had little titanium and less potassium, and more high-silica, low-alumina glass inclusions. The glass indicated that some of the deposits came from impact ejecta.

The Luna 24 mission closed out the first wave of lunar exploration. Fourteen years passed before the next flight to the Moon, the Japanese **Hiten (January 24, 1990)**.

September 15, 1976
Soyuz 22
U.S.S.R.

Soyuz 22 was an eight-day mission using the leftover back-up spacecraft from the **Soyuz 19–Apollo 18 (July 15, 1975)** joint U.S.-Soviet mission. The crew was Valeri Bykovsky (**Vostok 5–June 14, 1963; Soyuz 31–August 26, 1978**) and Vladimir Aksenov (**Soyuz T2–June 5, 1980**).

Like **Soyuz 16 (December 2, 1974)**, this flight was dedicated to science. The spacecraft's orbit was adjusted to give it an inclination 64.8 degrees from the earth's equator, an inclination not used by any Soviet manned mission since **Voskhod 2 (March 18, 1965)**. This orbit covered a wider range of latitudes for Earth resources observations.

Replacing the Soyuz docking port was an East German–built multispectral camera system, able to take six simultaneous photographs, four in visible light and two in the infrared. A single set of images covered 200,000 square miles with a resolution of 65 feet. This equipment was later installed on the next Soviet space laboratory, **Salyut 6 (September 29, 1977)**.

Extensive photography of the Soviet Union was undertaken during the eight-day flight. Twice a duplicate camera system was flown on a plane, taking simultaneous comparison photos.

Because a major NATO military exercise was also taking place during this flight, some Soviet observers believed that Soyuz 22 also did military reconnaissance. Whether this is true or not is unknown.

October 14, 1976

The next manned space flight, **Soyuz 23** (October 14, 1976), was a mission to **Salyut 5** (**June 22, 1976**).

October 14, 1976
Soyuz 23
U.S.S.R.

Flown by two rookie cosmonauts, Vyacheslav Zudov and Valery Rozdestvensky, Soyuz 23 failed to dock with **Salyut 5** (**June 22, 1976**) and therefore returned to Earth after only two days in space. The main antenna of the Soyuz's automatic docking system failed, and the cosmonauts could not accomplish a manual docking.

Soyuz 23's landing was unusual—at night in a blizzard on frozen Lake Tengiz. The two men crashed through the ice, unintentionally achieving the Soviet's first and only manned splashdown.

The next and last flight to Salyut 5 was **Soyuz 24** (February 7, 1977).

October 26, 1976
Ekran 1
U.S.S.R.

The Ekran ("screen" in Russian) satellites were Soviet geosynchronous communication satellites for providing television service to the remote northern and eastern parts of the Soviet Union. They supplemented the Molniya 2 (see **Molniya 2-01–November 24, 1971**), Molniya 3 (see **Molniya 3-01–November 21, 1974**), and Gorizont (see **Gorizont 1–December 19, 1978**) satellites.

Because the Ekran satellites transmitted their signal at very high power, the receiving terminal could be very small and simple, a distinct advantage in the remote and isolated regions of Russia and what was then the Soviet Union. A total of 20 Ekran satellites were launched through 1992, with usually two satellites kept in operation at all times.

In 1994 Russia began replacing the Ekran satellites. See **Gals 1** (**January 20, 1994**).

December 15, 1976
Cosmos 881 / Cosmos 882
U.S.S.R.

These two Cosmos satellites were the first test of the Merkur manned capsule, an alternative Soviet spacecraft to the Soyuz capsule for placing humans in orbit. During their short flight, lasting only a few hours, both capsules performed tests of the spacecraft's re-entry system. After another four unmanned tests during the next three years, the program was abandoned.

December 19, 1976
KH-11 1
U.S.A.

The KH-11 series of U.S. Air Force photo-reconnaissance satellites replaced both the wide-angle KH-9 and telephoto KH-7/KH-8 surveillance satellites. See **KH-9 1** (**June 15, 1971**) and **KH-7 1** (**July 12, 1963**) for details on these earlier programs.

KH-11 satellites are believed to use technology similar to the Hubble Space Telescope (see **STS 31–April 24, 1990**). Each carries a mirror telescope, an infrared scanner, and a multispectral scanner. Its mirror is thought to be about 77 inches in diameter, only slightly smaller than Hubble's 94.5 inches, and with a resolution thought to be anywhere between four to six inches. Some experts also believe that the mirror is not glass, like Hubble's, but a thin surface that unfolds after launch and is continuously adjusted by computer to maintain its perfectly correct curvature. This same technology is being considered for the successor to the Hubble Space Telescope, the Next Generation Space Telescope, scheduled for launch in the early 2000s.

The KH-11 1 satellite was also one of the first in-space telescopes to use CCDs (charge-coupled devices) to gather light. Invented in 1970, a standard CCD measures 800 by 800 pixels in a half-inch-square space. Each pixel collects light, and this information is recorded digitally and then transformed by computer into extremely high definition images.

KH-11 satellites, unlike their predecessors, also do not use recoverable capsules. All data is downlinked digitally, thereby allowing for almost instantaneous coverage should political events require it. For example, a KH-11 satellite located the American refugees being held in the embassy in Teheran in April 1980. Another KH-11 took photographs of the exterior of the space shuttle Columbia on its inaugural flight, **STS 1** (**April 12, 1981**), to assess the damage to its exterior tiles.

This first KH-11 functioned a little over two years. Later models operated over three years. Eight such satellites were launched through November 1988. The KH-11 series was superseded by the KH-12 series. See **STS 36** (**February 28, 1990**).

1977

February 7, 1977
Soyuz 24
U.S.S.R.

Soyuz 24 marked the last flight to **Salyut 5** (**June 22, 1976**). The mission lasted 18 days and was crewed by Viktor Gorbatko (**Soyuz 7–October 12, 1969**; **Soyuz 37–July 23, 1980**) and Yuri Glazkov.

No evidence of the atmospheric problems that had caused the abrupt end to **Soyuz 21** (**July 6, 1976**) were reported, though the two men did several engineering tests to monitor Salyut 5's atmosphere. Once, during a television broadcast, the cosmonauts partially purged the station's atmosphere, replacing this with air from stored tanks.

As with Soyuz 21, Soyuz 24 was dedicated mostly to military photography, so few details of the mission were reported to the public. The science research included Earth resource photographs of the Soviet Union, as well as infrared spectrograms of the amounts of water vapor, ozone, nitrogen oxide, and pollution in the upper atmosphere. The cosmonauts also repeated the biological experiments conducted during Soyuz 21.

The cosmonauts undocked and returned to Earth on February 26, 1977. Prior to leaving, they packed a second cargo capsule with film and experiments, which was recovered safely in the Soviet Union one day after the cosmonauts' landing. (Ironically, this same capsule, part of a secret military mission, was sold for $48,875 at auction at Sotheby's in New York on December 11, 1993.)

This last flight to the Salyut 5 laboratory was also the last flight to a first-generation Salyut station. The next Soviet orbital laboratory, **Salyut 6 (September 29, 1977)**, had significant improvements and increased capabilities.

February 23, 1977
Kiku 2 (ETS 2)
Japan
Kiku 2, Japan's first geosynchronous communications satellite, was also called Engineering Test Satellite (ETS) 2. It followed **Kiku 1 (September 9, 1975)** and continued tests of the Japanese N1 launch rocket.

Kiku 2 used the same design as **Skynet 1 (November 22, 1969)** and was orbited above Japan where it performed propagation experiments in the 1.7 GHz, 12 GHz, and 34.5 GHz bandwidths. See **ATS 6 (May 30, 1974)**.

The satellite operated until May 1978 and was followed by **Kiku 3 (February 11, 1981)**.

April 20, 1977
ESA-GEOS 1
Europe / U.S.A.
The European Space Agency Geosynchronous Satellite (ESA-GEOS) 1 was launched by NASA and built by ESA. It was intended to be Europe's first satellite in geosynchronous orbit, studying the earth's magnetosphere in the equatorial regions.

A failure of the launch rocket's third stage, however, placed the satellite in too low an orbit. Using the spacecraft's attitude-control thrusters, the orbit was adjusted to a perigee of 1,324 miles and an apogee of 23,922 miles. From this perch, ESA-GEOS 1 spent approximately 12 hours each day at altitudes sufficient to salvage a part of its mission, measuring small variations in the earth's electrical field.

ESA launched **ESA-GEOS 2 (July 14, 1978)** to complete ESA-GEOS 1's mission. See that mission for results.

June 17, 1977
SIGNE 3
France / U.S.S.R.
The Solar Interplanetary Gamma-Neutron Experiment (SIGNE) 3 satellite was one of a continuing series of joint French-Soviet satellites (see **Aureol 1–December 27, 1971**), whereby France built the satellite and the Soviets provided the rocket and launch facilities. SIGNE 3 carried two sensors, the first monitoring celestial gamma rays in the spectral range from 20 KeV to 10 MeV, and the second monitoring the Sun's ultraviolet output in two bands, 1,800–1,950 angstroms and 2,050–2,200 angstroms. The satellite provided data for less than a year.

The next solar research satellites were **ISEE 1/ISEE 2 (October 22, 1977)**. The next gamma-ray telescope was **HEAO 1 (August 12, 1977)**.

June 23, 1977
NTS 2
U.S.A.
NTS 2 was the U.S. Navy's fourth test flight of global positioning technology, following **NTS 1 (July 14, 1974)**. It was also the first built under the new Navstar program to create a global positioning system (GPS). Under this program, a person or moving vehicle anywhere on the earth always had no fewer than three satellites overhead and, using ground equipment synchronized to the satellites' clocks and positions, could obtain a positional fix within 30 feet in three dimensions at speeds up to 360 miles per hour. See **Timation 1 (May 31, 1967)**.

NTS 2 functioned for about six months, during which extensive tests were performed at the Yuma Proving Grounds in Arizona. The program then launched its first test cluster of satellites, beginning with **Navstar 1 (February 22, 1978)**.

July 14, 1977
Himawari 1
Japan
Himawari 1 ("sunflower" in Japanese) was Japan's first weather satellite and the second in a network of geosynchronous weather satellites that began with the American **GOES 1 (October 16, 1975)**. Upon completion, this network provided continuous and complete daily coverage of the world's weather patterns in the middle and tropical latitudes. See GOES 1 for details.

Positioned over Japan, Himawari 1 operated for 4.5 years. Japan has since launched four more Himawari weather satellites, the last in 1995, to maintain the system. The first three satellites each carried a scanner that produced images in both visible wavebands (0.5–0.75 microns) and infrared (10.5–12.5 microns). The last two satellites also included a sensor for monitoring solar activity, including flares and magnetic storms, by detecting high-energy solar particles such as protons, electrons, and alpha particles.

The fifth Himawari satellite also carried a second-generation international SARSAT/COSPAS search-and-rescue receiver-transmitter package, used to locate ships and aircraft in distress. See **Cosmos 1383 (June 30, 1982)**.

The next geosynchronous weather satellite to be added to the global network was the European **Meteosat 1 (November 23, 1977)**.

July 17, 1977
Cosmos 929
U.S.S.R.
This 200-day mission was the first flight test of the Heavy Cosmos (see **Cosmos 1267–April 25, 1981; Cosmos 1443–March 2, 1983**) modules that were flown attached to the **Salyut 6 (September 29, 1977)** and **Salyut 7 (April 19, 1982)** space laboratories. The spacecraft's total weight was about 20 tons, about as heavy as the Salyut stations.

During the flight, Cosmos 929 was maneuvered many times, its orbit raised and lowered more than a dozen times to test its power and thruster controls, as well as the ground control's ability to command the spacecraft. Its capsule returned to Earth and was recovered safely on August 16, 1977, after 30 days in orbit. The spacecraft itself was deorbited on February 2, 1978, burning up in the atmosphere.

August 3, 1977
Bion 4 (Cosmos 936)
U.S.S.R. / International

Bion 4 was part of the Soviet biological research program begun with **Bion 1 (October 31, 1973)**. It lasted 19.5 days, its experiments mostly follow-ups of research done on **Bion 3 (November 25, 1975)**, and including work by Soviet, American, French, Czechoslovakian, Bulgarian, Hungarian, East German, Polish, and Romanian scientists. On this flight, the research tested the effects of weightlessness on rats. Five test animals each were placed in two revolving centrifugal platforms (to create gravity artificially), with others left in zero gravity.

For the rats subjected only to zero gravity, scientists found a decrease in the water-soluble proteins in the rats' spinal cords. The rats in the centrifuges also lost proteins, but to a lesser degree. Weightlessness also caused muscle atrophy and bone loss, with the artificial gravity limiting the extent of these changes. Both groups recovered quickly, however, upon their return to Earth. The scientists concluded that the effect of artificial gravity on biological specimens was the same as natural gravity. **Bion 5 (September 25, 1979)**, the next in this program, made the first attempt to breed mammals in space.

August 12, 1977, 6:39 GMT
HEAO 1
U.S.A.

The High Energy Astronomical Observatory (HEAO) 1 was the first of three second-generation x-ray telescopes, designed by NASA to study the bewildering and fluctuating nature of the many x-ray objects discovered by **Uhuru (December 12, 1970)** and later confirmed by **Copernicus (August 21, 1972)**, ANS 1 **(August 30, 1974)**, Ariel 5 **(August 30, 1974)**, and OSO 8 **(June 21, 1975)**. Unlike these earlier satellites, HEAO 1 was designed with these x-ray sources in mind, most of which emitted x-rays in the energy range from 1 to 10 KeV. HEAO 1's instruments could detect emissions from 0.15 KeV to 10 MeV, and its goal was to map the entire x-ray sky, finding objects with .00001 the intensity of Scorpius X-1, the brightest known x-ray object.

During its 1.5-year life span, HEAO 1 found exploding x-ray stars both in and out of the Milky Way galaxy, increasing the number of known sources from 200 to nearly 1,500. Some bursters erupted once, and then faded away. Others repeated their bursts again and again, in a random pattern that at first defied explanation.

About 200 pulsars, thought to be the corpses of supernovae, were also discovered, exhibiting a baffling variety of features. Their pulse rates ranged from less than one second to almost 20 minutes. Many showed evidence that they were orbiting other objects. For eight x-ray pulsars, the pulse rate showed signs of gradual increase. This increase contrasted with the pulse rate decline previously seen in the radio pulsars.

HEAO 1 raised the number of black hole candidates, while finding that some of the previous candidates were merely x-ray pulsars. All seemed part of binary systems, with the x-ray source sucking matter from its orbiting companion.

Later analysis of HEAO 1's data revealed the existence of a number of supernovae remnants in the Cygnus region, 6,000 light years away, caused by a series of supernova explosions

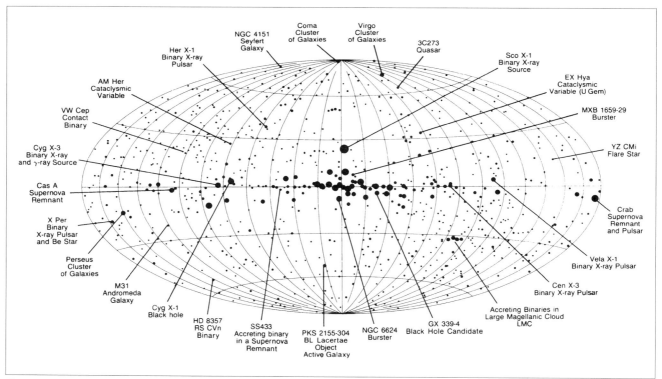

The HEAO 1 x-ray map of the sky. Compare with Uhuru (December 12, 1970) map. *NASA*

over the last 15 million years in a nearby region dubbed the Scorpius-Centaurus Association. As the waves of matter expanded outward from each explosion, they cleared the debris before them to create what scientists dubbed "a superbubble," a region of relatively empty space within which the Sun and our solar system now appeared to be traveling.

HEAO 1 also carried a gamma-ray burst detector to continue the monitoring of these mysterious bursts, first discovered by **Vela 5A/Vela 5B (May 23, 1969)**. Like earlier spacecraft (see **Solrad 11A/Solrad 11B–March 15, 1976**), HEAO 1 recorded bursts regularly, but the radiation usually faded so quickly no visible counterparts could be identified by other telescopes. Gamma-ray bursts remained a complete enigma. Gamma-ray burst monitoring continued on **ISEE 3 (August 12, 1978)**.

The next x-ray telescope was **Einstein (November 13, 1978)**.

August 12, 1977, 15:00 GMT
Enterprise, Flight 4
U.S.A.

Enterprise Flight 4 was the maiden flight of Enterprise, the first full scale prototype of NASA's space shuttle. Though Enterprise had flown three previous times attached to the back of a 747 carrier aircraft, this test was its first manned free flight, in which it glided to a landing from 23,000 feet altitude. The shuttle prototype was manned by Fred Haise (**Apollo 13–April 11, 1970**) and Gordon Fullerton (**STS 3–March 22, 1982; STS 51F–July 29, 1985**).

Despite having no engines and weighing 75 tons, Enterprise glided to a smooth landing on a dry lake bed at Edwards Air Force Base in California. Over the next several months, Enterprise was flown three more times, testing a variety of landing scenarios including its first landing on a concrete runway at Edwards Air Force Base.

These tests proved the reliability and feasibility of using unpowered lifting body concepts for returning spacecraft safely from space. Such concepts have become the foundation for all reusable Earth-to-orbit spaceships.

The first shuttle flight, **STS 1 (April 12, 1981)**, was the maiden flight of Columbia and the first orbital test of the shuttle design.

August 25, 1977
Sirio 1
Italy / U.S.A.

An experimental communications satellite, Sirio (Satellite Italiano Ricerca Industriale Orientata) 1 was built by Italy and launched by NASA. First placed over Europe in geosynchronous orbit, it was used for six years to test a variety of new engineering technologies, as well as to conduct propagation experiments in the 11 and 17 GHz frequencies. See **ATS 6 (May 30, 1974)** for details about these propagation experiments.

From 1983 until it was turned off in 1985, Sirio 1 was moved over the Indian Ocean for joint Italian-Chinese experiments.

September 18, 1977
Cosmos 954
U.S.S.R.

Cosmos 954 was the Soviet's seventeenth Radar Ocean Reconnaissance Satellite (RORSAT), designed to intercept the radar and electronic signals of ocean vessels. See **Cosmos 198 (December 27, 1967)** for details.

Because of their high power requirements, these satellites are powered with radioactive fuel. Previous RORSATs ejected their nuclear battery prior to deorbit. The battery compartment then used its own small rocket to lift itself into a higher parking orbit, where its nuclear fuel could not contaminate the atmosphere.

When Cosmos 954 lost attitude control after three months in orbit, ground controllers were unable to eject the battery compartment. Because the radioactive fuel had been shielded inside a separate compartment, it was not destroyed during re-entry. About 110 pounds of enriched uranium-235 was scattered across a 200-mile track in the Northwest Territory of Canada east of the Great Slave Lake. The clean-up cost the Canadian government over $1 million.

With this failure, Soviet engineers redesigned the RORSATs. See **Cosmos 1402 (August 27, 1982)**.

September 29, 1977
Salyut 6
U.S.S.R.

Salyut 6 was the Soviet Union's first second-generation space laboratory, following **Salyut 1 (April 19, 1971)**, **Salyut 3 (June 25, 1974)**, **Salyut 4 (December 26, 1974)**, and **Salyut 5 (June 22, 1976)**. The laboratory now included two docking ports, increased electrical power, refueling capabilities, and greater life-support capacity.

Of all the Soviet Salyuts, Salyut 6 was by far the most successful. Eighteen different manned missions were launched to Salyut 6 during its almost five years of operation, achieving a number of significant milestones. Human space flight was ex-

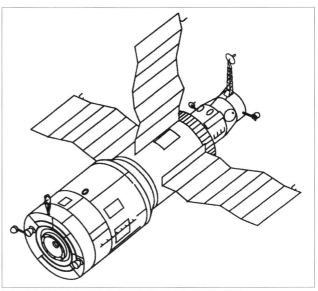

Salyut 6. Compare with Salyut 1 (April 19, 1971). *NASA*

tended to over six months, with the laboratory regularly holding two crews totaling four individuals.

In addition, 12 Progress freighters (see **Progress 1–January 20, 1978**) docked with the space station, bringing fuel and cargo weighing over 22 tons—more than the 20-ton weight of the station. Also docking with Salyut 6 was **Cosmos 1267 (April 25, 1981)**, a different module design that Western observers have since dubbed a Heavy Cosmos. This second module almost doubled the laboratory's total orbital weight.

The manned missions to Salyut 6 were as follows, the dates indicating the launch and return of their crews:

First unsuccessful flight
Soyuz 25–October 9, 1977 to October 11, 1977

First occupancy (97 days)
Soyuz 26–December 12, 1977 to March 16, 1978
Soyuz 27–January 10, 1978 to January 16, 1978
Soyuz 28–March 2, 1978 to March 10, 1978

Second occupancy (140 days)
Soyuz 29–June 15, 1978 to November 2, 1978
Soyuz 30–June 27, 1978 to July 4, 1978
Soyuz 31–August 26, 1978 to September 3, 1978

Third occupancy (175 days)
Soyuz 32–February 25, 1979 to August 19, 1979
Soyuz 33–April 10, 1979 to April 12, 1979 (emergency re-entry)

Fourth occupancy (185 days)
Soyuz 35–April 9, 1980 to October 11, 1980
Soyuz 36–May 26, 1980 to June 3, 1980
Soyuz T2–June 5, 1980 to June 9, 1980
Soyuz 37–July 23, 1980 to July 31, 1980
Soyuz 38–September 18, 1980 to September 26, 1980

Fifth occupancy (13 days)
Soyuz T3–November 27, 1980 to December 10, 1980

Sixth occupancy (74 days)
Soyuz T4–March 12, 1981 to June 10, 1981
Soyuz 39–March 22, 1981 to March 30, 1981
Soyuz 40–May 14, 1981 to May 22, 1981

Salyut 6 was deorbited on July 29, 1982. **Salyut 7 (April 19, 1982)** was the next and last Salyut station.

October 9, 1977
Soyuz 25
U.S.S.R.
Despite the unqualified success that the **Salyut 6 (September 29, 1977)** space laboratory achieved, this first Soyuz mission to the station was a failure. Much like **Soyuz 10 (April 23, 1971)**, **Soyuz 15 (August 26, 1974)**, and **Soyuz 23 (October 14, 1976)**, Soyuz 25 was unable to dock. The spacecraft, manned by Valery Ryumin (**Soyuz 32–February 25, 1979**; **Soyuz 35–April 9, 1980**; **STS-91–June 2, 1998**) and Vladimir Kovalyonok (**Soyuz 29–June 15, 1978**; **Soyuz T4–March 12, 1981**), returned to Earth after only two days in orbit.

Some obstruction, either in the forward docking port of the Salyut station or in the Soyuz orbital module, prevented the two spacecraft from achieving a hard dock. Since the orbital module was destroyed during re-entry, there was no way of knowing if its docking port had caused the problem. The next manned flight, **Soyuz 26 (December 10, 1977)**, docked with the station's aft docking port as a precaution.

October 22, 1977
ISEE 1 / ISEE 2
U.S.A. / Europe
The two International Sun-Earth Explorers, ISEE 1 and ISEE 2, continued the research into the earth's magnetosphere and its interaction with the Sun previously done by the IMP (see **IMP A–November 27, 1963**), OGO (see **OGO 1–September 5, 1964**), Intercosmos (see **Intercosmos 1–October 14, 1969**), and Prognoz (see **Prognoz 1–April 14, 1972**) satellites. The satellites were part of a joint NASA and European Space Agency (ESA) project, NASA building ISEE 1, ESA building ISEE 2, with NASA launching both.

Like the Imp, Prognoz, and HEOS (see **HEOS 1–December 5, 1978**) satellites, both ISEE satellites were placed in complementary eccentric orbits (perigee: about 260 miles; apogee: about 85,000 miles). Unlike the earlier satellites, both ISEE satellites were designed for a much longer service life, taking advantage of technology advances that allowed them to resist the Van Allen belt radiation. Both ISEE spacecraft operated for 10 years. Hence, replacement launches on a yearly basis were not necessary.

From their perch, the satellites repeatedly crossed in and out of the earth's magnetosphere, monitoring its changing shape and diameter as it was impinged by the solar wind. Their similar equipment and matched orbits allowed scientists to compile comparison data on magnetospheric phenomena and determine if some events were caused by local occurrences or by distant solar flares.

The two spacecraft found that the bow shock and the magnetopause, the dividing line between interplanetary space and the earth's magnetosphere, could change position quickly, depending on solar activity, flying back and forth past each spacecraft as the solar wind blew against it. Detailed analysis of the data also showed folds and pleats in the magnetopause, indicating that the boundary zone was a complex and chaotic region, with ripples and waves undulating along the magnetopause. Often the two spacecraft, separated by great distance, registered changes simultaneously, implying that such events were large-scale phenomena.

The third ISEE satellite, **ISEE 3 (August 12, 1978)**, complemented **Kyokko (February 4, 1978)** and **ESA-Geos 2 (July 14, 1978)**. See **Dynamics Explorer 1/Dynamics Explorer 2 (August 3, 1981)** for a summary of results.

November 23, 1977
Meteosat 1
Europe / U.S.A.
Similar in design to the American GOES satellites (see **GOES 1–October 16, 1975**), Meteosat 1 was a weather satellite built by the European Space Agency and launched by NASA. Placed in geosynchronous orbit over Europe, Meteosat 1 formed part of the global network of geosynchronous weather satellites begun with GOES 1 and continued with the Japanese **Himawari**

1 (**July 14, 1977**). See GOES 1 for details about this global network.

Meteosat 1 took hemispheric images in both visible and infrared wavelengths every 30 minutes. Images revealed cloud cover, water vapor, and cloud temperatures. Resolution was 1.5 miles for visible and 3 miles for infrared images.

Through the end of 1998, a total of seven Meteosat satellites have been launched. The first two satellites were considered test flights, followed by five operational missions, each with a service life of between two and five years. All carried instruments for imaging the earth in both visible (0.5–0.9 microns) and infrared (10.5–12.5 microns) wavebands. Because the second Meteosat satellite exceeded its life span, working until the 1990s, the third Meteosat satellite was available as a spare, and was therefore moved temporarily over the Western Atlantic in 1988 to replace a GOES satellite lost during launch.

As with the GOES satellites, Meteosat's wide-angle weather images work in conjunction with the close-up NOAA (see **NOAA 1–December 11, 1970**) images and extreme close-up Landsat (see **Landsat 1–July 23, 1972**; **Landsat 4–July 16, 1982**) images.

The next addition to this geosynchronous weather satellite network was the Indian **Insat 1A (April 10, 1982)**. With this satellite's launch, continuous coverage of the world's weather throughout the middle and tropical latitudes was permanently established.

December 10, 1977
Soyuz 26
U.S.S.R.

The crew of the Soyuz 26 completed the first long-term occupancy of the **Salyut 6 (September 29, 1977)** orbiting laboratory. Their spacecraft docked with the aft docking port, normally reserved for Progress freighters (see **Progress 1–January 20, 1978**). Then the cosmonauts, Georgi Grechko (**Soyuz 17–January 11, 1975**; **Soyuz T14–September 18, 1985**) and Yuri Romanenko (**Soyuz 38–September 18, 1980**; **Soyuz TM2–February 5, 1987**), performed the first Soviet space walk in almost nine years (since **Soyuz 4–January 14, 1969** and **Soyuz 5–January 15, 1969**), checking out the forward docking port to make sure it had not caused the docking failure on **Soyuz 25 (October 9, 1977)**. Grechko examined the port, found it clear of obstructions, and then used tools to confirm that its equipment functioned properly. During this space walk Romanenko, who was only supposed to wait in the hatch in case his help was needed, drifted out into space and had to be caught by Grechko.

The crew remained in orbit a total of 96 days, a new human space endurance record, exceeding that of the crew of **Skylab 4 (November 16, 1973)**. During their more than three months in space, Grechko and Romanenko were visited by **Soyuz 27 (January 10, 1978)** and **Soyuz 28 (March 2, 1978)**. They also performed the first refueling in space, using the Progress 1 freighter. See these flights for more information.

In order to do that refueling, however, it was necessary to remove their Soyuz 26 spacecraft from the rear docking port. To achieve this, the visiting crew of Soyuz 27 left their spacecraft in the front port and returned on Earth using Soyuz 26, becoming the first humans since the docking mission of **Soyuz 4 (January 14, 1969)** and **Soyuz 5 (January 15, 1969)** to enter space in one spacecraft and leave on another. See Soyuz 27.

Soyuz 26 initiated a number of experiments that were continued or revised during all six Salyut 6 occupancies.

In smelting and crystal research, the cosmonauts used a materials-processing furnace similar to—but more advanced than—that used by Valery Kubasov on **Soyuz 6 (October 11, 1969)**. Designed to operate in a vacuum, the furnace was placed in one of Salyut's waste disposal airlocks. The outer door was opened to allow the atmosphere to evacuate and the waste heat to dissipate into open space. Each test experimented with creating mixed alloys of elements that, under Earth gravity, did not mix well. Small capsules of aluminum-tungsten, molybdenum-gallium, copper-indium, aluminum-magnesium, and indium-antimonide were melted and solidified for later study on Earth. The results showed that even the slight accelerations caused by the cosmonauts' movements were enough to interfere with the pure mixing of the alloys.

The cosmonauts continued the routine of daily exercises and medical tests. Unlike previous Salyut crews, their day-night schedule was no longer adjusted so that the men were awake when the laboratory was over Earth-based ground stations, a schedule that was sometimes unnatural and tiring. Instead, the cosmonauts lived a normal 24-hour day in space.

In biological research, the behavior of two sets of tadpoles was studied. One group had been hatched on Earth, then brought to space. The second was hatched in space. While the space-born tadpoles tended to swim in spirals, the earth-born tadpoles swam about randomly, showing less ability to orient themselves to weightlessness.

During their last few weeks in space, the cosmonauts were visited by Soyuz 28, carrying the first person to fly in space who was neither Soviet nor American. See Soyuz 28 for more details.

As their return to Earth approached, Grechko and Romanenko increased their daily exercise schedule. Then, on March 16, 1978, after 96 days in space, they entered the Soyuz 27 descent module and landed in the Soviet Union. Their space flight remained the longest until **Soyuz 29 (June 15, 1979)**.

Upon return, the two men experienced physical fatigue, muscular pain, and mental difficulties in readapting to Earth gravity. During the transition, they kept trying to "swim" out of their beds in the morning, and even simple actions required thought. It took two weeks before the cosmonauts' weight, heart rate, and other physical conditions returned to their preflight numbers. Interestingly, Grechko thought that his readaption was easier than it had been on his first flight on Soyuz 17.

December 15, 1977
Sakura 1
Japan / U.S.A.

Built and owned by National Space Development Agency (NASDA) of Japan, Sakura 1 (sakura means "cherry blossom" in Japanese) was Japan's first operational geosynchronous communications satellite. It was built by Ford Aerospace and Communications under contract with Mitsubishi and launched by NASA.

The satellite, similar in design to **NATO 3A (April 22, 1976)**, was the first to use 20 GHz and 30 GHz bandwidths for normal communications links. These frequencies, first tested on **ATS 6 (May 30, 1974)**, were required because the traditional communications bandwidths, 4, 6, 12, and 14 GHz, were already heavily used in Japan. The satellite provided government and public telephone service both to Japan's remote outer islands and to its large urban areas. Capacity was between 192 to 672 voice circuits, depending on whether two, one, or no television signals were being broadcast.

Sakura 1 operated until 1983, when it was replaced with Japan's next generation of communications satellites. See **Sakura 2A (February 4, 1983)**.

1978

January 10, 1978
Soyuz 27
U.S.S.R.

This six-day mission carried Oleg Makarov (**Soyuz 12–September 27, 1973, Soyuz 18A–April 5, 1975; Soyuz T3–November 27, 1980**) and Vladimir Dzhanibekov (**Soyuz 39–March 22, 1981; Soyuz T6–June 24, 1982; Soyuz T12–July 17, 1984; Soyuz T13–June 6, 1985**) to **Salyut 6 (September 29, 1977)**, joining Georgi Grechko and Yuri Romanenko of **Soyuz 26 (December 10, 1977)** during their 96-day mission. Though Soyuz 27 remained in orbit until March 16, 1978, Makarov and Dzhanibekov returned on January 16th, using Soyuz 26 as their return vehicle. The purpose of this switch was twofold: to free the rear docking port for unmanned Progress freighters, and to provide Grechko and Romanenko with a fresh return vehicle.

With the arrival of Soyuz 27, the Salyut 6 laboratory consisted of three separately launched spacecraft, with a habitable volume of almost 3,900 cubic feet. In order to test this complex for any unexpected vibrational resonance, the four cosmonauts held onto the Salyut 6 treadmill and "bounced" up and down together, seeing if the spacecraft complex amplified this vibration (just as soldiers marching in unison across a bridge can cause resonance that can shake the bridge apart). No resonance was found, but this experiment was conducted whenever a new configuration was achieved in future space laboratories.

Soyuz 27 was the first manned spacecraft to resupply another manned mission in space. In addition to food, books, and letters from Earth, Soyuz 27 also brought a French experiment for testing how weightlessness affected the ability of cells to divide. Single celled paramecia were carefully unfrozen and allowed to divide. Comparisons showed that the in-space specimens were nearly identical to the ground controls, with only subtle changes in cell metabolism.

The Soyuz 27 cosmonauts entered Soyuz 26 on January 16th and returned to Earth, leaving Grechko and Romanenko to continue their record-breaking occupancy and freeing the rear port for the arrival of the first unmanned Progress freighter, **Progress 1 (January 20, 1978)**.

The next manned mission to Salyut 6 was **Soyuz 28 (March 2, 1978)**.

January 20, 1978
Progress 1
U.S.S.R.

Progress 1 automatically docked with **Salyut 6 (September 29, 1977)** on January 22, 1978, and spent 18 days attached to the space laboratory, performing the first refueling of a spacecraft in space. See **Soyuz 26 (December 10, 1977)** for further details.

Progress 1 was the first unmanned resupply freighter flown in space. Its design was that of a modified Soyuz spacecraft, carrying fuel, food, air, and equipment for the cosmonauts. Like Soyuz, the freighters had no solar panels and carried only two days of power in their batteries. Once docked to the Salyut laboratory, they used the station's solar panels to recharge their batteries and remained attached until their docking port was needed. Cosmonauts would unload their cargo and use the freighter's space to store garbage, which accumulated at a rate of about 40 to 70 pounds a day. When the freighter undocked and deorbited, both it and the garbage were destroyed during re-entry.

Through 1990, a total of 42 Progress freighters were launched. In 1989, a second-generation freighter, **Progress M1 (August 23, 1989)**, was introduced.

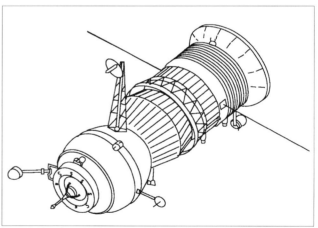

Progress freighter. *NASA*

January 26, 1978
IUE
U.S.A. / Europe / U.K.

The International Ultraviolet Explorer (IUE) was a joint project of the United States, the European Space Agency, and the United Kingdom. The United States provided two-thirds of the funding, with ESA and the United Kingdom splitting the rest. This ultraviolet telescope, a follow-up to **Copernicus (August 21, 1972)**, was placed in geosynchronous orbit where it observed the celestial sky in wavelengths from 1,150 to 3,200 angstroms, just beyond the visible wavelengths (4,000–7,000 angstroms). Using gyroscopes to stabilize the spacecraft (see **Miranda–March 9, 1974**), IUE was able to point at a single celestial object for periods as long as 14 hours, allowing it to record ultraviolet images of objects as faint as 17th magnitude. Spectrograms of objects as faint as 12th magnitude were also possible.

IUE was the first space telescope operated like a ground-based telescope. Astronomers could apply for telescope time, which was then scheduled by a review committee.

The telescope functioned for almost 20 years, until 1996, taking tens of thousands of images of thousands of celestial objects. IUE discovered that winds blew outward from many star types, from hot dwarfs to G and K type supergiants. It also found that these winds interacted with the interstellar material surrounding old supernovae, producing the beautiful optical planetary nebulae. The mass loss from these winds was also found to be high enough to directly affect the evolutional history of stars.

Moreover, IUE discovered that the stellar x-ray sources found by **Uhuru (December 12, 1970)**, **Ariel 5 (October 5, 1974)**, and **HEAO 1 (August 12, 1977)** were surrounded by hot circumstellar disks, more evidence that their x-ray emissions were caused by the infall of matter from a companion star onto a dense neutron star or black hole.

In its observations of galaxies, IUE found that the ultraviolet emissions from quasars and from Seyfert galaxies (by definition, galaxies with bright, compact nuclei) varied over time. Such variations were thought to occur, like the stellar x-ray sources, from the infall of the matter into a supermassive black hole. Evidence from IUE indicated that the central nucleus of one Seyfert galaxy contained an object several hundred million times more massive than the Sun.

IUE also detected the presence of a number of elements and molecules not previously seen outside the solar system. Gold was found in what astronomers called "type A peculiar" stars. The halo that surrounded the Milky Way was found to contain oxygen, sulfur, iron, silicon, and carbon.

In 1980, IUE made the first confirmation that comets are mostly water, finding that Comets Bradfield and Sergeant were similarly composed of hydrogen and oxygen, with measures of carbon, sulfur, and nitrogen. This discovery was later confirmed by the five probes sent to Halley's Comet in 1986, as well as by IUE itself (which made the first in-space observations of Halley's Comet upon its return). See **Vega 1, Halley's Comet Flyby (March 6, 1986)**, **Suisei, Halley's Comet Flyby (March 8, 1986)**, **Vega 2, Halley's Comet Flyby (March 9, 1986)**, **Sakigake, Halley's Comet Flyby (March 11, 1986)**, and **Giotto (March 13, 1986)**.

IUE also made detailed observations of supernovae, including Supernova SN1987A, the closest supernova since 1604. It found that these explosions play a much more significant role in enriching the universe with heavier elements (such as nitrogen, oxygen, carbon, and iron) than previously thought. IUE essentially proved that the Sun and the earth were formed from the essence of stars.

In solar system observations, the spacecraft discovered that Uranus has an aurora like the earth, and that Saturn's atmosphere contains acetylene.

In 1995, NASA transferred the operation of IUE to the European Space Agency, which after one year terminated the spacecraft's operation due to lack of funds. During that last year of operation, however, the spacecraft made observations of Comet Hale-Bopp, finding that the different ices in its nucleus were apparently clumped into isolated but distinct veins.

The over 100,000 spectra taken by IUE have since been reprocessed and are available on the Internet for any scientist to download and study.

February 4, 1978
Kyokko (EXOS A)
Japan

Dedicated to studying the aurora, Kyokko took ultraviolet images of auroral patterns every two minutes. It also carried equipment for measuring the energy, density, and temperature of charged particles in the aurora, as well as instruments to make ultraviolet spectra in five wavelengths ranging from 304 to 1,304 angstroms. The satellite operated in conjunction with **Jikiken (September 16, 1978)** and supplemented data from **ISEE 1/ISEE 2 (October 22, 1977)** and **ESA-GEOS 2 (July 14, 1978)**. See **Dynamics Explorer 1/Dynamics Explorer 2 (August 3, 1981)** for results.

February 9, 1978
FLTSATCOM 1
U.S.A.

The first in a cluster of six satellites, FLTSATCOM (Fleet Satellite Communication) 1 followed the experimental **LES 6 (September 26, 1968)** and **Tacsat 1 (February 9, 1969)** satellites, providing tactical communication links for U.S. military ships, submarines, aircraft, and ground stations. FLTSATCOM satellites worked in conjunction with the shuttle-launched LEASAT satellites (see **STS 41D–August 30, 1984**). Both systems were replaced by the UHF Follow-On satellites (see **UHF Follow-On 1–September 3, 1993**).

February 16, 1978
Ume 2 (ISS 2)
Japan

Ume 2—also called the Ionosphere Sounding Satellite (ISS) 2—was designed to study the ionosphere and how radio signals propagated through it. It also measured radio noise at five frequencies, from 2.5 to 25 MHz. See **ATS 6 (May 30, 1974)**.

February 22, 1978
Navstar 1
U.S.A.

Following **Timation 1 (May 31, 1967)** and **NTS 2 (June 23, 1977)**, Navstar 1 was the first in a four-satellite test cluster of global positioning system (GPS) satellites launched during 1978 by the U.S. military. Over the next two years, extensive tests in developing ground equipment and in refining satellite construction were performed. Over 40 different types of GPS ground instruments were tested. Location fixes of aircraft, helicopters, jeeps, ships, trucks, and even soldiers were achieved, with accuracy as good as 30 feet in three dimensions, even when moving at speeds up to 360 miles per hour.

Despite the program's success, the first two satellites experienced attitude control problems and clock failure, even though they each carried three redundant rubidium clocks.

Following these tests, the Navstar program began launching its first operational GPS cluster. See **Navstar 5** (**February 9, 1980**).

March 2, 1978
Soyuz 28
U.S.S.R. / Czechoslovakia

Soyuz 28 was the first manned launch involving crew members from two different countries. Soviet Alexei Gubarev (**Soyuz 17–January 11, 1975**) and Czechoslovakian Vladimir Remek docked with **Salyut 6 (September 29, 1977)** and spent eight days with the crew of **Soyuz 26 (December 10, 1977)**. Remek was the first non-Soviet and non-American to fly in space.

The joint mission focused on scientific research. In alloy tests, the two crews melted and mixed samples of lead, silver, glass, and copper chlorides in the materials processing furnace stored in Salyut 6's waste-disposal airlock. Other tests used plant life to recycle oxygen, carbon dioxide, and food. One biological experiment analyzed the ability of algae to grow in weightlessness, finding no difference between Earth- and space-grown cells. Another measured oxygen levels in human tissue samples. Micrometeoroid dust densities at 50 to 60 miles altitude were studied. Multispectral photographs of the earth were taken.

The crew of Soyuz 28 returned to Earth on March 10, 1978. The crew of Soyuz 26 returned on March 16th, completing the first occupancy of Salyut 6. The second occupancy began with **Soyuz 29 (June 15, 1978)**.

March 5, 1978
Oscar 8 / PIX 1
U.S.A. / International

• *Oscar 8*   Oscar 8 was the eighth in a series of privately financed amateur radio satellites and the third built by AMSAT (the Radio Amateur Satellite Corporation), an organization of ham radio operators from the United States, Australia, West Germany, Canada, and Japan. Oscar 8 was very similar to its two predecessors, **Oscar 6 (October 15, 1972)** and **Oscar 7 (November 15, 1974)**. During the period from 1978 to 1981 when both it and Oscar 7 functioned, the two satellites often worked together, relaying signals from amateur radio operators throughout the world. Oscar 8 operated until 1983. AMSAT's next Oscar satellite was **Oscar 10 (June 16, 1983)**. The next amateur communications satellite was **UoSat 1 (October 6, 1981)**.

• *PIX 1*   PIX (Plasma Interaction Experiment) 1 was a NASA experiment to find out if high voltage plasma events, which could short-circuit satellite electrical systems, could take place between the charged particles in the ionosphere and the higher voltage electrical systems of modern satellites. PIX 1 found that these events, caused by the impact of such charged particles, did occur, and that they could disable or damage satellite systems. A second test flight in 1983 studied how the size and area of solar panels affected these events.

April 4, 1978
Cosmos 1001
U.S.S.R.

This unmanned nine-day flight was the first test flight of the new Soyuz T manned spacecraft. A second 60-day test flight was flown in early 1979, followed then by the first official flight. See **Soyuz T1 (December 16, 1979)**.

April 7, 1978
Yuri 1
Japan / U.S.A.

Yuri 1 ("yuri" means lily in Japanese) was built by General Electric for the National Space Development Agency (NASDA) of Japan and launched by NASA. The satellite was the first use of direct broadcast technology, attempting to provide two television signals to areas of Japan that had poor or no reception. Home antennas could be as small as two feet across.

The satellite operated until 1980, when equipment failure ended its broadcasting ability. Through 1990, NASDA launched an additional three Yuri satellites, two of which were failures—one shortly after reaching orbit and the other at launch. The single success operated from 1986 until it was replaced in 1990 with a second-generation direct broadcast satellite that was able to carry three television channels.

With these satellites, more than a million Japanese homes were equipped with antennas, making this the first successful direct broadcast system in the world. The next such system was **Aussat A1** (see **STS 51I–August 27, 1985**).

April 26, 1978
HCMM
U.S.A.

Built by NASA, HCMM (Heat Capacity Mapping Mission) was the first, and (through December 1999) the only satellite dedicated solely to recording the temperature of the earth's surface and atmosphere. It used a scanner working in two spectral ranges, 0.55–1.11 microns and 10.5–12.5 microns, producing both visible and infrared images. The best resolution was a little less than one-third of a mile.

By noting how the temperatures of different features shifted differently throughout the day-night cycle, scientists produced detailed maps of surface features. Water, vegetation, and land features could be differentiated because they cooled at different rates. Urban areas stood out distinctly as hot regions.

The satellite operated for three years, until 1980. An archive of HCMM's images are available for public use. Later satellites, beginning with **Tiros N (October 13, 1978)**, used more advanced versions of HCMM's temperature-mapping scanner.

May 11, 1978
OTS 2
Europe / U.S.A.

This second Orbital Test Satellite (OTS) was built by ESA, the European Space Agency (its first OTS satellite having failed at launch). Placed in geosynchronous orbit over Europe by NASA, OTS 2 did communications tests, checking the satellite's design and performing propagation experiments at

12 GHz and 14 GHz, continuing the research of **Hermes** (**January 17, 1976**). See also **ATS 6** (**May 30, 1974**). OTS 2 also carried two pay-television cable channels, one for the United Kingdom and the other for France.

The satellite was the first for Eutelsat, an international consortium of 17 European nations established to provide business and public satellite communications services in Europe. Shareholders included the United Kingdom (15%), France (15%), Italy (10%), and West Germany (10%), with the remaining 50 percent divided among 12 other nations.

OTS 2 remained in use until 1984, at which time it was replaced by the first operational Eutelsat satellite cluster. See **Eutelsat 1** (**June 16, 1983**).

June 10, 1978
Chalet 1
U.S.A.

Placed in near geosynchronous orbit, Chalet 1 was the first of a new generation of U.S. Air Force signal and electronic surveillance satellites, replacing the Canyon satellites (see **Canyon 1–August 6, 1968**) and supplementing the Rhyolite satellites (see **Rhyolite 1–March 6, 1973**). It monitored and recorded the signals generated by Soviet and Chinese military spacecraft and equipment, beaming them back to American ground stations.

A total of five such satellites were successfully launched, the last reaching orbit in May 1989. Later versions, dubbed Magnum, were launched on the shuttle (see **STS 51C–January 24, 1985**).

June 15, 1978
Soyuz 29
U.S.S.R.

The crew of Soyuz 29, Vladimir Kovalyonok (**Soyuz 25–October 9, 1977**; **Soyuz T4–March 12, 1981**) and Alexander Ivanchenkov (**Soyuz T6–June 24, 1982**), completed the second occupancy of **Salyut 6** (**September 29, 1977**), spending 140 days in space, breaking the record of **Soyuz 26** (**December 10, 1977**) by over six weeks.

During this occupancy, two additional international crews docked with Salyut 6 for short visits, **Soyuz 30** (**June 27, 1978**) and **Soyuz 31** (**August 26, 1978**), with the crew of Soyuz 29 switching spacecrafts with the crew of Soyuz 31. See those flights for more information.

During their four-month flight, Salyut 6 was resupplied by three separate Progress freighters (see **Progress 1–January 20, 1978**). The first brought fuel and supplies, including a new instrument panel and a more advanced materials processing furnace. The second brought no fuel, but instead was stocked with air, supplies, and a guitar for Ivanchenkov. The third brought enough fuel and supplies to maintain the laboratory until the next manned occupancy. The fueling operations were controlled entirely on the ground, leaving the cosmonauts free to perform other tasks.

The cosmonauts continued the material and alloy tests begun by the crew of Soyuz 26, using both the old and new materials processing furnaces, to test aluminum, tin, and molybdenum. Attempts were made to manufacture gallium-arsenide crystals (used in many electrical diodes and solar cells—see **Sakura 3A–February 19, 1988**). Glass, suspended in weightlessness, was melted and remelted to test techniques of containerless manufacture. Crystals of germanium, indium-antimonide, and lead-telluride were manufactured.

Multispectral photographs of the Soviet Union specifically studied agricultural and mineral resources in the country's southern regions. In total, over 18,000 images were taken. The cosmonauts also recorded ocean color changes and directed fishing fleets to the location of fish and plankton. Their observations also revealed that the Ural Mountains began further south than had previously been recognized.

On July 29, 1978, the crew performed a two-hour space walk to recover a number of experiments attached to the outside of Salyut 6. With Kovalyonok acting as cameraman, Ivanchenkov removed a micrometeoroid detector and several test cartridges containing plastics, rubbers, and other materials being tested in the space environment for use in spacecraft construction.

Medical tests continued. For 16 hours a day, the cosmonauts wore the "penguin" suits first tested on **Soyuz 11** (**June 6, 1971**). A routine daily exercise program was followed, including a heavier workout schedule during the mission's last few weeks.

On November 2, 1978, the two men undocked from Salyut 6 and returned to Earth. Despite spending six weeks longer in space, Kovalyonok and Ivanchenkov recovered faster than the crew of Soyuz 26. For the first several days, they were weak and tired, with lower body weight and lower red blood cell and hemoglobin counts. Yet, they were able to walk on their second day on Earth, and by the sixth day, their major vital signs had returned to normal, though it still took almost four weeks for their muscular system to return to its preflight condition.

The third occupancy of Salyut 6 was **Soyuz 32** (**February 25, 1979**).

June 27, 1978, 1:12 GMT
Seasat
U.S.A.

Specifically designed by NASA to study the earth's oceans, Seasat was the first satellite to use radar imaging. All previous imaging satellites for observing either the earth or other planets had produced images in either ultraviolet (0.3–0.4 microns), visible (0.4–0.7 microns), or infrared (0.7–14 microns) wavelengths (see **Landsat 1–July 23, 1972**; **GOES 1–October 16, 1975**; **HCMM–April 26, 1978**). Seasat's primary imaging tool, a synthetic aperture radar, used a radar beam having a wavelength of 23 centimeters (about 9 inches). See *The Electromagnetic Spectrum* in the Introduction (page xii).

Synthetic aperture radar was designed to avoid the need for large orbiting antennas. A narrow beam swept the ground, which was reflected back to the spacecraft's small antenna in varying amounts and with measurable Doppler shifts, depending upon the undulations of the topography. Data-processing then measured these Doppler shifts and produced a synthesized image. Seasat's synthetic aperture radar created images about 62 miles wide with a resolution of about 80 feet.

Such radar could be oriented to look at the ground at a number of angles in order to simulate shadows of different lengths. In the case of Seasat, this "look" angle was set at 20

June 27, 1978

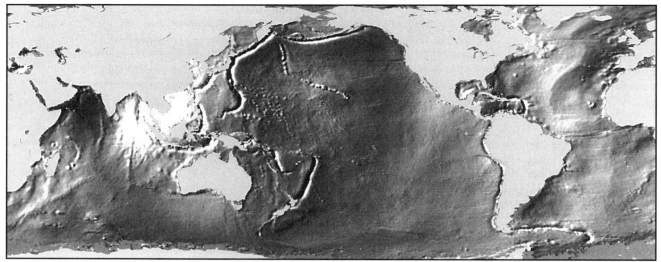
Topographical map of the ocean's surface, from Seasat. Note how sea floor features are reflected by surface topology. *NASA*

degrees and aimed to the starboard side of the spacecraft (the right-hand side when looking in the direction the spacecraft was moving). Later radar imaging spacecraft used different angles and looked either forward or backward, as well as sideways, depending on terrain. See, for example, **STS 41G (October 5, 1984)** and **Magellan (August 10, 1990)**.

The spacecraft also carried three other radar instruments. A radiometer monitored sea surface temperatures, wind speeds, rain, and ice conditions, working in five spectral bands from 8 to 45 millimeters. An altimeter attempted to measure altitude to within four-inch accuracy. A third sensor attempted to map the surface wind patterns across the entire globe.

Though the spacecraft's purpose was to scrutinize the earth's oceans, its sensors were also used to produce radar images of land features. Placed in a sun-synchronous orbit (see **Nimbus 1–August 28, 1964**), with radar that could see through clouds, Seasat was able to create radar images comparable to Landsat's photographic pictures.

Seasat functioned until October 1978, when a sudden power failure ended the mission. During that time, it produced radar images covering most of the Western Hemisphere. Its ocean images revealed strong currents on a planetary scale and mapped large-scale ocean eddies of high and low pressure. Later satellites (**Cosmos 1870-July 25, 1987** and **ERS 1–July 17, 1991**) confirmed this data.

The next spacecraft to use radar imaging was **Pioneer Venus Orbiter (December 4, 1978)**, which was on its way to Venus during Seasat's mission. The next use of radar imaging to observe the earth was on the space shuttle (see **STS 2– November 12, 1981**).

June 27, 1978, 15:27 GMT
Soyuz 30
U.S.S.R. / Poland

Soyuz 30 was the second manned international flight. Its crew comprised of Pyotr Klimuk (**Soyuz 13–December 18, 1973**; **Soyuz 18B–May 24, 1975**) of the Soviet Union and Miroslav Hermaszewski of Poland. The spacecraft docked with Salyut 6, joining the crew of **Soyuz 29 (June 15, 1978)** for eight days during the laboratory's second occupancy.

The two crews performed cardiovascular experiments, testing new equipment that monitored the men's heart rates while exercising. Polish materials experiments studied the uniformity of alloy formation in weightlessness, producing a mercury-cadmium-telluride alloy three to four times more homogeneous than any formed on Earth. Another test studied why the taste of foods seemed to change in weightlessness. Multispectral photography of Poland and the aurora was taken.

Soyuz 30 returned to Earth on July 5, leaving the Soyuz 29 crew to continue the second long-term occupancy of Salyut 6.

July 14, 1978
ESA-GEOS 2
Europe / U.S.A.

This European Space Agency Geosynchronous Satellite (ESA-GEOS) 2, launched by NASA and built by ESA, was intended to complete the mission of ESA's first geosynchronous satellite, **ESA-GEOS 1 (April 20, 1977)**, studying the earth's magnetosphere at the equatorial latitudes and geosynchronous altitudes and complementing data from **Kyokko (February 4, 1978)** and **ISEE 1/ISEE 2 (October 22, 1977)**.

While ESA-GEOS 2 successfully reached its planned orbit with all instruments working, less than a month later, it was apparently hit by a small object, which damaged a solar panel and caused a short-circuit with the spacecraft's electrical system. Despite this problem the spacecraft's experiments continued to gather data though 1983, mapping the composition and movement of the magnetosphere's plasma. See **Dynamics Explorer 1/Dynamics Explorer 2 (August 3, 1981)** for results.

After this mission, ESA was no longer dependent on NASA to launch its satellites—it was able to use Europe's own Ariane rocket, launched from French Guiana. See **CAT 1 (December 24, 1979)** and **Ariane 1 (June 19, 1981)**.

The next magnetospheric research satellite was **ISEE 3 (August 12, 1978)**.

The Sun and Stars

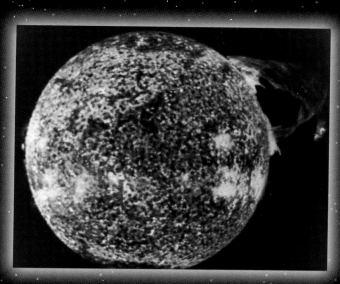

FIGURE 1. A coronal mass ejection erupting from the Sun, as photographed by astronauts on Skylab 2 (May 25, 1973). *NASA*

FIGURE 2. The Sun's corona, or atmosphere, taken by Solar Max (February 14, 1980). The colors indicate density, with purple the densest and yellow the least dense. The spikes, called streamers, can extend 10 million miles from the surface and are produced by streams of material flowing out of sunspot regions. *NASA*

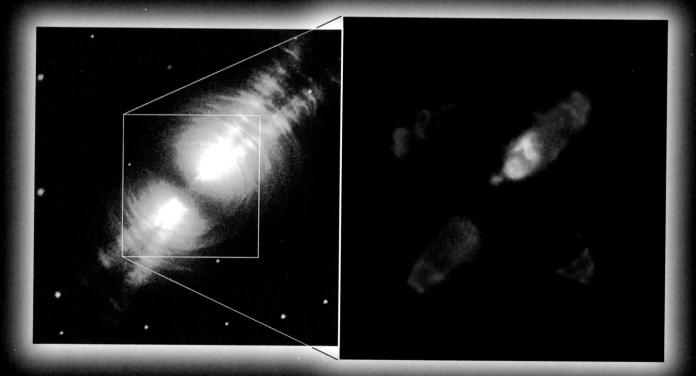

FIGURE 3. The Egg Nebula, as seen by the NICMOS infrared camera on the Hubble Space Telescope (see STS 82–February 11, 1997). The optical Hubble image is on the left, showing the star's bipolar jets. The infrared image on the right also reveals the red elliptical disk at right angles to the jets. *NASA*

The Gas Giants and Their Moons

FIGURE 4. Jupiter's Red Spot, with the shadow of the moon Io. See Pioneer 10 (December 3, 1973). *NASA*

FIGURE 5. Jupiter's dark bands and light zones, with an uplifted, lighter colored cloud prong. See Pioneer 10 (December 3, 1973). *NASA*

FIGURE 6. Pioneer 11 (December 2, 1974) looks down on Jupiter's chaotic northern latitudes. *NASA*

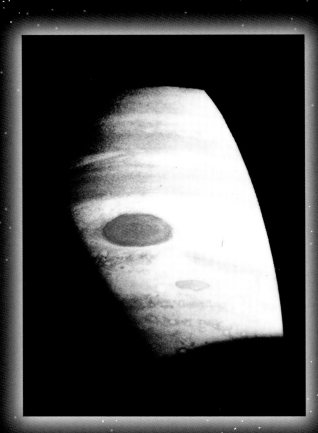

FIGURE 7. Jupiter's Red Spot up close. See Pioneer 11 (December 2, 1974). *NASA*

The Gas Giants and Their Moons

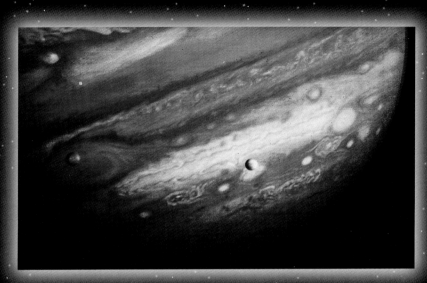

FIGURE 8. Jupiter's belts, zones, and Red Spot, as seen by Voyager 1 (March 5, 1979). *NASA*

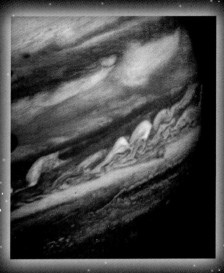

FIGURE 9. Jupiter's belts and zones as photographed by Voyager 2 (July 9, 1979). *NASA*

FIGURE 10. Turbulence around Jupiter's Red Spot, as seen by Voyager 1 (March 5, 1979). *NASA*

FIGURE 11. Jupiter's Red Spot, photographed by Voyager 2 (July 9, 1979). Compare with Figure 10, taken four months earlier. *NASA*

The Gas Giants and Their Moons

FIGURE 12. A false color infrared image of Jupiter's Red Spot, photographed by Galileo (December 7, 1995). Lighter colors indicate higher elevations, indicating that the Red Spot generally sits high in Jupiter's atmosphere. Compare with Figures 7, 10, and 11. *NASA*

FIGURE 13. Io—one of Jupiter's moons—photographed by Voyager 2 (July 9, 1979). Black spots are volcanoes. *NASA*

FIGURE 15. False color image of Saturn's rings taken by Voyager 2 (August 26, 1981). The colors indicate subtle variations in chemical composition from ring to ring. *NASA*

FIGURE 14. Photos of Saturn taken by Voyager 1 (December 12, 1980) and Voyager 2 (August 26, 1981) nine months apart show changes in the planet's bands and rings. *NASA*

The Gas Giants and Their Moons

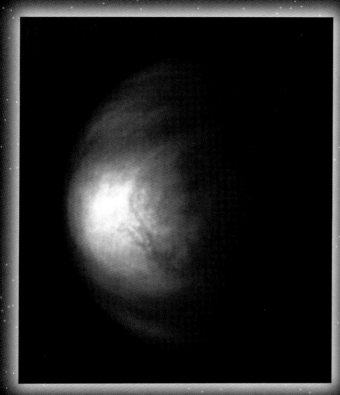

FIGURE 16. Uranus's clouds and bands photographed by Voyager 2 (January 24, 1986). The planet rotates on its side, the Sun shining on the planet's south pole for the last 40 years. *NASA*

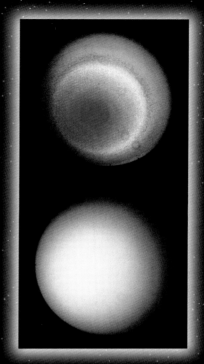

FIGURE 17. Images taken by Voyager 2 (January 24, 1986). On the bottom is a true color image of Uranus. On the top, false colors bring out the details of the different gases at the planet's south polar region. The "donut" shapes and pink edge on the false-color image are artifacts of the optics and processing. *NASA*

FIGURE 18. Neptune's clouds, storms, and Dark Spot, photographed by Voyager 2 (August 25, 1989). *NASA*

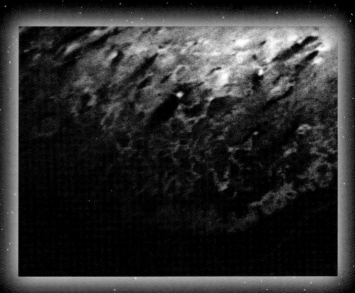

FIGURE 19. A close-up of the plumes and wind blown debris on Triton, Neptune's largest moon. See Voyager 2 (August 25, 1989). *NASA*

The Earth and Moon

FIGURE 20. The first global color photographs of Earth, taken November 8, 1967 (left), and November 10, 1967 (right), by ATS 3 (November 3, 1967). South America dominates the image, with Africa to the right and the southeastern United States to the upper left. Note the cloud movement over the two-day span. *NASA*

FIGURE 21. The Gemini 11 Agena Target (September 12, 1966) attached to Gemini 11 (September 12, 1966) by a 100-foot-long tether. *NASA*

FIGURE 22. The Apollo 11 (July 16, 1969) lunar module Eagle approaches the command module Columbia, with Earth in the background. *NASA*

FIGURE 23. Nepal and the snow-covered Himalayas as photographed by Landsat 1 (July 23, 1972). *NASA*

The Earth and Moon

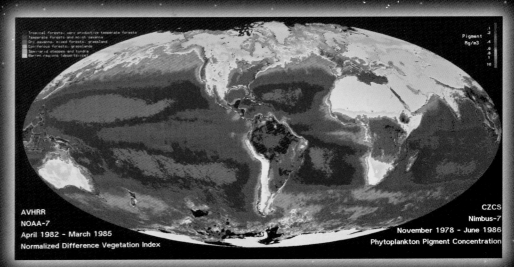

FIGURE 24. The global variation of phytoplankton and vegetation growth as seen by Nimbus 7 (October 13, 1978) from 1982 to 1985. On land, the darker colors signify tropical and very productive temperate forests, while bright yellows and reds indicate barren regions like deserts and icecaps. In the ocean, purple indicates less phytoplankton growth while red means more. *NASA*

FIGURE 25. False color image of the Washington, D.C., metropolitan area, taken by Landsat 4 (July 16, 1982) on November 2, 1982. The Capitol, the Mall, and Reagan National Airport are clearly visible. Red indicates forest or vegetation. *NASA*

FIGURE 26. The global topography of Earth's seafloor from Geosat (March 13, 1985). Pink indicates highest elevations. Dark blue indicates lowest elevations. *NASA*

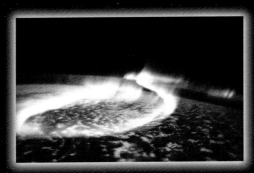

FIGURE 27. The aurora over Earth's northern hemisphere, photographed by the crew of STS 39 (April 28, 1991). *NASA*

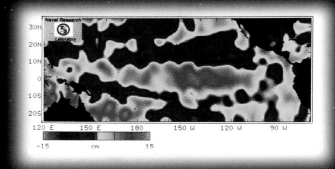

FIGURE 28. Variations in Earth's sea level as seen by TOPEX-Poseidon (August 10, 1992). The bright red strip along the Pacific equator measures a sea level rise of more than six inches, indicating the onset of El Niño. The silhouettes of North and South America can be seen on the right, with Australia in the lower left corner. *NASA*

Mars and Venus

FIGURE 29. Panorama of the Mars surface as seen by Viking 2 (August 7, 1976). *NASA*

FIGURE 30. The first color image of Venus's surface, taken by Venera 13 (March 1, 1982). *RIA-Novosti*

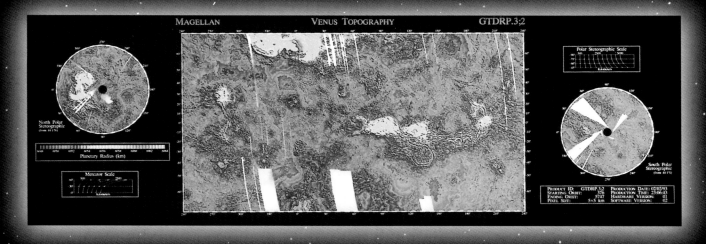

FIGURE 31. The global topography of Venus, as seen by Magellan (August 10, 1990). From deepest blue to pink indicates a rise in elevation of 10 miles. Compare with Pioneer Venus Orbiter (December 4, 1978). The Ishtar Terra plateau is at the top center, while the Aphrodite Terra plateau extends across the right side, south of the equator. Beta Regio is north of the equator near the left edge of the map. *NASA*

August 12, 1978
ISEE 3
U.S.A. / Europe

The International Sun-Earth Explorer (ISEE) 3 joined **ISEE 1/ISEE 2 (October 22, 1977)** in mapping the interaction of the solar wind and Earth's magnetosphere. All three spacecraft were part of a joint U.S. and European program that augmented data from **Kyokko (February 4, 1978)** and **ESA-GEOS 2 (July 14, 1978)**.

While the other spacecraft circled the earth in highly eccentric orbits, ISEE 3 was the first spacecraft placed in Lagrangian libration point L1. This position in space, one million miles closer to the Sun than the earth, was where the earth's and the Sun's gravitational fields balanced. Objects placed there tended to stay there. The L1 point was one of five such points surrounding the earth.

From this position, ISEE 3 had a continuous and unobstructed view of the solar wind and was able to detect changes in wind activity about an hour before they reached Earth. Scientists used this advance warning to correlate the data from ISEE 3 and from ISEE 1 and ISEE 2.

Over the next few years, all three spacecraft found that the northern and southern halves of the earth's magnetic field had somewhat different properties. While the northern field was fairly steady, the southern magnetopause was erratic, reacting violently to changes in solar wind activity.

The next magnetosphere research satellite was **Jikiken (September 16, 1978)**. See **Dynamics Explorer 1/Dynamics Explorer 2 (August 3, 1981)** for more results.

ISEE 3 also carried a gamma-ray burst detector, augmenting the monitoring by **Helios 2 (January 15, 1976)**, **Solrad 11A/Solrad 11B (March 15, 1976)**, and **HEAO 1 (August 12, 1977)** of the mysterious gamma-ray bursts first discovered by **Vela 5A/Vela 5B (May 23, 1969)**. Despite the attempts by all these satellites working together for many years to triangulate precisely the sky position of the bursts so that a visible counterpart could be identified, no such discovery was made. Gamma-ray bursts remained a mystery. Monitoring continued with **Ginga (February 5, 1987)**.

In 1982, NASA decided to send ISEE 3 on a new mission, to rendezvous with Comet Giacobini-Zinner on September 11, 1985. Because of this new project goal, the spacecraft's name was changed to International Cometary Explorer, or ICE. See **International Cometary Explorer (ICE)–September 11, 1985)** for more information.

August 26, 1978
Soyuz 31
U.S.S.R. / East Germany

This third international manned mission arrived at **Salyut 6 (September 29, 1977)** for an eight-day visit during the laboratory's second long-term occupancy by **Soyuz 29 (June 15, 1978)**. Soyuz 31 was manned by Soviet Valeri Bykovsky (**Vostok 5–June 14, 1963**, **Soyuz 22–September 15, 1976**) and East German Sigmund Jahn.

Multispectral photography was taken of East Germany using a camera system built by the Carl Zeiss camera factory in Jena. Other East German experiments included further experiments in weightless crystal growth and alloy mixing. Several alloy tests involved boiling a beryllium-thorium glass for use and study by the electronic optical industry.

In a more unusual test, Jahn repeated the phrase "226" in German each time he was in communication with Earth. By comparing how the tone, volume, and rate of his speech changed, scientists hoped to measure how the redistribution of body fluids to the upper body during weightlessness affected speech mannerisms.

The crew of Soyuz 31 also switched spacecraft with the Soyuz 29 crew, leaving their newer spacecraft behind and returning in the older Soyuz 29 descent module.

This flight was the last launched during the second occupancy of Salyut 6. The next Soyuz mission, **Soyuz 32 (February 25, 1979)**, began the station's third occupancy.

September 16, 1978
Jikiken (EXOS B)
Japan

Jikiken, called EXOS B prior to launch, was designed to study the earth's magnetosphere, working in conjunction with **Kyokko (February 4, 1978)**, **ESA-GEOS 2 (July 14, 1978)**, and all three ISEE satellites (see **ISEE 1/ISEE 2–October 22, 1978** and **ISEE 3–August 12, 1978**). Jikiken studied the plasma waves in the Van Allen belts and how those waves affected radio transmissions at certain low frequencies. See **Dynamics Explorer 1/Dynamics Explorer 2 (August 3, 1981)** for results.

The next magnetospheric research satellites were **Intercosmos 18/Magion 1 (October 24, 1978)**.

October 13, 1978
Nimbus 7 / Tiros N / CAMEO
U.S.A.

• *Nimbus 7* One of the most successful Earth research satellites ever launched, Nimbus 7 was also the last of NASA's Nimbus experimental environmental-meteorological research satellites. See **Nimbus 1 (August 28, 1964)** for a program description.

The spacecraft operated for more than a decade, with several sensors functioning into the early 1990s. Nimbus 7's instruments were the first to provide data on the earth's climate on a global basis, using many carefully refined spectral bands. Some sensors were like those flown on previous Nimbus flights, others were upgrades.

The Coastal Zone Color Scanner monitored chlorophyll (such as algae blooms), temperature, and sediment in coastal waters and the ocean *(see color section, Figure 24)*. It used five finely drawn spectral bands in visible light (0.433–0.453, 0.51–0.53, 0.4–0.56, 0.66–0.68, and 0.7–0.8 microns) and one infrared band (10.5–12.5 microns). Images covered a 1,000-mile swath with a resolution of about a half a mile. A second instrument, the scanning multichannel microwave radiometer, supplemented these data by measuring sea ice, glacial changes, and ocean surface conditions across five additional wavelengths.

These sensors functioned until 1986 and were not replaced until the launch of **SeaWiFS (August 1, 1997)**. They found that while seasonal changes in chlorophyll concentrations in the oceans were relatively constant in the southern hemisphere,

October 24, 1978

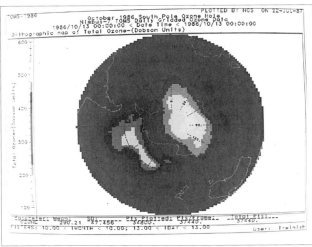

South Pole ozone hole as imaged by Nimbus 7. *NASA*

specific areas in the northern hemisphere (such as in the North Atlantic) showed dramatic seasonal swings.

The Earth Radiation Budget (ERB) sensor measured the balance of radiation entering and leaving the earth's atmosphere. This experiment operated for eight years, providing the first accurate measure of the solar constant, the amount of energy hitting the earth. Earthbound observations are incomplete because the atmosphere screens out many wavelengths, including ultraviolet radiation and x-rays. The data from ERB suggested that the Sun was actually a variable star; that its solar constant varied by about 0.1 percent through the 11-year solar cycle, with ultraviolet radiation actually varying by as much as 50 percent. Instrument uncertainties required more data to confirm this result. See **Solar Max (February 14, 1980)**.

The temperature-humidity infrared radiometer, flown on the previous two Nimbus satellites, measured infrared radiation in two wavebands. The 10.5–12.5 micron band was the standard infrared band, used by almost all infrared satellites to produce infrared images. The 6.5–7 micron band specifically measured water vapor in the upper atmosphere. Along with two sensors for monitoring water, gas, and temperatures in the stratosphere (between about 10 to 33 miles elevation), these instruments provided a long-term profile of the atmosphere's temperature and moisture, as well as global maps of the sea surface temperature. Through 1989, the data showed no indication of any global climate changes, either warming or cooling.

The Solar Backscatter Ultraviolet (SBUV) and Total Ozone Mapping (TOMS) sensors monitored 18 narrow spectral ultraviolet bands (0.25–0.34 microns), measuring the ultraviolet light radiated by the Sun and Earth. This data made possible the first detailed look at the ozone layer at high latitudes.

The instrument discovered that during the southern winter months, a large "hole" developed in the ozone layer over Antarctica. This discovery seemed to confirm the theories of many scientists predicting that the use of chlorofluorocarbons (CFCs) in aerosol spray cans would eventually damage the ozone layer. Because the database was so short, however, covering less than one solar cycle, and because the variation in solar ultraviolet radiation was seemingly so great over this same period of time, further research was necessary to confirm this hypothesis. Moreover, specific events in 1982—a volcanic eruption in Mexico and an El Niño event—might have also contributed to some of the ozone drop. See **SME (October 13, 1981)** for later results.

• **Tiros N** Placed in a sun-synchronous orbit (see **Nimbus 1–August 28, 1964**), Tiros N tested several new meteorological instruments. It was built and launched by NASA and operated by NOAA.

The satellite's prime sensor was the Advanced Very High Resolution Radiometer (AVHRR). AVHRR used five spectral wavebands in visible and infrared light (ranging from 0.55–0.68 microns to 10.5–11.5 microns) to measure cloud coverage, sea surface temperatures, atmospheric composition, and plant cover. Resolution could be less than two-thirds of a mile, depending on mode. Data could be recorded either onboard and transmitted later, or downloaded directly by local weather bureaus.

A second instrument, the Tiros Operational Vertical Sounder (TOVS), consisted of three separate sensors for measuring the composition, temperature, and humidity of the atmosphere's profile from the ground to 30 miles elevation. In addition to gathering daily weather data, TOVS confirmed the presence of the ozone hole seen by **Nimbus 7 (October 13, 1978)**.

This suite of instruments formed the backbone of the next generation of NOAA satellites, beginning in June, 1979. See **NOAA 1 (December 11, 1970)**.

• **CAMEO** On October 29, 1978, a barium cloud was released from the still-orbiting second stage of the Delta rocket that had been used to place Nimbus 7 and Tiros N in orbit. This NASA experiment, labeled CAMEO (Chemically Active Material Ejected into Orbit), created a bluish-white cloud more than 600 miles long at an altitude of about 600 miles, drifting southwestward. The cloud's movement was used by scientists to study the circulation patterns within the ionosphere.

October 24, 1978
Intercosmos 18 / Magion 1
U.S.S.R. / Czechoslovakia

This eighteenth launch in the Intercosmos series of satellites (see **Intercosmos 1–October 14, 1969**) was only unique in that it carried a second subsatellite, Magion 1, the first satellite built by Czechoslovakia. Magion 1 separated from Intercosmos 18 on November 14, 1978, and was designed to augment Intercosmos 18 by providing complementary proximity data of the ionosphere and magnetosphere.

Through 1996, a total of five Magion subsatellites were launched, the first three with an Intercosmos satellite, the last two with an Interball satellite (See **Interball 1–August 2, 1995**). All acted to augment data from the main satellite.

See **Dynamics Explorer/Dynamics Explorer 2 (August 3, 1981)** for results.

October 26, 1978
Radio Sputnik 1 / Radio Sputnik 2
U.S.S.R.

These two satellites were the Soviet equivalent to the amateur radio satellites launched under the Oscar series in the West (see **Oscar 1**–**December 12, 1961**). Both satellites were placed in a near polar low-Earth orbit and carried active repeaters for relaying ham radio messages. Through 1991, the Soviets launched 14 similar Radio Sputniks. The last, launched piggyback on a civilian experimental communications satellite, **Informator 1 (January 29, 1991)**, would also be called Oscar 21, the first to also participate in the Western program.

The Soviets also launched three Iskra satellites (Iskra means "spark" in Russian), all described as amateur radio communications satellites. One was launched piggyback with a weather satellite in 1981, and two were released from the airlock of **Salyut 7 (April 19, 1982)**. See **Iskra 2 (May 17, 1982)** and **Iskra 3 (November 18, 1982)**.

November 13, 1978
Einstein (HEAO 2)
U.S.A.

Built by NASA, Einstein (also called High Energy Astronomy Observatory 2, or HEAO 2) was a follow-up to **HEAO 1 (August 12, 1977)**. While HEAO 1 did an all-sky survey in x-rays, Einstein performed targeted observations on specific x-ray objects that emitted energy between 0.2 and 4 KeV. Unlike previous x-ray telescopes, Einstein did not simply detect the emission of x-rays; it produced detailed images of these bursts, allowing scientists to compare their shape to optical images and in some cases identify the visible counterpart to the x-ray object.

During its over-two-year life span, Einstein made x-ray images of what were then considered black hole candidates, Cygnus X-1, Hercules X-1, Scorpius X-1, and Circinus X-1, accumulating new data on the energy variability of these strange

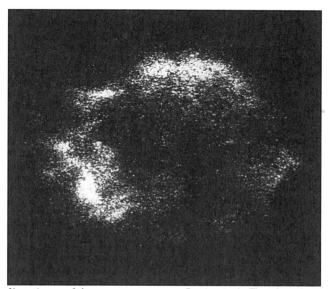

X-ray image of the supernova remnant Cassiopeia A. Thought to have occurred around 1657, this supernova was previously unknown because its visible light was obscured by dust. *NASA*

objects. (Circinus X-1 and Hercules X-1 are now believed to be neutron stars, not black holes.) Also studied were about 60 quasars, leading to the finding that these distant objects were some of the most powerful emitters of x-rays in the sky and apparently contribute over 20 percent of the sky's low-level background x-radiation.

The Crab Nebula was also photographed in x-rays, showing that its central pulsar beat with x-rays as well as radio pulses, and that these waves of energy interacted with the gaseous cloud surrounding the neutron star to shape it into the complex weave of gas seen from Earth. Other supernova remnants, such as Cassiopeia A and SN1006, also revealed the existence of gigantic x-ray emitting shell structures left over from the initial supernovae explosions. In many cases, these giant cloud shells were only visible in x-radiation.

Scientists also found that many young giant O, B, and A stars (such as the unpredictably variable giant Eta Carinae), as well as K and M stars, shone with more x-ray light than expected. This discovery required astronomers to rethink their theories about what heated the corona of stars.

Previous spacecraft had indicated that giant galaxy clusters were x-ray emitters. Einstein resolved these images to reveal that the emissions were coming from individual galaxies. Centaurus A, the closest galaxy whose nucleus radiated in radio waves, x-rays, and gamma rays, was found to contain several distinct x-ray sources, including an inner jet aligned with the galaxy's faintly visible optical jet. In fact, Einstein found that most Seyfert galaxies were also x-ray emitters. From these discoveries, astronomers created a new class of objects, called active galactic nuclei—galaxies whose nuclei emit copious amounts of ultraviolet, x-ray, and gamma radiation, and within which scientists suspected were hidden giant supermassive black holes.

Several nearby galaxies were also photographed. In the Andromeda galaxy were found about 50 x-ray sources, most clustered in the central bulge with a number tracing out the galaxy's spiral arms. Scientists believed these sources were similar to the Milky Way's population of x-ray bursters.

The next x-ray telescope was **Hakucho (February 21, 1979)**. The last HEAO satellite was **HEAO 3 (September 20, 1979)**. The next American x-ray telescope was **Rossi (December 30, 1995)**.

December 4, 1978
Pioneer Venus Orbiter
U.S.A.

This NASA spacecraft was the first American probe dedicated entirely to studying Venus since **Mariner 5, Venus Flyby (October 19, 1967)**. It was also the second radar-imaging satellite, following **Seasat (June 27, 1978)**. Launched on May 20, 1978, it reached Venus on December 4, 1978, and spent the next two years radar-mapping about 93 percent of the planet's shrouded surface. The satellite was then turned off until 1991, when its orbit had migrated to a position that allowed it to radar-map previously inaccessible polar regions. It finally ran out of fuel and was destroyed when it entered the Venusian atmosphere in August 1992.

The radar mapper revealed the surface of Venus to be flatter than Earth's, with few areas suggesting the earth's depres-

December 4, 1978

sions and continental shelves. While several plateaus and peaks resembling shield volcanoes were identified, the planet's surface generally appeared to be a rolling terrain with few distinct topographical features. Moreover, most of the surface features that could be seen had little in common with features on other planets in the inner solar system. Though a limited number of circular features were seen, they did not match the expected topography of impact craters (see **Mariner 10, Mercury Flyby #3–March 16, 1975**). The lack of craters implied a young surface, which further implied volcanism. Yet, while the data indicated a volcanic history, that history did not seem to resemble Earth's tectonic plate geology. Instead, the nature of the topography hinted at a process somewhere between Earth's moving plates and Mars's giant volcanoes and deep rift valleys.

Pioneer Venus Orbiter's maps showed that both **Venera 9 (October 22, 1975)** and **Venera 10 (October 25, 1975)** had landed on the skirt of an elevated region, Beta Regio, thought to be a giant shield volcano, located at 20°N 285°E on the map. The fact that these two Veneras had detected variations of basalt beneath them helped confirm this conclusion. **Venera 8 (July 22, 1972)**, meanwhile, had landed on rolling plains, far from any elevated regions, which matched the granitelike ancient crust material detected beneath it.

Where once scientists had asked whether the craters on the Moon were volcanic or impact in origin, scientists now wondered about the extent of volcanism on Venus. Though the data indicated that the planet's surface had been extensively shaped by volcanism, scientists simply had too little evidence to know for certain.

The spacecraft also carried sensors for studying the Venusian atmosphere over time. Based on this data, scientists learned that the atmosphere's cloud cover was mostly contained within a main cloud layer of sulfuric acid droplets sitting at 28 to 44 miles altitude. Though essentially opaque, the thick main layer was still quite tenuous, with several miles of visibility at any point within it—its opacity was more a function of its depth than its density. Above and below this were many other thin haze layers, with the top cloud layer at an altitude of 55 miles, sitting at temperatures between –60°F to –120°F. The clouds and haze seemed to fade out below 20 miles altitude. All the layers appeared to have a stability and structure enduring across years and great distances, as implied by images from **Mariner 10, Venus Flyby (February 5, 1974)**.

Nonetheless, from 1979 to 1986, instruments on Pioneer Venus Orbiter found that while the cloud cover over the poles was slowly brightening, there was a slow global decrease in the amount of atmospheric haze and sulfur dioxide. One theory proposed that when the orbiter arrived, a volcanic eruption had just taken place, saturating the atmosphere. The decrease indicated the slow return to earlier levels.

The spacecraft also measured other variations within the atmosphere. The northern latitudes were colder than the equator by about 90°F, with northern cloud tops floating at lower elevations. Both of these facts indicated circulation patterns not unlike those on Earth—air rose at the warm equator, flowed to the cold north and south, and sank, to begin the cycle over. Mariner 10 had indicated this circulation pattern; Pioneer Orbiter's data reinforced its existence.

The data still did not clarify what was causing the dark streaks in Venusian clouds when photographed in ultraviolet light. Some scientists had proposed sulfur dioxide as its cause—SO_2 could easily absorb ultraviolet light. Others proposed the dark streaks

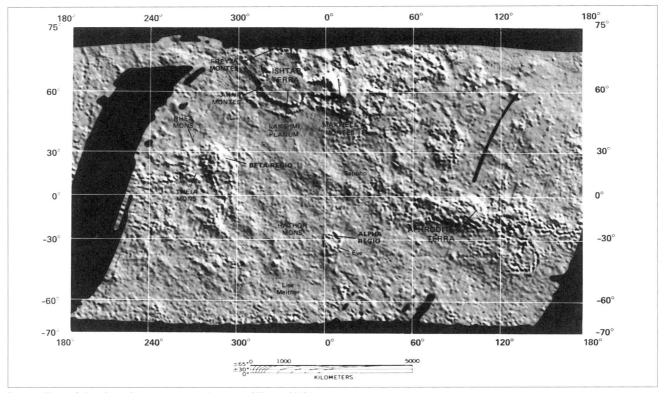

Pioneer Venus Orbiter's preliminary topographic map of Venus. *NASA*

were caused by the existence of an overlying haze layer that obscured the main cloud bank below.

As Pioneer Venus Orbiter continued its study of Venus, more spacecrafts arrived at the planet, first **Pioneer Venus Probe (December 9, 1978)** and followed by two more Soviet probes, **Venera 12, Venus Landing (December 21, 1978)** and **Venera 11, Venus Landing (December 25, 1978)**.

December 9, 1978
Pioneer Venus Probe
U.S.A.

Built by NASA and launched on August 8, 1978, Pioneer Venus Probe dropped five different spacecraft into the Venusian atmosphere. One large and one small probe entered on the planet's day side, while two other small probes landed on the night side. The Bus, which had carried the four probes to Venus, also cut across the atmosphere at a low angle.

In general, the data from these five probes agreed with the information from previous Venera spacecraft (see **Venera 6, Venus Landing–May 16, 1969**). Venus's atmosphere was incredibly hot and dense, which was apparently the case in many different places across the face of the planet. The atmospheric pressure and temperature increased steadily as the probes dropped into the main sulfuric cloud layer at about 45 miles altitude until they reached the surface or ceased operating, rising from about −10°F to 850°F. At the surface, the atmospheric pressure was approximately 1,350 pounds per square inch, almost 100 times greater than Earth's surface air pressure.

In fact, only minor differences were observed in temperature and pressure by all 14 Soviet and American probes that had studied Venus's atmosphere to this date. From the surface to the top of the main cloud bank at about 45 miles elevation, the atmosphere was stable, unaffected by the slow Venusian day and night cycle. Above this height, however, the atmosphere varied significantly between its day and night sides, with the flow moving away from the Sun.

The next Venus probes were **Venera 12, Venus Landing (December 21, 1978)** and **Venera 11, Venus Landing (December 25, 1978)**.

December 15, 1978
Anik B (Telesat 4)
Canada / U.S.A.

Built by RCA, Anik B was a geosynchronous communications satellite owned and operated by Telesat Canada (see **Anik A1–November 10, 1972**). The satellite augmented the capacity of the Anik A satellite cluster and performed atmospheric propagation tests at 12 GHz and 14 GHz (see **ATS 6–May 30, 1974**).

The next Telesat Canada satellite cluster included the **Anik D (August 26, 1982)** and **Anik C (STS 5–November 11, 1982)** satellites.

December 19, 1978
Gorizont 1
U.S.S.R.

Though this first Gorizont satellite was a partial failure (its fourth stage failed to fire, preventing the satellite from reaching geosynchronous orbit), its launch initiated the establishment of the third Soviet geosynchronous communications satellite cluster, following **Raduga 1 (December 22, 1975)** and **Ekran 1 (October 26, 1976)**. Gorizont satellites, however, were part of the Intersputnik consortium, the communist bloc's answer to Intelsat (see **Early Bird–April 6, 1965** for information about Intelsat). Intersputnik, officially formed in 1972, had 14 members in 1990 with an additional nine nonmember countries using its satellite system, including Algeria, Iraq, and Nicaragua. The members included the Soviet Union, Afghanistan, Bulgaria, Cuba, Czechoslovakia, East Germany, Hungary, Laos, Mongolia, North Korea, Poland, Romania, Vietnam, and Yemen.

Gorizont ("horizon" in Russian) provided television and telephone service to the Soviet Union and the members of the Intersputnik organization. Beginning in the mid-1980s, an agreement allowed U.S. users to also access the system. When completed, the Gorizont cluster covered every inhabited part of the globe except parts of North America. It also supplemented the Soviet elliptically orbiting **Molniya 1 (April 23, 1965)**, **Molniya 2 (November 24, 1971)**, and **Molniya 3 (November 21, 1974)** satellites.

Through 1996, 32 Gorizont satellites were launched. In 1994, Russia began replacing the cluster. See **Gals 1 (January 20, 1994)** and **Express 1 (October 13, 1994)**.

December 21, 1978
Venera 12, Venus Landing
U.S.S.R.

Launched on September 14th, Venera 12 managed to beam back data for 110 minutes after soft-landing on Venus. It was joined on the Venusian surface four days later by **Venera 11 (December 25, 1978)**. Neither spacecraft returned photographs.

Both spacecraft measured surface conditions similar to those seen by previous landers (see **Venera 9, Venus Landing–October 22, 1975** and **Venera 10, Venus Landing–October 25, 1975**). The surface air pressure was about 90 Earth atmospheres. The surface temperature was about 860°F.

The gamma-ray spectrometer, built in France, discovered traces of the inert element argon in Venus's atmosphere, at levels 200 to 300 times greater than found on Earth. Carbon monoxide was also measured. The Venusian atmosphere below 60 miles altitude was now known to be made up of about 96.5 percent carbon dioxide, with the remaining 3.5 percent almost all nitrogen. Traces of the noble gases (helium, neon, argon, and krypton) were found in quantities greater than found on Earth. Also detected was sulfuric acid, hydrogen fluoride, and hydrogen chloride. The amount of oxygen, however, was still unclear. Data from most of the landing probes showed very little oxygen in the air, with an inexplicable increase in oxygen at altitudes between 32 and 40 miles.

As the Venera 11 and Venera 12 spacecraft dropped through Venus's atmosphere, they both recorded several severe electromagnetic pulses, similar to the radio impulses produced on Earth by lightning. Venera 11 recorded an almost continuous series of short bursts, lasting 20 minutes, as it dropped from about 19 to 8 miles altitude. While these lightning bursts were clearly a discharge of electromagnetic energy, scientists were unsure of their origin.

As it descended, Venera 12—surprisingly—measured chlorine as the most abundant element, along with quantities of sulfur. Why chlorine was suddenly dominant puzzled the scientists.

December 25, 1978
Venera 11, Venus Landing
U.S.S.R.

Though launched on September 9th, Venera 11 arrived at Venus four days after **Venera 12** (**December 21, 1978**), which had been launched five days later. After successfully soft-landing on the surface, the spacecraft returned data for 95 minutes. See Venera 12 for a summary of results.

Following Venera 12 and Venera 11, three years passed before another probe was sent to any planet—the longest pause in exploration since the dawn of the space age. This gap in exploration ended with the launch in 1981 of two Venus probes, **Venera 13** (**March 1, 1982**) and **Venera 14** (**March 5, 1982**).

1979

January 30, 1979
SCATHA
U.S.A.

SCATHA (Spacecraft Charging at High Altitudes) was an engineering test launched by NASA to study the build-up of static electricity—sometimes as high as 20,000 volts—on spacecraft surfaces in geosynchronous orbit (see **Explorer 8–November 3, 1960** and **Gemini 4–June 3, 1965**). Engineers suspected that such build-ups had contributed to the failure and loss of data from several satellites.

February 12, 1979
Cosmos 1076 (Okean-E)
U.S.S.R.

Also called Okean-E, this Cosmos satellite was the first test of a Soviet satellite design dedicated to observing the sea ice conditions in the Arctic Ocean. Placed in a sun-synchronous orbit about 400 miles high, it was one of the first Soviet satellites to be placed in such an orbit. See **Meteor 1–March 26, 1969** for why a sun-synchronous orbit was not usually used by Soviet satellites.

Cosmos 1076 carried both visible and infrared scanners. A second test flight with an identical configuration was flown in early 1980. These two test satellites not only mapped sea ice changes but also detailed global surface water temperatures in the middle north latitudes. This information related directly to sea current changes in the Gulf Stream and North Atlantic.

In 1983, a third test flight, **Cosmos 1500** (**September 28, 1983**), included additional radar sensing equipment, and was put to use immediately observing ice conditions in the East Siberian Sea.

February 18, 1979
SAGE
U.S.A.

Refining measurements begun with **Nimbus 7** (**October 13, 1978**), NASA's Stratospheric Aerosol and Gas Experiment (SAGE) satellite made vertical maps of the ozone and aerosol levels in the stratosphere between 6 and 30 miles altitude. Such aerosols—ultramicroscopic solid or liquid particles suspended in the atmosphere—had never before been measured, and their quantity had a direct bearing on whether the earth's climate would heat or cool in the coming centuries.

SAGE was used to track the aerosol cloud from the Mount St. Helens eruption in 1980, as well as four other volcanic eruptions, showing how such clouds spread and dissipated through the atmosphere. The data suggested that the aerosol distribution and concentration between the northern and southern hemispheres might not be identical.

Through 1980, these volcanic events significantly increased the aerosol levels over the earth's poles. The satellite, which operated until 1981, also measured a slight overall increase in aerosol over that period. Whether this increase was because of the volcanic eruptions or a more permanent atmospheric change due to pollution remains an open scientific question, requiring many more years of data collection.

A similar sensor was flown on ERBS (see **STS 41G–October 5, 1984**).

February 21, 1979
Hakucho (CORSA B)
Japan

Hakucho ("swan" in Japanese) was Japan's first celestial x-ray telescope. Its detectors measured x-rays in the range from 0.1 to 100 KeV, continuing the monitoring of the x-ray sky by **HEAO 1** (**August 12, 1977**) and **Einstein** (**November 13, 1978**).

The next x-ray telescope was **Ariel 6** (**June 2, 1979**).

February 24, 1979
Solwind (P78-1)
U.S.A.

This U.S. Navy satellite, designed to evaluate a number of new engineering satellite designs, also carried several scientific sensors for monitoring the solar wind, solar flare activity, and the Sun's corona from 2.5 to 10 solar radii from the Sun's center. It continued the solar observations of **Solrad 11A/Solrad 11B** (**March 15, 1976**).

Though its service life was only planned for one year, Solwind operated until 1985. In 1979, it took the first picture of a comet colliding with the Sun—the first record of any such solar impact—and later analysis of its data found two more such Sun-grazers.

Solwind was destroyed on September 13, 1985 (while still functioning) as part of an American military anti-satellite (ASAT) test. See **MHV** (**September 13, 1985**) for details.

The next solar observation satellite was **Solar Max** (**February 14, 1980**).

**February 25, 1979
Soyuz 32
U.S.S.R.**

Soyuz 32 was crewed by Vladimir Lyakhov (**Soyuz T9–June 27, 1983**; **Soyuz TM6–August 29, 1988**) and Valery Ryumin (**Soyuz 25–October 9, 1977**; **Soyuz 35–April 9, 1980**; **STS 91–June 2, 1998**). This flight completed the third long-term occupancy of **Salyut 6 (September 29, 1977)**, setting a new space endurance record of 175 days, exceeding that of **Soyuz 29 (June 15, 1978)**.

Unlike the previous two occupancies, Soyuz 32 received no human visitors during its long mission. The only other manned mission during this period, **Soyuz 33 (April 10, 1979)**, was aborted when the spacecraft's main propulsion system failed after reaching orbit.

Three Progress freighters (see **Progress 1–January 20, 1978**) did dock with the spacecraft during Lyakhov's and Ryumin's six-month stay, bringing food, fuel, and air supplies. The first freighter also brought a third materials-processing furnace to replace the second, as well as a gamma-ray telescope, a chemical storage battery for back-up electricity, and a television receiver.

Also docking to Salyut 6 during this third occupancy was an unmanned Soyuz spacecraft, **Soyuz 34 (June 6, 1979)**, launched specifically to replace the three-month-old Soyuz 32, providing the cosmonauts a newer return spacecraft. Because Soyuz 32 returned to Earth unmanned, it was used by the cosmonauts as a cargo craft for doubling the quantity of photographic film and test specimens from the materials-processing furnaces normally returned to Earth.

During this mission it was revealed that near the end of the Soyuz 29 mission, a leak in Salyut 6's fuel tanks had been detected. Though Lyakhov and Ryumin of Soyuz 32 managed to locate and drain this leaking tank, Salyut 6's engines were not used to maintain its orbit for the remainder of its life. Instead, either Progress or Soyuz engines periodically lifted the station into higher orbits.

During their six-month stay, the cosmonauts continued or started experiments involving astronomy, biology, mineralogy, medicine, and Earth resources. They took numerous multispectral photographs of the earth and studied land geology and weather patterns. Materials processing experiments continued. Gamma-ray astronomy was performed. They also tested an electron beam gun for recoating the surface of mirrored surfaces in weightlessness and a vacuum.

The cosmonauts successfully hatched several quail eggs in an incubator, but the embryos developed much more slowly in space. One even had a body but no head. This experiment was repeated with more success on **Soyuz TM9 (February 11, 1990)**.

The cosmonauts also assembled and tested a 33-foot radio antenna in the rear docking port, the first in orbit since **RAE 2 (June 10, 1973)**. Such an antenna, when combined with an Earth-based radio antenna, created an extremely large radio antenna array. Because the antenna did not deploy properly, however, it produced less than acceptable results. In fact, a space walk on August 15 was required to free the antenna so that the rear docking port would be available for future Progress freighters. During this space walk, Ryumin and Lyakhov used wire cutters to free and dispose of the antenna, then retrieved a micrometeoroid detector as well as a number of other material samples being tested for use in future space operations. **Prognoz 9 (July 1, 1983)** was the next orbiting radio telescope.

Probably the most important results from Soyuz 32 were the medical results. The cosmonauts had followed a consistent routine of two hours exercise per day, and upon their return on August 19, 1979, they recovered as fast as the Soyuz 29 crew (who spent 140 days in space), and faster than the Soyuz 26 crew (96 days in space). Nonetheless, both men showed muscle and bone mass loss, and both had a slight difficulty speaking immediately after landing until the fluid distribution in their bodies readapted to gravity.

Salyut 6 remained unmanned for the next eight months, until **Soyuz 35 (April 9, 1980)** began the laboratory's fourth occupancy. During those eight months, an unmanned test Soyuz spacecraft, **Soyuz T1 (December 16, 1979)**, was docked with the station for 95 days.

**March 5, 1979
Voyager 1, Jupiter Flyby
U.S.A.**

Built by NASA and launched on September 5, 1977, Voyager 1 was the third spacecraft to fly past Jupiter, following **Pioneer 10 (March 3, 1972)** and **Pioneer 11 (April 6, 1973)**. It dipped to within 485,000 miles of the gas giant, and as it did so, it also flew close enough to the moons Io, Ganymede, Callisto, Europa, and Amalthea to take the first tantalizingly detailed pictures of these objects. The two most astonishing discoveries of the Voyager 1 flyby were that Jupiter itself had rings similar to Saturn's and that Io was a hellish home to the first active volcanism seen outside the earth.

Historic image of a volcano erupting on Jupiter's moon Io. *NASA*

March 5, 1979

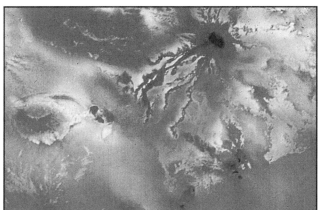
Lava flows off of a volcano on Io. Black spot in the upper right is the volcano peak. *NASA*

The first photograph showing a volcano erupting on Io revealed a several-mile-high plume silhouetted against the moon's horizon. Later, other plumes were spotted, along with evidence of three lava lakes, one the size of Hawaii with a temperature between 100°F and 200°F. Other photographs of Io showed a strange "Halloween" surface, with a shiny orange and white surface blistered by black volcanoes and sable lava flows with few craters. The salt flats postulated after Pioneer 10's flyby simply did not exist.

One of the most interesting aspects of this surprising discovery about Io was that one week before the flyby, three scientists had predicted it. Stanton Peale of the University of California and Patrick Cassen and Ray Reynolds of the Ames Research Center proposed that tidal heating as Io orbited Jupiter could cause its interior to churn and boil, thereby provoking repeated volcanic eruptions and a resurfacing of the planet's crust.

Jupiter's ring system was thin and flat, but nonetheless present. Its existence defied the theories of scientists who had predicted that such rings were not stable and would exist only rarely and for short periods.

Jupiter's Great Red Spot was found to be colder than the surrounding atmosphere, its counterclockwise rotation repeating every six days. White storm centers were once again vis-

Jupiter's moon Europa. *NASA*

Jupiter's moon Ganymede. *NASA*

ible to the south of the Spot *(see color section, Figure 8)*, as had been seen by the two Pioneer spacecraft. To its west was a region of vast turbulence *(see color section, Figure 10)*.

Voyager 1 found that the Jovian atmosphere contained aurorae, cloud-top lightning bolts and fireballs, and a much hotter day-side temperature (about 1,500°F) than seen on the previous two flybys. This change indicated that the atmosphere's weather could vary significantly over time. In the polar regions, east-to-west wind currents were measured, not the up- and down-welling convection cells indicated by Pioneer 11 photos.

The Jovian magnetosphere was as powerful and active as had been indicated by previous spacecraft. The flux tube flow-

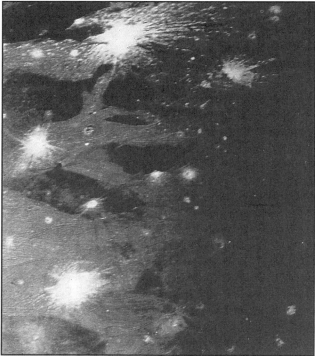
Patchwork patterns on Ganymede. *NASA*

April 10, 1979

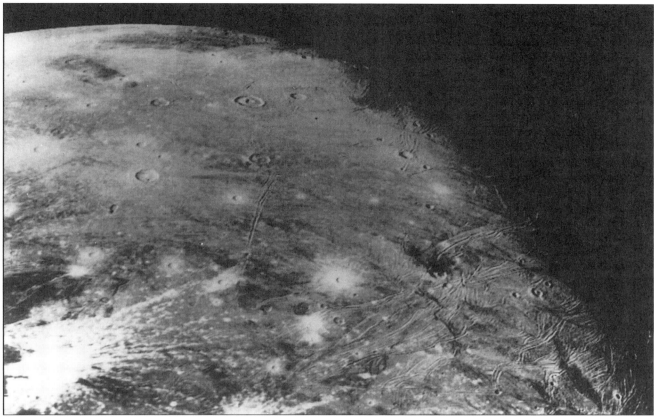
Grooves and craters on Ganymede. *NASA*

ing to and from Io and Jupiter was measured to have an electric current of more than a million amps. A hot plasma torus surrounding Io was discovered, with plasma electron densities over 2,700 per cubic centimeter. Jupiter's radiation belts contained ions of sulfur, oxygen, and sulfur dioxide, extending out for hundreds of thousands of miles.

Of Jupiter's moons, Europa revealed a white and brown surface covered with numerous intersecting lines. Surprisingly, it had no Moonlike craters.

Ganymede was covered with a patchy brown and white surface with many white spots (thought to be craters) and a complex weave of rays and grooved bands that baffled scientists.

Callisto, the only Jovian Moon that appeared anything like the earth's Moon, unveiled a cratered surface with its largest impact site, Valhalla, resembling a bull's-eye, surrounded by many frozen ripple marks. Amalthea was an irregularly shaped dark-reddish object 170 by 100 miles.

After passing Jupiter, Voyager 1 flew on to a rendezvous with Saturn (see **Voyager 1, Saturn Flyby–December 12, 1980**). The next spacecraft to reach Jupiter was **Voyager 2 (July 9, 1979)**.

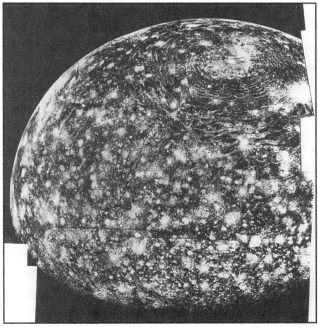
Jupiter's moon Callisto. *NASA*

April 10, 1979
Soyuz 33
U.S.S.R. / Bulgaria

This unsuccessful internationally manned space flight, manned by Nikolai Rukavishnikov (**Soyuz 10–April 23, 1973**; **Soyuz 16–December 2, 1974**) of the Soviet Union and Georgi Ivanov of Bulgaria, was unable to dock with **Salyut 6 (September 29, 1977)** due to a failure in Soyuz 33's main propulsion engines after reaching orbit. After two days in orbit, the landing was further complicated by the failure of the main capsule's retro-rockets, requiring use of its back-up retro-rockets. Although both men survived without injury, they were subjected to forces in excess of 8–10 g's during re-entry.

June 2, 1979
Ariel 6 (UK 6)
U.K. / U.S.A.

A follow-up to **Ariel 5 (October 5, 1974)**, Ariel 6 carried three instruments, two to study celestial x-ray sources and one to measure high energy heavy cosmic ray particles. It was built by the United Kingdom and launched by NASA. The launch was also the last successful NASA launch from its facility on Wallops Island, Virginia.

During Ariel 6's three-year life span, it made the first measurements of the relative abundance of the very heavy cosmic ray particles hitting the atmosphere, finding an unexpected high proportion of elements rare on Earth.

The spacecraft also continued the monitoring of a number of the brightest x-ray sources in the sky, begun by **Uhuru (December 12, 1970)**.

The next gamma-ray telescope was **HEAO 3 (September 20, 1979)**. The next x-ray telescope launched was on **Salyut 7 (April 19, 1982)**, followed by **Tenma (February 20, 1983)**.

June 6, 1979
Soyuz 34
U.S.S.R.

Originally intended to carry a two-man international crew to **Salyut 6 (September 29, 1977)**, Soyuz 34 was flown unmanned because of safety concerns related to the orbiting **Soyuz 32 (February 25, 1979)** spacecraft. Because of the failure of **Soyuz 33 (April 10, 1979)**, Soyuz 32 had spent more than 90 days in space, exceeding its safety limit. If a manned crew had lifted off on Soyuz 34, either its crew or the crew of Soyuz 32 would have had to return on Soyuz 32, something the Soviet space program felt was an unwarranted risk.

As a result, Soyuz 34 was sent to Salyut 6 unmanned. The crew of Soyuz 32, Vladimir Lyakhov and Valery Ryumin, then used it to return to Earth at the end of their 175-day mission in space. Soyuz 32, meanwhile, returned to Earth unmanned, packed with experiment results and film.

June 7, 1979
Bhaskara 1
India / U.S.S.R.

This satellite was the first non-American civilian satellite dedicated to Earth resources, following the Landsat satellites (see **Landsat 1–July 23, 1972**). Bhaskara 1 was also the second Indian satellite, following **Aryabhata (April 19, 1975)**. Designed and built in India, it was launched on a Soviet rocket from the U.S.S.R. Besides doing satellite engineering tests, Bhaskara 1 (named after two early Indian astronomers-mathematicians) conducted an Earth resources survey of India, filling a coverage gap of the Landsat satellites.

The spacecraft carried a television system resembling Landsat 1's, producing images in two wavebands, one visible (0.54–0.65 microns) and one near-infrared (0.75–0.85 microns). Both bands provided detailed observations of plant and vegetation changes. Each image covered an area 200 miles to a side, with resolution of about two-thirds of a mile. Bhaskara 1 also carried a radar scanner for observing the surrounding Indian Ocean. Sea currents and pollution were studied, as well.

Bhaskara 1 collected television images for about a year, and continued to gather additional data until it was turned off in March 1981. A second identical Bhaskara satellite was launched in November 1981, taking Earth resource images for about two months. The first fully operational long-term Indian Earth resource satellite was **IRS 1 (March 17, 1988)**. See also the second generation American **Landsat 4 (July 16, 1982)**, as well as the French **Spot 1 (February 22, 1986)**.

July 9, 1979
Voyager 2, Jupiter Flyby
U.S.A.

Built by NASA and launched on August 20, 1977, Voyager 2 was the fourth spacecraft to pass through the Jovian system, following **Pioneer 10 (March 3, 1972)**, **Pioneer 11 (April 6, 1973)**, and **Voyager 1 (March 5, 1979)**.

Pictures of Jupiter's Red Spot *(see color section, Figure 11)* found that in the four months since Voyager 1's flyby, the region had changed significantly. The white storm regions south

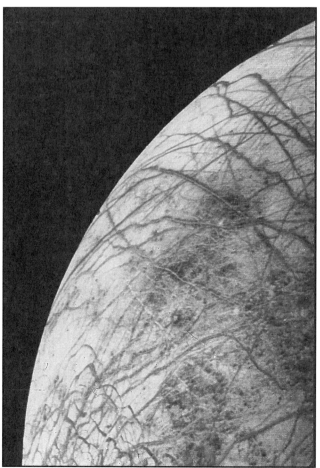

Europa. *NASA*

July 9, 1979

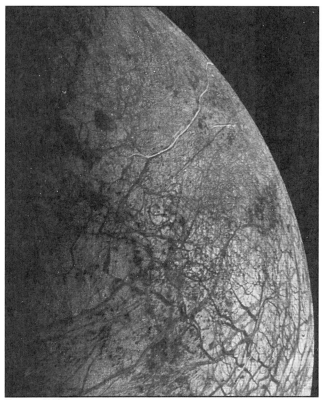

Europa. *NASA*

Ganymede—spots, patches, and grooves. *NASA*

of the Spot were gone, replaced by one large white spot to the southwest. The western region of turbulence had also changed considerably, now resembling wavelike ripples flowing past a rock in a stream. This gigantic storm was seen as one of many dotting Jupiter's mid-latitudes.

The atmospheric east-to-west jet streams identified by Voyager 1 were seen again, though some changes in velocity were measured *(see color section, Figure 9)*. An ultraviolet map of the planet's absorbing haze showed that the polar regions were darker than expected, indicating that most of the haze was in higher latitudes. And the cause of Jupiter's aurora was apparently linked to the charged particles in Io's magnetospheric torus, transmitted to Jupiter along the flux tube con-

Ganymede. *NASA*

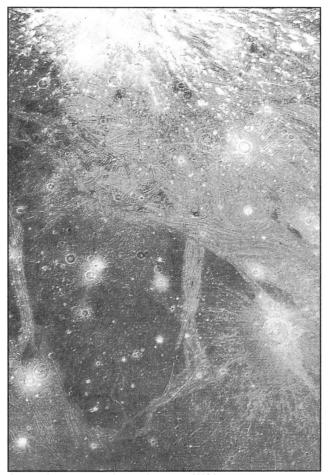

Ganymede—spots, patches, and grooves. *NASA*

167

September 1, 1979

Callisto. *NASA*

necting the satellite and planet. Jupiter's magnetosphere was confirmed to be a hot plasma, composed mostly of hydrogen, oxygen, and sulfur ions. Its size was vast, over 14 million miles wide from side to side, with a vast tail flowing away from the Sun.

Like its predecessor, Voyager 1, Voyager 2 not only took close-up pictures of Jupiter, it took even more detailed looks at the Jovian moons Io *(see color section, Figure 13)*, Europa, Ganymede, and Callisto. Six of the volcanoes seen erupting on Io by Voyager 1 were still gushing, though the largest, Pele, had become dormant. Many changes to Io's surface could be identified, providing scientists with a scale to measure the moon's resurfacing rate.

Europa was revealed to be a smooth, white ball covered with mysterious reddish lines. No craters were visible, nor was any vertical relief identifiable. The reddish lines looked almost like giant cracks in Europa's crust, like "a cracked egg," as one scientist described it. Some of the larger reddish lineaments had additional lines running down their center.

Ganymede revealed more details of the baffling grooved terrain, splattered with craters. One gigantic impact basin, dubbed Regio Galileo, resembled the multiringed Valhalla basin on Callisto. And the new pictures also uncovered a younger, smooth terrain along with the craters and grooves.

Callisto, however, continued to resemble the earth's Moon, its entire surface appearing heavily cratered, indicating an immense age exceeding several billion years.

Jupiter's rings were once again photographed, showing a bright, narrow central region only a few thousand miles across surrounded by a broad area of faint, diffuse material, possibly extending right down to the top of Jupiter's atmosphere. Careful inspection of these images revealed two new innermost Jovian satellites, orbiting only 35,000 and 36,000 miles above Jupiter's cloud tops. Named Metis and Adrastea, both were estimated to be less than 30 miles across.

The pictures from both Voyager spacecraft had reemphasized what **Viking 1 (June 19, 1976)** and **Viking 2 (August 7, 1976)** had demonstrated about Mars: the solar system's planets and moons included a much more bewildering array of shapes, sizes, and compositions than ever predicted by any scientist. As scientist Larry Soderblom noted after looking at the Voyager images of the Galilean satellites, "We're in a relatively high state of ignorance."

After this flyby, Voyager 2 was redirected towards Saturn (see **Voyager 2, Saturn Flyby–August 26, 1981**). The next spacecraft to reach Jupiter was Ulysses (**Ulysses, Jupiter Flyby–February 8, 1992**), on its way into a polar orbit of the Sun. The next spacecraft dedicated to studying Jupiter was Galileo (**Galileo, Jupiter Orbit–December 7, 1995**).

September 1, 1979
Pioneer 11, Saturn Flyby
U.S.A.

Launched on April 6, 1973, Pioneer 11 had flown past Jupiter on December 4, 1974 (see also **Pioneer 11–April 6, 1973; Pioneer 11, Jupiter Flyby–December 4, 1974**). Now, after 5.5 years in space, the spacecraft zipped past Saturn on September 1, 1979, flying just 22,000 miles from the outer edge of Saturn's rings on its inbound and outbound pass, and coming within 13,300 miles of the planet's cloud tops at its closest approach.

Pioneer 11 discovered that Saturn has a powerful magnetic field, though not as powerful as Jupiter's. Three times on its inward journey and nine times on its outward journey, the spacecraft crossed the magnetosphere's bow shock, indicating that, like Jupiter, the push and pull between the solar wind and magnetosphere caused the field to shrink and expand. Unlike Jupiter's (and most other known magnetic fields), Saturn's magnetic field was tilted less than one degree from the planet's axis. This lack of tilt contradicted several theories for explaining the origin of planetary magnetic fields—theories which held that the tilt in the internal dynamo of a molten core contributes directly to creating the field. Saturn's was not tilted, however, so these theories did not offer a good explanation for that planet.

While Saturn's magnetosphere carried trapped radiation particles like Jupiter's and the earth's, it also had several regions that were entirely clear of trapped particles, including the lower latitudes inside the rings. The rings had swept this area clear, making it almost free of particles. In general, Saturn's magnetosphere was much more stable than Jupiter's, less intense, and less powerful.

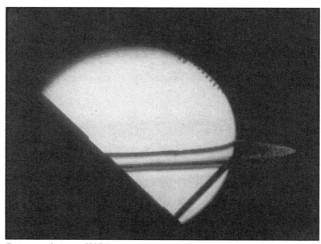
Saturn and rings. *NASA*

Saturn's rings, seen with sunlight shining through them. The distortion is an instrument effect. *NASA*

Saturn's rings were photographed for the first time with the Sun's light shining through them rather than reflected off their surface (as normally seen from Earth). How much light penetrated the rings and the gaps between allowed scientists to calculate their thickness and total mass and to map precisely the number of rings and gaps. For example, the width of Cassini's division was reduced from 4,000 to 2,600 miles. A new outer ring, labeled the F ring, was discovered, along with a new gap named the Pioneer Division.

The rings' total mass was estimated to be less than one three-millionth that of Saturn. Their thickness could not exceed 2.5 miles, with their particle size averaging less than half an inch in diameter, including evidence of a large population of particles about 100 microns in size, or about one ten-thousandth of a meter.

Saturn itself was photographed, revealing surprisingly little detail. Zones and belts like Jupiter's were seen, but they were narrower in width and lacked the extreme contrasts of light and dark.

Saturn's composition, inferred by careful measurements of the spacecraft's passage past the planet, was estimated to be 90 percent hydrogen and 10 percent helium, with trace amounts of water, ammonia, and methane. Like Jupiter, Saturn's interior was determined to be mostly liquid hydrogen, divided into two layers: a liquid hydrogen layer floating on top of an ocean of metallic hydrogen. At the planet's core was a solid, Earth-size sphere of iron-rich rocky materials weighing about 15 Earth masses.

Saturn was also found to radiate about three times the heat that it absorbs from the Sun. Scientists theorized that as the liquid helium in Saturn's interior slowly sank, it generated heat.

Upon leaving Saturnian space, Pioneer 11 headed out of the solar system, aimed at the constellation of Aquila. In this direction, the spacecraft was traveling upstream of the Sun's path through the galaxy, toward the heliosphere's bow shock. Scientists hoped that before its nuclear fuel ran out sometime around the mid-1990s, Pioneer 11 would pass out of the heliosphere and into interstellar space. See **Pioneer 11, Shutdown–September 30, 1995** for more details about what the spacecraft discovered over the next 16 years.

The next spacecraft to reach Saturn was **Voyager 1 (December 12, 1980)**.

September 20, 1979
HEAO 3
U.S.A.

The third and last High Energy Astronomy Observatory (HEAO), HEAO 3 was designed by NASA to complete the first all-sky survey in the high-energy cosmic and gamma-ray range of the electromagnetic spectrum, rather than the x-rays of the previous two HEAO satellites. It continued research of **COS B (August 9, 1975)**. The gamma-ray detector measured energy levels from 60 KeV to 10 MeV and from 325 MeV to 1,500 MeV, while the cosmic ray detector measured emissions from 1 GeV to 15 GeV.

During its two-year operational life, HEAO 3 discovered what many scientists believed was evidence of anti-matter in the center of the Milky Way. Several sources spread throughout the galaxy's core appeared to be emitting energy at precisely 5.11 KeV, which, according to theory, could only happen if matter and anti-matter electrons were colliding.

The spacecraft also detected gamma rays from several well known objects, such as the Crab Nebula and Cygnus X-1 (one of the most likely black hole candidates). The intensity of both sources fluctuated over time, indicating that violent local events caused their occurrence.

SS433 was also discovered by HEAO 3. This strange object, located in the Milky Way about 15,000 light years from Earth, emitted energy across the entire spectrum, from radio to gamma rays, and appeared to have two jets of matter blowing off it in opposite directions. Scientists believed it was either a black hole or a neutron star.

The next gamma-ray telescope was **Gamma** (**July 11, 1990**).

September 25, 1979
Bion 5 (Cosmos 1129)
U.S.S.R. / International

Bion 5 was part of the Soviet biological research program begun with **Bion 1** (**October 31, 1973**), with scientists from the U.S.S.R., Czechoslovakia, France, and the United States participating.

This mission, lasting 18.5 days, attempted the first breeding of mammals in space. In fact, this attempt was only the second flight ever to carry female mammals, following cosmonaut Valentina Tereshkova on **Vostok 6** (**June 16, 1963**).

Five female and two male rats were placed in a breeding chamber prior to flight. Though a control group of rats on the ground all produced offspring, none of the females flown in space successfully bred. Two did produce normal litters following their return, suggesting that some aspect of the environment of space had prevented them from breeding. The specific cause was nevertheless unclear: though weightlessness could have hindered embryo development, it could have just as easily interfered with copulation and insemination.

The next Bion satellite was **Bion 6** (**December 14, 1983**).

October 30, 1979
Magsat
U.S.A.

Built by NASA, Magsat gathered the first systematic global map of the earth's magnetic field and the anomalies within it caused by features in the earth's crust. This research followed up discoveries made on **OGO 6** (**June 5, 1969**) and was intended to provide geologists with a more detailed map of the earth's crust. This information could also be used to discover oil and mineral deposits, and to indicate the present and future state of the earth's magnetic field.

During Magsat's two-year life span, it found that the measured decline in the intensity of the earth's magnetic field in the 10 years since OGO 6 suggested that the field will reverse in polarity in approximately 1,200 years, much sooner than previously expected. Why this was happening, and what effect it will have on life on Earth, remains unknown. The previous reversal of Earth's magnetic field is believed to have occurred approximately 780,000 years ago.

The next satellite to map the earth's magnetic field was POGS (**April 11, 1990**).

November 1, 1979
Intercosmos 20
U.S.S.R. / East Europe

Unlike most of the satellites in the Intercosmos program (see **Intercosmos 1**–**October 14, 1969**), Intercosmos 20 did not do ionospheric research. Instead, it provided meteorological data, testing technology that made possible the real-time gathering of weather information from many different ocean and land stations, similar in design to equipment carried on the American NOAA satellites (see **NOAA 1**–**December 11, 1970**). The Soviet Union, Czechoslovakia, East Germany, Hungary, and Romania contributed to the satellite's instrumentation.

December 16, 1979
Soyuz T1
U.S.S.R.

Soyuz T1 was an unmanned 100-day mission, 95 days of which were spent docked to the **Salyut 6** (**September 29, 1977**) orbiting laboratory. The flight tested a new modified Soyuz capsule, allowing the Soviets to fly three-person missions for the first time since the death of Dobrovolsky, Volkov, and Patsayev on **Soyuz 11** (**June 6, 1971**).

The new spacecraft carried additional fuel and supply capacity, including the reinstatement of two solar panels to the outside of the descent module, allowing it to remain in inde-

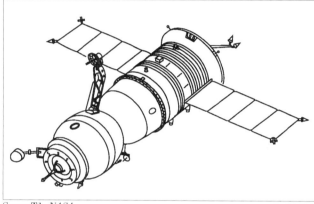

Soyuz T1. *NASA*

pendent orbit without docking for four days, twice as long as the earlier Soyuz crafts. Its propulsion system was also redesigned, as was its computer system.

The first manned mission using this more advanced Soyuz was **Soyuz T2** (**June 5, 1980**).

December 24, 1979
CAT 1
Europe

This launch was the first successful use of the Ariane 1 rocket, a three-stage launch vehicle developed by the European Space Agency as a direct competitor to American, Soviet, Chinese, and Japanese launch rockets. CAT 1 was a small test capsule for transmitting orbital telemetry to mission control in French Guiana.

Following one additional test flight in 1980, the first operational use of the Ariane rocket took place on June 19, 1981 (see **Ariane 1**–**June 19, 1981**).

1980

February 9, 1980
Navstar 5
U.S.A.
Navstar 5 was the first satellite in a planned global positioning system (GPS) eight-satellite cluster established by the U.S. military. It was the first such operational system flown, following the test cluster of satellites begun with **Navstar 1 (February 22, 1978)**. When completed, this cluster allowed a GPS sensor anywhere in the world to fix its position to within 30 feet in three dimensions, even if moving at 360 miles per hour. See **Timation 1 (May 31, 1967)** for a description of GPS.

The cluster was completed in October 1985, with seven satellites in the planned eight-satellite cluster reaching orbit. During this same period, beginning in October 1982, the Soviet Union began test flying its own GPS satellites, called Glonass. See **Cosmos 1413 through Cosmos 1415 (October 12, 1982)**.

In 1989, the United States inaugurated a second more accurate cluster of 24 GPS satellites. See **Navstar B2 1 (February 14, 1989)**.

February 14, 1980
Solar Max
U.S.A.
Solar Max, or the Solar Maximum Mission, was a solar observatory built by NASA that monitored solar flare activity as the Sun passed through the most active period of its solar cycle in 1980–81. It carried ultraviolet, x-ray, and gamma-ray sensors, as well as instruments for measuring the Sun's total irradiance. This satellite continued the monitoring of the Solrad satellites (see **Solrad 11A/Solrad 11B–March 15, 1976**) and the ISEE satellites (see **ISEE 1/ISEE 2–October 22, 1977**).

Solar Max was the first satellite to use the modular space-shuttle design, which allowed later shuttle flights to refurbish and modify satellites. It was launched with the intention that the shuttle would eventually retrieve it from orbit so that its main structure could be refurbished and it could be reused for later satellite projects.

During its first nine months of operation, the spacecraft observed about 2,000 solar flares, including some of the most powerful ever recorded *(see color section, Figure 2)*. One in May 1980 covered 2 billion square miles of the Sun's surface and generated core temperatures exceeding 100 million degrees Fahrenheit. From these observations, scientists were able to confirm that the solar flares were directly related to a twisting and unraveling of the Sun's magnetic field lines. Solar Max also found that variations in ultraviolet and x-ray emissions during solar flares were closely related.

The instrument measuring the Sun's total irradiance found that the Sun's brightness dipped by as much as 0.2 percent over periods of days, months, and years. When compared with longer-term data collected by **Nimbus 7 (October 13, 1978)**, these data proved that the Sun's overall output had steadily declined as it had gone from its active to its quiet state after 1981. This decline also corresponded to severe winter weather conditions in the United States in 1981–82 and suggested that solar flare activity and the subsequent cyclical variations in solar output could have a direct effect on the earth's climate. Further solar monitoring continued on ERBS (see **STS 41G–October 5, 1984**), as well as on three NOAA satellites from 1984 through 1988 (see **NOAA 1–December 11, 1970**).

In November 1980, three fuses in Solar Max's attitude control system burned out, making precise pointing of the satellite impossible. Although three of its instruments could still make coarse measurements, most of its research became impossible.

Because Solar Max was built using space shuttle module design, NASA decided to attempt an in-orbit repair, the first such repair ever tried. See **STS 41C (April 6, 1984)**.

The next solar observation satellite was **Hinotori (February 21, 1981)**.

April 9, 1980
Soyuz 35
U.S.S.R.
Soyuz 35 began the fourth occupancy of **Salyut 6 (September 29, 1977)**, following the 175-day occupancy of **Soyuz 32 (February 25, 1979)**. Crewed by Leonid Popov (**Soyuz 40–May 14, 1981; Soyuz T7–August 19, 1982**) and Valeri Ryumin (**Soyuz 25–October 9, 1977; Soyuz 32–February 25, 1979; STS 91–June 2, 1998**), the crew of Soyuz 35 set a space endurance record of 185 days—more than half a year—exceeding Soyuz 32 by 10 days. This record was not broken until **Soyuz T5 (May 13, 1982)** on **Salyut 7 (April 19, 1982)**.

Two weeks before the arrival of Soyuz 35, a Progress freighter (see **Progress 1–January 20, 1978**) docked with Salyut 6. Because of a leak in Salyut 6's fuel lines (see **Soyuz 32–February 25, 1979**), the freighter's engines were used to lift the station's orbit. The freighter also carried an additional electrical storage battery to supplement Salyut 6's solar panels. Before the Soyuz 35 mission ended, another three Progress tankers brought fuel, food, water, air, and replacement equipment (including a color television and Polaroid camera).

Ryumin, who had just flown on the Soyuz 32 mission, was enlisted for Soyuz 35 when cosmonaut Valentin Lebedev injured his knee and was temporarily grounded.

Unlike Ryumin's previous flight, during which the Soyuz 32 crew on Salyut 6 received no visits from other manned missions, Salyut 6 was visited by four different crews during Soyuz 35's six-month occupancy: **Soyuz 36 (May 26, 1980), Soyuz T2 (June 5, 1980), Soyuz 37 (July 23, 1980)**, and **Soyuz 38 (September 18, 1980)**. Soyuz T2 was the first manned flight of the new modified Soyuz spacecraft, while the other three missions included international crews. See the entries for these missions for more details.

Before, between, and after these visits, Popov and Ryumin spent most of their time maintaining the station, doing minor repairs and equipment checks. Their research focused on multispectral photography of the earth (taking more than 2,000 images and 40,000 spectrograms) and growing vegetables in the station's small greenhouse. They had found that while flowers in space tended to die before reaching maturity, vegetables like onions seemed to do better. Hence, this crew cultivated onions, peas, radishes, garlic, cucumbers, parsley, and dill, with only onions and garlic shoots prospering long enough to produce seeds. However, both an arabidopsis plant and an orchid became the first flowers to bloom in space. For more plant research, see **Soyuz T4 (March 12, 1981)**.

The Soyuz 35 crew returned to Earth on October 11, 1979, using the Soyuz 37 spacecraft. Like the Soyuz 32 crew, the two men recovered quickly, their heart rates returning to normal within days. Ryumin's recovery was especially significant since he had flown both Soyuz 32 and Soyuz 35 and had spent 12 out of 20 months in space. He was even able to walk unaided from the capsule to the lounge chairs provided for returning cosmonauts.

Salyut 6's fifth occupancy, **Soyuz T3 (November 27, 1980)**, was a short solo flight that performed further testing of the redesigned Soyuz T spacecraft.

May 26, 1980
Soyuz 36
U.S.S.R. / Hungary

An eight-day internationally manned flight, Soyuz 36 carried Soviet cosmonaut Valery Kubasov (**Soyuz 6–October 11, 1969; Soyuz 19–July 15, 1975**) and Hungarian cosmonaut Bertalan Farkas to **Salyut 6 (September 29, 1977)** during its fourth occupancy by **Soyuz 35 (April 9, 1980)**. Kubasov and Farkas returned to Earth using the older Soyuz 35 spacecraft, leaving the newer Soyuz 36 as an emergency return vehicle.

During their eight days in space, the two crews performed the usual regimen of experiments involving earth resource photography, human biology, and materials processing. Semiconductor alloys of gallium-arsenide, indium-antimonide, and gallium-antimonide were smelted in the station's furnace. One Hungarian experiment tested the effect of weightlessness on human interferon production and on the interferon in several blood samples.

The crew of Soyuz 36 returned to Earth on June 3, 1980. Two days later, **Soyuz T2 (June 5, 1980)** was launched, the second mission to the fourth occupancy of Salyut 6.

June 5, 1980
Soyuz T2
U.S.S.R.

Soyuz T2 was the first manned use of the redesigned Soyuz T spacecraft, previously flown unmanned on **Soyuz T1 (December 16, 1979)**. The crew, Yuri Malyshev (**Soyuz T11–April 3, 1984**) and Vladimir Aksenov (**Soyuz 22–September 15, 1976**), remained in space for only four days. After one day in orbit checking out the spacecraft's new solar panels, new computer system, and engine design, they docked with **Salyut 6 (September 29, 1977)**, performing additional tests on its new automatic docking systems. After another three days of tests with the **Soyuz 35 (April 9, 1980)** crew, Soyuz T2 undocked and returned to Earth.

The next mission to Salyut 6's fourth occupancy was **Soyuz 37 (July 23, 1980)**.

June 18, 1980
Meteor 1-30
U.S.S.R.

Unlike previous Meteor 1 weather satellites (see **Meteor 1-01–March 26, 1969**), Meteor 1-30 was designed for earth resource study rather than weather observations. It was placed in a sun-synchronous orbit (see **Nimbus 1–August 28, 1974**) and, like the American Landsat satellites (see **Landsat 1–July 23, 1972**), it carried several multispectral cameras for photographing the earth's resources, both agricultural and geological.

The Soviets launched one other earth resource satellite under the Meteor designation, **Meteor 1-31 (July 10, 1981)**, after which these earth resource satellites were launched under the Cosmos name until **Resurs F1 (May 25, 1989)**.

July 18, 1980
Rohini 1B
India

With this successful test launch of India's first orbital rocket, India became the seventh nation and eighth independent organization to place a satellite in orbit, following the Soviet Union (**Sputnik–October 4, 1957**), the United States (**Explorer 1–February 1, 1958**), France (**Asterix–November 26, 1965**), Japan (**Ohsumi–February 11, 1970**), China (**Mao 1–April 24, 1970**), the United Kingdom (**Prospero–October 28, 1971**), and the European Space Agency (**CAT 1–December 24, 1979**).

India made an additional seven test flights of the Rohini rocket through 1994, each time testing a progressively more powerful rocket.

The next nation to join the space community with its own orbital rocket was Israel. See **Ofeq 1 (September 19, 1988)**.

July 23, 1980
Soyuz 37
U.S.S.R. / Vietnam

Soyuz 37 docked with **Salyut 6 (September 29, 1977)**, joining the **Soyuz 35 (April 9, 1980)** crew during their 185-day fourth occupancy of the orbiting laboratory. The two-man international crew included Soviet Viktor Gorbatko (**Soyuz 7–October 12, 1969; Soyuz 24–February 7, 1977**) and Vietnamese Pham Tuan. As had the **Soyuz 36 (May 26, 1980)** crew, when the Soyuz 37 crew returned to Earth, they switched spacecraft, leaving their newer Soyuz 37 spacecraft for the Soyuz 35 crew and returning home in the older Soyuz 36 spacecraft.

During this eight-day mission, the four cosmonauts took multispectral earth resource photographs of Vietnam. Materials processing experiments produced a gallium-phosphide semiconductor crystal and a compound of bismuth-antimony-telluride.

The next manned mission to Salyut 6 during its fourth occupancy was **Soyuz 38 (September 18, 1980)**.

September 18, 1980
Soyuz 38
U.S.S.R. / Cuba

This eight-day mission was the fourth and last manned mission visiting the crew of **Soyuz 35 (April 9, 1980)** during their 185-day mission, the fourth occupancy of **Salyut 6 (September 29, 1977)**.

An internationally manned mission, the crew included Soviet Yuri Romanenko (**Soyuz 26–December 12, 1977**; **Soyuz TM2–February 5, 1987**) and Cuban Arnaldo Mendez. As during previous international flights, earth resource photographs of the country of the visiting astronaut were taken. A number of medical experiments designed by Cuban doctors studied changes in the cosmonauts' muscles, bones, immune systems, mineral and water balance, metabolism, hormone levels, and brain activity.

In the first crystal-growth experiments ever flown in space, four different mono-crystals of sucrose were grown, using different solutions. All four crystals grew faster than expected. Such research, producing large crystals of many different proteins that allowed scientists to map their structure easily, was later continued extensively on the American space shuttle (see **STS 51D–April 12, 1985**).

Materials processing experiments produced more semiconductor alloys of gallium-arsenide, tin-telluride, germanium-telluride, and other elements for later study on Earth.

Soyuz 38 returned to Earth on September 26, 1980. Two weeks later, the Soyuz 35 crew ended their 185-day mission, returning to Earth on October 11, 1980. Salyut 6 remained unoccupied for six weeks, until **Soyuz T3 (November 27, 1980)** arrived to begin the station's fifth manned occupancy.

November 15, 1980
SBS 1
U.S.A.

Owned and operated by Satellite Business Systems (SBS), this satellite was part of the fourth private communications satellite cluster launched to serve the U.S. domestic market, following **Westar 1 (April 13, 1974)**, **RCA Satcom 1 (December 13, 1975)**, and **Comstar 1 (May 13, 1976)**.

SBS was originally owned jointly by IBM, Aetna, and Comsat Corporation. Initially, the plan was to use these satellites to provide telephone and data services to large American corporations, linking together office sites to reduce long distance telephone charges. In the early 1980s, this market strategy did not work, and SBS sustained huge losses. In 1984, Comsat sold its share to IBM and Aetna. In 1985, IBM and Aetna sold SBS to MCI Communications Corporation, with IBM retaining some shares and ownership rights to three SBS satellites. In an attempt to cover some of its investment, MCI began offering long distance service to residential customers. This action was the first direct competition to AT&T's telephone service and eventually led to today's open market for telephone services.

Through September 1988, SBS placed five satellites in geosynchronous orbit over the United States. These satellites were the first to use the higher 12 and 14 GHz frequencies tested by **Hermes (January 17, 1976)**. A sixth satellite was launched in 1990 with a greater power capability and redundancy. In 1990, Hughes Communications purchased the last three satellites of this cluster from IBM to augment its own satellite cluster (see **Galaxy 1–June 28, 1983**).

Of the remaining three SBS satellites, all are operational as of December 1999 and are maintained in orbit by MCI as a backup to its fiber-optic network.

November 27, 1980
Soyuz T3
U.S.S.R.

This fifth occupancy of **Salyut 6 (September 29, 1977)** was the shortest, lasting only 13 days. Its main purpose was to reinitiate Soviet three-man flights into space. The crew, Leonid Kizim (**Soyuz T10B–February 8, 1984**; **Soyuz T15–March 13, 1986**), Oleg Makarov (**Soyuz 12–September 27, 1983**; **Soyuz 18A–April 5, 1975**; **Soyuz 27–January 10, 1978**), and Gennady Strekalov (**Soyuz T8–April 20, 1983**; **Soyuz T10A–September 27, 1983**; **Soyuz T11–April 3, 1984**; **Soyuz TM10–August 1, 1990**; **Soyuz TM21–March 14, 1995**), devoted their time either to testing the Soyuz T spacecraft or to performing maintenance and repair work on Salyut 6 (which had been in orbit for more than three years). The cosmonauts installed a new hydraulic unit for the laboratory's temperature control system and replaced several electrical components.

After their return to Earth on December 10, 1980, Salyut 6 remained unoccupied for three months until **Soyuz T4 (March 12, 1981)**, the orbiting laboratory's sixth and last manned occupancy.

December 6, 1980
Intelsat 5 F2
International / U.S.A.

Intelsat 5 F2 was the first satellite in the fifth generation of geosynchronous communication satellites owned by Intelsat, the Western international satellite consortium (see **Early Bird–April 6, 1965**). Intelsat 5's satellites supplemented and then replaced the Intelsat 4 and Intelsat 4A satellites (see **Intelsat 4 F2–January 26, 1971** and **Intelsat 4A F1–September 26, 1975**), updates required by the increasing demand for telephone and television communications, mostly across the Atlantic. Intelsat 5 satellites had twice the capacity of the Intelsat 4A satellites, between 12,000 and 36,000 simultaneous two-way voice circuits each, depending on the satellite's particular position and configuration. Working in conjunction with the earlier satellites, the capacity over the Atlantic was now more than 100,000 simultaneous two-way telephone calls at one time, a 400-fold increase in capacity in the 15 years since Early Bird, Intelsat's first satellite.

Intelsat 5 satellites were also the first by Intelsat that used the higher 11 and 14 GHz frequency bands, first tested by **Hermes (January 17, 1976)** and used operationally by **SBS 1 (November 15, 1980)**.

Through March 1984, eight Intelsat 5 satellites were launched, with three positioned over the Atlantic, three over the Pacific, and two over the Indian Ocean.

December 12, 1980

Despite the increased capacity, the demand continued to grow faster than predicted. Intelsat was soon forced to launch the Intelsat 5A satellites (see **Intelsat 5A F10–March 22, 1985**), a five-satellite addition to the Intelsat 5 satellite cluster.

December 12, 1980
Voyager 1, Saturn Flyby
U.S.A.

Having flown past Jupiter on March 5, 1979, Voyager 1 was the second spacecraft to reach Saturn, following Pioneer 11 (**Pioneer 11, Saturn Flyby–September 1, 1979**). Voyager 1's path past Saturn allowed it to get within 2,800 miles of the surface of Saturn's largest moon, Titan, the only moon in the solar system known to have an atmosphere. Moreover, the spacecraft took close-up pictures of four of Saturn's moons, as well as of the planet's famous rings.

Voyager 1's pictures of the rings revealed a plethora of baffling detail, including an almost uncountable number of rings within rings, visible spokes of brightness, and two shepherding moons that together cause one ring to be a thin wavy line. Two other rings appear braided and misshapen rather than circular. The overall appearance of the ring system resembles a vinyl record, with its innumerable grooves representing Saturn's many rings. Voyager 1's sensors confirmed that the rings were made of chunks of ice and snow, ranging in size from .0001 inch in diameter to 30 feet across.

The shepherd moons were two of six newly discovered satellites. Also discovered was a giant torus of hydrogen gas girdling the planet beyond its rings, trailing away from the moon

Over 95 different rings can be counted. Saturn's fourteenth satellite, discovered by Voyager 1, can be seen just outside the rings, at the upper right. *NASA*

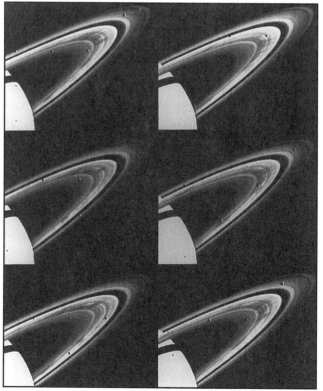
Montage showing rotation of the spokes in Saturn's rings. Sequence goes from bottom left to top right. *NASA*

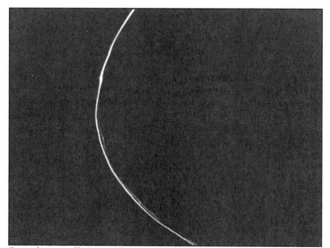
Saturn's outer F ring, with two strands that appear kinked or braided. *NASA*

Titan. Scientists believed that the Sun's ultraviolet radiation hitting Titan's atmosphere caused the methane within it to break into hydrogen, ethane, and acetylene. The hydrogen then escaped into space, creating the cloud. Titan's atmosphere, however, was found to be mostly nitrogen, not methane as scientists had expected. This large amount of nitrogen reduced the likelihood of methane oceans and rain, which had been postulated by scientists. Titan's atmosphere itself appeared to be about 500 miles thick, about six times thicker than Earth's atmosphere, with a temperature of about −300°F and an atmospheric pressure 50 percent greater than Earth's. The moon had no magnetic field, indicating that its core was not molten. Measurements of Titan's diameter also revealed that it is not

December 12, 1980

the solar system's largest moon, as had been thought. Instead, it was slightly smaller than Jupiter's Ganymede, 3,180 miles across versus 3,275.

Of Saturn's other moons, each provided surprises. Cracks or white streaks, as well as craters, could be seen on Mimas, Tethys, Dione, and Rhea. One crack on Mimas, along with a giant impact crater, covered nearly a quarter of the surface, indicating that the impact had almost split that moon. Mimas, Tethys, Dione, and Rhea appeared to be icy snowballs with rocky cores. Enceladus, however, was smooth and featureless—scientists posited that, like Io, tidal forces had heated its inte-

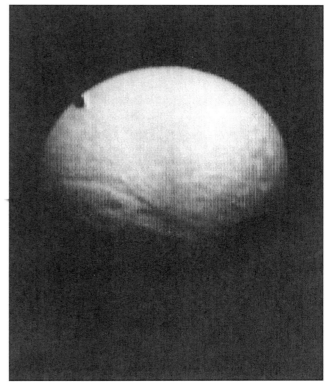

Saturn's moon Tethys. *NASA*

Saturn's moon Mimas. *NASA*

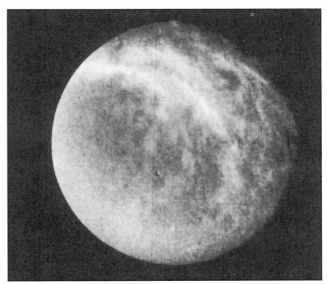

Saturn's moon Rhea. *NASA*

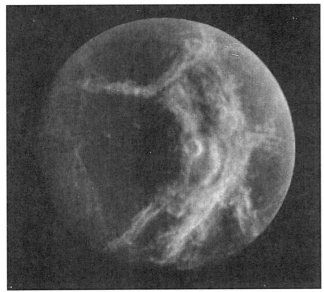

Saturn's moon Dione. This leading hemisphere, covered with white streaks that might be fractures, differs from the trailing side, which is cratered like Earth's moon. *NASA*

rior, causing icy volcanoes to spew watery lava and thus resurface it.

Saturn itself appeared to be a less colorful version of Jupiter, with parallel horizontal bands and gigantic storms running along the contact zones. Wind speeds of 900 to 1,000 miles per hour were measured, four times faster than winds on Jupiter. Several brownish ovals, similar to the giant storms seen on Jupiter, were also discovered. Temperatures at the cloud tops were about −300°F, and the bulk of the atmosphere was hydrogen, with helium comprising only 11 percent of the total.

February 6, 1981

Saturn, looking back after flyby. *NASA*

Like Jupiter, Saturn was surrounded by a powerful magnetic field that also generated radio signals—though the field was only one-third the strength of Jupiter's. Aurorae could be seen in Saturn's clouds. Data suggested that both the planet and its magnetosphere rotated with a period of 10 hours, 39 minutes, 26 seconds.

Having passed Saturn, Voyager 1 headed out of the solar system. Its path took it ahead of the Sun in its journey through the Milky Way. This upstream direction gives it the best chance to reach the heliopause, the dividing line between the Sun's magnetosphere and that of interstellar space, before its electrical power supply gives out sometime around the year 2020. See **Pioneer 11, Shutdown (September 30, 1995)** and **Pioneer 10, Shutdown (March 31, 1997)** for later details.

The next spacecraft to reach Saturn was **Voyager 2 (August 26, 1981)**.

1981

February 6, 1981
Intercosmos 21
U.S.S.R. / East Europe

Intercosmos 21 was part of the Intercosmos program (see **Intercosmos 1–October 14, 1969**), carrying experiments from the U.S.S.R., Czechoslovakia, East Germany, Hungary, and Romania. Unlike other Intercosmos satellites, it did not conduct ionospheric research. Instead, it tested an early design for the Soviet Okean oceanographic satellites (see **Cosmos 1500–September 28, 1983**) during its 17 months in orbit, carrying multichannel spectrometers for studying the ocean's colors and the transparency of the earth's atmosphere. Both phenomena had a direct bearing on measuring how much solar radiation reached the earth's surface. The ocean's color also indicated areas of pollution and high concentrations of marine biological life.

February 11, 1981
Kiku 3 (ETS 4)
Japan

Kiku 3, also called Engineering Test Satellite (ETS) 4, tested new larger and heavier satellite designs while transmitting telemetry of this first test flight of the Japanese N2 rocket. Kiku 3 was placed in elliptical orbit comparable to the Soviet Molniya satellites (see **Molniya 1-01–April 23, 1965**) and was tested in orbit for several months.

The next Japanese Engineering Test Satellite was **Kiku 4 (September 3, 1982)**.

February 21, 1981
Hinotori (Astro-A)
Japan

Called Astro-A prior to launch, Hinotori studied the x-ray radiation released by solar flares during the solar maximum, augmenting observations by **Solar Max (February 14, 1980)** before and after its repair (see **STS 41C–April 6, 1984**). It carried six instruments for detecting x-rays from 5 to 40 KeV, and gamma rays from 0.2 to 9 MeV.

March 12, 1981
Soyuz T4
U.S.S.R.

Soyuz T4 was the sixth and last manned occupancy of **Salyut 6 (September 29, 1977)**, crewed by Vladimir Kovalyonok (**Soyuz 25–October 9, 1977; Soyuz 29–June 15, 1978**) and Viktor Savinykh (**Soyuz T13–June 6, 1985; Soyuz TM5–June 7, 1988**).

During their 74-day mission, the Soyuz T4 crew was visited by two international flights, **Soyuz 39 (March 22, 1981)** and **Soyuz 40 (May 14, 1981)**. No Progress freighters (see **Progress 1–January 20, 1978**) were launched during the mission—the only freighter to supply this occupancy had docked with the station seven weeks prior to Soyuz T4's arrival.

The cosmonauts focused on repairs to Salyut 6 during their stay, including replacing a battery unit for powering the laboratory's solar panels and a pump for removing water condensation in the temperature control system. They also dismantled and reassembled the internal equipment of the station's front docking port, demonstrating that the equipment could be maintained in space.

The cosmonauts tended a greenhouse, attempting—as had previous crews—to grow plants and flowers from seed. While an arabidopsis plant flowered, they could not get an orchid to bloom, as had the crew of **Soyuz 35 (April 9, 1980)**. For later research with arabidopsis plants, see **Soyuz T5 (May 13, 1982)**.

For more details about Soyuz T4, see Soyuz 39 and Soyuz 40. Kovalyonok's and Savinykh's return to Earth on May 26, 1981, signaled the end of Salyut 6's manned operations, though a Heavy Cosmos module, **Cosmos 1267 (April 25, 1981)**, docked with the station on June 19, 1981.

March 22, 1981
Soyuz 39
U.S.S.R. / Mongolia

Soyuz 39 was a two-man, eight-day mission to **Salyut 6 (September 29, 1977)** that joined the crew of **Salyut T4 (March 12, 1981)** during the station's sixth manned occupancy. The eighth internationally manned flight in the Soviet-manned program, Soyuz 39 was crewed by Soviet Vladimir Dzhanibekov

(Soyuz 27–January 10, 1978; Soyuz T6–June 24, 1982; Soyuz T12–July 17, 1984; Soyuz T13–June 6, 1985) and Mongolian Jugderdemidiyn Gurragcha.

Multispectral earth resource images of Mongolia were taken. In a medical experiment to prevent space sickness, the cosmonauts wore special space collars that did not allow rapid head movements, with unclear results. Attempts were made to transmit holograms to and from Earth. Holographic images of crystal growth were also taken.

Soyuz 39 returned to Earth on March 30, 1981. The next manned flight to Salyut 6 was **Soyuz 40** (**May 14, 1981**).

April 12, 1981
STS 1, Columbia Flight #1
U.S.A.

STS (Space Transportation System) 1 was the first flight of the space shuttle Columbia and the first flight of a reusable manned spacecraft. Crewed by John Young (**Gemini 3–March 23, 1965**; **Gemini 10–July 18,1966**; **Apollo 10–May 18, 1969**; **Apollo 16–April 16, 1972**; **STS 9–November 28, 1983**) and Robert Crippen (**STS 7–June 18, 1983**; **STS 41C–April 6, 1984**; **STS 41G–October 5, 1984**), Columbia remained in space for just over two days, landing on a runway on a dry lake bed at Edwards Air Force Base in California. This short test flight, the first of four, was to ensure that the shuttle's systems worked as NASA had designed them.

As Young stated in a post-flight press conference, "[The shuttle] worked like a dream all the way through." Among the biggest concerns had been the shuttle's protective coating of tiles. Prior to the launch, many experts believed that numerous tiles would fall off, risking both the crew and the space-

Columbia during flight; note missing tiles. *NASA*

ship. However, although a few tiles covering the top of Columbia's engines did detach, none of the critical tiles protecting the shuttle's bottom fell off.

In total, Columbia weighed over 99 tons at landing, making it the largest object ever placed in orbit and exceeding the record set by **Skylab** (**May 14, 1973**). Columbia, the first spacecraft to test lifting-body aerodynamics from orbital altitudes, also set the record for the heaviest runway landing of an unpowered glider.

Columbia was the first of a planned fleet of four American shuttles, each of which had three decks, could carry as many as 10 passengers, and used an 80-20 nitrogen-oxygen atmosphere. Their cargo bays were so large that four Apollo capsules could fit inside.

The next shuttle flight was **STS 2** (**November 12, 1981**).

April 25, 1981
Cosmos 1267
U.S.S.R.

Cosmos 1267 was what Western observers dubbed a "Heavy Cosmos." Weighing over 40,000 pounds, almost as much as **Salyut 6** (**September 29, 1977**), Cosmos 1267 is thought to have resembled the Soviet military space laboratories **Salyut**

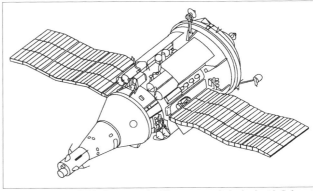
Cosmos 1267, with Merkur capsule. The spacecraft docked with Salyut 6 after the Merkur capsule had returned to Earth. *NASA*

Columbia at launch. *NASA*

May 14, 1981

3 (**June 25, 1974**) and **Salyut 5** (**June 22, 1976**). For seven weeks, Cosmos 1267 flew on its own, releasing a capsule on May 24, 1981, that returned to Earth and was recovered safely.

Three weeks later, on June 19, 1981, after the crew of **Soyuz T4** (**March 12, 1981**) had departed from Salyut 6, Cosmos 1267 docked with the station, creating an orbiting facility weighing more than 40 tons—a little less than half the weight of the American **Skylab** (**May 14, 1973**).

Over the next four months, this combined facility made several orbital maneuvers using Cosmos 1267's engines. Then the two spacecraft were allowed to drift, and after another nine months, on July 29, 1982, Cosmos 1267's engines were fired again, deorbiting both Salyut 6 and Cosmos 1267 over the Pacific Ocean.

The next Soviet orbiting space laboratory was **Salyut 7** (**April 19, 1982**).

May 14, 1981
Soyuz 40
U.S.S.R. / Romania

Soyuz 40 was the last manned mission to dock with **Salyut 6** (**September 29, 1977**). It was also the last flight of the Soyuz ferry in use since **Soyuz 12** (**September 27, 1973**). For the next five years, all Soviet manned flights used the Soyuz T spacecraft, first flown manned on **Soyuz T1** (**December 16, 1979**).

Soyuz 40, an eight-day internationally crewed flight, was manned by Soviet Leonid Popov (**Soyuz 35–April 9, 1980**; **Soyuz T7–August 19, 1982**) and Romanian Dumitru Prunariu. Docking with Salyut 6 on May 15th, they joined the crew of **Soyuz T4** (**March 12, 1981**) for six days of joint research. Studies were made of the increased blood flow to the cosmonauts' brains and the changes in their heart rates. In materials research, crystals of germanium were grown by capillary action. The cosmonauts also attempted (but failed) to grow a silicon thin-film mono-crystal that could be used in solar cells. See **STS 51** (**September 12, 1993**).

The Soyuz 40 crew returned to Earth on May 22, 1981, followed two days later by the crew of Soyuz T4, ending manned operations on Salyut 6. The next Soviet manned mission, **Soyuz T5** (**May 13, 1982**), docked with the next Soviet orbiting laboratory, **Salyut 7** (**April 19, 1982**).

June 19, 1981
Ariane 1
Apple / CAT 3
Europe / India

Ariane 1 was the third flight of the European Space Agency's Ariane rocket (see **CAT 1–December 24, 1979**) and the first to place operational satellites in orbit.

• *Apple* Designed and built by the Indian Space Research Organization (ISRO), the Ariane Passenger PayLoad Experiment (Apple) was India's first communications satellite, placed in geosynchronous orbit over India. During its two-year service life, Apple was used for communications and engineering experiments, despite the failure of one solar panel to deploy. These tests were followed by India's first operational satellite cluster, **Insat 1** (**April 10, 1982**), which provided both communications and weather data.

• *CAT 3* A small test satellite used for transmitting the Ariane rocket's launch telemetry, CAT 3 had battery power for only six orbits.

August 3, 1981
Dynamics Explorer 1 / Dynamics Explorer 2
U.S.A.

The identical NASA satellites Dynamics Explorer 1 and Dynamics Explorer 2 continued research into the lower ionosphere previously conducted by **Explorer 51** (**December 16, 1973**), **Explorer 54** (**October 6, 1975**), and **Explorer 55** (**November 20, 1975**). They also augmented **Kyokko** (**February 4, 1978**), **ESA–GEOS 2** (**July 14, 1978**), all three ISEE satellites (see **ISEE 1/ISEE 2–October 22, 1978** and **ISEE 3–August 12, 1978**), and **Jikiken** (**September 16, 1978**).

Both satellites were placed in a polar orbit with their apogees above the North Pole. Their different apogees (Dynamics Explorer 1: 15,000 miles; Dynamics Explorer 2: 600 miles) provided scientists with data for comparison from two different altitudes, which allowed them to map in great detail the interacting currents and flows of the magnetosphere, the ionosphere, and the atmosphere. Detailed real-time, in-space photographs of the aurora were taken, revealing that it was actually a flickering bright halo about 5,000 miles above the earth's poles.

Dynamics Explorer 1 operated through 1990. Dynamics Explorer 2 lasted until 1983. Combining data from almost 25 years of observations from many satellites, scientists developed a detailed view of the magnetosphere's structure.

The photographs also contained what some scientists called atmospheric holes—dark spots on the images believed to be caused by the infall of small chunks of ice or water, or "small comets," as these scientists called them. The rate of impact shown by the photographs indicated that enough water was

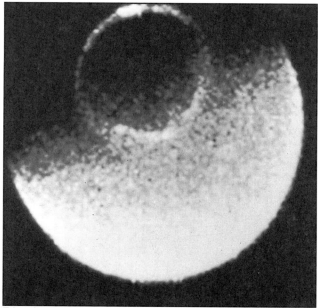

Taken September 15th, this ultraviolet image of the North Pole was the first to show the entire auroral halo. The mottled appearance along the terminator is caused by never-before-observed convection cells of upwelling nitrogen and oxygen in the atmosphere. *NASA*

August 26, 1981

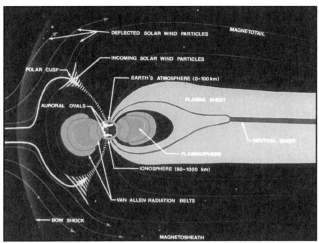

The Earth's magnetosphere as generally understood in the late 1990s. Compare with OGO 1 (September 5, 1964) diagram. *NASA*

being deposited from space to have filled Earth's oceans over the eons. Though photographs by **Polar (February 24, 1996)** also showed these atmospheric holes, most scientists remain skeptical, believing them to be instrument artifacts.

These two satellites later worked in conjunction with Sweden's **Viking (February 22, 1986)**. The next American satellite to study the ionosphere and magnetosphere was **CCE (August 16, 1984)**.

August 26, 1981
Voyager 2, Saturn Flyby
U.S.A.

Voyager 2 was the third visit to Saturn, following **Pioneer 11, Saturn Flyby (September 1, 1979)** and **Voyager 1, Saturn Flyby (December 12, 1980)**. The spacecraft dipped to within 64,000 miles of Saturn's cloud tops, and within 24,000 miles of the outside edge of its rings. It also took close-up images of the moons Iapetus, Hyperion, Titan, Dione, Mimas, Enceladus, Tethys, Rhea, and Phoebe. Unlike the two previous spacecraft, Voyager 2's trajectory viewed the sunlit side of Saturn's rings, providing the most spectacular pictures of the ringed planet yet taken.

The false-color images revealed thousands of individual rings and a wide range of colors, including green, orange, blue, purple, violet, and yellow *(see color section, Figure 15)*. No two rings appeared alike. By measuring the changing light of a bright star as it passed behind the rings, Voyager 2 was able to determine the density of the many rings, as well as the exact size of the many gaps between rings. Some rings appeared to be only 500 feet wide, and the data indicated that large wave patterns rippled for miles through them. Thousands of lightning strikes were detected, as well. These discharges of electricity, exceeding several million volts, appeared to be associated with the ring spokes seen by both Voyager spacecraft shortly after Saturn sunrise. Lightning and spokes also seemed connected to Saturn's magnetosphere and the acceleration of charged particles along magnetic lines of force and onto the rings.

Pictures of Saturn revealed ribbon-like clouds, bright and dark bands, and 1,000-mile-per-hour jet streams, including a cyclone as large as Earth. During the flyby, scientists were able to watch this storm's break-up as its size shrank by half *(see color section, Figure 14)*.

Of Saturn's moons, Voyager 2 found that they were as baffling and varied as suggested by earlier spacecraft photographs. Four more moons were discovered, raising the total to almost two dozen. All but Titan seemed to be composed mostly of ice. Cratered Tethys had one impact crater literally one-third the size of the entire moon.

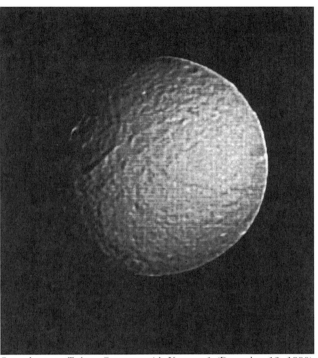

Saturn's moon Tethys. Compare with Voyager 1 (December 12, 1980) image. *NASA*

Hyperion was found to have an irregular shape, variously called "peanut-shaped" and "a hockey puck" by mission scientists. Tilted 45 degrees from its orbital plane, the moon should have been unstable and broken apart. Scientists suspected that a recent collision might have jiggled its rotational axis.

Enceladus looked like a snowball with giant grooves resembling glaciers running across a cratered icy surface. Scientists believed that tidal action was causing the ice to melt and flow periodically, not unlike lava on Earth.

Iapetus had always been known to have two very contrasting hemispheres, its leading face very dark and its trailing hemisphere very bright. When Voyager 2 flew by, scientists were able to measure Iapetus's density and found that it was mostly ice. Since ice could not produce the black coating on Iapetus's dark hemisphere, that material had to come from elsewhere. As the dark material had the same reflectivity as carbonaceous chondritic material found in the oldest meteorites, scientists postulated that as Iapetus circled Saturn, it swept the material out of space. Other scientists theorized that the material came from Phoebe, the next moon outward from Saturn.

September 19, 1981

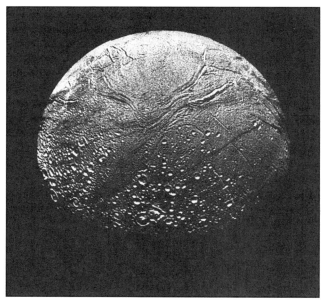
Saturn's moon Enceladus. *NASA*

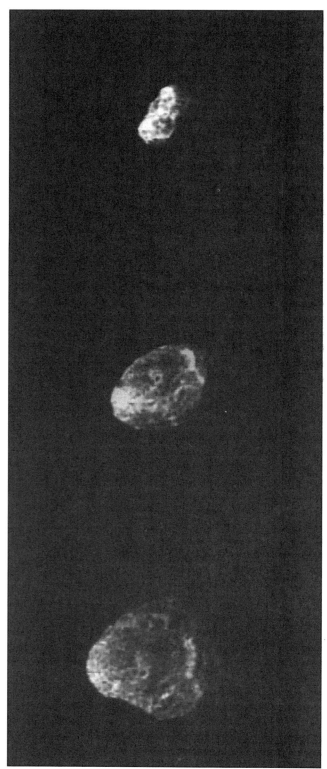
Montage of Saturn's moon Hyperion. *NASA*

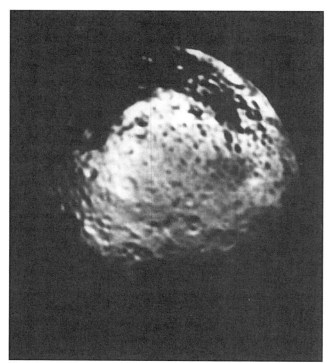
Saturn's moon Iapetus. *NASA*

As of December 1999, Voyager 2 is the last probe to reach Saturn. The next probe, **Cassini (October 15, 1997)**, is presently enroute, set to arrive in 2004.

Voyager 2, meanwhile, traveled on to a rendezvous with Uranus (see **Voyager 2, Uranus Flyby– January 24, 1986**).

September 19, 1981
China 9 through China 11
China

This triple launch—China 9, China 10, and China 11—was the first time China placed multiple satellites into orbit using a single rocket. Though described as ionospheric and atmospheric research satellites, Western experts believe all three conducted electronic surveillance for the Chinese military.

October 6, 1981
SME / UoSat 1 (Oscar 9)
U.S.A. / U.K.

• **SME** Solar Mesosphere Explorer (SME), a NASA satellite, was specifically designed to study the earth's mesosphere—the region from 20 to 50 miles elevation—during a period of declining solar activity. SME carried sensors for detecting the vertical profile of ozone, water vapor, nitrogen dioxide, temperature, and air pressure. It also carried two instruments for monitoring the Sun's ultraviolet and energy flux. Using these data, scientists hoped to find out how the earth's ozone layer fluctuated as the Sun's ultraviolet radiation varied, following up research conducted by **Nimbus 7 (October 13, 1978)**.

During its five years of operation, SME found that the ozone density at these elevations changed day to day and seasonally, and that the changes were caused by variations in temperature within the atmosphere. As temperatures dropped, ozone densities increased. Although specific solar events such as solar flares also affected ozone levels, the daily and yearly variations in solar ultraviolet output had much less influence than had temperature changes.

Later ozone layer research was conducted by Solar Max (see **STS 41C–April 6, 1984**).

• **UoSat 1 (Oscar 9)** Built by the University of Surrey, England, UoSat 1 was the ninth privately developed amateur communications satellite in the Oscar series, following **Oscar 8 (March 5, 1978)**. UoSat 1—carrying beacons transmitting at 7 MHz, 14 MHz, 21 MHz, 29.5 MHz, 145 MHz, and 435 MHz, and at 2.4 GHz and 10.4 GHz—did student propagation research on how changing atmospheric conditions affected radio signals. See also **ATS 6 (May 30, 1974)**.

The satellite also carried a magnetometer, a CCD camera (see **KH-11 1–December 19, 1976**), two geiger counters, and a speech synthesizer for performing other experiments. It operated at full capacity until 1989.

In 1984, the University of Surrey launched **UoSat 2 (March 1, 1984)** to replace UoSat 1. The next amateur communications satellite in the Oscar series was **Oscar 10 (June 16, 1983)**.

November 12, 1981
STS 2, Columbia Flight #2
U.S.A.

STS 2 was the second flight of the space shuttle Columbia, following **STS 1 (April 12, 1981)**. The flight was the first in history in which a reusable spacecraft was returned to space, and only the second time a spacecraft had been reorbited (see **Gemini 2–November 3, 1966**). Crewed by Joe Engle (**STS 51I–August 27, 1985**) and Richard Truly (**STS 8–August 30, 1983**), the mission had originally been scheduled to last five days, but it was ended after two for safety reasons when one of the shuttle's three fuel cells failed.

The astronauts performed the first tests of the shuttle's 50-foot-long remote manipulator arm, built by Canada. Columbia also carried the shuttle's first scientific payload, a set of earth observation multispectral cameras for scanning the ocean's color, air pollution levels, mineral deposits, and land features.

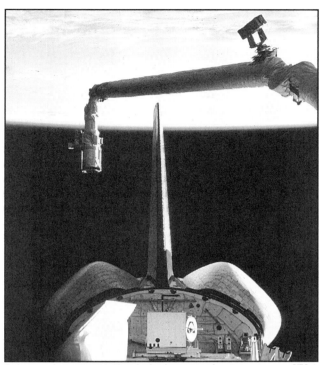

Robot arm being tested. Note lack of missing tiles, in contrast to STS 1 (April 12, 1981). *NASA*

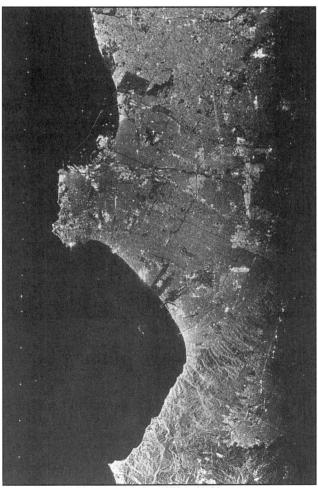

Infrared image of Los Angeles, from STS 2. *NASA*

The ocean color instrument mapped areas of heavy chlorophyll concentration in the China Sea between China, Korea, and Japan, indicating regions of rich aquatic life. The infrared radiometer identified several mineral clays that geologists used to find hidden oil, gold, silver, and copper deposits.

The most significant instrument in the package was a synthetic aperture radar, dubbed SIR-A (Shuttle Imaging Radar A), which was similar to that used on **Seasat (June 27, 1978)**. Since clouds are invisible to radar, the sensor was able to produce images regardless of the weather, scanning the earth's topography in swathes 30 miles wide and resolving objects 30 feet across.

During STS 2, more than 3.8 million square miles were radar imaged (including a single radar track more than 10,000 miles long), which produced excellent images of the topography of such remote areas as Indonesia and the Sahara. Geological structures such as thrust faults, folds, lineaments, and strike-and-dip, previously unidentifiable in these regions, were mapped in detail. In desert regions, scientists were stunned to find that the overlying sand was practically invisible to the radar, and that the radar easily revealed the underlying ancient drainage patterns and bedrock structures.

The next space flights to use radar imaging were the two Soviet missions to Venus, **Venera 15 (October 10, 1983)** and **Venera 16 (October 14, 1983)**. The next radar mission studying Earth was **Cosmos 1500 (September 28, 1983)**. The next shuttle radar mission was STS 41G (October 5, 1984).

The third flight of Columbia was STS 3 (**March 22, 1982**).

December 20, 1981
MARECS A / CAT 4
Europe

This mission was the second operational use of the European Space Agency's Ariane 1 rocket (see **Ariane 1–June 19, 1981**).

• *MARECS A* The Maritime European Communications Satellite (MARECS) A was built and launched by the European Space Agency (ESA) and was the first of two satellites to test maritime communications technology. After several months of experimentation by ESA, MARECS A was leased to Inmarsat (see **Marisat 1–February 19, 1976**) and placed in geosynchronous orbit above the Atlantic Ocean to replace Marisat 1. The satellite had capacity for 35 two-way voice channels plus a search-and-rescue channel transmitting from ship to shore.

One more MARECS satellite was launched in November 1984, which Inmarsat placed over the Atlantic Ocean, shifting MARECS A to the Pacific. In 1990, Inmarsat began replacing this Marisat/MARECS constellation. See **Inmarsat 2 F1 (October 30, 1990)**.

• *CAT 4* Battery powered with a life span of only 65 orbits, CAT 4 transmitted telemetry to ground stations, verifying the performance of the Ariane 1 rocket.

1982

February 26, 1982
Westar 4
U.S.A.

Built and owned by Western Union, Westar 4 was the first of a planned cluster of three satellites to replace Western Union's first-generation communications satellites—first launched in 1974 (see **Westar 1–April 13, 1974**). These geostationary satellites provided telephone and television service to the United States, including Alaska and Hawaii, and to the Caribbean. This second generation of Westar satellites was only slightly more capable. Their telephone capacity was the same, 1,200 one-way circuits, but they could broadcast twice the number of television channels, 24 instead of 12.

Western Union launched one other second-generation Westar satellite through 1982. A third, Westar 6, was deployed by the space shuttle Challenger in 1984 but failed to reach geosynchronous orbit when its upper stage malfunctioned. A later shuttle mission recovered it, allowing the insurance company to resell it as **Asiasat (April 7, 1990)**. See STS 41B (**February 3, 1984**) and STS 51A (**November 8, 1984**).

In 1988, Hughes Communications purchased Westar 3, Westar 4, and Westar 5 from Western Union, removing Western Union from the satellite business. See **Galaxy 1 (June 28, 1983)**.

March 1, 1982
Venera 13, Venus Landing
U.S.S.R.

Launched on October 30, 1981, Venera 13 was the fifteenth spacecraft to enter the atmosphere of Venus and return data, following **Venera 12 (December 21, 1978)** and **Venera 11 (December 25, 1978)**. It landed on the rolling Venusian plains just south of the equator and west of the suspected shield volcano Phoebe Regio, where it functioned for just over two hours—the longest of any spacecraft through December 1999. During its life span, Venera 13 used a drill to obtain and test a small rock sample and took two color panorama photographs—the first color images from the surface of Venus *(see color section, Figure 30)*. See **Venera 14 (March 5, 1982)** for results.

March 4, 1982
Venera 14, Venus Landing
U.S.S.R.

Venera 14 was essentially identical to **Venera 13 (March 1, 1982)**. It was launched on November 4, 1981, and arrived on Venus three days after Venera 13, landing on the same rolling plains only a few hundred miles to the southeast. On Venus's surface it functioned for almost one hour.

Both spacecraft sent back two color panoramas. Like the previous images taken by **Venera 9 (October 22, 1975)** and **Venera 10 (October 25, 1975)**, the camera was a fish-eye wide-angle lens that showed a close-up at the center of the frame and a long shot of the horizon in its upper left and right corners. Unlike the previous probes, Venera 13 and Venera 14

March 4, 1982

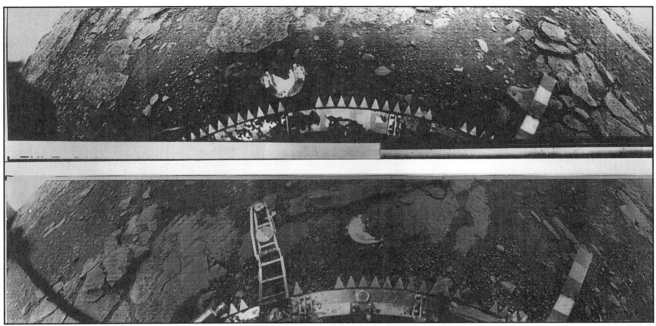

The surface of Venus, as photographed by Venera 13 *(see also color section, Figure 30)*. NASA

each carried two cameras in order to photograph the full 360 degrees surrounding them.

The panoramas resembled the previous Venus photographs, showing a flat, rocky terrain covered with scattered light-colored slabs resting on dark soil. As in Venera 9's pictures, some stones appeared layered. The only significant difference between the two sites was that the ground at Venera 14 was tiled with more flat rocks, revealing less of the granular soil underneath.

Because of the atmosphere's thickness, the horizon visible in the upper corners of the panoramas seemed much closer than it actually was. This illusion, thought to exist for any observer anywhere on the planet, made a visitor think he or she was standing on a small globe looking down to the horizon.

As V. I. Moroz of the U.S.S.R. Academy of Sciences noted, "In general, the neighborhood of each probe is reminiscent of the bottom of terrestrial oceans." The layered appearance of many stones suggested sedimentation, the dense atmosphere resembled the ocean, and the winds were akin to deep-water currents.

Like Venera 13, Venera 14 carried a small drill to obtain soil samples and test their composition. Placed in a chamber and examined by an x-ray spectrometer, each sample was found to resemble terrestrial basalt. Because of the nearness to Phoebe Regio to the west, these results implied that the two spacecraft had landed on the outer margin of that volcano's lava field.

The surface of Venus, as photographed by Venera 14. *NASA*.

As the two Venera spacecraft descended through the Venusian atmosphere, they recorded electromagnetic discharges resembling lightning flashes, very similar to the pulses detected by **Venera 12 (December 21, 1978)** and **Venera 11 (December 25, 1978)**. These charges faded out below 6 miles elevation, with no discharges detected on the surface. Though scientists still did not know what had caused them, some evidence indicated that they might be related to volcanic activity at the nearby shield volcanoes of Phoebe Regio, Beta Regio, and Alta Regio. See map from **Pioneer Venus Orbiter (December 4, 1978)**.

Atmospheric readings during descent revealed numerous thin-cloud layers (many less than 300 feet thick). Sulfur was the most abundant element in the clouds, with chlorine second; the opposite finding from that of Venera 12. Both spacecraft found that the main cloud layer, from 28 to 44 miles elevation, was actually three stable layers. Significant quantities of water vapor were detected in the center of these main cloud layers, between 30 and 36 miles altitude.

As found by Venera 11 and Venera 12, surprising quantities of certain isotopes of the noble gases were detected, including neon, argon, and krypton. This discovery puzzled scientists—while Venus seemed to contain the same amount of carbon dioxide and nitrogen as Earth, its allotment of inert gases was much greater. (With carbon dioxide and nitrogen, the only difference between the two planets seemed to be the amount trapped in the soil and the amount in the atmosphere. On Venus, carbon dioxide was in the atmosphere, while on Earth, it was in the soil.) If both planets formed from the same planetary nebula, the quantities of inert gases should have been the same. Some scientists proposed that the discrepancy existed because the Venusian atmosphere was enriched by the solar wind. Others discounted this theory, noting that the ratios didn't quite match, and that the data were still insufficient to theorize.

Both spacecraft carried seismographs, but during their short life span on the surface, they did not detect any Venusian quakes. Venera 14 did detect two very tiny micro-quakes, though these were insufficient to provide much information about the deep interior of the planet.

At the completion of Venera 14's flight, the theory that the planet's atmosphere resulted from a runaway greenhouse effect (see **Mariner 5, Venus Flyby–October 19, 1967**) was accepted by almost all scientists. Beyond this, however, the scientists had no explanation for why Earth and Venus, so close together in the solar system and of similar size, could be so different. Some theorists proposed that Venus might once have had a benign atmosphere like Earth, and that the planet's proximity to the Sun had caused its oceans to evaporate and then escape into space. The evidence to prove such theories remained weak and inconclusive, however.

Furthermore, while theories existed to describe the interior of the planet based on its known density and mass, nothing was known about its actual internal structure. Some scientists posited that the planet exhibited some plate tectonic activity, with more hot spots than the earth. The raised region, Ishtar Terra, was thought to be the planet's only continent, with regions like Phoebe, Beta, and other raised areas evidence of younger volcanic hot spots. Testing this hypothesis, however, would require additional space probes.

The overall weather and circulation patterns of the Venusian atmosphere were still only partly understood. The entire atmosphere seemed to rotate as a unit every four days. This flow was so strong that some scientists believed it contained about 15 percent of the planet's total angular momentum, and could actually change by several hours the length of Venus's long day as weather patterns evolved over time. Scientists also had evidence suggesting a flow from the subsolar point (the spot on the globe closest to the Sun) to the antisolar point. Other evidence indicated circulation patterns flowing from the equator to the poles and back again. No one, however, was able to fit all these patterns together into a general theory explaining Venus's weather.

The next spacecraft to reach Venus were the radar mappers **Venera 15 (October 10, 1983)** and **Venera 16 (October 14, 1983)**. The next landers were released from the Halley's Comet probes, Vega 1 and Vega 2 (see **Vega 1, Venus Flyby–June 10, 1985** and **Vega 2, Venus Flyby–June 14, 1985**).

March 22, 1982
STS 3, Columbia Flight #3
U.S.A.

Crewed by Jack Lousma (**Skylab 3–July 28, 1973**) and Gordon Fullerton (**Enterprise–August 12, 1977**; **STS 51F–July 29, 1985**), STS 3 was the third test flight of the space shuttle Columbia, following **STS 2 (November 12, 1981)**. This eight-day mission, which was longer than the first two missions combined, tested the spacecraft's ability to function in space for long periods. Columbia was allowed to orbit in a number of different orientations—nose-to-Sun, tail-to-Sun, bay-to-Sun, bay-away-from-Sun—to see how the shuttle absorbed or deflected the extreme temperatures of space. In general, Columbia handled the temperature ranges better than expected, keeping its interior temperature constant.

The Canadian-built robot arm was tested for the second time. Despite the temporary failure of a television camera on the arm's wrist, the arm was used for the first time to lift and return a payload from the cargo bay.

Besides evaluating the shuttle's ability to make an unprecedented third flight into space, STS 3 also carried a package of engineering and scientific experiments. One engineering test measured the electromagnetic wake created by the shuttle as it plowed its way through Earth's magnetosphere. Scientists needed to know the extent of this wake to accurately interpret scientific data that would be gathered on future shuttle missions. The experiment not only provided baseline data, but it found that the radio noise produced by the spacecraft's interaction with the magnetosphere was low enough not to interfere with future research.

This experiment produced one unexpected discovery: developed photographs of Columbia's tail during the night portion of one orbit revealed a faint glow 2 to 4 inches thick emanating from the shuttle's surfaces and most visible on the shuttle's leading edges. This glow was somehow caused by the interaction of the shuttle and the oxygen in the ionosphere.

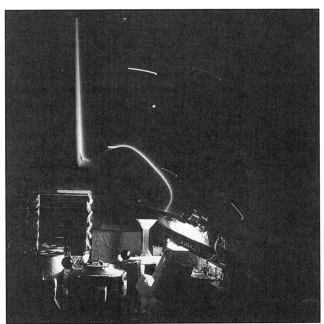

Looking into the shuttle's cargo bay. Note the glow on the rear tail surfaces. *NASA*

See **STS 39 (April 28, 1991)** for more research about this glow.

Another engineering test measured the build-up of static electricity on the shuttle's surface (see **Explorer 8–November 3, 1960**), finding that the build-up held to a relatively constant 1 to 2 volts. Several other sensors measured solar x-ray and ultraviolet radiation, as well as a micrometeoroid detector for measuring the impact rate in the shuttle's cargo bay.

A third engineering test investigated the use of passive heat pipes to maintain the temperatures within a canister. Rather than having each experiment carry its own heating system, an expensive and heavy option, this design showed that one heating system could control the temperature of a number of experiments.

This flight officially inaugurated the use of the shuttle as an orbiting laboratory. One experiment attempted to germinate 96 seedlings of pine, oat, and bean plants to see how the production of lignin, the structural material used by plants to give them rigidity, was affected by growth in weightlessness. Since lignin was indigestible to humans, the ability to grow plants without it could increase agricultural food production exponentially. Lignin production, however, remained unaffected by lack of gravity. While the plants did germinate and were able to orient their shoots upward toward the light, their roots failed to consistently orient downward. Some roots instead grew upward and out into the air, away from their water source. See **STS 61A (October 30, 1985)** for more plant root research.

Another experiment carried a cage of moths, houseflies, and honeybees to see if they would fly in zero gravity. Except for when the cage was rattled by the astronauts, the insects clung to its walls.

The monodisperse latex reactor experiment attempted to produce tiny 5- to 10-micron-wide latex particles that could be used in medical research, such as for measuring the size of pores in the intestine and eye. Scientists also hoped to use the latex particles to transport medicine directly to cancerous tumors. Although some technical problems reduced the results, scientists were still able to obtain quantities of particles 5 microns wide. Through 1984, this reactor made five shuttle flights, eventually producing quantities of perfect latex spheres all exactly the same size. See **STS 41B (February 3, 1984)**.

Because of bad weather at Edwards Air Force Base, the intended landing site, STS 3 landed at White Sands, New Mexico. Through December 1999, it was the first and only time this emergency landing site has been used by the shuttle program. Even so, high winds at White Sands caused the landing to be delayed one day, which extended the mission from seven to eight days.

The next shuttle mission, **STS 4 (June 27, 1982)**, completed the four-flight test program.

April 10, 1982
Insat 1A
India / U.S.A.

Following tests on India's first experimental communications satellite, Apple (see **Ariane 1–June 19, 1981**), the Indian Space Research Organization (ISRO) had Ford Aerospace and Communications build Insat 1A, India's first operational geosynchronous communications satellite. This satellite, launched by NASA, was also India's first weather satellite, and the first in a four-satellite cluster.

Insat 1 satellites were unusual in design. Because each satellite carried both communications and weather-monitoring payloads, the solar panel arrangement was asymmetrical, with all panels on one side so that the infrared radiometer had a clear view for weather images. A solar sail was used to balance the satellite against the pressure of the solar wind blowing against its solar panels.

The satellite's capacity for telephone and television communications equaled about 8,000 two-way voice circuits plus three television channels. Ground antennas could be as small as 12 feet across.

Insat 1A was part of the global network of geosynchronous weather satellites, begun with **GOES 1 (October 16, 1975)**. Weather images were produced in both visible light (0.55–0.75 microns) and infrared (10.5–12.5 microns). Resolution equaled 1.7 miles in visible light and 6.8 in the infrared, with a new image produced every 30 minutes. Each image covered the entire Indian subcontinent.

Through 1988, India launched four Insat 1 satellites. Insat 1A failed after five months when its solar sail failed to deploy and the satellite could not be oriented. Insat 1C lost half its broadcast capacity shortly after deployment due to a power failure. Insat 1B and Insat 1D functioned without problems.

In 1992, ISRO launched India's second-generation cluster of communications-weather satellites. See **Insat 2 (July 9, 1992)**.

April 19, 1982
Salyut 7
U.S.S.R.

As a replacement for the previous Soviet orbiting laboratory, **Salyut 6 (September 29, 1977)**, Salyut 7 was only slightly more sophisticated. It weighed essentially the same, about

43,000 pounds, provided about the same amount of habitable space (about 3,300 cubic feet), and included a similar assortment of telescopes, cameras, materials, and semiconductor furnaces, plant greenhouse experiments, and other equipment. In addition to a gamma-ray telescope, the station carried an x-ray telescope and a spectrometer. Additional exercise equipment was included, as well as more complete medical and biological facilities, a refrigerator, and water heater.

The most significant improvement concerned the front docking port, which had been strengthened to accommodate the greater weight of Heavy Cosmos modules, two of which would dock with the laboratory during its lifetime, **Cosmos 1443** (**March 2, 1983**) and **Cosmos 1686** (**September 27, 1985**). These modules, similar in design to the earlier military **Salyut 3** (**June 25, 1974**) and **Salyut 5** (**June 22, 1976**), were direct descendants of **Cosmos 1267** (**April 25, 1981**), which had docked with Salyut 6 following the completion of manned operations. These modules added about 1,700 cubic feet of habitable space to the laboratory, bringing the total volume to about 5,000 cubic feet (in contrast to the nearly 13,000 cubic feet of space of **Skylab–May 14, 1973**). The Heavy Cosmos modules also increased Salyut 7's available electrical power with their large solar panels. Each brought as much as five tons of cargo to Salyut 7, about triple the capacity of a Progress freighter (see **Progress 1–January 20, 1978**), and a recoverable capsule with which more than 1,000 pounds of experiments, specimens, and exposed film could be returned to Earth. See diagram with Cosmos 1686.

Because of several equipment problems, Salyut 7 did not have the same success as Salyut 6. The failure in 1983 of some of the station's solar panels reduced its electrical power, causing the inside temperature to drop to about 65°F with 100 percent humidity. In September 1983, a leak developed in one of the station's fuel lines, shutting down half the laboratory's 32 attitude control thrusters. Finally, in early 1985, while the station was unoccupied, radio control was lost, and Salyut 7 began drifting, lost power, and became crippled.

Despite these problems, a new space flight endurance record of 237 days (almost eight months) was set. Furthermore, the repair efforts to save the station (including 13 space walks) helped train Soviet cosmonauts in a variety of in-space construction tasks. This experience provided the skills needed for maintaining the next Soviet orbiting laboratory, **Mir (February 20, 1986)**. The manned missions to Salyut 7 are as follows, including the launch and return dates of crews:

First occupancy (211 days)
Soyuz T5–**May 13, 1982** to December 10, 1982
Soyuz T6–**June 24, 1982** to July 2, 1982
Soyuz T7–**August 19, 1982** to August 27, 1982

First unsuccessful flight
Soyuz T8–**April 20, 1983** to April 22, 1983

Second occupancy (149 days)
Soyuz T9–**June 27, 1983** to November 23, 1983
Soyuz T10A–**September 26, 1983** to September 26, 1983 (aborted during launch)

Third occupancy (237 days)
Soyuz T10B–**February 9, 1984** to October 2, 1984

Soyuz T11–**April 3, 1984** to April 11, 1984
Soyuz T12–**July 17, 1984** to July 29, 1984

Fourth occupancy (150 days for 1 cosmonaut)
Soyuz T13–**June 6, 1985** to September 25, 1985 and November 21, 1985
Soyuz T14–**September 17, 1985** to September 25, 1985 and November 21, 1985

Fifth occupancy (51 days)
Soyuz T15–**March 13, 1986**; launched to Mir March 13, 1986; occupied Salyut 7 from May 6, 1986 to June 25, 1986

Salyut 7 remained in orbit until February 7, 1991, when its orbit decayed and it burned up over Argentina. The next Soviet space laboratory, Mir, carried multiple docking ports to permit the assembly of a much larger orbiting research facility.

May 13, 1982
Soyuz T5
U.S.S.R.

Soyuz T5 was the first occupancy of **Salyut 7** (**April 19, 1982**). Its crew included Anatoly Berezovoi and Valentin Lebedev (**Soyuz 13–December 18, 1973**), who remained in orbit for 211 days. They returned in the **Soyuz T7** (**August 19, 1982**) spacecraft on December 10, 1982, setting a new space endurance record that surpassed the 185-day record of **Soyuz 35** (**April 9, 1980**).

During their mission, the cosmonauts released **Iskra 2** (**May 17, 1982**) and **Iskra 3** (**November 18, 1982**), small active repeater satellites for use in Soviet amateur radio communications. They were also visited twice by three-man missions: **Soyuz T6** (**June 24, 1982**) and **Soyuz T7** (**August 19, 1982**). Soyuz T7 included the first woman in space since Valentina Tereshkova (**Vostok 6–June 16, 1963**). See those missions for details.

Also docking with Salyut 7 during this occupancy were four Progress freighters (see **Progress 1–January 20, 1978**), which brought food, air, water, fuel, and replacement equipment.

Soyuz T5's research continued earlier experiments from **Salyut 6** (**September 29, 1977**). The cosmonauts managed a small greenhouse for testing how plants grew in space. Previous experiments had yielded poor production—except for onions, no plant grown in space had grown well (see **Soyuz 35–April 9, 1980**). The Soyuz T5 cosmonauts were more successful. With increased light and a better water dispenser, an arabidopsis plant not only bloomed but produced the first seeds in space. For later research with arabidopsis plants, see **STS 54** (**January 13, 1993**).

Materials research included continued alloy and semiconductor kiln experiments. The initial batch of cadmium-selenide and indium-antimonide semiconductor crystals weighed 4 pounds and exhibited a highly uniform structure. Later, crystals of cadmium and germanium-sulfide were also produced.

On July 30, 1982, Lebedev performed a 2.5-hour space walk, retrieving several experiments placed on the exterior of Salyut 7 prior to launch, including samples of amino acids and biopolymers being tested in the weightlessness, vacuum, and radiation of space. Lebedev also tested future in-space construction techniques. In one case, he tested the load-bearing capability of several materials. In another, he experimented with a variety of joint designs for assembling girders. A third test checked the dependability of bolts made from different materials.

Two hours of exercise per day on the treadmill and bicycle machine were now routine, as were regular checks of each cosmonaut's heart, blood, and brain activity.

Upon their return, Berezovoi and Lebedev recovered swiftly. More significant, however, were the psychological difficulties during the flight. As the mission wound down, the two men became increasingly irritable and hardly spoke to one another during the last few months. They also needed more sleep than previous crews. Based on this and other data, scientists determined that long-term missions usually operate best with crews of three or more.

The next attempt to occupy Salyut 7 was **Soyuz T8 (April 20, 1983)**.

May 17, 1982
Iskra 2
U.S.S.R.

Iskra 2 was deployed from **Salyut 7 (April 19, 1982)** by the **Soyuz T5 (May 13, 1982)** cosmonauts. Similar to the Oscar amateur radio satellite series (see **Oscar 1–December 12, 1961**), Iskra 2 had a small computer memory and a radio repeater for transmitting messages. It remained in orbit for two months before re-entering the atmosphere.

June 4, 1982
Cosmos 1374 (BOR-4)
U.S.S.R.

This half-day mission was the first orbital test of a Soviet unmanned space plane. Also called BOR-4, it was a subscale variation of the Soviet space shuttle **Buran (November 15, 1988)**, then under development. It also used American lifting body concepts tested in the 1960s. After orbiting Earth once, the spacecraft performed re-entry tests as it glided to a safe recovery in the Indian Ocean—the third ocean recovery in Soviet space history (see **Zond 8–October 20, 1970**).

Through 1987, the Soviets completed 13 flights of several subscale versions of this prototype, followed by a single unmanned test flight of Buran.

June 24, 1982
Soyuz T6
U.S.S.R. / France

This eight-day internationally crewed mission to **Salyut 7 (April 19, 1982)** was crewed by a three-man Soviet-French crew: Soviets Vladimir Dzhanibekov (**Soyuz 27–January 10, 1978; Soyuz 39–March 22, 1981; Soyuz T12–July 17, 1984; Soyuz T13–June 6, 1985**) and Alexander Ivanchenkov (**Soyuz 29–June 15, 1978**), and Frenchman Jean-Loup Chretien (**Soyuz TM7–November 26, 1988; STS 86–September 25, 1997**). It supported the first occupancy of Salyut 7 by **Soyuz T5 (May 13, 1982)**, placing five humans in an orbiting station for the first time. Chretien was also the first person not from a Communist-bloc country to fly on a Soviet spacecraft.

Soyuz T6's mission focused on medicine and biology. Three different experiments tested human vision in space, measuring the eye's ability to perceive details, depth of field, and color. Another experiment monitored the increased blood circulation to the brain in weightlessness. Antibiotics were tested on specific bacteria. The effect of cosmic radiation on living organisms was measured.

About 350 astronomical images were taken in both the visible and infrared, including three observations of the Crab Nebula.

Alloy and materials research continued, with experiments mostly testing the performance of the station's furnace, and how differing temperatures and Salyut 7's orientation affected the mixing and solidification of materials.

The Soyuz T6 crew returned to Earth on July 2, 1982. The next mission to Salyut 7 was **Soyuz T7 (August 19, 1982)**.

June 27, 1982
STS 4, Columbia Flight #4
U.S.A.

This fourth test flight of the space shuttle Columbia lasted seven days and was crewed by Ken Mattingly (**Apollo 16–April 16, 1972; STS 51C–January 24, 1985**) and Henry Hartsfield (**STS 41D–August 30, 1984; STS 61A–October 30, 1985**). Because the two-man crew of **Soyuz T5 (May 13, 1982)** and the three-man crew of **Soyuz T6 (June 24, 1982)** were already in orbit onboard **Salyut 7 (April 19, 1982)**, with Columbia's launch, seven humans from three countries (United States, Soviet Union, and France) were concurrently in space, tying a record set by the triple flight of **Soyuz 6 (October 11, 1969), Soyuz 7 (October 12, 1969), and Soyuz 8 (October 13, 1969)**. This record was broken during **STS 41C (April 6, 1984)**.

As with Columbia's first three flights (see **STS 3–March 22, 1982**), STS 4 was mostly dedicated to evaluating the shuttle's performance in space. More thermal tests were conducted to test the spacecraft's ability to maintain its interior temperature in the extremes of space. Detailed measurements were taken of the shuttle's wake through the ionosphere and magnetosphere.

Although no EVAs (extra-vehicular activities) were conducted, the astronauts donned spacesuits in anticipation of future space walks, confirming that the suits could be put on and off with ease. Also tested was the use of suction-cup footwear.

In addition to these engineering tests, STS 4 carried several science experiments, including what NASA dubbed its first "getaway specials." These self-contained canisters provided a low-cost method for placing scientific experiments in space. On STS 4, a single getaway special was packed with experiments by students from Utah State University, testing how oil and water mix and how shrimp and green algae grow in weightlessness.

Medical research included the shuttle's first private commercial cargo. This experiment, called CFES (Continuous Flow Electrophoresis System), was a prototype of a production unit for producing pure samples of biological materials—such as interferon, beta cells (which could act as a single-injection cure for diabetes), and growth hormones—by electrophoresis, research first conducted on **Apollo 14 (January 31, 1971)** and **Apollo 18 (July 15, 1975)**. Ortho Pharmaceuticals Corporation and McDonnell Douglas Astronautics were so satisfied with their results, having four times greater purity than possible on Earth, that they flew units on several later shuttle missions (see **STS 6–April 4, 1983**), hoping to sell the produced materials on the open market.

Another experiment attempted to make a photographic survey of thunderstorm lightning during both daytime and night. The images revealed that lightning not only occurred within clouds, as meteorologists had believed, but also linked separate clouds at high altitudes. Bolts were observed as long as 25 miles and occurring as far as 60 miles from storm centers.

STS 4 also carried a military payload, the first American manned mission to do so. Though both NASA and the Defense Department refused to release any information about this payload, reporters soon learned that it was mostly dedicated to testing the environment surrounding the shuttle in order to improve surveillance equipment for later flights. These tests took both infrared and ultraviolet measurements.

Columbia's landing at Edwards Air Force Base was also the shuttle's first on a concrete runway.

The next shuttle mission, STS 5 (**November 11, 1982**), officially began operations of the space transportation system and was the first to place commercial satellites in orbit.

June 30, 1982
Cosmos 1383 (COSPAS 1)
U.S.S.R.

Cosmos 1383 was the first component of an international search and rescue system, called SARSAT in the West and COSPAS in the Soviet Union. In a 1980 agreement among the United States, the Soviet Union, Canada, and France, a four-satellite constellation (two satellites from the United States and two from the Soviet Union) would be placed in polar orbit with the ability to automatically pick up emergency signals transmitted by ships or aircraft in distress. The satellites would then retransmit the signal to a number of SARSAT-COSPAS ground stations, which would then process the signal to identify the ship or plane, as well as pinpoint its location to within 11 miles. Later, more advanced transmitters could narrow locations to a little more than 2 miles.

By the early 1990s, 13 ground stations had been built in the United States, the Soviet Union, France, Canada, and other countries worldwide.

By the end of 1984, four satellites were in orbit, completing the cluster. To maintain the system through 1998, the Soviet Union/Russia launched both separate Cosmos satellites, as well as Nadezhda satellites (see **Nadezhda 1–July 4, 1989**), while the United States included the receiver-transmitters on seven NOAA satellites (see **NOAA 1–December 11, 1970**) and three GOES satellites (see **GOES 1–October 16, 1975**). Later, the system was expanded with receiver-transmitters on Japanese (see **Himawari 1–July 14, 1977**) and Indian (see **Insat 2–July 9, 1992**) satellites.

By 1984, the system was used to track competitors in the Paris-Dakar car rally across Africa, a dogsled expedition across the northern Canadian wilds, and a Russian ski expedition to the North Pole. Through 1989, the system's ability to quickly locate individuals in remote areas was estimated to have saved more than 1,200 lives.

July 16, 1982
Landsat 4
U.S.A.

The fourth Landsat earth resource satellite launched by NASA and the first of a second-generation design (see **Landsat 1–July 23, 1972**), Landsat 4 operated for 11 years, until 1993. It was joined by Landsat 5 in 1984.

Like the earlier Landsats, the second-generation satellites were placed in a sun-synchronous orbit (see **Nimbus 1–August 28, 1964**), able to photograph every point on the earth's surface every 16 days *(see color section, Figure 25)*. Because the new satellites used NASA's Tracking and Data Relay Satellites (see **STS 6–April 4, 1983**) instead of ground stations, an almost continuous stream of images could be transmitted. This capability made it possible for the satellites to photograph almost the entire surface of the earth between 81 degrees north and south latitudes. The only gaps, centered over the Indian subcontinent, were filled by two Indian earth resource satellites (see **Bhaskara 1–June 7, 1979**).

Landsat 4 carried the same multispectral scanner used on the first three Landsats, covering four spectral bands: red, green, and two near-infrared wavelengths. Images scanned a 115- by 115-mile area, resolving objects as small as 260 feet across.

Landsat 4 also carried a second scanning camera, called the thematic mapper. This instrument scanned in seven specific and narrow wavelengths, chosen by scientists based on what had been learned from earlier Landsat images. While the multispectral scanner's two visible bands covered wavelengths from 0.5 to 0.7 microns, the mapper used three filters: blue-green from 0.45 to 0.52 microns, green from 0.52 to 0.60 microns, and red from 0.63 to 0.69 microns. With these three filters, the mapper could penetrate clear water to map the seafloor on shallow coastlines, record changes in polluted or turgid water, and distinguish between healthy and unhealthy vegetation as well as soil and plant life. The red filter was specifically chosen because it recorded levels of chlorophyll, allowing scientists to discriminate easily between different plant species.

The infrared wavelengths were also refined. While the multispectral scanner had only two infrared filters, looking at wavelengths from 0.7 to 1.1 microns, the thematic mapper had four. The first, covering from 0.76 to 0.9 microns, was specifically for recording biomass and plant cell structures. Moreover, since this wavelength was completely absorbed by water (hence, water bodies appeared black in images), it could be used to easily map shorelines.

The second infrared filter covered 1.55 to 1.75 microns. It could record plant and soil moisture content, as well as differentiate between snow and clouds. It could also see through thin clouds, permitting observations at times when earlier Landsats would have been blind.

The third infrared filter, from 2.08 to 2.35 microns, discriminated between different rock types. Using this waveband, geologists could map the geology of a region and do mineral exploration.

These six bands, three visible and three infrared, all had resolutions of about 100 feet across.

The thematic mapper's seventh filter recorded heat radiation, working in the far-infrared from 10.4 to 12.5 microns. This waveband showed changes in radiant heat released by the ground. It could penetrate through smoke to locate forest fires, map volcanic eruptions precisely, and record day-night surface temperature changes. Its resolution was coarse, however, capturing nothing smaller than 400 feet across.

Researchers could produce images combining these spectral bands. Different combinations revealed different details. In the decades since their launch, researchers have used Landsat images for a myriad of discoveries. They have located oil and mineral resources in places as widespread as Chile, Nevada, Zaire, Spain, and Jordan; they have recorded volcanic eruptions from Mount Saint Helens in the United States to the Lascar volcano in Chile; they have discovered ancient caravan trails in west-central Oman, leading to the rediscovery of the lost city of Ubar; and they have mapped seasonal changes in ocean currents off the coast of California and sea ice changes off the east and west coasts of Greenland, as well as the Arctic Ocean north of Alaska and Canada.

After many years of delays and the loss of a sixth Landsat satellite at launch in April 1999, a seventh Landsat satellite joined the still-working Landsat 5 in orbit. See **Landsat 7 (April 15, 1999)**. The next earth resource satellite, however, was **Spot 1 (February 22, 1986)**.

August 19, 1982
Soyuz T7
U.S.S.R.

Soyuz T7 docked with **Salyut 7 (April 19, 1982)** for an eight-day mission, joining the crew of **Soyuz T5 (May 13, 1982)** during their 211-day flight. Soyuz T7's three-person crew included Leonid Popov (**Soyuz 35–April 9, 1980; Soyuz 40–May 14, 1981**), Alexander Serebrov (**Soyuz T8–April 20, 1983; Soyuz TM8–September 5, 1989; Soyuz TM17–July 1, 1993**), and Svetlana Savitskaya (**Soyuz T12–July 17, 1984**), the first woman to fly in space since Valentina Tereshkova in **Vostok 6 (June 16, 1963)**. Savitskaya lost a little more than two pounds during the flight, but showed no other ill effects, adapting quickly to weightlessness and recovering easily back on Earth.

When the crew of Soyuz T7 returned to Earth, they used the older Soyuz T5, leaving the newer Soyuz T7 spacecraft for the return of the Soyuz T5 crew. The next Soviet manned space flight was **Soyuz T8 (April 20, 1983)**.

August 26, 1982
Anik D1 (Telesat 5)
Canada / U.S.A.

This geosynchronous communications satellite was built by Spar Aerospace (of Canada) and Hughes Aircraft (of the United States) for Telesat Canada. Launched by NASA, it was the first of a two-satellite cluster replacing the Anik A (see **Anik A1–November 10, 1972**) and Anik B (**Anik B1–December 15, 1978**) satellites servicing Canada and its northernmost provinces. Though similar to earlier satellites, Anik D satellites had greater power output, which allowed smaller ground stations to use them.

This two-satellite cluster was further augmented by the Anik C satellites (see **STS 5–November 11, 1982**). Both were eventually replaced by the Anik E satellites (see **Anik E2–April 4, 1991**).

August 27, 1982
Cosmos 1402
U.S.S.R.

Cosmos 1402 was the Soviet Union's 24th Radar Ocean Reconnaissance Satellite (RORSAT). Designed to intercept radar and electronic signals from ocean vessels, these satellites are powered with radioactive fuel. See **Cosmos 198 (December 27, 1967)**.

Cosmos 1402 was also the second RORSAT to re-enter the atmosphere uncontrolled, following **Cosmos 954 (September 18, 1977)**. Unlike Cosmos 954, which had scattered radioactive debris and fuel across the northern Canada, Cosmos 1402 ejected its fuel prior to its deorbit in January 1983. Unprotected by the spacecraft's body, the fuel burned up in the atmosphere two weeks later, thereby posing less of a threat to health and safety on Earth.

September 3, 1982
Kiku 4 (ETS 3)
Japan

Kiku 4, also called Engineering Test Satellite (ETS) 3, was an experimental flight to test several Japanese-built satellite components, including an attitude control system maneuverable in three dimensions, deployable solar panels, and a thermal control system for maintaining the satellite's temperature.

The next Japanese Engineering Test Satellite was **Kiku 5 (August 27, 1987)**.

September 9, 1982
Conestoga
U.S.A.

Conestoga was the first privately financed rocket. Built by Space Services, Inc., it was 37 feet high and took a 1,000-pound payload on an almost 11-minute suborbital flight 326 miles downrange of its launch site on Matagorda Island off the Texas coast.

Although Space Services did not follow this success with an orbital flight, its Conestoga rocket foreshadowed the many private launch vehicles being built in the 1990s. See **Pegsat (April 5, 1990)**.

October 12, 1982
Cosmos 1413 through Cosmos 1415
U.S.S.R.

Cosmos 1413, Cosmos 1414, and Cosmos 1415 were the first satellite test flights of a Soviet navigational system, comparable to the American global positioning system (GPS) first tested with **Timation 1 (May 31, 1967)** and made operational with **Navstar 5 (February 9, 1980)**. Called Glonass, the Soviet system uses a constellation of 24 satellites, launched in groups of three. GPS ground receivers calculate position by measuring precisely the time it takes for signals to travel from the satellites to the receiver. This travel time is then used to obtain the distance, which is then used to triangulate position. Glonass receivers can obtain ground fixes within 330 feet on a horizontal plane, somewhat better than early American civilian GPS receivers.

Though similar in design and concept, the Soviet Glonass and American GPS systems are not compatible, requiring different ground receivers to process each system's signal and obtain a positional fix. Several manufacturers have attempted to develop ground receivers that use the signals from both systems, thereby doubling the available orbiting satellites and increasing the receiver's accuracy. The success of these products, as of 1995, has been inconclusive.

After a number of test flights, launches of the first constellation began in 1983, becoming operational in 1988. A second, more advanced satellite cluster soon followed and is now in operation. Through 1998, the Soviet Union/Russia launched more than 80 Glonass satellites, averaging about two launches per year, with each launch usually placing three satellites in orbit. This launch pace was apparently needed to maintain the operating cluster of 24 satellites, since many Glonass satellites had short life spans, functioning less than three years. In the late 1990s, the status of the Glonass system was hindered by satellite failures, lack of funds, and a slower launch pace. As of December 1999, only 14 Glonass satellites were thought to be operational. The Russian government had attempted in early 1999 to arrange a deal with the European Union to sell half its ownership of Glonass in exchange for an infusion of cash. This arrangement, however, did not take place.

October 28, 1982
RCA Satcom 5 (Aurora 1)
U.S.A.

RCA Satcom 5 was a satellite developed and built by the Radio Corporation of America (RCA). Prior to launch, it was sold to RCA Alascom and was renamed Aurora 1. Aurora 1 was dedicated to providing long-distance communications within Alaska, as well as from Alaska to the rest of the United States. The satellite was also part of RCA's communications satellite cluster servicing the U.S. telephone and television market (see **RCA Satcom 1–December 13, 1975**). Satcom 5 was also the first of a second generation of improved RCA satellites, with an increase in capacity from 1,000 one-way voice circuits to 6,000. Its power output also doubled.

RCA launched two more of these second-generation satellites through September 1983. It then followed this with a third-generation satellite, its Ku-band satellites using the 12/14 GHz frequencies first tested on **Hermes (January 17, 1976)**.

Shortly after the first launch of the Ku-band satellites (see **STS 61B–November 26, 1985**), RCA and its entire space division was purchased by General Electric, which later replaced RCA's older Satcom satellites (see **GE Satcom C1–November 20, 1990** and **Aurora 2–May 29, 1991**).

October 30, 1982
DSCS III
U.S.A.

The third-generation American military communication satellite, Defense Satellite Communication System (DSCS) III replaced the DSCS II satellite cluster (see **DSCS II 1/DSCS II 2–November 3, 1971**). Eleven of these satellites were launched, two from the space shuttle (**STS 51J–October 3, 1985**). Unlike the earlier DSCS II satellites, which provided strategic military communications links between large ground antennas, DSCS III satellites also provided communications between smaller mobile ground and ship terminals, thereby supplementing the tactical communication systems of **FLTSATCOM (February 9, 1978)**, **LEASAT (STS 41D–August 30, 1984)**, and **UHF Follow-On (September 3, 1993)**.

The DSCS III satellite cluster is being replaced, as of 1999, by the Milstar satellites (see **Milstar 1–February 7, 1994**).

November 11, 1982
STS 5, Columbia Flight #5
SBS 3 / Anik C3 (Telesat 6)
U.S.A. / Canada

STS 5, a five-day flight of the space shuttle Columbia (its fifth space flight), was the first shuttle mission to deploy commercial communications satellites. It was also the first spacecraft to place more than three humans in space, carrying a crew of four: Vance Brand (**Apollo 18–July 15, 1975; STS 41B–February 3, 1984; STS 35–December 2, 1990**), Robert Overmyer (**STS 51B–April 29, 1985**), Joe Allen (**STS 51A–November 8, 1984**), and William Lenoir. While in orbit, the astronauts held a televised press conference, holding up a sign reading "Satellite Deployment by the Ace Moving Co., Fast and Courteous Service. We Deliver."

The two satellites deployed from Columbia used their own separate thruster engines to lift them into geosynchronous orbit.

• *SBS 3* Deployed on November 11th, SBS (Satellite Business Systems) 3, a geosynchronous satellite, was part of the satellite cluster owned and operated today by MCI Communications. See **SBS 1 (November 15, 1980)**.

• *Anik C3 (Telesat 6)* Deployed on November 12th, Anik C3 was built by Hughes Aircraft for Telesat Canada (see **Anik A1–November 10, 1972** and **Anik B–December 15, 1978**). This geosynchronous communications satellite was the first of a three-satellite cluster providing service to Canada using the 12/14 GHz frequencies. Similar to the SBS satellites (see **SBS 1–November 15, 1980**), each had a capacity of 1,344 two-way voice circuits or 16 television channels. The Anik C cluster augmented the Anik D satellites (see **Anik D1–August 26, 1982**), with both Anik C and Anik D satellites replacing the Anik A and Anik B cluster.

Through April 1985, Telesat Canada launched three Anik C satellites, covering the southern urban areas of Canada. The Anik D satellites, meanwhile, were configured to cover the entire country.

Telesat Canada later replaced both the Anik C and Anik D satellites with the Anik E satellites (see **Anik E2–April 4, 1991**).

Columbia's experiments included three student getaway specials and a continuing series of engineering tests of the shuttle's thermal and structural systems. One getaway special experiment studied convection currents in zero gravity. In gravity, heated substances rise, causing convection circulation patterns (such as in thunderstorms, jet streams, and the bubbles in boiling water). This experiment found that while a heated drop of oil does not expand upward as it floats in water, it forms strange geometric shapes, circular at the top and bottom and tapered at the waist. See **STS 65 (July 8, 1994)** for later bubble research.

Although the health of all crew members was reviewed prior to and during the mission in order to observe their adaptation to weightlessness, this American flight was the first in which the crew's life signs were not continuously monitored. The symptoms for space sickness were well known, along with effective treatment. Moreover, research during the first four shuttle flights had shown that daily exercise lasting 30 minutes on the treadmill, along with the drinking of additional fluids prior to re-entry, had reduced the weight loss and mitigated the other physical changes experienced by earlier astronauts. Hence, for short-term shuttle missions lasting less than a month, the effect of zero gravity was no longer a concern.

The next shuttle flight was **STS 6 (April 4, 1983)**.

November 18, 1982
Iskra 3
U.S.S.R.

Iskra 3 was released by the **Soyuz T5 (May 13, 1982)** cosmonauts on **Salyut 7 (April 19, 1982)**. Described as a small satellite for amateur radio experiments similar to the Oscar satellites (see **Oscar 8 (March 5, 1978)**), it remained in orbit for two months and then burned up in the atmosphere.

1983

January 26, 1983
IRAS
U.S.A. / U.K. / Netherlands

The InfraRed Astronomical Satellite (IRAS) was the first infrared space telescope, providing an all-sky survey of astronomical infrared objects in four wavebands (the lower two bands covering 8–30 microns and the upper two bands covering 40–120 microns). An additional experiment built by scientists from the Netherlands obtained infrared spectra across three wavebands, from 7.4 to 23 microns, 41 to 62.5 microns, and 84 to 114 microns. At these wavelengths, the telescope was able to look through galactic dust clouds and see the bright objects hidden behind.

During its 10-month life span, IRAS observed more than 96 percent of the sky, producing a catalog of nearly a quarter-million objects, including 20,000 new galaxies. It found many previously unknown cloud regions of star formation, in the Milky Way as well as in other nearby galaxies. The Milky Way

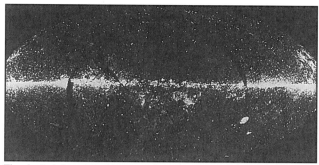

The sky in infrared, looking away from the center of the Milky Way. The bright region at the edges of the image is the Milky Way's core, where Sagittarius A* (pronounced A-star) is located. This object is suspected to be a supermassive black hole. The other bright regions are molecular clouds where star formation is still taking place. *NASA*

appeared more active in star formation than previously thought (producing approximately one new star per year), while the nearby Andromeda Galaxy was less so, despite the discovery of a ring of material circling Andromeda's outer perimeter.

Within these gigantic molecular clouds, IRAS also identified many protostars, newly born stars as young as 100,000 years old, their thermonuclear flames just igniting. About 50 or so bright stars, including Vega, Fomalhaut, Beta Pictoris, and Epsilon Eridani, seemed to have dust disks, indicating the presence of a forming solar system. A series of asymmetri-

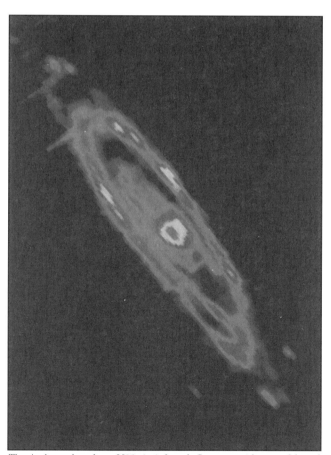

The Andromeda galaxy, M31, in infrared. Compare with ground-based image in the Introduction, page xvii. *NASA*

cal gas shells was found around the gas giant Betelgeuse, extending for some 4.5 light years and indicating that the star had undergone a series of eruptions in its past.

IRAS also observed several colliding galaxies, tearing each other apart as they flew through each other. It found that while spiral galaxies were filled with dust and star formation regions, elliptical galaxies had almost no dust or star formation. IRAS also showed that at the 60-micron wavelength, the sky was dotted with many previously unknown galaxies glowing in infrared.

IRAS also discovered five comets, the first such discoveries by infrared detection. Moreover, it found a small Earth-crossing asteroid orbiting the Sun along the same track as the Geminid meteor shower, suggesting that the asteroid and meteor shower were the remains of a dead comet. Within the asteroid belt between Mars and Jupiter, IRAS identified three hidden bands of dust, as well as evidence of thousands of previously unknown asteroids.

COBE (**November 18, 1989**) was the next satellite dedicated to infrared observations. IRAS's specific research, however, was continued with **ISO** (**November 17, 1995**).

February 4, 1983
Sakura 2A
Japan

Sakura 2A, built and launched by Japan's National Space Development Agency (NASDA), was the first of a two-satellite communications constellation replacing **Sakura 1** (**December 15, 1977**). This launch was also the first operational use of Japan's N2 rocket, first tested on **Kiku 3** (**February 11, 1981**).

Placed in geosynchronous orbit over Japan, the Sakura 2 satellites (the second was launched in August 1983) provided government and public telephone service between Japan's main and remote islands. Their capacity was essentially the same as Sakura 1, although both satellites incorporated several technological changes to improve their reliability and efficiency. They functioned until 1988, when they were replaced with Japan's third-generation communications satellite. See **Sakura 3A** (**February 19, 1988**).

February 20, 1983
Tenma (Astro-B)
Japan

Tenma (a legendary Japanese winged horse) was Japan's second x-ray telescope, following **Hakucho** (**February 21, 1979**). It did an all-sky survey and monitored the output of a number of transient x-ray objects, continuing observations from previous x-ray telescopes like **Ariel 6** (**June 2, 1979**).

The next x-ray telescope was **Exosat** (**May 26, 1983**).

March 2, 1983
Cosmos 1443
U.S.S.R.

On March 10, 1983, just prior to the planned occupancy of **Soyuz T8** (**April 20, 1983**), Heavy Cosmos 1443 docked with Salyut 7 (**April 19, 1982**), adding a second habitable module to the station and increasing its total habitable space by one-third, from 3,300 to 5,000 cubic feet. It also brought more than three tons of cargo to the station. The Salyut 7 complex then weighed over 51 tons, not including cargo.

This module remained docked with Salyut 7 until August 14, 1983, providing support during the second occupancy of the station by the crew of **Soyuz T9** (**June 27, 1983**). Nine days after separation, Cosmos 1443's recoverable capsule returned to Earth, carrying about 800 pounds of film and equipment. Cosmos 1443 was deorbited on September 19, 1983, burning up in the atmosphere. In 1993, the recovered capsule was sold at a Sotheby's auction in New York for $552,500.

The next and last Heavy Cosmos module to dock with Salyut 7 was **Cosmos 1686** (**September 27, 1985**).

March 23, 1983
Astron
U.S.S.R. / France

A joint project of the Soviet Union and France (see **Aureol 1–December 27, 1971**), Astron carried an ultraviolet telescope-spectrometer and an x-ray spectrometer to search for ultraviolet and x-ray sources in interstellar and intergalactic space. In its first year of operation, the satellite charted the constellations of Taurus, Leo, and Orion, acquiring ultraviolet images of over 100 stars, more than 30 galaxies, and dozens of nebulae. See **Rosat** (**June 1, 1990**).

April 4, 1983
STS 6, Challenger Flight #1
TDRS 1
U.S.A.

STS 6 was the sixth flight in the American space shuttle program and was also the inaugural flight of the second space shuttle, Challenger. Lasting five days, it was crewed by Paul Weitz (**Skylab 2–May 25, 1973**), Karol Bobko (**STS 51D–April 12, 1985**; **STS 51J–October 3, 1985**), Don Peterson, and Story Musgrave (**STS 51F–July 29, 1985**; **STS 33–November 22, 1989**; **STS 44–November 24, 1991**; **STS 61–December 2, 1993**; **STS 80–November 19, 1996**). Musgrave and Peterson performed the first shuttle space walk, which was also the first American space walk in almost a decade, since **Skylab 4** (**November 16, 1973**). On April 7th, they spent 3.5 hours inside Challenger's open cargo bay, testing the usefulness of their spacesuits and the design of the shuttle's handholds and several newly designed tools for working in space.

STS 6 carried a total payload (excluding the shuttle's own weight) of almost over 23 tons, the most ever lifted into orbit. That payload included:

- *TDRS 1* Deployed on April 5th, TDRS (Tracking and Data Relay Satellite) 1 was developed by NASA to provide communications with its orbiting spacecraft. Previously, the space agency had used an extensive network of ground stations, which were expensive and did not provide a full coverage of every possible orbit. TDRS 1, the first of a six-satellite cluster, replaced these ground stations. All signals to and from the TDRS ground station in White Sands, New Mexico, and an orbiting spacecraft are relayed from satellite to satellite until they reach their destination. All six satellites were deployed by the space shuttle, and as of December 1999 are operational, with three in use and three orbiting as spares.

The TDRS satellites were originally designed in cooperation with Western Union, and were therefore built with a separate and independent capacity for domestic communications. When NASA and Western Union parted ways prior to TDRS 1's launch (both decided they were better off with separate satellites: see **Westar 4–February 26, 1982**), the TDRS satellites were left with this unused capacity. In 1990, Columbia Communications won the bid to utilize this capacity, marketing it in competition with Intelsat (see **Early Bird–April 6, 1965**). Thus, Columbia became the second private company following Pan American Satellite (see **Pan American Satellite 1–June 15, 1988**) to provide international video, voice, and data communications.

Other experiments on STS 6 included the privately owned Continuous Flow Electrophoresis System (CFES), last flown on STS 4 (**June 27, 1982**), which tested electrophoresis methods for separating hemoglobin and a polysaccharide, as well as albumins from rat and egg tissues and a cell culture of proteins. It confirmed the results of the previous flight—that in weightlessness, it was possible to produce pharmaceutical materials in great quantities and four times purer than on Earth. With this success, McDonnell Douglas purchased space on later shuttle flights to produce materials for commercial sale. See STS 8 (**August 30, 1983**).

STS 6 also carried several getaway specials, including a Japanese experiment proposed by two high school students to produce the first snowflakes in zero gravity (see **STS 8–August 30, 1983**). A second getaway special tested methods for packaging plant seeds in weightlessness.

The next shuttle mission was STS 7 (**June 18, 1983**).

April 20, 1983
Soyuz T8
U.S.S.R.

Although Soyuz T8 was intended to be the second occupancy of **Salyut 7** (**April 19, 1982**), it failed to dock with the station—the first such failure since **Soyuz 33** (**April 10, 1979**). The crew included Vladimir Titov (**Soyuz T10A–September 27, 1983**; **Soyuz TM4–September 21, 1987**; **STS 63–February 3, 1995**; **STS 86–September 25, 1997**), Gennady Strekalov (**Soyuz T3–November 27, 1980**; **Soyuz T10A–September 27, 1983**; **Soyuz T11–April 3, 1984**; **Soyuz TM10–August 1, 1990**; **Soyuz TM21–March 14, 1995**), and Alexander Serebrov (**Soyuz T7–August 19, 1982**; **Soyuz TM8–September 5, 1989**; **Soyuz TM17–July 1, 1993**). Serebrov was the first man to fly on successive missions, having flown on the previous Soviet manned flight.

The docking failure occurred because Soyuz T8's automatic docking radar antenna had been ripped off during launch. Although the cosmonauts attempted to dock manually, without this system they had no way of accurately judging their distance from the laboratory. They canceled a second docking attempt and returned to Earth after two days in space.

The next mission, **Soyuz T9** (**June 27, 1983**), completed Salyut 7's second occupancy.

May 26, 1983
Exosat
Europe / U.S.A.

Launched by NASA and built by the European Space Agency, Exosat was an x-ray telescope that studied the sky for energy emissions between 0.4 to 50 KeV until April 1986. With an accuracy of 2 arc seconds, Exosat allowed scientists to identify the visible counterparts for many of the x-ray sources discovered by previous x-ray telescopes (see **Uhuru–December 12, 1970** and **Einstein–November 13, 1978**).

The next x-ray telescope was **Ginga** (**February 5, 1987**).

June 16, 1983
Eutelsat 1 (ECS 1) / Oscar 10
Europe / International

• *Eutelsat 1 (ECS 1)* Also called ECS (European Communications Satellite) 1, Eutelsat 1 was a satellite owned and operated by Eutelsat (see **OTS 2–May 11, 1978**). The satellite, placed in orbit by an Ariane rocket, inaugurated the consortium's first operational geosynchronous satellite cluster to provide both public and business communication services throughout Europe. This first-generation satellite was based upon the designs developed on OTS 2 and had a capacity for 1,800 telephone calls or one television signal with multiple audio channels.

Through July 1988, Eutelsat successfully launched four Eutelsat satellites. The consortium's original plan had called for two operating satellites, but when both satellites first opened for business, Eutelsat received requests for three times their capacity, and the company was forced to build and launch two more satellites. Then, in 1990, it launched the more advanced Eutelsat 2 cluster (see **Eutelsat 2 F1–August 30, 1990**).

• *Oscar 10* Oscar 10 was the fourth satellite built by AMSAT (the Radio Amateur Satellite Corporation), an organization of ham radio operators (see **Oscar 6–October 15, 1972**). West German AMSAT members played the central role in designing and building Oscar 10. The satellite was a significant improvement over the previous AMSAT satellites (see **Oscar 8–March 5, 1978**), designed for a high-altitude orbit and long life. It was placed in a high eccentric orbit similar to the Soviet Molniya satellites (see **Molniya 1-01–April 23, 1965**), its apogee chosen to provide Earth coverage to the largest number of AMSAT members. Because the satellite failed to enter its optimum orbit, however, it spent several hours per day within the Van Allen belts, thereby damaging and shortening the service life of its solid-state memory. It functioned through 1990.

The next Oscar satellite was **UoSat 2** (**March 1, 1984**).

June 18, 1983
STS 7, Challenger Flight #2
SPAS 1 / Palapa B1 / Anik C2 (Telesat 7)
U.S.A. / Germany / Indonesia / Canada

STS 7 was the second flight of the space shuttle Challenger, the second spacecraft ever to go into space, return to Earth, and return to space successfully. The six-day flight was crewed by Robert Crippen (**STS 1–April 12, 1981**; **STS 41C–April 6, 1984**; **STS 41G–October 5, 1984**), Fred Hauck (**STS 51A–November 8, 1984**; **STS 26–September 29, 1988**), Sally

Ride (**STS 41G–October 5, 1984**), John Fabian (**STS 51G– June 17, 1985**), and Norman Thagard (**STS 51B–April 29, 1985; STS 30–May 4, 1989; STS 42–January 22, 1992; Soyuz TM21–March 14, 1995**). Ride was the third woman to fly into space, following Valentina Tereshkova (**Vostok 6–June 16, 1963**) and Svetlana Savitskaya (**Soyuz T7–August 19, 1982**), and the first American woman. With a crew of five, this flight set a record for the most humans lifted into space by one spacecraft.

STS 7 released three satellites, including the first satellite to be retrieved from orbit.

• *SPAS 1* The German-built scientific satellite SPAS (Shuttle PAllet Satellite) 1 was deployed from Challenger's cargo bay on June 22nd, the first such satellite released using the shuttle's robot arm. It carried a number of instruments for studying the environment surrounding the shuttle, as well as for taking the first photographs of the shuttle while in orbit. Although it was allowed to fly free for a period of hours, the satellite was also used to test the robot arm. The astronauts used the arm five times to release and then retrieve the satellite, testing its ability to make precise motions and respond to commands.

SPAS 1 was then restored to the shuttle's cargo bay and returned to Earth, which was the first time an orbiting satellite safely returned from space without using its own shielding to protect it from the heat of re-entry.

• *Palapa B1* Owned by Indonesia, Palapa B1 was the first satellite of a second-generation cluster to provide communications services to Indonesia and its neighboring countries, replacing Indonesia's first two Palapa satellites (see **Palapa 1–July 8, 1976**). Similar to Anik C3 (see **STS 5–November 11, 1982**), Palapa B satellites were twice the size of Indonesia's earlier satellites, with double the television signal capacity (24 versus 12 channels) and increased two-way voice capacity.

Through 1992, Indonesia orbited five Palapa B satellites. One, **Palapa B2R (April 13, 1990)**, required two launches to reach geosynchronous orbit (see **STS 41B–February 3, 1984** and **STS 51A– November 8, 1994**).

In 1996, Indonesia began replacing the Palapa B satellites with its third-generation communications satellite. See **Palapa C1 (February 1, 1996)**.

• *Anik C2 (Telesat 7)* See **STS 5 (November 11, 1982)** for details about the Anik C geosynchronous communications satellites.

The shuttle also carried seven getaway specials and the Continuous Flow Electrophoresis System (CFES). The CFES experiment had flown three times previously (see **STS 4–June 27, 1982**). On this flight, its owner, McDonnell Douglas, designed it to produce larger quantities of pure pharmaceutical materials, as well as further test the equipment's design. See **STS 8 (August 30, 1983)**.

Of the getaway specials, the most interesting was an attempt by 300 high school students to study the effect of weightlessness on a colony of 150 carpenter ants. Unfortunately, the experiment failed when the ants, stored hidden in their canister, died of dehydration prior to launch. Other experiments conducted soldering and desoldering tests, observed how newly sprouted radish seeds responded to weightlessness, and studied how oil and water mixed in zero gravity.

Almost a dozen other experimental packages tested the effect of weightlessness on alloys, crystals, and a variety of other materials. One test grew crystals 10 times larger than was possible on Earth. Another observed the behavior of lead droplets and gas bubbles within a molten mix of cesium and chloride.

The next shuttle flight was **STS 8 (August 30, 1983)**.

June 27, 1983: 9:12 GMT
Soyuz T9
U.S.S.R.

Soyuz T9's mission was the second occupancy of **Salyut 7 (April 19, 1982)**. The crew, Vladimir Lyakhov (**Soyuz 32– February 25, 1979; Soyuz TM6–August 29, 1988**) and Alexander Alexandrov (**Soyuz TM3–July 22, 1987**), remained in space for 149 days.

It took Lyakhov and Alexandrov almost six weeks to unload the more than three tons of cargo from **Cosmos 1443 (March 2, 1983)**. During this time, they also performed extensive Earth observations using Salyut 7's multispectral cameras and telescopes. Twenty-thousand multispectral Earth observation photographs were taken, studying the agricultural regions and crop production of the Soviet Union. The exposed film magazines were stored in Cosmos 1443's recoverable capsule, along with parts from Salyut 7, including a failed computer memory circuit board and an air regenerator that had been replaced with a newer unit. Ground engineers inspected this equipment to determine how weightlessness and space affected their durability.

On July 27, 1983, Salyut 7 was hit by a small object, creating a quarter-inch crater in one of its windows. No air escaped, however, although the cosmonauts both heard the "crack" of impact.

On August 14, 1983, Cosmos 1443 undocked, and nine days later its capsule returned to Earth safely, carrying a total of almost 800 pounds of film and experiment samples. Cosmos 1443 was then allowed to burn up in the atmosphere on September 19, 1983.

With the front docking port clear, Lyakhov and Alexandrov entered their Soyuz T9 spacecraft and switched docking ports, a maneuver that was becoming routine. This action freed the rear docking port, allowing Progress freighters (see **Progress 1–January 20, 1978**) to dock with the laboratory. In the remaining three months of their mission, two freighters brought the standard supplies and fuel. During refueling operations in late August, however, a break in one of the station's fuel lines occurred, shutting down half of Salyut 7's attitude thrusters and requiring four space walks over the next year by other crews to attempt a repair.

Lyakhov and Alexandrov completed two space walks of their own, each just less than three hours long. On November 1st and 2nd, the cosmonauts installed additional solar panels—brought to Salyut 7 by Cosmos 1443—to the perimeter of the station's main panels. Because several older solar panels had failed prematurely, the station's power supply was limited, forcing the cosmonauts to keep its temperature at 65°F and its humidity at 100 percent. These panels, originally intended to supplement the laboratory's electrical supply, helped alleviate the problem.

Like the previous occupancies, Lyakhov and Alexandrov also conducted medical, biological, mineralogical, and astronomical research. In electrophoresis experiments, they purified an anti-gene of the influenza virus for study on Earth. In their greenhouse, they experimented with new techniques for growing plants in space, testing the use of electrical stimulation to improve plant growth, producing radishes and tomato sprouts. The results were mixed, at best. See **STS 51 (September 12, 1993)**.

Because of the launchpad failure of **Soyuz T10A (September 27, 1983)**, the Soyuz T9 cosmonauts received no visitors during their five months in orbit. They returned to Earth on November 23, 1983, using the same spacecraft that had put them in orbit. This five-month flight by a Soyuz spacecraft would remain a record until six-month missions became routine on **Mir (February 20, 1986)**. See **Soyuz TM16 (January 24, 1993)**.

The next manned mission to Salyut 7 was **Soyuz T10B (February 8, 1984)**, which set a new in-space endurance record of almost seven months.

June 27, 1983, 15:37 GMT
HiLat (P83-1)
U.S.A.

This Air Force experimental satellite continued propagation experiments started on **P76-5 (May 22, 1976)**. It carried a beacon that transmitted across five different frequency bands, from 138 to 1,239 MHz. From its high latitude orbit of 82 degrees inclination, HiLat (for HIgh LATitude) measured the atmospheric effects on radio transmissions in order to improve military communications.

HiLat was followed by **Polar Bear (November 14, 1986)**, which studied the atmosphere's effects on the same frequencies directly above the north and south poles.

June 28, 1983
Galaxy 1
U.S.A.

Owned and operated by Hughes Communications Inc. (formerly Hughes Aircraft), Galaxy 1 inaugurated the fifth private communications satellite cluster serving the U.S. domestic market, following **Westar 1 (April 13, 1974)**, **RCA Satcom 1 (December 13, 1975)**, **Comstar 1 (May 13, 1976)**, and **SBS 1 (November 15, 1980)**.

Through February 1994, Hughes launched seven satellites, all using the 4/6 GHz frequencies and with a capacity of 1,000 voice circuits or 24 television signals. To further augment its cluster, in 1990, Hughes purchased three satellites from Satellite Business Systems (see **SBS 1–November 15, 1980**) and three satellites from Western Union (see **Westar 4–February 26, 1982**). With the retirement of its first two satellites, Hughes had in 1992 a cluster of six satellites, three of which operated at the higher capacity 11/14 GHz frequencies.

Just as RCA used its Satcom satellites to sell broadcast distribution to cable channels, so did Hughes. The growth of this cable business was so precipitous that, late in 1992, Hughes began launching a new series of satellites using both the lower and higher GHz frequencies. See **Galaxy 7 (October 28, 1992)**.

July 1, 1983
Prognoz 9
U.S.S.R. / Czechoslovakia / France

Unlike the previous Prognoz satellites (see **Prognoz 1–April 14, 1972**), Prognoz 9 did more than monitor the Sun's solar flare output and the interaction of the solar wind and Earth's magnetosphere. Placed in an eccentric orbit (perigee: 240 miles, apogee: 450,000 miles) with an apogee four times greater than early Prognoz satellites, Prognoz 9 carried a radio telescope, a magnetometer, and three spectrometers (two working in x-ray frequencies). The radio telescope was the first successfully placed in space since **RAE 2 (June 10, 1973)**. Its goal was to produce a map of the sky in radio wavelengths, looking for the residual radiation left from the "big bang." One x-ray spectrometer studied the Sun's x-ray radiation. The second x-ray spectrometer searched for x-ray bursts from interstellar sources. The third spectrometer measured solar wind fluxes. The experiments were built by the Soviet Union, Czechoslovakia, and France.

The next radio telescope was **Halca (February 12, 1997)**. The next Prognoz satellite, **Prognoz 10 (April 26, 1985)**, was also the last in that program.

July 28, 1983
Telstar 3A
U.S.A.

Telstar 3A was the first satellite built and launched entirely by AT&T in more than 20 years, since **Telstar 2 (May 7, 1963)**. It was also the first of three satellites to supplement and replace the Comstar satellites (see **Comstar 1–May 13, 1976**), providing telephone service to the United States domestic market.

The Telstar 3 satellites were similar in design to the Anik C satellites (see **STS 5–November 11, 1982**) and had significantly greater capacity than the Comstar satellites—7,800 one-way voice circuits instead of 1,200. They were built to have a service life of 10 years, which they exceeded.

In 1993, AT&T began launching its next-generation communications satellite. See **Telstar 401 (December 16, 1993)**.

August 30, 1983
STS 8, Challenger Flight #3
Insat 1B
U.S.A. / India

This third flight of the space shuttle Challenger carried a crew of five: Richard Truly (**STS 2–November 12, 1981**), Daniel Brandenstein (**STS 51G–June 17, 1985; STS 32–June 9, 1990; STS 49–May 7, 1992**), Guion Bluford (**STS 61A–October 30, 1985; STS 39–April 28, 1991; STS 53–December 2, 1992**), Dale Gardner (**STS 51A–November 8, 1984**), and William Thornton (**STS 51B–April 29, 1985**). Bluford was the first black man to enter space, while Thornton was the oldest at 54 years of age. The six-day mission included the first night launch and night landing. One satellite was deployed.

• *Insat 1B* Deployed on August 8th, this Indian geosynchronous communications-weather satellite was part of that country's first-generation satellite cluster. See **Insat 1A (April 10, 1982)**.

After releasing this satellite, the astronauts spent five hours using what NASA dubbed the Payload Flight Test Article (PFTA) to make further tests of the shuttle's robot arm. PFTA weighed almost 7,500 pounds and was the heaviest object manipulated by the arm to date. PFTA's shape was also made intentionally awkward—a large dumbbell that barely fit within the shuttle's cargo bay—in order to test the robot arm's ability to manipulate it accurately.

Thornton, a doctor, had been added to the crew list to study the cause of space sickness. By the time of this flight, accumulated knowledge from U.S. and Soviet missions had found that about half the individuals who flew in space experienced some form of space sickness or nausea, lasting from several minutes to several hours. During the mission, Thornton kept careful records of both his and other crew members' reactions to the environment of space, measuring the redistribution of fluid from the legs toward the head, which caused the face to appear puffy (see **Soyuz 11–June 6, 1971**). This fluid redistribution was suspected to be a cause of some of the headaches and nausea experienced by astronauts while in orbit. See **STS 55 (April 26, 1993)** for later research.

McDonnell Douglas's Continuous Flow Electrophoresis System (CFES) experiment (see **STS 4–June 27, 1982**) flew once again. After four shuttle flights to test the facility's operation, CFES carried its first live cell samples, including human kidney, rat pituitary, and dog pancreas cells. The rat pituitary cells showed that in microgravity there was a 20-fold decrease in growth hormone production, which scientists believed might be the cause of the muscle wasting seen in space. A follow-up experiment occurred on **STS 46 (July 31, 1992)**.

STS 8 also carried six live white rats in a newly designed animal cage, testing the cage's operation. Future missions (see **STS 51B–April 29, 1985**) used this enclosure to transport research animals into space.

The getaway specials included a larger repeat of the Japanese experiment to produce snowflakes in space, flown previously on **STS 6 (April 4, 1983)**. Other specials measured the damage caused by cosmic rays to integrated circuits and the effect of the space environment on photographic film. The erosion of carbon and osmium while exposed in space was also studied.

To measure the effect of the earth's atmosphere on the shuttle and these materials, Challenger's orbit was lowered to 139 miles, with its nose pointed down and its open cargo bay facing to the front, to expose both the shuttle and its experiments to the maximum atmospheric effect.

The next shuttle mission was **STS 9 (November 28, 1983)**.

September 27, 1983
Soyuz T10A
U.S.S.R.

Soyuz T10A, manned by Gennady Strekalov (**Soyuz T3–November 27, 1980**; **Soyuz T8–April 20, 1983**; **Soyuz T11–April 3, 1984**; **Soyuz TM10–August 1, 1990**; **Soyuz TM21–March 14, 1995**) and Vladimir Titov (**Soyuz T8–April 20, 1983**; **Soyuz TM4–December 21, 1987**; **STS 63–February 3, 1995**; **STS 86–September 25, 1997**), was the second launch that, when aborted, required the use of the rocket's emergency escape system (see **Soyuz 18A–April 5, 1975**). Ninety seconds before blast-off, a fire broke out at the base of the launchpad. Ten seconds later, the cosmonauts activated the emergency escape rocket, and they and the Soyuz T10A capsule were ejected more than 3,000 feet into the air. The descent module then separated, parachuting them safely to the ground. The main rocket, meanwhile, exploded on the launchpad, setting fires that took 20 hours to extinguish.

The next Soviet manned launch was **Soyuz T10B (February 8, 1984)**, which achieved the third occupancy of **Salyut 7 (April 19, 1982)** and set a new in-space endurance record.

September 28, 1983
Cosmos 1500 (Okean-OE)
U.S.S.R.

Designed to observe sea ice conditions in the Arctic Ocean, Cosmos 1500 (also called Okean-OE) was the third flight in this Soviet program and the first operational flight. See **Cosmos 1076 (February 12, 1979)**. Like the previous two Okean satellites, Cosmos 1500 carried several multispectral scanners, each able to image in both visible and infrared light. Scan widths ranged from 280 to 1200 miles, with resolutions as small as 1,200 feet.

Cosmos 1500 also carried a radar package able to "photograph" sea ice conditions in any weather, similar in concept to the radar instruments used on **Venera 15 (October 10, 1983)** and **Venera 16 (October 14, 1983)**. The radar, called side-looking because it imaged the surface at an angle in order to see the profile of surface features, scanned a width of 280 miles with a resolution of 0.75 to 1.0 mile. This instrument could map ice development and determine ice sheet thickness and sea currents. Unlike the synthetic-aperture radar of **Seasat (June 27, 1978)**, Cosmos 1500 radar was real-aperture, requiring a much larger antenna to produce its image.

Cosmos 1500 was immediately put to use imaging the serious ice conditions unfolding in 1983 in the East Siberian Sea. A warm summer followed by early freezing had trapped 70 ships, including seven icebreakers. One ship had been lost and several had been damaged by the ice. The satellite was used to guide the icebreakers to thinner ice so that the imprisoned ships could be freed.

Through December 1999, the Soviet Union/Russia launched 10 Okean satellites, establishing a working cluster of at least two working satellites at any time that covered both the northern Arctic and the southern Antarctic. Sea routes through the ever-changing ice sheets were routinely revised. In 1985, for example, these satellites, used in conjunction with observations by the **Soyuz T13 (June 6, 1985)** cosmonauts on **Salyut 7 (April 19, 1982)**, helped the icebreaker *Vladivostok* free a trapped ship in the Antarctic.

In 1995, a modified version of the Okean satellites, **Sich 1 (August 31, 1995)**, was launched. The next spacecraft to carry radar imaging equipment to study Earth was **STS 41G (October 5, 1984)**. The next spacecraft to carry radar imaging was **Venera 15, Venus Orbit (October 10, 1983)** and **Venera 16, Venus Orbit (October 14, 1983)**.

October 10, 1983
Venera 15, Venus Orbit
U.S.S.R.
Launched on June 2, 1983, Venera 15 was the Soviet Union's first radar mapper sent to Venus, following the American radar mapper, **Pioneer Venus Orbiter (December 4, 1978)**. By combining its data with its twin, **Venera 16 (October 14, 1983)**, radar maps of previously unobserved parts of the Venusian surface were produced. See Venera 16 for details.

October 14, 1983
Venera 16, Venus Orbit
U.S.S.R.
Venera 16 was launched on June 7, 1983, five days after its companion orbiter, **Venera 15 (October 10, 1983)**. Both spacecraft were designed to provide radar maps of Venusian regions previously unmapped, resolving features approximately one mile across. Though this resolution was about equal to that of Earth-based radar telescopes like the Arecibo and Goldstone radio antennas, the two Venera craft mapped the north polar regions of Venus, inaccessible to instruments on Earth.

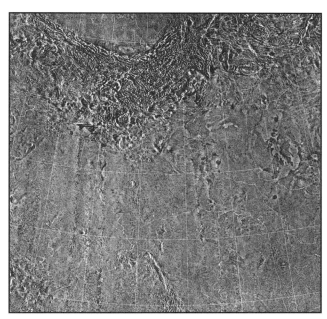

The middle latitudes. The Ishtar Plateau, with its tessera ridges lines, can be seen near the top. Compare with Pioneer Venus Orbiter (December 4, 1978) map. *RIA Novosti*

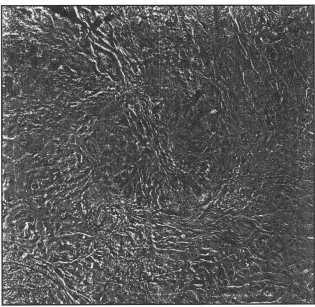

The north pole of Venus. The northern edge of the Ishtar Plateau can be seen in the lower left corner. *RIA Novosti*

Both Venera spacecraft operated for more than a year. Terrain features included smooth and hilly plains, lava flows, calderas, scarps, ridges, impact craters, and a number of new surface types. Newly discovered geological formations included tessera (radar bright areas with many criss-crossing ridges and fractures) and coronae (raised oval features exhibiting evidence of surface deformation or volcanism). The geological origin of these formations was unknown.

The next Venus probes included two landers released by the Soviet Halley's Comet probes, Vega 1 (**Vega 1, Venus Flyby–June 10, 1985**) and Vega 2 (**Vega 2, Venus Flyby–June 14, 1985**).

November 28, 1983
STS 9, Columbia Flight #6
U.S.A. / Europe / International
STS 9, the sixth flight of the space shuttle Columbia, lasted over 10 days and was crewed by John Young (**Gemini 3–March 23, 1965; Gemini 10–July 18, 1966; Apollo 10–May 18, 1969; Apollo 16–April 16, 1972; STS 1–April 12, 1981**), Brewster Shaw (**STS 61B–November 26, 1985; STS 28–August 8, 1989**), Robert Parker (**STS 35–December 2, 1990**), Owen Garriott (**Skylab 3–July 28, 1973**), Byron Lichtenberg (**STS 45–March 24, 1992**), and Ulf Merbold (**STS 42–January 22, 1992; Soyuz TM20–October 4, 1994**). This six-person crew—the most passengers launched into space by a single spacecraft—also included the first non-American (German astronaut Merbold) to fly on an American spacecraft. It was also the first shuttle flight to be placed in a high-inclination orbit, tilted from the equator by 57 degrees. This orbit allowed the researchers onboard to observe a larger area of the earth's surface.

STS 9 was also the first flight of Spacelab, built by the European Space Agency. Spacelab was a module designed to fit inside the shuttle's cargo bay and provide a shirt-sleeve laboratory environment for scientific research. Connected to the shuttle's interior by a tunnel, the module filled about two-thirds of the cargo bay, with the remaining one-third reserved for the sensors used by the scientists in their research.

Although much of the mission's work was devoted to verifying Spacelab's operation, this flight still carried 38 separate experiments involving more than 70 investigators worldwide. To maximize the research time while in space, the six-man crew was divided into two 12-hour shifts.

Three different furnaces, as well as a number of other instruments, studied how different materials were affected by the absence of gravity. Both organic and inorganic crystals were grown, formation of metal alloys impossible on Earth was attempted, liquid and gas flow was studied, and extensive soldering and brazing tests were performed. One test produced a very homogeneous aluminum-zinc alloy, which could not occur on Earth. Another produced what scientists called Marangoni convection, caused not by gravity but by surface tension when silicon oil suspended between two disks began flowing. A third grew the first silicon crystal in space.

Several different multispectral cameras took photographs of the earth's surface and atmosphere, mapping land and cloud formations. Two telescopes studied the celestial sky in optical and ultraviolet wavebands, while a third instrument obtained x-ray spectra of transient x-ray bursts. Three sensors measured the Sun's energy output from infrared to ultraviolet wavelengths.

Sixteen different experiments investigated space sickness and the effect of weightlessness on life. Tests showed that individuals in space learn to depend less on their sense of balance and more on what they see and touch to orient themselves. Space sickness apparently occurred during the period of adaptation.

The next shuttle mission was STS 41B (**February 3, 1984**). The next Spacelab mission was STS 51B (**April 29, 1985**).

December 14, 1983
Bion 6 (Cosmos 1514)
U.S.S.R. / International

This biological research satellite, part of a Soviet program begun with **Bion 1** (**October 31, 1973**), carried 2 monkeys, 18 white rats, and 3 guppies into space for five days, then returned them to Earth for study.

The monkeys, the first ever placed in space by the Soviet Union and the first primates orbited since **Biosat 3** (**June 29, 1969**), were implanted with sensors built by the United States to monitor their blood circulation. Upon their return to Earth, the data showed that one monkey had adapted to weightlessness better than the other, with the second monkey dying less than three days after recovery from a strangulated bowel. Whether this condition was caused by the space flight or had existed before was not clear.

Of the rats, five were pregnant females used for testing the effect of weightlessness on embryos. Four of five produced normal offspring upon return to Earth. The fifth female rat delivered a litter of dead fetuses. All the space rats experienced less weight gain than the earthbound control rats. Liver mass, hemoglobin, and other fluids also showed a decrease.

The next Bion satellite was **Bion 7** (**July 10, 1985**).

1984

February 3, 1984
STS 41B (STS 10), Challenger Flight #4
Westar 6 / Palapa B2
U.S.A. / Indonesia

This fourth flight of Challenger lasted eight days and was crewed by Vance Brand (**Apollo 18–July 15, 1975**; **STS 5–November 11, 1982**; **STS 35–December 2, 1990**), Robert Gibson (**STS 61C–January 12, 1986**; **STS 27–December 2, 1988**; **STS 47–September 12, 1992**; **STS 71–June 27, 1995**), Bruce McCandless (**STS 31–April 24, 1990**), Robert Stewart (**STS 51J–October 3, 1985**), and Ronald McNair (**STS 51L–January 28, 1986**). Because some in NASA were superstitious and were averse to flying a mission with a "13" designation, the mission numbering system was changed. In the new, very arcane system—used for only two years—the first digit represented the year ("4" standing for "1984"), the second digit indicated the launch site ("1" for Florida and "2" for California), and the letter was the year's mission ("B" for the second flight of the government's fiscal year).

Two satellites were deployed, both of which failed to reach their planned orbit.

• **Westar 6**   Intended as part of Western Union's cluster of geosynchronous communications satellites (see **Westar 4–February 26, 1982**), Westar 6 was deployed on February 4th. However, its upper-stage rocket, intended to lift the satellite into geosynchronous orbit, failed to fire, and the satellite was left stranded in a useless low orbit. The satellite was recovered and returned to Earth during **STS 51A** (**November 8, 1984**). The insurance companies, after taking ownership, had it refurbished and sold. The satellite was eventually relaunched into orbit as **Asiasat 1** (**April 7, 1990**).

• **Palapa B2**   Palapa B2 was intended to be part of Indonesia's second-generation cluster of geosynchronous communications satellites (see **STS 7–June 18, 1983**). After deployment on February 4th, however, Palapa B2's upper stage failed just like Westar 6's, leaving the satellite stranded in a low, useless orbit. Like Westar 6, Palapa B2 was recovered from orbit and returned to Earth during **STS 51A** (**November 8, 1984**). The insurance companies refurbished the satellite and resold it to Indonesia, which relaunched it successfully as **Palapa B2R** (**April 13, 1990**).

Ironically, STS 41B also tested techniques and equipment for recovering and repairing orbiting satellites. This work was in anticipation of the next shuttle flight, **STS 41C** (**April 6, 1984**), which would snag the research satellite **Solar Max** (**February 14, 1980**) from orbit, repair it, and then release it again. These tests included the first untethered space walks, in which McCandless and Stewart used the Manned Maneuvering Unit (MMU) to fly freely as far as 320 feet away from Challenger.

The Shuttle Pallet Satellite (SPAS), first flown on **STS 7** (**June 18, 1983**), was held by the robot arm and used as a dummy test satellite around which McCandless and Stewart flew. Because of a balky "wrist" on the robot arm, a docking and dummy repair of the SPAS while being held by the robot arm could not be done. The repair, which simulated what was planned for Solar Max, was instead performed by the astronauts while SPAS rested in the cargo bay.

Both men also tested a specially designed foot restraint attached to the end of the robot arm. Standing on this device, the robot arm became the equivalent of an orbiting cherry-picker, from which an astronaut could be safely maneuvered to a work area.

Stewart also performed tests for transferring fuel from a storage tank to a satellite and from satellite to satellite. He found that he could make the pump and pipe connections with-

February 8, 1984

Orbiting satellite Bruce McCandless. *NASA*

out leaks. Later shuttle refueling tests were conducted on **STS 41G (October 5, 1984)**.

Also released in conjunction with these space walk tests was a 6.5-foot-diameter balloon to be used as a dummy satellite with which Challenger could practice rendezvous maneuvers planned in the rendezvous with Solar Max. Unfortunately, the balloon tore apart as it was inflating, forcing a scrubbing of the test. The crew did, however, sight and track one balloon fragment as far as eight miles away.

Five getaway specials studied liquid currents, wave motions, capillary action, the growth of radish seeds, and the circulation of gas inside an arc lamp. The last found that the gases burned more symmetrically in space, and that more energy was consumed in heat than in light, making the arc lamp less bright.

STS 41B also carried two 35-mm motion picture cameras, shooting film used by a consortium of four planetariums. Film footage of the shuttle's interior, as well as the space walks, was used in a movie about the shuttle program.

A small acoustical furnace was tested for use in future shuttle flights, using sound waves to suspend materials during melting and processing and anticipating the development of containerless furnaces. See **STS 61C (January 12, 1986)**.

A NASA-built automatic processing laboratory for separating proteins by using electrical charges, called Isoelectric Focusing Experiment, was similar to the privately financed electrophoresis experiment flown on four previous shuttle flights (see **STS 4–June 27, 1982**). It tested methods for producing purer biological samples than was possible on Earth.

The Monodisperse Latex Reactor (MLR) made its fifth shuttle flight (see **STS 3–March 22, 1982**), continuing experiments in the production of perfectly formed micron-sized latex spheres for use in medical research. On this flight, spheres 18 and 30 microns in diameter were produced. Based on these results, NASA obtained a contract with the National Bureau of Standards to produce spheres 10, 30, and 100 microns in diameter on four later shuttle flights. The spheres were to be used for calibrating scientific instruments.

STS 41B completed its mission by making the first runway landing at the Kennedy Space Center in Florida.

The next shuttle mission was **STS 41C (April 6, 1984)**.

February 8, 1984
Soyuz T10B
U.S.S.R.

The crew of Soyuz T10B docked with **Salyut 7 (April 19, 1982)** and remained in orbit for 237 days—almost eight months—surpassing the 211-day record of **Soyuz T9 (June 27, 1983)**. The crew included Leonid Kizim (**Soyuz T3–November 27, 1982**; **Soyuz T15–March 13, 1986**), Vladimir Solovyov (**Soyuz T15–March 13, 1986**), and Oleg Atkov. Atkov was a doctor, a specialist in heart diseases.

During this third occupancy of Salyut 7, five Progress freighters (see **Progress 1–January 20, 1978**) resupplied the station. Two short-term Soyuz missions, **Soyuz T11 (April 3, 1984)** and **Soyuz T12 (July 17, 1984)**, also visited. Each of these manned missions carried crews of three, meaning that for the first time, Salyut 7 carried six humans.

For this long-endurance mission, the cosmonauts were required to perform at least 2.5 hours of exercise per day. Previous flights had shown, however, that over time cosmonauts became resistant to this tedious schedule. As a doctor, one of Atkov's research goals was to find ways to safely vary the exercise regimen to eliminate the resistance. He also monitored the redistribution of fluids in the body during weightlessness, making frequent check-ups of each crew member.

Possibly the most important part of this third occupancy was its large number of space walks. Besides one space walk by two crew members of Soyuz T12, the crew of Soyuz T10B performed six separate space walks, five dedicated to repairing the fuel line leak that had occurred during a Progress refueling on Soyuz T9 and had disabled 16 of Salyut 7's 32 attitude control thrusters. The first five space walks were planned before launch. The sixth took place only after the two cosmonauts received in-orbit training from one of the Soyuz T12 cosmonauts.

On April 23rd, Kizim and Solovyov spent more than four hours in space preparing their work area and installing a ladder, tools, and the needed repair equipment on the outside of Salyut 7. On April 26th, they began actual work, opening a panel near the laboratory's engines, where they cut aside the insulation protecting the fuel lines and installed a new valve on the leaking fuel line, testing it to make sure it worked. This space walk lasted almost five hours. On April 29th, the two men returned to the same work area for almost three hours. They installed a replacement fuel line, tested it to prove it was airtight, then replaced the panel protecting the propulsion system.

Ground tests soon indicated that these repairs had not worked. Fuel was still leaking from the line. On May 3rd, Kizim and Solovyov went outside again, installing and testing a second fuel line, as well as identifying the exact location of the leak.

On May 18th, Kizim and Solovyov performed their last scheduled space walk, spending a little more than three hours to install supplementary solar panels to a second main panel of Salyut 7, similar to the work the cosmonauts on Soyuz T9 had done. These additional panels increased the power output of this main panel by 50 percent.

By this time, mission control had decided that the leak could be sealed only if that particular fuel line were pinched off from the rest of the system. Vladimir Dzhanibekov, who commanded the visiting Soyuz T12 mission, trained on the ground to do this repair. Once in orbit, however, Kizim and Solovyov requested the right to finish the job they had started, and the decision was made to let Dzhanibekov instruct them on what they needed to do. On August 8th, Kizim and Solovyov performed their sixth space walk, pinching the line after removing the heat insulation that surrounded it. Along the way, they also dismantled an older solar panel section for return to Earth for analysis.

While the fuel line repairs were not successful (a fresh leak soon reappeared), the space walk experience was invaluable. Kizim and Solovyov spent a total of about 45 hours in space, more than all other Soviet space walks combined. Their increased expertise is illustrated by the fact that it took them half the time of the Soyuz T9 cosmonauts to install the supplementary solar panels.

The two men increased their daily workouts significantly as their return date approached, procedures that were routine. On October 2, 1983, after 237 days in space, Kizim, Solovyov, and Atkov returned to Earth, using the Soyuz T11 spacecraft left by its crew. Initially, the three men had difficulty readapting themselves to Earth gravity, having lost about 15 percent of their bone mass, significant muscle strength, and blood volume. Doctors equated their condition to that of patients after long periods of bed rest.

Atkov was also about 2 inches taller due to the lengthening of his spine. The accumulated number of long endurance space missions had found that about two-thirds of all cosmonauts and astronauts experienced some form of temporary back pain once back on Earth, possibly because of this spine lengthening. See **STS 42 (January 22, 1992)**.

After three weeks, all three men were back to normal. Their endurance record in space lasted until the flight of Yuri Romanenko on **Soyuz TM2 (February 5, 1987)**.

The next occupancy of Salyut 7 docked with and reactivated a crippled and dead space station. See **Soyuz T13 (June 6, 1985)**.

February 14, 1984
Ohzora (EXOS C)
Japan

Ohzora carried eight instruments for measuring the ionosphere, including the ionospheric plasma over the South Atlantic anomaly, and worked in conjunction with **CCE (August 16, 1984)**.

March 1, 1984
UoSat 2 (Oscar 11)
U.K. / U.S.A.

UoSat 2 was the second satellite built by the University of Surrey, England, following **UoSat 1 (October 6, 1981)**. It was also the eleventh privately developed amateur communications satellite in the Oscar series (see **Oscar 10–June 16, 1983**). It was launched piggyback with the fifth American Landsat satellite (see **Landsat 4–July 16, 1982**).

UoSat 2's primary payload was a digital store-and-forward memory bank, being tested for its use relaying messages from one part of the globe to another. Because the satellite was in low-Earth orbit, it was almost never visible simultaneously to two communications stations. The memory bank stored the message from the sending station, then forwarded it to the receiving station when that station next appeared over the horizon. See **GLOMR (STS 51D–April 29, 1985)** for further research in this field.

The satellite also carried a speech synthesizer, a CCD camera (see **KH-11 1–December 19, 1976**), as well as a number of other experiments for use by students at the university.

UoSat 2 operated through the early 1990s, when the university replaced it with **UoSat 3/UoSat 4 (January 22, 1990)**. The next Oscar satellite was **Oscar 12 (August 12, 1986)**, the first Japanese amateur satellite.

April 3, 1984
Soyuz T11
U.S.S.R. / India

Soyuz T11 was a short, eight-day internationally crewed mission in support of the 237-day third occupancy of **Salyut 7 (April 19, 1982)** by **Soyuz T10B (February 8, 1984)**. The crew included Yuri Malyshev (**Soyuz T2–June 5, 1980**), Gennady Strekalov (**Soyuz T3–November 27, 1980**; **Soyuz T8–April 20, 1983**; **Soyuz T10A–September 27, 1983**; **Soyuz TM10–August 1, 1990**; **Soyuz TM21–March 14, 1995**), and Indian Rakesh Sharma. With the three crewmen of Soyuz T10B already onboard, a Soviet orbiting laboratory supported six humans for the first time. With the launch a few days later of the American shuttle flight **STS 41C (April 6, 1984)**, carrying a five-person crew, a record 11 humans were in orbit.

Sharma, the first Indian in space, tested the use of yoga in weightlessness. The cosmonauts also filled out questionnaires to measure any changes in their balance and reflexes. Forty percent of the Indian subcontinent was photographed using multispectral photography. Details of a major forest fire in Burma were radioed to Earth.

On April 11, 1984, the crew of Soyuz T11 boarded Soyuz T10B and returned to Earth, leaving their fresher spacecraft as a lifeboat. The next visiting mission to the third occupancy of Salyut 7 was **Soyuz 12 (July 17, 1984)**.

April 6, 1984
STS 41C (STS 11), Challenger Flight #5
Solar Max / LDEF
U.S.A.

A seven-day flight, STS 41C was Challenger's fifth. It was crewed by Robert Crippen (STS 1–April 12, 1981; STS 7–June 18, 1983; STS 41G–October 5, 1984), Francis Scobee (STS 51L–January 28, 1986), Terry Hart, James van Hoften (STS 51I–August 27, 1985), and George Nelson (STS 61C–January 12, 1986; STS 26–September 29, 1988). With this launch, a total of 11 humans were in space concurrently, including the three crewmen from Soyuz T10B (February 8, 1984) and the three crewmen from Soyuz T11 (April 3, 1984)—which exceeded the record of seven set by Soyuz 6 (October 11, 1969), Soyuz 7 (October 12, 1969), and Soyuz 8 (October 13, 1969). This record held for six years until STS 35 (December 2, 1990).

STS 41C deployed one satellite and performed the first in-orbit repair of another.

• *Solar Max*   First launched on February 14, 1980, to study solar flare activity, the NASA satellite Solar Max became disabled after nine months of operation when three fuses in its attitude control system failed (see **Solar Max–February 14, 1980**). Soon thereafter, the satellite's one remaining useful instrument, the coronagraph-polarimeter, suffered electrical failure as well. Because Solar Max had been built using shuttle modular design and was originally intended to be retrieved from orbit by a shuttle, its use to test in-orbit repair techniques made perfect sense.

On the third day in space, Crippen and Scobee piloted Challenger to within 200 feet of a slowly tumbling Solar Max. George Nelson and James van Hoften then began a space walk, with Nelson planning to grab the satellite using the Manned Maneuvering Unit (MMU)—first tested on **STS 41B (February 3, 1984)**. After Nelson stopped the satellite's tumble, the shuttle would ease over and the robot arm would grab it.

Nelson's attempt failed. The docking jaws attached to his MMU failed to close. After three docking attempts, followed by Nelson's attempt to slow Solar Max's tumble by grabbing onto its solar panels by hand, Nelson was called back to the shuttle. Then the crew tried four times to snatch the spinning satellite directly with the robot arm—to no avail.

With the satellite tumbling wildly, it could not be rescued unless ground controllers could gain control. After many of the satellite's systems were turned off and a new computer program was uploaded, Solar Max finally stabilized.

The next day, the shuttle crew made another attempt to grab the satellite with the robot arm and finally succeeded. Once the satellite was secured in Challenger's cargo bay, Nelson and van Hoften spent the flight's sixth day replacing Solar Max's attitude control module and the electronics of the coronagraph-polarimeter. After another day of check-out, the satellite was released to resume its research.

During the satellite's second life, which lasted five years, Solar Max recorded over 12,500 solar flares, including one of the most intense just two weeks after the repair mission. The satellite discovered 10 comets colliding with the Sun. Its gamma-ray spectrometer recorded the gamma rays released by the supernova SN1987, which took place on February 24,

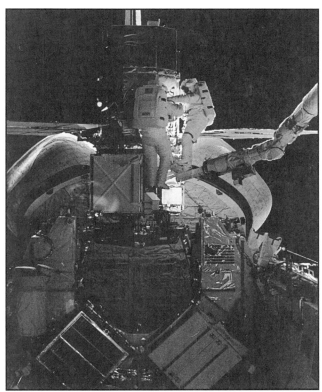

Nelson and van Hoften work on Solar Max, attached to the robot arm. *NASA*

1987 within the Large Magellanic Cloud—the closest supernova event since 1604.

Overall, Solar Max's data showed that from 1980 to 1985, the Sun's total output decreased as the Sun's cycle went from maximum to minimum. This discovery strongly suggested that past climate changes on the earth had been at least partly triggered by solar variation, and that any models describing climate change must consider the Sun to correctly predict future changes. Further climate monitoring continued with ERBS (see **STS 41G–October 5, 1984**).

The satellite's data also suggested that coronal mass ejections, first discovered by **Skylab 2 (May 25, 1973)**, were linked to the complex loops of the Sun's magnetic field lines. Similarly, Solar Max found that both solar flare activity and the Sun's magnetic field were linked. The mass ejections, which seemed to occur somewhat regularly, also seemed related to solar flares.

Following up ozone data from **SME (October 6, 1981)**, Solar Max's ultraviolet spectrometer was used to measure the ozone concentration in the earth's upper atmosphere from 50 degrees north and south latitudes, finding small *increases* in ozone levels during its last five years of operation. It also found that the ozone variations were greater in the southern hemisphere. See **STS 34 (October 18, 1989)** for more ozone research.

Though NASA's original plans had called for the satellite's retrieval by a later shuttle flight, the Challenger explosion on **STS 51L (January 28, 1986)** prevented this. NASA had to make a choice between recovering Solar Max or LDEF, and chose LDEF (see below).

The next satellite dedicated to solar observations was **Ulysses** (see **STS 41–October 6, 1990**).

- *LDEF* The passive satellite LDEF (Long-Duration Exposure Facility) carried 57 experiments to study how the environment of low-Earth orbit affected a wide range of materials, from tomato seeds to paints. Deployed on April 7th, LDEF was originally planned to remain in orbit for about 10 months. Instead, program delays and the Challenger accident (see **STS 51L–January 28, 1986**) caused LDEF to remain in space almost six years, until 1990. See **STS 32 (January 9, 1990)**.

STS 41C carried only one other experiment, a colony of 3,300 bees that—after a few hours of disorientation—began building their hive as if nothing was unusual. The honeycomb was nearly normal, and upon return to Earth, the bees quickly began gathering pollen and nectar.

The flight also carried the first IMAX 70-mm cameras to fly in space, shooting film for the movie *The Dream Is Alive*.

The next shuttle flight was **STS 41D (August 30, 1984)**.

April 8, 1984
STW (Shiyan Tongxin Weixing) 1
China

Shiyan Tongxin Weixing 1 ("experimental communications satellite" in Chinese) was the first Chinese satellite placed in geosynchronous orbit. An earlier attempt in January 1984 had been stranded in low-Earth orbit when its upper stage failed to fire.

Through 1991, China successfully placed six STW satellites in geosynchronous orbit. The first two were used largely for communications experiments. The remaining four were used for telephone communications and television broadcasts to the mountainous interior of China, where standard ground communications were rare. Typical ground antennas are about 20 feet across, more than 2,000 of which had been installed by the end of the 1980s.

In 1994, China began replacing these satellites with a second-generation communications satellite. See **DFH 3A (November 29, 1994)**.

May 23, 1984
Spacenet 1
U.S.A.

Designed by General Telephone and Electronics (GTE) and the Southern Pacific Communications Company (SPCC), the Spacenet satellites formed the sixth private cluster of geosynchronous communications satellites for serving the United States domestic market. The earlier private systems were **Westar 1 (April 13, 1974)**, **RCA Satcom 1 (December 13, 1975)**, **Comstar 1 (May 13, 1976)**, **SBS 1 (November 15, 1980)**, and **Galaxy 1 (June 28, 1983)**.

In 1983, GTE bought out SPCC, and through 1988, GTE launched three additional Spacenet satellites on its own. Combined with its four GStar satellites (see **GStar 1–May 7, 1985**) and the purchase of American Satellite Corporation's (ASC) second satellite (see **STS 51I–August 27, 1985**), GTE had, in 1990, a cluster of eight satellites in orbit.

The Spacenet satellites were the first to use the 4/6 and 12/14 GHz frequencies in a single satellite (see **ATS 6–May 30, 1976** and **Hermes–January 17, 1976**). Each Spacenet satellite also had a capacity of 3,600 voice circuits or one television signal. As of December 1999, two Spacenet satellites are still in operation, owned by GE Americom.

Though these satellites did provide cable television distribution (similar to both the RCA Satcom and Hughes Galaxy systems), GTE also used them to directly compete with AT&T, offering long distance telephone service to both business and residential customers. This service was given the name GTE Sprint, becoming one of the largest American long distance telephone companies.

July 17, 1984
Soyuz T12
U.S.S.R.

Soyuz T12, a visiting flight to **Salyut 7 (April 19, 1982)** during the **Soyuz T10B (February 8, 1984)** residency, lasted 12 days—four days longer than other visiting missions. Its crew consisted of Vladimir Dzhanibekov (**Soyuz 27–January 10, 1978**; **Soyuz 39–March 22, 1981**; **Soyuz T6–June 24, 1982**; **Soyuz T13–June 6, 1985**), Svetlana Savitskaya (**Soyuz T7–August 19, 1982**), and Igor Volk. Volk was a skilled military test pilot and part of the Soviet program to build a reusable space shuttle, similar to the American shuttle. This flight gave him experience in space. See **Buran (November 15, 1988)**.

Savitskaya, on her second space flight, set a record of 12 days in space for a woman. This record lasted until **STS 50 (June 25, 1992)**. She also became the first woman to walk in space. On July 25, 1984, she and Dzhanibekov exited Salyut 7 for a four-hour EVA (extra-vehicular activity), during which she performed a number of tasks with a purpose similar to those performed by Buzz Aldrin on **Gemini 12 (November 11, 1966)**—to demonstrate and prove in-space construction techniques. Savitskaya's tasks, however, were more sophisticated. She used a newly developed portable electron beam-welding tool to cut, weld, and solder a variety of metal plates (see **Soyuz 6–October 11, 1969**), while Dzhanibekov photographed and described her actions. She welded titanium and stainless steel, soldered using tin and lead, and cut plates of titanium and steel. She also applied a silver coat to a surface of anodized aluminum. Dzhanibekov then traded places with her so that he could also test the equipment. Just before ending their EVA, the two cosmonauts retrieved several exterior experiments testing the effect of space on sample materials.

The original plans for Soyuz T12 called for Dzhanibekov to perform one more space walk, finishing the repairs on Salyut 7's leaking fuel line (see **Soyuz T9–June 27, 1983**). Once in orbit, however, Kizim and Solovyov of Soyuz T10B insisted that they be allowed to finish the job, and the last few days of this joint flight were used as a training session to teach the Soyuz T10B cosmonauts what they needed to do to complete the repairs.

On July 29, 1984, Soyuz T12 returned to Earth. The next flight to Salyut 7 began its fourth occupancy. **Soyuz T13 (June 6, 1985)** also reactivated a crippled and dead space station.

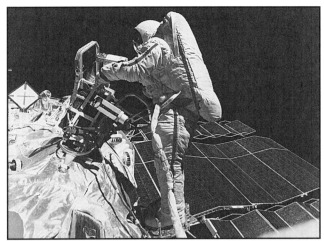
Svetlana Savitskaya. Note welding tool in her right hand. *RIA Novosti*

August 4, 1984
Telecom 1A
France / Europe

Telecom 1A, a French satellite launched by the European Space Agency, was the first in a three-satellite cluster dedicated specifically to the communication needs of France and its overseas territories. It replaced the Symphonie satellites (see **Symphonie 1–December 18, 1974**), its design an outgrowth of **Eutelsat 1 (June 16, 1983)**.

The French launched three Telecom 1 geosynchronous satellites through 1988. Each had three payloads. The first payload broadcast in the 4/6 GHz band and was for communications with French Guiana, the Caribbean, and other territories. It had a capacity of about 1,000 voice circuits. The second used the 7/8 GHz band and was used by the French military to relay signals from oceangoing vessels as well as other military operations in and out of France. This band could also operate in conjunction with NATO communications. The third payload used the 12/14 GHz band and was for business and television communications, both in and out of France, with the capability to relay data from ground antennas about 10 feet across.

In 1991, the French replaced the Telecom 1 satellites with its next generation communications satellites. See **Telecom 2A (December 16, 1991)**.

August 16, 1984
CCE / UKS / Ion Release Module (IRM)
U.S.A. / U.K. / West Germany

The three satellites CCE, UKS, and Ion Release Module (IRM), launched together from Cape Canaveral, were part of a joint U.S., U.K., and West German project to study how the particles in the solar wind are captured and transported into the earth's magnetosphere. See **Dynamics Explorer 1/Dynamics Explorer 2 (August 3, 1981)** for earlier results. The West German contribution, the Ion Release Module (IRM), carried 16 canisters, eight filled with a lithium gas and eight with barium gas. When released, the gas in each canister reacted with sunlight, creating clouds of either barium or lithium ions, which the satellites observed as the ions traveled along the earth's magnetic field lines.

To maximize the observations, the American Charge Composition Explorer (CCE) was placed in an eccentric orbit 700 by 31,000 miles with an inclination of 4.8 degrees, while IRM and the United Kingdom Satellite (UKS), still linked, entered an orbit of about 620 by 70,000 miles with an inclination of 27 degrees. UKS and IRM then separated, maintaining a distance of about 120 miles from each other, with IRM periodically releasing gas charges over the next seven months.

The first two releases of lithium found that less than 1 percent of the solar wind penetrated the earth's magnetosphere. On December 25th, a large release of barium took place on the sunward side of the earth, creating what scientists dubbed an artificial comet—a 6,000-mile tail blown by the solar wind to trail away from IRM, which was visible to many observers on Earth. Scientists found that the solar wind caused the tail to grow faster than expected and to disperse within 15 minutes. A subsequent barium release confirmed these results.

Additional releases in the spring of 1985 took place within the earth's magnetosphere, on its anti-sunward side. Although the cloud was visible from Earth, the instruments on all three satellites failed to detect the presence of ions from the clouds, contradicting all theories of magnetospheric behavior.

This mission's end began a period of fewer satellites launched to research the ionosphere and the magnetosphere. Six years passed before the next American mission, **Pegsat (April 5, 1990)**, which, like CCE, used gas canisters. **Ionosonde (December 18, 1986)** began the Soviet slowdown in such research.

The next magnetospheric satellite was **Prognoz 10 (April 26, 1985)**.

August 30, 1984
STS 41D (STS 12), Discovery Flight #1
SBS 4 / LEASAT 2 / Telstar 3B
U.S.A.

STS 41D was the maiden flight of the third space shuttle, Discovery, taking place after two months of delays and two scrubbed launches. The mission lasted six days and was crewed by Henry Hartsfield (**STS 4–June 27, 1982**; **STS 61A–October 30, 1985**), Michael Coats (**STS 29–March 14, 1989**; **STS 39–April 28, 1991**), Richard Mullane (**STS 27–December 2, 1988**; **STS 36–February 28, 1990**), Steven Hawley (**STS 61C–January 12, 1986**; **STS 31–April 24, 1990**; **STS 82–February 11, 1997**), Judith Resnik (**STS 51L–January 28, 1986**), and Charles Walker (**STS 51D–April 12, 1985**; **STS 61B–November 26, 1985**). STS 41D deployed two commercial and one military satellite.

• *SBS 4* Deployed on August 30th, the SBS (Satellite Business Systems) 4 geosynchronous satellite was the fourth in its six-satellite cluster. It is owned and operated by MCI Communications. See **SBS 1 (November 15, 1980)**.

• *LEASAT 2* Built by Hughes Communications, Inc., and leased by the U.S. military, LEASAT 2 was the first of five military communications satellites for providing tactical communications links between small mobile terminals, such as ships, submarines, and aircraft. Deployed on August 31st, it was placed in geosynchronous orbit, where LEASAT satellites worked in conjunction with the FLTSATCOM satellites (see **FLTSATCOM 1–February 9, 1978**).

These LEASAT military satellites were specifically designed for launch by the shuttle, and a total of five were deployed through 1990 (see also **STS 51A–November 8, 1984**; **STS 51D–April 12, 1985**; **STS 51I–August 27, 1985**; and **STS 32–January 9, 1990**). The third was unique in that it was repaired in orbit after failing to turn on immediately after deployment. The entire constellation was replaced by the next generation of tactical communications satellites. See **UHF Follow-On 1 (September 3, 1993)**.

• *Telstar 3B* Deployed on September 1st, Telstar 3B was part of AT&T's geosynchronous telephone satellite network. See **Telstar 3A (July 28, 1983)**.

Among the onboard experiments, McDonnell Douglas's Continuous Flow Electrophoresis System (CFES) (see **STS 4–June 27, 1982**) was tested again. Flown on four previous missions, this drug-processing unit was able to separate biological hormones better than was possible on Earth. For this flight, CFES was monitored and maintained by Charles Walker, the first astronaut flown at commercial expense. Because the unit had problems the first day, Walker and the ground crew spent considerable time tinkering with its operation, eventually running it almost exclusively in manual mode. Although Walker produced about 83 percent of what was planned for the experiment, later ground analysis of the hormone showed that it had been contaminated by toxins and was therefore useless. See **STS 51D (April 12, 1985)** for CFES's next flight.

In a technology demonstration, the astronauts deployed from Discovery's cargo bay a 102- by 13-foot solar panel, testing the operation of sizable and collapsible structures necessary for building large space stations. Although most of the panels carried dummy cells, a full working array this size would have been capable of producing 12.5 kilowatts of power. Resnik, who handled the multiple deployments, reported that the structure opened and closed reliably and was very stable, showing only about 1 to 2 inches of oscillation as the shuttle performed orbital maneuvers.

STS 41D also tested the Large Format Camera, installed in the cargo bay and capable of taking high resolution pictures of the ground for cartography and topographic research. See **STS 41G (October 5, 1984)**.

Four getaway special experiments included a student experiment sponsored by Rockwell International that successfully grew indium crystals larger than was possible on Earth.

Two motion picture cameras were flown, including an IMAX camera, producing footage for two different documentary films, including the IMAX film *The Dream Is Alive*.

The flight's one serious anomaly occurred when a chunk of ice developed near the front of one of the shuttle's wings. Because there was concern the ice could damage the wing during re-entry, Commander Hartsfield broke it off using the robot arm.

The next shuttle mission was **STS 41G (October 5, 1984)**.

October 5, 1984
STS 41G (STS 13), Challenger Flight #6
ERBS
U.S.A. / Canada

This sixth flight of Challenger lasted eight days. It was manned by Robert Crippen (STS 1–April 12, 1981; STS 7–June 18, 1983; STS 41C–April 6, 1984), Jon McBride, David Leetsma (STS 28–August 8, 1989; STS 45–March 24, 1992), Sally Ride (STS 7–June 18, 1983), Kathryn Sullivan (STS 31–April 24, 1990; STS 45–March 24, 1992), Paul Scully-Power, and Marc Garneau (STS 77–May 19, 1996). Garneau was the first Canadian astronaut. The seven crew members constituted the largest number of people launched into orbit by a single vehicle, a record until **STS 61A (October 30, 1985)**.

STS 41G's prime mission was Earth observation from space. The mission deployed one satellite.

• *ERBS* Deployed on October 5th, ERBS (Earth Radiation Budget Satellite) was NASA's follow-up of the ERBS sensor first flown on **Nimbus 7 (October 13, 1978)**. It continued the long-term measurement of the earth's climate and its radiation budget, produced by the Sun and absorbed by the earth. Its instruments, including SAGE (Stratospheric Aerosol and Gas Experiment) 2, a follow-up of **SAGE (February 18, 1979)**, worked in conjunction with NOAA satellite data (see **NOAA 1–December 11, 1970**).

ERBS's measurements further confirmed results from Nimbus 7 and **Solar Max (February 14, 1980, STS-41C–April 6, 1984)**, showing that the Sun's total output varied with the solar cycle. The data also showed that about 30 percent of the Sun's radiant energy was reflected back into space, 20 percent was absorbed by the atmosphere, and 50 percent was absorbed by the earth's surface. Global monthly maps revealed how different land areas, such as deserts and jungle areas, absorbed different amounts of solar energy, as well as illustrated how that energy was redistributed through the atmosphere. These details were essential for scientists to begin developing computer models that might someday predict climate changes.

ERBS operated through the end of the 1980s. Further solar output monitoring continued with **UARS (STS 48–September 12, 1991)** and **EURECA (STS 46–July 31, 1992)**.

STS 41G also carried a variety of other sensors for studying the earth's surface and atmosphere. The Large Format Camera, flown previously on **STS 41D (August 30, 1984)**, was again installed in the cargo bay. Over both flights, the camera snapped more than 4,500 images of the ground, of which about 60 percent were cloud free. Stereo images of Mount Everest in Nepal were taken, as well as other remote regions from Peru to Vietnam. Hurricane Josephine off the eastern coast of the United States was also imaged. Despite its success, this was the camera's last flight. Although NASA had hoped researchers would be attracted by its capabilities and would pay for further flights, the cost (about $20 million per flight) was apparently too high.

The mission also included the second use on the space shuttle of a radar imaging scanner, following STS 2 (**November 12, 1981**). Dubbed SIR-B (Shuttle Imaging Radar), this synthetic-aperture radar (see **Seasat–June 27, 1978**) imaged a swathe 20–25 miles wide and was able to resolve objects as small as 55 feet across, depending on look angle. Unlike the previous radar instruments used on Seasat and STS 2, the look angle on STS 13 could be adjusted, thereby changing the highlights and shadows of radar images. Each day of the mission's eight-day flight, the look angle was adjusted, and as the spacecraft repeatedly flew over the same regions its radar images were correspondingly modified. Scientists then evaluated the images to see how these changing angles enhanced image interpretation. They found that different types of features required different look angles to be properly identified. This discovery suggested that the interpretations of the radar mapping performed by **Pioneer, Venus Orbiter (December 4, 1978), Venera 15 (October 10, 1983),** and **Venera 16 (October 14, 1983)** probably contained inaccuracies, and that radar mapping of Venus planned by **Magellan (August 10, 1990)** required multiple "look angles" to guarantee an accurate map.

Because of a temperamental antenna and the temporary shutdown of one of NASA's TDRS communications satellites (see **STS 6–April 4, 1983**), only about 20 percent of the planned radar data was obtained. Despite this failure, coverage included Bangladesh, Hawaii, Argentina, and Mt. Shasta in California. Forest cover in West Germany was penetrated to map the ground. Nevada was imaged to a depth of 3 feet, revealing previously unknown fault and tectonic features. Radar images of the rift valley in Kenya and Sahara Desert found evidence of ancient human settlements, as well as buried stream patterns.

The next radar imaging mission was **Cosmos 1870 (July 25, 1987)**. The next shuttle mission to employ radar imaging was **STS 59 (April 9, 1994)**.

Of STS 41G's many Earth observing instruments, one was crew member Paul Scully-Power, a trained oceanographer. He spent his eight days in orbit studying ocean currents and eddies, his observations proving that the eddies were as important as the currents for making long-term weather predictions. He also showed the U.S. Navy how they could better hide their submarines from satellite detection.

The astronauts tested what NASA dubbed its Orbital Refueling System, continuing research from **STS 41B (February 3, 1984)**. The system could pump up to 550 pounds of fuel into a retrieved satellite. During the mission, Sullivan and Leetsma used the shuttle's computers to pump fuel remotely from one tank to another. Later they space-walked to the system and tested tools for attaching a fuel pump to a dummy satellite tank, which they later used to transfer more fuel from tank to tank. This space walk by Sullivan was the first by an American woman. Later shuttle refueling tests were conducted on **STS 53 (December 12, 1992)**.

The IMAX camera made its third flight on the shuttle, gathering additional footage for the film *The Dream Is Alive.*

The next shuttle mission was **STS 51A (November 8, 1984)**.

November 8, 1984
STS 51A (STS 14), Discovery Flight #2
LEASAT 1 / Anik D2
U.S.A. / Canada

STS 51A was Discovery's second flight, making it the third reusable spacecraft to fly into space twice. The flight lasted eight days and was crewed by Fred Hauck (**STS 7–June 18, 1983; STS 26–September 29, 1988**), David Walker (**STS 30–May 4, 1989; STS 53–December 2, 1992; STS 69–September 7, 1995**), Joe Allen (**STS 5–November 11, 1982**), Anna Fisher, and Dale Gardner (**STS 8–August 30, 1983**). This mission proved the shuttle's ability to deploy and retrieve orbiting satellites. It deployed two communications satellites.

• *LEASAT 1*   Built and leased from Hughes Communications, Inc., LEASAT 1 was the second of five military LEASAT communications satellites. See **STS 41D–August 30, 1984** for program information.

• *Anik D2*   Anik D2 was owned and operated by Telesat Canada (see **Anik A1–November 10, 1972**). See **Anik D1 (August 26, 1982)** for more details about this geosynchronous communications satellite.

STS 51A also made the first retrieval from orbit of two satellites: Palapa B2 and Westar 6, deployed by **STS 41B (February 3, 1984)** and stranded in low-Earth orbit when their upper booster rockets malfunctioned. First, Fred Hauck and David Walker guided Discovery to a rendezvous with each satellite. Then, Joe Allen and Dale Gardner used the manned maneuvering units first tested on STS 41B to fly to each and stabilize their spin. Anna Fisher then used the robot arm to grasp and place the satellites in the shuttle's cargo bay. In one case, it was necessary for Allen and Gardner to manually position the satellite in the cargo bay so that Fisher could grab it with the robot arm. Upon return to Earth, both satellites were refurbished, sold by the insurance companies that owned them, and relaunched. See **Asiasat 1 (April 7, 1990)** and **Palapa B2R (April 13, 1990)**.

In addition to these satellite operations, STS 51A carried two experiments, the Diffusive Mixing of Organic Solutions (DMOS) experiment and several handheld self-contained radiation monitoring units for measuring the background radiation in various sections of the shuttle. DMOS was a private unit developed by the 3M company to study methods for producing commercial products in space using organic and polymer chemistry. During the flight, DMOS produced large organic crystals impossible to produce on Earth. 3M followed this experiment with tests on **STS 51I (August 27, 1985)** and **STS 61B (November 26, 1985)**.

The next shuttle mission was **STS 51C (January 24, 1985)**.

1985

January 7, 1985
Sakigake
Japan

Sakigake ("pioneer" in Japanese) was Japan's first interplanetary spacecraft, flying about 4.3 million miles from Halley's Comet on March 11, 1986, passing between the Sun and the comet to study the interaction of the solar wind with the comet's coma (see **Sakigake, Halley's Comet Flyby–March 11, 1986**).

As Japan's first interplanetary probe, Sakigake carried as many engineering tests as scientific experiments, verifying the design of its operational components for use in future spacecraft. Its success cleared the path for the launch of Suisei (**Suisei, Halley's Comet Flyby–March 8, 1986**).

January 24, 1985
STS 51C (STS 15), Discovery Flight #3
Magnum 1
U.S.A.

STS 51C ushered in the most active year in space shuttle history, through 1999: nine missions were launched in 1985. STS 51C, Discovery's third flight, lasted three days and was crewed by Ken Mattingly (**Apollo 16–April 16, 1972**; **STS 4–June 27, 1982**), Loren Shriver (**STS 31–April 24, 1990**; **STS 46–July 31, 1992**), James Buchli (**STS 61A–October 30, 1985**; **STS 29–March 13, 1989**; **STS 48–September 12, 1991**), Ellison Onizuka (**STS 51L–January 28, 1986**), and Gary Payton.

The mission was the first American manned space flight that was almost entirely military in nature. Hence, it was also the first American space flight for which all details were to be kept secret. However, the mission's main purpose—to deploy a surveillance satellite—was revealed by the press well before launch.

• *Magnum 1*   A variation of the U.S. Air Force Chalet series of satellites (see **Chalet 1–June 10, 1978**), Magnum 1 listened and intercepted electronic signals, including radio, voice communications, radar, and missile telemetry. Three Magnum satellites were launched by the space shuttle (see **STS 33–November 22, 1989**; **STS 38–November 15, 1990**) before the military ceased using the shuttle to deploy its satellites. Later versions were also given the Trumpet designation (see **Trumpet 1–May 3, 1994**).

STS 51C did carry one civilian biological experiment. Blood samples from both healthy and ill donors (including those with heart disease, diabetes, cancer, and hypertension) were flown to investigate the effect of weightlessness on blood flow and the structure of red cells. It was found that the healthy and diseased blood cells both returned undeformed from space, an unexpected result suggesting that any deformed blood cells found in the astronauts from previous missions were caused not from weightlessness directly but from the changes in calcium and bone production. The healthy state of the cells also indicated that weightlessness could provide a therapy for curing some diseases. These tests continued on STS 26 (September 29, 1988).

The next shuttle mission was STS 51D (**April 12, 1985**).

February 8, 1985
Brasilsat A1 / Arabsat 1A
Brazil / Arab States / Europe

Brasilsat A1 and Arabsat 1A were satellites launched by the European Space Agency's Ariane rocket.

• *Brasilsat A1*   Developed by Spar Aerospace of Canada for Embratel, the Brazilian space agency, Brasilsat A1 was Brazil's first geosynchronous communications satellite. The satellite was similar in design to **Anik D** (**August 26, 1982**), using the 4/6 GHz bandwidths, and provided telephone service throughout the remote Brazilian backcountry.

Embratel built one additional Brasilsat A satellite through 1986. In 1994, it replaced these two satellites with a second-generation satellite cluster. See **Brasilsat B1** (**August 10, 1994**).

• *Arabsat 1A*   Built by Aerospatiale and Ford Aerospace and Communications for the Arab Satellite Communications Organizations (Arabsat), Arabsat 1 was the first geosynchronous satellite dedicated to providing Arab nations with telephone and television service. Formed in 1976 by 22 Arab states, Arabsat launched three Arabsat 1 satellites through 1992. Each had a capacity for 8,000 voice circuits and eight television signals.

Political disagreements and economics limited the satellites' productivity. In 1989, only one-third of their capacity was utilized, mostly for telephone communications, with their television broadcasting capability completely unused because of internal disagreements within Arabsat over political and religious issues.

In 1996, Arabsat launched a second-generation two-satellite cluster. The new satellites provided television and telephone service, carried a larger capacity, and used both the 4/6 GHz and 12 GHz frequencies. As of December 1999, one of the older satellites and both newer satellites were operating, with a third-generation satellite launched in February 1999.

March 13, 1985
Geosat
U.S.A.

Geosat (GEOdetic SATellite), built by the U.S. Navy, used a radar altimeter to map precisely the shape of the earth above the ocean, measuring subtle changes in sea level. This research followed work done by **GEOS 2** (**January 11, 1968**), **GEOS 3** (**April 9, 1975**), and **Seasat** (**June 27, 1978**).

Geosat operated for 19 months, measuring the ocean's average sea level to within 1 inch. The U.S. Navy needed this data so that the routing used by its submarine-launched ballistic missiles above the earth took into consideration the fluctuations in gravity caused by depressions and bulges in sea level.

In the early 1990s, the data from this satellite were declassified, and because the changes in sea level were direct reflections of the topography of the sea floor, scientists used the data to produce extremely detailed maps of the ocean floors.

Seafloor image from Geosat (March 13, 1985) of the Atlantic and Caribbean. Note ribbed pattern surrounding the spreading center between the plates of North America and Africa. *NOAA*

The maps doubled the known number of sea floor volcanoes, and revealed that spreading centers between tectonic plates were complex regions *(see color section, Figure 25)*. The plate's edges were considerably smeared as they rubbed against each other, and, in total, the earth's plates appeared less rigid and more malleable than scientists had expected.

The data also showed that the Pacific was generally higher in elevation than the Atlantic, and that the Atlantic was overall a denser ocean.

The next American geodetic satellite was **TOPEX–Poseidon (August 10, 1992)**. The next geodetic satellite was **Ajisai (August 12, 1986)**.

March 22, 1985
Intelsat 5A F10
International / U.S.A.

The exponential increase in demand for international telephone and television service forced Intelsat (see **Early Bird–April 6, 1965**) to supplement its Intelsat 5 satellite cluster (see **Intelsat 5 F2–December 6, 1980**) with the Intelsat 5A satellites. Though similar to the Intelsat 5 satellites, the new satellites had a slightly larger capacity, between 15,000 and 45,000 two-way voice circuits per satellite. Through January 1989, Intelsat launched five Intelsat 5A satellites, positioning two over the Atlantic, one over the Pacific, and two over the Indian Ocean. These satellites increased the system's capacity by almost 50 percent.

Nonetheless, by the end of the 1980s, Intelsat was forced to upgrade its service once again. See **Intelsat 6A F2 (October 27, 1989)**.

April 12, 1985
STS 51D (STS 16), Discovery Flight #4
Anik C1 / LEASAT 3
U.S.A. / Canada

STS 51D was Discovery's fourth flight. It lasted seven days and carried a crew of seven: Karol Bobko (**STS 6–April 4, 1983; STS 51J–October 3, 1985**), Don Williams (**STS 34–October 18, 1989**), Jeffrey Hoffman (**STS 35–December 2, 1990; STS 46–July 31, 1992; STS 61–December 2, 1993; STS 75–February 22, 1996**), David Griggs, Rhea Seddon (**STS 40–June 5, 1991; STS 58–October 18, 1993**), Charles Walker (**STS 41D–August 30, 1984; STS 61B–November 26, 1985**), and Jake Garn. Garn was a U.S. senator, the first untrained non-astronaut launched into space. STS 51D deployed two satellites.

- *Anik C1* See Anik A1 (**November 10, 1972**) and **STS 5 (November 11, 1982)** for information about this geosynchronous communications satellite that was owned and operated by Telesat Canada.

- *LEASAT 3* Built and leased from Hughes Communications, Inc., LEASAT 3 was the third of five military LEASAT communications satellites. See **STS 41D–August 30, 1984** for program information.

LEASAT 3 was released from Discovery's cargo bay on April 13. The computer program for firing its booster engines failed to run, however, leaving the satellite stranded in an unusable low-Earth orbit. In an improvised attempt to flip the switch that would activate the program, Hoffman and Griggs did an unscheduled space walk to attach two jury-rigged devices to the robot arm, dubbed alternatively a "fly swatter" and a "lacrosse stick." Then Bobko and Williams piloted the shuttle close to LEASAT 3 so that Seddon could use the "modified" robot arm to activate the computer. Despite four tries, Seddon was unable to flip the switch. Fuel limitations then forced an end to the rescue attempt. LEASAT 3 remained in its useless orbit until **STS 51I (August 27, 1985)**.

Among STS 51D's scientific experiments, McDonnell Douglas's Continuous Flow Electrophoresis System (CFES) made its sixth shuttle flight, with Charles Walker making his second flight to run this commercial production unit for producing pure biological specimens for sale as pharmaceuticals. See **STS 4 (June 27, 1982)** and **STS 41D (August 30, 1984)**. This flight used better sterilization methods, as well as modified equipment, to try and avoid the impurity problems seen on STS 41D. STS 51D also tested additional electrophoresis equipment, which attempted to separate cells and protein.

STS 51D was the first shuttle flight to do protein crystal growth research, following tests during **Soyuz 38 (September 18, 1980)**. In space, it was hoped that protein crystals could be grown much larger than possible on Earth. From these crystals, scientists could more easily study their molecular structure. STS 51D successfully produced a large lysozyme crystal, demonstrating that the technology was feasible. See **STS 51F (July 29, 1985)**.

In medical research, STS 51D carried French-built equipment to take daily ultrasound images of an astronaut's heart to document how its pumping action changed in weightlessness, continuing research from **Soyuz T6 (June 24, 1982)**. See **STS 61G (June 17, 1985)** for later research. Moreover, Senator Garn had small microphones attached to his belly to record the sounds of his digestive tract. Unfortunately for him but fortunately for the scientists, Garn experienced an especially rough case of space sickness, lasting more than two days.

The flight also included an informal study of how ordinary toys, such as yo-yos, paper airplanes, gyroscopes, tops, wind-up cars and mice, and Slinkys, behaved in zero gravity. The yo-yo, for example, could be thrown very slowly and still return, but it could not be made to "sleep" at the end of the line. The Slinky, meanwhile, could no longer "climb" down stairs, but when stretched out and vibrated either in and out or from side to side, the wave patterns produced more precisely reflected mathematical curves than in tests conducted on Earth.

The next shuttle mission was STS 51B (**April 29, 1985**).

April 16, 1985
Foton 1 (Cosmos 1645)
U.S.S.R.

Foton 1, a 12-day unmanned mission, was the first in a new series of Soviet recoverable research satellites, most of which performed materials processing research. Using the same Vostok capsule design as the Bion satellites (see **Bion 1–October 31, 1973**), each mission remained in space for a few weeks, testing a wide range of substances in weightlessness—from crystals to polymers. The capsule then re-entered the atmosphere and was recovered so that scientists could study the results.

Through December 1999, a total of 12 Foton satellites were launched, with the first three given the Cosmos designation. Beginning with **Foton 5 (April 26, 1989)**, the satellites were launched on a commercial basis, whereby European researchers paid the Soviet/Russian launch facilities to placed their experiments in orbit. The last four Foton satellites carried biological experiments for the European Space Agency. See **Foton 6 (April 11, 1990)**.

April 26, 1985
Prognoz 10 (Intercosmos 23)
U.S.S.R. / Czechoslovakia

Carrying experiments devised by the Soviet Union and Czechoslovakia, Prognoz 10 resumed the solar monitoring work of earlier Prognoz satellites (see **Prognoz 1–April 14, 1972**). Like its predecessors, the satellite was placed in an eccentric orbit, with its apogee at about 125,000 miles so that it crossed the bow shock of the earth's magnetosphere. Because Prognoz 10 carried experiments from two countries, it was also listed as part of the Soviet Intercosmos program. See **Intercosmos 1 (October 14, 1969)**.

Prognoz 10 functioned for only a few weeks, however, and ended the Prognoz program, beginning a four-year period in which few satellites to study the magnetosphere or Sun were launched. See **Ionosonde (December 18, 1986)**.

April 29, 1985
STS 51B (STS 17), Challenger Flight #7
NUSAT / GLOMR
U.S.A. / Europe

The seventh flight of Challenger, STS 51B lasted seven days and carried a crew of seven: Robert Overmyer (STS 5–November 11, 1982), Fred Gregory (STS 33–November 22, 1989; STS 44–November 24, 1991), Norman Thagard (STS 7–June 18, 1983; STS 30–May 4, 1989; STS 42–January 22, 1992; Soyuz TM21–March 14, 1995), William Thornton (STS 8–August 30, 1983), Don Lind, Taylor Wang, and Lodwijk van den Berg. This was the second flight of Spacelab (see **STS 9–November 28, 1983**). As before, the crew was divided into two shifts in order to operate around the clock, performing research in materials processing, life sciences, Earth observations, and astrophysics.

The shuttle's two getaway specials were both small deployable satellites.

• *NUSAT*   A joint project of Weber State College and Utah State University, NUSAT (Northern Utah Satellite) tested the low-cost fast development of lightweight communications satellites. It was designed to help the Federal Aviation Administration calibrate its air traffic control radar facilities, and it operated for 20 months before re-entering the atmosphere in 1986.

• *GLOMR*   This American military test satellite also demonstrated that a satellite could be built cheaply in less than one year. Moreover, GLOMR (Global Low Orbiting Message Relay) tested digital communications technologies, including store-and-forward transmission of data, first tested on **UoSat 2 (March 1, 1984)**, as well as the use of unattended ground stations. On this flight, however, a battery problem prevented the canister door from opening. The satellite could not be released and was returned to Earth. It was later deployed on **STS 61A (October 30, 1985)**.

Meanwhile, in Spacelab, the round-the-clock lab work continued. In life sciences research, STS 51B was the first shuttle to use its animal enclosures (see **STS 8–August 30, 1983**) for research, carrying 2 squirrel monkeys and 24 rats, and it was the first American spacecraft since **Biosatellite 3 (June 29, 1969)** to conduct biological research on animals in space. STS 51B also continued research from **Bion 6 (December 14, 1983)**. Although problems with the cages caused the release of animal feces and food throughout Spacelab, the animals survived the mission, with all 24 rats showing significant loss of muscle tone—up to 40 percent—and a 10 to 15 percent loss of bone mineral.

Other life sciences research included tests for improving the shuttle's waste management system. Biofeedback methods that might help the astronauts control their bodily processes and thereby reduce the effects of space sickness were also tested.

All three materials-processing experiments attempted to grow crystals in space. The crystals, which would have been impossible to grow on Earth, had applications as heat sensors and in other high technology equipment. In one case, a red mercuric oxide crystal grew much faster and 20 times larger than expected. The size of a sugar cube, this crystal's value was estimated at millions of dollars; it was later sliced apart and used in x-ray and gamma-ray spectrometers.

Two triglycine sulfate crystals were also grown. In this case, the growth process itself was photographed, illustrating how ground- and space-grown crystals develop differently.

Earth observations included some of the most spectacular photographs ever taken of the earth's aurora, showing how it sits on top of the atmosphere and encircles the poles (see **Dynamics Explorer 1/Dynamics Explorer 2–August 3, 1981). During these observations, the astronauts watched as Challenger flew directly through the aurora three times.

The next shuttle mission was STS 51G (**June 17, 1985**).

May 7, 1985
GStar 1
U.S.A.

This geosynchronous communications satellite, owned and operated by General Telephone and Electronics (GTE) Sprint, formed part of GTE Sprint's satellite cluster begun with **Spacenet 1** (**May 23, 1984**), providing long distance telephone service and cable television distribution in the United States. Through November 1990, GTE Sprint launched four GStar satellites, each having a capacity for 5,400 voice circuits.

June 6, 1985
Soyuz T13
U.S.S.R.

Soyuz T13 was a rescue mission. In February 1985, all contact with the then-unmanned **Salyut 7 (April 19, 1982)** was lost. Without ground control to maintain the laboratory's orientation, the station began to drift, placing its solar panels in darkness and draining its electrical batteries. This power loss, in turn, caused the Salyut 7's temperature control system to turn off, causing its water pipes to freeze.

The crew members of Soyuz T13, Vladimir Dzhanibekov (**Soyuz 27–January 10, 1978**; **Soyuz 39–March 22, 1981**; **Soyuz T6–June 24, 1982**; **Soyuz T12–July 17, 1984**) and Viktor Savinykh (**Soyuz T4–March 12, 1981**; **Soyuz TM5–June 7, 1988**), were specifically trained to dock with the crippled laboratory and repair and reactivate it.

After making a visual inspection of the laboratory's exterior, the cosmonauts docked manually with it, using a handheld laser range-finder to judge distances. (This additional piece of equipment prevented the same problems that **Soyuz T8–April 20, 1983** faced when its radar system failed.)

Wearing full spacesuits, Dzhanibekov and Savinykh entered the station on June 8th. They found its atmosphere "stuffy and cold," though breathable. For a week, Dzhanibekov and Savinykh commuted every 40 minutes from their Soyuz T13 capsule to the freezing interior of Salyut 7, wearing portable breathing apparatus and warm clothing. First, they reoriented the laboratory so that its solar panels could once again charge the station's batteries. Then, they reactivated Salyut 7's life support systems, along with its communications systems; tested the station's control systems and reactivated them; and turned on the station's television system.

By June 21st, when a Progress freighter (see **Progress 1–January 20, 1978**) arrived packed with replacement parts, Salyut 7 was once again functioning. By July 2nd, Dzhanibekov and Savinykh were able to begin a normal schedule of research, using the laboratory's multispectral cameras to photograph agricultural regions of the Soviet Union. All told, this occupancy photographed about 6 million square miles.

The remainder of the fourth occupancy followed a more routine schedule of science and maintenance. On August 2nd, Dzhanibekov and Savinykh performed a five-hour space walk, installing the third set of supplementary solar panels to the station's last main panel (see **Soyuz T9–June 27, 1983** and **Soyuz T10B–February 8, 1984**). They also recovered a French micrometeoroid sensor for catching particles from Comet Giacobini-Zinner as it made its close approach to the Sun the previous October and replaced this with a new detector for the impending visit of Halley's Comet in March 1986. They also retrieved samples that were being tested in the environment of space.

During August, the cosmonauts used the station's multispectral cameras to help free an Antarctic expedition ship trapped in ice. They did this work in conjunction with the Soviet Okean satellites (see **Cosmos 1500–September 28, 1983**), collecting information about ice thickness and sea currents so that an icebreaking ship could reach the trapped vessel.

This crew received one additional Progress-type freighter (see **Progress 1–January 20, 1978**), followed by **Soyuz T14 (September 17, 1985)**. This manned mission not only switched spacecraft, as was routine, but also rotated crews for the first time.

While Dzhanibekov returned to Earth with Georgi Grechko on September 26th after 112 days in orbit, Savinykh remained on Salyut 7 with two of Soyuz T14's crew: Alexander Volkov and Vladimir Vasyutin. These three men had originally been

Soyuz T13 leaves the repaired Salyut 7 and the docked Soyuz T14. *RIA Novosti*

teamed up for Soyuz T13's planned long-term occupancy. Volkov and Vasyutin had been bumped when Salyut 7 failed. Finally, they were reunited to complete their scheduled mission.

June 10, 1985
Vega 1, Venus Flyby
U.S.S.R. / International

Vega 1, launched on December 15, 1984, was one of two identical probes (see **Vega 2, Venus Flyby–June 14, 1985**) built by the Soviet Union to fly past Halley's Comet. These spacecraft were the most complex unmanned probes ever built by the Soviets. They involved more international cooperation than any previous Soviet project, using experiments and equipment built in communist bloc countries such as Bulgaria, Hungary, Poland, Czechoslovakia, and East Germany, as well as capitalist countries such as the United States, Austria, France, and West Germany. This project was also the first in which both East and West Germany cooperated.

The Vega spacecraft were also the first Soviet craft to use a planet to redirect their flight, using concepts first developed by **Mariner 10 (November 3, 1973)**. As the two spacecraft passed Venus in June, they each deployed a lander and a balloon, then used Venus's gravitational pull to redirect their flight toward a rendezvous with Halley's Comet in March 1986.

The Vega 1 lander and balloon entered the Venusian atmosphere on June 10, 1985. The lander, similar to the Soviet Venera probes (see **Venera 14–March 5, 1982**), was designed to survive on the surface long enough to obtain a rock sample and test its composition. Because Vega 1's drill was accidentally deployed early, however, it could not get its rock sample.

Both the balloon and the lander did gather data as they descended through the atmosphere. See **Vega 2, Venus Flyby (June 14, 1985)** for results from Vega 1's Venus research.

Vega 1 went on to its rendezvous with Halley's Comet on March 6, 1986 (see **Vega 1, Halley's Comet Flyby–March 6, 1986**).

June 14, 1985
Vega 2, Venus Flyby
U.S.S.R. / International

Identical in design to Vega 1 (**Vega 1, Venus Flyby–June 10, 1985**), Vega 2 was launched on December 21, 1984. During its June 14, 1985, Venus flyby, it deployed a lander and a balloon and used the planet's gravity to redirect the main probe to a rendezvous with Halley's Comet on March 9, 1986.

The two French-built balloons—one from Vega 1 and the other from Vega 2—remained aloft for almost two days after their deployment into Venus's atmosphere. Each drifted in the middle cloud layer about 33 miles above the planet's surface, in winds averaging about 145 miles per hour. Each balloon entered the atmosphere at about midnight local time, 7 degrees to the north and south of the equator, respectively. Both traversed almost 100 degrees of longitude during their life span. Each drifted slowly away from the equator and poleward, although the southern balloon did so at a faster rate. Scientists were surprised to discover a 12°F difference in temperature between the two balloons, showing that large temperature variations in the Venusian atmosphere were possible.

The data from the balloons, widely separated, proved that Venus's atmosphere flowed in a retrograde direction (opposite the rotation of the planet), and did so in a planetwide manner. Both balloons also recorded evidence of vertical winds sometimes exceeding 6 miles per hour, indicating the presence of large-scale planet-sized waves of wind extending throughout the atmosphere.

The data gathered by both landers as they descended through the atmosphere matched the findings of earlier spacecraft (see **Pioneer Venus Probe–December 9, 1978; Venera 12–December 21, 1978**). However, the Vega 1 and Vega 2 landers and balloons registered previously unseen and surprisingly high levels of water vapor in the middle cloud layer, at between 32 and 38 miles altitude. Below this, the water vapor levels matched the numbers from previous probes. The landers also made the first direct detection of sulfuric acid in these clouds, proving unquestionably the nature of Venus's clouds. The probes also detected chlorine, sulfur, and, for the first time, phosphorus in the clouds.

Since both landers were aimed at Venus's night side, neither was equipped with cameras. They were the first to land on the large raised continent-sized plateau named Aphrodite Terra, located on the opposite hemisphere from all previous Venera landing sites. Though the Vega 1 lander failed to get a rock sample, Vega 2 was more successful. Its sample resembled basalt, as had earlier samples tested by **Venera 13 (March 1, 1982)** and **Venera 14 (March 5, 1982)**. Once again, the evidence indicated that Venus's geology was strongly shaped by volcanic activity.

As of December 1999, the Vega 1 and Vega 2 landers remain the last spacecraft to touch the surface of Venus and return data. Only one other probe, the radar mapper **Magellan (August 10, 1990)**, has been sent to Venus, and no other Venus spacecraft is presently being planned. Vega 2 continued on to Halley's Comet, rendezvousing with it on March 9, 1986 (see **Vega 2, Halley's Comet Flyby–March 9, 1986**).

June 17, 1985
STS 51G (STS 18), Discovery Flight #5
Morelos 1 / Arabsat 1B / Telstar 3C / Spartan 1
U.S.A. / Mexico / Arab States / France

STS 51G, the fifth flight of the space shuttle Discovery, lasted seven days and was crewed by Daniel Brandenstein (STS 8–August 30, 1983; STS 32–June 9, 1990; STS 49–May 7, 1992), John Creighton (STS 36–February 28, 1990; STS 48–September 12, 1991), John Fabian (STS 7–June 18, 1983), Steven Nagel (STS 61A–October 30, 1985; STS 37–April 5, 1991; STS 55–April 26, 1993), Shannon Lucid (STS 34–October 18, 1989; STS 43–August 2, 1991; STS 58–October 18, 1993; STS 76–March 22, 1996), Frenchman Patrick Baudry, and Saudi Arabian Salman Abdul Aziz Al Sa'ud. This shuttle flight deployed three communications satellites and one temporary free-flyer.

- *Morelos 1* Named in honor of one of the heroes of Mexican independence, Morelos 1 was Mexico's first geosynchronous satellite. It was deployed on June 17th and was the first of two satellites launched to provide telephone and television service for Mexico. The satellites, built by Hughes Communications, were similar in design to Anik C3 (see **STS 5–No-**

vember 11, 1982) and **Telstar 3A (July 28, 1983)**. Ground antennas range from 10 to 36 feet across, depending on bandwidth.

In 1993, the Mexican government launched **Solidaridad 1 (November 20, 1993)**, the first of two geosynchronous satellites replacing the Morelos satellites.

• *Arabsat 1B* Deployed on June 18th, Arabsat 1B was the second geosynchronous satellite in the Arab States' first communications cluster. See **Arabsat 1A (February 8, 1985)**.

• *Telstar 3C* Deployed on June 19th, the satellite Telstar 3C was part of AT&T's geosynchronous telephone satellite network. See **Telstar 3A (July 28, 1983)**.

• *Spartan 1* The maiden flight of Spartan 1 (which stood for Shuttle Pointed Autonomous Research Tool for Astronomy) demonstrated its usefulness as an inexpensive research satellite. Very simply designed, the NASA-built satellite carried no communications or power-generating equipment. For three days, from June 20th to June 22nd, Spartan 1 gathered x-ray data (0.5–KeV) on the Perseus cluster of galaxies, the center of the Milky Way, and the x-ray object Scorpius X-2.

Following this flight, Spartan 1 was retired, its experience used to design later Spartan spacecraft. Three different Spartans flew on later shuttle missions, with a fourth lost in the Challenger accident (see **STS 51L–January 28, 1986**). Spartan 201 carried sensors for studying the Sun and flew five different times through Ocrober 1998. See **STS 56 (April 8, 1993)** for its first mission. Spartan 204 (also called the Spartan Service Module) was used in a variety of ways, its purpose modified with each flight. On its first mission, **STS 63 (February 3, 1995)**, it made ultraviolet astronomical observations. On its second flight, **STS 77 (May 19, 1996)**, it performed engineering experiments. Spartan 206 flew once, on **STS 72 (January 11, 1996)**, doing technology tests.

Other STS 51G experiments included six getaway specials that covered research such as the sloshing of liquids within a tank and the testing of a heating system using capillary action instead of moving pumps to circulate fluid.

The Automated Directional Solidification Furnace (ADSF) was intended for regular shuttle use, investigating how different materials mixed and solidified in weightlessness. On this flight, a sample of manganese and bismuth was melted to produce an alloy that had a significantly higher magnetic strength than possible from a ground-produced alloy. ADSF next flew on **STS 61C (January 12, 1986)**.

Continuing research from **Soyuz T6 (June 24, 1982)** and **STS 51D (April 12, 1985)**, the French Echocardiograph Experiment took regular ultrasonic images of the hearts of both Baudry and Lucid. Another French experiment tested how the body's posture, orientation, and movement adapted to the loss of gravity. Baudry and Al Sa'ud measured changes in these functions each day using sensors, cameras, and tape recorders.

In a military tracking test, a laser beam successfully hit an 8-inch target on Discovery and was able to track the spacecraft for 2.5 minutes as it flew overhead at more than 17,000 miles per hour.

The next shuttle flight was **STS 51F (July 29, 1985)**.

July 10, 1985
Bion 7 (Cosmos 1667)
U.S.S.R.

Bion 7, a biological research satellite, was part of a Soviet program begun with **Bion 1 (October 31, 1973)**; it followed **Bion 6 (December 14, 1983)**. It continued the life sciences research performed on **STS 51B (April 29, 1985)**, placing two monkeys, 10 rats, 1,500 flies, and 10 newts in orbit for seven days, after which the capsule was successfully recovered.

The flies were launched in their larval stage to see how weightlessness influenced their emergence as flies.

The monkeys and rats were studied to see how blood circulation and balance changed in weightlessness. Unlike in Bion 6, neither monkey died upon return to Earth and both readapted quickly to the earth environment.

To study how wounds healed in space, the 10 newts each had a foot and an eye lens removed prior to launch. A control group of identically prepared newts remained on Earth. The experiment found that both earthbound and spacebound newts grew new limbs and eye lenses at the same rate.

The next Bion satellite was **Bion 8 (September 29, 1987)**.

July 29, 1985
STS 51F (STS 19), Challenger Flight #8
U.S.A. / Europe

The eighth flight of Challenger, STS 51F, lasted eight days and was crewed by Gordon Fullerton (**Enterprise–August 12, 1977; STS 3–March 22, 1982**), Story Musgrave (**STS 6–April 4, 1983; STS 33–November 22, 1989; STS 44–November 24, 1991; STS 61–December 2, 1993; STS 80–November 19, 1996**), Roy Bridges, Anthony England, Karl Henize, Loren Acton, and John-David Bartoe. Although one of the shuttle's main engines shut down prematurely during launch due to the failure of two thermometers, the spacecraft was still able to make orbit and complete its mission.

STS 51F was the third mission of the European Space Agency's Spacelab, following **STS 9 (November 28, 1983)** and **STS 51B (April 29, 1985)**. As with the previous two missions, the crew was divided into two teams operating in a round-the-clock schedule, conducting 13 different experiments in the fields of biology, Earth observation, and astronomy.

For more than a day, the astronauts conducted detailed observations of the Sun. One instrument recorded the Sun's outer layers in ultraviolet radiation, another measured its helium and hydrogen abundance, and a third mapped the Sun's magnetic field activity.

Other instruments made celestial observations. An x-ray telescope attempted to map the x-ray emissions (2–20 KeV) of a number of galaxy clusters to find out whether the x-ray emissions came from individual galaxy nuclei or from the intergalactic gas in between. An infrared telescope studied the Milky Way's structure in four discreet wavebands from 4 to 120 microns, while a cosmic ray detector collected cosmic rays from 50 to 2,000 GeV.

The Plasma Diagnostics Package (PDP), a free-flying package of sensors, studied the wake and bow shock produced by the shuttle as it plowed its way through the earth's ionosphere and magnetic field. It was released with the robot arm on August 1st and recovered the next day.

In the life sciences, tests continued on crystal growth equipment, following **STS 51D (April 12, 1985)**. On this flight, the crystals had direct practical applications—large and pure mercury iodide crystals grown in space could be used in x-ray and gamma-ray detectors, both in hospitals and telescopes, at a much-reduced cost because they operated at room temperature, unlike other detector designs. As in earlier tests, the experiment proved that in space, large, pure crystals could be grown and at a rate much faster than expected. See **STS 26 (September 29, 1988)** for later research, and **STS 42 (January 22, 1992)** for a follow-up on the mercury iodide crystal research.

During STS 51F, the first astronaut conversations with ham radio operators took place, beginning a practice that has become a routine for almost all shuttle flights.

The next shuttle mission was **STS 51I (August 27, 1985)**.

August 27, 1985
STS 51I (STS 20), Discovery Flight #6
Aussat A1 (Optus A1) / ASC 1 / LEASAT 4
U.S.A. / Australia

The sixth flight of Discovery, STS 51I, lasted seven days and was crewed by Joe Engle (**STS 2–November 12, 1981**), James van Hoften (**STS 41C–April 6, 1984**), Richard Covey (**STS 26–September 29, 1988**; **STS 38–November 15, 1990**; **STS 61–December 2, 1993**), Mike Lounge (**STS 26–September 29, 1988**; **STS 35–December 2, 1990**), and William Fisher. One of the most demanding of all shuttle flights, STS 51I released three satellites and rescued a fourth.

• *Aussat A1* Deployed on August 27th soon after Discovery reached orbit, the satellite Aussat A1 (also called Optus A1) was Australia's first geosynchronous communications satellite. It was built by Hughes Communications for Aussat Proprietary Limited, a government-owned company created for this purpose. The satellite's design was similar to **Anik C3 (STS 5–November 11, 1982)** and **Telstar 3A (July 28, 1983)**.

Australia launched three Aussat A satellites through 1987 that provided direct broadcast services to homes; cable television distribution; long-distance telephone service from remote areas; as well as service for New Zealand, Papua New Guinea, and the southwest Pacific, servicing approximately one million people. Home antennas as small as four feet across could receive one television channel and three radio stations.

Because of the increasing demand for service, in 1992, Aussat supplemented the Aussat A satellites with a second-generation satellite system. See **Aussat B1 (August 14, 1992)**.

• *ASC 1* Designed by Contel American Satellite Company, ASC (American Satellite Company) 1 was the first satellite in a planned seventh cluster of privately owned geosynchronous communications satellites for serving the United States domestic market, following **Westar 1 (April 13, 1974)**, **RCA Satcom 1 (December 13, 1975)**, **Comstar 1 (May 13, 1976)**, **SBS 1 (November 15, 1980)**, **Galaxy 1 (June 28, 1983)**, and **Spacenet 1 (May 23, 1984)**.

The satellite was deployed from Discovery on August 27th, almost immediately after Aussat A1, marking the first time that two satellites were deployed by a shuttle on one day. It subsequently entered geosynchronous orbit over the eastern United States. ASC 1 was similar to the Spacenet satellites. It had a capacity for 1,000 voice circuits or one television signal, with a design life of about 10 years. The satellite was mostly used by government agencies (such as NASA and the Department of Defense) and large corporations.

Additional Contel ASC launches had been planned for 1986 but were delayed until 1991 due to a series of launch failures (including the Challenger explosion on **STS 51L–January 28, 1986**). By that time, Contel ASC had merged with General Telephone and Electronics (GTE); thus, its second satellite was renamed Spacenet 4 and was incorporated into that cluster.

ASC 1 represented the end of the first wave of private investment in space. This first wave had seen seven different companies successfully launch satellites into orbit, followed by shake-outs and mergers reducing the private players to five—Hughes Communications, AT&T, General Electric, MCI Communications, and GTE Sprint. All together, these companies orbited over 50 geosynchronous satellites through 1998, all providing service to the U.S. market and revolutionizing the telephone, television, and radio industries.

The second wave of new private investment in space began with the launch of the Iridium cluster of low-Earth-orbiting communications satellites. See **Iridium 4 through Iridium 8 (May 5, 1997)**.

• *LEASAT 4* Built and leased from Hughes Communications, Inc., LEASAT 4 was deployed on August 29th. It was the fourth of five military LEASAT communications satellites. See **STS 41D–August 30, 1984** for program information. Unlike the other satellites in the system, LEASAT 4 failed due to a faulty cable after only a few weeks in orbit.

Having successfully deployed three satellites, the astronauts then proceeded to rescue LEASAT 3, first deployed by **STS 51D (April 12, 1985)**. The satellite had been stranded dormant in low-Earth orbit because the computer program for operating its final-stage booster had been shut off accidentally during launch.

STS 51I's rescue attempt took three days. On the first day, Engle and Covey rendezvoused with the satellite, pulling to within 35 feet of it. On the second day, van Hoften and Fisher performed a seven-hour space walk to grab the satellite and secure it in the Discovery's cargo bay. Van Hoften, attached to the end of the robot arm, manually grabbed the satellite and held it. Lounge, inside the shuttle, then used the robot arm to guide van Hoften and the satellite into position in the cargo bay so that Fisher could install a new electronics package to bypass LEASAT 3's balky system. On the third day, the satellite was released, with van Hoften manually giving it a spin of about 3 revolutions per minute in order to stabilize it.

STS 51I carried one experiment, developed by the 3M company, producing vapor transport thin films with highly ordered crystalline structures. Follow-up research continued on **STS 26 (September 29, 1988)**.

The next shuttle flight was **STS 51J (October 3, 1985)**.

September 11, 1985
International Cometary Explorer (ICE)
U.S.A.
Originally launched as the **International Sun-Earth Explorer (ISEE) 3 (August 12, 1978)**, NASA renamed this probe the International Cometary Explorer (ICE) when its goal and orbit were changed so that, after five lunar flybys, it escaped Earth orbit and entered solar orbit to cut through the tail of Comet Giacobini-Zinner less than 5,000 miles from the comet's nucleus. This flyby, on September 11, 1985, was the first close approach to a comet, passing close to the dividing line between the comet's coma head and its tail.

Since it carried no imaging system, ICE could not take pictures of the comet. Its original design, to measure the solar wind and its interaction with the earth's magnetic field, was now used to investigate how the solar wind pulled and pushed at the dust cloud surrounding Comet Giacobini-Zinner's coma cloud. Scientists were surprised to discover significantly less cometary dust within a dust cloud that was three times larger than expected. The cloud itself appeared asymmetrical.

The spacecraft did show that the Sun's magnetic field lines draped around the comet, as predicted. It failed to definitively confirm a bow shock at the comet's head, although this region of collision between the solar wind and the comet's dust cloud was turbulent and complex. The spacecraft also detected many high-speed electrically charged particles at much greater distances from the comet than expected. These charged particles had been ionized by the Sun's ultraviolet radiation after escaping from the comet's nucleus.

The International Cometary Explorer ceased operations in May 1997. However, its present course will bring it back to Earth in the year 2014, where an additional gravity assist from the Moon will once again place it in Earth orbit. From this position, the spacecraft will actually be retrievable and, if recovered, has been promised to the Smithsonian's Air and Space Museum for display.

September 13, 1985
MHV
U.S.A.
The Miniature Homing Vehicle (MHV) was the United States military's answer to the Soviet anti-satellite (ASAT) system (see **Cosmos 252–November 1, 1968**). This 12-inch-diameter heat-seeking interceptor in the nosecone of a two-stage rocket was launched from the belly of an F-15 fighter at 50,000 feet. Its target was **Solwind P78-1 (February 24, 1979)**, a solar and gamma-ray research satellite orbiting at approximately 200 miles altitude.

Unlike the Soviet co-orbital ASAT, where the interceptor reaches orbit and then makes a rendezvous with its target, MHV was a direct-ascent ASAT. As it accelerated to speeds exceeding 30,000 miles per hour, its eight infrared telescopes and laser gyro guided the rocket precisely to its target, scoring a direct hit, shattering Solwind P78-1 into about 150 pieces.

September 17, 1985
Soyuz T14
U.S.S.R.
Soyuz T14 docked with **Salyut 7 (April 19, 1982)** after it was reactivated and repaired by Vladimir Dzhanibekov and Viktor Savinykh, the crew of **Soyuz T13 (June 6, 1985)**. This eight-day visiting mission brought Georgi Grechko (**Soyuz 17–January 11, 1975; Soyuz 26–December 10, 1977**), Alexander Volkov (**Soyuz TM7–November 26, 1988; Soyuz TM13–October 2, 1991**), and Vladimir Vasyutin to the laboratory, reuniting Savinykh with his original crewmates, Volkov and Vasyutin. These three men had trained together as the original long-term mission crew of Soyuz T13 and had been split up after the failure of Salyut 7 in February 1985. When the eight-day mission was over, Grechko and Dzhanibekov returned to Earth, the original Soyuz T13 crew proceeding with their original long-term mission.

One day after Dzhanibekov's and Grechko's return to Earth, **Cosmos 1686 (September 27, 1985)** was launched, the second Heavy Cosmos module to dock with Salyut 7, bringing with it over 3 tons of fuel and 5.5 tons of cargo. Unlike earlier Heavy Cosmos modules, **Cosmos 1267 (April 25, 1981)** and **Cosmos 1443 (March 2, 1983)**, Cosmos 1686 had no recoverable capsule. The absence of a capsule allowed the super-freighter to bring additional supplies to the recovering station, including a number of astronomical telescopes and research equipment.

Attempts to grow plants in space continued, with the cosmonauts succeeding in growing cotton. Tests were done on several older pieces of Salyut 7's solar panels, which were removed from the laboratory's exterior by Kizim and Solovyov on **Salyut T10B (February 8, 1984)**.

In the middle of November, Vladimir Vasyutin began to feel ill and experienced inflammation and a fever as high as 104°F. By November 21st, his illness had become so severe that it became necessary to quickly mothball Salyut 7 and return to Earth. Vasyutin was immediately taken to a hospital where he spent a month recovering.

Both Dzhanibekov and Savinykh of Soyuz T13 noted upon return to Earth that the station's condition was significantly affected by its period of deep freeze, indicating that the laboratory's useful life was nearing its end. Nonetheless, because many experiments had been left unfinished or unresolved due to Vasyutin's illness, a fifth and final occupancy was scheduled, **Soyuz T15 (March 13, 1986)**. This manned flight was the first and only (though December 1999) mission to occupy two orbiting laboratories, Salyut 7 and **Mir (February 20, 1986)**.

September 27, 1985
Cosmos 1686
U.S.S.R.
Cosmos 1686, a Heavy Cosmos module, docked with **Salyut 7 (April 19, 1982)** on October 2nd, bringing more than 3 tons of fuel and 5.5 tons of cargo. The cargo was used by both the **Soyuz T14 (September 17, 1985)** and **Soyuz T15 (March 13, 1986)** crews.

October 3, 1985

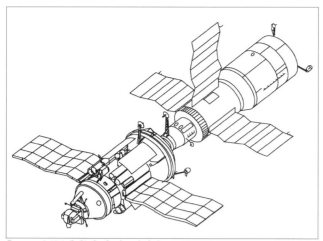

Cosmos 1686 *(left)* docked with Salyut *(right)*. NASA

Both Salyut 7 and Cosmos 1686 were destroyed when they re-entered the atmosphere together over Argentina on February 7, 1991.

October 3, 1985
STS 51J (STS 21), Atlantis Flight #1
U.S.A.

The mission STS 51J was the first for the space shuttle Atlantis, the fourth in the American fleet of reusable spaceships. STS 51J was the second American military space flight, following **STS 51C (January 24, 1985)**. It lasted four days and was crewed by Karol Bobko (**STS 6–April 4, 1983; STS 51D–April 12, 1985**), Ron Grabe (**STS 30–May 4, 1989; STS 42–January 22, 1992; STS 57–June 21, 1993**), David Hilmers (**STS 26–September 29, 1988; STS 36–February 28, 1990; STS 42–January 22, 1992**), Robert Stewart (**STS 41B–February 3, 1984**), and William Pailes.

The flight was the first and last shuttle mission to deploy satellites for the Defense Satellite Communication System III (DSCS). See **DSCS III (October 30, 1982)** for program details. Two satellites were released, each using a booster engine to reach geosynchronous orbit.

The next shuttle mission was **STS 61A (October 30, 1985)**. The next military shuttle flight was **STS 27 (December 2, 1988)**.

October 24, 1985
Meteor 3-01
U.S.S.R.

In 1988 and 1989, a third-generation Soviet weather satellite, Meteor 3-01, began replacing the Meteor 2 satellites (see **Meteor 2-01–July 11, 1985**). In order to guarantee coverage of the entire earth, Meteor 3 satellites were placed in a higher orbit—at an altitude of 750 miles instead of 550 miles. They were distributed in three different orbital planes separated by 120 degrees. Each satellite's scans covered a wider swathe in the equatorial regions, thereby eliminating any observational gaps. The satellites carried two cameras operating in both visible and infrared (9.65–18.7 microns) wavelengths. Besides cloud cover, the satellites could monitor ground and sea-surface temperature, snow and ice cover, and humidity.

From 1985 until 1994, a total of six Meteor 3 satellites were launched. These satellites were then augmented in geosynchronous orbit by **Elektro 1 (October 31, 1994)**.

October 25, 1985
Altair 1 (Cosmos 1700)
U.S.S.R.

Altair 1 was a geosynchronous communications satellite used by the Soviet space program and military, supplementing the Raduga satellite cluster (see **Raduga 1–December 22, 1975**). Through October 1995, two more Altair satellites were launched, then augmented by two Luch satellites (see **Luch 1–December 16, 1994**). All five satellites provided continuous communications between **Mir (February 20, 1986)** and the ground, much like NASA's TDRS satellites (see **STS 6–April 4, 1983**). In April 1999, however, the last of these satellites failed. As of December 1999, no clear plans exist to launch a replacement.

October 30, 1985
STS 61A (STS 22), Challenger Flight #9
GLOMR
U.S.A. / West Germany / Netherlands

STS 61A was the ninth space flight for the shuttle Challenger. It lasted seven days and was manned by a crew of eight. This crew size was a new record for the most humans sent into space by one vehicle, exceeding the record of seven first set by **STS 41G (October 5, 1984)**. As of December 1999, this record has only been matched but not exceeded. The crew included Henry Hartsfield (**STS 4–June 27, 1982; STS 41D–August 30, 1984**), Steven Nagel (**STS 51G–June 17, 1985; STS 37–April 5, 1991; STS 55–April 26, 1993**), James Buchli (**STS 51C–January 24, 1985; STS 29–March 13, 1989; STS 48–September 12, 1991**), Guion Bluford (**STS 8–August 30, 1983; STS 39–April 28, 1991; STS 53–December 2, 1992**), Bonnie Dunbar (**STS 32–January 9, 1990; STS 50–June 25, 1992; STS 71–June 27, 1995; STS 89–January 22, 1998**), Reinhard Furrer, Wubbo Ockels, and Ernst Messerschmid. Furrer and Messerschmid were Germans, and Ockels was from the Netherlands.

Only one satellite was deployed, using a getaway special canister.

• *GLOMR*   GLOMR, a military experimental communications satellite, tested several new digital technologies, including store-and-forward transmission of data (see **UoSat 2–March 1, 1984**) and the use of unattended ground stations. The first attempt to orbit the satellite from **STS 51B (April 17, 1985)** failed due to a battery problem. After deployment, GLOMR participated in orbital tests through the end of 1986.

Similar tests using equipment comparable to GLOMR's were performed on several other lightweight military satellites. See **Macsat 1 and 2 (May 9, 1990)**, **Microsat 1 through Microsat 7 (July 17, 1991)**, and **Secs (April 5, 1990)**.

STS 61A was the fourth flight of Spacelab, following **STS 51F (July 29, 1985)**. It carried 76 experiments funded by the West German government. To maintain a 24-hour operation, six of the eight crew members were divided into two teams, each of which worked a 12-hour shift. The remaining two crew members alternated between crews.

A number of experiments were performed to study the effects on the human body of acceleration in zero gravity. A sled attached to 12-foot-long rails ran down the center of Spacelab. Each day, four astronauts took a jolting sled ride to see if the experience increased or decreased their space sickness. Instead, the rides produced no effect at all.

Another experiment grew corn seeds, confirming what had been seen on **STS 3 (March 22, 1982)**: while the plant shoots grew correctly out of the soil, the roots developed in all directions, both in and out of the soil. When returned to Earth, however, the roots reoriented themselves and began growing downward again, into the ground. They apparently needed gravity to orient themselves correctly. See **STS 42 (January 22, 1992)** for later research.

Of Spacelab's 76 experiments, 48 studied materials processing or how different kinds of materials behaved in space, either when melted, solidified, or mixed. For example, the behavior of fluids was studied, and crystals of silicon, cadmium, proteins, and other semiconductor materials were grown. The results from this work mostly suggested design changes to improve later experiment performance.

The next flight of the German Spacelab configuration was STS 55 (April 26, 1993). The next shuttle mission was STS 61B (November 26, 1985).

November 26, 1985
STS 61B (STS 23), Atlantis Flight #2
Morelos 2 / Aussat A2 (Optus A2) / RCA Satcom Ku2 / OEX
U.S.A. / Mexico / Australia

This space flight was Atlantis's second, making it the fourth spaceship to fly into space and later return. The mission lasted just under seven days and was crewed by Brewster Shaw (**STS 9–November 28, 1983; STS 28–August 8, 1989**), Bryan O'Connor (**STS 40–June 5, 1991**), Mary Cleave (**STS 30– May 4, 1989**), Jerry Ross (**STS 27–December 2, 1988; STS 37–April 5, 1991; STS 55–April 26, 1993; STS 74–November 12, 1995; STS 88–December 4, 1998**), Charles Walker (**STS 41D–August 30, 1984; STS 51D–April 12, 1985**), Sherwood Spring, and Rodolfo Neri Vela. Vela was the first Mexican to fly in space. The mission was the ninth shuttle mission in 1985, completing the busiest shuttle year through 1999. STS 61B deployed four satellites.

• *Morelos 2* Deployed on November 27th, Morelos 2 was the second in Mexico's first cluster of geosynchronous communications satellites. See **STS 51G (June 17, 1985)**.

• *Aussat A2 (Optus A2)* Deployed on November 28th, Aussat A2 (also called Optus A2) was part of Australia's first geosynchronous communications satellite cluster. See **STS 51I (August 27, 1985)**.

• *RCA Satcom Ku2* Developed and owned by the Radio Corporation of America (RCA), RCA Satcom Ku2 was the first launch of the company's third-generation geosynchronous communications satellites, following the RCA Satcom 5 satellites (see **RCA Satcom 5–October 28, 1982**). The Satcom Ku2 satellites had twice the bandwidth and used the 12/14 GHz frequencies in the Ku-band (first tested on **Hermes–January 17, 1976**). Deployed on November 28th, this satellite helped augment the company's telephone and television broadcast capabilities for the U.S. market.

In 1986, RCA was purchased by General Electric (GE), which then took over the launching and operation of these satellites. GE subsequently placed in orbit one additional Satcom Ku-band satellite and replaced RCA's older-generation satellites (see **GE Satcom C1–November 20, 1990**).

• *OEX* Released on November 30th, OEX (Orbiter Experiment) was used as a target to test a new autopilot software package. Twice during the mission, the program maintained precise station-keeping between Atlantis and OEX.

STS 61B also tested two methods for building large structures in space. One method, called ACCESS, involved astronauts Ross and Spring in two space walks to assemble and then disassemble a 45-foot-high structure extending out of the cargo bay. Both men completed their tasks in about half the time scheduled. A second method, called EASE, required one astronaut to be in a foot restraint while the second moved about freely as both assembled the sections. Alternating tasks, the two men were able to erect and then take apart a 12-foot-high structure eight times, two more times than was planned.

McDonnell Douglas's Continuous Flow Electrophoresis System (CFES) made its fourth flight into space (see **STS 4– June 27, 1982**). As he had on **STS 41D (August 30, 1984)** and **STS 51D (April 12, 1985)**, Charles Walker operated the unit, which, on this mission, made the first in-space attempt to mass produce about one liter of raw hormone material that was then submitted to the Food and Drug Administration for testing. The material, erythropoietin, helps the body produce red blood cells and can be used by people with anemia. Tests on **Apollo 18 (July 15, 1975)** had indicated that electrophoresis could produce this material in space.

The 3M Company's Diffusive Mixing of Organic Solutions (DMOS) was also flown. DMOS had first been tested on **STS 51A (November 8, 1984)**. On STS 61B, the unit studied how the growth process took place and also produced four crystalline organic compounds. Based on this research, 3M decided that further work required an in-space period of at least 30 days.

An IMAX camera flew for the fourth time, shooting footage for the film *The Dream Is Alive*. Also onboard was a handheld, large-format camera used to take extensive pictures of Africa, including Ethiopia and Somalia, looking for surface evidence of underground water.

The one getaway special, designed by Telesat Canada, tested a method for putting metallic tungsten deposits on surfaces and, hence, for producing mirrors in space. The experiment also attempted to produce metallic tungsten crystals as the vapor cooled.

Besides overseeing the release of Mexico's satellite, astronaut Vela also conducted several life sciences experiments, testing whether acupuncture could mitigate the effects of space sickness and attempting to grow wheat, amaranth, and lentil seeds.

The next shuttle mission was **STS 61C (January 12, 1986)**.

1986

January 12, 1986
STS 61C (STS 24), Columbia Flight #7
GE Satcom
U.S.A.

STS 61C, the seventh flight of Columbia, lasted six days. It was crewed by Robert Gibson (STS 41B–February 3, 1984; STS 27–December 2, 1988; STS 47–September 12, 1992; STS 71–June 27, 1995), Charles Bolden (STS 31–April 24, 1990; STS 45–March 24, 1992; STS 60–February 3, 1994), George Nelson (STS 41C–April 6, 1984; STS 26–September 29, 1988), Steven Hawley (STS 41D–August 30, 1984; STS 31–April 24, 1990; STS 82–February 11, 1997), Franklin Chang-Diaz (STS 34–October 18, 1989; STS 46–July 31, 1992; STS 60–February 3, 1994; STS 75–February 22, 1996; STS 91–June 2, 1998), Robert Cenker, and Bill Nelson. Bill Nelson, who was a congressman from Florida, was the second American elected official to fly into space, following Jake Garn on STS 51D (April 12, 1985).

STS 61C deployed one satellite.

- **GE Satcom** Developed and owned by General Electric (GE), the satellite GE Satcom was part of the company's third-generation geosynchronous communications satellites. See **RCA Satcom 5 (October 28, 1982)** and **STS 61B (November 26, 1985)** for more details.

This satellite was the last privately financed communications satellite deployed by NASA's shuttle program. Following the Challenger explosion (STS 51L–January 28, 1986), the Reagan administration ruled that NASA was no longer to be involved in commercial launches. All subsequent private American communications satellites have been put in orbit using either privately financed rockets or the subsidized rocket programs of other countries. See **Marco Polo 1 (August 27, 1989)** for the subsequent first commercial launch on an American rocket.

STS 61C's onboard experiments included an equipment package of three materials-processing facilities, with most of the research consisting of checking out the design of the equipment; small failures limited results from two of three facilities. The acoustic levitator (previously tested on STS 41B–February 3, 1984) functioned successfully throughout the flight, observing how bubbles of glycerol water and silicone oil behave in weightlessness. Its next flight was STS 50 (June 25, 1992). The Automated Directional Solidification Furnace (ADSF), a package of four furnaces flown previously on STS 51G (June 17, 1985), worked for a total of only 2 hours and 20 minutes, confirming that as liquids solidify in weightlessness, there is an absence of convection. Thus, the resulting alloys were much more evenly mixed. ADSF's next flight was STS 26 (September 29, 1988).

Another experiment attempted to isolate the reasons for the limited shelf life (usually about 35 days) of human blood. Over time, the blood's components separate, with the heavier constituents settling to the bottom of the container. Identical blood samples, half flown in space and half kept on Earth as a control, were compared to see if weightlessness increased or reduced the blood's shelf life.

A package of NASA experiments tested several new engineering designs, including a camera to photograph the particle environment surrounding the shuttle, a heating system to use capillary action rather than pumps, and a test to see the effects of the space environment on coated mirrors.

Thirteen getaway specials included experiments by high school and college students from across the United States, investigating such topics as the behavior of liquids and solids in space, the effect of weightlessness on brine shrimp eggs hatched in space, the formation of paper in space, and the observation of the celestial ultraviolet background radiation between 350 and 6,000 angstroms.

One Department of Agriculture experiment confirmed earlier research showing that gypsy moth eggs hatched sooner in weightlessness, and that weightlessness shortened the moths' development, producing sterile moths. This research continued on STS 68 (September 30, 1994). Another experiment included engorged dog ticks to see how zero gravity affected them.

The next shuttle flight was STS 51L (January 28, 1986), the Challenger accident.

January 24, 1986
Voyager 2, Uranus Flyby
U.S.A.

Voyager 2, launched on August 20, 1977, had already made close approaches to Jupiter on July 9, 1979, and to Saturn on August 26, 1981 (see **Voyager 2, Jupiter Flyby–July 9, 1979** and **Voyager 2, Saturn Flyby–August 26, 1981**). On January 24, 1986, it made the first close approach to Uranus, passing about 50,000 miles from the planet's cloud tops *(see color section, Figure 6)*.

As it traversed the region surrounding the gas giant, it discovered that Uranus had a substantial magnetic field, which was comparable to the one found around Earth and Saturn. This discovery proved that the planet's interior, with a core of rocky ice and rock surrounded by a thick ocean of liquid ("molten") water, methane, and ammonia, was able to act like a conducting dynamo. Surprisingly, the magnetic field was tilted 60

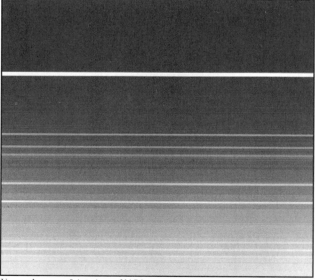

Uranus's many faint rings. *NASA*

degrees from the planet's axis of rotation, the largest tilt of any magnetic field in the solar system. This tilt caused the field's magnetotail to corkscrew away from the Sun, as if a garden hose were being twirled as it spewed water. The data also proved that the planet's rotation was about 17 hours long.

Because Uranus's axis of rotation was tilted 90 degrees, the planet rotated on its side as it circled the Sun, each pole spending half the year pointed at the Sun and half pointed away. The sunlit pole during Voyager 2's flyby was labeled the southern pole.

The spacecraft found that the planet's atmosphere were unlike Saturn's and Jupiter's in that it had few distinct features. Uranus's atmosphere of hydrogen and helium was cold, −350°F, or only 60°K above absolute zero. It was aqua blue in

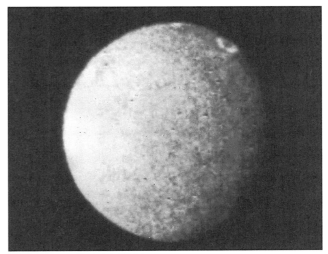
Uranus's moon Umbriel. *NASA*

Uranus's moon Titania, with craters and ridgeline indicating past tectonic activity. *NASA*

color, with faint bands encircling the planet's equator similar to those seen on the larger gas giants. A faint patch of brown could be seen at the sunlit pole. The existence of bands on Uranus, rotating on its side with the sunlight hitting one pole directly, suggested that such features are caused by a planet's rotation and not its absorption of sunlight *(see color section, Figure 17)*.

Wind speeds of 300 miles per hour were measured, but unlike all other known atmospheres in the solar system, they blew in the same direction as the planet's rotation. The winds were also strongest at the equator, also unlike any other planet.

Even more baffling was that Uranus's north pole, hidden in darkness for the last 40 years, was the warmest spot on the planet.

The planet's faint rings, first discovered in 1977 by Earth observations when a bright star passed behind them, were seen to be complex and many—similar to Saturn's rings, though much fainter and less dense. The outermost ring, dubbed Ep-

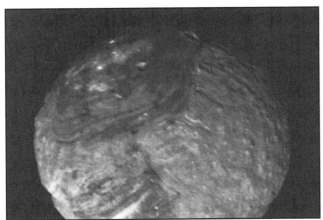
Uranus's moon Miranda. *NASA*

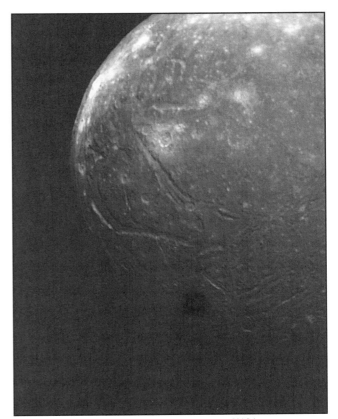
Uranus's moon Ariel, with grooves and fractures. *NASA*

January 28, 1986

silon, was inexplicably made up of rocks, fist-sized or larger, with no evidence of dust particles.

In its passage, Voyager 2 discovered seven more moons orbiting the planet, doubling the number known. These new objects were all between 20 and 30 miles in diameter and generally lurked in or near Uranus's ring system. Voyager 2 also took close-up pictures of the planet's five largest moons: Miranda, Ariel, Umbriel, Titania, and Oberon. The pictures revealed moons that were a mixture of rock and ice, much like most of Saturn's moons, with heavily cratered surfaces interspersed with gigantic cracks, canyons, and fault systems. All the moons showed evidence of violent internal geological activity, increasing in evidence at closer distances to the gas giant. Miranda, the innermost moon, revealed a strange patchwork of old cratered terrain, freshly melted water-lava plains, and terraced streaks reminiscent of earth-like glaciers.

As of December 1999, Voyager 2 is the only spacecraft to have reached this distant world, with no other missions presently planned.

Voyager 2 continued outward, flying past Neptune on August 25, 1989 (see **Voyager 2, Neptune Flyby–August 25, 1989**).

January 28, 1986
STS 51L (STS 25), Challenger Flight #10
U.S.A.

STS 51L is commonly referred to as the Challenger accident. Seventy-three seconds after launch, a leak in one of the spacecraft's two solid rocket boosters caused the destruction of Challenger, killing all seven crew members.

The flight was Challenger's tenth. The crew included Francis Scobee (**STS 41C–April 6, 1984**), Michael Smith, Ellison Onizuka (**STS 51C–January 24, 1985**), Judith Resnik (**STS 41D–August 30, 1984**), Ron McNair (**STS 41B–February 3, 1984**), Greg Jarvis, and Christa McAuliffe. McAuliffe was part of the Teacher-in-Space program and was the first ordinary civilian to fly into space. Their seven deaths not only crippled the American space program for more than two years but also doubled the number of casualties involved in the testing and flight of spacecraft. See **Apollo 1 (January 27, 1967)**, **Soyuz 1 (April 23, 1967)**, and **Soyuz 11 (June 6, 1971)**.

Also lost in the explosion was NASA's second Tracking and Data Relay Satellite (TDRS), intended to provide continuous communications between the ground and the shuttle (see **STS 6–April 4, 1983**), as well as a Spartan free-flying satellite (see **STS 51G–June 17, 1985**) designed to observe Halley's Comet.

The explosion occurred when an O-ring in the right solid rocket booster failed. The cold outdoor temperature of 36°F on launch day made the O-ring stiff, inflexible, and unable to maintain its seal. The subsequent leak of burning solid fuel ignited the hydrogen and oxygen fuel in the large External Fuel Tank, causing the explosion that destroyed the spacecraft.

Based on these findings, the shuttle program was overhauled, its solid rocket boosters redesigned. More than two years passed before the next shuttle flight. See **STS 26 (September 29, 1988)**.

A secondary decision by the Reagan administration ruled that the shuttle would no longer be used to launch private commercial satellites. This decision has had far-reaching im-

The Challenger crew *(back row, left to right)*: El Onizuka, Christa McAuliffe, Greg Jarvis, Judy Resnik; *(front row, left to right)* Mike Smith, Dick Scobee, Ron McNair. *NASA*

plications. While it caused a temporary delay in the development of an American commercial space industry and increased the launch business for the European Space Agency's Ariane rocket (see **Pan American Satellite 1/Meteosat 3/Oscar 13–June 15, 1988**), as well as the Japanese, and later Russian and Chinese space industries (see **Jindai/Ajisai/Fuji 1–August 12, 1986**; **IRS 1A–March 17, 1988**; **Asiasat 1–April 7, 1990**), it has also caused the creation of new private companies competing for the business of placing satellites in orbit inexpensively. See **Pegsat (April 5, 1990)**.

February 20, 1986
Mir
U.S.S.R.

Mir was the Soviet Union's sixth successful orbiting laboratory. Somewhat heavier than the previous station, **Salyut 7 (April 19, 1982)**, Mir weighed 22 tons and was actually the core module of a planned larger facility. In place of the front transfer compartment and docking port of previous Salyut stations, Mir carried a multidocking adapter, outfitted with five separate docking ports.

Soviet plans called for four separate modules to be launched over the next few years: **Kvant 1 (March 31, 1987)**, **Kvant 2 (November 26, 1989)**, **Kristall (May 31, 1990)**, **Spektr (May 20, 1995)**, and **Priroda (April 23, 1996)**. These would increase the laboratory's habitable space from its initial 3,000 cubic feet to just under 13,000 cubic feet and its total weight to about 110 tons, 10 tons heavier than **Skylab (May 14, 1973)** and about the same as an average space shuttle plus cargo. Because of the collapse and break-up of the Soviet Union in 1991, however, the launch of the last two modules was delayed, and in the end, Spektr and Priroda were completed and launched only because of funds from the United States as part of the Mir-Shuttle program (see **Soyuz TM21–March 14, 1995**).

Mir's solar panel array was also redesigned from Salyut 7. The panels were larger and had a greater ability to adjust their position to maximize the sunlight hitting their surface. They

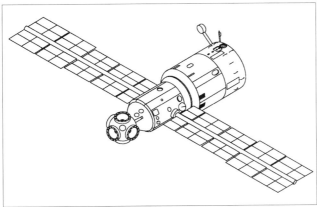

Mir's core module, with five docking ports at bow and one at stern. Compare with Salyut 6 (September 29, 1977) diagram. *NASA*

were also designed to have additional sections added during later manned flights. See **Soyuz TM2 (February 5, 1987)** and **Soyuz TM4 (December 21, 1987)**.

With this station, the Soviet/Russian space program attempted to maintain a permanent human presence in space, and except for two periods—July 17, 1986–February 4, 1987 and April 27, 1989–September 4, 1989—Mir was continuously occupied by at least one person during most of its 14-year existence. The last occupancy, which began on September 5, 1989, lasted almost a full decade, ending on August 27, 1999, with the return of the crew of **Soyuz TM29 (February 20, 1999)**.

Unlike previous Soviet stations (which depended on a limited network of ground stations and ships), communications were relayed from a cluster of orbiting communications satellites, beginning with two Altair satellites (see **Altair 1–October 25, 1985**) and later with three Luch satellites (see **Luch 1–December 16, 1994**).

Mir carried an allotment of scientific equipment comparable to that of the Salyut stations, including one furnace for growing crystals, two for doing alloy and materials experiments, and a fourth for testing semiconductor production. Also onboard was equipment for coating surfaces with silver in order to maintain the station's telescopes. Photographic equipment included the full array of multispectral cameras and spectrometers.

Because Mir's goal was to achieve a permanent human presence in space, crews rotated in and out of the station. Sometimes long-term occupancies overlapped, while in other cases individual crew members stayed onboard and other crews came and went. The manned missions to Mir are as follows, with crew members listed according to the spacecraft they launched on, and the numbers in parentheses indicating the approximate number of days each remained in space.

First occupancy (54 days)
Soyuz T15–March 13, 1986 to May 5, 1986

Second occupancy (21 days)
Soyuz T15–June 26, 1986 to July 16, 1986.

Third occupancy (2 years, 84 days)
Soyuz TM2–February 5, 1987
  Romanenko to December 29, 1987 (326 days)
  Levykin to July 29, 1987 (176 days)
Soyuz TM3–July 22, 1987
  Viktorenko to July 29, 1987 (8 days)
  Faris to July 29 1987 (8 days)
  Alexandrov to December 29, 1987 (161 days)
Soyuz TM4–December 21, 1987
  Levchenko to December 29, 1987 (8 days)
  Titov to December 21, 1988 (366 days)
  Manarov to December 21, 1988 (366 days)
Soyuz TM5–June 7, 1988
  A. Solovyov to June 17, 1988 (10 days)
  Savinykh to June 17, 1998 (10 days)
  A. P. Alexandrov to June 17, 1998 (10 days)
Soyuz TM6–August 29, 1988
  Lyakhov to September 7, 1988 (10 days)
  Mohmand to September 7, 1988 (10 days)
  Polyakov to April 27, 1989 (242 days)
Soyuz TM7–November 26, 1988
  Volkov to April 27, 1989 (153 days)
  Krikalev to April 27, 1989 (153 days)
  Chretien to December 21, 1988 (26 days)

February 20, 1986

Fourth occupancy (8 days less than 10 years)

Soyuz TM8–September 5, 1989
    Viktorenko to February 19, 1990 (166 days)
    Serebrov to February 19, 1990 (166 days)
Soyuz TM9–February 11, 1990
    A. Solovyov to August 9, 1990 (179 days)
    Balandin to August 9, 1990 (179 days)
Soyuz TM10–August 1, 1990
    Manakov to December 10, 1990 (132 days)
    Strekalov to December 10, 1990 (132 days)
Soyuz TM11–December 2, 1990
    Afanasyev to May 26, 1991 (176 days)
    Manarov to May 26, 1991 (176 days)
    Akiyama to December 10, 1990 (8 days)
Soyuz TM12–May 18, 1991
    Artsebarsky to October 10, 1991 (145 days)
    Krikalev to March 25, 1992 (313 days)
    Sharman to May 26, 1991 (8 days)
Soyuz TM13–October 2, 1991
    Volkov to March 25, 1992 (175 days)
    Aubakirov to October 10, 1990 (8 days)
    Viehboeck to October 10, 1990 (8 days)
Soyuz TM14–March 17, 1992
    Viktorenko to August 10, 1992 (146 days)
    Kaleri to August 10, 1992 (146 days)
    Flade to March 25, 1992 (8 days)
Soyuz TM15–July 27, 1992
    A. Solovyov to February 1, 1993 (189 days)
    Avdeyev to February 1, 1993 (189 days)
    Tognini to August 10, 1992 (14 days)
Soyuz TM16–January 24, 1993
    Manakov to July 22, 1993 (196 days)
    Poleshchuk to July 22, 1993 (196 days)
Soyuz TM17–July 1, 1993
    Tsibliyev to January 14, 1994 (195 days)
    Serebrov to January 14, 1994 (195 days)
    Haignere to July 22, 1993 (22 days)
Soyuz TM18–January 8, 1994
    Afanasyev to July 9, 1994 (183 days)
    Usachev to July 9, 1994 (183 days)
    Polyakov to March 22, 1995 (439 days)
Soyuz TM19–July 1, 1994
    Malenchenko to November 4, 1994 (127 days)
    Musabayev to November 4, 1994 (127 days)
Soyuz TM20–October 4, 1994
    Viktorenko to March 22, 1995 (169 days)
    Kondakova to March 22, 1995 (169 days)
    Merbold to November 4, 1994 (32 days)
Soyuz TM21–March 14, 1995
    Dezhurov to July 7, 1995 (116 days)
    Strekalov to July 7, 1995 (116 days)
    Thagard to July 7, 1995 (116 days)
STS 71–June 27, 1995
    Gibson to July 7, 1995 (10 days)
    Precort to July 7, 1995 (10 days)
    Baker to July 7, 1995 (10 days)
    Dunbar to July 7, 1995 (10 days)
    Harbaugh to July 7, 1995 (10 days)
    A. Solovyov to September 11, 1995 (76 days)
    Budarin to September 11, 1995 (76 days)
Soyuz TM22–September 3, 1995
    Gidzenko to February 29, 1996 (179 days)
    Avdeyev to February 29, 1996 (179 days)
    Reiter to February 29, 1996 (179 days)
STS 74–November 12, 1995
    Cameron to November 20, 1995 (8 days)
    Halsell to November 20, 1995 (8 days)
    Ross to November 20, 1995 (8 days)
    McArthur to November 20, 1995 (8 days)
    Hadfield to November 20, 1995 (8 days)
Soyuz TM23–February 21, 1996
    Onufrienko to September 2, 1996 (193 days)
    Usachev to September 2, 1996 (193 days)
STS 76–March 22, 1996
    Chilton to March 30, 1996 (9 days)
    Searfoss to March 30, 1996 (9 days)
    Godwin to March 30, 1996 (9 days)
    Clifford to March 30, 1996 (9 days)
    Sega to March 30, 1996 (9 days)
    Lucid to September 26, 1996 (188 days)
Soyuz TM24–August 17, 1996
    Korzun to March 2, 1997 (197 days)
    Kaleri to March 2, 1997 (197 days)
    Andre-Deshays to September 2, 1996 (16 days)
STS 79–September 16, 1996
    Readdy to September 26, 1996 (10 days)
    Wilcutt to September 26, 1996 (10 days)
    Akers to September 26, 1996 (10 days)
    Apt to September 26, 1996 (10 days)
    Walz to September 26, 1996 (10 days)
    Blaha to January 22, 1997 (129 days)
STS 81–January 12, 1997
    Baker to January 22, 1997 (10 days)
    Jett to January 22, 1997 (10 days)
    Wisoff to January 22, 1997 (10 days)
    Grunsfeld to January 22, 1997 (10 days)
    Ivins to January 22, 1997 (10 days)
    Linenger to May 24, 1997 (133 days)
Soyuz TM25–February 10, 1997
    Tsibliyev to August 14, 1997 (185 days)
    Lazutkin to August 14, 1997 (185 days)
    Ewald to March 2, 1997 (20 days)
STS 84–May 15, 1997
    Precourt to May 24, 1997 (9 days)
    Collins to May 24, 1997 (9 days)
    Clervoy to May 24, 1997 (9 days)
    Noriega to May 24, 1997 (9 days)
    Tsang Lu to May 24, 1997 (9 days)
    Kondakova to May 24, 1997 (9 days)
    Foale to October 6, 1997 (145 days)
Soyuz TM26–August 5, 1997
    A. Solovyov to February 19, 1998 (198 days)
    Vinogradov to February 19, 1998 (198 days)
STS 86–September 25, 1997
    Wetherbee to October 6, 1997 (11 days)
    Bloomfield to October 6, 1997 (11 days)
    Parazynski to October 6, 1997 (11 days)

Titov to October 6, 1997 (11 days)
Chretien to October 6, 1997 (11 days)
Lawrence to October 6, 1997 (11 days)
Wolf to January 31, 1998 (129 days)

STS 89–January 22, 1998
Wilcutt to January 31, 1998 (9 days)
Edwards to January 31, 1998 (9 days)
Reilly to January 31, 1998 (9 days)
Anderson to January 31, 1998 (9 days)
Dunbar to January 31, 1998 (9 days)
Sharipov to January 31, 1998 (9 days)
Thomas to June 12, 1998 (142 days)

Soyuz TM27–January 29, 1998
Musabayev to August 25, 1998 (207 days)
Budarin to August 25, 1998 (207 days)
Eyharts to February 19, 1998 (22 days)

STS 91–June 2, 1998
Precourt to June 12, 1998 (10 days)
Gorie to June 12, 1998 (10 days)
Lawrence to June 12, 1998 (10 days)
Chang-Diaz to June 12, 1998 (10 days)
Kavandi to June 12, 1998 (10 days)
Ryumin to June 12, 1998 (10 days)

Soyuz TM28–August 13, 1998
Padalka to February 28, 1999 (200 days)
Avdeyev to August 27, 1999 (380 days)
Baturin to August 25, 1998 (12 days)

Soyuz TM29–February 20, 1999
Afanasyev to August 27, 1999 (188 days)
Haignere to August 27, 1999 (188 days)
Bella to February 28, 1999 (8 days)

With the return of Soyuz TM29, Mir was unoccupied for the first time in almost 10 years. As of December 1999, Russian plans are to deorbit the station sometime in the spring of 2000. See Soyuz TM29 for details.

The next space station, the International Space Station, is under construction. See **Zarya (November 20, 1998)**.

February 22, 1986
Viking / Spot 1
Sweden / France / Europe

• *Viking* Sweden's first satellite, Viking studied the earth's aurora and the electromagnetic environment of the polar regions. Launched on an Ariane rocket, it worked in conjunction with the two American satellites **Dynamics Explorer 1/Dynamics Explorer 2 (August 3, 1981)**. The satellite operated until May 1987, helping to provide simultaneous observations of both poles. The next Swedish auroral research satellite was **Freja (October 6, 1992)**. See also **Sampex (July 3, 1992)**.

• *Spot 1* Developed by the French space program in cooperation with Belgian and Swedish scientists, Spot 1 was the first civilian high-resolution photo-reconnaissance satellite, similar to the KH satellites in the American Keyhole surveillance program (see **Discoverer 36–December 12, 1961**; **KH-4 9032–April 18, 1962**; and **KH-9 1–June 15, 1971**). The spacecraft provided the same earth resource data as the Landsat satellites (see **Landsat 1–July 23, 1972** and **Landsat 4–July 16, 1982**), but at much higher resolution. For security and political reasons, the resolution of the Landsat cameras had always been limited to no better than about 100 feet. Spot 1's cameras can see objects as small as 60 feet across, and its black-and-white pictures resolve objects as small as 30 feet.

Unlike Landsat 4's seven spectral filters, however, Spot 1 only has filters in two visible bands (0.5–0.59 microns and 0.61–0.68 microns) and one near-infrared band (0.79–0.89 microns). It lacks the mid-infrared bands for studying geological features and tropical moisture changes, and the thermal band for recording surface heat variations. And though it has a better resolution, each image covers less than one-tenth the area. Because Spot 1 carries two duplicate imaging systems, however, it is able to take three-dimensional stereo images.

For reasonable fees (from $250 to $1,790 in 1986), anyone can purchase any picture in Spot 1's inventory. Furthermore, on a fee basis, the satellite can be ordered to take a requested picture. For example, its cameras were rented by American news organizations to photograph the Chernobyl nuclear reactor immediately after the reactor meltdown in April 1986.

Spot 1 was augmented with the launch of two identical Spot satellites through 1993. A fourth satellite, **Spot 4 (March 24, 1998)**, carries a second-generation imaging system.

The next Earth resource satellite was **Momo (February 19, 1987)**.

March 6, 1986
Vega 1, Halley's Comet Flyby
U.S.S.R. / International

First launched on December 15, 1984, Vega 1 was the first of a two-satellite Soviet-led international mission to explore both Venus and Halley's Comet. Vega 1, along with its twin, Vega 2, had flown past Venus on June 10, 1985, each probe deploying a lander and a balloon, then using Venus's gravity to redirect their flight towards Halley's Comet. See **Vega 1, Venus Flyby (June 10, 1985)** and **Vega 2, Venus Flyby (June 14, 1985)** for details.

On March 6th, Vega 1 passed within 5,525 miles of the comet's nucleus, cutting across the leading sunward part of its large dust cloud or coma. As it passed, its photographs revealed an irregular potato-shaped nucleus about 5 by 9 miles in size, and that the heated material being released to form the coma cloud did not emanate outward from the nucleus's entire surface, but from discrete sources. Jets could be seen erupting from the daylight side of the nucleus. None were seen coming from the night side. These images implied that the comet's surface was covered with a thin frozen mantle, less than one half inch thick, with a temperature of about –100°F, surrounded by a coma cloud at between 80°F and 260°F. The jets or plumes were the equivalent of small "volcanoes" of water vapor and dust.

Combining data from the Vega 2 flyby of the comet **(Vega 2, Halley's Comet Flyby–March 9, 1986)**, scientists estimated that the nucleus rotated once every 53 hours, give or take three hours. One side appeared more active than the other, since Vega 2 saw a different hemisphere when it made its close approach, thereby observing fewer dust plumes.

The overall brightness of the comet's nucleus was very low, similar to the dark C type asteroids, considered to be the most ancient objects in the solar system, formed in its earliest history. The dense heart of the coma cloud itself appeared to be about 100,000 miles across.

The comet's interior, made mostly of the primitive substances thought to make up the C type asteroids—ice, carbon, silicon, sodium, magnesium, calcium, and iron—was heated by the Sun. This material was thrown out from the nucleus in jets on the daylight side to form the coma and then was turned back by the solar wind to trail away from the Sun, forming the comet's tail. In general, the data confirmed results from **IUE (January 26, 1978)**, proving that comets were like dirty snowballs.

March 8, 1986
Suisei, Halley's Comet Flyby
Japan

Launched on August 18, 1985, Suisei ("comet" in Japanese) was the second of two interplanetary probes sent by Japan to study Halley's Comet; the other probe was **Sakigake (January 7, 1985, March 11, 1986)**. Suisei flew past Halley's at a distance of less than 94,000 miles, passing between the Sun and the comet in order to study the collusion of the solar wind and the comet's coma and how that interaction caused the comet's tail. During its flyby, Suisei produced ultraviolet images of the comet's coma cloud for measuring the cloud's hydrogen gas.

Suisei found that the hydrogen cloud had a shell structure and varied in brightness from day to day. Like Vega 1 (**Vega 1, Halley's Comet Flyby–March 6, 1986**) and Vega 2 (**Vega 2, Halley's Comet Flyby–March 9, 1986**), Suisei found that the brightest sections corresponded to one side of the nucleus, indicating that discrete regions of eruption caused the comet's cloud, with the remainder of the comet's surface coated by a hard mantle. Suisei's data indicated that the eruptions were coming from two strong sources and four weak ones, concentrated on about two-thirds of the comet's surface.

Suisei also found that the comet's release of water vapor increased sharply near perihelion and then continued at a much higher rate afterward.

The spacecraft passed through the bow shock between the solar wind and the coma at about 280,000 miles. There, it measured a sudden and distinct drop in solar wind speed—from about one million miles per hour to about 120,000 miles per hour. The wind direction also changed, as the solar wind was forced to slide around the comet's coma cloud.

Suisei's estimate of the nucleus's rotation was more accurate than that of Vega 1 and Vega 2, measuring it as about 52.9 hours long.

March 9, 1986
Vega 2, Halley's Comet Flyby
U.S.S.R. / International

Identical to Vega 1, Vega 2 had been launched on December 21, 1984, and had flown past Venus on June 14, 1985 (see **Vega 2, Venus Flyby–June 14, 1985**), dropping off a lander and a balloon before heading toward its rendezvous with Halley's Comet. See **Vega 1, Halley's Comet Flyby (March 6, 1986)** for results.

March 11, 1986
Sakigake, Halley's Comet Flyby
Japan

Launched on January 7, 1985, Sakigake was Japan's first interplanetary probe (see **Sakigake–January 7, 1985**). It successfully flew within 4.3 million miles of Halley's Comet, passing between the Sun and the comet in order to monitor solar wind during the comet's passage. Throughout the close flybys of Halley's Comet by other spacecraft, Sakigake provided a database of the solar wind's velocity and direction just prior to the wind's impact with the coma cloud.

During its own distant flyby, Sakigake detected discrete bursts of x-ray radiation emanating from the comet. Scientists suspected that these bursts were produced by fluctuations of the solar wind as it hit the comet's coma cloud. Sakigake also found that the neutral sheet of the Sun's magnetic field, the equatorial dividing line between its oppositely charged northern and southern hemispheres, undulated up and down past Sakigake several times during this period. Scientists thought that this turbulence was a distant reflection of the collision between the solar wind and the coma cloud at the bow shock.

Using these data, scientists estimated that the bow shock was between 250,000 and 320,000 miles in front of the comet nucleus during Sakigake's closest approach, confirming the measurements from **Suisei (March 8, 1986)**.

March 13, 1986
Giotto, Halley's Comet Flyby
Europe

Launched on July 2, 1985, Giotto was a project of the European Space Agency. Its goal was to fly as close to the nucleus of Halley's Comet as possible, and it used information from the previous flybys of **Vega 1 (March 6, 1986)**, **Suisei (March 8, 1986)**, **Vega 2 (March 9, 1986)**, and **Sakigake (March 11, 1986)** to adjust its course accordingly.

On March 13, it flew only 370 miles from the nucleus of Halley's Comet. Fourteen seconds before closest approach, it was hit by a large particle that shook the spacecraft, damaged some instruments, and caused a 32-minute period when data were transmitted back to Earth only intermittently. This impact also prevented the spacecraft from photographing the comet on its outward journey.

Giotto's pictures confirmed what the Vega 1 and Vega 2 images had shown—Halley's nucleus was a single elongated potato-like object. Giotto found, however, that it appeared larger than indicated by Vega 1 and Vega 2, about 7 by 10 miles in size. Two distinct dust jets were spotted emanating from the daylight hemisphere, and most of the coma seemed to come from these specific eruptions. Other surface features that Giotto identified included circular structures reminiscent of impact craters and hill-like mounds. As noted by Vega 1 and Vega 2, the surface was much darker than expected.

The spacecraft's sensors confirmed that water was the dominant component of Halley's Comet. The dust cloud was mostly hydrogen, carbon, nitrogen, oxygen, and other basic elements, with many molecules of water, carbon, carbon dioxide, and

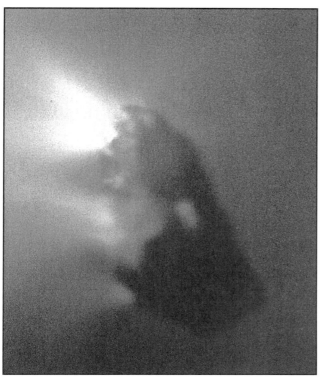

The nucleus of Halley's Comet. Both craters and mountains can be seen, as well as specific areas where jets are emanating. *ESA*

other materials also present. Although the cloud's make-up was close to what had been predicted, it did not precisely match the carbonaceous chondrite make-up of most meteorites, since it contained eight times as much carbon. This result surprised the scientists, leaving them puzzled as to the origin of meteorites and as to why comets and carbonaceous chondrite meteorites, both thought to be primitive solar system objects, were not alike.

Giotto located the bow shock at about 600,000 miles from the nucleus, about twice as far as measured by the Japanese spacecraft. It also found that the Sun's magnetic field lines managed to penetrate inward to a distance of 3,000 miles, where what scientists dubbed the "contact surface" formed. Inside this zone, Giotto was within the comet's atmosphere, where no particles from the solar wind could penetrate, and the spacecraft recorded no evidence of the Sun's magnetic field.

More than a dozen years passed before the next mission to a comet. See **Stardust (February 7, 1999)**.

March 13, 1986
Soyuz T15
U.S.S.R.

Soyuz T15, a 125-day mission, was crewed by Leonid Kizim (**Soyuz T3–November 27, 1980**; **Soyuz T10B–February 8, 1984**) and Vladimir Solovyov (**Soyuz T10B–February 8, 1984**). This was the first crew to occupy **Mir (February 20, 1986)** and the last crew to occupy **Salyut 7 (April 19, 1982)**, making them the first humans to perform operations on two different space laboratories.

The flight's initial goal was to activate Mir. The cosmonauts docked with the station and spent six weeks checking out its systems. They tested the stability of the laboratory's natural resonance by shaking the station in unison, much as had been done by the **Soyuz 27 (January 10, 1978)** cosmonauts on **Salyut 6 (September 29, 1977)**. They also tested the use of a geosynchronous satellite, **Altair 1 (October 25, 1985)**, for communications. Two Progress freighters (see **Progress 1–January 20, 1978**) docked with the station during their initial occupancy, testing Mir's automatic docking system and its ability to be refueled. The laboratory's atmosphere was tested repeatedly to make sure its carbon dioxide scrubbers were working properly.

On May 5th, after six weeks on Mir, Kizim and Solovyov undocked and maneuvered to **Salyut 7 (April 19, 1982)**, where they then spent another seven weeks wrapping up the experiments left behind because of the hasty departure of the Soyuz T13/Soyuz T14 crew (see **Soyuz T14–September 17, 1985**).

Kizim and Solovyov performed two space walks while occupying Salyut 7. During both, they experimented with assembling and using a 50-foot truss structure, not unlike that tested on **STS 61B (November 26, 1985)**. On the first EVA (extra-vehicular activity), they tested unfolding and folding the truss, then retrieved a French micrometeoroid detector for observing space particles from Halley's Comet, as well as samples being tested in the environment of space.

On the second EVA, they once again assembled the frame, this time attaching a detector at its far end to measure the truss's vibrations. They then proceeded to use the welding tool first tested during space walks on **Soyuz T12 (July 17, 1984)**.

For another three weeks, Kizim and Solovyov alternated between making Earth observations and packing experiment samples, film, and equipment in their Soyuz spacecraft.

On June 25th, they undocked from Salyut 7 and returned to Mir for another three weeks in space, bringing with them 20 experiments from the older laboratory. During their second occupancy of Mir, the astronauts continued checking out the new station and installed a new computer system. They also tested the laboratory's medical equipment, including its exercise equipment. They then returned to Earth on July 16, 1986, after 126 days in space.

This mission was the last manned visit to Salyut 7, which burned up in the atmosphere over Argentina on February 7, 1991. The next mission to Mir was **Soyuz TM2 (February 5, 1987)**.

May 21, 1986
Soyuz TM1
U.S.S.R.

Soyuz TM1, a nine-day unmanned flight, tested a modified Soyuz spacecraft designed to operate with **Mir (February 20, 1986)**. During its flight, it docked automatically with the laboratory and stayed linked for seven days during a period when the crew of **Soyuz T15 (March 13, 1986)** was occupying **Salyut 7 (April 19, 1982)**.

The Soyuz TM had an improved power system, a better altimeter for its descent module, and an enhanced propulsion system. It could carry 440 additional pounds of cargo into orbit and return to Earth with three times more cargo (about 330 pounds) than earlier Soyuz models.

The next Soyuz TM flight, **Soyuz TM2 (February 5, 1987)**, was manned.

August 12, 1986
Jindai (MABES) / Ajisai (EGP) / Fuji 1 (Oscar 12; JAS 1)
Japan

Jindai, Ajisai, and Fuji 1, a launch of three satellites, was the first test flight of Japan's H1 rocket, designed to provide Japan with a competitive launch capability able to put 3.5 tons into low-Earth orbit and a little more than 1 ton in geosynchronous orbit.

- *Jindai*   Also called the MAgnetic Bearing flywheel Experimental System (MABES), Jindai tested the use of a magnetic bearing flywheel in weightlessness. It also gathered telemetry for the test launch of the H1 rocket.

- *Ajisai*   Also called the Experimental Geophysical Payload (EGP), Ajisai (a Japanese flower) was a geodetic satellite. By reflecting ground-based lasers off the satellite, the locations of Japan's remote islands were more precisely determined. Moreover, the data helped link the local geodetic map system of Japan to that of the rest of the world.

- *Fuji 1 (Oscar 12)*   Also called JAS 1 (for Japanese Amateur Satellite), Fuji 1 was built by Japanese amateur radio enthusiasts and was the 12th in the Oscar series of satellites (see **Oscar 1–December 12, 1961**), following **UoSat 2 (March 1, 1984)**. Riding piggyback on the H1 rocket, Fuji 1 carried several channels for allowing ham radio operators to relay their messages worldwide. A battery failure ended operations in 1989. Japanese amateur radio operators then followed with a second Fuji satellite in 1990 and a third in 1996.

The next Oscar satellite was **Oscar 13 (June 15, 1988)**.

November 14, 1986
Polar Bear
U.S.A.

Polar Bear, an Air Force satellite, performed propagation experiments in the polar regions, testing the effect of the atmosphere on five frequencies ranging from 138 MHz to 1,239 MHz, the same frequencies tested on **HiLat (June 27, 1983)**. The U.S. Air Force used the data from both satellites to improve communications in the high latitudes.

December 18, 1986
Ionosonde (Cosmos 1809)
U.S.S.R. / Poland

The satellite Ionosonde carried experiments built by Polish and Soviet scientists for monitoring the earth's ionosphere and magnetic field at polar latitudes, working in conjunction with **Dynamics Explorer 1/Dynamics Explorer 2 (August 3, 1981)** and **Prognoz 10 (April 26, 1985)**. Ionosonde gathered data until the summer of 1987.

More than two years passed before the next ionospheric research satellite, **Akebono (February 21, 1989)**, was launched. This was the longest break in such research since the launch of **Sputnik (October 4, 1957)**. Almost three years passed before the next Soviet solar-magnetosphere satellite, **Intercosmos 24 (September 28, 1989)**, was launched.

1987

February 5, 1987, 6:28 GMT
Ginga (Astro-C)
Japan

Ginga ("galaxy" in Japanese) carried an x-ray telescope (covering the 1–20 KeV energy range) and a gamma-ray burst detector (1–400 KeV). It followed **Exosat (May 26, 1983)** and was the only operational x-ray telescope in orbit at launch. The satellite focused on studying the x-ray emissions of active galactic nuclei, as well as transient x-ray bursters. Launched only 19 days before the eruption of SN1987A in the Large Magellanic Cloud, the first supernova explosion visible to the naked eye since 1604, Ginga was able to observe the powerful x-rays emitted by it. Its data helped map the nature of the explosion's gas cloud.

Ginga operated for four years, during which it made about 1,000 observations of about 350 different celestial objects. Among its discoveries were two very bright x-ray transients, named GS 2023+338 and GS 2000+25, as well as numerous weak transient x-ray pulsars along the galactic plane.

Ginga also carried a gamma-ray burst detector. Although monitoring of gamma-ray bursts, which had been discovered by **Vela 5A/Vela 5B (May 23, 1969)**, had now continued for almost 20 years (see **ISEE 3–August 12, 1978**), no data yet explained their origin. See the next x-ray telescope, **Granat (December 1, 1989)**, for later gamma-ray burst research.

The next Japanese x-ray telescope was **ASCA (February 20, 1993)**.

February 5, 1987, 21:38 GMT
Soyuz TM2
U.S.S.R.

Soyuz TM2 was the second flight to **Mir (February 20, 1987)** following **Soyuz T15 (March 13, 1986)**, and the first dedicated exclusively to the orbiting laboratory. Its crew included Yuri Romanenko (**Soyuz 26–December 10, 1987**; **Soyuz 38–September 18, 1980**) and Alexander Leveykin.

Romanenko remained on Mir for 326 days, breaking the record of 237 days set by the crew of **Soyuz T10B (February 9, 1984)**. Leveykin, however, had difficulty adapting to the weightlessness. He soon began having heart problems and returned to Earth with the crew of **Soyuz TM3 (July 22, 1987)** on July 29, 1987, after 176 days in space. He was replaced by Alexander Alexandrov from Soyuz TM3.

During Romanenko's 11 months in space, seven different Progress freighters (see **Progress 1–January 20, 1978**) brought the station fuel, supplies, and additional equipment for experiments. A new furnace for growing crystals in space was brought in March, along with a wide-angle camera for multispectral Earth photography. In November, a second crystal furnace arrived, along with a sensor for detecting high-energy particles from deep space. This same freighter performed a second docking, testing new control software designed to reduce fuel consumption.

Also launched during Romanenko's occupancy was the second permanent module to the Mir complex, **Kvant 1 (March 31, 1987)**. The module rendezvoused with the laboratory on April 5th but could not dock. After several days of discussion, a second docking attempt was tried on April 9th, this time achieving a soft but not a hard dock. To find out what the problem was, Romanenko and Leveykin performed an unscheduled space walk on April 11th, working their way to the stern docking port of Mir where Kvant 1 hung. There, they discovered a fabric sheet entangled with the docking system. Apparently, this material had been attached to the last Progress freighter to use the port, and when it had undocked, the fabric had been left behind. The cosmonauts pulled the material free, and ground controllers were then able to bring Kvant 1 in for a hard docking.

Using Kvant 1's x-ray telescope-spectrometers, the two cosmonauts made measurements of SN1987a, a supernova that had exploded in the Large Magellanic Cloud in February 1987—the nearest to occur since the development of the telescope by Galileo. Their observations provided measurements of the high-energy spectrum of the explosion's initial blast wave.

On June 12th and June 16th, Romanenko and Leveykin conducted space walks to install extensions to Mir's solar panel arrays, brought to the station by Kvant 1.

On **July 22, 1987**, **Soyuz TM3** was launched, bringing to Mir three more astronauts, including Alexandrov—who replaced Leveykin.

February 19, 1987
Momo 1 (MOS 1)
Japan

Momo ("peach blossom" in Japanese) 1, also called MOS (Marine Observation Satellite) 1, was Japan's first Earth observation satellite. It was placed in a sun-synchronous orbit (see **Nimbus 1–July 28, 1964**), carrying sensors similar to, though less sensitive than, the French satellite **Spot 1 (February 22, 1986)**. Momo 1's multispectral radiometers scanned swathes either 60, 115, 200, or 930 miles wide in infrared and visible wavelengths. The narrowest swathe could resolve objects about 150 feet across.

The data were used to study sea color and temperature changes, as well as to monitor agriculture, forestry, aquatic life, sea ice, and ocean currents surrounding Japan.

The next Earth resource satellite was India's **IRS 1A (March 17, 1988)**. Japan also replaced Momo 1 in 1990 with a second Momo satellite.

March 31, 1987
Kvant 1
U.S.S.R.

Kvant 1 was the first add-on module to **Mir (February 20, 1986)**. It weighed about 20 tons, bringing the weight of the complex to about 42 tons, excluding Soyuz and Progress spacecraft. Kvant 1 mostly studied high-energy astronomical phenomena. It carried both an x-ray telescope-spectrometer and a high-energy scintillation telescope-spectrometer operating in the energy range from 2 to 30 KeV. Several of these instru-

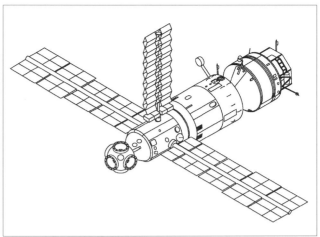

Kvant 1 docked with Mir's aft docking port. Compare with Mir (February 20, 1986). The third solar panel attached to Mir's top was brought to the complex by Kvant 1, and installed during later manned missions. *NASA*

ments were developed in cooperation with the European Space Agency.

In its initial rendezvous and docking with Mir on April 5th, it failed to dock, and a space walk was required by the crew of Soyuz TM2 to solve the problem. See **Soyuz TM2 (February 5, 1987)**.

The module was built with two docking ports, front and rear. It linked with Mir's rear docking port where Progress freighters (see **Progress 1–January 20, 1978**) had previously docked to refuel the station. Subsequent tankers docked in Kvant 1's second port, their fuel being pumped through Kvant 1 and into Mir.

Kvant 1 also carried a new system for orienting Mir, using 10 gyroscopes similar in concept to those first tested on **PAC 1 (August 9, 1969)** and used on **Skylab (May 14, 1973)**. Instead of firing rocket thrusters to adjust the laboratory's orientation, the gyroscopes spun to hold the station in position by the angular momentum. Kvant 1 also brought a set of add-on solar panels to be attached to Mir's main panels. See Soyuz TM2 for details about their installation.

The next add-on module to dock with Mir was **Kvant 2 (November 26, 1989)**.

July 22, 1987
Soyuz TM3
U.S.S.R. / Syria

This eight-day internationally crewed mission brought Soviets Alexander Viktorenko (**Soyuz TM8–September 5, 1989; Soyuz TM14–March 17, 1992; Soyuz TM20–October 4, 1994**) and Alexander Alexandrov (**Soyuz T9–June 27, 1983**) and Syrian Mohammed Faris to **Mir (February 20, 1986)**. They joined Yuri Romanenko and Alexander Leveykin from **Soyuz TM2 (February 5, 1987)**.

Five men occupied Mir for the first time. Experiments during this joint flight included using the Mir furnaces to grow crystals of gallium antimonide and to produce and study an alloy of aluminum and nickel. Multispectral images of Syria were taken. Routine exercise and medical tests continued.

One experiment continued electrophoresis research begun on **Soyuz T9 (June 27, 1983)**, attempting to produce very pure samples of interferon and anti-influenza medication. This work was similar to electrophoresis research being done on American space shuttle flights (see **STS 4–June 27, 1982**).

Viktorenko, Faris, and Leveykin returned to Earth on July 29th in the Soyuz TM2 spacecraft, leaving their fresher Soyuz TM3 craft in which Romanenko and Alexandrov could return. Leveykin was brought home early because his heart had begun beating irregularly during the last few weeks. Though he recovered fully, it took his body longer than usual to return to normal.

With the end of the joint flight on July 29th, and with five more months in their mission, Romanenko and Alexandrov settled down to the normal Soviet five-day work week in space, with two hours a day reserved for exercise.

They continued the materials processing experiments, testing the synthesis of a polyacrylamide gel for use in biology work. They studied the dust layer at 60 miles altitude. They tended their small greenhouse of onions and other vegetables.

In December, the two cosmonauts increased their daily exercise period to 2.5 hours per day in anticipation of their return to Earth. Then, **Soyuz TM4 (December 21, 1987)** was launched to Mir, carrying a crew of three and making possible the first transfer of the station's operation from two entirely different crews.

Romanenko and Alexandrov returned to Earth on December 29, 1987, having spent 326 and 161 days in space, respectively. Romanenko's condition upon landing was good—he was able to sit up in the rescue helicopter and actually walk away from it. One day later, he completed a 100-meter run, demonstrating that the human body can recover from long periods in weightlessness.

July 25, 1987
Cosmos 1870
U.S.S.R.

Cosmos 1870 was dedicated to providing radar images of the Soviet Union, much of which was unobservable in visible wavelengths because of heavy cloud cover. Radar imagery, for example, allowed researchers to track ice conditions in the Arctic Ocean north of Russia. Cosmos 1870 followed the **STS 41G (October 5, 1984)** radar mission and was the second orbiting radar mission, following **Seasat (June 27, 1978)**, to be placed permanently in orbit.

The spacecraft's initial design had been similar to the three military Salyut stations: the failed Salyut 2, **Salyut 3 (June 25, 1974)**, and **Salyut 5 (June 22, 1976)**. A bureaucratic competition between two different factions within the Soviet space program, however, caused the Almaz ("diamond" in Russian) design to be abandoned. Although the government ordered the junking of already-built models, two nearly completed station modules were carefully placed in storage and were resurrected later as radar imaging satellites. Hence, Cosmos 1870 weighed 20 tons, the same as the Salyut stations. Instead of carrying passengers, however, it was reconfigured to carry two synthetic-aperture radar sensors (see Seasat) for radar mapping the earth's surface. Both radars produced images 25 miles wide and as much as 150 miles long, resolving objects as small as 80 feet across. The image length was obtained by literally rolling the spacecraft on its axis as it took a radar scan. Though this technique consumed a great deal of fuel, Cosmos 1870's large size permitted it to launch with a gigantic fuel supply—over three tons. In addition, the spacecraft used gyroscopes for attitude control, similar to those developed for **Mir (February 20, 1986)**. See **Kvant 1 (March 31, 1987)**.

During its two years of operation, Cosmos 1870 made radar observations of large-scale ocean waves and currents, geological formations, and agricultural output of the Soviet Union.

The next orbiting radar imager was **Magellan (August 10, 1990)**, which produced a detailed radar map of Venus. The next radar imaging satellite aimed at Earth was **Almaz 1 (March 31, 1991)**, which was also the last Soviet radar imaging satellite.

August 5, 1987
China 20
China / France

China 20, the 20th satellite launched by China, carried research experiments for the French company Matra. The launch was the first time a Chinese satellite was used by another country for private commercial work. After five days in orbit, the capsule re-entered the earth's atmosphere and was recovered successfully. Matra's experiments studied the effects of zero gravity on photosynthesis in algae. The Chinese experiments studied whether an yttrium-barium-copper alloy could be produced in weightlessness.

Matra continued its experiments on the Soviet Union's **Foton 5 (April 26, 1989)**. The next Chinese international commercial research satellite was **China 23 (August 5, 1988)**.

August 27, 1987
Kiku 5 (ETS 5)
Japan

Kiku 5, also called ETS (Engineering Test Satellite) 5, was an experimental communications satellite placed into geosynchronous orbit over Japan. This flight was not only the second test flight of the H1 Japanese rocket and the first test of its upper stage, it was also the first Japanese working communications satellite using an attitude-control system maneuverable in three directions. It also used deployable solar panels, tested on the previous ETS mission, **Kiku 4 (September 3, 1982)**.

Kiku 5 performed tests of maritime communications from Japanese fishing vessels and aircraft over the Pacific Ocean. Further tests of mobile receivers/transmitters on the ground were also performed. The satellite completed its main research in March 1989, though an experimental transponder is still available for propagation experiments in December 1999.

The next successful Japanese Engineering Test Satellite was **Kiku 7 (November 28, 1997)**. The next use of the H1 rocket placed the first of Japan's third-generation communications satellites, **Sakura 3A (February 19, 1988)**, into geosynchronous orbit.

September 29, 1987
Bion 8 (Cosmos 1887)
U.S.S.R. / International

Bion 8, a biological research satellite, was part of a Soviet program begun with **Bion 1 (October 31, 1973)**, following **Bion 7 (July 10, 1985)**. It carried 2 monkeys, 12 white rats, and a number of fish and insect specimens into space for 13 days. Researchers from the Soviet Union, the United States, and five other countries, plus the European Space Agency, participated.

Of the two monkeys, only one lost weight. Nonetheless, the results from all the mammal specimens indicated that adolescent vertebrates experienced serious negative effects from exposure to weightlessness. For example, the strength of the rats' bones was reduced anywhere from 27 to 40 percent, while their heart muscles showed significant atrophy and degeneration. The rats also showed increased cholesterol, triglycerides, and organ weight in the liver, as well as a reduced immune response.

The next Bion satellite was **Bion 9 (September 15, 1989)**.

December 21, 1987
Soyuz TM4
U.S.S.R.

This nine-day mission carried Vladimir Titov (**Soyuz T8–April 20, 1983; Soyuz T10A–September 27, 1983; STS 63–February 3, 1995; STS 86–September 25, 1997**), Musa Manarov (**Soyuz TM11–December 2, 1990**), and Anatoly Levchenko to **Mir (February 20, 1986)**, who joined Yuri Romanenko (from **Soyuz TM2–February 5, 1987**) and Alexander Alexandrov (from **Soyuz TM3–July 22, 1987**) as those two men finished up their long-endurance flight.

Titov and Manarov were to set a new space endurance record of 366 days, or one year and one day in space, breaking the record of 326 days that had just been set by Romanenko.

Levchenko, however, remained in space for only nine days to become familiar with weightlessness. He had been trained in a separate program to fly the Soviet Union's space shuttle (see **Buran–November 15, 1988**), and after the difficulties that Alexander Leveykin had experienced with zero gravity during **Soyuz TM2 (February 5, 1987)**, the Soviets decided

Musa Manarov with his children. *RIA Novosti*

that Levchenko's training required at least one short flight into space prior to his piloting their shuttle.

For nine days, Romanenko and Alexandrov prepared Titov and Manarov for the transfer of Mir. They then returned to Earth on December 29, 1987, bringing Levchenko with them. See **Soyuz TM3 (July 22, 1987)** for details about Romanenko and Alexandrov's return to Earth.

With the departure of Romanenko, Alexandrov, and Levchenko, Titov and Manarov settled down to the normal work routine of a Soviet long-term space flight. They exercised about two hours a day; performed experiments in materials processing, earth resources, medicine, and biology; and unloaded supplies from five Progress freighters (see **Progress 1–January 20, 1978**). They were also visited by three Soyuz crews, **Soyuz TM5 (June 7, 1988)**, **Soyuz TM6 (August 29, 1988)**, and **Soyuz TM7 (November 26, 1988)**, the last of which took control of the station from them.

Titov and Manarov also performed three space walks during their year in space. The first, on February 26, 1988, finished the solar panel installation work begun by Romanenko and Leveykin of Soyuz TM2 while also replacing a faulty panel with a more advanced design. The last two space walks, on June 30 and October 20, 1988, performed repairs to Mir's x-ray telescope. Titov and Manarov also retrieved exterior sample packages for study on Earth.

Manarov, Titov, and Chretien (of Soyuz TM7) returned to Earth on December 21, 1988. See both Soyuz TM6 and Soyuz TM7 for further details.

Vladimir Titov in Soyuz capsule. *RIA Novosti*

1988

February 19, 1988
Sakura 3A
Japan

Built and launched by Japan's National Space Development Agency (NASDA), Sakura 3A was the first of a two-satellite constellation. It was Japan's third generation of geosynchronous communications satellites, replacing the Sakura 2 satellites (see **Sakura 2A–December 15, 1977**). Sakura 3A was also the first operational use of the H1 rocket, previously tested with **Kiku 5 (August 27, 1987)**.

Sakura 3 satellites had increased reliability and capacity; telephone conversation capacity had been increased from 700 to 4,000 two-way conversations. Electrical power generation was also increased by 50 percent with the use of gallium-arsenide solar cells rather than silicon cells—the first time any communications satellite had used this technology.

While the previous two generations had provided just public and government telephone service, 20 percent of the Sakura 3 satellites' capacity was allocated to testing business communication concepts.

The Sakura satellites were replaced in 1995 by the N-Star satellites. See **N-Star A (August 29, 1995)**.

March 17, 1988
IRS 1A
India / U.S.S.R.

Following the launch of two short-term experimental earth resource satellites in 1979 and 1981 (see **Bhaskara 1–June 7, 1979**), India developed the fully operational India Remote Sensing (IRS) satellites, of which IRS 1A was the first. Built in India but launched on a Soviet rocket, IRS 1A was also the first official Soviet commercial launch.

IRS 1A took photos in four spectral bands, three visible and one near-infrared, matching the three visible and the shortest infrared bands of **Landsat 4 (July 16, 1982)**. IRS 1A's images covered a 92- by 92-mile area with a resolution of about 240 feet.

Through December 1999, a total of six IRS satellites have been launched. All were placed in sun-synchronous orbits (see **Nimbus 1–August 28, 1964**), photographing the entire Earth's surface every 22 to 24 days. Later satellites carried tape recorders, allowing for data acquisition (even when the satellite was out of contact with ground stations), increased resolution (as low as 75 feet in visible light), and a second infrared band replacing one visible band. The lost visible band was compensated for by the addition of two more camera systems, a panchromatic system and a wide field sensor.

The panchromatic camera covered the upper two-thirds of the visible spectrum, from 0.5 to 0.75 microns, taking color photographs covering a 50-mile swathe with a resolution of less than 33 feet, the best resolution of any commercial color remote-sensing camera to this date.

The wide field system takes two simultaneous images, one in visible and one in infrared wavelengths. The images cover a wide area, 500 miles across, with a resolution of less than 620 feet, and are used to monitor global changes in vegetation and plant life.

The next Earth resource satellite was **JERS 1 (February 11, 1992)**. See **IRS-P4 Oceansat (May 26, 1999)** for India's next-generation Earth resource satellite.

March 25, 1988
San Marco D/L
Italy / West Germany / U.S.A.

San Marco D/L was the eighth and last satellite launched from the Italian San Marco launch platform floating off the coast of Kenya (see **San Marco 2–April 26, 1967**). The satellite included experiments from Italy, West Germany, and the United States, and remained in orbit until December 1988. This satellite was also the first scientific satellite in which the United States participated after the Challenger accident (see **STS 51L–January 28, 1986**).

During its eight months of operation, San Marco D/L studied the earth's ionosphere at equatorial latitudes, including wind and temperature variations, along with atmospheric density changes in low-Earth orbit, continuing research last done by **Explorer 55 (November 25, 1975)**. It also carried sensors for monitoring the Sun's ultraviolet output.

Pion 1 and 2 (**Pion 1/Pion 2–May 25, 1989**) were the next satellites to monitor atmospheric density changes. For the next satellite launch from a floating ocean platform, see **DemoSat (March 27, 1999)**.

June 7, 1988
Soyuz TM5
U.S.S.R. / Bulgaria

Soyuz TM5, a 10-day internationally crewed flight to **Mir (February 20, 1986)**, was manned by Soviets Anatoly Solovyov (**Soyuz TM9–February 11, 1990; Soyuz TM15–July 27, 1992; STS 71–June 27, 1995; Soyuz TM26–August 5, 1997**) and Viktor Savinykh (**Soyuz T4–March 12, 1981; Soyuz T13–June 6, 1985**), and Bulgarian Alexander P. Alexandrov (not to be confused with Soviet cosmonaut Alexander Alexandrov of **Soyuz TM2/Soyuz TM3–February 5, 1987**). They joined Vladimir Titov and Musa Manarov of **Soyuz TM4 (December 21, 1987)** during their year-long occupancy of Mir, by then almost half completed.

During the eight-day joint flight, the five cosmonauts conducted more than 40 experiments focusing on three research areas. In life sciences, they experimented with several new exercise routines to improve an individual's condition while weightless. They grew protein crystals and attempted to produce extremely pure hormones by electrophoresis. This work was similar to that performed on the space shuttle (see **STS 4–June 27, 1982**). In materials processing, an alloy of aluminum-copper-iron was attempted. Finally, Earth resource research included three multispectral scans of Bulgaria.

Solovyov, Savinykh, and Alexandrov left their Soyuz TM5 spacecraft behind and returned to Earth in the older Soyuz TM4 spacecraft, landing in the Soviet Union on June 17, 1988.

The next spacecraft to visit Titov and Manarov was **Soyuz TM6 (August 29, 1988)**.

June 15, 1988
Pan American Satellite 1 / Meteosat 3 / Oscar 13
U.S.A. / Europe

Three satellites—Pan American Satellite 1, Meteosat 3, and Oscar 13—were launched commercially by the European Space Agency's (ESA's) Ariane rocket.

• *Pan American Satellite 1*   Pan American Satellite 1 was the first privately owned international communications satellite, built by RCA Astro Electronics and owned by Pan American Satellite. It was placed in geosynchronous orbit above the Atlantic Ocean and focused on providing telephone and television communication between the United States and South America, competing directly with **Intelsat** (see **Early Bird–April 6, 1965**).

Pan American Satellite launched an additional seven communications satellites through December 1999, placing four over the Atlantic, one over the Pacific, and one over the Indian Ocean—with two spares. Later satellites carried capacity for as many as 320 radio and 120 direct television channels.

Later satellites, launched in the late 1990s, were also the first commercial satellites to use a xenon ion engine to maintain their orientation in space. See **Pan American Satellite 5** (**August 28, 1997**).

• *Meteosat 3*   The third Meteosat satellite launched (see **Meteosat 1–November 23, 1977**), Meteosat 3 was part of the global network of geosynchronous weather satellites begun with **GOES 1** (**October 16, 1975**).

Meteosat 3, built by ESA, was stationed over Europe for most of its service life. Because of the launch failure of a U.S. GOES satellite in 1986, however, and because the second Meteosat satellite exceeded its life expectancy, Meteosat 3 was available as a spare. It was, therefore, initially moved over the Western Atlantic in 1988 to replace the lost GOES satellite.

• *Oscar 13*   Oscar 13 was the fifth satellite built by AMSAT (see **Oscar 6–October 15, 1972**). As with the previous AMSAT satellite, **Oscar 10** (**June 16, 1983**), West German AMSAT members played the central role in designing and building Oscar 13. Resembling Oscar 10, but with greater capacity, it was placed in an elliptical orbit similar to that used by the Soviet Molniya communications satellites (see **Molniya 1-01–April 23, 1965**). Its apogee was chosen to provide the widest Earth coverage to the largest number of AMSAT members. It operated through 1996, when its orbit decayed and it burned up in the atmosphere.

The next Oscar satellites were a package of six satellites: **Oscars 14 through Oscar 19** (**January 22, 1990**).

July 7, 1988
Phobos 1
U.S.S.R. / Europe

The first Soviet spacecraft sent to Mars in 15 years (following the failures of **Mars 4, Mars Flyby–February 10, 1974; Mars 5, Mars Orbit–February 12, 1974; Mars 6, Mars Landing–March 12, 1974; and Mars 7, Mars Flyby–March 9, 1974**),

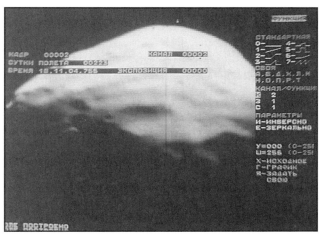

The Martian moon Phobos, just prior to the loss of signal from Phobos 2. *RIA Novosti*

Phobos 1 was one of two identical probes—its twin was **Phobos 2** (**July 12, 1988**). Each carried two landers and an orbiter. The landers were aimed at the Martian moon Phobos. Both spacecraft were built in cooperation with the Soviet Union's Eastern bloc and the European Space Agency.

Phobos 1 functioned normally for two months. Then, an upload of incorrect software caused the spacecraft to lose its lock on the Sun, causing its solar panels to drift and thereby drain its batteries. Contact was never restored.

July 12, 1988
Phobos 2
U.S.S.R. / Europe

Identical in design to **Phobos 1** (**July 7, 1988**), Phobos 2 also carried one orbiter and two landers. It entered Mars orbit on January 29, 1989. According to the probe's main mission, it would pass in this orbit within 175 feet of the Martian moon Phobos, where it would then release its two landers.

As it made this final approach to Phobos on March 27, 1989, however, a malfunction of its onboard computer caused the ground to lose contact with the spacecraft.

Prior to this failure, the spacecraft did detect a surprisingly high number of ionized particles leaving Mars's atmosphere in a stream flying away from the planet on its night side. These particles, made up of hydrogen and oxygen atoms, had been ionized by the Sun's ultraviolet radiation. Phobos 2 found that the rate of escape for these atoms was large enough to have contributed significantly to the depletion of the Martian atmosphere over time.

Since Mars seemed to lack a magnetic field to direct this flow, however, scientists were puzzled about what caused it. The goal of the Japanese probe **Nozomi** (**July 3, 1998**) is to solve this mystery.

The next attempt to visit Mars was **Mars Observer** (**September 25, 1992**).

August 5, 1988
China 23
China / West Germany

Similar to **China 20** (**August 5, 1987**), China 23 was a recoverable research satellite carrying experiments by foreign commercial companies. In this case, the West German company Intospace paid China to send 104 vials of protein crystals to test several crystal growth processes. Molecular maps of the grown crystals were later used to improve methods for making medicines such as interferon.

Although this satellite was the last Chinese recoverable research satellite to carry foreign experiments, the Chinese have continued their program, launching five more satellites through December 1999. The satellites conducted either materials research, remote Earth sensing, or surveillance work. Most were successfully recovered after 1–3 weeks in orbit.

August 29, 1988
Soyuz TM6
U.S.S.R. / Afghanistan

Soyuz TM6 was the second international visit to **Mir** (**February 20, 1986**), following **Soyuz TM5** (**June 7, 1988**). It carried Soviets Vladimir Lyakhov (**Soyuz 32–February 25, 1979**; **Soyuz T9–June 27, 1983**) and Valery Polyakov (**Soyuz TM18–January 8, 1994**), as well as Abdul Ahad Mohmand of Afghanistan for an eight-day mission. They joined Vladimir Titov and Musa Manarov, who had been occupying Mir for eight months since their launch on **Soyuz TM4** (**December 21, 1987**). Polyakov was a physician, and his job was to remain on Mir and monitor the health of Titov and Manarov as they completed a full year in space.

Most of the research during Soyuz TM6 involved extensive multispectral observations of Afghanistan, gathering remote-sensing data on oil, gas, and ore deposits.

On September 6th, Lyakhov and Mohmand entered the Soyuz TM5 capsule to return to Earth, leaving Polyakov and their fresher Soyuz TM6 behind. After undocking and ejecting the Soyuz orbital module, however, a computer failure caused their first retro-rocket burn to misfire. The cosmonauts waited two orbits and tried again manually. The burn once again misfired. The two men then had to wait an additional 24 hours before trying a third time in a cramped descent module with no food and only enough battery power to last 48 more hours. The third attempt on September 7th worked, with Lyakhov and Mohmand landing safely in the Soviet Union.

In Mir, Titov and Manarov continued their year in space, with Polyakov there to monitor their health. Their next visitors were the crew members of **Soyuz TM7** (**November 26, 1988**).

September 6, 1988
Feng Yun 1A
China

Feng Yun 1A ("wind cloud" in Chinese) was China's first weather satellite. It was placed in a sun-synchronous orbit (see **Nimbus 1–August 28, 1964**) and took downloadable multispectral images across five wavebands from 0.58 to 12.5 microns, observing clouds patterns, ocean color, and ice and snow conditions. Scans were 1,800 miles wide, and objects as small as 2.5 miles across were resolvable.

Because of attitude control problems, Feng Yun 1A operated for only 38 days. It was followed by a second, identical weather satellite in September 1990, which functioned for more than a year. A third weather satellite, **Feng Yun 2B** (**June 10, 1997**), was China's first geosynchronous meteorological satellite.

September 19, 1988
Ofeq 1
Israel

Ofeq 1 ("horizon" in Hebrew) was Israel's first satellite, launched on its own rocket, making it the eighth country to place a spacecraft in orbit around the earth, following India (see **Rohini 1B–July 18, 1980**). The launch took place south of Tel-Aviv on the Mediterranean coast. The satellite was an engineering test of both rocket and satellite. It operated for four months, re-entering the atmosphere in January 1989.

The next Israeli satellite was **Ofeq 2** (**April 3, 1990**).

September 29, 1988
STS 26, Discovery Flight #7
TDRS 3
U.S.A.

STS 26 was the first shuttle flight after the Challenger accident (**STS 51L–January 28, 1986**) and the seventh flight of the shuttle Discovery. This four-day mission focused on verifying the safety of the redesigned shuttle and solid rocket boosters. Its crew included Fred Hauck (**STS 7–June 18, 1983**; **STS 51A–November 8, 1984**), Richard Covey (**STS 51I–August 27, 1985**; **STS 38–November 15, 1990**; **STS 61–December 2, 1993**), Mike Lounge (**STS 51I–August 27, 1985**; **STS 35–December 2, 1990**), George Nelson (**STS 41C–April 6, 1984**; **STS 61C–January 12, 1986**), and David Hilmers (**STS 51J–October 3, 1985**; **STS 36–February 28, 1990**; **STS 42–January 22, 1992**). This was the first American all-veteran crew since **Apollo 11** (**July 16, 1969**).

STS 26 deployed one satellite.

- **TDRS 3** TDRS (Tracking and Data Relay Satellite) 3 was developed by NASA to provide communications with its orbiting spacecraft. See **STS 6** (**April 4, 1983**). TDRS 3 replaced the satellite lost in the Challenger accident.

STS 26 also carried a large number of scientific experiments. This research was now the shuttle's main focus, since after Challenger's explosion, the shuttle no longer could be used to launch private commercial satellites. See STS 51L.

The vapor transport experiment developed by 3M and flown previously on **STS 51I** (**August 27, 1985**) once again tested methods for producing a variety of thin organic films. The Automated Directional Solidification Furnace (ADSF), which had flown twice before (see **STS 51G–June 17, 1985** and **STS 61C–January 12, 1986**), again attempted to perform melting and solidification experiments in its four furnaces. As had been done on **STS 51C** (**January 24, 1985**), blood samples from individuals ill with heart disease, hypertension, diabetes, and cancer were flown to see the effect of weightlessness on the red blood cells.

Two experiments tested methods for separating different biological samples. The Isoelectric Focusing Experiment used electrophoresis to attempt to separate samples of proteins, hemoglobin, and albumen using electricity, normally impossible on Earth. This work was similar to McDonnell Douglas's research begun on **STS 4 (June 27, 1982)**. The Phase Partitioning Experiment, tested first on **STS 51D (April 12, 1985)**, studied how different biological samples that naturally separated on Earth behaved in zero gravity.

In protein crystal growth experiments, nine different proteins were used to grow 60 different crystals. Previous tests (see **Soyuz 38–September 18, 1980**, and **STS 51F–July 29, 1985**) had shown that in space, the crystals grew much faster and larger than on Earth. Once produced, scientists used these larger crystals to study the structure of the molecules, which in turn helped them understand how the proteins interact in biology. Samples included enzymes related to AIDS, fungicides, and emphysema, as well as samples of interferon and the first totally synthetic peptide.

The experiment was successful in producing three different crystals that were purer and larger than ever seen before: gamma-interferon (a protein which stimulates the immune system and is used to treat some cancers), isocitrate lyase (a plant enzyme), and elastase (an enzyme that had been linked to the destruction of lung tissue in emphysema patients). All three crystals were returned to Earth, where scientists used them to better map each crystal's structure.

The next protein crystal growth experiment was **STS 29 (March 13, 1989)**.

The Earth Limb Radiance test took photographs of the earth's horizon line from orbit. Satellites routinely used the earth's limb as a reference for orientation. The photos provided satellite builders better baseline data for designing the sensors that did this work.

STS 26 also carried two student experiments. One tested how the strength of a titanium-aluminum alloy was affected by being melted and resolidified in space. The second experimented with growing lead iodide crystals.

The next shuttle flight was **STS 27 (December 2, 1988)**.

October 28, 1988
TDF 1
France / West Germany / Scandinavia

TDF 1 (named for Telediffusion de France, France's national broadcasting company) was part of a joint program among France, West Germany, and five Scandinavian countries to develop satellites for broadcast television signals directly to home use. The program successfully launched two satellites for France, one for West Germany, and one for the Nordic countries. The German satellite was known as TV-Sat, while the Nordic satellite was called Tele-X.

All four satellites were essentially the same. The German and French satellites could carry five television stations simultaneously, while Tele-X could carry three. As of December 1999, three still operate, though two have exceeded their expected service life.

November 15, 1988
Buran
U.S.S.R.

Buran ("blizzard" in Russian) was the Soviet version of the American space shuttle. This unmanned orbital test mission

Buran on the launchpad. *RIA Novosti*

was its only flight. Weighing 80 tons, it made two orbits of the earth before it made an unpowered glider landing in the Soviet Union.

Looking almost identical in design to the U.S. shuttle, Buran had been built for use with **Mir (February 20, 1986)**. Due to lack of funds, however, the program was canceled in 1993. Although three more shuttles had been planned, none were ever finished.

November 26, 1988
Soyuz TM7
U.S.S.R. / France

Soyuz TM7 carried Soviets Alexander Volkov (**Soyuz T14–September 17, 1985; Soyuz TM13–October 2, 1991**) and Sergei Krikalev (**Soyuz TM12–May 18, 1991; STS 60–February 3, 1994; STS 88–December 4, 1998**), and French cosmonaut Jean-Loup Chretien (**Soyuz T6–June 24, 1982; STS 86–September 25, 1997**) to **Mir (February 20, 1986)**. They joined Vladimir Titov and Musa Manarov from **Soyuz TM4 (December 21, 1987)** and Valery Polyakov from **Soyuz TM6 (August 29, 1988)** for 26 days, during which Titov and Manarov completed their record-setting, year-long mission in space before handing over occupancy of Mir to Volkov and Krikalev.

During this joint flight, Volkov and Chretien performed one space walk, during which they assembled and disassembled a 12- by 12- by 19-foot test structure, work similar to that done on **STS 61C** (**November 26, 1985**) and **Soyuz T15** (**March 13, 1986**). Getting the structure to unfold required a kick from one of the men. With this EVA (extra-vehicular activity), Chretien became the first non-American and non-Soviet to walk in space.

In another experiment, integrated circuits were exposed to the vacuum of space. The test found that unless the circuits were carefully shielded, the space environment easily damaged them.

Chretien's flight was announced by the Soviets as the last "free" guest cosmonaut mission. Future foreign visitors to Mir were expected to pay commercial rates for the space laboratory's use. This decision was an effort by the Soviets to make their space program a profitable operation.

Titov, Manarov, and Chretien returned to Earth on December 21, 1988, using new re-entry software developed because of the difficulties experienced by **Soyuz TM6** (**August 29, 1988**). A software bug, however, caused their landing to be delayed by two orbits.

Chretien's 26-day flight was the longest mission for any non-Soviet and non-American. For Titov and Manarov, this completed one year and one day in space, a new endurance record, exceeding the record set by Yuri Romanenko of **Soyuz TM2** (**February 5, 1987**). After landing, both men recovered quickly from their year in space. An extensive program of exercise returned them to normal within two months. While Titov had lost eight pounds during the flight, Manarov had gained five. Both had developed significant muscle atrophy in their legs, their calves reduced by nearly 20 percent. This loss, however, was not the worst seen during the Soviet long-term space flight program.

Polyakov, Volkov, and Krikalev, meanwhile, remained on Mir. They returned on April 27, 1989, in the Soyuz TM7 capsule, with Polyakov completing eight months in orbit and Volkov and Krikalev five months. During this period, three Progress freighters (see **Progress 1–January 20, 1978**) brought them several tons of supplies.

The cosmonauts performed 200 experiments. They grew a variety of semiconductor crystals; performed electrophoresis experiments, separating biological samples; and took extensive Earth observation images of the Soviet Union. And using the x-ray telescope on **Kvant 1** (**March 31, 1987**), they recorded observations of several x-ray transients, including the supernova SN1987A in the Large Magellanic Cloud.

While the original plans had called for a replacement crew to take over Mir from Polyakov, Volkov, and Krikalev, a funding shortage caused delays in the production of the next Soyuz spacecraft. The space laboratory remained unoccupied until the arrival of **Soyuz TM8** (**September 5, 1989**). That flight began almost a decade-long period in which there was a continual human presence in space.

December 2, 1988
STS 27, Atlantis Flight #3
Lacrosse 1
U.S.A.

STS 27, the third flight of the shuttle Atlantis and second flight after the Challenger accident (see **STS 51L–January 28, 1986** and **STS 26–September 29, 1988**), was also the third American manned space flight classified as a secret military mission, following **STS 51J** (**October 3, 1985**). The flight lasted four days and was manned by Robert Gibson (**STS 41B–February 3, 1984**; **STS 61C–January 12, 1986**; **STS 47–September 12, 1992**; **STS 71–June 27, 1995**), Guy Gardner (**STS 35–December 2, 1990**), Richard Mullane (**STS 41D–August 30, 1984**; **STS 36–February 28, 1990**), Jerry Ross (**STS 61B–November 26, 1985**; **STS 37–April 5, 1991**; **STS 55–April 26, 1993**; **STS 74–November 12, 1995**; **STS 88–December 4, 1998**), and William Shepherd (**STS 41–October 6, 1990**; **STS 52–October 22, 1992**). It deployed one military satellite.

• *Lacrosse 1*   A U.S. Air Force surveillance satellite, Lacrosse 1 was also a radar imaging satellite. It used side-looking radar similar with greater resolution (able to resolve objects as small as 3 feet) than, radar equipment flown on **Seasat** (**June 27, 1978**) and STS 41G (**December 5, 1984**). Because the images it produced used radar frequencies, it could see through cloud cover as well as produce images at night.

Through December 1999, the U.S. Air Force is reported to have launched a total of three such satellites.

The next shuttle mission was **STS 29** (**March 13, 1989**). The next military shuttle mission was STS 28 (**August 8, 1989**).

December 11, 1988
Skynet 4B / Astra 1A
U.K. / Luxembourg / Europe

The satellites Skynet 4B and Astra 1A were both launched by the European Space Agency's Ariane rocket.

• *Skynet 4B*   Skynet 4B was the first in a new cluster of British military satellites replacing **Skynet 2** (**November 23, 1974**). Placed in geosynchronous orbit, this constellation provides strategic communication links between both ship- and shore-based antennas, both wide and narrow beam. While initially designed for global strategic communication using large ground-based antennas, the technology has advanced so much that ground antennas as small as 2 feet can be used, allowing soldiers in the field to utilize it for tactical applications. Through December 1999, five satellites have been launched and are in use.

• *Astra 1A*   Built by a Luxembourg-subsidized company, Societe Europeenne des Satellites (SES), Astra 1A inaugurated a satellite cluster providing cable television distribution and direct broadcasting services to Europe. Through December 1999, SES has launched eight Astra satellites, primarily serving the United Kingdom, France, West Germany, Denmark, the Netherlands, Switzerland, Austria, and northern Italy. In these countries, SES's satellite signals can be picked up by ground antennas as small as two feet across. Larger antennas work in surrounding countries.

1989

January 10, 1989
Cosmos 1989
U.S.S.R.

Cosmos 1989 was the first official geodetic satellite launched by the Soviet Union. Similar to the previous geodetic satellite, **Ajisai (August 12, 1986)**, Cosmos 1989 was a laser reflecting satellite, a 4-foot-diameter sphere covered with 2,000 reflecting quartz prisms. Ground lasers bounced off the satellite could determine their relative position to within a few inches.

February 14, 1989
Navstar B2 1
U.S.A.

This satellite was the first launch in the United States military's second-generation global position system (GPS) satellite cluster. See **Timation 1 (May 31, 1967)** for a description of the GPS concept.

Larger and heavier than the previous generation (see **Navstar 5–February 9, 1980**), improvements included supplemental shielding against radiation, along with greater system redundancy. These modifications prolonged the satellite's service life from 4 to 7.5 years. (The short life span results from the satellites' orbit within the Van Allen radiation belts, which slowly destroys its solar panels and electronics.)

In March 1994, the United States completed this second cluster, comprising 21 active satellites and 3 spares. Through October 1999, an additional 8 satellites have been launched, and the cluster remains in operation as of December 1999, used by both the military and the general public.

February 21, 1989
Akebono (EXOS D)
Japan

Akebono studied the earth's aurora from an eccentric polar orbit (perigee: 186 miles, apogee: 5,000 miles). It was the fourth ionospheric satellite launched by Japan, following **Ohzora (February 14, 1984)**. Akebono was also the first satellite launched in more than two years, since **Ionosonde (December 18, 1986)**, to do such research. It carried eight sensors, including a television scanner in both visible and ultraviolet wavelengths, and gathered data for one week.

The next magnetospheric-ionospheric satellites were **Intercosmos 24/Magion 2 (November 28, 1989)**.

March 6, 1989
JCSat 1
Japan

Owned and operated by Japan Communications Satellite, JCSat 1 was the first privately owned Japanese geosynchronous satellite. Its design resembled **Intelsat 6 (October 27, 1989)** and the sixth SBS satellite (see **SBS 1–November 15, 1980**). It used the 12/14 GHz frequencies and was capable of broadcasting 32 television signals.

Through December 1999, the company has launched six JCSat satellites, all positioned over Japan, providing cable television distribution, direct broadcast capability, and digital voice transmissions. To receive the signal, ground antennas can range from 4 to 36 feet in diameter.

March 13, 1989
STS 29, Discovery Flight #8
  TDRS 4
U.S.A.

STS 29, the eighth flight of Discovery, lasted five days and was crewed by Michael Coats (STS 41D–August 30, 1984; STS 39–April 28, 1991), John Blaha (STS 33–November 22, 1989; STS 43–August 2, 1991; STS 58–October 18, 1993; STS 79–September 16, 1996), James Buchli (STS 51C–January 24, 1985; STS 61A–October 30, 1985; STS 48–September 12, 1991), Robert Springer (STS 38–November 15, 1990), and James Bagian (STS 40–June 5, 1991). The mission deployed one satellite.

• *TDRS 4* Deployed on March 13, TDRS (Tracking and Data Relay Satellite) 4 provided communications between NASA and its orbiting spacecraft. See **STS 6 (April 4, 1983)**.

Discovery's onboard experiments included protein crystal growth experiments, following results from **STS 26 (September 29, 1988)**. Protein crystals were grown using enzymes relating to emphysema, the kidneys, and interferon, as well as antibiotics, growth hormones, and a number of other body functions. Once again, gamma-interferon crystal growth was equal or superior to any previous ground results and had morphologies unlike anything seen before in Earth-grown crystals. The results this time gave scientists clues to why the crystals grew so quickly. For instance, the creation of a slight vacuum when the experiment's syringe plunger was pulled back seemed to accelerate the growth process.

The experiment also grew the first virus crystals, producing 200 crystals of the tobacco mosaic virus. Also grown was a crystal of lectin, a plant seed protein. Both crystals were greatly improved over Earth-grown crystals, and the research allowed scientists to better determine each crystal's structure.

An improved facility for growing protein crystals was next flown on **STS 32 (January 9, 1990)**.

Chromex (CHROMosome and plant cell EXperiment) attempted to see whether two kinds of plants, weeds and daylilies, could regenerate their roots in space, following up research on **STS 3 (March 22, 1982)** and **STS 61A (October 30, 1985)**. Both plants had had their roots removed prior to launch, and the results showed that although the plant roots did regrow, abnormalities were found in 3 to 30 percent of each plant's root chromosomes. Also, root tissue production in space was 40 to 50 percent greater than on the ground. Chromex next flew on **STS 41 (October 6, 1990)**.

Of the two student experiments, one attempted to find out how bone fractures healed in weightlessness. Four rats were flown with small bone sections removed to see how their bodies replaced the lost bone. The second student experiment tested how chicken embryos developed in space. Thirty-two fertilized chicken eggs were flown, half fertilized two days before launch and half nine days before launch. The eggs fertilized two days before launch were all dead upon return to Earth, while the nine-day eggs hatched successfully. The results indicated that, at least in the first week of life, the chicken embryos required gravity to develop.

For the first time since STS 61B (**November 26, 1985**), the shuttle carried an IMAX 70-mm camera, shooting film for *The Blue Planet*, the second IMAX film shot in space.

This shuttle mission was the first to use a laptop computer. The laptop was considered state-of-the-art at the time, with eight megabytes of RAM memory and a hard drive of 20 megabytes.

The next shuttle mission was STS 30 (**May 4, 1989**).

April 26, 1989
Foton 5
U.S.S.R. / France

Foton 5 was a recoverable Vostok-type capsule and part of a series of Soviet research satellites designed to study the behavior of materials in space (see **Foton 1–April 16, 1985**). It was also the first satellite to be launched under a new Soviet commercial program, whereby foreign researchers could buy experimental space on the capsule. On this flight, the French firm Matra paid several hundred thousand dollars for the right to use the Foton capsule to test the behavior of liquids in zero gravity.

Foton 5 also tested the growing of semiconductor crystals, as well as the separation of proteins by electrophoresis. This second experiment was similar to McDonnell Douglas's experiments flown on a number of space shuttle flights (see **STS 4–June 27, 1984**).

The next Foton satellite was **Foton 6 (April 11, 1990)**.

May 4, 1989
STS 30, Atlantis Flight #4
Magellan
U.S.A.

STS 30 was the fourth flight of the shuttle Atlantis. The mission lasted four days and was crewed by David Walker (STS 51A–November 8, 1984; STS 53–December 2, 1992; STS 69–September 7, 1995), Ron Grabe (STS 51J–October 3, 1985; STS 42–January 22, 1992; STS 57–June 21, 1993), Norman Thagard (STS 7–June 18, 1983; STS 51B–April 29, 1985; STS 42–January 22, 1992; Soyuz TM21–March 14, 1995), Mary Cleave (STS 61B–November 26, 1985), and Mark Lee (STS 47–September 12, 1992; STS 64–September 9, 1994; STS 82–February 11, 1997). This flight marked the first time that the shuttle launched a planetary probe.

• *Magellan*   Deployed on May 5, 1989, Magellan's mission was to go into orbit around Venus and provide a high-resolution radar map of the planet's surface. Built by NASA, Magellan was a follow-up to the **Pioneer Venus Orbiter (December 4, 1978)** radar mapping mission. The spacecraft reached Venus on August 10, 1990, where it spent four years mapping 98 percent of the planet's surface (see **Magellan, Venus Orbit–August 10, 1990**).

STS 30's onboard experiments included a test of what was called the floating zone process to produce crystals of indium. A sample rod was melted by a heater that was drawn along its length. As the melted indium cooled, it reformed into a single crystal that acted as a seed onto which the crystal's lattice structure grew. Any impurities in the lattice remained in the melted zone, leaving the solidifying region pure. This flight proved the successful functioning of the experiment's equipment, and a larger test was subsequently flown on STS 32 (**January 9, 1990**).

The astronauts also devoted a significant amount of time using 35-mm cameras to take photographs of lightning storms over Africa. This mission was also the first to use a fax machine to transmit information.

The next shuttle flight was STS 28 (**August 8, 1989**).

May 25, 1989
Resurs F1 / Pion 1 / Pion 2
U.S.S.R.

• *Resurs F1*   Resurs F1 was the first Soviet earth resource satellite under the official Resurs name. See **Meteor 1-30 (June 18, 1980)**.

Through September 1999, the Soviets/Russians have launched 25 Resurs satellites. Most were similar to the Soviet short-term recoverable surveillance satellites—remaining in orbit for about two weeks and then returning to Earth where their multispectral images could be developed and analyzed.

Some, however, operated for longer periods. They were placed in sun-synchronous orbits (see **Nimbus 1–August 28, 1964**) where they scanned the ground across eight wavelengths, from the visible to infrared wavelengths (0.5–12.4 microns), covering swathes about 375 miles wide.

• *Pion 1 and Pion 2*   Pion 1 and Pion 2 were two passive satellites deployed from Resurs F1 on June 9th. Both reentered the earth's atmosphere about a month later. By observing the slow decay of their orbits, Soviet scientists obtained data on the atmosphere's density at approximately 145 miles elevation. The next satellite to monitor atmospheric density was **Atmosphere 1 (September 3, 1990)**.

May 31, 1989
Cosmos 2024
U.S.S.R.

Identical to **Cosmos 1989 (January 10, 1989)**, Cosmos 2024 was a geodetic laser reflecting satellite, a 4-foot-diameter sphere covered with 2,000 reflecting quartz prisms. Ground-based lasers bounced off the satellite could determine their relative position to within a few inches. Both this and Cosmos 1989 were used by Soviet scientists to more precisely determine Earth's shape and the topographic location of the Soviet Union upon it.

June 5, 1989
Kopernikus 1 / Superbird A
West Germany / Japan / Europe

• *Kopernikus 1*   Developed and built in West Germany, Kopernikus 1 was the first of a three-satellite cluster designed to provide telephone and television broadcast service to West Germany and West Berlin. The third satellite was placed in orbit by October 1992, where today all three satellites serve all of reunified Germany. The satellites also transmit in the 20 and 30 GHz frequency band for propagation tests at these experimental frequencies, continuing work started by **ATS 6 (May 30, 1974)**. Although the signal at these frequencies can be seriously impaired by rain, the continuing demand for bandwidth encouraged further research, since such frequencies can carry more signal data than the 4/6 GHz and 12/14 GHz bands combined. See also **Olympus 1 (July 12, 1989)**.

• *Superbird A* Owned and operated by Space Communications Corporation (SCC), Superbird A was the first satellite in the second private communications cluster built by a Japanese company, following **JCSat 1 (March 6, 1989)**. Through December 1999, this company has successfully placed four Superbird satellites in geosynchronous orbit, three of which still operate today. Each has the capacity to broadcast 30 television signals simultaneously. The satellites provide cable television distribution, data transmissions, and direct broadcasting capability to private homes, as well as one channel for Japanese military communications.

July 4, 1989
Nadezhda 1
U.S.S.R.

Nadezhda 1 carried a second-generation version of the international SARSAT/COSPAS search-and-rescue receiver-transmitter package, developed to quickly locate ships and aircraft in distress. See **Cosmos 1383 (June 30, 1982)**.

Through July 1994, the Soviet Union/Russia launched four Nadezhda satellites.

July 12, 1989
Olympus 1
Europe

Built and launched by the European Space Agency (ESA), Olympus 1 was both an operational and experimental communications satellite placed in geosynchronous orbit above Europe. Olympus 1 carried four different payloads, each providing a different communications service. The first payload provided two television direct broadcast channels, one reserved for Italian use and the second for technology experiments. The second payload used the 12/14 GHz frequencies and was devoted to business services and data transmissions to small antennas (10 feet across).

The third payload did propagation experiments in the 20 and 30 GHz frequencies, continuing research started by **ATS 6 (May 30, 1974)** and renewed on **Kopernikus 1 (June 5, 1989)**. During Olympus's service life, more than 100 experiments were performed, including a data relay experiment with EURECA (see **STS 46–July 31, 1992** and **STS 57–June 21, 1993**). See **Italsat 1 (January 15, 1991)** for the first commercial use of these frequencies.

Olympus's fourth payload also used the 20 and 30 GHz frequencies and was reserved for data and voice transmissions to demonstrate the feasibility of satellite communications at these experimental frequencies.

August 8, 1989, 12:37 GMT
STS 28, Columbia Flight #8
USA 40 / USA 41
U.S.A.

Even though it was numbered 28, STS 28 was actually the 30th shuttle flight. Following the Challenger accident (see **STS 51L–January 28, 1986**), NASA changed the mission numbering system, giving each flight its own permanent number, regardless of flight sequence.

STS 28 was Columbia's eighth flight into space, and its first in 3.5 years, since **STS 61C (January 12, 1986)**. The mission was also the fourth American manned military space flight, following **STS 27 (December 2, 1988)**. It lasted five days and was crewed by Brewster Shaw (**STS 9–November 28, 1983**; **STS 61B–November 26, 1985**), Richard Richards (**STS 41–October 6, 1990**; **STS 50–June 25, 1992**; **STS 64–September 9, 1994**), David Leetsma (**STS 41G–October 5, 1984**; **STS 28–August 8, 1989**; **STS 45–March 24, 1992**), James Adamson (**STS 43–August 2, 1991**), and Mark Brown (**STS 48–September 12, 1991**). The mission deployed two classified Air Force satellites.

• *USA 40* Very little is known about USA 40. Some experts believed it to have been a photo-surveillance satellite, either an upgraded KH-11 or the first KH-12. See **KH-11 (December 19, 1976)** and **KH-12 (see STS 36–February 28, 1990)**. Others dubbed USA 40 "SDS 2," believing it to have been the first of five satellites in a second-generation U.S. Air Force communications satellite cluster, replacing the SDS 1 satellites. See **SDS 1 (June 2, 1976)**.

• *USA 41* USA 41 was deployed on August 8th. Little is known about it, as well, though some experts believed it to have been a ferret-type electronics satellite, designed to detect the electronic signals. See **Ferret 1 (June 2, 1962)**. Other experts thought it to be a research satellite in the strategic defense initiative.

Ground observations concluded that one of these two satellites was tumbling out of control within two months of its deployment.

The next shuttle flight was **STS 34 (October 18, 1989)**. The next military shuttle mission was **STS 33 (November 22, 1989)**.

August 8, 1989, 23:25 GMT
Hipparcos
Europe

Hipparcos, a European Space Agency astronomical satellite, was flown to obtain precise measurements of the parallax of more than 120,000 nearby stars, thereby allowing astronomers to determine their distance very accurately. With this knowledge, scientists could better estimate the absolute brightness of most stars, which would enable them to estimate the distance of stars throughout the universe more exactly.

A failure in the spacecraft's booster engine left it stranded in an inappropriate eccentric orbit. In order to compile a catalog of more than 100,000 stars, the data gathering required a significant amount of additional time. For more than three years, the satellite collected its data, then scientists spent another four years analyzing this information, finally releasing their results in early 1997.

According to their conclusions, the universe was generally larger than previously thought. Cepheid variable stars, one of astronomy's prime measuring sticks for estimating the distance of galaxies, averaged about 10 percent farther away than predicted. This meant that their absolute brightness was correspondingly 10 percent brighter than astronomers had believed. Using this new absolute brightness as a yardstick meant that the Cepheids in other galaxies were generally farther away than thought, indicating that the universe itself was larger and older, as well.

As of December 1999, astronomers still have not completely digested the conclusions presented by Hipparcos, nor have they completely succeeded in fitting its results into their theories about the age of the universe, its expansion, and its size.

August 23, 1989
Progress M1
U.S.S.R.

Progress M1 was a second-generation unmanned freighter for the **Mir (February 20, 1986)** space laboratory, replacing the **Progress 1 (January 20, 1978)** design.

The Progress M freighter carried solar panels, allowing the freighter to fly independent of Mir for up to 30 days. The

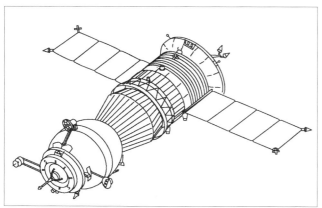

Progress M. Compare with Progress 1 (January 20, 1978). *NASA*

additional power capacity also permitted these freighters to remain docked with the station for as long as three months. Their improved automatic docking system, dubbed "Kurs," also saved the station's fuel, since the freighter did all the maneuvering.

Finally, three later Progress M freighters carried a small recoverable capsule that was capable of returning up to 440 pounds of cargo from Mir. The first successful recovery took place in 1991, with two additional capsules returning in 1993 and 1994.

Through December 1999, a total of 42 Progress M freighters have been launched to Mir. This particular freighter brought supplies to Mir in anticipation of its reoccupancy by **Soyuz TM8 (September 5, 1989)**.

August 25, 1989
Voyager 2, Neptune Flyby
U.S.A.

Voyager 2, launched on August 20, 1977, had previously made close approaches to Jupiter, Saturn, and Uranus (see **Voyager 2, Jupiter Flyby–July 9, 1979; Voyager 2, Saturn Flyby–**

August 26, 1981; and **Voyager 2, Uranus Flyby–January 24, 1986)**. With this Neptune flyby, it made its last flyby of another planet, passing within 3,000 miles of Neptune's cloud tops. Along the way, the spacecraft also flew within 25,000 miles of Neptune's moon Triton.

Images revealed that Neptune, a deep-blue planet, had the same horizontal light and dark bands that had been seen on all the other gas giants, though much more distinct than seen on Uranus. It also had a "Great Dark Spot," a huge hurricane-like storm more than 6,000 miles wide and reminiscent of Jupiter's Red Spot *(see color section, Figure 18)*. The atmosphere itself appeared very turbulent, with layers of clouds racing across the planet's face, changing so quickly that scientists were unsure if they were seeing the same clouds from day to day. At the poles, the winds blew in a westerly direction at about 700 miles per hour, while at the equator they blew easterly at 1,500 miles per hour. Temperatures of around –300°F were measured. These numbers indicated that the planet generated almost three times the heat it received from the Sun.

Six new moons were discovered, ranging in size from 30 to 125 miles in diameter and raising the known total from two to eight. Voyager also proved that the planet was encircled by rings, similar to those of Jupiter, Saturn, and Uranus. Unlike the other ring systems, however, Neptune's rings were unevenly distributed around the planet, with both thin and thick regions.

Close-up photos of Triton revealed a mottled surface of pink and white, with many cracks and ridges and only a few craters. Much of its surface appeared young and newly coated. The moon showed evidence of recurrent and very recent periods when its icy surface had melted and refrozen, with "molten" ice flowing up through cracks to cover the surface. In one photo, as many as 50 dark streaks—looking like wind-blown plume deposits—were seen. In several photos, two active geysers were identified, making Triton only the third world in the solar system known to have active volcanism, following Earth and Jupiter's moon Io.

Triton's atmosphere was mostly nitrogen and was much thinner than predicted. Its surface temperature was very cold, around –400°F, only 38°K above absolute zero. Its size was smaller than expected, only 1,700 miles across compared to the predicted 2,500. The error was because of the moon's ability

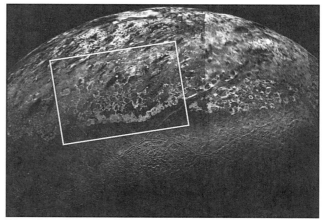

Triton. See color section (Figure 19) for detail. *NASA*

to reflect more light than expected, which made it appear larger from Earth.

Voyager 2 also discovered a magnetic field around Neptune, comparable in strength to Earth's. The field was tipped about 50 degrees from the planet's axis. By measuring this field's rotation, scientists determined the planet's day to be just over 16 hours long.

Voyager 2 continued on its journey out of the solar system, joining **Pioneer 10 (March 3, 1972)**, **Pioneer 11 (April 6, 1973)**, and its sister ship **Voyager 1 (March 5, 1979)** in their interstellar journeys. See **Pioneer 11, Shutdown (September 30, 1995)** and **Pioneer 10, Shutdown (March 31, 1997)** for more information.

August 27, 1989
Marco Polo 1
U.K. / U.S.A.

Owned and operated by British Satellite Broadcasting (BSB), Ltd., the geosynchronous communications satellite Marco Polo 1 was the first of two satellites providing radio and television broadcasting capability to Great Britain; both with broadcast distribution, as well as direct broadcasting to residential homes. A second Marco Polo satellite was launched in 1990.

Because of stiff competition from the Astra satellites (see **Astra 1A–December 11, 1988**), the Marco Polo satellites have had difficulty getting customers. In 1990, BSB merged with Sky Television, which had been leasing space on the Astra satellites to broadcast to the British market. The new company, British Sky Television, decided to continue the direct broadcasting using the Astra satellites, leaving the Marco Polo satellites for cable television channel distribution. Eventually, it sold the second Marco Polo satellite to Telenor of Norway, which renamed it Thor 1 and used it as part of its own Scandinavian direct broadcasting service. See **Thor 2 (May 21, 1997)**.

This mission was the first commercial launch using an American rocket following the Challenger accident (see **STS 51L–January 28, 1986**). It was also the first in which the satellite manufacturer—in this case, Hughes Space and Communications, Inc. (formerly Hughes Communications)—had to certify that the spacecraft was in orbit and functioning before it could transfer operations to its owner.

September 5, 1989
Soyuz TM8
U.S.S.R.

Soyuz TM8 was the first manned flight to **Mir (February 20, 1986)** in more than 10 months, since **Soyuz TM7 (November 26, 1988)**. The station itself had been unoccupied since the return of the Soyuz TM7 crew on April 27, 1988. The launch of Soyuz TM8, however, began a 10-year occupancy of Mir, ending with **Soyuz TM29 (February 20, 1999)**.

Because of declining cash reserves in the crumbling Soviet Union, however, the number of manned flights to Mir from this time on never matched the pace during the earlier **Salyut 6 (September 29, 1977)** and **Salyut 7 (April 19, 1982)** stations. Until the American space shuttle began its joint flights to Mir in 1995 (see **STS 71–June 27, 1995**), the station saw few multiple-crew visits. Instead, one crew generally remained onboard for a period of time and then transferred occupancy to the next crew.

Furthermore, the Soviet Union had begun to make an effort to earn a profit from each mission. For example, they announced that although the cost of Soyuz TM8 was 80 million rubles, its net profit was 25 million rubles.

Soyuz TM8 was crewed by Alexander Viktorenko (**Soyuz TM3–July 22, 1987**; **Soyuz TM14–March 17, 1992**; **Soyuz TM20–October 4, 1994**) and Alexander Serebrov (**Soyuz T7–August 19, 1982**; **Soyuz T8–April 20, 1983**; **Soyuz TM17–July 1, 1993**). During their stay on Mir, Viktorenko and Serebrov received three Progress freighters, two of which were the second-generation M type (see **Progress M1–August 23, 1989**). The second M freighter also carried an American experiment for growing protein crystals in space, developed by the commercial company Payload Systems, Inc., which was using the freighter as a paying customer.

Also docking with Mir during their occupancy was the station's second permanent module, **Kvant 2 (November 26, 1989)**. During their more than six months in orbit, Viktorenko and Serebrov performed five space walks. In January, they conducted three EVAs (extra-vehicular activities) to place and retrieve a number of scientific experiments on the outside of Mir, including several placed there by the Soviet-French cosmonauts from **Soyuz TM7 (November 26, 1988)**. In early February, they made two space walks to test a maneuvering unit similar to that used by Bruce McCandless and Robert Stewart on **STS 41B (February 3, 1984)**. With this unit, Viktorenko and Serebrov were able to move as much as 150 feet away from the station, although they remained tethered to Mir at all times. During Viktorenko's flight, he used a sensor to measure the radiation surrounding the station. Although both men said that the unit worked well, it was never used again.

Viktorenko and Serebrov returned to Earth on February 19, 1990, with more than 200 pounds of cargo, including the American protein crystal experiments. The station, meanwhile, was left in the hands of its next crew: **Soyuz TM9 (February 11, 1990)**.

September 15, 1989
Bion 9 (Cosmos 2044)
U.S.S.R. / International

Bion 9, a biological research satellite, was part of a Soviet program begun with **Bion 1 (October 31, 1973)**. It carried two monkeys, rats, newts, fish, fruit flies, ants, worms, cells, and seeds on a two-week space flight. Researchers from the Soviet Union, the United States, the European Space Agency, and five other countries participated in a total of 30 different experiments.

The flight lasted 14 days. In one experiment, scientists tested whether a rat's injured leg muscle healed in space, finding that the healing process seemed hindered. In another experiment, half the samples of stick insect larvae failed to hatch. Those that did hatch, however, developed normally. Fruit-fly larvae produced similar results, although in this case the spaceborn flies also lived a shorter life, their development apparently retarded.

The previous Bion satellite was **Bion 8 (September 29, 1987)**. The next was **Bion 10 (December 29, 1992)**.

September 28, 1989
Intercosmos 24 / Magion 2
U.S.S.R. / Czechoslovakia / U.S.A. / East Europe

Intercosmos 24 and Magion 2 were part of the Soviet Intercosmos program (see **Intercosmos 1–October 14, 1969**), including experiments from the Soviet Union, the United States, Bulgaria, Czechoslovakia, East Germany, Hungary, Poland, and Romania. Both satellites were the first ionospheric satellites since **Akebono (February 21, 1989)**, and the first from the Soviet Union and Eastern Europe in almost three years, since **Ionosonde (December 18, 1986)**.

After deploying the Czechoslovakian subsatellite Magion 2 on October 3rd, the two satellites worked in conjunction to study how low-frequency transmissions and loop antennas worked within the earth's magnetosphere. Intercosmos 24 transmitted a signal; Magion 2 then studied how it propagated through the ionosphere and the earth's magnetic field.

The next satellite to study the ionosphere was **Pegsat (April 5, 1990)**.

October 18, 1989
STS 34, Atlantis Flight #5
Galileo
U.S.A.

STS 34 was the fifth flight of the space shuttle Atlantis, lasted five days, and carried a crew of five. The crew included Don Williams (**STS 51D–April 12, 1985**), Michael McCulley, Shannon Lucid (**STS 51G–June 17, 1985; STS 43–August 2, 1991; STS 58–October 18, 1993; STS 76–March 22, 1996**), Ellen Baker (**STS 50–June 5, 1992; STS 71–June 27, 1995**), and Franklin Chang-Diaz (**STS 61C–January 12, 1986; STS 46–July 31, 1992; STS 60–February 3, 1994; STS 75–February 22, 1996; STS 91–June 2, 1998**). The flight deployed one planetary probe.

- *Galileo* Deployed on October 18th, Galileo was the first spacecraft to orbit Jupiter and release a probe into the gas giant's atmosphere. Initially scheduled for launch in 1986, the Challenger accident (see **STS 51L–January 28, 1986**) caused a three-year delay and a complete redesign of Galileo's mission. In NASA's original plans, Galileo would have been placed in Earth orbit by the space shuttle, then a Centaur upper stage would have been used to send it on a direct path to Jupiter. After the Challenger accident, the use of the Centaur upper stage on the shuttle was considered too risky, and a different method was devised to reach Jupiter. To acquire the necessary speed, Galileo made one flyby of Venus and two of Earth. Along the way, it passed close to two different asteroids, Gaspra and Ida. Finally, after more than six years in space, Galileo entered Jupiter orbit, simultaneously sending its atmospheric probe into Jupiter's cloud tops (see **Galileo, Gaspra Flyby–October 29, 1991; Galileo, Ida Flyby–August 28, 1993**; and **Galileo, Jupiter Orbit–December 7, 1995**).

STS 34's onboard experiments included an attempt to find out how the plant growth hormone indoleacetic acid, which corn plants use to guide their growth upward, changed in zero gravity. A student experiment also tested how ice crystals form in zero gravity.

The 3M Company's fifth experiment on the shuttle (see **STS 26–September 29, 1988**) studied how a variety of polymers changed in microgravity as they were melted and recrystalized. Surprisingly, the samples showed no obvious differences from processed samples on Earth.

The Shuttle Solar Backscatter Ultraviolet (SSBUV) sensor was similar to sensors for measuring ozone layer changes on the NOAA satellites (see **NOAA 1–December 11, 1970**). SSBUV was designed to provide scientists with reliable comparison data so that they could correct and calibrate other already orbiting instruments. There were concerns that the calibration on the previous satellites had drifted so much that their measurements were unreliable (see for example **Aeros 2–July 16, 1974**). To improve the data's dependability, SSBUV was flown on seven later shuttle missions through 1996. Further ozone research continued on **Meteor 3-05 (August 15, 1991)**.

An IMAX camera filmed several thousand feet of film of the earth for use in the documentary film *The Blue Planet*. Another payload camera took spectacular photos of lightning discharges over Australia, Africa, Indonesia, and Brazil.

The next shuttle flight was **STS 33 (November 22, 1989)**.

October 27, 1989
Intelsat 6A F2
International / Europe

Intelsat 6A F2 was the first satellite in the sixth constellation of communications satellites built and owned by Intelsat (see **Early Bird–April 6, 1965**) and launched on an Ariane rocket. The Intelsat 6A cluster supplements the Intelsat 5 and Intelsat 5A satellite cluster (see **Intelsat 5 F2–December 6, 1980** and **Intelsat 5A F10–March 22, 1985**). Intelsat 6A satellites have more than double the capacity, carrying 15,000 to 24,000 two-way voice circuits under normal conditions and a maximum capacity of 120,000 voice circuits per satellite under certain conditions. Each satellite also carries three television circuits. They use technology developed on the SBS satellites (see **SBS 1–November 15, 1980**), Anik C (see **STS 5–November 11, 1982**), the LEASAT satellites (see **STS 41D–August 30, 1984**), and a number of other satellites.

Through October 1991, Intelsat launched four Intelsat 6A satellites, placing two over the Atlantic, one over Europe, and one over the Indian Ocean. When combined with the still functioning Intelsat 5 and 5A satellites, the capacity between Europe and North America was more than a quarter of a million telephone calls at any moment.

This capacity was soon insufficient. See **Intelsat 7 (October 22, 1993)**. Intelsat also launched **Intelsat K (June 10, 1992)** to handle television and business communications over the Atlantic Ocean.

November 18, 1989
COBE
U.S.A.

Built by NASA, the COsmic Background Explorer (COBE) measured the diffuse background infrared radiation of the sky from 1 micron to 1 centimeter. This faint radiation, discovered in 1964, was the strongest evidence that the universe as we know it was created in a single "big bang" between 10 and 20 billion years ago. The radiation, predicted by the "big bang,"

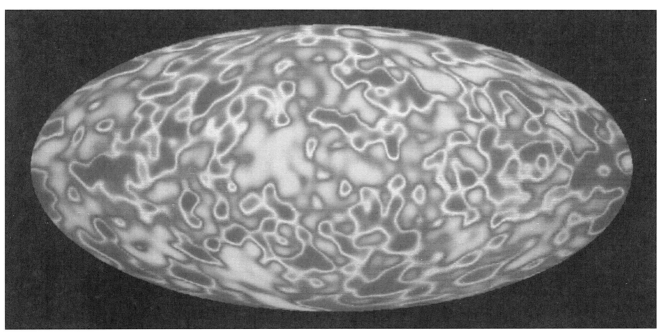

COBE: Map of the background microwave radiation. The light and dark areas indicate the initial uneven distribution of matter following the Big Bang. *NASA*

has a temperature of only 3°K and has the same brightness in all directions. Therefore, it was not caused by any single celestial object.

In order for galaxies to have formed, cosmologists believed that very slight undetected variations in the radiation had to exist. COBE, 100 times more sensitive than any previous infrared space telescope, attempted to discover these variations.

COBE operated until 1993, making eight complete sweeps of the celestial sky. From the data, scientists created a whole sky map, showing that there were faint but distinct variations in the background radiation, depending on direction. This result agreed with predictions proposed by the "inflationary big bang" theory, which postulated the universe's expansion had been accelerated in its first trillionth of a second. It also provided scientists with the first evidence of how matter had differentiated into the clumps that eventually became galaxies. Why this variation formed, however, remains a mystery.

COBE's sensors also measured the infrared radiation of the galaxy, helping astronomers pinpoint the concentrations of dust within the Milky Way.

Four years of additional analysis of the COBE data also revealed that an unexpected but measurable infrared glow diffused the universe. Scientists believed that this glow was caused by the warming of the intergalactic dust by all the stars that have existed since the "big bang." Larger than expected, this glow indicated that more starlight and hence mass had existed through time than had been predicted by most theorists.

The next infrared space telescope was **ISO** (**November 17, 1995**).

November 22, 1989
STS 33, Discovery Flight #9
Magnum 2
U.S.A.

STS 33, the ninth flight of the space shuttle Discovery, was a five-day classified military mission, the fifth in the shuttle program, following **STS 28** (**August 8, 1989**). The crew consisted of Fred Gregory (STS 51B–April 29, 1985; STS 44– November 24, 1991), John Blaha (STS 29–March 13, 1989; STS 43–August 2, 1991; STS 58–October 18, 1993; STS 79–September 16, 1996), Story Musgrave (STS 6–April 4, 1983; STS 51F–July 29, 1985; STS 44–November 24, 1991; STS 61–December 2, 1993; STS 80–November 19, 1996), Kathryn Thornton (STS 49–May 7, 1992; STS 61– December 2, 1993; STS 73–October 20, 1995), and Manley Carter. The mission deployed one satellite.

• *Magnum 2* Magnum 2 was a military surveillance satellite for intercepting electronic signals and voice communications. See **STS 51C** (**January 24, 1985**) and **Chalet 1** (**June 10, 1978**).

The next shuttle mission was **STS 32** (**January 9, 1990**). The next military shuttle mission was **STS 36** (**February 8, 1990**).

November 26, 1989
Kvant 2
U.S.S.R.

Kvant 2 was the second permanent module to be added to **Mir** (**February 20, 1986**), following **Kvant 1** (**March 31, 1987**). Its docking, however, was delayed until December 12th due to solar panel and maneuvering problems.

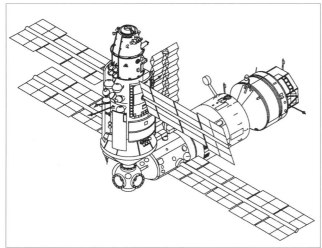

Mir Core module with Kvant 1 and Kvant 2 attached. Compare with diagram on March 31, 1987. *NASA*

Kvant 2 added almost 2,000 cubic feet to the Mir station, increasing its total interior space to 7,000 cubic feet. It also added a shower and sink to the laboratory. Kvant 2 was divided into two airtight chambers, with one room designed as an airlock to ease access into space. See **Soyuz TM9 (February 11, 1990)**.

Its cargo included a manned maneuvering unit (MMU), which the Soviets nicknamed a "space motorcycle." Similar to that used by the astronauts on **STS 41B (February 3, 1984)**, the two cosmonauts performed two space walks in February 1990 to test this unit's capability. See **Soyuz TM8 (September 5, 1989)**.

Its scientific cargo included a Japanese incubator containing quails' eggs to study how the eggs developed over a six-month period in space. See **Soyuz TM9 (February 11, 1990)** for results. Kvant 2 also carried two multispectral cameras and an experiment for studying the behavior of fluids in weightlessness.

After initially docking along Mir's main axis, Kvant 2 was transferred to its permanent position on a lateral port on December 8th. This transfer was done with the use of a short robot arm attached to Mir's multidocking adapter, which manually maneuvered the module from one docking port to another.

Mir's next module was **Kristall (May 31, 1990)**.

December 1, 1989
Granat (Astron 2)
U.S.S.R. / France

Granat (originally called Astron 2) was an x-ray and gamma-ray telescope for studying high-energy transient objects both inside and outside the Milky Way galaxy. It followed **Ginga (February 5, 1987)**, the previous x-ray telescope.

The spacecraft's main camera, built by France, could produce images of the sky in the energy range from 35 to 1,300 KeV. Its other x-ray instruments extend the lower end of this range to 2 KeV.

X-ray observations of the center of the Milky Way increased the plethora of x-ray objects known to exist there, including a kind of x-ray object that scientists called x-ray bursters—binary systems made up of an ordinary star and a neutron star. As the ordinary star orbited the neutron star, the neutron star ripped matter from its outer atmosphere, which then built up into a disk-like cloud surrounding the neutron star. Scientists believed that periodically the cloud collapsed, crashing into the neutron star and causing the gigantic thermonuclear explosions that produced the x-ray bursts.

Granat's discoveries also added to the number of black hole candidates. For example, GRS 1915+105, sometimes labeled a microquasar, periodically ejects matter at such great speed that, because of its orientation relative to the earth, the material appears to travel faster than the speed of light. The repeating nature of these ejections have prompted scientists to nickname this object astronomy's "Old Faithful."

Granat also carried a gamma-ray burst sensor, observing energy bursts between 0.1 to 100 MeV, augmenting the monitoring of **Ginga (February 5, 1987)**. During the satellite's first eight months of operation, this instrument detected over 50 bursts coming from all across the sky. The cause of these events, first discovered by **Vela 5A/Vela 5B (May 23, 1969)**, remained a mystery, however. See the **Compton Gamma Ray Observatory (STS 37–April 5, 1991)** for results.

The next x-ray telescope was **ROSAT (June 1, 1990)**. The next high-energy telescope was **Gamma (July 11, 1990)**.

1990

January 9, 1990
STS 32, Columbia Flight #9
   LEASAT 5 / LDEF
U.S.A.

STS 32 was Columbia's ninth space flight. The mission lasted 11 days (the longest shuttle mission to date) and was crewed by Daniel Brandenstein (**STS 8–August 30, 1983; STS 51G–June 17, 1985; STS 49–May 7, 1992**), James Wetherbee (**STS 52–October 22, 1992; STS 63–February 3, 1995; STS 86–September 25, 1997**), Bonnie Dunbar (**STS 61A–October 30, 1985; STS 50–June 25, 1992; STS 71–June 27, 1995; STS 89–January 22, 1998**), David Low (**STS 43–August 2, 1991; STS 57–June 21, 1993**), and Marsha Ivins (**STS 46–July 31, 1992; STS 62–March 4, 1994; STS 81–January 12, 1997**). STS 32 deployed one satellite and retrieved another.

• *LEASAT 5*   Built and leased from Hughes Communications, Inc., LEASAT 5 was the last of five military LEASAT communication satellites. See **STS 41D–August 30, 1984** for program information.

• *LDEF*   LDEF, which stands for Long Duration Exposure Facility, was first deployed on **STS 41C (April 6, 1984)**. It carried 57 different experiments testing a wide variety of materials to exposure to the space environment. The original plans had been to recover LDEF after 10 months in orbit. Program delays and the Challenger accident (see **STS 51L–January 28, 1986**) delayed that retrieval for six years.

After rendezvousing and grabbing LDEF with the shuttle's robot arm on January 13th, astronauts Marsha Ivins, David Low, and Bonnie Dunbar made a three-hour photographic survey of the satellite from inside Columbia, documenting how the samples in each panel had been affected by six years in space. All told, they took about 800 pictures.

LDEF showed significant signs of wear, appearing to be in worse condition than expected. Some thin film specimens, such as test balloon material, had become completely eroded. The thermal covers to a number of other experiments were either peeled away, discolored, or showed evidence of meteoroid impacts.

Other experiments onboard STS 32 included the second test flight (see **STS 30–May 4, 1989**) of a floating zone furnace. The apparatus took a sample rod of material, such as indium, and drew a heater along it. On STS 32, the equipment was used to produce large and pure crystals of indium while testing how the small changes in motion produced by the shuttle and its crew affected the stability of the molten region of each sample and the crystal formation itself. The experiment found that the molten region generally was four times larger than possible on Earth, indicating that less power was needed than expected to melt each sample. The tests also found that shuttle engine firings and crew movements caused some variations in the crystal structure, as well as surface irregularities. In one case, the molten region developed an arch because of an engine burn.

In continuing protein crystal growth experiments (see **STS 26–September 29, 1988** and **STS 29–March 13, 1989**), 120 different protein crystals using 24 different samples of enzymes, antibiotics, and growth hormones involving such diseases as emphysema, cancer, diabetes, arthritis, and AIDS were flown. For the first time, an attempt was made to grow crystals in a controlled cold environment of 39°F, rather than in the ambient cabin temperature of 71°F. Although a power failure prevented most of the cooled samples from developing, the small deformed cooled crystals that did grow still appeared more pure than Earth-grown crystals. And, as in previous tests, the uncooled crystals that successfully grew were larger and more pure than any grown on Earth. Of the 12 uncooled samples, 8 produced superior crystals. Returned to Earth, these were x-rayed and their structure mapped to help scientists better design drugs to either block or enhance each substance's effect on the body. See **STS 31 (April 24, 1990)** for more protein crystal growth experiments.

Like the floating zone furnace, the crystal growth equipment had become sensitive enough to detect the small motions caused by shuttle engine firings and crew activity, important data for further refining future operations. These effects caused scientists to abandon use of the term "zero gravity," calling the weightless orbital environment "microgravity" instead.

Another life sciences experiment tested whether the day-night growth cycle in pink bread mold persisted in microgravity. It found that the growth cycle persisted; weightlessness had no effect.

The Latitude/Longitude Locator was a camera system for more precisely identifying the location of photographed targets. On **STS 41G (October 5, 1984)**, oceanographer Paul Scully-Power had seen some unusual oceanographic events but was unable to determine their location accurately enough for subsequent research. This new system was an attempt to solve the problem.

An IMAX camera filmed footage for the second in-space documentary, *The Blue Planet*.

The next shuttle mission was **STS 36 (February 28, 1990)**.

January 22, 1990
UoSat 3 (Oscar 14) / UoSat 4 (Oscar 15) /
Oscar 16 / Oscar 17 / Oscar 19 /
Webersat (Oscar 18)
U.K. / International / U.S.A. / Europe

• *UoSat 3 (Oscar 14) and UoSat 4 (Oscar 15)*   Built by the University of Surrey, England, the two satellites UoSat 3 and UoSat 4 were also part of the Oscar series of privately developed amateur communications satellites (see **Oscar 13–June 15, 1988** for the previous satellite). Originally intended as a single satellite to replace **UoSat 2 (March 1, 1984)** and launched piggyback on a NASA rocket, the university was forced to redesign the package into two satellites when the Challenger accident (see **STS 51L–January 28, 1986**) eliminated its launch opportunity. The payloads would fit on the Ariane rocket only as two satellites.

UoSat 3 continued the store-and-forward memory storage experiments started on UoSat 2, using a 4-megabyte memory bank. It also carried several instruments for measuring radiation in the ionosphere. UoSat 4 was dedicated to demonstrating the feasibility of several new technologies, including a CCD camera (see **KH-11 1–December 19, 1976**) and solar cells made of gallium-arsenide, indium-phosphorus, and silicon. Unfortunately, this satellite's transmitters did not work after its first day in orbit, and although the satellite seemed to function, no data was ever received from it. The University of Surrey replaced this failure with **UoSat 5 (July 17, 1991)**.

• *Oscar 16/Oscar 17/Oscar 19*   Oscars 16, 17, and 19 were built by AMSAT (the Radio Amateur Satellite Corporation), a worldwide organization of ham radio operators (see **Oscar 6–October 15, 1972**). Oscar 17 was aimed at promoting scientific education in the schools, while Oscars 16 and 19 provided digital memory for relaying messages by store-and-forward, tested earlier on **UoSat 2 (March 1, 1984)**.

• *Webersat (Oscar 18)*   Built by Weber State College in Utah, Webersat was similar in design to Oscars 16, 17, and 19. It also carried an Earth-imaging package including a CCD camera and CCD spectrometer (see **KH-11 1–December 19, 1976**).

January 24, 1990
Hagoromo / Hiten (Muses A)
Japan

• *Hagoromo*   Hagoromo, an unmanned probe, was the first attempt to send a spacecraft to the Moon in 14 years, since **Luna 24 (August 9, 1976)**. It also made Japan the third nation to do so, after the United States and the Soviet Union. Hagoromo was a small sphere, only 14 inches in diameter, intended merely to test its deployment engineering. It was released from Hiten on March 19, 1990, as Hiten approached the Moon at the apogee of its eccentric Earth orbit. The maneuver successfully placed Hagoromo in a 14,000- by 5,500-mile orbit around the Moon, though all radio contact was lost immediately after release.

• *Hiten (Muses A)*   Hiten (the Japanese name of a Buddhist angel who plays music in heaven) was an engineering test demonstrating that the Moon's gravity could be used to modify and adjust the path of a spacecraft. Hiten proved this, and Japan later used this same technique for **Geotail (July 24, 1992)** and **Nozomi (July 4, 1998)**.

As Hiten made its first close approach to the Moon, it released the probe Hagoromo into lunar orbit. Later in Hiten's mission, the spacecraft was maneuvered into lunar orbit as well, and after circling the Moon for over a year, it crashed onto its surface. Because Hiten carried very sensitive tracking equipment, its course could be calculated precisely, thereby allowing scientists to better map the variations in the Moon's gravitational field.

The next lunar mission was **Clementine (January 25, 1994)**.

February 7, 1990
Orizuru (DEBUT)
Japan

Also called DEBUT (for Deployable Boom and Umbrella Test), Orizuru ("folded paper crane" in Japanese) was an engineering satellite. It tested the use of a retractable 5-foot boom and a large unfolding umbrella for aerodynamically adjusting a satellite's orbit.

February 11, 1990
Soyuz TM9
U.S.S.R.

Soyuz TM9 was crewed by Anatoly Solovyov (**Soyuz TM5–June 7, 1988**; **Soyuz TM15–July 27, 1992**; **STS 71–June 27, 1995**; **Soyuz TM26–August 5, 1997**) and Alexander Balandin. They arrived on **Mir (February 20, 1986)** on February 13th, where they and the crew from **Soyuz TM8 (September 5, 1989)** began a six-day period of station transfer. With the return to Earth of the Soyuz TM8 crew on February 19th, Solovyov and Balandin settled in for a planned seven-month stay on Mir.

Almost immediately thereafter, they discovered that their Soyuz TM9 re-entry module had been damaged. A layer of insulation protecting the module had been ripped off during launch, with one strip six feet long still hanging loose. While the spacecraft could land without the damaged blankets, the loose strips interfered with the capsule's attitude control system and had to be removed prior to re-entry. After several months of planning, the two men performed an unscheduled space walk on July 17th to remove the excess material. They exited Mir through the airlock in **Kvant 2 (November 26, 1989)**. After six hours, they completed their work. In attempting to re-enter Mir through the Kvant 2 airlock, however, Solovyov and Balandin discovered that they could not shut the outer hatch door. The two men were forced to use the module's inner chamber as an airlock, losing an additional 200 cubic feet of air and leaving the outer chamber in vacuum with its hatch open.

In a second space walk on July 26th to fix the stuck hatch door, the two men were able to clear an obstruction and force the door closed. Nonetheless, this hatch would cause problems on later flights. See **Soyuz TM10 (August 1, 1990)**.

During their occupancy, the two men were resupplied by three Progress freighters (see **Progress M1–August 23, 1989**). They also received Mir's third permanent module, **Kristall (May 31, 1990)**.

As with Soyuz TM8, Soyuz TM9 attempted to turn a profit. Although the cost of the mission was estimated at about $133 million, payments by a number of Soviet companies to produce crystals in the Kristall module was expected to earn $175 million. A total of 23 different crystals were grown, including zinc oxide, gallium arsenide, and several protein crystals.

Because Kristall's launch was delayed by more than two months, however, much of the intended research was not done by this crew. In fact, in an attempt to complete the delayed research, Solovyov and Balandin did not follow their exercise program thoroughly during their last weeks in space. Consequently, the two men had more trouble readapting to Earth gravity than previous crews.

The cosmonauts took multispectral photography of the earth covering approximately 8 million square miles, resolving objects as small as 15 feet. In a Japanese incubator first brought to Mir by Kvant 2, several quail eggs successfully hatched, producing healthy chicks, which was an improvement from results on **Soyuz 32 (February 25, 1979)**. See **Soyuz TM21 (March 14, 1995)** for the next quail egg experiment.

On August 3rd, **Soyuz TM10 (August 1, 1990)** docked with Mir. For the next six days, the two crews transferred operations from Soyuz TM9 to Soyuz TM10. Then, Soyuz TM9 returned safely to Earth on August 9, 1990, its crew having spent 179 days in space.

February 14, 1990
LACE / RME
U.S.A.

LACE and RME, two military experimental satellites, tested the use of lasers and mirrors to destroy enemy missiles. Ground lasers were fired at the Relay Mirror Experiment (RME) satellite, which used a 2-foot-diameter mirror to redirect the light beam at a target with an aiming error of less than one-fifty-seven-millionth of a degree.

The second satellite, called the Low-power Atmospheric Compensation Experiment (LACE), carried 210 different sensors for measuring the laser beam's accuracy and any distortion it might experience as it traveled through the atmosphere. LACE also carried a camera designed to take ultraviolet images of rocket plumes, testing technology for future surveillance satellites. Over the next few years, the spacecraft regularly observed the thruster and rocket firings of the space shuttle, using these data to calibrate its sensors.

February 28, 1990
STS 36, Atlantis Flight #6
KH-12 1
U.S.A.

STS 36, the sixth flight of the shuttle Atlantis, was also the sixth classified military mission in the shuttle program, following **STS 33 (November 22, 1989)**. It lasted four days and carried a crew of five, including John Creighton (**STS 51G–June 17, 1985**; **STS 48–September 12, 1991**), John Casper (**STS 54–January 13, 1993**; **STS 62–March 4, 1994**; **STS 77–May 19, 1996**), David Hilmers (**STS 51J–October 3, 1985**; **STS 26–September 29, 1988**; **STS 42–January 22, 1992**), Richard Mullane (**STS 41D–August 30, 1984**; **STS 27–December 2, 1988**), and Pierre Thuot (**STS 49–May 7, 1992**; **STS 62–March 4, 1994**). The mission deployed one surveillance satellite. In order to place this satellite in a high inclination orbit passing over most of the Soviet Union, Atlantis was launched into an orbit with the highest inclination of any shuttle mission—62 degrees.

• *KH-12 1* Deployed on February 28th, the surveillance satellite KH-12 1 replaced the KH-11 satellites (see **KH-11 1–December 19, 1976**). It had a longer orbital service life and a greater ability to take high-resolution pictures—capable of resolving objects as small as four *inches*. Some experts believed the satellites also used advanced radar technology to produce images at night and through clouds.

Because of a failure in the satellite's booster rockets, KH-12 1 broke apart less than a month after launch.

Furthermore, the Challenger accident (see **STS 51L–January 28, 1986**) caused a rethinking of the military's dependence on the shuttle and its KH-12 satellite program, which had been designed to be launched and maintained by the shuttle. Subsequent launches used standard disposable rockets and, through December 1999, only two more KH-12 satellites are known to have been orbited.

The next shuttle mission was **STS 31 (April 24, 1990)**. The next classified military shuttle mission was **STS 38 (November 15, 1990)**.

April 3, 1990
Ofeq 2
Israel

A follow-up of Israel's first satellite, **Ofeq 1 (April 19, 1988)**, Ofeq 2 was launched from a site in central Israel. Like its predecessor, Ofeq 2 was a short-term satellite for testing Israel's launch and satellite technology. It remained in orbit for three months, providing ground stations information on its orbit and operation.

The next Israeli satellite was **Ofeq 3 (April 5, 1995)**, Israel's first operational surveillance satellite.

April 5, 1990
Pegsat / SECS
U.S.A.

Pegsat and SECS, two small experimental satellites, were the first objects placed in orbit by the Pegasus launch system, which was the first newly designed American rocket to put a satellite in orbit since the development of the Saturn 5 (see **Apollo 4–November 9, 1967**). Unlike the previous attempt to build a new launch system, **Conestoga (September 9, 1982)**, Pegasus rockets were launched not from the ground but from the air, first from a B-52 and later from an L1011 aircraft. Developed privately by Orbital Sciences Corporation, Pegasus was a three-stage winged rocket that, once released at about 43,000 feet altitude, automatically piloted itself into an orbital trajectory.

• *Pegsat* The satellite Pegsat provided telemetry of the aerodynamics of Pegasus' launch, detailing the attitude, temperature, pressure, and stresses on both rocket and satellite as it was released by the B-52 and accelerated to orbital velocity.

Once placed in a polar orbit 414 by 250 miles, Pegsat released two canisters of barium and strontium gas above northern Canada so that scientists could study the behavior of the earth's magnetic field at high latitudes, continuing ionospheric research by **Intercosmos 24/Magion 2 (September 28, 1989)**. Pegsat was the first American ionospheric satellite in six years, since **CCE (August 16, 1984)**. Ground observations noted that on the first discharge the gas was accelerated upward along magnetic field lines to elevations exceeding 3,100 miles. On the second release nine days later, however, the gas did not move upward but remained close to Earth at orbital altitude.

This gas release program continued with **CRRES (July 25, 1990)**.

• *SECS* The U.S. Navy Small Experimental Communications Satellite (SECS) was similar to GLOMR (see **STS 61A–October 30, 1985**), as well as **UoSat 3/UoSat 4 (January 22, 1990)**, testing digital communication store-and-forward technologies.

April 7, 1990
Asiasat 1
Hong Kong / China

Owned by the Asia Satellite Telecommunications Company, based in Hong Kong, Asiasat 1 had one of the most interesting histories of any previous communication satellite. Originally built by Western Union and named Westar 6, the satellite was slated to go into geosynchronous orbit in 1984, supplementing that company's second generation of satellites serving the U.S. domestic telephone and television market (see **Westar 4–February 26, 1982**). Deployed from the space shuttle (see **STS 41B–February 3, 1984**), Westar 6 was stranded in a useless low-Earth orbit when its upper stage failed to fire.

Because of this failure, the insurance company, Lloyd's of London, paid off Western Union's policy and took ownership, immediately working out a rescue plan with NASA. On a subsequent shuttle flight (see **STS 51A–November 8, 1984**), the satellite was recovered and returned to Earth. It was then resold to Asia Satellite Telecommunications Company, which refurbished it.

When launched into orbit, Asiasat 1 became the first satellite to be placed in orbit twice. Asiasat 1's history was further complicated as this second launch was also China's first commercial launch. Because of concerns in the United States about technology transfers to China, the satellite was kept under constant supervision by U.S. officials during its transport to China and preparation for launch.

Today, Asiasat provides telephone and broadcast communications for a number of Asian and Southeast Asian nations, including Hong Kong, China, Thailand, South Korea, and Mongolia. Because of Asiasat 1's success, Asia Satellite Telecommunications has since augmented it with two satellites, one launched in 1995 and a second in 1999. All three satellites were placed in geosynchronous orbit above Asia and are still in use as of December 1999.

A third Asiasat satellite, Asiasat 3, was initially stranded in low-Earth orbit in 1997 when the upper stage of its Russian Proton booster failed and was therefore never used by Asia Satellite Telecommunications. Instead, the insurance company took possession, and the satellite's builder, Hughes Space and Communication, used the Moon's gravity to place the satellite in a more useful orbit. See **HGS 1 (December 25, 1997)**.

April 11, 1990, 16:00 GMT
POGS / TEX / SCE
U.S.A.

• *POGS* The Polar Orbiting Geomagnetic Survey (POGS) was a U.S. Air Force satellite for updating the maps of the earth's magnetic field, last surveyed by **Magsat (October 30, 1979)**. POGS also carried an experimental solid-state recorder, testing whether modern high-speed miniaturized integrated circuits could function dependably in space.

• *TEX* The Transceiver EXperiment (TEX) measured, for the U.S. Air Force, how a variety of frequencies propagated through the upper atmosphere. It used a variable-power transmitter to test how atmospheric interference could be mitigated depending upon signal power.

• *SCE* The Selective Communications Experiment (SCE) also conducted U.S. Air Force propagation experiments using a variable-power transmitter, testing different frequencies than TEX.

April 11, 1990, 17:00 GMT
Foton 6
U.S.S.R. / France

Called Foton 3 when launched, Foton 6 was actually the sixth satellite in a Soviet commercial program using recoverable satellites for materials research. See **Foton 1 (April 16, 1985)** and **Foton 5 (April 26, 1989)** for program details.

Foton 6, which remained in orbit for 16 days, carried both Soviet and French microgravity experiments, testing methods for growing crystals in weightlessness. Both protein and semiconductor crystals were grown. The French experiment grew protein albumin crystals for medical research.

With this launch began a six-satellite program of French, British, German, and European research on Foton satellites. Through September 1999, six more Fotons were launched, each remaining in orbit from 14 to 18 days. Research included testing a variety of materials in the space environment, as well as crystal growth and cellular biological experiments.

Of these satellites, only Foton 9 failed, becoming seriously damaged after landing. A helicopter that was transporting the capsule to a local airfield was forced to drop it due to very bad weather conditions.

April 13, 1990
Palapa B2R
Indonesia / U.S.A.

Palapa B2R was the third satellite in Indonesia's second-generation cluster of geosynchronous communications satellites (see **STS 7–June 18, 1983**). It was also the second satellite to be placed in orbit twice, following **Asiasat 1 (April 7, 1990)**. Like Asiasat 1, Palapa B2R had first been deployed by **STS 41B (February 3, 1984)**, only to be stranded in low orbit when its upper stage failed to fire.

The insurance companies paid off Indonesia's claim and arranged with NASA to retrieve the satellite on a subsequent shuttle flight (see **STS 51A–November 8, 1984**). Palapa B2R was then refurbished and sold back to Indonesia to be relaunched on a Delta rocket.

April 24, 1990
STS 31, Discovery Flight #10
Hubble Space Telescope
U.S.A.

The tenth flight of the shuttle Discovery, STS 31 lasted five days and was crewed by Loren Shriver (**STS 51C–January 24, 1985; STS 46–July 31, 1992**), Charles Bolden (**STS 61C–January 12, 1986; STS 45–March 24, 1992; STS 60–February 3, 1994**), Steven Hawley (**STS 41D–August 30, 1984; STS 61C–January 12, 1986; STS 82–February 11, 1997**), Bruce McCandless (**STS 41B–February 3, 1984**), and Kathryn Sullivan (**STS 41G–October 5, 1984; STS 45–March 24, 1992**). Its main mission, the deployment of NASA's Hubble

Space Telescope, required that Discovery's orbit reach an altitude of 383 miles, higher than any previous shuttle mission.

• *Hubble Space Telescope* Deployed on April 25th, the Hubble Space Telescope heralded in the beginning of a golden period of space-based astronomical research. Throughout the 1990s, more than a dozen orbiting telescopes covering almost the entire electromagnetic spectrum, from the infrared (**ISO–November 17, 1995**) to the ultraviolet (**EUVE–June 7, 1992**) to x-rays (**Rossi–December 30, 1995**) to gamma radiation (see **Compton/STS 37–April 5, 1991**), were placed in orbit. Unlike earlier space-based telescopes, which flew for only short periods and whose researchers designed and operated the equipment during that period, most of these newer telescopes had long life spans and were operated much like ground-based telescopes—in which astronomers applied to a scientific review committee for observation time.

Hubble was the first optical telescope able to operate successfully above the cloudy interference of the earth's atmosphere. First conceived in the early 1970s, and then delayed for four years because of the Challenger accident (see **STS 51L–January 28, 1986**), Hubble's primary mirror, 94.5 inches across, was the 31st largest in the world. Its instrumentation at launch allowed it to take photographs from the ultraviolet to the near-infrared (1,150–11,000 angstroms) and included lenses for both wide-angle and telephoto images. Scientists estimated that the telescope would be able to see stars as dim as 28th magnitude, between 25 to 50 times fainter and with a resolution seven to eight times better than previously possible. Furthermore, Hubble carried two different spectrographs, able to produce spectrums of many faint objects simultaneously.

Scientists expected the Hubble Space Telescope to revolutionize astronomy and humanity's perspective of the universe. And though that expectation was eventually realized, Hubble did not begin its life auspiciously. An engineering mistake during the manufacture of its primary mirror put the telescope's images out of focus. Although some research was possible using the telescope's spectrographs, most of the scientific results during its first three years of operation were disappointing at best.

Fortunately, the telescope had been built with the space shuttle in mind. Scheduled inspections and upgrades every three years by a manned shuttle crew were part of its design. In order to make the telescope usable, a corrective set of lenses was devised and installed on **STS 61 (December 2, 1993)**, during the first maintenance visit. See that flight for Hubble's repair, as well as a summary of some of its discoveries during its first six years of operation.

Onboard experiments included more protein crystal growth research, continuing work from **STS 32 (January 9, 1990)**. See **STS 26 (September 29, 1988)** for a research outline. On STS 31, 60 samples attempted to grow 12 different protein crystals at controlled cold temperatures. Eight crystals were successfully grown, larger and more uniform than possible on Earth. These results matched the previous crystal growth experiments, and included crystals for studying emphysema, arthritis, and the production of sugar from proteins. The results also indicated that in space about 20 percent of all samples would produce superior crystals. Shuttle protein crystal research continued on **STS 37 (April 5, 1991)**.

Two instruments monitored the shuttle's environment, with one measuring the radiation dosage within the cabin during orbit and the other measuring the amount of contamination experienced within the shuttle's cargo bay prior to and during launch. These instruments flew routinely on later shuttle flights, allowing scientists to develop baseline data.

The single student experiment tested to see how arcs of electricity behaved in a weightless environment.

Two IMAX cameras were carried, one in the shuttle's cargo bay and one in the crew compartment. Not only did these cameras document the release of Hubble, they also took footage of Japan, the Andes, the Amazon River basin, and San Francisco Bay. Because of the shuttle's high altitude, the view was especially spectacular. Footage was used in both the second and third IMAX space films, *The Blue Planet* and *Destiny in Space*.

Also on STS 31 was a new experiment to study how very thin polymer membranes could be produced in the weightlessness of space. These membranes were used as filters by many industries in desalination, the filtration of food products, and the dialysis of the kidneys and blood. Researchers hoped that in microgravity, more uniform membranes could be produced. The experiment found that the space-produced membranes were significantly different than Earth-produced ones, although minor acceleration effects on the membranes following their creation apparently skewed the results. The next shuttle mission, **STS 41 (October 6, 1990)**, carried hardware to try and counter this.

May 9, 1990
Macsat 1 / Macsat 2
U.S.A.

Two U.S. military experimental satellites, called Multiple Access Communications Satellites (Macsats), were similar to GLOMR (see **STS 61A–October 30, 1985**). They were extremely lightweight and quickly built, and they were designed to test digital communication store-and-forward technologies.

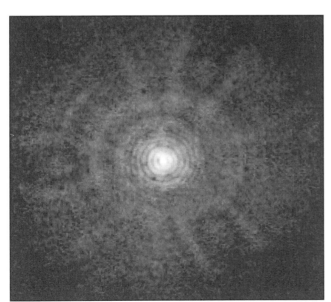

This image of a single star shows the spherical aberration—the multiple halos—that prevented Hubble images from focusing sharply. *NASA*

May 31, 1990
Kristall
U.S.S.R.

Kristall was the third permanent module added to **Mir (February 20, 1986)**, following **Kvant 1 (March 31, 1987)** and **Kvant 2 (November 26, 1989)**. The module increased the station's interior space by 1,400 cubic feet, for a total of approximately 8,400 cubic feet. With this module, the entire Mir complex weighed about 70 tons.

Kristall was divided into two compartments. The main section was a laboratory for performing materials processing experiments. It contained four furnaces, producing both metal and protein crystals. On this initial launch, the compartment was also packed with food cargo.

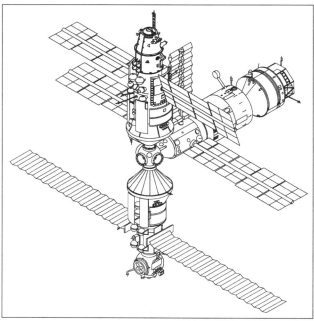

Mir Core with Kvant 1 (aft port), Kvant 2 (top port), and Kristall (bottom port). Compare with diagram on November 26, 1989. *NASA*

The second compartment was a multidocking adapter with two androgynous docking units, increasing Mir's docking ports from six to eight. One of these docking ports was designed for use by the Soviet space shuttle (see **Buran–November 15, 1988**), while the other was for housing an x-ray telescope that Buran was supposed to bring to the station. Neither Buran nor the x-ray telescope ever flew, though Buran's port was used once by the American space shuttle on **STS 71 (June 27, 1995)**. A third opening on Kristall's multidocking adapter housed several multispectral cameras and a gamma-ray telescope and spectrometer.

One day after docking with the front docking port on Mir's multidocking adapter, a robot arm manually moved Kristall to a lateral port, opposite Kvant 2, so that Mir was shaped like the letter T. This configuration balanced the station, thereby reducing the engine burns necessary to keep it oriented properly.

See **Soyuz TM9 (February 11, 1990)** for more details.

June 1, 1990
Rosat
Germany / U.S.A. / U.K.

The ROentgen SATellite (ROSAT), built by West Germany (with additional instruments from the United Kingdom) and launched by NASA, was designed to produce images in the extreme ultraviolet (60–300 angstroms, or 0.04–0.2 KeV)—the first spacecraft to do so—and the low or soft x-ray range (6–100 angstroms, or 0.1–2.4 KeV). Rosat followed and supplemented the still operating high-energy x-ray telescope **Granat (December 1, 1989)**.

Rosat's cameras were unique in that they not only covered a greater range with more sensitivity, they were also able to resolve and focus these emissions into detailed images, which allowed astronomers to pinpoint precisely each object's position in the sky and thereby find its optical counterpart.

The satellite produced the most complete all-sky survey in these wavelengths to date, increasing the number of known x-ray objects from about 1,000 to more than 60,000. Of these, 40,000 were immediately linked to visible objects: 20,000 were identified as objects within the Milky Way, most of which were thought to be neutron stars, while the other 20,000 were extragalactic, thought to be supermassive black holes in the centers of galaxies.

The x-ray emitting stars included a wide assortment of objects, from ordinary stars with hot turbulent atmospheres to black hole candidates. Most significant were the x-ray binary systems, which consisted of a normal star circling an x-ray emitting object, either a neutron star (a dead star whose matter has collapsed into a tiny superdense object less than 15 miles across) or a black hole.

Rosat's observations over time further confirmed results from Granat about x-ray bursters. Astronomers believed that these bursts were caused by the in-fall of material onto the surface of a neutron star.

Of the extra-galactic sources, Rosat initially proved that at least 45 percent of the background x-ray radiation could be attributed to individual quasars. Scientists believed that better instruments would show that most of the remainder came from quasars as well.

In the ultraviolet, Rosat more than doubled the number of known white dwarfs, from 50 to over 115. It also made a more complete map of the gas in the Milky Way, confirming data from **IRAS (January 26, 1983)** that the clouds were distributed unevenly, with giant spheres of empty space surrounded by vast waves of gas. A large number of these clouds appeared to be supernova remnants, the bubbles of gas blown out from the star's explosion. The evidence also indicated that the Sun itself might be traveling through one such region of bubbles.

Following the completion of this six-month survey, Rosat operated through 1999 like most ground-based telescopes, made available to astronomers for guest observations of specific celestial objects.

The next space telescope observing in soft x-rays was flown on **STS 35 (December 2, 1990)**. **EUVE (June 7, 1992)** was the next permanently orbiting extreme ultraviolet–soft x-ray telescope.

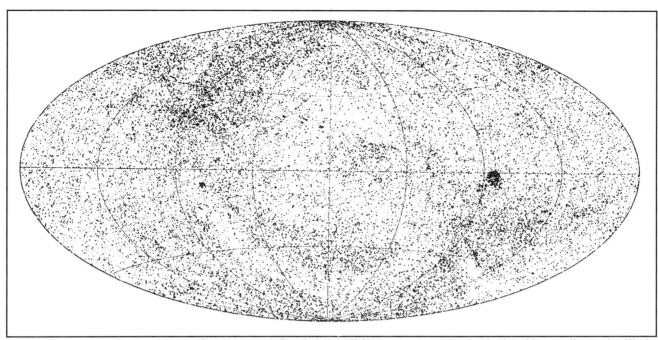

Rosat all-sky survey in galactic coordinates. Because x-rays in Rosat's 0.1–2.4 KeV sensor range are absorbed by the cold interstellar gas that fills the plane of the Milky Way, the survey does not show an increase in density along that plane, as had previous x-ray surveys by Uhuru (December 12, 1970) and HEAO 1 (August 12, 1977). *NASA*

June 8, 1990
USA 59
U.S.A.

USA 59 was the first satellite in a next-generation cluster of the Navy Ocean Sensing System (NOSS) used by the U.S. Navy for obtaining data on ocean weather and wave patterns, as well as for naval reconnaissance. See **NOSS 1 (April 30, 1976)** for the previous generation. Some of the information provided by these satellites was similar to the NOAA weather satellites (see **NOAA 1–December 11, 1970**). Through 1996, three NOSS satellites (now called NROSS for Navy Remote Ocean Sensing System) were launched, each a minicluster of three subsatellites floating several hundred feet apart, connected to each other by fine wires. See **Triad (September 2, 1972)**.

July 11, 1990
Gamma
U.S.S.R. / France

A joint Soviet-French project, Gamma was designed to study specific celestial objects in x-rays (2–25 KeV) and gamma rays (50–5,000 MeV), following up research by **Granat (December 1, 1989)** but with the ability to observe celestial radiation at much higher energies. Gamma operated for about a year, making observations of the Large Magellanic Cloud and supernova SN1987a, as well as the central regions of the Milky Way.

The next high-energy telescope was the Compton Gamma Ray Observatory (see **STS 37–April 5, 1991**).

July 16, 1990
Badr A
Pakistan / China

Badr A was built by the Pakistan Amateur Radio Society and carried digital store-and-forward equipment similar to that on UoSat 2 (**March 1, 1984**). It was used to relay ham radio messages until its orbit decayed in December 1990.

Badr A was placed in orbit by the first test flight of the Chinese Long March 2 rocket, the middle-sized rocket in a family of three launch vehicles being developed by the Chinese to compete in the global commercial satellite launch industry.

July 25, 1990
CRRES
U.S.A.

The Combined Release and Radiation Effects Satellite (CRRES) was a joint U.S. Air Force and NASA project to study how particles in the Van Allen radiation belts traveled through the earth's magnetic field. As part of its research program, the spacecraft released a series of 24 chemical canisters through early 1991, producing gas clouds so that scientists could observe their development in the belts. CRRES also investigated how the radiation in the Van Allen belts affected computer software and hardware, studying different methods for protecting and prolonging their life span.

Both studies continued research from **Pegsat (April 5, 1990)**. They found that certain miniaturized electronics were badly damaged by the Van Allen belt radiation. Computer chips produced with silicon and gallium arsenide were particularly hard hit, as were the computer's memory chips. Unexpectedly, the worst damage was caused by protons from the Sun rather than high-energy intergalactic cosmic rays.

The next ionospheric-magnetospheric satellites were **Intercosmos 25/Magion 3 (December 18, 1991)**.

August 1, 1990
Soyuz TM10
U.S.S.R.

The eight-day Soyuz TM10 mission was crewed by Gennady Manakov (**Soyuz TM16–January 24, 1993**) and Gennady Strekalov (**Soyuz T3–November 27, 1980; Soyuz T8–April 20, 1983; Soyuz T10A–September 27, 1983; Soyuz T11–April 3, 1984; Soyuz TM21–March 14, 1995**). It took over occupancy of **Mir (February 20, 1986)** from the **Soyuz TM9 (February 11, 1990)** crew. For six days, the two crews occupied the station, transferring operations. Then, the Soyuz TM9 crew returned to Earth, leaving Manakov and Strekalov to operate Mir.

Their mission, similar to the last two Soyuz flights (see **Soyuz TM7–November 26, 1988**), was to make a profit from space. For example, the surface area of their launch rocket had been treated as a billboard and sold to private commercial firms for advertising.

During their stay, the two men were resupplied twice by Progress freighters (see **Progress M1–August 23, 1989**). The first brought cables and equipment to prepare for the arrival of a Japanese journalist on **Soyuz TM11 (December 2, 1990)**. The second was the first operational use of the Progress recoverable capsule, able to return 330 pounds to Earth. Though the capsule was successfully deorbited on November 28th, its beacon signal was never detected, and the capsule with all its experimental specimens was lost.

Soyuz TM10 continued the commercial materials research started by the Soyuz TM9 crew, using **Kristall (May 31, 1990)** to grow experimental semiconductor and protein crystals. The loss of the recoverable capsule and its specimens, however, destroyed most of the results from this research.

On October 29th, the two men performed a space walk, during which they placed micrometeorite detectors on the outside of Mir, as well as made an attempt to repair the balky **Kvant 2 (November 26, 1989)** hatch that had given the Soyuz TM9 crew problems. Because of a stripped screw, the repairs could not be completed, so completion of the task was left to the next Soyuz crew. See **Soyuz TM11 (December 2, 1990)**.

On December 4th, Manakov and Stekalov were joined by the crew of Soyuz TM11: Musa Manarov, Victor Afanasyev, and Toyohiro Akiyama. Akiyama was a Japanese journalist and the first civilian to reach orbit who was not part of a government space program. See Soyuz TM11 for details.

On December 10, 1990, Manakov, Strekalov, and Akiyama returned to Earth, leaving Mir in the hands of the Soyuz TM11 crew. Manakov and Strekalov had spent 132 days in space.

Four years later, the Soyuz TM10 capsule became the first manned capsule to be sold to the public. The Russians, in an attempt to raise cash, had the capsule auctioned off at Sotheby's in December 1993 for $1,652,500. Also auctioned were Manakov's and Strekalov's spacesuits ($34,500 and $85,000, respectively),

August 10, 1990
Magellan, Venus Orbit
U.S.A.

Magellan, first deployed by the space shuttle Atlantis (see **STS 30–May 4, 1989**), was a follow-up of the **Pioneer Venus Orbiter (December 4, 1978)**. It orbited Venus to provide a high-resolution radar map of the planet's surface.

The spacecraft reached Venus on August 10, 1990, where it spent the next four years mapping 98 percent of the planet's surface *(see color insert, Figure 31)*. Magellan's side-looking radar could see objects as small as 250 feet. Magellan's polar orbit was positioned so that every 243 days, Venus's entire surface rotated under the spacecraft, and during its life span the planet's surface passed under Magellan six times. During the first three cycles, Magellan radar-mapped the surface, twice using its side-looking radar looking leftward, and once looking rightward. The two left-looking images were taken at slightly different orientations in order to create stereo 3-D images.

In Magellan's fourth and fifth cycles, it obtained gravity data, mapping the density changes in the planet's interior. The sixth cycle performed a "windmill" experiment. The spacecraft dipped down to 85 miles, using its solar panels as sails to measure the thickness of the planet's atmosphere. Shortly thereafter, on October 12, 1994, the spacecraft's mission ended as it burned up in Venus's atmosphere.

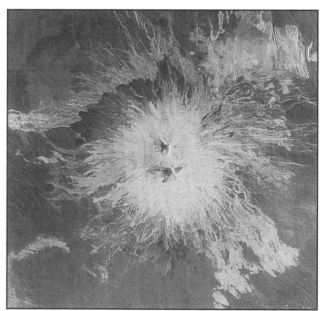

The Sapas Mons volcano on Venus. Note lava flows from peak and on flanks. *NASA*

The radar maps revealed an alien and volcanic surface, with more than 85 percent of the surface covered with lava flows. Magellan discovered over 1,100 volcanoes with diameters greater than 12 miles. The diameters of 167 volcanoes were larger than 60 miles. Scientists estimated that the surface had about a million volcanoes. The volcanoes were also not randomly distributed on the surface, but were clustered in specific areas, mostly in the equatorial latitudes, indicating the presence of long-lasting volcanic hot spots. The surface had relatively few craters, and most of these appeared young, less than half a billion years old.

Magellan confirmed and refined some of the surface features first seen by **Venera 15 (October 10, 1983)**, **Venera 16 (October 14, 1983)**, and **Pioneer Venus Orbiter**. About

August 10, 1990

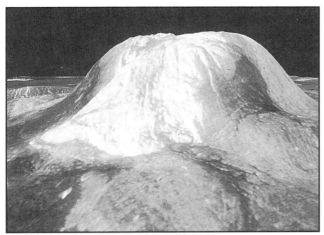
Computer-generated image of two-mile-high Gula Mons volcano. NASA

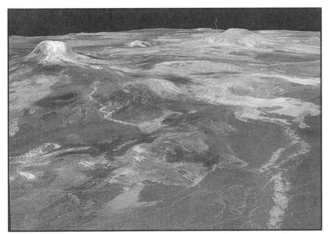
Computer-generated image looking southwest at Gula Mons on horizon left and Sif Mons (about 6,300 feet high) on horizon right. Note the lava flows coming off Gula Mons and across the plain. NASA

Overhead radar image of Gula Mons (bottom right of image). Lava flows from the computer image above can be seen moving northward from the volcano. NASA

70 percent of the planet was lowland plains. Two large highlands, Ishtar and Aphrodite Terrae (equivalent to Earth's continents), were covered mostly with tessera, regions of complex, thin intersecting ridgelines. These features were generally the oldest terrain. What geological event had caused them was entirely unknown.

Lava channels, similar to the Moon's rilles, extended for thousands of miles. The longest, Baltis Vallis, runs for over 4,200 miles, the longest such channel known in the solar system. Because ordinary lava cannot flow such distances without solidifying—even in Venus's hot and thick atmosphere—some scientists theorized that these channels might have been caused by rivers of silicate, sulfur, or carbonite.

Coronae, discovered by Venera 15 and Venera 16, were now thought to be a special type of Venusian volcanic activity. These circular features generally resembled volcanic calderas, though they were not on the top of volcanoes. Often, they were surrounded by many concentric ridgelines. Sometimes they were a depression, sometimes a plateau. Scientists believed that they formed when a blob of magma rises from the planet's interior, applying pressure to the surface and causing a bulge, with lava leaking across its top. Eventually, the magma cools, and the bulge turns into a depression surrounded by the circular ridgelines.

Magellan found that the mountaintops of Venus generally had a high radar reflectivity. On Earth, compounds of the volatile metals—copper, zinc, tin, lead, arsenic, antimony, and bismuth—are often vented from volcanoes. Scientists theorized that on Venus such materials were vented, then formed aerosol hazes in its hot atmosphere. These metallic clouds would eventually settle on the planet's mountaintops, falling as metallic snow.

Scientists now believed that Venus's global geology was generally vertical, lava moving up through the crust, rather than the horizontal plate tectonics of Earth. Magellan's gravity data and surface maps of numerous volcanoes indicated a much thinner crust than Earth's. The crust was also not broken into a collection of shifting plates, but appeared fixed as a single unit across the entire globe. Scientists suspected that the lack of surface oceans weighing down upon the lithosphere allowed it to stay unbroken. Conversely, scientists also concluded that the earth's plate tectonics were somehow directly linked to the existence of its vast oceans.

Since Venus had almost no craters older than 500 million years in age, scientists believed volcanic resurfacing had recoated most of the planet's surface. Only the few highland tessera regions remained from before this time period. Whether the resurfacing had taken place in a single global event half-a-billion years ago, or in a series of volcanic flows over a long period of time, is still unknown. Nor is it clear today how much volcanic activity still occurs on Venus.

The atmosphere could be summarized as being very hot (averaging 900°F), very dense (almost 100 times denser than Earth's and equal to more than 1,300 pounds of pressure per square inch), very dry, and made up almost entirely of carbon dioxide. The planet's clouds were 75 percent sulfuric acid and 25 percent water vapor, forming three layers between 30 and 42 miles altitude. Although scientists knew that the atmosphere seemed to circulate in planetary-sized waves in the opposite direction of the planet's rotation, its overall planetary flow patterns were not clearly understood—the data were still too sketchy. The most popular theory said that the temperature difference between the day and night sides drove the upper atmosphere nightward. At noon, the atmosphere rose and flowed to the night side, where it sank and flowed back to the day side.

The scientists also did not have a clear understanding why Venus's clouds appeared yellow. The atmosphere's structure below 6 and above 70 miles elevation remained unknown. On the surface, the ground's make-up was largely a mystery, limited to the few samples tested by the Venera landers. And though the planet's size and density closely resemble Earth's, the answer to why it had evolved so differently remains largely unexplained.

Magellan was the last planetary mission to Venus through December 1999, and no future mission to Venus was planned at that date.

August 30, 1990
Eutelsat 2 F1
Europe

Owned and operated by Eutelsat (see **OTS 2–May 11, 1978**) and launched on an Ariane rocket, Eutelsat 2 F1 inaugurated the consortium's second-generation satellite cluster. See **Eutelsat 1 (June 16, 1983)**. Placed in geosynchronous orbit over Europe, it was the first of five satellites launched through January 1994, later supplemented by a sixth more advanced satellite in October 1998. Eutelsat 2 satellites operated with three times the power, had more efficiency, greater redundancy, and better spectrum coverage. Aware of the increasing demand for satellite broadcast television, they also had a greater capacity for television signal distribution.

Despite this, increased demand for television signal capacity forced Eutelsat in 1995 to again supplement its Eutelsat 2 satellites. See **Hot Bird 1 (March 28, 1995)**.

September 3, 1990
Atmosphere 1
China

Atmosphere 1, a Chinese research project, placed two inflatable balloons in orbit to study the changing density of the earth's atmosphere between 250 to 600 miles. Similar to the earlier American balloon satellites, such as **Explorer 19 (December 19, 1963)**, the two satellites remained in orbit until March and July of 1991.

Until the launch of Starshine (see **STS 96–May 27, 1999**), these satellites are the last specifically dedicated to atmospheric density research. By 1990, earlier research (see **Explorer 19** and **Explorer 55–November 20, 1975**) was sufficient to allow reasonably accurate predictions of atmospheric changes, and any further data could be acquired by monitoring the normal orbital changes of most other satellites. While Starshine is dedicated to atmospheric density study, its main focus is education.

October 5, 1990
Fanhui Shu Weixing 12
China

With the recoverable satellite Fanhui Shu Weixing 12, China performed its first in-space biological research. The satellite remained in orbit for eight days, carrying a collection of animals and plants.

October 6, 1990
STS 41, Discovery Flight #11
Ulysses
U.S.A. / Europe

STS 41 was Discovery's eleventh flight. It lasted four days and was crewed by Richard Richards (STS 28–August 8, 1989; STS 50–June 25, 1992; STS 64–September 9, 1994), Robert Cabana (STS 53–December 2, 1992; STS 65–July 8, 1994; STS 88–December 4, 1998), Bruce Melnick (STS 49–May 7, 1992), William Shepherd (STS 27–December 2, 1988; STS 52–October 22, 1992), and Thomas Akers (STS 49–May 7, 1992; STS 61–December 2, 1993; STS 79–September 16, 1996). STS 41 carried one interplanetary probe.

• *Ulysses* Deployed on October 6th, Ulysses was a solar research satellite augmenting observations of Solar Max (see **STS 41C–April 6, 1984**). It was also the first spacecraft to fly above the ecliptic of the solar system, flying over the Sun's polar regions. A joint project of NASA and the European Space Agency, Ulysses was first aimed at Jupiter in order to use that giant planet's gravitational field to swing the spacecraft into a polar orbit. See also **Ulysses, Jupiter Flyby–February 8, 1992; Ulysses, First Solar Orbit–April 17, 1998**.

The next satellite dedicated to solar observations was **Yohkoh (August 30, 1991)**.

Onboard experiments included the polymer membrane experiment, which hoped to produce thin films for use as filters in industry and medicine. This mission tested hardware redesigned to try and dampen the minor vibrations caused by shuttle and crew activity and observed on the experiment's first flight, **STS 31 (April 24, 1990)**. Despite the improvements, the shape and structure of the polymer membranes were still affected by the slightest vibrations from the shuttle or crew. See **STS 43 (August 2, 1991)** for further attempts to eliminate these vibrations.

A private commercial study, designed by Genentech, Inc., of South San Francisco, California, used 16 rats to investigate how certain medical treatments counteracted the effects of weightlessness. Since these effects were similar to other Earth-based medical disorders, the company hoped their results would improve the treatment of bone disease, heart conditions, and the immune system. Eight rats received protein injections, while eight others did not. Because of unexpected temperature fluctuations within the rats' enclosure, however, the results were uncertain. Research continued on **STS 48 (September 12, 1991)**.

Chromex (CHROMosome and Plant Cell Experiment) made its second flight, following **STS 29 (March 13, 1989)**, attempting to isolate the cause of the chromosomal abnormalities and tissue growth differences found in the weed and daylily roots grown in space. For results, see **STS 51 (September 12, 1993)**.

Intelsat (see **Early Bird–April 6, 1965**) tested an array of solar panel materials to measure their durability in space. These materials had been used on the second Intelsat 6 communications satellite (see **Intelsat 6A F2–October 27, 1989**). This satellite had been stranded in low-Earth orbit due to the failure of its booster rocket. Before committing to a planned shuttle rescue mission in 1992 (see **STS 49–May 7, 1992**), Intelsat needed to know the decay rate of the satellite's solar panels.

A fire and combustion experiment took a small piece of ashless filter paper in a pressurized chamber and set it on fire, testing how materials burned in space. The fire, rather than forming the familiar pointed flame, instead took the shape of a ball. Because there was no convection, as the fire burned it became increasingly isolated from the surrounding atmosphere. The nearby oxygen became depleted and was replaced by a spherical cloud of soot encircling the flame and causing it to wither. Such fire and combustion experiments were flown repeatedly on later shuttles. See **STS 47 (September 12, 1992)** and **STS 54 (January 13, 1993)**.

The next shuttle mission was **STS 38 (November 15, 1990)**.

October 30, 1990
Inmarsat 2 F1
International / U.S.A.

Launched from Cape Canaveral, this satellite was the first in a four-satellite constellation owned by Inmarsat (see **Marisat 1–February 19, 1976**). It supplemented and then replaced the MARECS constellation (see **MARECS A–December 20, 1981**), tripling its capacity to 250 two-way voice circuits, as well as one shore-to-ship channel and four ship-to-shore channels. The satellites also provide mobile communications capability. The first satellite was placed over the Indian Ocean, the second and fourth over the Atlantic, and the third over the Pacific.

The constellation was designed to last into the twenty-first century. Increasing capacity demands, however, forced Inmarsat to launch an additional five geosynchronous satellites from 1996 to 1998. See **Inmarsat 3 F1 (April 3, 1996)**.

November 15, 1990
STS 38, Atlantis Flight #7
  ## Magnum 3
U.S.A.

STS 38, a five-day mission, was the seventh flight of the space shuttle Atlantis and was crewed by Richard Covey (**STS 51I–August 27, 1985; STS 26–September 29, 1988; STS 61–December 2, 1993**), Frank Culbertson (**STS 51–September 12, 1993**), Robert Springer (**STS 29–March 13, 1989**), Carl Meade (**STS 50–June 25, 1992; STS 64–September 9, 1994**), and Charles Gemar (**STS 48–September 12, 1991; STS 62–March 4, 1994**). This flight was the seventh classified military shuttle mission, following **STS 36 (February 28, 1990)**. It released one surveillance satellite.

- *Magnum 3* Deployed on November 15th, Magnum 3 was a variation of the Chalet satellites (see **Chalet 1–June 10, 1978**), designed to listen and intercept electronic signals, including radio, voice communications, radar, and missile telemetry. See also **STS 51C (January 24, 1985)**.

The next shuttle mission was **STS 35 (December 2, 1990)**. The next military shuttle mission was **STS 39 (April 28, 1991)**.

November 20, 1990
Satcom C1
U.S.A.

Owned and operated by General Electric (GE), Satcom C1 began the replacement of that company's RCA Satcom 5 communications satellites (see **RCA Satcom 5–October 28, 1982**), which provided telephone and television service to the U.S. market. Through 1992, GE launched three Satcom C satellites, all only slightly more advanced than the RCA Satcom 5 satellites, and all stationed over the United States. In December 1999, they were still in use.

December 2, 1990, 6:49 GMT
STS 35, Columbia Flight #10
U.S.A.

STS 35, Columbia's tenth flight, was a nine-day mission; it was the Spacelab's fifth flight and the first since **STS 61A (October 30, 1985)**. The seven-person crew consisted of Vance Brand (**Apollo 18–July 15, 1975; STS 5–November 11, 1982; STS 41B–February 3, 1984**), Guy Gardner (**STS 27–December 2, 1988**), Jeffrey Hoffman (**STS 51D–April 12, 1985; STS 46–July 31, 1992; STS 61–December 2, 1993; STS 75–February 22, 1996**), Mike Lounge (**STS 51I–August 27, 1985; STS 26–September 29, 1988**), Robert Parker (**STS 9–November 28, 1983**), Ronald Parise (**STS 67–March 2, 1995**), and Sam Durrance (**STS 67–March 2, 1995**).

The 7-person Columbia crew, combined with the 5 men onboard Mir (the crews of **Soyuz TM10–August 1, 1990** and **Soyuz TM11–December 2, 1990**), made 12 humans in space, a new record, which exceeded the previous record of 11 set during **STS 41C (April 6, 1984)** and was not broken until **Soyuz TM21 (March 14, 1995)**.

STS 35 was the first shuttle mission dedicated entirely to astronomy. It carried three ultraviolet telescopes and one x-ray telescope in the shuttle's cargo bay. The ultraviolet instruments covered wavelengths from 420 to 3,200 angstroms, while the x-ray telescope observed from 0.3 to 12 KeV and supplemented observations by **Rosat (June 1, 1990)**.

During the nine-day flight, 130 astronomical objects were studied. Images of a supernova remnant, dubbed the Cygnus Loop, revealed new details of its expanding bubble of gas and found that the shock wave's temperature and velocity were higher than predicted.

Several nearby galaxies were mapped for the first time in ultraviolet light. Of the active galactic nuclei observed (see **Einstein–November 13, 1978**), STS 35 found further evidence that these high-powered emitters of radiation were supermassive black holes in the centers of galaxies.

Other observations indicated that the dust grains in interstellar space had a variety of shapes, including sand-like particles (with angular edges) and soot-like particles (spherical in shape). Astronomers were surprised to find that this variety was not evenly distributed throughout space.

The data also failed to find evidence that neutrinos have mass. Scientists had hoped that these ghost-like particles, which traveled at the speed of light, carried no charge, and apparently had no mass, could supply the missing mass needed to cause the universe to contract, rather than expand forever. STS 35's telescopes, however, found no evidence of neutrino mass.

Images of Jupiter and its moon Io refined measurements made by **Voyager 1 (March 5, 1979)** and **Voyager 2 (July 9, 1979)** of the magnetic flux tube between the two bodies and the powerful magnetic torus that surrounds Io.

The next shuttle flight was **STS 37 (April 5, 1991)**. The next astronomical shuttle flight was **STS 67 (March 2, 1995)**. The next x-ray telescope, **ASCA (February 20, 1993)**, used the same mirror design as this shuttle flight.

December 2, 1990, 8:13 GMT
Soyuz TM11
U.S.S.R.

Soyuz TM11 carried a crew of three: Musa Manarov (**Soyuz TM4–December 21, 1987**), Victor Afanasyev (**Soyuz TM18–January 8, 1994; Soyuz TM29–February 20, 1999**), and Toyohiro Akiyama. Akiyama was a Japanese journalist and the first civilian not part of a government program to orbit Earth. His berth on the spacecraft had been purchased for between $10 and $28 million by the Tokyo Broadcasting System. During his eight-day flight, he made daily broadcasts back to Japan, describing the experience to his Earth-bound audience. He also participated in research to study the effect of weightlessness on his body.

On December 4th, Soyuz TM11 docked with **Mir (February 20, 1986)**, and Manarov, Afanasyev, and Akiyama joined the station's 2-man crew from **Soyuz TM10 (August 1, 1990)**. With the launch of STS 35 and its crew of 7, there were 12 human beings in space, a new record (see **STS 35–December 2, 1990** for details).

The two Soyuz crews did joint research on Mir for six days. The men observed the behavior of six Japanese tree frogs, finding that the frogs' posture and behavior was unique in weightlessness. Most preferred to hold onto some surface object, remaining passive and stationary. Their posture was unusual with their heads tilted backwards, backs arched back, and hind legs unfolded. When jumping, the frogs had difficulty coordinating their movements and limbs. When left floating, they acted as if they were in the middle of a leap, stretching their limbs outward as if anticipating a landing, and then grabbing the first object they could find. Upon return to Earth, the frogs readapted within a few hours, but a density loss in their vertebrae was noted.

On December 10th, Akiyama returned to Earth with the crew of Soyuz TM10. Manarov and Afanasyev then settled down to a 176-day stay on Mir. During their occupancy, they were visited by two Progress freighters (see **Progress M1–August 23, 1989**). The first brought new life-support equipment, along with the disassembled parts of the Strela boom, a 46-foot-long crane that, when installed on the Mir core module, could move large equipment on the station's exterior.

Manarov and Afanasyev made four space walks during their mission, the first to complete repairs to the balky hatch on **Kvant 2 (November 26, 1989)** that had given the crews of **Soyuz TM9 (February 11, 1990)** and Soyuz TM10 so much difficulty. In January 1991, they performed two space walks to install the Strela crane and then used it to relocate the solar panels attached to **Kristall (May 31, 1990)**. By moving them to the outside of **Kvant 1 (March 31, 1987)**, they were better exposed to sunlight.

The second Progress freighter had serious docking difficulties. On March 21, 1991, it missed the station by about 1,500 feet. On March 23rd, a second attempt also failed, with a near-collision when the freighter missed Mir by only 15 feet. In both attempts, ground controllers aborted the docking when they realized that the freighter's Kurs automatic docking system was guiding it incorrectly.

The freighter was then placed in a station-keeping orbit near Mir. On March 26th, Manarov and Afanasyev undocked their Soyuz TM11 spacecraft from the station's multidocking adapter and attempted their own automatic docking to the rear docking port on **Kvant 1 (March 31, 1987)**, which the Progress tanker had tried to use. When their attempt also failed, they took over manually and successfully docked. In a subsequent space walk, the cosmonauts discovered that the antenna for the Kurs docking system on the Kvant 2 module was missing. In fact, it was later learned it had been accidentally kicked off during a January space walk. Until a replacement was installed during **Soyuz TM12 (May 18, 1991)**, all automatic dockings took place on Mir's forward port.

On March 28th, the freighter was maneuvered to the multidocking adapter normally used by the manned Soyuz craft, using its Kurs automatic docking system to dock there successfully.

Manarov and Afanasyev remained on Mir until May 26th. Prior to their return to Earth, they were joined on Mir by the crew of Soyuz TM12.

1991

January 8, 1991
NATO 4A
NATO / U.S.A.

Replacing the NATO 3 satellites (see **NATO 3A–April 22, 1976**), NATO 4A is one of two geosynchronous satellites providing NATO with military communications, both narrow- and wide-beam coverage on land and sea. It works in conjunction with the American **DSCS III (October 30, 1982)** and the British **Skynet 4 (December 11, 1988)** military satellites. Built for a seven-year design life, both satellites are still in use as of December 1999.

January 15, 1991
Italsat 1
Italy / Europe

Launched on an Ariane rocket, this geosynchronous communications satellite was built by Italy to provide that country with high-volume telephone service, up to 12,000 simultaneous telephone calls, using what had been the experimental 20 and 30 GHz frequencies (first tested on **ATS 6–May 30, 1974** and **Sirio–August 25, 1977** and retested on **Kopernikus–June 5, 1989** and **Olympus 1–July 12, 1989**). Italsat 1 also performed the propagation experiments in the high frequencies of 20, 30, 40, and 50 GHz.

In August 1996, Italy launched a second Italsat satellite to supplement its communications capacity. Both satellites are still in operation as of December 1999.

January 29, 1991
Informator 1
U.S.S.R.

Informator 1, an experimental communications satellite, was a prototype for a cluster of 12 medium altitude satellites planned for launch beginning in 1997. These satellites, dubbed Gonets, were to be more advanced versions of the Soviet Strela store-and-forward military communications satellites (see **Strela 1–April 25, 1970**) that are similar to the American Iridium satellites (see **Iridium 4 through Iridium 8–May 5, 1997**).

As of December 1999, only one launch of three satellites had taken place. See **Gonets D1 (February 14, 1997)**.

March 31, 1991
Almaz 1
U.S.S.R.

Originally designed as a Salyut space station (see **Salyut 3–June 25, 1974**), Almaz 1 was reconfigured as a platform for performing radar imaging of the earth. This satellite was the third orbiting radar imager, following **Seasat (June 27, 1978)** and **Cosmos 1870 (July 25, 1987)**. See Cosmos 1870 for the background of Almaz 1's change from Salyut to radar imager.

The spacecraft operated for 19 months, gathering radar images 25 by 220 miles, with a resolution as good as 35 feet. It also had the ability to adjust its radar look angle to change each image's pseudo shadows. Moreover, Almaz 1 carried a radiometer for scanning the earth's surface temperature, accurate to within 1°F. This scanner took swathes 375 miles wide with a resolution of 6 to 20 miles. In addition to monitoring crop yields in Russia and ocean currents and wave patterns, the data from Almaz 1 were used to track ice formation in the Arctic Ocean.

The next radar imaging satellite was **ERS 1 (July 17, 1991)**.

April 4, 1991
Anik E2
Canada / Europe

Launched on an Ariane rocket and owned and operated by Telesat Canada (see **Anik A1–November 10, 1972**), Anik E2 is one of two communication satellites in geosynchronous orbit to serve Canada, replacing the Anik C (see **STS 5–November 11, 1982**) and **Anik D (August 26, 1982)** satellites. A second Anik E satellite was launched in September 1991 as a back-up to Anik E2. Both have a planned service life of more than 12 years.

The Anik E satellites were augmented by **Nimiq 1 (May 21, 1999)**, which provided direct broadcast capabilities.

April 5, 1991
STS 37, Atlantis Flight #8
Compton Gamma Ray Observatory
U.S.A.

The eighth flight of the space shuttle Atlantis, STS 37 lasted six days and was crewed by Steven Nagel (**STS 51G–June 17, 1985**; **STS 61A–October 30, 1985**; **STS 55–April 26, 1993**), Kenneth Cameron (**STS 56–April 8, 1993**; **STS 74–November 12, 1995**), Jerome Apt (**STS 47–September 12, 1992**; **STS 59–April 9, 1994**; **STS 79–September 16, 1996**), Linda Godwin (**STS 59–April 9, 1994**; **STS 76–March 22, 1996**) and Jerry Ross (**STS 61B–November 26, 1985**; **STS 27–December 2, 1988**; **STS 55–April 26, 1993**; **STS 74–November 12, 1995**; **STS 88–December 4, 1998**). STS 37 deployed one astronomical satellite, the second of the United States' great observatories after the Hubble Space Telescope (see **STS 31–April 24, 1990** and **STS 61–December 2, 1993**).

- *Compton Gamma Ray Observatory* Named in honor of the American physicist Arthur Holly Compton, the Compton Gamma Ray Observatory was deployed from the shuttle on April 7th. This new space telescope, built by NASA, was the first to provide an all-sky continuous survey in the x-ray *and* gamma-ray regions of the electromagnetic spectrum, from 10 KeV to 30 GeV (30,000,000 KeV), building on data provided by **HEAO 3 (September 20, 1979)**, **Kvant 1 (March 31, 1987)**, and **Gamma (July 11, 1990)**. To cover this wide energy range, Compton carried four instruments.

BATSE (Burst And Transient Source Experiment) looked for the sudden and temporary soft gamma-ray sources first identified by **COS B (August 9, 1975)**. BATSE also searched for gamma-ray bursts, those mysterious objects first discovered more than two decades earlier by **Vela 5A/Vela 5B (May 23, 1969)**. The instrument's observation range was from 60 to 600 KeV.

OSSE (Oriented Scintillation Spectrometer Experiment) studied the continuous emissions of x-ray and gamma-ray sources in the energy range 100 KeV to 10 MeV.

COMTEL (for COMpton TELescope) performed an all-sky survey of the x-ray sky, from 0.1 to 10 MeV, finding and mapping both galactic and extragalactic sources.

EGRET (Energetic Gamma-Ray Experiment Telescope) made observations of specific gamma-ray point sources emitting radiation in the 20 MeV to 30 GeV energy range.

After deployment, Compton's main high-gain antenna failed to open. To correct this failure, Jerry Ross and Jerome Apt performed an unscheduled space walk. On April 7th, Ross climbed out to the satellite's base, still attached to the shuttle's robot arm, and manually pulled the antenna free.

In the years since, Compton has produced a wealth of data about celestial objects pulsing in the high-energy part of the electromagnetic spectrum. This information nicely complemented the spectacular optical pictures produced by the Hubble Space Telescope.

April 5, 1991

Jerry Ross, with Compton Gamma Ray Observatory silhouetted by Earth behind him. He has just released the observatory's antenna. *NASA*

Compton identified the location of the sky's many x-ray and gamma-ray sources, coming from either quasars or the centers of a variety of galaxy types. For example, through March 1999, EGRET had discovered more than 70 galaxies emitting energy in the 20 MeV to 30 GeV energy range, with COMTEL finding an additional nine emitting radiation from 0.1 to 10 MeV. These data, along with other Compton evidence, provided further proof that supermassive black holes lurked in the nucleus of all these distant objects.

Many other gamma-ray sources were found in the Milky Way. For example, a pulsar known as Geminga was discovered as the closest pulsar to the earth—only 180 light years away—and that it probably had erupted as a supernova no more that 340,000 years ago. This blast was probably the brightest supernova ever seen by humans and likely swept clear what scientists now dub the "Local Bubble," a region of rarefied gas within which the solar system presently travels.

Compton also discovered a number of x-ray bursters, including the first pulsar that also burst, a combination that scientists had considered impossible (see **Rossi–December 30, 1995**). Compton also increased the number of known pulsars emitting energy in both radio and gamma rays from two to six.

Compton's most significant discovery, however, was its demonstration that the mysterious gamma-ray bursts occurred all across the sky, were not concentrated along the plane of the Milky Way, and therefore had to originate from outside it. This discovery contradicted what most scientists had believed and meant that the bursts were tremendously powerful events, producing more energy in seconds than the Sun will produce in its entire life span.

What caused these colossal explosions, however, remained unknown. Scientists proposed several theories, including the merging of neutron star binaries and the swallowing of a neutron star by a black hole. Direct evidence, however, did not exist, as no gamma-ray burst had yet been linked to any visible event in the sky despite combined monitoring by Compton, **Granat (December 1, 1989)**, **Wind (November 1, 1994)**, and other satellites over more than 20 years. See **BeppoSax (April 30, 1996)** for the first visible counterpart.

The Compton Gamma Ray Observatory was designed so that future shuttle missions could refuel it, thereby extending its service life indefinitely. As of December 1999, however, no such mission has been necessary, or is planned.

The next x-ray telescope was BeppoSax. The next gamma-ray telescope was **Minisat 1 (April 21, 1997)**. The third of the United States Great Observatories was the Chandra X-Ray Observatory (see **STS 93–July 23, 1999**).

In addition to their unscheduled EVA (extra-vehicular activity), Ross and Apt performed two other space walks, testing several construction techniques anticipated during the assembly of the American space station Freedom, later replaced by the International Space Station (see **Zarya–November 20, 1998** and **STS 88–December 4, 1998**). These tasks included using ropes to move about the cargo bay and working while secured to foot restraints at the end of the shuttle's robot arm.

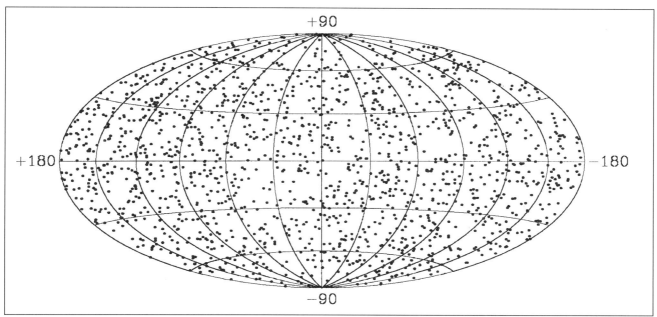

Compton Gamma Ray Observatory: The sky distribution of 1,637 gamma ray bursts from 1991 to 1998. The equator matches the galactic plane of the Milky Way. *NASA*

The two men also tested three different rail transportation systems—an electrically powered cart, a cart powered by a hand pump, and a cart that the astronaut merely pulled along with his hands. According to Ross, the hand cart worked best: "There is no significant advantage to the more complex carts."

STS 37 also tested a prototype space radiator, intended for later use on the space station. While circulating liquid to regulate temperature, the equipment experienced problems with air bubbles blocking the liquid flow.

Protein crystal growth research (see **STS 26–September 29, 1988**) continued from **STS 31** (**April 24, 1990**). On STS 37, the experiment used new hardware designed to produce larger quantities of protein crystals and focused only on growing crystals from bovine insulin. The experiment was also repeated on a 1991 shuttle flight. The insulin crystals produced on both flights, while very pure and well ordered, were not much larger than comparable Earth-grown crystals. A new suite of 11 protein crystals were flown on **STS 48** (**September 12, 1991**).

Another package of 61 experiments tested protein crystal growth, manufacturing techniques, fluid behavior in microgravity, and other life sciences research. For example, one test successfully hatched 5 out of 44 shrimp in space, but they died soon after hatching for causes believed to be related to weightlessness—although this was not confirmed.

The next shuttle mission was **STS 39** (**April 28, 1991**).

April 28, 1991
STS 39, Discovery Flight #12
U.S.A.

STS 39 was the twelfth flight of Discovery. It lasted eight days and was crewed by Michael Coats (**STS 41D–August 30, 1984; STS 29–March 14, 1989**), Blaine Hammond (**STS 64–September 9, 1994**), Guion Bluford (**STS 8–August 30, 1983; STS 61A–October 30, 1985; STS 53–December 2, 1992**), Gregory Harbaugh (**STS 54–January 13, 1993; STS 71–June 27, 1995; STS 82–February 11, 1997**), Richard Hieb (**STS 49–May 7, 1992; STS 65–July 8, 1994**), Donald McMonagle (**STS 54–January 13, 1993; STS 66–November 3, 1994**), and Charles Veach (**STS 52–October 22, 1992**). This military shuttle flight, following **STS 38** (**November 15, 1990**), was the first flown unclassified. It researched the environment of the earth's magnetosphere and ionosphere and the shuttle's interaction with it.

In order to study this environment, STS 39 employed three main tools. It temporarily deployed SPAS (Shuttle Pallet Satellite) 2, previously flown on **STS 7** (**June 18, 1983**) and **STS 41B** (**February 3, 1984**). The shuttle separated from and trailed behind this passive satellite by about 7 miles so that SPAS 2 could take photographs and make measurements of the shuttle environment in infrared, visible, and ultraviolet wavelengths. The shuttle then released three small subsatellites, each of which produced small short-lived gas clouds that SPAS 2 could image and sense. The shuttle also released four additional plumes of gas (xenon, neon, carbon dioxide, and nitrous oxide) from the shuttle's cargo bay.

The tests were designed to give the U.S. military baseline data on the shape of rocket plumes in space, as well as their spectral characteristics, so that future satellite sensors could be designed to detect them.

These experiments discovered that the glow first seen on the shuttle's skin on **STS 3** (**March 22, 1982**) was caused by a chemical reaction between the atmosphere's oxygen molecules and the small amounts of nitric oxide released from the shuttle's exhaust gases, producing nitrogen dioxide and light. During STS 39's experiments, the glow intensified (and did so significantly) only when plumes of nitrous oxide were released.

In addition to these artificial aurora experiments, Discovery actually flew through the earth's aurora at one point, during which the crew took spectacular pictures *(see color section, Figure 27)*. "It's just like flying through a curtain of light," noted Michael Coats.

255

A variety of smaller onboard experiments ranged from measuring the contamination levels in the cargo bay during launch to observing how the perceived cloud cover of the earth changed depending on the angle a photograph is taken.

The next shuttle mission was STS 40 (**June 5, 1991**). The next military shuttle mission was STS 44 (**November 24, 1991**).

May 18, 1991
Soyuz TM12
U.S.S.R. / U.K.

The launch crew of Soyuz TM12 included Anatoly Artsebarski, Sergei Krikalev (**Soyuz TM7–November 26, 1988**; **STS 60–February 3, 1994**; **STS 88–December 4, 1998**), and Helen Sharman, the first British citizen to fly in space and the first woman to occupy **Mir (February 20, 1986)**. Like the previous flight of **Soyuz TM11 (December 2, 1990)**, in which a Japanese journalist had been a paid passenger, Sharman was intended to be a paying customer, with her airfare provided by a British consortium. When the consortium failed to come up with the money, however, Sharman's participation continued at Soviet expense, but with a greatly limited suite of experiments.

Soyuz TM12 docked with Mir on May 20th, joining Afanasyev and Manarov from Soyuz TM11 as they wrapped up 176 days in space. For the next six days, the old crew briefed Artsebarski and Krikalev on the station's operation while also performing several joint experiments. Then, on May 26th, Afanasyev, Manarov, and Sharman returned to Earth in the Soyuz TM11 spacecraft, leaving the station in the hands of Krikalev and Artsebarski.

Krikalev remained on Mir until March 25, 1992. He was visited twice by manned crews during this period: **Soyuz TM13 (October 2, 1991)** and **Soyuz TM14 (March 17, 1992)**. The original plans had called for both Artsebarski and Krikalev to return following the arrival of Soyuz TM13, but the late replacement of a Soviet cosmonaut for a less experienced cosmonaut from Kazakhstan on Soyuz TM13 made it necessary for Krikalev to remain on Mir for two consecutive crew shifts.

Krikalev's flight was even more remarkable in that he had taken off from the Soviet Union but returned to a new nation, the Russian Federation. During his 10-month stay in space, the Soviet Union had collapsed and newly independent states had formed.

Before the arrival of Soyuz TM13, Mir was resupplied by two Progress freighters (see **Progress M1–August 23, 1989**). The second carried a recoverable capsule that successfully returned 330 pounds of cargo to Earth on September 30th, part of which included multispectral photography of Europe, Kazakhstan, and the central Asian republics of the Soviet Union.

Artsebarski and Krikalev performed six separate space walks in July, replacing the missing Kurs antenna on Kvant 2 that had caused the docking problems during Soyuz TM11. They also built a 46-foot-long girder on **Kvant 2 (November 26, 1989)** using special thermomechanical joints. When heated, the material "remembered" its previous shape and reverted to it, thereby creating a solid joint. This girder, dubbed Sofora, should not be confused with the Strela crane (see **Soyuz TM11–December 2, 1990**). Sofora was, instead, a platform for a thruster engine for orienting Mir, later installed on its top by the crew of **Soyuz TM15 (July 27, 1992)**.

On June 17th and August 15th, the men deployed two small satellites from Mir's airlock, but both satellites failed.

See Soyuz TM13 for later events on Mir.

May 29, 1991
Aurora 2
U.S.A.

Aurora 2, owned and operated by General Electric (GE), replaced **RCA Satcom 5 (October 28, 1982)**, which provided telephone and television service to the remote small villages of Alaska and linked these locations to the rest of the United States. By the late 1980s, more than 200 small terminals had been built, providing at least two television stations to every Alaskan village of 20 people.

General Electric had purchased RCA in 1986 and placed Aurora 2 in geosynchronous orbit to replace the aging first satellite (which had been renamed Aurora 1 when GE purchased it). The two satellites were similar, although the second has increased transmitter power and equipment redundancy. Aurora 2 is expected to have a service life of about 12 years and is still in use as of December 1999.

June 5, 1991
STS 40, Columbia Flight #11
U.S.A.

STS 40 was Columbia's eleventh space flight. It lasted nine days and was crewed by Bryan O'Connor (**STS 61B–November 26, 1985**), Sidney Gutierrez (**STS 59–April 9, 1994**), James Bagian (**STS 29–March 13, 1989**), Tamara Jernigan (**STS 52–October 22, 1992**; **STS 67–March 2, 1995**; **STS 80–November 19, 1996**), Rhea Seddon (**STS 51D–April 12, 1985**; **STS 58–October 18, 1993**), Francis Gaffney, and Millie Hughes-Fulford. This mission was also the sixth shuttle flight to carry Spacelab and the first entirely dedicated to life sciences research.

Eighteen different experiments researched the effects of weightlessness on both the crew and 29 laboratory rats. Six experiments studied the astronauts' cardiovascular systems. Four studied bone loss in both humans and rats. Others investigated how the body metabolizes protein in space, regulates its fluid and kidney function, maintains its balance and muscle tone, and performs a number of other functions. Blood, saliva, and urine samples were taken, as well as ultrasound images of the astronauts' hearts.

Results from these tests showed that the astronauts experienced a 25 percent muscle loss, more than had been expected, and that they recovered only half that loss after nine days back on Earth. These results indicated that a heavy and dedicated exercise regimen, as introduced by the Soviets on their Salyut and Mir missions (see, for example, **Soyuz T10B–February 8, 1984**), was essential if humans were to spend extended periods in space.

The research also found that weightlessness contributes to a loss in red blood cells. Young blood cells were seen to die off before they could mature and enter the bloodstream from the bone marrow (where they are produced).

Other parts of the astronauts' bodies adapted to space faster than predicted, however. Their hearts beat faster with a higher blood pressure, and the number of the nervous systems' synapses in the inner eye varied greatly, seemingly adapting with ease to the environment of weightlessness.

The lung, however, was found to function exactly as it did on Earth—where blood flow is greater near the lung's base, with more air near its top. To the scientists' surprise, the lack of gravity had no effect on this uneven distribution.

The first jellyfish to fly in space (2,478 of them) were studied to find out what role gravity played in their development and growth. The jellyfish were launched in their polyp stage and matured into their ephyrae stage while in orbit. Upon their return to Earth, they continued their normal development, showing no negative effects from the short period without gravity.

STS 40 also carried 12 getaway specials, the first launched since before the Challenger accident (see **STS 51L–January 28, 1986**). These canisters mostly focused on materials research, testing a variety of new processing techniques, including the production of seamless hollow ball bearings, ultralight metal foams, and pure gallium arsenide semiconductor crystals. Other private getaway canisters performed soldering experiments and tested the effect of the space environment on computer floppy disks and 57 different plant seeds.

The crew also took photographs and videos of the giant ash plume of the Mount Pinatubo volcano, which began erupting during the mission.

The next shuttle mission was **STS 43 (August 2, 1991)**. The next life sciences shuttle mission was **STS 58 (October 18, 1993)**.

June 29, 1991
Rex 1
U.S.A.

Rex (Radiation Experiment) 1 was a U.S. Air Force experimental satellite for testing the operation of modern miniaturized communications equipment in polar orbit at an altitude of over 500 miles. Such an orbit exposed the equipment to higher than normal radiation doses.

The Air Force flew a second experimental Rex test satellite in March 1996, doing further tests on the effect of high polar orbit on communications equipment.

July 4, 1991
Losat X
U.S.A.

Losat (LOw Altitude SATellite) X was a military research satellite that tested newly designed surveillance sensors for observing ground-launched rocket plumes. It gathered data for five months before re-entering the atmosphere.

July 17, 1991, 1:46 GMT
ERS 1 / Orbcomm X / SARA / Tubsat 1 / UoSat 5
Europe / U.S.A. / France / Germany / U.K.

All five satellites were launched on an Ariane rocket.

- *ERS 1*   The European Remote-sensing Satellite (ERS) 1 was the European Space Agency's first radar satellite and the fourth orbiting radar satellite, following **Almaz 1 (March 31, 1991)**. Placed in a sun-synchronous orbit (see **Nimbus 1–August 28, 1964**), the spacecraft studied the ocean, its currents, temperatures, and wind speeds, as well as ice formation and coastal mapping. ERS 1 carried a synthetic-aperture radar scanner (see **Seasat–June 27, 1978**) working in the 5.7-cm waveband. Depending on the instrument's configuration, it could make radar images of land features 62 miles wide with a resolution of 80 feet, discern ocean wave patterns including their direction and size, and detail surface wind speeds and direction. The spacecraft also carried a radar altimeter for precisely measuring the elevation of surface features. This was used to monitor changes in ocean levels due to tides, currents, winds, and the very shape of the earth. As with previous radar satellites, ERS 1's archive of data is available for public purchase.

   The next radar mission was the Japanese satellite, **JERS 1 (February 11, 1992)**. The next European Space Agency radar satellite was **ERS 2 (April 21, 1995)**.

- *Orbcomm X*   Built by Orbital Sciences Corporation, Orbcomm X was a prototype of the low-Earth-orbit store-and-forward communications satellite the company hoped to launch for mobile telephone communications. It was intended to test technology similar to **Microsat 1 through Microsat 7 (July 17, 1991)**, **UoSat 2 (March 1, 1984)**, and a number of other satellites. Although Orbcomm X failed after one day in orbit, Orbital Sciences completed these tests with a second satellite in 1993. The concept itself was eventually launched with the **Iridium 4 through Iridium 8 (May 5, 1997)** and **Orbcomm FM5 through Orbcomm FM12 (December 23, 1997)** clusters.

- *SARA*   A 40-pound satellite built in France, SARA studied the radio emissions that periodically came from Jupiter (see **Pioneer 10, Jupiter Flyby–December 3, 1973**).

- *Tubsat 1*   Built by the Technical University of Berlin (TUB), Tubsat tested store-and-forward technologies for communications satellites. It also participated in wildlife studies by relaying data accumulated from small transmitters attached to wildlife. Using the Doppler shift produced by each transmitter, the animal's location and movements across time could be studied.

   The next satellite built by the university was **Tubsat 2 (January 25, 1994)**.

- *UoSat 5*   Built by the University of Surrey, England, UoSat 5 replaced **UoSat 4 (January 22, 1990)** and was the 22nd satellite in the Oscar series of amateur communications satellites (see **Oscar 1–December 12, 1961**). It carried several technology demonstration tests, including an experimental CCD camera (see **KH-11 1–December 19, 1976**) and solar cells made of gallium arsenide, indium phosphide, and silicon.

The next Oscar satellite was **Oscar 23 (August 10, 1992)**.

July 17, 1991, 17:33 GMT
Microsat 1 through Microsat 7
U.S.A.

Seven satellites—Microsat 1 through Microsat 7—were placed in orbit by a Pegasus launch vehicle; they were a U.S. Air Force test of a low-Earth-orbit cluster of communications satellites. The concept of satellite communications in low-Earth orbit had first been conceived in the early days of space satellite communications and demonstrated by **Telstar 1 (July 10, 1962)** and **Telstar 2 (May 7, 1963)**. The use of a cluster of such satellites was proven by **IDCSP 1 through IDCSP 7 (June 16, 1966)**. The large cost of launching many satellites, however, had favored the development instead of satellites orbiting in geosynchronous altitude—in the 1960s and 1970s, the expense of lifting a single satellite to high orbit was far less than the cost of orbiting many low altitude satellites.

The advances in computer and digital technology in the 1980s, the shrinking of the size and weight of communications equipment, and the growing congestion at geosynchronous orbit had renewed interest in the low-Earth-orbit idea. The seven Microsats tested the feasibility of such a cluster, using miniaturized digital microprocessors to store-and-forward any transmitted data. Other test flights (see **GLOMR/STS 51B–April 29, 1985**; **UoSat 3–January 22, 1990**; **Macsat 1/Macsat 2–May 5, 1990**; **Informator 1–January 29, 1991**; and **Orbcomm X–July 17, 1991**) supplemented this research, leading to the new low-Earth-orbit clusters just being completed and coming into operation in 1998 and 1999. See **Iridium 4 through Iridium 8 (May 5, 1997)**, **Orbcomm FM5 through Orbcomm FM12 (December 23, 1997)**, and **Globalstar 1 through Globalstar 4 (February 14, 1998)**.

August 2, 1991
STS 43, Atlantis Flight #9
TDRS 5
U.S.A.

STS 43, a nine-day mission, was the ninth for the space shuttle Atlantis. It was crewed by John Blaha (**STS 29–March 13, 1989**; **STS 33–November 22, 1989**; **STS 58–October 18, 1993**; **STS 79–September 16, 1996**), Michael Baker (**STS 52–October 22, 1992**; **STS 68–September 30, 1994**; **STS 81–January 12, 1997**), James Adamson (**STS 28–August 8, 1989**), David Low (**STS 32–January 9, 1990**; **STS 57–June 21, 1993**), and Shannon Lucid (**STS 51G–June 17, 1985**; **STS 34–October 18, 1989**; **STS 58–October 18, 1993**; **STS 76–March 22, 1996**). The mission released one satellite.

• *TDRS 5*   Deployed on August 2nd, TDRS (Tracking and Data Relay Satellite) 5 provided communications for NASA's orbiting spacecraft. See **STS 6 (April 4, 1983)**.

Onboard experiments included the third attempt to grow thin polymer membrane films (see **STS 31–April 24, 1990**), continuing tests conducted on **STS 41 (October 6, 1990)**. On this flight, the thin membrane was created while the equipment was allowed to float in the air, thereby reducing the effect of vibrations from crew movements. The next four flights used the identical equipment and repeated the experiment, but each time varied how the equipment was attached to the shuttle or the methods used for preventing jitter and vibration. This provided comparison data so that scientists could learn exactly how the shuttle environment affected membrane production. See **STS 51 (September 12, 1993)** for further results.

The next shuttle flight was **STS 48 (September 12, 1991)**.

August 15, 1991
Meteor 3-05
U.S.S.R. / U.S.A.

The Soviet weather satellite Meteor 3-05, part of the Meteor 3 satellite cluster (see **Meteor 3-01–October 24, 1985**), was unusual in that it also carried an American instrument for measuring the ozone layer. The Total Ozone Mapping Sensor (TOMS) had first flown on **Nimbus 7 (October 13, 1978)**. An identical instrument was included on Meteor 3-05 in order to maintain continuous monitoring of the ozone layer's annual variations to find out if the use of chlorofluorocarbons (CFCs) by northern hemisphere industrial nations was damaging it. TOMS's data also augmented **SME (October 6, 1981)** and **Solar Max (see STS 41C–April 6, 1984)**.

The TOMS instrument operated until the end of 1994. In 1992, it measured a small increase in the size of the southern ozone hole, discovered by Nimbus 7 and by this time known to appear over Antarctica each year in the late winter months. Over the next three years, however, the hole's size remained stable, showing no increase. In the northern hemisphere, meanwhile, no ozone hole was detected at all. This dichotomy was puzzling to scientists, since chlorofluorocarbons are used mostly in the northern hemisphere and it seemed that they should have produced a greater effect there.

The monitoring of the ozone layer continued with **UARS** (see **STS 48–September 12, 1991**).

August 30, 1991
Yohkoh (Solar-A)
Japan / U.S.A. / U.K.

Yohkoh ("sunbeam" in Japanese) was a Japanese-led solar research satellite that also carried instruments from the United States and the United Kingdom. Following **Solar Max (STS 41C–April 6, 1984)**, it studied the Sun's high-energy solar flares in what scientists call both soft and hard x-rays. The soft x-ray sensor observed energy emissions across the lower part of the x-ray electromagnetic spectrum, from 0.1 to 4 KeV. The hard x-ray sensor covered the waveband's high end, from 20 to 80 KeV. Yohkoh also carried a spectrometer for detecting x-rays and gamma rays from 3 KeV to 20 MeV.

Yohkoh found that the Sun's corona, the vast and very hot atmosphere of particles streaming away from the Sun and visible on Earth only during eclipses, was much more active than previously thought. The spacecraft's ability to take rapid, sequential pictures showed that the corona changes continually, its structure of loops and arcs pulsing violently as they adjust to changes in the Sun's magnetic field lines. Within solar flare regions, the corona is ever-expanding, contributing significantly to loss of the Sun's mass.

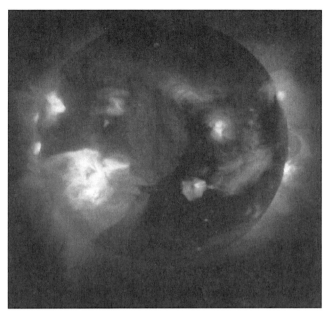

The Sun in soft x-rays on 25 October 1991. *NASA*

Using Yohkoh's image archive, scientists in 1999 determined that coronal mass ejections generally erupt from S-shaped surface patterns caused by twisted magnetic lines of force. Eruptions could now be predicted, allowing satellite companies to better protect their orbiting equipment.

As of December 1999, Yohkoh continues to send back spectacular x-ray images of the Sun. Its sensors work in conjunction with instruments on **Soho (December 2, 1995)**, providing a very extensive picture of the Sun's daily activity.

September 12, 1991
STS 48, Discovery Flight #13
UARS
U.S.A.

The thirteenth flight of Discovery, STS 48 was a five-day mission carrying a crew of five: John Creighton (**STS 51G–June 17, 1985**; **STS 36–February 28, 1990**), Kenneth Reightler (**STS 60–February 3, 1994**), Mark Brown (**STS 28–August 8, 1989**), Charles Gemar (**STS 38–November 15, 1990**; **STS 62–March 4, 1994**), and James Buchli (**STS 51C–January 24, 1985**; **STS 61A–October 30, 1985**; **STS 29–March 13, 1989**). STS 48 deployed one Earth observation satellite.

• *UARS* Deployed on September 15, the Upper Atmosphere Research Satellite (UARS) was a follow-up of **Nimbus 7 (October 13, 1978)**, **Solar Max (February 14, 1980)** and **ERBS** (see **STS 41G–October 5, 1984**). Built by NASA, it carried 10 instruments, 4 to study the chemical make-up of the earth's upper atmosphere, 2 to measure high-altitude wind speeds, and 4 to measure any variations in the Sun's energy output and its direct impact on the atmosphere.

UARS measurements of the earth's ozone layer matched the patterns seen by the TOMS instrument on **Meteor 3-05 (August 15, 1991)**—no ozone hole in the northern hemisphere, with a slightly larger ozone hole in the southern hemisphere in 1992. UARS, however, also detected high concentrations of chlorine monoxide over both Northern Europe and Antarctica during the winter months. This chemical was thought to be a component of the chemical process in which chlorofluorocarbons (CFCs) deplete ozone. UARS also detected hydrogen fluoride, a chemical whose presence could only be caused by CFCs. This evidence was an indirect confirmation of the theory that CFCs—chemicals manufactured and used by humans—might cause ozone layer depletion. For further ozone layer research, see **Sampex (July 3, 1992)**.

UARS's instruments also detected huge, violent, and continent-sized windstorms in the region between 30 and 60 miles altitude, a region never before studied. Winds of 200 miles per hour stretched for thousands of miles.

STS 48's onboard experiments included protein crystal growth research (see **STS 26–September 29, 1988**). While the previous two missions (**STS 37–April 5, 1991** and **STS 43–August 2, 1991**) had only grown protein crystals from bovine insulin, this flight attempted to grow crystals from 11 different proteins in 60 different samples. This research continued on STS 42 (**January 22, 1992**).

A private commercial study continued research from **STS 41 (October 6, 1990)**, using eight young 30-day-old rodents to see how their muscles developed in the absence of gravity.

The behavior of sloshing fluids was tested inside two differently shaped tanks: flat-bottomed and spherical. This engineering test—anticipating the need to build large structures in space—measured the stresses produced on the tanks when the fluids moved.

STS 48 also tested a prototype electronic still camera, set to automatically capture high-resolution digital images and

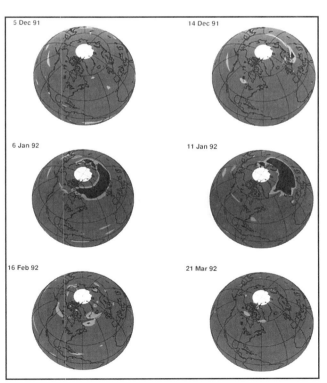

The rise and fall of chlorine monoxide over northern Europe during the dark winter months. *NASA*

downlink them directly to Earth. See **STS 42–January 22, 1992** for later tests.

As the third shuttle flight launched since the eruption of Mount Pinatubo and the first since the igniting of the Kuwaiti oil fires, the astronauts were struck by how dirty the atmosphere seemed. "Of all the flights I've been on, this is by far the worst the atmosphere has looked during this time of season," noted James Buchli. For a comparison of this reaction with that of the next shuttle crew, see **STS 44 (November 24, 1991)**.

The next shuttle flight was **STS 44 (November 24, 1991)**.

October 2, 1991
Soyuz TM13
U.S.S.R. / Kazakhstan / Austria

Soyuz TM13 carried an international crew: Alexander Volkov (**Soyuz T14–September 17, 1985; Soyuz TM7–November 26, 1988**) of the Soviet Union, Toktar Aubakirov of Kazakhstan, and Franz Viehbock of Austria. While Aubakirov was officially a citizen of the Soviet Union, he actually represented his native republic, which would soon declare its independence. Viehbock, the first Austrian to fly in space, was a paid passenger—Austria paid the Soviets $7 million for him to occupy **Mir (February 20, 1986)**.

During the eight-day flight, six of which were spent on Mir with Anatoly Artsebarski and Sergei Krikalev of **Soyuz TM12 (May 18, 1981)**, 15 scientific experiments in medicine and biology were conducted, and extensive multispectral photographs were taken of Kazakhstan and Austria.

Aubakirov and Viehbock, neither of whom was trained for a long-term mission on Mir, had originally been scheduled to fly on separate missions. Money shortages, as well as political pressure to put a Kazakh in space (though all previous Soviet manned missions had been launched from Kazakhstan, no native-born Kazakh had ever been in space), forced the Soviets to combine the missions into one flight. With only one Soyuz TM13 cosmonaut (Volkov) trained for a long-term Mir occupancy, one member of the crew of Soyuz TM12, Krikalev, was forced to remain in space through a second consecutive crew shift.

During the subsequent residency of Krikalev and Volkov, two Progress freighters (see **Progress M1–August 23, 1989**) visited Mir with supplies. Despite one aborted docking attempt and a delay in final separation from Mir because of attitude control problems, the first freighter successfully delivered supplies as well as a recoverable capsule that returned to Earth filled with over 300 pounds of experimental specimens, exposed film, and other cargo. The second cargo ship had no technical problems, but its launch was almost delayed because of a threatened strike by unpaid employees at mission control.

Krikalev and Volkov continued the routine of most Mir long-term missions. They produced semiconductor crystals of gallium arsenide with one of the station's furnaces, smelted samples of germanium and cadmium sulfide, made x-ray observations of the black hole candidate Cygnus X-1, and gathered data on the environment surrounding the orbiting laboratory. They also continued the daily regimen of at least two hours of exercise.

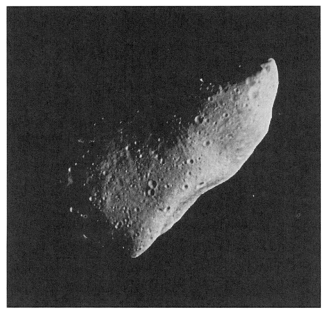

The asteroid Gaspra. *NASA*

The next mission to Mir was **Soyuz TM14 (March 17, 1991)**.

October 29, 1991
Galileo, Gaspra Flyby
U.S.A.

Launched by **STS 34 (October 18, 1989)** on a mission to Jupiter, Galileo's route through space took it within 1,000 miles of the asteroid Gaspra, floating in the asteroid belt between Mars and Jupiter. Galileo's images of Gaspra, the first close-ups ever taken of an asteroid, revealed a dark, gray, irregular-shaped, and crater-pocked object, 7.5 by 10 miles across and rotating once every seven hours.

Other flyby data indicated that the asteroid might have a magnetic field. As the spacecraft flew past, it recorded changes in solar wind direction a few hundred miles from Gaspra, suggesting the presence of a small magnetosphere. Because the wind change might have also been caused by normal solar wind activity coincidentally occurring during the spacecraft's rendezvous, scientists were uncertain of the field's presence.

Galileo continued on its roundabout course towards Jupiter, zipping past Earth and then flying past the asteroid Ida on August 28, 1993. See **Galileo, Ida Flyby (August 28, 1993)**.

November 24, 1991
STS 44, Atlantis Flight #10
DSP 16
U.S.A.

STS 44 was the tenth flight of the shuttle Atlantis. It lasted seven days and was the second unclassified military shuttle flight, following **STS 39 (April 28, 1991)**. The crew was Fred Gregory (**STS 51B–April 29, 1985; STS 33–November 22, 1989**), Terrence Henricks (**STS 55–April 26, 1993; STS 70–July 13, 1995; STS 78–June 20, 1996**), Mario Runco (**STS 54–January 13, 1993; STS 77–May 19, 1996**), James Voss

(STS 53–December 2, 1992; STS 69–September 7, 1995), Story Musgrave (STS 6–April 4, 1983; STS 51F–July 29, 1985; STS 33–November 22, 1989; STS 61–December 2, 1993; STS 80–November 19, 1996), and Tom Hennen. STS 44 released one military satellite.

- *DSP 16* Deployed from STS 44 on November 25th, this Defense Support Program (DSP) satellite was a U.S. Air Force early-warning satellite. See **IMEWS 1 (May 5, 1971)**.

The remainder of the shuttle mission focused on Earth observation. Tom Hennen tested a new U.S. Army reconnaissance system dubbed "Terra Scout." He used the system to identify and describe specific predefined Earth objects viewed also by an earthbound observer using conventional techniques. Their results were then compared to see which method was more accurate.

Another military experiment, dubbed "Military Man in Space," tested the ability of a spaceborne astronaut to participate in air, ground, and naval operations. The astronauts observed ground operations, including armored vehicle and flying aircraft formations, with Mario Runco at one point able to identify a 16-foot boat and a 747 commercial airliner near Hawaii. This research was similar to research performed on **Gemini 5 (August 21, 1965)**.

Because one of the shuttle's navigational units malfunctioned, the mission was shortened from 10 days to 7 days, becoming only the second shuttle mission ever shortened (see **STS 2–November 12, 1981**) and the first in over a decade.

Unlike the previous shuttle flight (**STS 48–September 12, 1991**), in which the crew had commented on the dirtiness of the earth's atmosphere, this crew noted a much clearer atmosphere only five months after the eruption of Mount Pinatubo. Moreover, the Kuwaiti oil fires had only been extinguished a few weeks prior to launch, yet the astronauts could see a marked decline in atmosphere smoke, even while they were in orbit. Fred Gregory later noted, "I equated [Earth] to a cat cleaning itself. When it became dirty it just licked itself clean."

The next shuttle mission was **STS 42 (January 22, 1992)**. The next military shuttle mission was **STS 53 (December 2, 1992)**.

December 16, 1991
Telecom 2A
France / Europe
Owned and operated by France and launched on an Ariane rocket, Telecom 2A was the first in that country's second-generation cluster of geosynchronous communications satellites, replacing its Telecom 1 satellites (see **Telecom 1A–August 4, 1984**). Telecom 2 satellites produced more than three times the electrical power and twice the capacity of the earlier satellites. Through December 1999, France orbited a total of four Telecom 2 satellites, all of which are still in use.

December 18, 1991
Intercosmos 25 / Magion 3
U.S.S.R. / Czechoslovakia / East Europe
Intercosmos 25 was the last satellite launched under the Soviet banner in the Intercosmos program (see **Intercosmos 1–October 14, 1969** for program information). It continued ionospheric research of **CRRES (July 25, 1990)**, carrying experiments from scientists from the Soviet Union, Bulgaria, Czechoslovakia, East Germany, Hungary, Poland, and Romania. Similar to **Prognoz 10 (April 26, 1985)**, Intercosmos 25 monitored solar activity and the resulting fluctuations in the earth's magnetic field and ionosphere, working with Magion 3, the third Czechoslovakian satellite. After separating on December 28th, the two satellites worked in conjunction.

The next ionospheric research satellite was **Sampex (July 3, 1992)**. **Intercosmos 26 (March 2, 1994)** was the last Intercosmos satellite.

December 19, 1991
Raduga 28
U.S.S.R.
Raduga 28, a military geosynchronous communications satellite (see **Raduga 1–December 22, 1975**), is unique only because it was the last spacecraft placed in orbit under the hammer-and-sickle flag of the Soviet Union. Following this flight, the Soviet Union ceased to exist, its space program falling under the control of the Russian Federation, the Soviet Union's largest republic.

1992

January 22, 1992
STS 42, Discovery Flight #14
U.S.A. / International
Discovery's fourteenth flight, STS 42 lasted eight days and was crewed by Ron Grabe (STS 51J–October 3, 1985; STS 30–May 4, 1989; STS 57–June 21, 1993), Stephen Oswald (STS 56–April 8, 1993; STS 67–March 2, 1995), William Readdy (STS 51–September 12, 1993; STS 79–September 16, 1996), Norman Thagard (STS 7–June 18, 1983; STS 51B–April 29, 1985; STS 30–May 4, 1989; Soyuz TM21–March 14, 1995), David Hilmers (STS 51J–October 3, 1985; STS 26–September 29, 1988; STS 36–February 28, 1990), Roberta Bondar, and German Ulf Merbold (STS 9–November 28, 1983; Soyuz TM20–October 4, 1994). For this flight, the Spacelab module was reconfigured and given the name the International Microgravity Laboratory, the first of several such flights (scientists usually refer to the flight as IML-1). More than 60 experiments were flown, proposed by more than 200 scientists from 13 countries researching subjects from the human body's response to weightlessness to the behavior of fluids. Research highlights follow.

A follow-up of the plant-root experiment from **STS 61A (October 30, 1985)** studied how roots use gravity to orient themselves. In this case, a centrifuge was used to periodically subject lentil roots to 1 g of gravity; it was found that this short exposure helped them maintain correct orientation. Another plant-root experiment studied two different strains of thale cress, one whose roots grew downward in gravity and a mutant whose roots grew in all directions. In both cases, the addition of gravity aided the roots' ability to orient, but without gravity, the plants grew faster.

January 24, 1992

Protein crystal growth research (see **STS 26–September 29, 1988**) continued from **STS 48 (September 12, 1991)**; this time, attempts were made to grow crystals from 15 proteins in more than 120 different samples. Several experiments tested how variations in temperature and methods could improve crystal grow rate, size, and purity. For example, techniques for growing mercury iodide crystals, first tried on **STS 51F (July 29, 1985)**, attempted to grow the crystals twice as fast by carefully raising the experiment's temperature during the growth process. See **STS 49 (May 7, 1992)** for later research.

Another experiment, called GOSAMR and designed by the 3M Company, tested the behavior of gelatins and colloids in weightlessness. Such substances are used in the production of the advanced ceramic composites that are used as polishing and abrasive materials.

In an attempt to explain why almost two-thirds of all humans in space experienced back pain (see **Soyuz T10B–February 8, 1984**), careful measurements were taken of the astronauts' spinal columns, showing how the column expanded and straightened due to the lack of gravity. This research continued on **STS 52 (October 22, 1992)**.

STS 42 also carried 10 getaway-special canisters, as well as two student experiments. One canister flew brine shrimp to see how weightlessness affected their behavior. Another included three furnaces for testing the solidification of alloys. A third carried an Australian ultraviolet telescope, intended to make observations of several specific deep space objects, including the hot gas in the Magellanic Clouds.

The IMAX camera also flew, taking footage of the interior of Spacelab for the third IMAX space film, *Destiny in Space*. Also flown was a handheld digital camera, able to take high-resolution electronic pictures comparable to film (see **STS 48–September 12, 1991** for earlier tests). Instead of film, a CCD was placed on the film plane of a 35-mm Nikon F4 camera, giving it the capability of downlinking its images directly to Earth (see **KH-11 1–December 19, 1976**).

The next shuttle mission was **STS 45 (March 24, 1992)**. The next mission dedicated to microgravity research was **STS 62 (March 4, 1994)**. The second International Microgravity Laboratory flight was **STS 65 (July 8, 1994)**.

January 24, 1992
Cosmos 2176
Russia

Cosmos 2176, a surveillance early-warning satellite, is significant only because it was the first spacecraft launched under the flag of the Russian Federation, following the fall of the Soviet Union.

February 8, 1992
Ulysses, Jupiter Flyby
U.S.A.

Launched from the space shuttle (see **STS 41–October 6, 1990**), Ulysses was designed to study the Sun's magnetic field and the solar wind over its poles. To reach this polar orbit, the spacecraft used the gravitational field of Jupiter, swinging past the gas giant on this date to be flung into an orbit carrying it over the Sun's southern pole in 1994 and its northern pole in 1997.

In its flight past Jupiter, Ulysses found that the planet's magnetosphere was larger than recorded during Pioneer 10's flyby (**Pioneer 10, Jupiter Flyby–December 3, 1973**), extending outward from the planet by more than 4.6 million miles. This expansion indicated that the pressure from the solar wind was less than before, and that the interaction between the planet's magnetosphere and the wind could vary significantly. The magnetosphere's bow shock and outer structure, however, were found to resemble Earth's quite closely, and, like Earth, the solar wind was able to follow Jupiter's magnetic field lines to impact directly onto its polar regions (see **OGO 4–July 28, 1967**).

The spacecraft made detailed measurements of the plasma torus that surrounded Io and interacted with Jupiter, indicating that the amount of gas within it was comparable to that seen by **Voyager 1, Jupiter Flyby (March 5, 1979)** and **Voyager 2, Jupiter Flyby (July 9, 1979)**, although the torus appeared less extended at the equator, implying a cooler temperature than had been estimated previously. Its make-up was mostly oxygen and sulfur ions ejected from Io's volcanoes (produced at a rate of about a ton a second), though solar wind ions and ions from Jupiter were also detected.

Ulysses also found that Jupiter's strong magnetic field generally swept the space surrounding the planet clean of dust. Most of the dust detected came from interplanetary space and not from Jupiter or its moons. Furthermore, on March 10th, the spacecraft passed through an intense dust stream, its trajectory resembling what would be found if Ulysses had passed through a comet's dust tail. No known comet was nearby, however, which left the source of the stream completely unknown. See **Galileo (December 7, 1995)** for further data about the dust rings around Jupiter.

Ulysses then moved below the ecliptic of the solar system, swinging into a polar orbit that took it above each pole at a distance of between 170 and 300 million miles. See **Ulysses, First Solar Orbit (April 17, 1998)** for Ulysses' later discoveries.

February 11, 1992
JERS 1
Japan

JERS (Japanese Earth Resource Satellite) 1 was Japan's first Earth resource satellite, following satellites launched by the United States (**Landsat 1–July 23, 1972**), India (**Bhaskara 1–June 7, 1979**), the Soviet Union (**Meteor 1-30–June 18, 1980**), and France (**Spot 1–February 22, 1986**).

JERS 1 carried both radar and multispectral imaging systems. The multispectral system took photographs in seven spectral ranges. The two visible and first two infrared wavebands were identical to Landsat 4's (see **Landsat 4–July 16, 1982**), with the three highest infrared bands subdividing Landsat 4's 2.08- to 2.35-micron band into three narrow ranges: 2.01–2.12, 2.13–2.25, and 2.27–2.4 microns. Researchers hoped that these subdivisions would allow JERS 1 to provide more detailed and specific information about geological strata. Unfortunately, blurring and other interference patterns made many of these images unusable.

The swathe width of the radar-imaging system was 50 miles, with resolution as small as 60 feet. It used a synthetic-aperture radar (see **Seasat–June 27, 1978**) working in the 23.5-centimeter waveband. Its 55-degree look angle, which produced the pseudo shadows of the radar image, was intermediate between that of **ERS 1 (July 17, 1991)**, which had a steep 67-degree angle, and both **Almaz 1 (March 31, 1991)** and the previous shuttle radar mission (**STS 2–November 12, 1981**). Hence, images of a region could be compared to reveal different features.

JERS 1 operated until 1998, taking nearly half a million radar images and more than 300,000 optical images.

The next radar instrument flew on two space shuttle missions: STS 59 (**April 9, 1994**) and STS 68 (**September 30, 1994**).

March 17, 1992
Soyuz TM14
Russia / Germany

Soyuz TM14 carried Alexander Viktorenko (**Soyuz TM3–July 22, 1987; Soyuz TM8–September 5, 1989; Soyuz TM20–October 4, 1994**), Alexander Kaleri (**Soyuz TM24–August 17, 1996**), and German Klaus-Dietrich Flade. Flade was a paying customer; his airfare was about $12 million, paid by the unified German government. This space flight was also the first manned mission under the flag of the Russian Federation, the Soviet Union having ceased to exist.

After two days of maneuvers, the spacecraft docked with **Mir (February 20, 1986)**, where its three crewmen joined Sergei Krikalev of **Soyuz TM12 (May 18, 1991)** and Alexander Volkov of **Soyuz TM13 (October 2, 1991)**. Because of cash shortages and political troubles on Earth, Krikalev had been forced to remain in space for two consecutive crew shifts. During his 10-month space flight, an attempted coup had failed in Moscow and the Soviet Union had fallen.

During the six days of joint occupancy, the five men conducted a number of medical and biological experiments. Viktorenko and Kaleri were also briefed on the station's operation.

On March 25th, Krikalev, Volkov, and Flade boarded Soyuz TM13, undocked from Mir, and returned to Earth. Krikalev had completed 313 days in space, a record exceeded only by the 366 days flown by Vladimir Titov and Musa Manarov on **Soyuz TM4 (December 21, 1987)**.

On Mir, the Soyuz TM14 crew, Viktorenko and Kaleri, settled down to a four-month crew shift. During their occupancy they were visited by two Progress freighters (see **Progress M1–August 23, 1989**), both bringing supplies and replacement parts. The second needed two attempts to dock safely with Mir; its first docking attempt was aborted due to a problem in the software of the Kurs automatic docking system. Neither freighter carried a recoverable capsule.

The two men did one space walk during their occupancy, replacing two gyroscopes on the station's exterior.

Their shift ended with a six-day joint occupancy with **Soyuz TM15 (July 27, 1992)**.

March 24, 1992
STS 45, Atlantis Flight #11
U.S.A. / International

STS 45, a nine-day mission, was the eleventh flight of Atlantis. It was crewed by Charles Bolden (**STS 61C–January 12, 1986; STS 31–April 24, 1990; STS 60–February 3, 1994**), Brian Duffy (**STS 57–June 21, 1993; STS 72–January 11, 1996**), Kathryn Sullivan (**STS 41G–October 5, 1984; STS 31–April 24, 1990**), David Leetsma (**STS 41G–October 5, 1984; STS 28–August 8, 1989**), Michael Foale (**STS 56–April 8, 1993; STS 63–February 3, 1995; STS 84–May 15, 1997; STS 103–December 19, 1999**), Byron Lichtenberg (**STS 9–November 28, 1983**), and Dirk Frimout, a Belgian.

As had been done on **STS 42 (January 22, 1992)**, the Spacelab module was reconfigured, this time to conduct international studies of the earth and its atmosphere by the United States, France, Germany, Belgium, the United Kingdom, Switzerland, the Netherlands, and Japan. The lab included a package of 12 instruments, 4 to measure the Sun's energy output, 7 to study the magnetosphere and the composition of the upper atmosphere, and 1 to conduct ultraviolet observations of selected galaxies, quasars, and the celestial background ultraviolet radiation. See **STS 56 (April 8, 1993)** for results.

One onboard experiment attempted to learn the exact cause of bone loss during space flight. It found that weightlessness significantly impaired the body's ability to produce osteoblasts, the chemical the body uses to mineralize bone fibers. This research continued on STS 56.

STS 45 carried one getaway special, an experiment to grow a crystal of gallium arsenide, 1 inch in diameter and 3.5 inches long. Such crystals were used extensively in the semiconductor and electronics industry. This experiment was a reflight of a successful experiment first flown on **STS 40 (June 5, 1991)**, this time carrying additional sensors for analyzing the growth process.

The next shuttle flight was STS 49 (**May 7, 1992**).

May 7, 1992
STS 49, Endeavour Flight #1
Intelsat 6 F3
U.S.A. / International

STS 49 was the maiden flight of the space shuttle Endeavour, built as a replacement for Challenger (see **STS 51L–January 28, 1986**). With its flight, Endeavour became the fifth spacecraft to fly into space and return to Earth in a condition allowing it to fly again. The nine-day mission was crewed by Daniel Brandenstein (**STS 8–August 30, 1983, STS 51G–June 17, 1985, and STS 32–June 9, 1990**), Kevin Chilton (**STS 59–April 9, 1994; STS 76–March 22, 1996**), Bruce Melnick (**STS 41–October 6, 1990**), Thomas Akers (**STS 41–October 6, 1990; STS 61–December 2, 1993; STS 79–September 16, 1996**), Richard Hieb (**STS 39–April 28, 1991; STS 65–July 8, 1994**), Kathryn Thornton (**STS 33–November 22, 1989; STS 61–December 2, 1993; STS 73–October 20, 1995**), and Pierre Thuot (**STS 36–February 28, 1990; STS 62–March 4, 1994**). STS 49 retrieved and redeployed one stranded communications satellite.

June 7, 1992

• *Intelsat 6 F3* Initially launched on March 14, 1990, Intelsat 6 F3 had been intended as the second geosynchronous communications satellite in Intelsat's (see **Early Bird–April 6, 1965**) sixth satellite cluster (see **Intelsat 6A F2–October 27, 1989**). Due to a failure of its second stage, however, Intelsat 6 F3 was stranded in low-Earth orbit. STS 49's job, for which Intelsat paid nearly $150 million, was to capture the satellite and attach a replacement solid rocket booster that would be able to lift it to geosynchronous orbit.

The rendezvous required both spacecraft (Endeavour and Intelsat 6 F3) to adjust their orbits, which was the first time such a dual maneuver had been done. In the first attempt to capture the satellite, Pierre Thuot attached himself to the end of the robot arm, and Bruce Melnick guided the arm close to the satellite so he could fasten a special capture bar to it. Unfortunately, each time Thuot tried to attach the bar, the satellite swung away from him.

A second space walk the next day had the same result. For more than three hours, Daniel Brandenstein flew Endeavour in close formation with Intelsat 6 F3 while Thuot and Melnick attempted four times without success to attach the capture bar.

Brandenstein and NASA then came up with a daring plan for capturing the satellite. Three astronauts would stand in a circle in the cargo bay, and Brandenstein would maneuver Endeavour so that the satellite would literally fly into their hands.

Left to right: Hieb, Akers, and Thuot grab Intelsat 6 F3. *NASA*

With Brandenstein at the controls, Thuot, Richard Hieb, and Thomas Akers positioned themselves in the cargo bay, and at Hieb's command, they reached up and grabbed the satellite. For a half-hour more, Thuot and Akers held on while Hieb attached the capture bar. Then Melnick grabbed it with the robot arm and docked the satellite. The new booster engine was then attached to it.

Once the astronauts were back inside Endeavour, the satellite was released. The next day, its booster fired, which successfully placed Intelsat 6 F3 in geosynchronous orbit.

The space walk by Thuot, Hieb, and Akers lasted 8.5 hours and remains the longest EVA in history, through December 1999.

STS 49 included one more space walk, by Akers and Kathryn Thornton. For several hours, the two astronauts tested a variety of construction techniques being planned for the planned American space station Freedom (which was later replaced by the International Space Station; see **Zarya–November 20, 1998**). They constructed a 15- by 15- by 15-foot truss framework and climbed along it to test its strength. Then, with Melnick controlling the robot arm, the astronauts attached a two-ton structure (normally placed in the cargo bay to hold instruments and experiments) to the top of the framework. The exercise simulated the tasks necessary to link the station's various modules and trusses.

STS 49 also continued protein crystal growth research (see **STS 26–September 29, 1988**). The equipment attempted to grow a large batch of crystals from bovine insulin. Previous flights (see **STS 42–January 22, 1992**) had shown that the growth rate—and, therefore, the final size of the crystals—could be increased by very slowly lowering the temperature from 104°F to 71°F as the crystals grew. This research continued on the next shuttle flight, which was **STS 50** (**June 25, 1992**).

June 7, 1992
EUVE
U.S.A.

Built by NASA, the Extreme Ultra-Violet Explorer (EUVE) produced an all-sky survey in the 60–740 angstrom range of the ultraviolet spectrum. It followed and supplemented instruments on the Hubble Space Telescope (see **STS 31–April 24, 1990**) and **Rosat** (**June 1, 1990**).

Astronomers had previously thought that this part of the electromagnetic spectrum would be completely blocked by interstellar gas, making it pointless to build a space telescope to observe it. Observations by **Copernicus** (**August 21, 1972**) had shown, however, that the interstellar medium was very patchy, with dense cloudy regions interspersed with bubbles and tunnels of emptiness. The solar system itself was now known to be traveling through a particularly empty region dubbed the Local Bubble (see **Compton/STS 37–April 5, 1991**). Hence, an extreme ultraviolet telescope might be able to pick out details previously unseen.

EUVE's initial sky survey increased the number of known sources emitting radiation in these wavelengths from 10 (4 of which were the Sun, Jupiter, Saturn, and its moon Titan) to about 100. It observed an outburst of extreme ultraviolet radiation from what astronomers now call "cataclysmic variables," a closely orbiting binary star system where periodically the gravitational force of the white dwarf sucks matter from its partner, causing a gigantic thermonuclear explosion. It also detected solar flare events on several red dwarf stars.

EUVE startled scientists when it detected one extreme ultraviolet source about 2 billion light years away, inside the nucleus of another galaxy. Belonging to the class of galaxies known as BL Lac objects, or blazars, these objects vary in brightness across a wide range of wavelengths, sometimes fluctuating radically in time spans as short as a single day. The data implied that a single object, possibly a supermassive black hole, was the radiation source.

The spacecraft discovered ionized helium in the nearby interstellar medium. It also helped map the nearby galactic clouds that surround the Sun.

EUVE was still in operation as of December 1999, available to astronomers to make observations, much like any ground-based telescope. Its research was further supplemented by **ASCA (February 20, 1993)**, **ALEXIS (April 25, 1993)**, and **Fuse (June 24, 1999)**.

EUVE also carried a small technology package for measuring the contamination produced by the venting of satellite materials over time. These data put realistic limits on the amount of contamination produced, and thereby allow engineers to redesign future satellites for less cost.

June 10, 1992
Intelsat K
International / U.S.A.

Intelsat K, owned by the international consortium Intelsat (see **Early Bird–April 6, 1965**), was dedicated to television and business communications between Europe and North America. Launched from Cape Canaveral, it augmented the already functioning Intelsat 5/Intelsat 5A and Intelsat 6A constellations (see **Intelsat 5 F2–December 6, 1980**; **Intelsat 5A F10–March 22, 1985**; **Intelsat 6A F2–October 27, 1989**).

Intelsat K was originally developed by General Electric for the American company Crimson Satellite. When Crimson Satellite abandoned its plans, the satellite was sold to Intelsat. It carries capacity for 32 television channels or a maximum of 65,000 two-way voice circuits. Ground antennas as small as 4 feet across could be used to relay transmissions.

The next Intelsat satellite was the first in its seventh generation of communication satellites, **Intelsat 7 F1 (October 22, 1993)**.

June 25, 1992
STS 50, Columbia Flight #12
U.S.A.

STS 50 was the longest shuttle mission to date, lasting almost 14 days and exceeding the previous record of 11 days set by **STS 32 (January 9, 1990)**. The mission was Columbia's twelfth flight, crewed by Richard Richards (**STS 28–August 8, 1989**; **STS 41–October 6, 1990**; **STS 64–September 9, 1994**), Kenneth Brown, Bonnie Dunbar (**STS 61A–October 30, 1985**; **STS 32–January 9, 1990**; **STS 71–June 27, 1995**; **STS 89–January 22, 1998**), Carl Meade (**STS 38–November 15, 1990**; **STS 64–September 9, 1994**), Ellen Baker (**STS 34–October 18, 1989**; **STS 71–June 27, 1995**), Lawrence DeLucas, and Eugene Trinh. The 14 days in space by Dunbar and Baker set a record for the longest space flight by a woman, breaking the 12-day record set by Svetlana Savitskaya on **Soyuz T12 (July 17, 1984)**. This new record lasted until **STS 65 (July 8, 1994)**.

STS 50 carried what NASA had dubbed the United States Microgravity Laboratory (USML-1). This facility used a reconfigured Spacelab module in the shuttle's cargo bay to conduct 32 different experiments, from growing protein crystals to studying the dynamics of candle flames in weightlessness.

Protein crystal growth research (see **STS 26–September 29, 1988** and **STS 49–May 7, 1992**) took advantage of the flight's length to grow very large crystals, with a total of almost 900 different samples tested and a 50 percent success rate. The first malic enzyme crystals were grown; Lawrence DeLucas manually mixed the protein and a precipitating agent solution to generate the growth. Other crystals helped scientists understand and develop drugs to treat AIDS and other immune-system disorders. See **STS 52 (October 22, 1992)** for later protein crystal research.

A new experiment to grow zeolite crystals resulted in larger crystals then ever before seen. And a world-record 6.5-inch-long gallium arsenide semiconductor crystal was produced.

Another experiment studied the behavior of fluid drops when held within a containerless box of sound waves (see **STS 41B–February 3, 1984** and **STS 61C–January 12, 1986**). Acoustics were used to move, break apart, and then rejoin these drops. The sound waves were also successfully used to rotate the droplets and make them oscillate. This research continued on **STS 47 (September 12, 1992)**.

A newly designed high-tech greenhouse, called Astroculture, was also tested, investigating methods for circulating air, water, and light to plants while also providing their roots an artificial soil in which to grow. Previous experiments on a number of different American and Soviet/Russian space flights had failed to find any consistently successful method for growing plants in space (see for example **Soyuz 35–April 9, 1980**). This greenhouse's next flight was **STS 57 (June 21, 1993)**.

The next shuttle mission was **STS 46 (July 31, 1992)**. The next U.S. Microgravity Laboratory flight was **STS 73 (October 20, 1995)**.

July 3, 1992
Sampex
U.S.A.

The goal of the Sampex (Solar Anomalous and Magnetospheric Particle Explorer) was to study high-energy particles in a polar orbit about 300 to 400 miles altitude. These particles included cosmic rays coming from outside the solar system, ionized particles from the Sun, and other energized electrons in the earth's magnetosphere. It continued research by **CRRES (July 25, 1990)** and **Intercosmos 25 (December 18, 1991)** and signaled the renewal of such work after a general period of inactivity (see **CCE–August 16, 1984** and **Ionosonde–February 21, 1986**).

Sampex was the first in a NASA series of small research satellites that were meant to be built and launched frequently and at low cost. For example, two of Sampex's instruments had previously flown as getaway specials on earlier shuttle flights.

Sampex found that the Van Allen radiation belts, previously thought to have two layers (see **OGO 1–September 5, 1964**), also had a third belt close to the earth. This belt trapped many interstellar high-energy particles and cosmic rays, as well as many of the heavier ionized particles from the Sun, bouncing the particles back and forth along the earth's magnetic field lines before they finally leaked out into the atmosphere or into space. Because the layer seemed to vary greatly in intensity over time, and because its particles could act to deplete ozone, some scientists thought that it could have contributed to the fluctuations seen in the ozone layer in recent years by satellites like **Meteor 3-05 (August 15, 1991)** and **UARS** (see **STS 48–September 12, 1991**). See **STS 56 (April 8, 1993)** for further ozone layer research.

Sampex also found that high-energy particles from both the Sun and Jupiter reached Earth, depending on the activity level of the Sun—when the Sun was quiet, Jupiter's particles bombarded Earth.

As of December 1999, Sampex is still in operation. The next satellite to study the earth's magnetosphere was **Geotail (July 24, 1992)**, initiating a new international project.

July 9, 1992
Insat 2A
India / Europe

Built in India under the direction of the Indian Space Research Organization (ISRO) and launched on an Ariane rocket, Insat 2A was a second-generation satellite to replace India's first communications-weather satellites (see **Insat 1A–April 10, 1982**). Through December 1999, India launched five Insat 2 satellites, placing them in geosynchronous orbit in two positions above India.

Similar to the Insat 1 satellites, Insat 2 satellites are almost three times heavier and generate four times the electrical power from solar panels that are 50 percent larger. Like the Insat 1 satellite, the Insat 2 satellites are asymmetrical, with their solar panels on one side and a solar sail on the other to balance each satellite against the pressure of the solar wind. Each satellite's capacity to provide both telephone and television communications is about the same, as is its visible and infrared radiometer for producing weather images of the Indian subcontinent.

Insat 2A also carried a second-generation international SARSAT/COSPAS search-and-rescue receiver/transmitter package, used to locate ships and aircraft in distress. See **Cosmos 1383 (June 30, 1982)**.

July 24, 1992
Geotail
Japan / U.S.A.

Geotail was a joint Japanese and American project to study the tail of the earth's magnetosphere, which trails away from the Sun beyond the Moon's orbit. The spacecraft was also part of a new international coordinated effort by the United States, Japan, the European Space Agency, and Russia to study the dynamics of the solar wind with the earth's magnetosphere. Subsequent satellites in the program were **Wind (November 1, 1994)**, **Interball 1 (August 2, 1995)**, **Soho (December 2, 1995)**, **Polar (February 24, 1996)**, and **Interball 2 (August 29, 1996)**. **Sampex (July 3, 1992)** also contributed data.

For the first two years of Geotail's operation, it was placed in a long eccentric orbit extending out into the tail by almost 900,000 miles. Later, the orbit was adjusted inward so that the spacecraft could study the tail's features closer to the earth. As of December 1999, it is still in operation.

The data indicated that the earth's plasma sheet, the region that separates the earth's magnetosphere from the solar wind, is composed mostly of hydrogen ions with traces of helium and oxygen ions. The uneven distribution of particles moving at different velocities fit the theoretical models that scientists had developed—they could now use these models to extrapolate the plasma sheet's formation process and how it has fluctuated in time with the Sun's solar wind.

This research was further augmented by **Freja (October 6, 1992)**, which studied the aurora.

July 27, 1992
Soyuz TM15
Russia / France

Soyuz TM15 included Anatoly Solovyov (**Soyuz TM5–June 7, 1988**; **Soyuz TM9–February 11, 1990**; **STS 71–June 27, 1995**; **Soyuz TM26–August 5, 1997**), Sergei Avdeyev (**Soyuz TM22–September 3, 1995**; **Soyuz TM28–August 13, 1998**), and Michel Tognini. Tognini, the second Frenchman to fly in space following Jean-Loup Chretien (see **Soyuz T6–June 24, 1982**; **Soyuz TM7–November 26, 1986**), was a paying customer; his airfare of $12 million was paid by the French government. According to the Russian-French deal, Tognini's flight was the first of three planned visits to **Mir (February 20, 1986)**. See **Soyuz TM17 (July 1, 1993)** and **Soyuz TM24 (August 17, 1996)** for the subsequent flights.

After two days of maneuvers, Soyuz TM15 docked with Mir, where Alexander Viktorenko and Alexander Kaleri of **Soyuz TM14 (March 17, 1992)** had been occupying the station for the last four months.

For the next 12 days, the two crews performed a range of commercial experiments in biology, medicine, and fluid and materials research. Much of Tognini's work was to set up long-term experiments to be retrieved by the later French flights. Then, on August 10th, Viktorenko, Kaleri, and Tognini returned to Earth in Soyuz TM14, leaving Solovyov and Avdeyev on Mir. Viktorenko and Kaleri had completed 146 days in space. Tognini returned with about 22 pounds of experiment results.

Solovyov and Avdeyev settled down to six months in space. Their flight's primary goal was maintenance to prolong Mir's orbital life. They replaced several gyroscopes that controlled the station's orientation in space. They performed four space walks, during which they installed a new rocket thruster to the end of the 46-foot Sofora boom that had been constructed by the cosmonauts on **Soyuz TM12 (May 18, 1991)**. This thruster improved their ability to control Mir's roll and thereby to aim its solar panels more accurately at the Sun. During these space walks, they also repositioned a Kurs docking antenna to the androgynous docking port on **Kristall (May 31, 1990)** in anticipation of future Russian shuttle flights.

During these EVAs (extra-vehicular activities), they also removed the hammer-and-sickle emblem of the Soviet Union that had been painted on the side of the space station.

During their stay, they took over 7,000 multispectral photographs of the earth, covering 1.35 million square miles of Europe, America, and the former Soviet Union. Other research produced slightly more than 1.5 pounds of gallium arsenide, cadmium sulfide, and zinc oxide crystals, used as semiconductors in the computer and electronics industry.

Two Progress freighters resupplied the station (see **Progress M1–August 23, 1989**). The first contained a recoverable capsule, which was used to return an additional 22 pounds of experimental results produced during Michel Tognini's flight. The second carried an experimental solar mirror, about 66 feet in diameter, part of a Russian experiment to see if such an orbiting mirror could be used to light portions of the earth. See **Soyuz TM16 (January 24, 1993)**.

On November 20th, the two cosmonauts also released from Mir's airlock a small subsatellite, called Mak 2, for studying the earth's ionosphere. The satellite remained in orbit for about three months before burning up in the earth's atmosphere.

On January 26, 1993, Solovyov and Avdeyev were joined by their replacement crew on **Soyuz TM16 (January 24, 1993)**.

July 31, 1992
STS 46, Atlantis Flight #12
EURECA / TSS 1
U.S.A. / Europe / Switzerland / Italy

The twelfth flight of Atlantis, the eight-day STS 46 mission was crewed by Loren Shriver (STS 51C–January 24, 1985; STS 31–April 24, 1990), Andrew Allen (STS 62–March 4, 1994; STS 75–February 22, 1996), Jeffrey Hoffman (STS 51D–April 12, 1985; STS 35–December 2, 1990; STS 61–December 2, 1993; STS 75–February 22, 1996), Franklin Chang-Diaz (STS 61C–January 12, 1986; STS 34–October 18, 1989; STS 60–February 3, 1994; STS 75–February 22, 1996; STS 91–June 2, 1998), Marsha Ivins (STS 32–January 9, 1990; STS 62–March 4, 1994; STS 81–January 12, 1997), Claude Nicollier (STS 61–December 2, 1993; STS 75–February 22, 1996; STS 103–December 19, 1999), and Franco Malerba. Nicollier was the first Swiss astronaut; Malerba was the first Italian. STS 46 released one satellite.

• *EURECA* Deployed on August 2nd, EURECA (the EUropean REtrievable CArrier) was a European Space Agency research satellite. EURECA was designed to remain in orbit for between six months and a year, to be retrieved by a later space shuttle mission. This reusable satellite took advantage of this long period in space to carry out 15 different experiments in the life sciences and materials processing fields. Protein and semiconductor crystals were grown, different types of experimental gallium arsenide solar cells were tested, and the effect of space on biological samples was studied.

One experiment studied the mechanics of adhesion, how solid bodies stick to each other. A variety of spherical projectiles, weighing from 0.66 to 1.1 pounds, were fired almost 200,000 times at a target surface at differing speeds. The resulting rebounds were measured to see how weightlessness changed their bounce and the friction of impact. The experiment found that, in space, slight variations in surface roughness significantly affect the impact results.

Another instrument measured the Sun's output in the wavelengths from 170 to 3,200 microns to learn whether the Sun was a variable star and by how much it varied, continuing research by **Nimbus 7 (October 13, 1978)** and **ERBS** (see STS 41G–October 5, 1984). It found that the Sun's output varied with the passage of dark sunspots and bright faculae across its face, causing overall brightness variations as the number of these features changed throughout the solar cycle. See **STS 56 (April 8, 1993)** for further solar variation research.

EURECA also performed data relay tests in conjunction with the geosynchronous experimental communications satellite **Olympus 1 (July 12, 1989)**. This research tested the 23–28 GHz frequency band for relaying data to and from geosynchronous and low orbit satellites and worked in conjunction with NASA's TDRS communications satellites (see **STS 6–April 4, 1983**).

EURECA remained in orbit for 11 months and was retrieved by **STS 57 (June 21, 1993)**.

• *TSS 1* The technology experiment TSS (Tethered Satellite System) 1, a joint U.S. and Italian project, tested the concept of using long tethers to connect two spacecraft while simultaneously generating electricity. Theory held that as the long copper tether swept through the earth's magnetic field, it would produce an electrical current, which the spacecraft could then tap for power. Previous tether experiments (see **Gemini 11–September 12, 1966**, and **Gemini 12–November 11, 1966**) had not looked for this current.

The plan called for TSS 1, weighing 1,139 pounds, to be unreeled almost 14 miles by a copper cord less than 0.1 inch in diameter. The two spacecraft would then perform several hours of experiments, testing their maneuverability and the generation of electricity.

Unfortunately, the tether kept jamming, and the astronauts were only able to unreel TSS 1 about 850 feet. Because the jamming problems also made it difficult to reel the satellite in, the experiment was cut short to avoid the risk of losing TSS 1.

The next tethered satellite test occurred on **Seds (March 30, 1993)**. The next flight of TSS was **STS 75 (February 22, 1996)**.

Other STS 46 experiments, placed in the shuttle's cargo bay, evaluated how different materials stood up to exposure to space, many of which were being considered as construction materials for future space station projects. Some specifically measured the effect of the atomic oxygen flow that circulated around the shuttle while in orbit. One tested electroplating using pure nickel.

Atlantis also carried the IMAX camera in its cargo bay, shooting footage for the film *Destiny in Space*, the third IMAX film shot in space.

The next shuttle flight was **STS 47 (September 12, 1992)**.

August 10, 1992
TOPEX-Poseidon / S80-T / Uribyol (Kitsat 1)
U.S.A. / France / South Korea / U.K. / Europe

- *TOPEX-Poseidon* TOPEX-Poseidon was launched on an Ariane rocket. This joint project of the United States and France (TOPEX was the American name, Poseidon the French) was the first radar satellite dedicated solely to observing the earth's oceans since **Seasat (June 27, 1978)**. It also followed up research of **Geosat (March 13, 1985)**. Its radar equipment could measure the distance from the satellite to the ocean surface to an accuracy of 1.2 inches, thereby producing accurate maps of the oceans' currents and eddies.

The spacecraft produced immediate dividends, including the first detailed images of the development and growth of an El Niño event in the winter of 1992–93 *(see color section, Figure 28)*.

The satellite showed the enormous strip of warm water—hundreds of miles wide and about 1,000 feet deep—building up along the equator in the Pacific Ocean and flowing toward South America. The event resulted in wet weather in the western United States and a warm winter in the east. In early 1996, the satellite successfully predicted a second El Niño event for the 1997–98 winter.

TOPEX-Poseidon also detected subtle changes in the global sea level as the seasons changed. For example, in 1993, it was able to measure a drop of about a foot in sea level in both the Gulf Stream off of North America and in the ocean east of Japan, caused by a cooling of the ocean by cold winter air blowing off the nearby continents. A corresponding rise was detected in the southern hemisphere.

Over time, TOPEX-Poseidon will measure changes in the global sea level, showing whether climate change is causing the oceans to rise, as predicted by some scientists. Although the first four years of results indicated a very tiny global sea level rise of somewhere between 0.04 and 0.12 inch per year, the uncertainties were so large that scientists were unsure if any increase had occurred at all. As of December 1999, TOPEX-Poseidon remains in operation.

- *S80-T* A French experimental communications satellite, S80-T did propagation tests at several frequencies, evaluating their use with state-of-the-art message pagers using mobile receivers.

- *Uribyol* Also called Kitsat 1, Uribyol ("our star" in Korean) was a demonstration South Korean communications satellite. It was built by South Korean students at the University of Surrey in England as part of a $12 million student exchange program. The satellite carried a system to transfer electronic mail, a camera to take pictures of the earth, and a sensor for measuring cosmic rays.

August 14, 1992
Aussat B1 (Optus B1)
Australia / China / U.S.A.

Also called Optus B1, the Australian geosynchronous communications satellite Aussat B1 augmented the Aussat A satellites (see **STS 51I–August 27, 1985**). Built by Hughes Communications and launched on a Chinese Long March rocket, Aussat B1 was Hughes's first three-axis-stabilized satellite, with solar panels deployed on opposite sides of the main payload and attitude jets for stabilizing the spacecraft's orientation. All of Hughes's previous satellites, such as Anik C3 (see **STS 5–November 11, 1982**) and Telstar 3A (**July 28, 1983**), had been spin-stabilized—a spinning drum within which the main payload rested on a despun platform (see **DATS–July 1, 1967**).

While the main communications payload of this satellite is similar to the earlier satellites, Aussat B satellites generate more than three times the electricity, have more bandwidth, and have a more efficient design. The satellites include a 28 GHz beacon for performing propagation experiments at this frequency, renewing experiments first performed on **Comstar 1 (May 13, 1976)**.

Australia successfully launched one additional Aussat B satellite in 1994, once again using a Chinese Long March rocket. As of December 1999, three Aussat satellites still operate.

September 10, 1992
Hispasat 1A
Spain / U.S.A.

Hispasat 1A was launched on a U.S. rocket and was owned and operated by Hispasat SA, a semi-private company formed by the Spanish government and several Spanish aerospace and television companies. Hispasat 1A was the first satellite devoted to providing cable television, direct broadcast television, and telephone service to Spain and the Canary Islands, as well as communications for the Spanish military both in and out of Spain. The satellite was also used to broadcast Spanish events celebrating the 500th anniversary of Columbus's discovery of America, as well as the 1992 summer Olympic games from Barcelona.

A second Hispasat satellite was launched in 1993. Both are still in operation as of December 1999, though identical problems with their military transponders disabled these functions in January 1998.

September 12, 1992
STS 47, Endeavour Flight #2
U.S.A. / Japan

STS 47 was Endeavour's second flight, making it the fifth reusable spacecraft to return to space. The eight-day mission was crewed by Robert Gibson (**STS 41B–February 3, 1984**; **STS 61C–January 12, 1986**; **STS 27–December 2, 1988**; **STS 71–June 27, 1995**), Curtis Brown (**STS 66–November 3, 1994**; **STS 77–May 19, 1996**; **STS 85–August 7, 1997**; **STS 95–October 29, 1998**; **STS 103–December 19, 1999**), Mark Lee (**STS 30–May 4, 1989**; **STS 64–September 9, 1994**; **STS 82–February 11, 1997**), Jan Davis (**STS 60–February 3, 1994**; **STS 85–August 7, 1997**), Jerome Apt (**STS 37–April 5, 1991**; **STS 59–April 9, 1994**; **STS 79–September 16, 1996**), Mae Jemison, and Mamoru Mohri. Lee and Davis were the first husband and wife to fly into space on the same spacecraft, while Jemison was the first black woman. Mohri was the second Japanese person to fly in space, following Toyahiro Akiyama on **Soyuz TM11 (December 2, 1990)**.

STS 47 carried what was called Spacelab-J, a Spacelab configuration of 43 experiments, 34 by Japan, 7 by the United

States, and 2 joint tests. More than half the experiments involved research into the behavior of metals, alloys, and fluids, as well as testing methods for growing semiconductor crystals. Five different furnaces were used. The acoustical tests from **STS 50 (June 25, 1992)** were repeated, in which studies were made of how sound waves could position, rotate, and shape floating drops of mineral oil. Other experiments studied the flow of melted glass in weightlessness and attempted to make an evenly mixed alloy of indium and aluminum, impossible on Earth.

The behavior of boiling liquids in weightlessness was examined, using 10 pounds of liquid freon. The lack of gravity prevented convection, the process whereby the bubbles of gas rise to the liquid's surface as it boils, then dissipate into the air. Instead, the gas bubbles grew much larger than they did on Earth, and while they did tend to migrate to the surface of liquid bubble, they remained there, forming a layer of vapor that enclosed the liquid and caused its temperature to rise.

Among the life sciences experiments, one took unfertilized frog eggs and fertilized them while in orbit to watch them develop. On Earth, the eggs grow with an up, down, right, and left symmetry directly related to gravitational pull. In space, the eggs developed normally and were apparently unaffected by the lack of gravity. Other frogs, however, that had been fertilized on Earth and hatched as tadpoles during the flight behaved abnormally, some swimming in a variety of wild random patterns, from circles to somersaults. More than half also died.

STS 47 included the first electrophoresis experiments since **Foton 5 (April 26, 1989)** and the first on the shuttle since STS 26 (see **STS 4–June 27, 1982**). Prior to the Challenger accident (see **STS 51L–January 28, 1986**), electrophoresis research by McDonnell Douglas had been a regular payload on shuttle missions. When commercial satellite launches on the shuttle ceased after the accident, that research stopped as well. On STS 47, two different experiments attempted to separate several biological mixtures, including salmonella bacteria.

Endeavour also carried an Israeli scientific experiment, investigating how the oriental hornet's ability to build its combs was affected by weightlessness. On Earth, the hornets' sense of gravity forced them to build their combs a certain way. In space, however, the hornets appeared disorganized and unable to work normally.

The astronauts continued fire experiments from **STS 41 (October 6, 1990)**, igniting several samples and filming the resulting fires. Further fire tests were conducted on **STS 54 (January 13, 1993)**.

STS 47 carried nine getaway specials, with experiments from five countries. Most studied how different kinds of materials behaved in space. One investigation mixed flour, water, and yeast and then baked it to see whether the bread yeast could cause the bread to rise as it did on Earth. Unfortunately, the experiment failed when the researchers found that the ingredients had been mixed improperly while in space.

The next shuttle flight was **STS 52 (October 22, 1992)**.

September 25, 1992
Mars Observer
U.S.A.

The first American probe to Mars in 17 years, since **Viking 2 (August 7, 1976)**, Mars Observer was intended by NASA as an orbital mapper to study the red planet's atmosphere, surface, and geological make-up.

Though the spacecraft functioned well during its cruise to Mars, all contact was lost on August 21, 1993, three days before orbital insertion. Although the cause of this failure is still unknown, some scientists suspect that a ruptured fuel line caused the spacecraft to spin out of control. This was the first American planetary mission to fail since Mariner 8 had fallen in the ocean during its launch in 1971 (see **Mariner 9, Mars Orbit–November 14, 1971**).

The next missions to reach Mars was **Mars Pathfinder (July 4, 1997)** and **Mars Global Surveyor (September 12, 1997)**.

October 6, 1992
China 35 / Freja
China / Sweden

• *China 35* Also called FSW-1 4, China 35 was part of a series of Chinese science missions in which a recoverable capsule was sent into orbit for periods of about a week and then recovered. See **China 23 (August 8, 1988)** for program information.

On this flight, plant seeds from the mission were subsequently distributed to 1,500 elementary schools, where they were planted and successfully grown.

• *Freja* A Swedish satellite, Freja also carried instruments built by U.S., Canadian, and German scientists and continued research by Sweden's **Viking (February 22, 1986)**, studying the earth's aurora and how that light show was caused by the impact of the Sun's solar wind on the earth's magnetosphere. Its data also supplemented the international program begun by **Geotail (July 24, 1992)** to study the magnetosphere.

Freja operated for four years. Its orbital inclination took the spacecraft through the auroral halo for long periods during each orbit, traveling along magnetic field lines directly connected to electromagnetic events, a perspective not previously viewed. Its instrumentation was also more precise, resolving hot plasma features as small as 600 feet across and cold plasma features smaller than 30 feet. Scientists used these data to create more accurate models for predicting when and how solar events affected the electromagnetic turbulence that caused the aurora.

The next satellite to study the aurora, **Astrid (January 24, 1995)**, was also Sweden's next satellite. The next satellite to study the ionosphere was **Intercosmos 26 (March 2, 1994)**.

October 22, 1992
STS 52, Columbia Flight #13
Lageos 2 / CTA
U.S.A. / Italy / Canada

STS 52, a 10-day flight and Columbia's thirteenth, was crewed by James Wetherbee (**STS 32–January 9, 1990**; **STS 63–February 3, 1995**; **STS 86–September 25, 1997**), Michael

Baker (STS 43–August 2, 1991; STS 68–September 30, 1994; STS 81–January 12, 1997), Charles Veach (STS 39–April 28, 1991), Tamara Jernigan (STS 40–June 5, 1991; STS 67–March 2, 1995; STS 80–November 19, 1996), William Shepherd (STS 27–December 2, 1988; STS 41–October 6, 1990), and Canadian Steven MacLean. It released two research satellites.

• *Lageos 2* LAser GEOdynamic Satellite (Lageos) 2 was an Italian geodetic satellite, a twin of **Lageos 1 (May 4, 1976)**. A 2-foot-wide sphere weighing about 900 pounds and covered with laser reflectors, Lageos 2 was deployed from Columbia on October 23rd, where it used its booster engine to place itself in orbit about 3,600 miles high and circling the earth's middle latitudes. Ground-based stations then measured tiny motions in the earth's crust by firing lasers at the satellite and measuring the changing distances. With both Lageos satellites, scientists were able to double the accuracy of their measurements, more precisely monitoring changes in the earth's rotation, volcanic activity, and the shifting of its tectonic plates. See Lageos 1 for the geological results.

The data compiled from these two satellites were so accurate that in 1998 scientists announced that they had detected evidence that the earth's mass and rotation was actually dragging space around it, something that Einstein had predicted in his theory of general relativity. Ground lasers had found that both spacecraft's orbits had inexplicably shifted about six feet per year in the direction of the earth's rotation. According to Einstein, this shift was caused because as the earth spun, it pulled the space in which the satellites were orbiting along with it, thereby causing their orbits to shift.

The next geodetic satellite was **Stella (September 26, 1993)**.

• *CTA* The Canadian Target Assembly (CTA) was actually part of an experiment package designed to aid astronauts in gauging distance, size, and motion in space—astronauts had found that their sense of perspective was unreliable, due to the lack of reference points. CTA, a passive satellite 4 by 7 by 1.5 feet in dimensions, was covered with a known pattern of dots. A video camera onboard the shuttle followed CTA as it was deployed, and a computer program monitored the dot pattern to give Steve MacLean, the Canadian astronaut, a precise description of the satellite's location and orientation.

During the mission, MacLean did several tests with CTA as it was maneuvered about on the robot arm. On the ninth day of the mission, CTA was released so that MacLean could track it as it slowly moved away. He found that the program accurately measured the distance to the satellite.

STS 52's onboard experiments included two American experiments studying the behavior of solids and fluids. One investigated the transition of helium from its normal liquid state to that of a superfluid, a transition difficult to study on Earth. More than 40 times, the helium sample was passed through the transition phase. The second experiment tested how three samples of tin-bismuth alloy mixed as they were melted and then resolidified.

In a Canadian experiment package, more than 350 different material samples were exposed to space for 30 hours to try and understand how and why the atomic oxygen flow surrounding the shuttle damaged them. Another experiment studied the solidification process in 40 samples, combining various amounts of gold, silver, bismuth, manganese, and lead.

A Canadian medical experiment investigated the cause of back pain that about two-thirds of all astronauts and cosmonauts had experienced during and after space flight, caused by the elongation of the spine in weightlessness (see **Soyuz T10B–September 8, 1984** and **STS 42–January 22, 1992**). Steve MacLean, a doctor, kept track of changes in his height, as well as recorded the precise location of the back pain. Later back research occurred on **STS 65 (July 8, 1994)**.

A private commercial study, developed jointly by NASA and Merck & Co., investigated the ability of a drug to retard or stop bone loss in weightlessness. It was hoped that the protein could eventually become a treatment for osteoporosis in women. The same drug was also being tested on Earth in wide-scale clinical studies.

Protein crystal growth experiments (see **STS 26–September 29, 1988**) continued. As on **STS 49 (May 7, 1992)**, the research attempted to grow much larger batches of bovine insulin protein crystals by precisely controlling the temperature of the crystals as they developed. To do this more effectively, the equipment on this flight used a preprogrammed computer set prior to launch to monitor and adjust the experiment's temperature as required. These temperature tests were repeated on one shuttle flight in 1993, and then human insulin crystals were grown on **STS 60 (February 3, 1994)**.

In another crystal growth experiment, large semiconductor crystals of cadmium telluride were grown. Previously, the largest such crystal had been about the size of a pencil eraser. This experiment tried to produce crystals the size of a dime.

A third crystal growth experiment continued the zeolite crystal growth research done on **STS 50 (June 25, 1992)**. Zeolites were a porous mineral that, if formed uniformly as membranes, could be used as sieves in the gasoline and nuclear waste industries.

In technology tests, new methods for controlling the tank pressure of such fluids as liquid hydrogen, oxygen, and nitrogen were studied, as well as methods for circulating liquid in pipes to control the shuttle's temperature.

The next shuttle flight was **STS 53 (December 2, 1992)**.

October 28, 1992
Galaxy 7
U.S.A.

Galaxy 7 was the first in a second generation of geosynchronous communications satellites owned and operated by Hughes Communications, Inc., providing cable television service to the American market. See **Galaxy 1 (June 28, 1983)**.

Each satellite in this second generation transmits and receives in both the 4/6 GHz frequencies, as well as the 12/14 GHz frequencies first tested by **Hermes (January 17, 1976)** and used by **SBS 1 (November 15, 1980)**. The satellites also use the three-axis design first adopted by Hughes for **Aussat B1 (August 14, 1992)**. They generate four times the power of the earlier Galaxy satellites and have twice the television channel capability.

Through Dedcember 1999, Hughes has launched seven of these satellites.

November 16, 1992
Resurs-500
Russia

The Resurs-500 satellite (see **Meteor 1-30–June 18, 1980** and **Resurs F1–May 25, 1989**) is of special interest only because its descent module carried promotional and goodwill materials to the United States, celebrating the 500th anniversary of the European discovery of the New World by Christopher Columbus. Instead of being recovered in Russia, the capsule splashed down in the Pacific off the coast of the state of Washington, where a Russian ship recovered it and brought it to Seattle for a week of exhibitions and other public events.

November 21, 1992
MSTI 1
U.S.A.

MSTI 1, a U.S. Air Force satellite, was the first of three MSTI satellites testing infrared imaging equipment used to detect the heat from missile plumes. This technology was similar in concept to the early Midas surveillance satellites (see **Midas 2–May 24, 1960**) but more advanced. During its four-day mission, it took infrared images of a number of remote islands in the eastern Pacific Ocean.

The next MSTI satellite was **MSTI 2 (May 9, 1994)**.

December 2, 1992
STS 53, Discovery Flight #15
USA 89
U.S.A.

As of December 1999, STS 53 remains the last manned military space flight by the United States. The fifteenth flight of Discovery, it lasted seven days and was manned by David Walker (**STS 51A–November 8, 1984**; **STS 30–May 4, 1989**; **STS 69–September 7, 1995**), Robert Cabana (**STS 41–October 6, 1990**; **STS 65–July 8, 1994**; **STS 88–December 4, 1998**), Guion Bluford (**STS 8–August 30, 1983**; **STS 61A–October 30, 1985**; **STS 39–April 28, 1991**), James Voss (**STS 44–November 24, 1991**; **STS 69–September 7, 1995**), and Michael Clifford (**STS 59–April 9, 1994**; **STS 76–March 22, 1996**). The mission's classified cargo was an Air Force satellite.

- *USA 89*   Deployed on December 2nd, USA 89 was part of the Air Force's communications satellite network. See USA 40 on **STS 28 (August 8, 1989)**.

A second deployment of six small spheres, intended to test the ability of ground facilities to track orbital debris, did not take place because an equipment failure prevented the release of the spheres. This test successfully flew on **STS 60 (February 3, 1994)**.

STS 53 also conducted several technological experiments, from trying out new laser-sensing equipment to testing the refueling of tanks in space. In refueling tests (see **STS 41G–October 5, 1984**), colored water was transferred eight times between two small tanks, with cameras and sensors recording the events. Later shuttle refueling tests occurred on **STS 57 (June 21, 1993)**.

One experiment, making its first shuttle flight, tested methods for producing what scientists called microencapsulated products. In this process, an active material is encapsulated as tiny spheres within a polymer base. As the base deteriorates or is destroyed, it allows the release of the material. This technology is used in time-released drugs or with ubiquitous products such as scratch-and-sniff magazine ads and carbonless copy paper. On Earth, gravity prevents the production of large numbers of uniform beads within the polymer, limiting its use as an antibiotic. In space, scientists hoped the lack of gravity would solve this problem. On this flight, the yield of drugs was too small. The next flight test flew on **STS 70 (July 13, 1995)**.

The next shuttle mission was STS 54 (**January 13, 1993**).

December 29, 1992
Bion 10 (Cosmos 2229)
Russia / International

Bion 10, a biological research satellite, was part of a program begun with **Bion 1 (October 31, 1973)**. Following **Bion 9 (September 15, 1989)**, it carried two monkeys, insects, reptiles, plants, and cell and tissue cultures on a planned two-week space flight. Researchers from Russia, the United States, the European Space Agency, and seven other countries participated in a total of two dozen different life sciences experiments.

The flight, which lasted 12 days, was forced to land 2 days early when temperatures in the capsule began rising uncontrollably. Several experiments were destroyed by the heat, and the malfunctions caused one of the two monkeys to go without food for the mission's last three days.

The next Bion mission was **Bion 11 (December 24, 1996)**.

1993

January 13, 1993
STS 54, Endeavour Flight #3
TDRS 6
U.S.A.

The third space flight of Endeavour, STS 54 was a six-day mission crewed by John Casper (**STS 36–February 28, 1990**; **STS 62–March 4, 1994**; **STS 77–May 19, 1996**), Donald McMonagle (**STS 39–April 28, 1991**; **STS 66–November 3, 1994**), Mario Runco (**STS 44–November 24, 1991**; **STS 77–May 19, 1996**), Gregory Harbaugh (**STS 39–April 28, 1991**; **STS 71–June 27, 1995**; **STS 82–February 11, 1997**), and Susan Helms (**STS 64–September 9, 1994**; **STS 78–June 20, 1996**). The flight released one satellite.

- *TDRS 6*   Deployed on January 13, Tracking and Data Relay Satellite (TDRS) 6 provided NASA with communications to its spacecraft. See **STS 6 (April 4, 1983)**.

During the mission, astronauts Greg Harbaugh and Mario Runco performed a four-hour space walk, practicing techniques needed for the construction of structures in space. In one case, to simulate moving large objects in space, they took turns carrying each other with one hand while working their way through the cargo bay.

Onboard experiments included an x-ray spectrometer, studying the soft x-ray background radiation of the sky, from 42 to 84 angstroms. Its observations supplemented the sky-survey work done by **Rosat (June 1, 1990)** and anticipated observations of **Alexis (April 25, 1993)**.

Plant research with the Chromex facility continued. Results from **STS 29 (March 13, 1989)** and **STS 41 (October 6, 1990)** had shown that in space, plant seeds developed very poorly. On this flight, the seed production of *Arabidopsis thaliana* (a mouse-ear cress plant, so named because its leaves resemble the round ears of a mouse) was studied, following up earlier research on **Salyut 6 (September 29, 1977)** during **Soyuz 35 (April 9, 1980)** and **Soyuz T4 (March 12, 1981)** and **Salyut 7 (April 19, 1982)** during **Soyuz T9 (June 27, 1983)**. Thirty-six seeds were planted just prior to launch so that their early growth in weightlessness could be observed. For results, see **STS 51 (September 12, 1993)**.

A new biological facility contained 28 different commercial experiments, covering immune system disorders, cancer, bone disorders, and agriculture research. Other tests investigated methods for recycling waste products in space, grew both protein and RNA crystals, and attempted to form tiny magnets using bacteria.

The astronauts also conducted fire tests in space (see **STS 41–October 6, 1990**), burning two small pieces of Plexiglas. The fire produced a flash of bright orange, which quickly changed to a steady blue ball-shaped flame similar in shape to previous tests. See **STS 75 (February 22, 1996)** for more flame experiments.

In a follow-up to the "Toys-in-Space" experiment on **STS 51D (April 12, 1985)**, the astronauts played horseshoes and basketball, with Susan Helms hitting a ringer and Harbaugh hitting a bank shot through the hoop. Mario Runco ran a toy car on a banked racetrack, finding that if he ran the car too slowly it rose off the track and floated away.

The next shuttle mission was **STS 56 (April 8, 1993)**.

January 24, 1993
Soyuz TM16
Russia

The eight-day mission Soyuz TM16, crewed by Gennady Manakov (**Soyuz TM10–August 1, 1990**) and Alexander Poleshchuk, spent six days docked with **Mir (February 20, 1986)** and the crew of **Soyuz TM15 (July 27, 1992)**, Anatoly Solovyov and Sergei Avdeyev. Unlike most previous Soyuz flights (which used the docking ports on the Mir core module), Soyuz TM16 docked using the androgynous docking port on the **Kristall (May 31, 1990)** module, developed for the Soviet shuttle (see **Buran–November 15, 1988**). This docking port was to be used by the American shuttle in the newly agreed-to Shuttle-Mir joint flights (see **STS 71–June 27, 1995**).

Crew transfer operations from the Soyuz TM15 crew to the Soyuz TM16 crew took six days. Then, on February 1st, Solovyov and Avdeyev undocked from Mir and returned to Earth. Their 188-day flight, while not a human record, was a record for the longest in-orbit space flight for a Soyuz capsule.

Manakov and Poleshchuk then began a six-month crew shift on Mir. Their first task involved testing a 66-foot-diameter unfolding mirror, brought to Mir by a Progress cargo ship during Soyuz TM15, to reflect sunlight onto the earth's night side and illuminate a swathe about 2.5 miles across. From inside the station, Manakov and Avdeyev attached the mirror to the freighter's docking port. Then, on February 4, 1993, ground controllers undocked the freighter from Mir, unfolded the mirror, and aimed it at the earth. Its beam of light, about as bright as two or three full moons, swept across a strip running from the Atlantic Ocean, across Europe, and into Russia. On the ground, the light beam appeared to be a quick flash of light in the sky. On Mir, the cosmonauts watched the faint patch of light move across the earth's surface. A second such mirror test was attempted on **Soyuz TM28 (August 13, 1998)**.

During their occupancy, Manakov and Poleshchuk were visited by three Progress freighters (see **Progress M1–August 23, 1989**). Besides bringing several tons of supplies, the first also brought gyroscopes, which the cosmonauts installed within the **Kvant 2 (November 26, 1989)** module in order to better control Mir's orientation. The second, after undocking from Mir after more than four months of joint flight, was kept in orbit another seven months to test its long-term durability in space. The third brought equipment, including a replacement handle used to control the manually operated Strela crane attached to the outside of Mir and installed during **Soyuz TM11 (December 2, 1990)**. It also carried a recoverable capsule for returning cargo to Earth.

The cosmonauts performed two space walks during their mission. On the first, in April, they installed solar panel motors to Kvant 2 to be used when solar panels from Kristall were transferred there by a later crew (see **Spektr–May 20, 1995**). It was during this EVA that Poleshchuk reported that the handle to the Strela crane was missing. It was thought that the handle had probably become detached and floated away. Without this handle, they could not use the crane, which prevented the installation of a second solar panel motor to Kvant 1. Their second space walk was therefore delayed until the replacement handle arrived. They then used it and the crane to install the motor.

Soyuz TM16's shift ended with the arrival of **Soyuz TM17 (July 1, 1993)**.

February 9, 1993
SCD 1
Brazil / U.S.A.

SCD 1 (Data Collection Satellite 1 in Portuguese) was the first satellite built by Brazil (**Brasilsat A1–February 8, 1985** had been built by Canada). SCD 1 was placed in orbit by Orbital Science's Pegasus rocket (see **Pegsat–April 5, 1990**), that rocket's first successful commercial launch.

SCD 1 was a weather and Earth resource satellite, gathering daily environmental data about Brazil and automatically transmitting this information to 30 ground stations. It operated for about two years and was then replaced by a second SCD satellite in October 1998.

February 20, 1993
ASCA (Astro-D)
Japan

The fourth Japanese x-ray telescope, following **Ginga (February 5, 1987)**, ASCA (Advanced Satellite for Cosmology and

Astrophysics) also continued astronomical research of **STS 35** (**December 2, 1990**) and **Rosat** (**June 1, 1990**). Its x-ray telescope, covering an energy range from 0.4 to 10 KeV, was the first to produce x-ray images using CCD detectors (see **KH-11 1–December 19, 1976**).

ASCA's observations have shown that the abundance of iron in the coronae of many active stars is significantly less than that of the Sun. In galaxy clusters, however, ASCA detected an abundance of heavy elements, evidence that many supernovae had previously occurred in these regions.

The spacecraft also confirmed that three strange objects called Soft Gamma-Ray Repeaters were pulsars with superstrong magnetic fields. Scientists dubbed these objects magnetars, their magnetic fields so strong (about 1,000 trillion times stronger than Earth's) that they heat the neutron star's surface to about 18 million degrees Fahrenheit and periodically cause starquakes in the pulsars' crust. These quakes in turn cause the resulting bursts of soft gamma-ray energy.

ASCA also detected evidence that the origin of interstellar cosmic rays came from the expanding gas clouds that surrounded supernovae. As the shock wave of each supernova crashed into the interstellar medium, the interaction produced the cosmic rays.

The telescope is still in operation as of December 1999.

The next ultraviolet–x-ray telescope was **Alexis** (**April 25, 1993**).

March 30, 1993
Seds
U.S.A.

Seds (Small Expendable-tether Deployer System) was built by the Air Force to test the use of tethers as a low-cost method for deploying satellites into higher or lower orbits. The tests continued tether technology research attempted on **STS 46** (**July 31, 1992**). Once in orbit, the deploying satellite released a 57-pound payload by gradually unwinding a 12-mile-long tether made of polyethylene fiber (not unlike ordinary ski rope). Deployed downward toward the earth, the attached spacecraft flew in formation for about 14 minutes, after which the tether was cut as planned.

The next tether test occurred on **STS 75** (**February 22, 1996**).

April 8, 1993
STS 56, Discovery Flight #16
Spartan 201
U.S.A.

Discovery's sixteenth flight, STS 56, was a nine-day mission, manned by Kenneth Cameron (**STS 37–April 5, 1991**; **STS 74–November 12, 1995**), Stephen Oswald (**STS 42–January 22, 1992**; **STS 67–March 2, 1995**), Kenneth Cockrell (**STS 69–September 7, 1995**; **STS 80–November 19, 1996**), Michael Foale (**STS 45–March 24, 1992**; **STS 63–February 3, 1995**; **STS 84–May 15, 1997**; **STS 103–December 19, 1999**), and Ellen Ochoa (**STS 66–November 3, 1994**). During this scientific research flight, one passive satellite was deployed and then retrieved.

• *Spartan 201* The retrievable free-flying spacecraft Spartan 201, using design concepts first tested during **STS 51G** (**June 17, 1985**), was dedicated by NASA to solar research, using a coronal spectrometer and a visible light coronagraph for studying the solar wind and the Sun's corona. Through October 1998, it has flown five times.

Spartan 201 was deployed on April 11 and retrieved on April 13. The observations during these two days indicated that the slow wind from the Sun might be coming from openings in the Sun's magnetic field associated with long pointed structures called helmet streamers. The data also confirmed data from **Helios 2** (**January 15, 1976**) that the fast wind comes from coronal holes. An observed coronal hole near the North Pole was seen to eject protons at high random speeds, indicating much higher temperatures than expected—in the millions of degrees.

The next flight of Spartan 201 was **STS 64** (**September 9, 1994**). The next solar observatory was **Soho** (**December 2, 1995**).

Onboard experiments included the second flight of the atmospheric research package previously flown on **STS 45** (**March 24, 1992**). On STS 56, this package included seven instruments for studying the atmosphere and the Sun. Atmospheric sensors measured hydrogen, nitrogen, chlorine, and fluorine, as well as pollutants, temperature changes, air pressure fluctuations, and ozone layer variations. Solar instruments measured the Sun's energy output, monitoring the short- and long-term variations in the solar constant.

The ozone data, augmenting data from other satellites like **Meteor 3-05** (**August 15, 1991**), **UARS** (see **STS 41C–April 6, 1984**), and **Sampex** (**July 3, 1992**), indicated that since STS 45 there had been a 10 percent decline in total ozone in the middle northern latitudes, thought by scientists to have been caused not by CFCs (chlorofluorocarbons) but mostly by the aerosols released during the eruption of Mount Pinatubo in June 1991. Other instruments confirmed predictions about the amount of daily ozone variations due to the day-night cycle. See **STS 66** (**November 3, 1993**) for further ozone layer research.

The data also proved that there was an overall upward motion of the atmosphere at the equator and a downward motion at the poles, as had been believed by meteorologists for decades.

The solar instruments confirmed data from **Solar Max** (**February 14, 1980**) that the solar constant fluctuated depending on the Sun's activity, declining by about 0.1 percent during the quiet period of the Sun's 11-year cycle. Furthermore, the variation in ultraviolet radiation was much greater than expected, dropping almost 50 percent during the Sun's quiet period. Because the collision of the Sun's ultraviolet radiation against the atmosphere's oxygen was what produced the ozone layer, these results had a direct bearing on any theories relating to ozone layer depletion.

A set of automatic minilabs carried 30 different experiments, researching crystal growth, bone development, drug processing, and food production. Protein crystals for studying AIDS, breast cancer, and other genetic diseases were grown. Following up discoveries from **Apollo 18** (**July 15, 1975**), protein crystals of urokinase, an enzyme directly related to the spread of breast cancer, were successfully produced, though they were not large enough for later study on Earth. A follow-up experiment was conducted on **STS 67** (**March 2, 1995**).

Discovery also carried several laboratory rats to study how weightlessness affected their bone strength, following up research from STS 45. The experiment confirmed that during space flight there was the expected 40 percent drop in the production of osteoblast, the chemical the body used to produce bone mineral. What surprised the scientists was that within 24 hours after landing, osteoblast production increased by 300 percent. It appeared that the rats' bones, weakened by weightlessness, almost instantly responded to the return of gravity to start producing bone. This research continued on STS 59 (**April 9, 1994**).

The next shuttle flight was STS 55 (**April 26, 1993**).

April 25, 1993
Alexis
U.S.A.

Alexis (Array of Low-Energy X-ray Imaging Sensors) was an Air Force test satellite, carrying two research instruments. Its military sensor studied the effect of lightning and low-frequency electromagnetic impulses from nuclear explosions on the transmission of radio signals through the ionosphere.

Its scientific sensor observed the celestial sky in the energy range from about 60 to 100 eV, or from 130 to 190 angstroms. This part of the spectrum could be called either the highest part of the ultraviolet range, the extreme ultraviolet, or the lowest part of the x-ray range, the soft x-ray. This region was also being observed by **EUVE** (**June 7, 1992**) and **ASCA** (**February 20, 1993**).

Because one of the satellite's two solar panels was damaged during launch, it took six weeks to establish communications with the spacecraft. The damage further caused Alexis to wobble in its orbit, requiring additional complex ground analysis to successfully interpret its data. Despite these problems, Alexis downloaded more than 650 megabytes of information through 1998, which when fully analyzed will produce an all-sky map in the soft x-rays. It has also monitored the sky for bursts of extreme ultraviolet energy from what scientists have now labeled EUV transients.

The next permanent x-ray telescope was the **Rossi X-ray Timing Explorer** (**December 30, 1995**). The next extreme ultraviolet telescope was Orfeus-Spas (see **STS 51**–**September 12, 1993**).

April 26, 1993
STS 55, Columbia Flight #14
U.S.A. / Germany

STS 55 was Columbia's fourteenth flight, lasted 10 days, and was manned by Steven Nagel (**STS 51G**–**June 17, 1985**; **STS 61A**–**October 30, 1985**; **STS 37**–**April 5, 1991**), Terrence Henricks (**STS 44**–**November 24, 1991**; **STS 70**–**July 13, 1995**; **STS 78**–**June 20, 1996**), Jerry Ross (**STS 61B**–**November 26, 1985**; **STS 27**–**December 2, 1988**; **STS 37**–**April 5, 1991**; **STS 74**–**November 12, 1995**; **STS 88**–**December 4, 1998**), Charles Precourt (**STS 71**–**June 27, 1995**; **STS 84**–**May 15, 1997**; **STS 91**–**June 2, 1998**), Bernard Harris (**STS 63**–**February 3, 1995**), and Germans Ulrich Walter and Hans Schlegel. During this flight, Columbia's accumulated time in space exceeded 100 days, with the entire shuttle program exceeding more than a year in space.

STS 55 was the second German Spacelab flight (see **STS 61A**–**October 30, 1985**); its launch was delayed five years due to the Challenger accident (see **STS 51L**–**January 28, 1986**). The flight carried 90 different experiments, covering both materials processing and life sciences.

Materials processing experiments included the growth of large semiconductor crystals, such as a gallium arsenide crystal almost an inch long. Alloy research studied how nickel percolated through liquid alloys of copper-aluminum and copper-gold. The solidification of a number of different alloys was also studied, including copper-manganese and aluminum-lithium. Also studied was the behavior of liquids in microgravity.

Life sciences studies included over two dozen experiments investigating the body's reaction to weightlessness The studies focused on the dramatic shift of fluids moving from the legs into the upper body, causing swollen faces (see **STS 8**–**August 30, 1983**), dehydration, and a significant strain on the cardiovascular system, especially upon return to Earth. To try and alleviate this, one experiment combined injections of a saline solution with the use a lower-body negative pressure device, similar to suits first tested on **Soyuz 11** (**June 6, 1971**) and **Skylab 2** (**May 25, 1973**). These suits temporarily force fluids into the lower body. This experiment found that by subjecting the lower body to four hours of lower pressure, combined with increased salt intake, the effects of microgravity were mitigated if the individual returned to Earth within the next 24 hours.

Another experiment tried to understand the mechanics that caused the lightheadedness felt by astronauts when they first stood up after landing. Scientists suspected that the fluid shift in weightlessness caused a change in the body's system for regulating blood pressure, requiring several days to readjust to a 1g environment.

During the flight, Bernard Harris, who was a doctor, also discovered that not only does the body's fluid get redistributed but the heart shifts position as well.

Further human medical research continued on **STS 58** (**October 18, 1993**).

Other life sciences experiments studied how weightlessness affected the development of tadpoles and perch. Both responded badly to microgravity, dying in large numbers during the flight.

In a technology test, a German-built robot arm, called Rotex, was tested inside of Spacelab. Unlike the large robot arm used to deploy satellites, Rotex was small and was designed to help the astronauts do basic tasks in weightlessness.

The next shuttle flight was STS 57 (**June 21, 1993**).

May 12, 1993
Arsene
France / Europe

Built by students from 30 different French engineering schools and launched on an Ariane rocket, Arsene provided a communications satellite link for amateur radio operators, much like the Amsat Oscar satellites (see **Oscar 8**–**March 5, 1978**). Arsene was operated by the French Amateur Club de l'Espace.

June 21, 1993
STS 57, Endeavour Flight #4
EURECA
U.S.A.

A 10-day flight, STS 57 was Endeavour's fourth flight. It was crewed by Ronald Grabe (**STS 51J–October 3, 1985**; **STS 30–May 4, 1989**; **STS 42–January 22, 1992**), Brian Duffy (**STS 45–March 24, 1992**; **STS 72–January 11, 1996**), David Low (**STS 32–January 9, 1990**; **STS 43–August 2, 1991**), Janice Voss (**STS 63–February 3, 1995**; **STS 83–April 4, 1997**; **STS 94–July 1, 1997**), Peter Wisoff (**STS 68–September 30, 1994**; **STS 81–January 12, 1997**), and Nancy Sherlock (**STS 70–July 13, 1995**; **STS 88–December 4, 1998**). STS 57 retrieved one satellite.

- *EURECA*   The EUropean REtrievable CArrier (EURECA) satellite had been deployed 11 months earlier by **STS 46 (July 31, 1992)**. STS 57 retrieved it, returning its experiments to Earth. The retrieval required an almost six-hour space walk by David Low and Nancy Sherlock in order to manually close a balky antenna on the satellite.

STS 57 was the first flight of Spacehab, a commercially developed pressurized crew module carried in the shuttle's cargo bay. Similar in concept to Spacelab (see **STS 9–November 28, 1983**), Spacehab added about 1,100 cubic feet of habitable space to the shuttle for experiments and working space. Built by McDonnell Douglas and Alenia and leased to NASA, private companies could lease space in the module to conduct research. Over the next two years, an additional six flights of Spacehab took place.

Spacehab carried 22 experiments in medicine, protein crystal growth, biology, and materials processing. Bacteria used in agriculture and pharmaceutical manufacturing were grown. A variety of polymer membranes and protein crystals were grown, using a number of experimental techniques. An attempt to produce perfectly round and stronger ball bearings was carried out. Additional tests on laboratory rats were conducted to study the loss of muscle tissue and bone mineralization in space.

The second flight of Astroculture (see **STS 50–June 25, 1992**) tested new space greenhouse technology, specifically the use of LEDs for providing plants with light. Though developed for plants in space, these LEDs were soon used in cancer treatment on Earth—the LEDs were inserted in light-sensitive tumor-treating drugs and then used to activate them. Plant research continued on **STS 51 (September 12, 1993)**. Astroculture's next flight was **STS 60 (February 3, 1994)**.

Endeavour also conducted two technology experiments for refueling tanks in space, including one flown previously on **STS 53 (December 2, 1992)** that allowed the astronauts to transfer colored water between two tanks. A second test, continuing research from **STS 52 (October 22, 1992)**, tried a method for moving superfluid helium between tanks to see whether the fuel transfer equipment could operate without leaks and efficiently manage the supercold (−454°F) helium. Some helium transfers were done while Endeavour did somersaults in orbit, as well as during thruster firings. Later refueling tests occurred on **STS 77 (May 19, 1996)**.

A new facility tested methods to grow non-organic crystals for use in lasers. Such crystals, impossible to grow on Earth, would increase the number of industrial uses for lasers. This flight's engineering results were used on the next test, **STS 59 (April 9, 1994)**.

STS 57 also carried 12 getaway specials. One student experiment took 1,000 photos of Earth, their targets purposely chosen to match pictures taken during **Skylab (May 14, 1973)**, in order to study surface changes over the ensuing two decades. Others tested new techniques for growing gallium-arsenide crystals.

The next shuttle flight was **STS 51 (September 12, 1993)**.

July 1, 1993
Soyuz TM17
Russia / France

Soyuz TM17 brought Vasily Tsibliyev (**Soyuz TM25–February 10, 1997**), Alexander Serebrov (**Soyuz T7–August 19, 1982**; **Soyuz T8–April 20, 1983**; **Soyuz TM8–September 5, 1989**), and Frenchman Jean-Pierre Haignere (**Soyuz TM29–February 20, 1999**) to **Mir (February 20, 1986)**. It docked at the station on July 3rd to join the crew of **Soyuz TM16 (January 24, 1993)**, which included Gennady Manakov and Alexander Poleshchuk.

For the next 19 days, the five men conducted a dozen experiments together, including the completion of one long-term attempt to grow semiconductor crystals that had been installed on Mir in August 1992 during Michel Tognini's two-week flight on **Soyuz TM15 (July 27, 1992)**. As with Tognini's flight, Haignere's airfare to Mir, about $12.5 million, had been paid for by the French government. Because Haignere's flight was the second in three planned French flights to Mir, much of his work involved maintaining or setting up experiments that were supposed to be retrieved during the third flight. Delays in the Russian program due to cash shortages, however, delayed this third flight, so that it did not take place until **Soyuz TM24 (August 17, 1996)**.

On July 22nd, Manakov, Poleshchuk, and Haignere returned to Earth in Soyuz TM16. Manakov and Poleshchuk had completed 196 days in space.

Tsibliyev and Serebrov then began a six-month tour of duty on Mir. During their stay, they were resupplied by two Progress freighters (see **Progress M1–August 23, 1989**). The second included a recoverable capsule, which returned to Earth on November 21st carrying about 22 pounds of experiment results. Among these were several protein crystals as well as a sample package of different materials that had been attached to the outside of Mir for the last two years.

The two men made five space walks during their flight, totaling more than 20 hours outside. The first two EVAs (extra-vehicular activities) completed the construction of a 15-foot girder to the outside of **Kvant 2 (November 26, 1989)**. The last three EVAs were mostly devoted to making a visual inspection of the station's exterior following the Perseid meteor shower; a number of small but minor impact sites were found.

The cosmonauts took extensive multispectral photographs of the earth's oceans. Other experiments included the growing of more protein and semiconductor crystals.

August 28, 1993

Though originally scheduled to return to Earth in November, the mission was extended through January when the necessary Soyuz booster rockets were not delivered on time. Russian subcontractors, complaining of lack of payment, had refused to meet their construction deadlines. These same financial difficulties also limited the amount of Russian research conducted during the mission. Instead, priority was given to German, American, and French experiments, for which a profit could be earned.

On January 10th, Tsibliyev and Serebrov were joined on Mir by the crew of **Soyuz TM18 (January 8, 1994)**.

August 28, 1993
Galileo, Ida Flyby
U.S.A.

Launched by STS 34 (October 18, 1989) on a mission to Jupiter, Galileo's route through space had already sent it past the asteroid Gaspra on October 29, 1991 (see **Galileo, Gaspra Flyby–October 29, 1991**). With the Gaspra flyby, Galileo had taken the first close-up images of an asteroid, revealing a dark, gray, irregular-shaped object.

On this date, Galileo flew within 1,500 miles of the asteroid Ida, taking five images as it zipped past. The pictures revealed a pock-marked irregular surface, about 32 miles in length and similar in appearance to Gaspra.

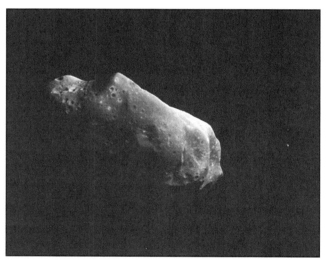

Ida, with its moon Dactyl to the right. *NASA*

The images also revealed a one-mile-diameter moon orbiting Ida. This unexpected discovery, named Dactyl, was also irregularly shaped and appeared to orbit the larger asteroid every 24 hours. Spectrometer readings of the two objects indicated that they were comprised of different materials, with Ida predominantly olivine and Dactyl a mix of olivine, orthopyroxene, and clinopyroxen. These differences indicated that the two asteroids had formed separately, and that the smaller asteroid was not a piece broken off from the larger.

Just as it had in its rendezvous with Gaspra, Galileo detected evidence of a magnetic field around Ida. In both cases, the data were sparse and results could have also been caused by the coincidental occurrence of normal solar wind activity.

Galileo continued on towards Jupiter, entering orbit around the planet on **December 7, 1995**, while also dropping a probe into the gas giant's cloud tops (see **Galileo, Jupiter Orbit–December 7, 1995**).

August 31, 1993
Temisat
Italy / Germany / Russia

Temisat, a small Italian weather satellite built by a German company and launched by Russia, tested new technology for relaying data. Weather information was gathered at about 50 ground stations in the Mediterranean, stored, then transmitted to the satellite when it was overhead, where it became available for all 50 ground stations to download.

September 3, 1993
UHF Follow-On 1
U.S.A.

UHF Follow-On 1 was the first of 10 U.S. Navy satellites launched through November 1999 providing tactical communications for ships, submarines, and aircraft at sea. This satellite constellation replaced the FLTSATCOM (**FLTSATCOM 1–February 9, 1978**) and LEASAT (**STS 41D–August 30, 1984**) military communications satellites.

September 12, 1993
STS 51, Discovery Flight #17
ACTS / Orfeus-Spas
U.S.A. / Germany

The seventeenth flight of Discovery, STS 51, was a 10-day mission. It was crewed by Frank Culbertson (**STS 38–November 15, 1990**), William Readdy (**STS 42–January 22, 1992; STS 79–September 16, 1996**), James Newman (**STS 69–September 7, 1995; STS 88–December 4, 1998**), Daniel Bursch (**STS 68–September 30, 1994; STS 77–May 19, 1996**), and Carl Walz (**STS 65–July 8, 1994; STS 79–September 16, 1996**). The mission released one communications satellite and an astronomical satellite.

• *ACTS* The experimental communications satellite ACTS (Advanced Communication Technology Satellite) was the first built by NASA in almost 20 years, since **ATS 6 (May 30, 1974)**. ACTS tested several new engineering components for geosynchronous satellites, including narrow beam scanning aimed at specific ground stations. It also carried three beacons for performing propagation experiments in the 20 and 30 GHz waveband. It was deployed on September 12 and then used its own booster to move to geosynchronous orbit over the United States.

Like ATS 6, ACTS was utilized by several colleges and medical facilities to test methods for instantaneous video communications links. Two-way video transmissions were used for education as well as to provide medical services to remote facilities. ACTS is still operational as of December 1999.

• *Orfeus-Spas* A passive astronomical satellite built by Germany, Orfeus-Spas carried an ultraviolet telescope to study the celestial sky in the waveband from 40 to 125 nanometers. It also carried two ultraviolet spectrometers. The telescope supplemented observations by **ASCA (February 20, 1993)** and **Alexis (April 25, 1993)**.

Orfeus-Spas also carried a remote-controlled IMAX camera for filming the satellite's deployment and retrieval. Working in conjunction with an onboard shuttle IMAX camera, the footage was used in the third IMAX space film, *Destiny in Space*.

Orfeus-Spas was deployed on September 13 and retrieved on September 20th. During its seven days of observations, it gathered data on several unusual celestial objects, including three white dwarf stars, one of three known binary systems that contained a soft gamma-ray repeating burster and the active galactic nuclei of galaxy PKS 2155+304. The telescope also studied the interstellar medium of the Milky Way, getting measurements of its density, content, and structure in the galactic regions surrounding the Sun.

Orfeus-Spas next flew on **STS 80 (November 19, 1996)**. Before this, the next permanently orbiting x-ray telescope, the **Rossi X-ray Timing Explorer (December 30, 1995)**, was launched. Moreover, **STS 67 (March 2, 1995)** flew three ultraviolet telescopes in its cargo bay.

On September 16th, Carl Walz and Jim Newman performed a seven-hour space walk, testing several tools designed for use in the repair of the Hubble Space Telescope (see **STS 31–April 24, 1990**), scheduled for **STS 61 (December 2, 1993)**. The new tools included a battery-powered socket wrench and a portable foot restraint that was to be attached directly to the telescope. The two men also practiced other EVA (extra-vehicular activity) techniques necessary for construction in space.

Other experiments included a package of sample composite plastic and metallic materials that were placed in the cargo bay to see how the ionospheric flux of atomic oxygen that the shuttle routinely flew through affected them.

Plant research with the Chromex facility (see **STS 29–March 13, 1989**) continued, repeating the arabidopsis seed experiment from **STS 54 (January 13, 1993)** as well as testing root growth in a strain of wheat. The four flights had found that plants grown in space had difficulty producing seed embryos, indicating that weightlessness might make many plants infertile. The difficulty, however, might also have been caused by poor circulation of water and air in weightlessness. Further plant research continued on **STS 60 (February 3, 1994)**. The next Chromex flight was **STS 68 (September 30, 1994)**.

The polymer membrane processing experiment (see **STS 31–April 24, 1990**) made its ninth shuttle flight. As had been done on the previous five flights, the same hardware and materials were flown to introduce as few variables into the experiment as possible, allowing the researchers to refine the technology to produce a wider range of porous membranes. The cumulative knowledge showed that the tiny vibrations and energy fields produced onboard the shuttle made it impossible to produce satisfactory membranes. Most subsequent thin-film research (see **STS 60–February 3, 1994**) placed its equipment on a free-flying satellite, Wake Shield 1, to eliminate these problems. For other polymer membrane research, see **STS 63 (February 3, 1995)**.

Discovery's 10-day mission ended on September 22nd with the first night landing at the Kennedy Space Center in Florida. The next shuttle mission was **STS 58 (October 18, 1993)**.

September 26, 1993
Stella / Healthsat 2 / Itamsat / Uribyol 2 / Posat 1 / Eyesat 1
France / U.K. / Italy / South Korea / Portugal / U.S.A. / Europe

• *Stella* Launched on an Ariane rocket, the French geodetic satellite Stella, a dense 128-pound sphere covered with laser reflectors, worked with its twin **Starlette (February 6, 1975)** to increase the accuracy of measurements of the earth's shape and crust and the movement of tectonic plates and ocean tides. Together, the two satellites made measurements accurate to within one half inch, supplementing geodetic research being performed by **Lageos 1 (May 4, 1976)** and **Lageos 2** (see **STS 52–October 22, 1992**).

The next geodetic satellite was **Geo-IK (November 29, 1994)**.

• *Healthsat 2* A joint U.S. and U.K. project, Healthsat 2 was a microsatellite that helped doctors in remote areas of China and Africa relay medical information to and from hospitals and other health centers. It was part of a series of educational/research satellites built by the University of Surrey, following **UoSat 3 (January 22, 1990)**. In 1995, Healthsat 2 was used to fight that year's outbreak of the Ebola virus.

• *Itamsat* The Italian microsatellite Itamsat was designed to receive and retransmit amateur radio communications, similar in concept to the Amsat Oscar satellites (see **Oscar 6–October 15, 1972** and **Oscar 13–June 15, 1988**).

• *Uribyol 2* Also called Kitsat 2, Uribyol 2 was South Korea's second satellite, following **Uribyol 1 (August 10, 1992)**. It was built entirely in South Korea and carried two cameras and other instruments for weather observation and communication tests. As of December 1999, it is still operational, and has been augmented by **Uribyol 3 (May 26, 1999)**.

• *Posat 1* The POrtuguese SATellite (Posat) 1 was Portugal's first satellite. Similar to Uribyol 2, it carried two cameras and other instruments for making weather observations and communications tests.

• *Eyesat 1* A U.S. microsatellite, Eyesat 1 performed store-and-forward communications tests, relaying environmental data to and from ground-based stations. It was similar in concept to **Temisat (August 31, 1993)**.

October 18, 1993
STS 58, Columbia Flight #15
U.S.A.

Columbia's fifteenth flight, STS 58 was a 14-day mission, the longest shuttle flight to date, exceeding **STS 50 (June 25, 1992)** by a few hours. Its crew included John Blaha (**STS 29–March 13, 1989**; **STS 33–November 22, 1989**; **STS 43–August 2, 1991**; **STS 79–September 16, 1996**), Richard Searfoss (**STS 76–March 22, 1996**; **STS 90–April 17, 1998**), Rhea Seddon (**STS 51D–April 12, 1985**; **STS 40–June 5, 1991**), William McArthur (**STS 74–November 12, 1995**), David Wolf (**STS 86–September 25, 1997**), Shannon Lucid (**STS 51G–June 17, 1985**; **STS 34–October 18, 1989**; STS

43–August 2, 1991; STS 76–March 22, 1996), and Martin Fettman.

The flight carried Spacelab in the shuttle's cargo bay, and focused on life sciences research, continuing work from the first life sciences Spacelab flight, **STS 40 (June 5, 1991)**, as well as **STS 55 (April 26, 1993)**. The crew included two physicians (Seddon and Wolf), a biochemist (Lucid), and a veterinarian (Fettman). Seddon had also flown on STS 40. The research studied how the human cardiovascular system, its sense of balance, and its metabolism changed in space. During launch and the first two days of flight, Shannon Lucid and Martin Fettman wore catheters inserted through veins in their arms in order to study the shifting of fluids from the legs to the body's upper half while in weightlessness. During the flight, the astronauts took turns working out on an exercise bicycle and using a lower-body negative pressure device to pull fluid into their legs.

In a test to measure how the body absorbs calcium while in orbit, each astronaut swallowed calcium tablets and then had blood drawn. The experiment hoped to better understand how and why the body experiences bone loss while in space.

STS 58 also carried 48 laboratory rats, the most ever flown on a single mission. In order to study how tissue samples changed in space without being subjected to re-entry, six of the rats were killed and dissected by Fettman while still in orbit, their spleen, bone marrow, whole blood, and other tissue samples preserved for later study on Earth. This dissection was the first ever performed in space. Unlike rat muscles from **Bion 8 (September 29, 1987)** and **Bion 9 (September 15, 1989)**, which had exhibited tissue damage when the rats were dissected after their return to Earth, the preserved tissue samples removed *during* this shuttle flight showed no tissue damage. This result suggested that it was the stress of re-entry that caused the damage, and that weightlessness only caused atrophy in muscles. See **Bion 11 (December 11, 1996)** for later results.

The next shuttle mission was STS 61 (**December 2, 1993**). See **STS 78 (June 20, 1996)** for the next shuttle life sciences mission.

October 22, 1993
Intelsat 7 F1
International / Europe

Launched on an Ariane rocket, Intelsat 7 F1 was the first satellite in the seventh generation of communications satellites owned by Intelsat (see **Early Bird–April 6, 1965**). The Intelsat 7 cluster supplemented the Intelsat 5/Intelsat 5A and Intelsat 6A clusters (see **Intelsat 5 F2–December 6, 1980; Intelsat 5A F10–March 22, 1985; Intelsat 6A F2–October 27, 1989**).

Once again, the new satellite significantly increased transmission capacity. Intelsat 7 satellites could transmit simultaneously from 18,000 to 112,500 telephone calls (depending on satellite configuration), as well as three color television channels. And while the contract for these satellites required a service life of just less than 11 years, they carry fuel for 19 years of service.

Through 1996, Intelsat launched nine Intelsat 7 satellites, placing them over the Atlantic, Pacific, and Indian Oceans to replace the aging Intelsat 5/Intelsat 5A cluster. Combined with the Intelsat 6A cluster, Intelsat capacity over the Atlantic could exceed more than half a million phone calls at any moment.

Increased demand nonetheless required Intelsat to augment its Intelsat 7 constellation. See **Intelsat 801 (March 1, 1997)**.

November 20, 1993
Solidaridad 1
Mexico / Europe

Launched on an Ariane rocket, Solidaridad 1 was the first satellite in Mexico's second cluster of geosynchronous communications satellites, replacing the Morelos satellites (see **Morelos 1–June 17, 1985**). In October 1994, Mexico added a second Solidaridad satellite, increasing its telephone and television service to Mexico as well as its land, sea, and air mobile communications capacity.

The next Mexican communications satellite system, **Satmex 5 (December 6, 1998)**, was privately owned.

December 2, 1993
STS 61, Endeavour Flight #5
Hubble Space Telescope, Repair
U.S.A. / Switzerland

The fifth flight of Endeavour, STS 61 was an 11-day mission. It was manned by Richard Covey (STS 51I–August 27, 1985; STS 26–September 29, 1988; STS 38–November 15, 1990), Kenneth Bowersox (STS 73–October 20, 1995; STS 82–February 11, 1997), Story Musgrave (STS 6–April 4, 1983; STS 51F–July 29, 1985; STS 33–November 22, 1989; STS 44–November 24, 1991; STS 80–November 19, 1996), Jeffrey Hoffman (STS 51D–April 12, 1985; STS 35–December 2, 1990; STS 46–July 31, 1992; STS 75–February 22, 1996), Kathryn Thornton (STS 33–November 22, 1989; STS 49–May 7, 1992; STS 73–October 20, 1995), Thomas Akers

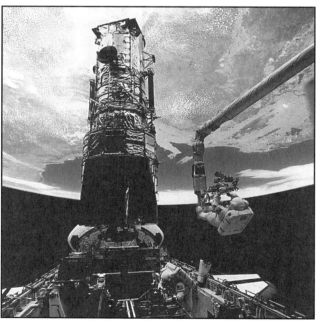

Story Musgrave, on the end of the robot arm, is being hoisted to the top of Hubble to install protective covers to the telescope's magnetometers. Jeffrey Hoffman, meanwhile, works in the cargo bay. *NASA*

December 2, 1993

The double nucleus of M31, the Andromeda Galaxy. This image covers only a tiny area in the center of the nucleus seen in the ground-based telescope picture in the Introduction. The second nucleus, separated by five light years from the first, is believed to be either a remnant of a second galaxy now merged with Andromeda, or an asymmetrical clustering of orbiting stars on one side of a supermassive black hole. *NASA*

(STS 41–October 6, 1990; STS 49–May 7, 1992; STS 79–September 16, 1996), and a Swiss astronaut, Claude Nicollier (STS 46–July 31, 1992; STS 75–February 22, 1996; STS 103–December 19, 1999).

STS 61 was dedicated to the repair and maintenance of the Hubble Space Telescope, placed in orbit by **STS 31 (April 24, 1990)**. The mission's primary goal was to add a specially designed corrective lens to try and bring the telescope's main mirror into focus. Video cameras and an IMAX camera in Endeavour's cargo bay captured the whole story for the film *Destiny in Space*, the third IMAX film shot in space.

After Claude Nicollier captured the 13-ton Hubble with the shuttle's robot arm, two teams of astronauts performed five different space walks, the most ever on a shuttle flight, to install or replace a variety of parts.

In the first EVA (extra-vehicular activity), Story Musgrave and Jeffrey Hoffman spent six hours replacing three failed gyroscopes, two electronic control units, and eight fuses. During the repair, the two men improvised a way to close the gyroscopes' stuck cover.

In the second EVA, Kathryn Thornton and Tom Akers removed and replaced Hubble's two solar panels. The original panels had expanded and contracted whenever they crossed in and out of sunlight, causing the telescope to wobble significantly and making accurate pointing difficult at best. The warping on one panel was so extreme that its shape had become permanently and visibly deformed.

In the third EVA, Musgrave and Hoffman replaced the telescope's main camera with the new Wide Field Planetary Camera (WFPC), lovingly referred to by astronomers as "wifpic." Designed to correct the focus problem in Hubble's mirror, this camera takes the telescope's optical images.

In the fourth EVA, Thornton and Akers removed the telescope's high-speed photometer and replaced it with COSPAR, the Corrective Optics Space Telescope Axial Replacement unit. This instrument corrected the focus problem for Hubble's three other instruments: its faint-object camera and two spectrographs.

In the fifth EVA, Hoffman and Musgrave completed the installation of the replacement solar panels by attaching eight new connector cables to each panel's drive units. The new solar panels were then unrolled, and the telescope released back into space.

All the repairs worked perfectly, making the Hubble Space Telescope one of the most productive astronomical instruments ever. As astronomer James Crocker of the Space Telescope Science Institute noted shortly thereafter, "We nailed it—got the prescription just right."

• *Hubble Space Telescope*   Over the next three years, the Hubble Space Telescope produced a stunning series of spectacular images.

The core of many galaxies were suddenly unveiled as regions of violent star birth and death. In the Andromeda galaxy, the nearest large spiral to the Milky Way, the nucleus was home to a turbulent cluster of star formation 1,000 light years across.

Other images revealed details about active galactic nuclei that confirmed the presence of supermassive black holes (see **Einstein–November 13, 1978**). Giant jets of matter and energy could be seen rocketing outward from these nuclei.

In the Orion nebula, Hubble quickly imaged protoplanetary disks of material surrounding young stars *(see back cover)*. Called proplyds by astronomers, these stars were less than a

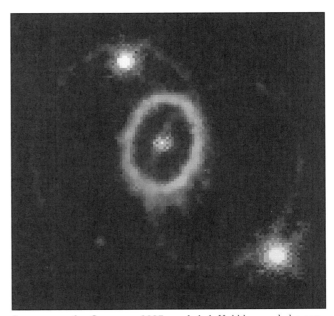

Seven years after Supernova 1987a exploded, Hubble revealed a complex ring structure. Astronomers believe they are showing us the impact zones between two different clouds. As the fast wind from the supernova rockets material outward in all directions, it first collides with the dense cloud at the star's elliptic, shown by the bright inner ring. This collision redirects the fast wind poleward, where it is now impacting an outer cloud to form the wider secondary rings above the star's north and south poles. *NASA*

few hundred thousand years old. Around another star, Beta Pictoris, the thin disk of matter suggested by **IRAS (January 26, 1983)**, was clearly visible and much thinner than expected—indicative of a planet-forming solar system.

Images of supernovae remnants throughout the Milky Way revealed graceful weaves of glowing filaments, shock waves of gas colliding against the interstellar matter around them. The pictures allowed astronomers to develop the first detailed theories describing how these shock waves develop.

The Crab Nebula was shown to be a strange, violent object, with waves and ripples of energy coursing outward from its central pulsar at stunning speeds. Several pictures taken only months apart showed remarkable, unexpected, and completely inexplicable changes to the inner parts of the nebula's inner cloud.

Hubble took the equivalent of a long 10-day exposure of a blank spot of the sky near the Big Dipper, and found that even there thousands of galaxies were hidden. This image gave astronomers a tiny glance at events taking place in the first 10 percent of the universe's life span, a period shortly after the formation of galaxies.

For the first time ever, planetary nebulae could be seen for what they were, vast clouds of material expanding outward from exploding stars *(see back cover)*. Astronomers could begin to make sense of these delicate and beautiful objects. The theory of bipolar jets was developed: Surrounding the equator of most aging stars was a disk of material, ejected during a period of its evolution when a slow wind blew from the star. When the star aged, a fast wind of ejected material began blowing. As this new wind blasted outward, it was blocked at the ecliptic by the slow wind's older disk. Finding the path of least resistance, the fast wind was funneled towards the star's poles, hence forming two north and south jets.

The space telescope's close-up pictures of supernovae SN1987a, the first nearby supernovae in 400 years, also showed this bipolar flow, revealing a ring structure that indicated the formation of a giant hourglass shape in space.

Closer to home, Hubble snapped pictures of Mars's weather, Jupiter's aurora, Saturn's rings, and the atmosphere and rings of Uranus and Neptune. Some of these images were equal to pictures taken previously by spacecraft like **Pioneer 10 (December 3, 1973)**, **Pioneer 11 (December 4, 1974/September 1, 1979)**, **Voyager 1 (March 5, 1979/December 12, 1980)**, and **Voyager 2 (July 9, 1979/August 26, 1981/January 24, 1986/August 25, 1989)**. Other images took the first pictures of objects in the Kuiper belt, a region of comet-like asteroids just beyond the orbit of Pluto. Pluto—known from ground-based observations in 1978 to have a moon, Charon, that was almost as massive as Pluto—itself was photographed, revealing changing light and dark areas on its surface that could be craters, basins, or drifting frost regions.

Hubble's other instruments provided astronomers with detailed spectrographs of thousands of objects, clarifying, proving, and shattering many theories about star formation and the universe's birth and age.

In 1997, STS 82 (**February 11, 1997**) made the second servicing mission to the Hubble Space Telescope. The next shuttle mission was STS 60 (**February 3, 1994**).

December 16, 1993
Telstar 401
U.S.A.

Telstar 401 was owned and operated by AT&T and used for providing video and broadcast service to the United States market. This satellite supplemented and replaced the Telstar 3 satellites (see **Telstar 3A–July 28, 1983**) and operated through January 1997. It was replaced in 1995 with a second Telstar 4 satellite.

Not only did both satellites have increased capacity and bandwidth (using the 4/6 GHz bands and 12/14 GHz frequencies), the Telstar 4 satellites also had the highest power-generating capacity of any satellite to date, able to produce 6,800 watts of power from both its solar cells and nickel-hydrogen batteries. This satellite was also the first to use electric arc jets (a form of ion engine) to maintain its position in space, rather than liquid fuel rockets (see **SERT 2–February 4, 1970**).

A comparison between Telstar 401 with AT&T's first satellite, **Telstar 1 (July 10, 1962)**, shows how much the private

The rotating surface of Pluto, photographed by the Hubble Space Telescope. The tile pattern is an artifact of the image enhancement technique. Compare with the image in the Introduction, made using a ground-based telescope. *NASA*

satellite communications industry had advanced in 31 years. Telstar 1 weighed 170 pounds; Telstar 401 weighed 4,212 pounds. Telstar 1 was a sphere less than 3 feet across. Telstar 401 was a 13- by 7- by 8-foot box with two solar panels stretching 40 feet on either side. Telstar 1 generated 15 watts of power; Telstar 401 generated 6,800 watts. And with 48 transponders covering two bandwidths, the spacecraft's capacity was 48 times greater than Telstar 1's.

Furthermore, by 1993, AT&T's private communications satellites no longer focused on providing telephone service. Instead, video conferencing, radio and television broadcasting, and many other services were offered, from 15-minute one-time conferences to up to 48 channels of regular television broadcasts.

In 1996, AT&T's satellites were purchased by Loral Skynet, which subsequently launched three more Telstar satellites, in May 1997, February 1999, and September 1999, in order to increase its capacity to meet the continually increasing demand.

December 18, 1993
DBS 1 / Thaicom 1A
U.S.A. / Thailand / Europe

• *DBS 1* Built by Hughes Communications, Inc., and launched on an Ariane rocket, the Direct Broadcast Satellite (DBS) 1 was first to provide direct broadcast television to the United States, following the first direct broadcast systems in Japan (see **Yuri 1–April 7, 1978**) and Europe (see **Astra 1A–December 11, 1988**). Placed in geosynchronous orbit above the United States, it was the first of a three-satellite cluster, providing U.S. customers access to the DirecTV television service. Once a customer installed an 18-inch satellite dish, they had access to 150 channels of television programming. This service has become cable television's biggest competition. Within a year, Hughes had signed up 250,000 subscribers, adding an additional 3,000 per day and by December 1999 had more than 7.5 million subscribers.

In 1994 and 1995, Hughes launched two more DBS satellites, providing added capacity and an orbiting spare satellite in case of failure. It would later also purchase the Primestar satellite network (see **Tempo 2–March 8, 1997**). Then, in October 1999, an additional DBS satellite was added to the combined cluster. See **DirecTV 1R (October 10, 1999)**.

• *Thaicom 1A* A communications satellite built by Hughes Space and Communications, Thaicom 1A was Thailand's first. It was placed in geosynchronous orbit slightly eastward of Thailand and transmitted signals on the 4/7 GHz and 11/17 GHz wavebands. Through April 1997, Thailand supplemented their telephone and television capacity with two more communications satellites. All three are still in operation as of December 1999.

1994

January 8, 1994
Soyuz TM18
Russia

On January 10th, Victor Afanasyev (**Soyuz TM11–December 2, 1990**; **Soyuz TM29–February 20, 1999**), Yuri Usachev (**Soyuz TM23–February 21, 1996**), and Valery Polyakov (**Soyuz TM6–August 29, 1988**) arrived at **Mir (February 20, 1986)**, joining the crew of **Soyuz TM17 (July 1, 1993)**: Vasily Tsibliyev and Alexander Serebrov. Polyakov, a doctor, began what would become the longest space flight to date, lasting 439 days, or one year, two months, and two weeks. This record broke the one-year record set by Vladimir Titov and Musa Manarov from **Soyuz TM4 (December 21, 1987)**. To accomplish this, Polyakov remained on Mir through three complete crew shifts.

Unlike previous Mir crew transfers, the switch from the Soyuz TM17 to the Soyuz TM18 crew was done quickly, in only four days. On January 14th, Tsibliyev and Serebrov undocked from Mir. Before firing their retro-rockets to return to Earth, they did a short inspection tour of the station, taking photographs of the docking port on the **Kristall (May 31, 1990)** module that would be used by the American space shuttle in the first Shuttle-Mir mission (see **STS 71–June 27, 1995**).

During this maneuver, however, the Soyuz TM17 spacecraft bumped into the Mir complex. The Soyuz TM18 crew inside the station felt no impact, and the station showed no outward signs of damage. Nonetheless, the incident required the Soyuz TM18 crew to perform one unscheduled space walk to inspect the outside of Mir. Tsibliyev and Serebrov, meanwhile, returned to Earth, landing safely about 60 miles from their target area.

During the six-month residency of Afansyev and Usachev, three Progress freighters (see **Progress M1–August 23, 1989**) resupplied Mir. The January and March ships brought about

Valeri Polyakov. One year, two months, two weeks in space. *RIA Novosti*

four tons worth of supplies. They were unloaded, filled with garbage from the station, and sent on their way, burning up upon re-entry into the atmosphere. The May tanker brought with it a recoverable capsule for returning cargo to Earth.

Polyakov's flight focused on medical research, studying the consequences of very long exposure to weightlessness on the human body, with physician Polyakov himself being the guinea pig. Other than the short EVA (extra-vehicular activity) to check for damage from Soyuz TM17's collision, no other space walks took place.

Polyakov also continued several medical and biological experiments installed during Klaus-Dietrich Flade's 10-day space flight on **Soyuz TM14 (March 17, 1992)**. For doing this research, Russia was paid slightly more than $700,000 by Germany.

On July 2nd, **Soyuz TM19 (July 1, 1994)** arrived at Mir, bringing replacement cosmonauts for Afanasev and Usachev.

January 20, 1994
Gals 1
Russia

Gals ("tack" in Russian) 1 was the first direct broadcast television satellite launched by Russia, following systems in Japan (see **Yuri 1–April 7, 1978**), Europe (see **Astra 1A–December 11, 1988**), and the United States (**DBS 1–December 18, 1993**). Gals 1 was placed in geosynchronous orbit and helped replace the older **Ekran (Ekran 1–October 26, 1976)** and **Gorizont (Gorizont 1–December 19, 1978)** television systems. As of December 1999, a total of two Gals satellites have been launched.

These satellites, along with Russia's Express satellites (see **Express 1–October 13, 1994**), were the first to use a xenon ion engine to maintain their orbital position. This design, originally developed in the United States in the 1960s, was an advance from the ion engines used on **Sert 2 (February 4, 1970)**, **Cosmos 699 (December 24, 1974)**, and AT&T's **Telstar 401 (December 16, 1993)** and was similar to the engine later used on **Pan American 5 (August 28, 1997)** and **Deep Space 1 (October 24, 1998)**.

January 25, 1994, 0:25 GMT
Tubsat 2
Germany / Russia

Launched piggyback with a Russian weather satellite, Tubsat 2 was the second satellite built by the Technical University of Berlin (see **Tubsat 1–July 17, 1991**). It tested the engineering of a three-axis attitude control system. It also carried instruments for making meteorological measurements.

The university's next satellites were **Tubsat N/Tubsat N1 (July 7, 1998)**.

January 25, 1994, 16:34 GMT
Clementine / ISAS
U.S.A.

• *Clementine* Though flown as a technology test of several new lightweight cameras and control systems for the U.S. military, Clementine was also given a scientific mission and ended

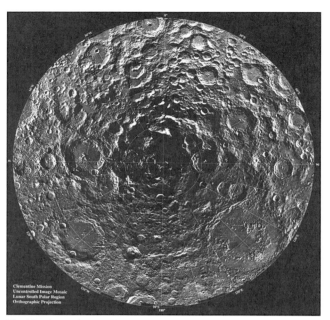

The Moon's south pole. The dark areas, including Aitken Basin, have the lowest relative elevation. *NRL*

up discovering the existence of ice on the Moon. The spacecraft was also the first American spacecraft to orbit the Moon in more than two decades, since **RAE 2 (June 10, 1973)**. It carried four cameras, taking multispectral images from the ultraviolet to the near-infrared, and it was able to resolve objects as small as 45 feet.

During its 73 days in lunar orbit, Clementine took just under two million digital images of the Moon, producing mosaics of several regions not previously photographed by earlier missions (see **Lunar Orbiter 4–May 5, 1967** and **Lunar Orbiter 5–August 1, 1967**).

The data also produced a more detailed map of the lunar sphere, discovering the existence of ancient, very deep impact basins rendered invisible on the surface due to subsequent cratering. The largest, Aitken Basin near the south pole, was 1,500 miles wide and about 7 miles deep, making it the largest and deepest—relative to its planet's size—in the solar system. Because of its polar location, many of the crater floors in this basin never saw sunlight. Other instruments on Clementine indicated that ice existed in these crater floors.

Clementine's initial flight plan had included a flyby of an asteroid. This second goal was canceled, however, when one of Clementine's attitude thrusters malfunctioned, using up its fuel supply.

The next lunar mission was **Lunar Prospector (January 7, 1998)**.

• *ISAS* InterStage Adapter Satellite (ISAS) was actually the booster stage that propelled Clementine out of Earth orbit and toward the Moon. This military satellite was inserted into an elliptical Earth orbit that sent it regularly through the Van Allen belts, where two radiation detectors monitored changing energy fluxes and a third sensor monitored the effect of this environment on several metal oxide semiconductors.

February 3, 1994, 12:10 GMT
STS 60, Discovery Flight #18
Wake Shield 1 / BremSat 1 / ODERACS 1A through ODERACS 1F
U.S.A. / Russia / Germany

STS 60, the eighteenth flight of Discovery, was an eight-day mission. It was manned by Charles Bolden (**STS 61C–January 12, 1986**; **STS 31–April 24, 1990**; **STS 45–March 24, 1992**), Kenneth Reightler (**STS 48–September 12, 1991**), Franklin Chang-Diaz (**STS 61C–January 12, 1986**; **STS 34–October 18, 1989**; **STS 46–July 31, 1992**; **STS 75–February 22, 1996**; **STS 91–June 2, 1998**), Jan Davis (**STS 47–September 12, 1992**; **STS 85–August 7, 1997**), Ronald Sega (**STS 76–March 22, 1996**), and Sergei Krikalev (**Soyuz TM7–November 26, 1988**; **Soyuz TM12–May 18, 1991**; **STS 88–December 4, 1998**). Krikalev was the first Russian to be launched on an American spacecraft, part of an agreement between the United States and Russia that eventually led to a partnership in the construction of the International Space Station.

STS 60 deployed two satellites and tested a third at the end of the robot arm.

• *Wake Shield 1* The free-flying satellite Wake Shield 1 continued NASA research into the production of thin very uniform films in microgravity. Previous experiments in producing polymer membranes had flown on a number of shuttles (see **STS 31–April 24, 1990** and **STS 51–September 12, 1993**). Wake Shield was designed to grow thin crystalline semiconductor films for use in modern advanced electronics and to do so in as pure a vacuum with as little gravitational vibration as possible. Its mission on STS 60 was merely to demonstrate that the technology could work.

Wake Shield is a 12-foot-diameter stainless steel disk, behind which is housed the hardware for growing the films. As it is deployed from the shuttle, the Shield is positioned so that it acts to block the few atoms of the thin ionosphere flying past the orbiting satellite, hence creating in its wake a super vacuum. The production equipment for growing the semiconductor films is attached to the shield so that it flies inside this wake.

Furthermore, because Wake Shield flies independent of the shuttle, it is isolated from the small accelerations and vibrations caused by shuttle thruster firings and the movement of astronauts. During its several days of free flight, scientists hoped that Wake Shield would provide one of the purest environments possible for the production of materials.

On this flight, however, a failed attitude control sensor on Wake Shield prevented its release by the robot arm. Instead, the Shield remained attached to the robot arm, where sensors measured the extent of the vacuum, and five gallium arsenide semiconductor films were grown. The experiment found that close to the shuttle the vacuum was significantly contaminated. Better results required Wake Shield to fly several miles away.

Wake Shield's next flight was **STS 69 (September 7, 1995)**.

• *BremSat 1* The satellite BremSat 1, a getaway special, was built by the University of Bremen in Germany. It was deployed on February 9th and contained experiments for studying the flux of atomic oxygen and dust particles in the earth's ionosphere. It transmitted data to Earth for a little over one year before re-entering the atmosphere, during which it also measured the temperature and air pressure prior to burning up.

• *ODERACS 1A through ODERACS 1F* The six Orbital Debris Radar Calibration Spheres (ODERACS) 1A through 1F were used by ground stations to calibrate their radar sensors in order to better track orbital debris in space. The six spheres were made of three pairs, 2, 4, and 6 inches in diameter, with the surface of one of each pair polished and the other dull. A failed battery had prevented the spheres' release in its first flight on **STS 53 (December 2, 1992)**.

All six spheres were released on February 9th, and radar facilities in Massachusetts quickly located them. Monitoring of their flight continued for the next 200 days. Scientists used these observations to refine their mathematical models for predicting atmospheric drag and the location of small orbiting objects. A second ODERACS test took place on **STS 63 (February 3, 1995)**.

STS 60 was the second flight of Spacehab (see **STS 57–June 21, 1993**), packed with 43 different experiments in life sciences, materials processing, and engineering.

Astroculture continued tests for improving space greenhouse technology (see **STS 50–June 25, 1992** and **STS 57–September 12, 1993**), this time exploring methods for maintaining the temperature and humidity within the greenhouse as well as using the plant byproducts to supply the astronauts' drinking and cooking water. See **STS 64 (September 9, 1994)** for more greenhouse research.

Protein crystal growth research (see **STS 26–September 29, 1988**) continued. After several missions to refine the experiment's equipment (see **STS 52–October 22, 1992**), STS 60 successfully grew for the first time large high-quality protein crystals of human insulin. These crystals were used by the pharmaceutical industry to make detailed maps of the insulin protein's structure, which in turn were used to develop today's improved insulin drugs. Further protein crystal growth research occurred on **STS 65 (July 8, 1994)**. Further insulin research occurred on **STS 72 (January 11, 1996)**.

Other experiments attempted to alleviate the suppression of the immune system of rats caused by weightlessness. If successful, this treatment could be applied to curing human immune system failures like AIDS. Biological experiments also tested methods for producing insecticides and drugs, as well as separating cells and cell fragments.

The next shuttle flight was **STS 62 (March 4, 1994)**.

February 3, 1994, 22:20 GMT
Ryusei / Myojo
Japan

This dual launch of Ryusei and Myojo was the first test flight of Japan's new H2 rocket, able to place large communication satellites in geosynchronous orbit and compete directly with the larger rockets of the United States, Russia, and the European Space Agency.

• *Ryusei* Also called the Orbital Re-entry Experiment Vehicle (OREX), Ryusei was the first test flight of a Japanese-built re-entry spacecraft. After one orbit, the capsule re-entered the atmosphere, releasing its parachutes and splashing down successfully in the mid-Pacific. Because the recovery cost was so high, however, no recovery attempt was made.

- *Myojo*   Also called the Vehicle Evaluation Payload (VEP), Myojo was designed to monitor the performance of the H2 rocket, transmitting telemetry of its liftoff and orbital insertion.

February 7, 1994
Milstar 1
U.S.A.

Milstar 1 was the first in a satellite cluster replacing the American military's **DSCS III (October 30, 1982)** satellites. A constellation of six Milstar satellites is planned, which when completed will have the capability of handling 10,000 callers simultaneously, directly linking many highly mobile small terminals on vehicles, ships, aircraft, and submarines. Through December 1999, a total of two Milstar satellites have been launched successfully.

February 8, 1994
Shijian 4
China

Shijian 4 was China's first satellite to study the earth's magnetosphere and ionosphere. It was placed in a 135- by 22,450-mile orbit, inclined 28 degrees, where the satellite studied the radiation of the Van Allen belts and their effect on satellite equipment. **Shijian 5 (May 10, 1999)** continued this research.

February 12, 1994
Cosmos 2268 through Cosmos 2273
Russia

This routine six-satellite launch of Strela military communications satellites 95 through 100 (see **Strela 1–April 25, 1970**) is only unique because of the official description provided by the Russian government: "The space object serves the Ministry of Defense of the Russian Federation." This was the first time that the Russian government admitted openly that one of its spacecraft was launched in connection with military defense.

March 2, 1994
Intercosmos 26 (Coronas I)
Russia / International

Intercosmos 26, the last satellite in the Intercosmos program (see **Intercosmos 1–October 14, 1969**), conducted solar and ionospheric research, carrying an optical-ultraviolet ray coronograph, an ultraviolet ray radiometer, and ultraviolet, x-ray and gamma-ray spectrometers. Its data supplemented **Freja (October 6, 1992)**.

Because of the break-up of the Soviet Union, this flight was the first Intercosmos mission including nations outside the former communist bloc—France and the United Kingdom contributed experiments. Moreover, Russia and the Ukraine as well as the Czech and Slovak Republics were now considered separate countries. Other participants included Poland and Bulgaria.

This program later continued under the Interball name. See **Interball 1 (August 29, 1996)**. The next satellite to study the magnetosphere was **Wind (November 11, 1994)**.

March 4, 1994
STS 62, Columbia Flight #16
U.S.A.

STS 62 was Columbia's sixteenth space flight. It lasted just under 14 days and was crewed by John Casper (**STS 36–February 28, 1990**; **STS 54–January 13, 1993**; **STS 77–May 19, 1996**), Andrew Allen (**STS 46–July 31, 1992**; **STS 75–February 22, 1996**), Charles Gemar (**STS 38–November 15, 1990**; **STS 48–September 12, 1991**), Marsha Ivins (**STS 32–January 9, 1990**; **STS 46–July 31, 1992**; **STS 81–January 12, 1997**), and Pierre Thuot (**STS 36–February 28, 1990**; **STS 49–May 7, 1992**).

Installed in Columbia's cargo bay were five instruments, running automatically. Several studied fluid behavior and how materials melted and solidified in weightlessness. Others monitored the tiny vibrations and accelerations experienced in the cargo bay as the shuttle periodically fired its thrusters or the astronauts moved about. Accumulating evidence from all shuttle flights indicated that these tiny movements had to be monitored and recorded in order to fully understand the data from the scientific experiments (see **STS 32–January 9, 1990** and **STS 51–September 12, 1993**). These sensors therefore flew on almost all subsequent science shuttle missions.

The cargo bay also carried seven automatically running technology experiments. These tests monitored radiation levels, examined different methods for shielding both humans and sensitive electronic equipment, studied various temperature control systems, and tested how advanced high-voltage solar arrays interacted with the atomic oxygen of the ionosphere. Three identical sets of 264 different samples were exposed to space for different lengths of time. During the mission, the shuttle was also turned to face different directions as it traveled through the ionosphere, in order to test different samples under different conditions. Comparing the identical samples later allowed researchers to measure the detailed effects of the spacecraft's travel.

The cargo bay also carried extra fuel cells, providing the shuttle with additional electrical power while also permitting it to stay in orbit longer.

Housed within Columbia's habitable middeck were six more experiments, including an array of protein crystal growth experiments testing a variety of new growth techniques. The astronauts also assembled a small-scale structure of trusses to test its ability to bear stress in zero gravity and hence its practicality as a large space station structure.

Finally, astronauts Marsha Ivins, Pierre Thuot, and Charles Gemar tested a new guidance and magnetic grappling system at the end of the shuttle's robot arm.

The next shuttle flight was **STS 59 (April 9, 1994)**.

March 13, 1994
STEP 0 (TAOS) / DarpaSat
U.S.A.

This mission was the first launch of the Taurus rocket, built by Orbital Sciences, Inc., and launched out of Vandenberg Air

Force Base in California. While Orbital's Pegasus rocket (see **Pegsat–April 5, 1990**) could orbit about 1,000 pounds, the Taurus rocket could orbit 3,000.

• *STEP 0* The Space Test Experiment Program (STEP) 0 (also called TAOS) was the first in a four-satellite Air Force program, testing new technologies for satellite construction and design. It carried 10 satellite subsystem experiments for improving the ability of satellites to operate independently of ground control. Though designed to last 12 months, STEP 0 operated more than four years, through March 1999. The program's remaining four satellites were launched, but all had technical failures that prevented them from achieving their goals.

• *DarpaSat* A military satellite, DarpaSat tested the use of a global positioning system (GPS) receiver to locate a satellite's position.

April 9, 1994
STS 59, Endeavour Flight #6
U.S.A. / Germany / Italy

The 11-day flight of STS 59 was Endeavour's sixth. It was manned by Sidney Gutierrez (**STS 40–June 5, 1991**), Kevin Chilton (**STS 49–May 7, 1992**; **STS 76–March 22, 1996**), Linda Godwin (**STS 37–April 5, 1991**; **STS 76–March 22, 1996**), Jerome Apt (**STS 37–April 5, 1991**; **STS 47–September 12, 1992**; **STS 79–September 16, 1996**), Michael Clifford (**STS 53–December 2, 1992**; **STS 76–March 22, 1996**), and Thomas Jones (**STS 68–September 30, 1994**; **STS 80–November 19, 1996**).

Endeavour carried what NASA dubbed the Space Radar Laboratory, a package of advanced radar imaging sensors, built jointly by the United States, Germany, and Italy. This flight (also dubbed SIR-C for Shuttle Imaging Radar C) was the third time the shuttle had been used to obtain radar images of portions of the earth (see **STS 41G–October 5, 1984**) and was the first half of a two-flight shuttle radar program, completed on **STS 68 (September 30, 1994)**.

This synthetic aperture radar (see **Seasat–June 27, 1978**) was more advanced than previous designs. It could be aimed, allowing greater flexibility in choosing observation targets and also permitting easy changes to the look angle that produces the radar image's shadows and highlights. It also scanned the earth in three wavelengths (3, 6, and 23 cm), thereby making it possible to produce false color radar images that distinguish between different surface features.

The radar scans studied vegetation changes in Michigan, North Carolina, Brazil, and Central Europe; the hydrology of Oklahoma, Austria, Brazil, and Italy; the ocean currents of the Gulf Stream and the Indian Ocean; and the geology of the Sahara Desert, Death Valley in California, the Andes Mountains in Chile, and Mount Pinatubo in the Philippines. The images also revealed a previously unknown chain of ancient impact craters across the Sahara Desert in Chad, as well as the underlying geological formations that guide the Nile River through the desert.

The images, when combined with previous scans from Seasat (**June 27, 1978**), ERS 1 (**July 17, 1991**), JERS 1 (**February 11, 1992**), STS 2 (**November 12, 1981**), and STS 41G (**October 5, 1984**) plus the later scans from STS 68, allowed scientists to see changing surface details spanning almost two decades. Further supplementing this database were the next two radar satellites: **ERS 2 (April 21, 1995)** and **Radarsat (November 4, 1995)**.

Onboard experiments included bone and muscle cell research begun on **STS 45 (March 24, 1992)** and continued on **STS 56 (April 8, 1993)**. Chicken embryos, cultured muscle cells, and rat fetus cells were exposed to weightlessness to study their development. Newly designed equipment allowed scientists on Earth to observe the experiments in real time, with the ability to adjust the experiment's operation. See **STS 69 (September 7, 1995)** for follow-up research.

Research to produce crystals for lasers (see **STS 57–June 21, 1993**) continued. On this flight, six different ovens grew a variety of different crystals that could increase the frequency range of lasers as well as act as high-speed laser switches. This facility flew on the shuttle one more time in 1995.

The three getaway special payloads studied the freezing and crystallization of water in space, the conductivity of liquids, and the growth behavior of slime mold in weightlessness. This last experiment was especially interesting in that the slime mold's growth rate *decreased* in weightlessness, the opposite of what scientists had expected.

The next shuttle mission was **STS 65 (July 8, 1994)**.

May 3, 1994
Trumpet 1
U.S.A.

Using a large deployable mesh antenna, Trumpet 1 was the first in a new generation of U.S. National Reconnaissance Office ferret-type satellites for monitoring foreign electronic and radio broadcasts, replacing satellites like **Rhyolite 1 (March 6, 1973)**, **Chalet 1 (June 10, 1978)**, and **Magnum 1** (see **STS 51C–January 24, 1985**). The satellite was placed in a highly eccentric orbit similar to that used by the **Molniya 1-01 (April 23, 1965)** satellites, with its apogee above Russia. Through 1999, three Trumpet satellites have been launched.

May 9, 1994
MSTI 2
U.S.A.

Miniature Sensor Technology Integration (MSTI) 2 was the second of three U.S. Air Force research satellites (see **MSTI 1–November 21, 1992**) doing low-cost tests of advanced infrared imaging technology to track and image the launch of ballistic missiles. During its six-month mission, MSTI 2 took three million infrared images, successfully spotting and tracking a test launch of a Minuteman 3 rocket launched from Vandenberg Air Force Base in California. Also tracked were a Sergeant rocket and a number of static engine firings.

MSTI 2 was also significant in that it was placed in orbit by the last Scout booster, first developed in the early 1960s.

MSTI 2 also was of interest because twice it passed close to a manned spacecraft. On January 12, 1996, it flew within 6 miles of **STS 72 (January 11, 1996)**. On September 17, 1997, it came within 1,500 feet of hitting **Mir (February 20, 1986)**. See **Soyuz TM26 (August 5, 1997)**.

The next MSTI satellite was **MSTI 3 (May 17, 1996)**.

June 17, 1994
STRV-1A / STRV-1B
U.K. / Europe

Launched on an Ariane rocket, the Space Technology and Research Vehicle (STRV) satellites A and B were built for the British military. Each satellite was a cube that was 20 inches on a side and weighed 110 pounds. Each tested new solar cell designs while also measuring the static charge build-up on their surfaces.

July 1, 1994
Soyuz TM19
Russia

Soyuz TM19 brought Yuri Malenchenko of Russia and Talgat Musabayev (**Soyuz TM27–January 29, 1998**) of Kazakhstan to **Mir (February 20, 1986)**. After one day of maneuvers, they joined the three-man crew from **Soyuz TM18 (January 8, 1994)**: Victor Afanasyev, Yuri Usachev, and Valery Polyakov. The two crews remained together on the station for seven more days, after which Afanasyev and Usachev returned to Earth in Soyuz TM18. Their crew member Polyakov remained in space, continuing his record-setting 439-day mission.

During the residency of the Soyuz TM19 crew, Mir was resupplied by one Progress freighter (see **Progress M1–August 23, 1989**). This freighter, arriving in September, had problems docking with the station when its automatic "Kurs" docking system failed to bring the spacecraft into its docking port. In two separate docking attempts, the Progress freighter touched the station four times but did not dock.

The freighter carried supplies as well as 600 pounds of scientific equipment for the joint Russian-European experiments to be performed on **Soyuz TM20 (October 3, 1994)** and the Mir-Shuttle missions beginning with **Soyuz TM21 (March 14, 1995)**. If it failed to dock with Mir, the station's food supplies could only last another two weeks. And since the Russians had no spare Progress tankers available, Mir would have to be abandoned and the international flights canceled.

After several days of analysis, a third attempt succeeded, this time using the Kurs system to bring the spacecraft to within 500 feet. Afterward, Malenchenko took over, manually guiding the freighter into its docking port.

Because of these docking problems, Malenchenko and Musabayev performed a space walk on September 9th to inspect the station for any damage that the failures might have caused. They found none.

They conducted a second space walk four days later, inspecting other parts of the exterior of Mir, including a number of electrical cable links, solar batteries, and the 46-foot-long Sofora girder. The girder was built by the **Soyuz TM12 (May 18, 1991)** cosmonauts, and an attitude thruster had been attached to it by the **Soyuz TM15 (July 27, 1992)** cosmonauts.

On October 6th, Malenchenko, Musabayev, and Polyakov were joined by the crew of Soyuz TM20.

July 8, 1994
STS 65, Columbia Flight #17
U.S.A. / International

The seventeenth space flight for Columbia, the 15-day mission of STS 65 was the longest shuttle flight to date, exceeding the previous record set by STS 58 (October 18, 1993). Its crew was Robert Cabana (**STS 41–October 6, 1990; STS 53–December 2, 1992; STS 88–December 4, 1998**), James Halsell (**STS 74–November 12, 1995; STS 83–April 4, 1997; STS 94–July 1, 1997**), Richard Hieb (**STS 39–April 28, 1991; STS 49–May 7, 1992**), Donald Thomas (**STS 70–July 13, 1995; STS 83–April 4, 1997; STS 94–July 1, 1997**), Carl Walz (**STS 51–September 12, 1993; STS 79–September 16, 1996**), Leroy Chiao (**STS 72–January 11, 1996**), and Chiaki Mukai (**STS 95–October 29, 1998**). Mukai was the first Japanese woman to fly in space. She spent 15 days there, which was the longest flight to that date for a woman. Her flight broke the 14-day record set by Bonnie Dunbar and Ellen Baker on **STS 50 (January 25, 1992)**. Mukai's record lasted less than a year, broken by Elena Kondakova's 169-day flight on **Soyuz TM20 (October 4, 1994)**.

STS 65 was the second flight of the International Microgravity Laboratory (dubbed by scientists IML-2), first flown on **STS 42 (January 22, 1992)**. More than 200 scientists from the United States, Europe, Germany, France, Canada, and Japan participated in more than 80 experiments covering the life sciences, materials processing, and technological development.

Several electrophoresis experiments continued research from **STS 47 (September 12, 1992)**, testing the use of electricity to separate mixtures of biological materials. STS 65 studied methods for isolating rat pituitary cells, the DNA of the nematode worm, and a number of other proteins used in the pharmaceutical industry.

Columbia also carried an aquarium of newts, jellyfish, goldfish, guppies, and salamanders, testing their adaptation to space. Four Japanese medaka fish laid five eggs while in space and were apparently unbothered by weightlessness, swimming normally.

Protein crystal growth research (see **STS 26–September 29, 1988**) continued. One experiment took more than 5,000 images of the crystal growing process itself so that scientists could unravel its inner workings. Another attempted to grow 60 crystals from six different types of proteins. These crystals were used in drug research relating to diseases such as AIDS and osteoporosis. Further protein crystal growth research occurred on **STS 68 (September 30, 1994)**.

The laboratory conducted six different experiments on the behavior of bubbles and particles within fluids (see **STS 5–November 11, 1982**). Since the formation of bubbles within alloys as they solidified often affected their purity and uniformity, scientists needed to better understand why they formed and how. Later bubble research occurred on **STS 78 (June 20, 1996)**.

A new centrifuge-microscope developed by the German Space Agency was used in eight experiments. It was designed to simulate differing levels of gravity for a number of life forms, including jellyfish, slime mold, cress plants, some simple plant forms, and single celled animals, to see how they grew in a centrifugally formed artificial gravity field.

Research on the changes in weightlessness to the spinal column continued from **STS 52 (October 22, 1992)**, confirming that the spine lengthened and straightened in zero gravity. For example, Richard Hieb's height increased from 6 feet and 3 inches to 6 feet and 4 inches during the two-week flight.

The next shuttle flight was **STS 64 (September 9, 1994)**.

July 21, 1994
Apstar 1
China / U.S.A.

The Apstar satellites were geosynchronous communications satellites built by Hughes Aircraft for a Hong Kong–based consortium. Apstar 1 was launched by China and positioned over that country to serve the Asian market.

The Chinese successfully launched two more Apstar satellites in 1996 and 1997, placing one over the western edge of China and the other two on its eastern edge. Of the three, the first two used the 4/6 GHz wavebands. All three were launched in cooperation with Hong Kong, anticipating that territory's return from the United Kingdom to China. All three are in operation as of December 1999.

August 3, 1994
APEX
U.S.A.

APEX (Advanced Photovoltaic and Electronic Experiments) was an Air Force engineering satellite testing the operation of new solar power panel designs in the radiation and plasma of the Van Allen belts.

August 10, 1994
Brasilsat B1 / Turksat 1B
Brazil / Canada / Turkey / France / Europe

• *Brasilsat B1* Built by Spar Aerospace of Canada for Embratel (the Brazil Space Agency) and launched on an Ariane rocket, Brasilsat B1 was the first satellite in Brazil's second-generation geosynchronous communications satellites, replacing the Brasilsat A satellites (see **Brasilsat A1–February 8, 1985**). The satellite not only provides telephone and television broadcast signals, it also provides a channel for Brazilian military use. Through December 1999, Embratel launched two additional Brasilsat B satellites; all three are still in use.

• *Turksat 1B* Built by Aerospatiale of France and launched on a Ariane rocket, Turksat 1B was Turkey's first satellite. Placed in geosynchronous orbit above that country, it provides television and radio service to Turkey, Europe, and Asia, using the 11/14 GHz frequencies. In order to meet increasing demand for television service, Turkey launched a second Turksat satellite in 1996. Both are still in operation as of December 1999.

September 9, 1994
STS 64, Discovery Flight #19
Spartan 201
U.S.A.

STS 64 was Discovery's nineteenth flight into space. It lasted 11 days and was crewed by Richard Richards (**STS 28–August 8, 1989**; **STS 41–October 6, 1990**; **STS 50–June 25, 1992**), Blaine Hammond (**STS 39–April 28, 1991**), Susan Helms (**STS 54–January 13, 1993**; **STS 78–June 20, 1996**), Carl Meade (**STS 38–November 15, 1990**; **STS 50–June 25, 1992**), Mark Lee (**STS 30–May 4, 1989**; **STS 47–September 12, 1992**; **STS 82–February 11, 1997**), and Jerry Linenger (**STS 81–January 12, 1997**). The mission deployed and retrieved one satellite.

• *Spartan 201* Deployed from Discovery on September 13, Spartan 201 spent two days using its two telescopes to observe the Sun, its corona, and the solar wind that emanated from it. This flight was Spartan 201's second, following **STS 56 (April 8, 1993)**. See **STS 51G (June 17, 1985)** for Spartan program information.

Spartan 201's observations were coordinated with those being taken at this time by Ulysses (**Ulysses, First Solar Orbit–April 17, 1998**) as it made its first passage over the Sun's southern pole. See that date for results.

Spartan 201's next flight was **STS 69 (September 7, 1995)**. A second Spartan spacecraft, Spartan 204, flew on **STS 63 (February 3, 1995)** and was dedicated to astronomical observations.

STS 64's other research included the Lidar In-Space Technology Experiment (LITE). This experiment gave a laser instrument called a lidar (LIght Detection And Ranging) its first in-space flight test. The laser beam was aimed earthward, with the instrument measuring the amount of reflection off of clouds and other suspended particles in the air. The lidar could reveal atmospheric details with a resolution of about 50 feet. The lidar took images of a variety of targets, which were then compared with ground data. Some images captured Saharan dust storms as they rose and drifted westward over the Atlantic Ocean. Others spotted a volcano in New Guinea as well as penetrated through a Pacific typhoon.

Another onboard experiment tested the use of a robot in the shuttle's cargo bay to handle and monitor six different materials processing experiments. Called Romps (for Robot-Operated Materials Processing System), the robot was designed to move samples to experimental furnaces in accordance with a programmed schedule. The six experiments attempted to produce pure semiconductor crystals while simultaneously testing new production technologies, of which the robot was one.

Astronauts Mark Lee and Carle Meade made the first untethered space walks in more than 10 years, since **STS 51A (November 8, 1984)**. They also made the first use of a maneuvering unit since **Soyuz TM8 (September 5, 1989)**, successfully testing a lighter backpack called Safer (for Simplified Aid for EVA Rescue). Not as powerful as the Manned Maneuvering Unit used on **STS 41B (February 3, 1984)**, **STS 41C (April 6, 1984)**, and STS 51A to retrieve satellites, Safer was designed as an emergency rescue device, carried by astronauts in the event they were thrown free of the shuttle and needed a way to get back safely. The unit was also rechargeable, using nitrogen from the shuttle's atmosphere supplies.

In another test, a package of sensors placed on the end of the shuttle's robot arm measured the environment near the shuttle while it fired its thrusters. This information provided scientists with baseline data on the effects of engine firings when the shuttle was docked to either **Mir (February 20, 1986)** or the International Space Station (see **STS 88–December 4, 1998**).

BRIC (for Biological Research in Canisters) made its first shuttle flight. A small self-contained canister, this facility was used for a variety of plant and insect experiments on seven shuttle flights through December 1999. On STS 64, it tested an artificially created soil to grow grass from seeds. BRIC's next flight was **STS 68 (September 30, 1994)**.

The 10 getaway specials included school experiments from all across the United States, as well as research from Chinese, Japanese, Canadian, Spanish, and American scientists. The experiments studied fluid behavior, grew many different types of alloy crystals, and observed the effect of weightlessness on lichen.

September 30, 1994
STS 68, Endeavour Flight #7
U.S.A.

The 11-day flight of STS 68 was Endeavour's seventh, crewed by Michael Baker (STS 43–August 2, 1991; STS 52–October 22, 1992; STS 81–January 12, 1997), Terrence Wilcutt (STS 79–September 16, 1996; STS 89–January 22, 1998), Thomas Jones (STS 59–April 9, 1994; STS 80–November 19, 1996), Daniel Bursch (STS 51–September 12, 1993; STS 77–May 19, 1996), Peter Wisoff (STS 57–June 21, 1993; STS 81–January 12, 1997), and Steven Smith (STS 82–February 11, 1997; STS 103–December 19, 1999). STS 68 was also Endeavour's second mission in 1994 and the second use that year of the Space Radar Laboratory, also dubbed SIR-C. See **STS 59 (April 9, 1994)**. Scientists used the two sets of radar images, taken six months apart, to compare seasonal changes in the earth's land and ocean surface.

The radar images showed, for example, new volcano eruptions on Russia's Kamchatka peninsula, the detection of an intentionally placed oil spill north of Denmark (designed to test the radar), and the discovery of a dark-green line in the Pacific several miles wide and hundreds of miles long, caused by the presence of microscopic plants held in position by the clash of two sea currents.

The next radar missions were **ERS 2 (April 21, 1995)** and **Radarsat (November 4, 1995)**.

Other onboard experiments included the Chromex greenhouse (see **STS 29–March 13, 1989**), once again studying why some plant species had difficulty producing seed embryos in space, continuing research from **STS 51 (September 12, 1993)**. On this flight, seedlings of the arabidopsis mouse-ear cress plant were successfully pollinated and grown in space, the first time such pollination had been achieved in weightlessness. These results indicated that earlier failures occurred because of poor air and water circulation. Chromex's next flight was **STS 63 (February 3, 1995)**.

This radar image of southern Egypt shows ancient buried drainage patterns near an uninhabited region near the Safsaf Oasis. On the ground, these patterns are invisible, obscured by desert sands. *NASA*

A BRIC canister (see **STS 64–September 9, 1964**) carried gypsy moths, continuing research from **STS 61C (January 12, 1986)** to learn why microgravity caused a shortening in their early development. Because this shortening resulted in sterile moths, the data might be used to control the moth population on Earth.

Protein crystal growth research (see **STS 26–September 29, 1988**) continued from **STS 65 (July 8, 1994)**, testing another six proteins from 60 different samples and refining the technology needed for producing large well-ordered crystals. These protein crystals were used in research producing better drugs, such as insulin, interferon, and human growth hormones. Further protein crystal growth research took place on the next shuttle mission, which was **STS 66 (November 3, 1994)**.

Of the five getaway special payloads, several student packages were dedicated to astronaut Ronald McNair, who had died on **STS 51L (January 28, 1986)**. McNair had been involved in a program to involve students in space science, and these experiments were an outgrowth of his work. One studied the milkweed bug in space, while a second used weightlessness to grow large-sized salt crystals.

One getaway special carried a half-million stamps commemorating the 25th anniversary of the **Apollo 11 (July 16, 1969)** lunar landing.

October 3, 1994
Soyuz TM20
Russia / Europe

Soyuz TM20 brought Alexander Viktorenko (**Soyuz TM3–July 22, 1987**; **Soyuz TM8–September 5, 1989**; **Soyuz TM14–March 17, 1992**), Elena Kondakova (**STS 84–May 15, 1997**), and Ulf Merbold (**STS 9–November 28, 1983**; **STS 42–January 22, 1992**) to **Mir (February 20, 1986)**. Merbold, a German, was being flown as part of the EuroMir project between the European Space Agency (ESA) and Russia, whereby ESA paid the Russians $50 million for the right to orbit two Germans to Mir and conduct long-term experiments on the station. The crew of Soyuz TM20 arrived at Mir on October 6th, joining Valery Polyakov (from **Soyuz TM18–January 8, 1994**), Yuri Malenchenko, and Talgat Musabayev (from **Soyuz TM19–July 1, 1994**). Polyakov was in the 10th month of his record-setting 14.5-month mission.

During the docking, the automatic docking system failed (see Soyuz TM19), and Viktorenko had to make a harder than normal manual docking.

For a month, the four men and one woman conducted research together. Merbold's flight, the longest by a German, included 20 planned experiments studying space sickness, the shift of his body fluids in space, and the effect of weightlessness on the cardiovascular system, the muscles and bones, and the system of balance. The research also included four experiments studying how materials melted and solidified in space, as well as three experimental engineering tests.

A week into the joint mission, however, Mir experienced a two-day power failure, almost forcing its abandonment. The power loss was so serious that radio communications with Earth were even lost for a time. Using the battery power on Soyuz TM19 to regain contact with Earth, the cosmonauts then reactivated the station by firing Soyuz TM19's thrusters to orient Mir so that its solar panels were in sunlight. The power failure caused Merbold's furnace to malfunction, forcing him to cancel his four experiments in materials processing.

On November 4th, Soyuz TM19 returned from space with Malenchenko, Musabayev, and Merbold. Malenchenko and Musabayev had spent 127 days in space; Merbold, 32 days. The day before they returned, however, the three men undocked from Mir and then redocked with it using the automatic "Kurs" docking system that had malfunctioned during their own docking as well as when an unmanned freighter ship had arrived in September (see Soyuz TM19). The docking test went smoothly, and cleared the way for the first American shuttle docking with Mir (see **STS 71–June 27, 1995**).

The next EuroMir manned flight to Mir was the 179-day flight of Thomas Reiter on **Soyuz TM22 (September 3, 1995)**.

Viktorenko, Kondakova, and Polyakov then settled into several more months in space, with Polyakov continuing his marathon flight of 439 days. On January 9th, he broke the record of 366 days set by Vladimir Titov and Musa Manarov on **Soyuz TM4 (December 21, 1987)**.

Moreover, Kondakova's flight set a new endurance record for a woman in space. As the first Russian woman to fly in space in 10 years (see **Soyuz T12–July 17, 1984**), Kondakova's 169-day flight was more than 11 times longer than the 15-day record that had been set by Japanese astronaut Chiaki Mukai on **STS 65 (July 8, 1994)**.

Because of the earlier power failure, however, much of the research program remained limited. Several of the station's batteries could no longer store power from Mir's solar panels, nor could Russia afford to build and launch replacements. Until 1996, Mir crews had to carefully ration their electrical use. Then, the American space shuttle brought the failed batteries back to Earth for refurbishment and returned them to Mir (see **STS 71–June 27, 1995** and **STS 76–March 22, 1996**).

During the stay of the Soyuz TM20 crew, the station was visited by two Progress freighters (see **Progress M1–August 23, 1989**). The first arrived in November, bringing two tons of cargo. The second arrived in February, its cargo including more than 200 pounds of American research equipment for use by Norman Thagard, a member of the crew of **Soyuz TM21 (March 14, 1995)** and the first American to occupy Mir.

On January 10, 1995, Viktorenko, Kondakova, and Polyakov successfully repeated the docking test performed by the Soyuz TM19 crew, moving about 500 feet from the station in their Soyuz TM20 spacecraft, then allowing the Kurs docking system to automatically dock them. No problems were reported, making the docking difficulties during Soyuz TM19 and Soyuz TM20 more puzzling.

In November, Mir developed a leak in its coolant system (also used to regenerate the station's atmosphere). This failure forced the cosmonauts to switch to a back-up system, which used lithium perchlorate candle canisters to generate oxygen and remove carbon dioxide from the air. While the leak was located and later sealed, an additional 20 candle canisters were brought to Mir by STS 74 (**November 12, 1995**).

On March 16, 1995, the crew of Soyuz TM21 joined Viktorenko, Kondakova, and Polyakov on Mir.

October 13, 1994
Express 1
Russia

The Russian communications satellite Express 1 was the first of a next-generation geosynchronous communications satellite cluster replacing the Gorizont satellites (see **Gorizont 1–December 19, 1978**). With 33 percent more capacity, Express 1 was placed over the Atlantic Ocean to provide telephone and television service to Russia, Europe, and the United States. Like the Russian Gals satellites (see **Gals 1–January 20, 1994**), the Express satellites used a xenon ion engine to maintain their position in space.

In 1996, the Russians launched a second Express satellite, placing it over the Indian Ocean. Both satellites still operate as of December 1999.

October 31, 1994
Elektro
Russia

The geosynchronous weather satellite Elektro, placed over the Indian Ocean, was Russia's contribution to the network of satellites providing hourly global images of the world's cloud cover in the middle and equatorial regions. See **GOES 1 (October 16, 1975)**.

November 1, 1994
Wind
U.S.A. / International

Built by NASA, Wind was part of a joint international program to monitor and study the solar wind and its interaction with the earth's magnetosphere. It joined **Geotail (July 24, 1992)**, already in orbit. Later satellites in the program were **Interball 1 (August 2, 1995)**, **Soho (December 2, 1995)**, **Polar (February 24, 1996)**, and **Interball 2 (August 29, 1996)**. Also supplementing this research were **Sampex (July 3, 1992)** and **Freja (October 6, 1992)**. Wind's eight sensors included one French and one Russian instrument. The Russian experiment, a gamma-ray burst detector, was the first ever carried on an American spacecraft.

The spacecraft was first placed in an extremely eccentric orbit (perigee: 30,000 miles; apogee: about 1 million miles). By repeatedly flying past the Moon, the spacecraft's orbital apogee was kept on the earth's day side, thereby allowing the satellite to study the nature of the solar wind just prior to and during its impact with the earth's magnetosphere, the region called the bow shock and magnetopause. This research was complementary to Geotail's work, which studied the magnetosphere's tail on the earth's night side.

For nine months, Wind and Geotail worked together, monitoring the passage of solar wind particles past the earth. Several coronal mass ejections were detected, with the blasts moving past Wind and reaching Geotail about an hour later.

During this part of Wind's operating life, it also detected oxygen, silicon, and aluminum in the Moon's very tenuous, almost nonexistent atmosphere.

Wind was then maneuvered into an L-1 halo orbit, similar to that first used by **ISEE 3 (August 12, 1978)**. From this perch, Wind continued to monitor the incoming solar wind until November 1998, providing a one-hour warning to other spacecraft of impending changes in wind speed and flux. During this period, it also allowed scientists to obtain data on the ion tail of Comet Hale-Bopp as that comet swung past the Sun in the spring of 1997. For further information, see both Soho and Polar.

With the launch of **ACE (August 25, 1997)**, Wind was then moved into an extremely eccentric Earth orbit that took it over 300,000 miles above the elliptical plane of the solar system, where it could sample areas of interplanetary space never previously visited. As of December 1999, the spacecraft continues to return data from this orbit.

The next magnetospheric satellite was **Astrid 1 (January 24, 1995)**.

November 3, 1994
STS 66, Atlantis Flight #13
Crista-Spas 1
U.S.A. / Germany / France

STS 66 was Atlantis's thirteenth space flight. It lasted 11 days and was manned by Don McMonagle (**STS 39–April 28, 1991**; **STS 54–January 13, 1993**), Curtis Brown (**STS 47–September 12, 1992**; **STS 77–May 19, 1996**; **STS 85–August 7, 1997**; **STS 95–October 29, 1998**; **STS 103–December 19, 1999**), Ellen Ochoa (**STS 56–April 9, 1993**), Joseph Tanner (**STS 82–February 11, 1997**), Scott Parazynski (**STS 86–September 25, 1997**; **STS 95–October 29, 1998**), and French astronaut Jean-Francois Clervoy (**STS 84–May 15, 1997**; **STS 103–December 19, 1999**).

STS 66 focused on studying the earth's atmosphere as well as the influence of the Sun upon it. The astronauts deployed one retrievable satellite.

- *Crista-Spas 1* The German Earth observation satellite Crista-Spas 1 used the same structural design as Orfeus-Spas, flown on **STS 51 (September 12, 1993)**. But instead of doing astronomical research, Crista-Spas studied the earth's atmosphere. It was released on November 4th and retrieved eight days later. During the retrieval, Commander McMonagle practiced new rendezvous techniques to be used when the shuttle docked with **Mir (February 20, 1986)** in 1995. See STS 71 (**June 27, 1995**).

Crista-Spas 1 measured the composition of the middle atmosphere, from 24 to 72 miles elevation, obtaining data on the amount of nitric oxide, so that scientists could better model the atmosphere's structure.

Crista-Spas 1 next flew on **STS 85 (August 7, 1997)**.

The shuttle's onboard instruments included four solar and two atmospheric sensors, part of the atmospheric research package flown previously on **STS 45 (March 17, 1992)** and **STS 56 (April 8, 1993)**. Scientists compared the autumn data from this flight to the spring data of the previous two flights. The data also augmented information from **Meteor 3-05 (August 15, 1991)**, **UARS (see STS 41C–April 6, 1984)**, and **Sampex (July 3, 1992)**.

The results from these missions indicated that the ozone hole seen over the southern pole was somehow isolated from other parts of the ozone layer, and that its existence was independent of changes in ozone at other latitudes. This result was unexpected and raised questions about previous theories on the causes of ozone depletion. See **TOMS (July 2, 1996)** for more ozone layer research.

Other onboard experiments included two protein crystal growth experiments (see **STS 26–September 29, 1988**), growing crystals of albumin, malic enzymes, and other drug related proteins. The experiments tested the use of better engineering methods for producing large, well-ordered crystals for later study on Earth. See **STS 67 (March 2, 1995)** for further protein crystal research.

Ten laboratory rats were flown to study the effects of weightlessness on their nerves, bones, muscles, skin, balance, and immune system, a follow-up of research from **STS 58 (October 18, 1993)**. One experiment studied how the tendon's bone attachment was affected by microgravity.

Another experiment studied the effect of weightlessness on 10 pregnant rats, measuring how the mothers maintained the necessary fluid balance for the fetuses at great cost to themselves. After a short period of disorientation, the rats successfully adjusted. Upon return to Earth, all 10 rats gave birth to a total of 122 pups, proving that despite spending half their pregnancy in weightlessness, healthy offspring could be produced. Nonetheless, though born on Earth, the infant rats showed a temporary lack of balance soon after birth, requiring a short period of time to adjust to gravity. This same experiment was repeated on a shuttle flight in 1995, confirming its results.

The next shuttle mission was STS 63 (**February 3, 1995**).

November 29, 1994, 2:54 GMT
Geo-IK
Russia

Much like the previous geodetic satellite, **Stella** (**September 26, 1993**), the Russian geodetic research satellite Geo-IK weighed more than 3,300 pounds so that its orbit would remain extremely stable. Its data augmented the previous two Russian geodetic satellites, **Cosmos 1989** (**January 10, 1989**) and **Cosmos 2024** (**May 31, 1989**).

November 29, 1994, 10:21 GMT
Orion 1
U.S.A.

Orion 1, owned by Orion Satellite Corporation, was placed at geosynchronous orbit above the Atlantic to provide business communications among North America, Europe, and northern Africa.

While this satellite was not the first privately owned international communications satellite (see **Pan American Satellite 1–June 15, 1988**), Orion Satellite Corporation's application in 1983 to the Federal Communications Commission in the United States for permission to launch it in direct competition with Intelsat (see **Early Bird–April 6, 1965**) established the precedent for others to apply and be approved. In 1985, the FCC approved three licenses, including Pan American Satellite 1. Orion's was approved a few months later.

Orion Satellite Corporation was later sold to Loral Skynet, which launched a second Orion satellite in October 1999. Both satellites, still in operation, were then renamed as Telstar satellites to link them with the rest of Loral's satellite cluster. See **Telstar 401** (**December 16, 1993**).

November 29, 1994, 17:02 GMT
DFH (Dong Fang Hong) 3A
China

DFH, a second-generation geosynchronous communications satellite, replaced China's STW satellites (see **STW 1–April 8, 1984**). The name means "The East is Red." It is thought to use both 4/6 and 12/14 GHz frequencies.

In May 1997, China launched one additional DFH satellite. In addition, in August 1996 and May 1998, China launched two other geosynchronous communications satellites built by Lockheed Martin and Hughes Space and Communications. All four satellites are still in operation as of December 1999.

December 16, 1994
Luch 1
Russia

Though officially the first Luch satellite, the geosynchronous communication satellite Luch 1 was actually the fourth in the program. Previous launches, considered experimental, were given either the Cosmos or Altair name (see **Altair 1–October 25, 1985**). Through 1995, a total of five have been launched, replacing the Raduga satellites (**Raduga 1–December 22, 1975**). They mainly provide military communications and ground-to-**Mir** (**February 20, 1986**) communications.

December 26, 1994
Radio-Rosto
Russia

Similar to the Oscar satellites (see **Oscar 6–October 15, 1972**), the Russian amateur radio satellite Radio-Rosto was built by a group of amateur ham radio operators from the town of Kaluga, about 110 miles south of Moscow. Radio-Rosto was a prototype for a proposed constellation of six larger satellites, called Radio-M. As of December 1999, however, this constellation had not been launched.

1995

January 15, 1995
Express RV
Japan / Germany / Russia

The scientific mission Express RV was the first time a Russian re-entry capsule was launched on a foreign (Japanese) rocket. It contained experiments by Germany and Japan. Due to a failure of the second stage rocket, however, the capsule was placed in too low an orbit and re-entered the atmosphere after only a few hours. Having lost contact with the capsule, German scientists assumed it had landed in the Pacific Ocean and was lost. The project was written off as a failure.

One year later, the capsule was found in Africa. Its parachutes had automatically opened, and it had landed safely in a remote part of northeast Ghana. Not knowing to whom the spacecraft belonged, Ghanaian authorities had shipped it to a nearby airport for storage.

January 24, 1995
FAISAT / Astrid 1
U.S.A. / Sweden / Russia

• *FAISAT*   Built by the U.S. company Final Analysis, Inc., FAISAT was an experimental communications satellite testing digital store-and-forward technology, similar to research on **UoSat 2** (**March 3, 1984**), **Macsat 1 and 2** (**May 9, 1990**), **Informator 1** (**January 29, 1991**), and **Microsat 1 through Microsat 7** (**July 17, 1991**). Faisat's launch was unique in that it was the first time an American private satellite was placed in orbit by a Russian rocket.

Through December 1999, one additional FAISAT satellite was launched. Both satellites tested technology that first looked for open channels in the 100 to 460 MHz wavebands, then used those channels to broadcast its signal. These tests were in anticipation of Faisat's proposed 26-satellite private communications constellation providing two-way paging and tracking information to subscribers. One customer, for instance, was a utility company that planned to use the satellite constellation to read water meters.

February 3, 1995

- *Astrid 1* A Swedish research satellite, Astrid 1 studied the neutral particles in the earth's magnetosphere. It operated for three months, gathering data as well as ultraviolet images of the earth's aurora, following up observations by Sweden's previous satellite, **Freja (October 6, 1992)**, and supplementing the international program begun with **Geotail (July 24, 1992)**.

  The next satellite to study the aurora was **Polar (February 24, 1996)**. The next magnetospheric satellites were **Interball 1/Magion 4 (August 2, 1995)**.

February 3, 1995
STS 63, Discovery Flight #20
ODERACS 2A through ODERACS 2F / Spartan 204
U.S.A. / Russia

STS 63 was Discovery's twentieth flight into space. The mission lasted eight days and was crewed by James Wetherbee (STS 32–January 9, 1990; STS 52–October 22, 1992; STS 86–September 25, 1997), Eileen Collins (STS 84–May 15, 1997), Bernard Harris (STS 55–April 26, 1993), Michael Foale (STS 45–March 24, 1992; STS 56–April 8, 1993; STS 84–May 15, 1997; STS 103–December 19, 1999), Janice Voss (STS 57–June 21, 1993; STS 83–April 4, 1997; STS 94–July 1, 1997), and Vladimir Titov (**Soyuz T8–April 20, 1983; Soyuz T10A–September 27, 1983; Soyuz TM4–September 21, 1987; STS 86–September 25, 1997**). Titov was the second Russian to fly on the space shuttle, following Sergei Krikalev on **STS 60 (February 3, 1994)**. Titov's flight, however, was part of a new U.S.-Russian joint agreement that included a series of seven shuttle dockings to **Mir (February 20, 1986)**—later increased to nine—combined with several long-term visits by American astronauts. These Mir-Shuttle missions, in which the United States paid Russia $400 million to use Mir for scientific research and astronaut training, also provided training in joint operations necessary for the construction of the International Space Station (see **Zarya–November 20, 1998**).

To initiate this program, STS 63 made the first shuttle rendezvous with Mir, moving within 40 feet of the Soviet/Russian station. The maneuvers tested the techniques necessary for future dockings as well as demonstrated that the shuttle's engine firings did not damage Mir.

During the rendezvous, the cosmonauts on Mir, Alexander Viktorenko, Elena Kondakova (both from **Soyuz TM20–October 3, 1994**), and Valery Polyakov (from **Soyuz TM18–January 8, 1994**), could be seen in the station's windows. They took pictures of the shuttle and waved stuffed toys at Discovery's crew. The shuttle astronauts returned the favor, taking video and film images of the Mir station.

STS 63 also released a number of satellites.

- *ODERACS 2A through ODERACS 2F* The six Orbital Debris Radar Calibration spheres (ODERACS 2A through 2F) were deployed from Discovery on February 4th, continuing radar calibration tests begun in the first ODERACS test on **STS 60 (February 3, 1994)**.

- *Spartan 204* This spacecraft used the design first tested on STS 51G (**June 17, 1985**). Unlike Spartan 201 (see **STS 64–September 9, 1994**), which was the first operational Spartan spacecraft and was dedicated to solar research, Spartan 204 (also called the Spartan Service Module) was reconfigured from flight to flight. On STS 63, it conducted astronomical observations. First, it spent one day attached to the robot arm studying the airglow that surrounded the shuttle as it traveled through the earth's ionosphere (see **STS 39–April 28, 1991**). Then, it was released on February 7th for two days of astronomy, studying a variety of celestial objects in the far ultraviolet.

  The next Spartan flight was **STS 69 (September 7, 1995)**, with the third flight of Spartan 201. The next flight of Spartan 204 was **STS 77 (May 19, 1996)**.

One space walk conducted by Mike Foale and Bernard Harris practiced techniques for maneuvering large objects (in this case Spartan 204) by hand inside the cargo bay. The space walk also tested the insulation of their spacesuit gloves and boots. When the astronauts were in darkness for an extended period of time—something labeled "a coldsoak"—both men felt their fingers and feet become cold. See **STS 69 (September 7, 1995)** for later tests.

For the first time since **STS 61 (December 2, 1993)**, the shuttle carried an IMAX camera in its cargo bay, shooting 10 minutes of 70-mm film for the documentary *Mission to Mir*, the fourth IMAX film shot in space.

The cargo bay also carried Spacehab, a commercially built module for performing scientific research (see **STS 57–June 21, 1993**). Twenty different experiments in microgravity were conducted.

The Chromex greenhouse (see **STS 29–March 13, 1989**) once again studied why some plant species had problems producing seed embryos in space. Results on **STS 68 (September 30, 1994)** had indicated that the difficulties might be a result of poor circulation of air and water caused by weightlessness, not from weightlessness itself. Using a new air exchange system, Chromex attempted to grow superdwarf wheat seedlings. Other plant experiments included the Astroculture facility, continuing experiments for developing better space greenhouse technologies (see **STS 50–June 25, 1992**). This flight, incorporating what had been learned on previous flights, was the first to include plants. It tested the ability of wheat seedlings to grow in the enclosed automatic chamber. A BRIC canister (see **STS 64–September 9, 1994**) also carried soybean seeds to see if they could germinate in space. Plant research continued on **STS 70 (July 13, 1995)**.

Also tested was a new design for growing polymer membranes in space, which was different than that first flown on **STS 31 (April 24, 1990)** and flown a number of times thereafter. The polymers grown on STS 63 had applications in the contact lens industry.

The Coca-Cola company sponsored an experiment to study how the astronauts' taste changed in space, designing and building a soda dispenser to provide the astronauts with samples of both diet and regular Coke. No one noticed a difference in taste. A second shuttle test, flown in 1996, also studied the behavior of carbonated liquids.

Other experiments continued research into the effect of weightlessness on bones and muscles using rat and chicken tissue samples (see STS 45–March 24, 1992) as well as two different bacteria, including *E. coli*. Protein crystal growth experiments (see STS 26–September 29, 1988) continued from STS 66 (November 3, 1966), with STS 63 carrying more than 370 different protein samples, the largest collection of samples flown to date. One sample grew crystals of interferon used in ground research to develop drugs to treat hepatitis B and C. See the next shuttle flight, STS 67 (March 2, 1995), for results.

By the time STS 63 was launched, the technology for growing crystals had been refined enough so that almost every subsequent American manned space flight, including those to Mir (February 20, 1986), was used to produce protein crystals. Though not all samples produced crystals, the success rate and the medical knowledge gained by studying the grown crystals justified the many flights.

March 2, 1995
STS 67, Endeavour Flight #8
U.S.A.

The eighth flight of Endeavour, STS 67 set a new shuttle endurance record, lasting more than 16.5 days, exceeding the previous record set by STS 65 (July 8, 1994). The crew was Stephen Oswald (STS 42–January 22, 1992; STS 56–April 8, 1993), William Gregory, John Grunsfeld (STS 81–January 12, 1997; STS 103–December 19, 1999), Wendy Lawrence (STS 86–September 25, 1997; STS 91–June 2, 1998), Tamara Jernigan (STS 40–June 5, 1991; STS 52–October 22, 1992; STS 80–November 19, 1996), Ronald Parise (STS 35–December 2, 1990), and Samuel Durrance (STS 35–December 2, 1990). With the launch of Soyuz TM21, 13 humans were in space, a new record. See **Soyuz TM21 (March 14, 1995)** for details.

STS 67 was the second flight of the astronomy payload first flown on STS 35 (December 2, 1990); its three ultraviolet telescopes covered wavelengths from 825 to 3,200 angstroms. Many of STS 35's same celestial targets were observed, supplementing research by Orfeus-Spas on STS 51 (September 12, 1993) as well as ASCA (February 20, 1993) and Alexis (April 25, 1993).

The next ultraviolet telescope was the **Midcourse Space Experiment (April 24, 1996)**.

Onboard experiments included two more protein crystal growth experiments (see STS 26–September 29, 1988) and continued work from STS 63 (February 3, 1995), attempting to grow over 370 crystals from eight different proteins. The grown crystals aided scientists in understanding the microscopic structure of the HIV virus, the *E. coli* bacteria, and enzymes relating to diabetes and the production of energy by the body from glucose. Another crystal growth experiment continued research from STS 56 (April 8, 1993), attempting to grow crystals of urokinase, an enzyme linked to the spread of breast cancer. The experiment produced urokinase crystals 50 microns across, about half the hoped-for size but still the largest ever grown. Scientists were able to partly map the enzyme's structure using these crystals. The experiment was repeated on STS 80 (November 19, 1996). See also the next shuttle flight, STS 71 (June 27, 1995), for protein crystal research on Mir (February 20, 1986).

March 14, 1995
Soyuz TM21
Russia / U.S.A.

Soyuz TM21 brought Vladimir Dezhurov, Gennady Strekalov (Soyuz T3–November 27, 1980; Soyuz T8–April 20, 1983; Soyuz T10A–September 27, 1983; Soyuz T11–April 3, 1984; Soyuz TM10–August 1, 1990) and Norman Thagard (STS 7–June 18, 1983; STS 51B–April 29, 1985; STS 30–May 4, 1989; STS 42–January 22, 1992) to Mir (February 20, 1986). Thagard was the first American to be launched by a Russian rocket and the first to visit Mir. His flight was part of a series of joint Shuttle-Mir missions (see STS 63–February 3, 1995) and the cooperative effort leading to the construction of the International Space Station (see **Zarya–November 20, 1998**)

The launch of Soyuz TM21 also set a new record for humans in space. Its crew of three joined the crew of seven from STS 67 (March 2, 1995), the crew of one from Soyuz TM18 (January 8, 1994), and the crew of two from Soyuz TM20 (October 3, 1994) for a total of 13 humans in space, exceeding the previous record of 12 (see STS 35–December 2, 1990). As of December 1999, this remains the record.

On March 16th, Soyuz TM21 docked with Mir. For the next six days, the station was occupied by six humans. Because of battery failures the station had limited power, and because of limited funds the station's interior space was cramped—Russia had been unable to complete and launch the station's last two modules (see **Spektr–May 20, 1995** and **Priroda–April 23, 1996**).

On March 22nd, Soyuz TM20 returned to Earth, carrying Alexander Viktorenko, Elena Kondakova, and Valery Polyakov. Viktorenko and Kondakova had completed 169 days in space, with Kondakov's flight being a new female endurance record, shattering the previous 15-day record set by Chiaki Mukai on STS 65 (July 8, 1994). Kondakov's record lasted until Shannon Lucid's flight (see STS 76–March 22, 1976).

Polyakov's record flight of 439 days (bettering a year by more than 2.5 months), proved that human beings could survive in space long enough to travel to Mars. Upon return, his health was not much different than that of other cosmonauts after a long flight. He took his first steps only hours after touchdown, and was completely readapted to gravity within two months.

Onboard Mir, Vladimir Dezhurov, Gennady Strekalov, and Norman Thagard continued the station's occupancy. While Dezhurov and Strekalov mostly worked to maintain the station, Thagard devoted himself to setting up and running a number of scientific experiments, including medical research into the long-term effects of weightlessness on the human body.

During their stay on Mir, the three men were resupplied by one Progress freighter (see **Progress M1–August 23, 1989**). In addition to food and fuel supplies, this freighter brought a small German geodetic satellite dubbed GFZ 1. This satellite was covered with laser reflectors much like previous geodetic satellites (see **Starlette–February 6, 1975; Lageos 1–May 4, 1976; Cosmos 1989–January 10, 1989; Cosmos 2024–**

May 31, 1989). The cosmonauts deployed it from Mir on April 19th, 1995. It remained in orbit until June 1999.

The freighter also brought a quail egg experiment to be returned to Earth with Thagard. Previous missions (**Soyuz 32–February 25, 1979** and **Soyuz TM9–February 11, 1990**) had hatched quail eggs while in space with mixed results. This flight, using eggs that had been fertilized on Earth, studied their development to better understand the influence of weightlessness.

Also arriving at Mir during their occupancy was Spektr, the station's fourth permanent module. In anticipation of the docking of the American space shuttle (see **STS 71–June 27, 1995**), Dezhurov and Strekalov did five space walks to reconfigure the station complex, repositioning solar panels and modules to allow room for the shuttle to dock with the station. See Spektr for further details about the Soyuz TM21 mission.

March 18, 1995
Space Flyer Unit
Japan / U.S.A.

The Space Flyer Unit (SFU) was a scientific reusable satellite capable of being launched on either a Japanese rocket or the American space shuttle. On this flight, it was launched from Japan for retrieval by the shuttle (see **STS 72–January 11, 1996**).

During its 10 months in orbit, SFU verified the satellite's design, testing advanced solar panel and propulsion systems. SFU also carried IRTS (Infrared Telescope in Space), observing the infrared sky from 1 to 1,000 microns. During IRTS's three-week service life, it surveyed about 10 percent of the celestial sky.

SFU also carried three small furnaces for conducting materials processing experiments, producing semiconductor crystals and testing new weightless furnace technology. One biology experiment studied the long-term effect of newt cell division in space.

SFU was recovered on January 13, 1996, by STS 72, which was the first time a spacecraft launched by one nation was retrieved by a spacecraft from another.

March 28, 1995
Hot Bird 1
Europe / France

The geosynchronous communications satellite Hot Bird 1, built by the French company Aerospatiale for Eutelsat (see **OTS 2–May 11, 1978**), was positioned over Europe to provide direct broadcast television to European and Mediterranean countries. The first of five Hot Bird satellites launched through October 1998, these satellites supplemented the telephone service provided by Eutelsat's Eutelsat 2 satellite cluster (see **Eutelsat 2 F1–August 30, 1990**).

A later-generation geosynchronous Eutelsat satellite was added to the cluster in April 1999 and positioned over Europe.

April 3, 1995
Orbcomm FM1 / Orbcomm FM2 / Microlab 1
U.S.A.

• *Orbcomm FM1 and Orbcomm FM2* These two satellites were built, launched, and operated by Orbital Sciences Corporation. They were the first test satellites in a 26-satellite low-Earth-orbit communications cluster—eventually increased to 36 satellites—for providing paging services throughout the world.

Onboard computer problems with both satellites prevented satellite communications soon after launch. After several weeks of reprogramming, however, both satellites were recovered, initiating a period of satellite testing. Both satellites were then made available to commercial companies for paging and e-mail service.

Orbital Sciences' full constellation began launching in 1997. See **Orbcomm FM5 through Orbcomm FM12 (December 23, 1997)**.

• *Microlab 1* A research mini-satellite, Microlab 1 carried two payloads designed to measure lightning and large-scale storm changes as well as atmospheric temperature and humidity.

April 5, 1995
Ofeq 3
Israel

Ofeq 3 was Israel's first operational surveillance satellite and its third overall, following **Ofeq 2 (April 3, 1990)**. Operating for about a year, Ofeq 3 was placed in a retrograde orbit allowing it to transmit high-resolution pictures of Syria, Iran, and Iraq.

April 7, 1995
AMSC 1
U.S.A.

American Mobile Satellite Corporation (AMSC) 1 is a geosynchronous satellite built by a consortium of eight private companies, including Hughes Communications, Singapore Telecommunications, and AT&T Wireless, to provide mobile telephone communications in areas of North America where cellular technology is unavailable. The satellite works in partnership with **MSAT 1 (April 20, 1996)**, which provides similar service for Canada. Each satellite acts as a back-up for the other.

April 21, 1995
ERS 2
Europe

Resembling the European Space Agency's first radar satellite, **ERS 1 (July 17, 1991)**, ERS (Earth Remote-sensing Satellite) 2 carried a synthetic aperture radar scanner (see **Seasat–June 27, 1978**) as well as a wide beam radar scanner. The satellite's close-up radar images were 62 miles wide with a resolution of 80 feet. Other instruments measured ocean topography, wave heights, and sea ice changes to an accuracy of 4 inches. Wind speeds were also monitored.

ERS 2 also carried a sensor for measuring global changes in the atmosphere's ozone content, continuing the monitoring begun with the TOMS ozone monitoring equipment on **Meteor 3-05 (August 15, 1991)**. ERS 2's ozone monitoring was later supplemented by **TOMS (July 2, 1996)**.

For almost a year, ERS 1 and ERS 2 were operated in conjunction. Because the orbits of the two satellites allowed them to observe the same surface locations a day apart, three-dimensional stereoscopic images could be produced by combining images. The satellites could also trace surface changes precisely over time. Furthermore, the radar images could be compared with earlier images brought back from **STS 41G (October 5, 1984)**, **STS 59 (April 9, 1994)**, and **STS 68 (September 30, 1994)** in order to compile a better understanding of long-term surface changes. These radar pictures are available for commercial purchase.

One result, accurate to within 4 inches, indicated that the Antarctic ice shield had not changed during five years of data gathering, from 1991 to 1996. This evidence suggested that the world's climate was not warming as predicted by some scientists.

The next radar instrument was **Radarsat (November 4, 1995)**.

May 20, 1995
Spektr
Russia / U.S.A.

Spektr was the fourth permanent module added to **Mir (February 20, 1986)**, following **Kvant 1 (March 31, 1987)**, **Kvant 2 (November 26, 1989)**, and **Kristall (May 31, 1990)**. Originally scheduled for launch in late 1991, with its prime mission to test military technology for monitoring ballistic missile launches, cash shortages in Russia had prevented its completion. When the United States and Russia agreed to a joint Shuttle-Mir program, with the United States paying Russia $400 million for the right to use Mir, Russia used these funds to redesign Spektr as a science module holding much of NASA's research equipment.

Spektr carried a Belgian-made spectrometer for monitoring atmospheric gases such as ozone, carbon dioxide, freon, and sulfur. It also carried two freezers for storing biological samples as well as a centrifuge, all for the use of the American astronauts occupying Mir.

Spektr added about 2,200 cubic feet to Mir, bringing its habitable volume to about 10,600 cubic feet. Its weight of about 20 tons raised the weight of Mir to approximately 90 tons. With the addition of Soyuz and Progress spacecraft, the full complex weighed just over 100 tons, about the same as an average American space shuttle (see **STS 1–April 12, 1981**), without cargo.

Spektr docked with Mir on June 1. Two days later, cosmonauts Vladimir Dezhurov and Gennady Strekalov (from **Soyuz TM21–March 14, 1995**) repositioned the module using the robot arm attached to Mir's multidocking adapter, moving it from the main docking port along Mir's axis to a side port. They then moved the Kristall module from a side port to the main port, thereby placing Kristall along the station's main axis. The main docking port on Kristall's own multidocking adapter had been designed as the port for the Russian space shuttle (see **Buran–November 15, 1988**). The American space shuttle would use it.

The two men then performed a series of space walks to remove the solar panels on the Kristall module (where they might interfere with a shuttle docking) and attach them to Spektr. In total, Mir had 11 solar panel arrays: three on Mir, two on Kvant 1, two on **Kvant 2 (November 26, 1989)**, and four on **Spektr (May 20, 1995)**.

These maneuvers completed several years of preparation, both on the ground and on Mir, to make the station ready for the arrival of the American space shuttle. See **STS 71 (June 27, 1995)**.

Mir's last module was **Priroda (April 23, 1996)**.

June 27, 1995
STS 71, Atlantis Flight #14
U.S.A. / Russia

STS 71 was Atlantis's fourteenth space flight, and the 100th American manned mission into space, beginning with Alan Shepard's 15-minute suborbital flight on **Freedom 7 (May 5, 1961)** 35 years earlier. In contrast, STS 71 lasted just under 10 days, orbited Earth 152 times, and lifted off with a crew of seven. The crew included Robert Gibson (**STS 41B–February 3, 1984**; **STS 61C–January 12, 1986**; **STS 27–December 2, 1988**; **STS 47–September 12, 1992**), Charles Precourt (**STS 55–April 26, 1993**; **STS 84–May 15, 1997**; **STS 91–June 2, 1998**), Ellen Baker (**STS 34–October 18, 1989**; **STS 50–June 5, 1992**), Gregory Harbaugh (**STS 39–April 28, 1991**; **STS 54–January 13, 1993**; **STS 82–February 11, 1997**), and Bonnie Dunbar (**STS 61A–October 30, 1985**; **STS 32–January 9, 1990**; **STS 50–June 25, 1992**; **STS 89–**

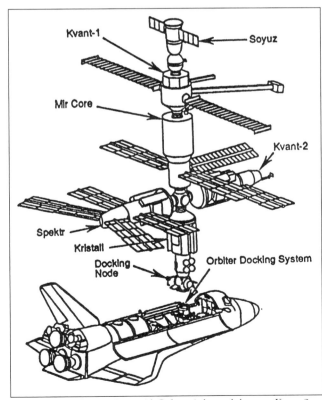

Mir Core, with Kvant 1 (now with Sofora girder and thruster, Kvant 2, Spektr, Kristall, and Space Shuttle). Compare with diagram on May 31, 1990. *NASA*

January 22, 1998), and Russians Nikolai Budarin (**Soyuz TM27–January 29, 1998**) and Anatoly Solovyov (**Soyuz TM5–June 7, 1988; Soyuz TM9–February 11, 1990; Soyuz TM15–July 27, 1992; Soyuz TM26–August 5, 1997**).

Atlantis also made the first American docking with the Soviet-built/Russian-owned space station **Mir (February 20, 1986)**, during which it also made the first American-Russian crew transfer from Mir; it left Budarin and Solovyov behind and returned Vladimir Dezhurov, Gennady Strekalov, and Norman Thagard to Earth. These three men had been launched to Mir on **Soyuz TM21 (March 14, 1995)** and had spent 116 days in space. Thagard's flight broke the American in-space record of 84 days set by the **Skylab 4 (November 16, 1973)** crew more than two decades earlier.

Atlantis docked with the **Kristall (May 31, 1990)** module of Mir on June 29th using the docking port originally intended for the Soviet-built shuttle **Buran (November 15, 1988)**. Mir and Atlantis then remained docked for five days, creating the largest manmade object ever placed in orbit, weighing more than 200 tons.

During this five-day period, the operation of Mir was transferred from Dezhurov, Strekalov, and Thagard of Soyuz TM21 to Budarin and Solovyov of STS 71. About three tons of equipment and supplies were moved from Atlantis into Mir, including a new, larger docking port to be used in the next shuttle docking (see **STS 74–November 12, 1995**). Other cargo transferred ranged from water and air sampling equipment to hard drive replacements for laptop computers. Water that normally collected in the shuttle and was ejected into space before re-entry was instead transferred to Mir to restock its water supply. Oxygen and nitrogen were also bled into Mir's atmosphere. A protein crystal growth experiment (see **STS 26–September 29, 1988** and **STS 63–February 3, 1995**) and plant growing equipment were also transferred. The crystal growth experiment contained 46 different proteins and was packaged to remain on Mir for pick-up by the next shuttle docking mission (see **STS 74**), testing several new engineering techniques for growing crystals.

Cargo removed from Mir and returned to Earth included one of the failed NiCad batteries that prevented the station from storing power (see **Soyuz TM20–October 3, 1994**). This battery was refurbished and returned to Mir by **STS 76 (March 22, 1996)**. Also removed from Mir were more than 100 packages that had been retrieved from the outside of the station during Dezhurov's and Strekalov's space walks.

On July 4th, Atlantis undocked from Mir. Also undocking was the Soyuz TM21 spacecraft, flown by Budarin and Solovyov. For a short while, both shuttle and Soyuz capsule were flying in formation with a temporarily unoccupied Mir. Due to a computer failure on the station at that very moment, however, the station began to drift, and the cosmonauts were forced to hurriedly redock with it.

During these joint operations, an IMAX camera shot footage for the film *Mission to Mir*, the fourth IMAX space film.

On July 7th, Atlantis returned to Earth, landing at Cape Canaveral. The return of the Soyuz TM21 crew on Atlantis posed a slight immigration problem, however. Neither Dezhurov nor Strekalov had the proper visas for entering the United States. To make their arrival conform to the immigration laws, the U.S. State Department went to the Immigration and Naturalization Services and applied for a visa waiver for "aliens from outer space." The visas were obtained without difficulty.

Budarin and Solovyov remained on Mir, the first Russian crew to have been brought to a Russian orbiting station in something other than a Soyuz spacecraft. During the remainder of their flight, they were visited by one Progress freighter (see **Progress M1–August 23, 1989**), bringing them the usual food and fuel supplies. They also performed three space walks, doing routine repairs and maintenance to the station's exterior.

On September 5th, **Soyuz TM22 (September 3, 1995)** docked with Mir, bringing a new replacement crew.

July 7, 1995
Cerise / Helios 1A / UPM-LBSAT 1
France / Spain / Italy / Europe

- *Cerise*   A surveillance satellite, Cerise was launched on an Ariane rocket for France's military and monitored the high-frequency communications of other nations.

On July 24th, Cerise was hit by a piece of orbital debris from the third stage of an old Ariane rocket, first launched in 1986. The impact sent the satellite tumbling. It was the first such impact directly observed by ground control. Despite the impact, the spacecraft remained operable, and after reprogramming its computers Cerise was reoriented properly.

- *Helios 1A*   A French military surveillance satellite, Helios 1A was a follow-up of the civilian Spot satellites (see **Spot 1–February 22, 1986**). It used the more advanced technology eventually launched on **Spot 4 (March 23, 1998)**; its cameras were believed to be capable of resolving objects less than 3 feet across. The satellite was used by the militaries of France, Italy, and Spain, all three of which had contributed to its construction.

A second Helios surveillance satellite was launched in December 1999 to augment the reconnaissance work of these three nations.

- *UPM/LBSAT 1*   The Spanish micro-satellite UPM/LBSAT 1 conducted communications and microgravity research for the Universidad Politecnia de Madrid.

July 13, 1995
STS 70, Discovery Flight #21
TDRS 7
U.S.A.

STS 70 was Discovery's 21st mission. Its launch, however, was delayed by five weeks because woodpeckers had poked almost 200 holes into the orange insulation foam on the shuttle's external tank.

The flight lasted just under nine days and was crewed by Terrence Henricks (**STS 44–November 24, 1991; STS 55–April 26, 1993; STS 78–June 20, 1996**), Kevin Kregel (**STS 78–June 20, 1996; STS 87–November 19, 1997**), Nancy Sherlock Currie (**STS 57–June 21, 1993; STS 88–December 4, 1998**), Donald Thomas (**STS 65–July 8, 1994; STS 83–April 4, 1997; STS 94–July 1, 1997**), and Mary Ellen Weber. STS 70 deployed one satellite.

- **TDRS 7** Deployed on July 13, Tracking and Data Relay Satellite (TDRS) 7 was used by NASA to provide communications with its orbiting spacecraft, completing NASA's six-satellite cluster. See **STS 6 (April 4, 1983)**.

STS 70 also carried the usual assortment of scientific experiments, including several using both live rats and rat tissue samples. Protein crystal growth research (see **STS 26–September 29, 1988** and **STS 63–February 3, 1995**) attempted this time to grow large crystals of alpha interferon protein, used in the treatment of hepatitis B and C.

Plant research continued from STS 63 in two BRIC canisters (see **STS 64–September 9, 1994**). One studied how the hormone system and muscle formation of the tobacco hornworm were affected by weightlessness, while a second used daylily plant cells to observe how changes in water availability affected cell reproduction. Scientists were beginning to suspect that the inability of water to flow downward in weightlessness was the cause of most plant growth problems in space. Plant research continued on **STS 73 (October 20, 1995)**.

STS 70 also carried the second attempt to produce microencapsulated drugs, previously tested on **STS 53 (December 2, 1992)**. With microencapsulated drugs, tiny spheres of medicine are surrounded by a polymer substance, which slowly dissolves to release the drug over time. On Earth, some drugs could not be encapsulated. For this experiment, newly designed equipment was tested. See **STS 95 (October 29, 1995)** for later research.

Another experiment conducted technology tests of equipment for growing cell cultures in space, using colon cancer cells to test the system's ability to circulate fluids so that the cultures could grow.

The next shuttle flight was **STS 69 (September 7, 1995)**.

August 2, 1995
Interball 1 / Magion 4
Russia / Czech Republic

The two satellites Interball 1 and Magion 4 were part of a four-satellite international scientific program led by Russia, continuing the research previously performed by Prognoz and Intercosmos satellites (see **Prognoz 10–April 25, 1985** and **Intercosmos 26–March 2, 1994**) and supplementing data obtained by the international program begun with **Geotail (July 24, 1992)**.

Interball 1, also called the Interball Tail Probe, was placed in an orbit that allowed it to study the tail of the earth's magnetic field, much as had been done by Geotail. Its 20 instruments conducted almost two dozen experiments, monitoring the solar wind, solar flares, the Sun's magnetic field, as well as the effect of each on the earth's magnetotail.

Magion 4 was deployed from Interball 1 on August 3rd. This Czech research satellite carried 10 sensors of its own. The two satellites, working in conjunction, provided scientists with comparison data over a wider region.

The next launch in this international program was **Soho (December 2, 1995)**. The second launch in the Russian Interball program included **Interball 2/Microsat/Magion 5 (August 29, 1996)**.

August 5, 1995
Mugunghwa 1 (Koreasat 1)
South Korea / U.S.A.

The first of two South Korea geosynchronous communications satellites, Mugunghwa 1 was built by Lockheed Martin for South Korea and launched from Cape Canaveral on a Delta-2 rocket. The satellite was initially stranded in low-Earth orbit when the final stage of the launch rocket failed. By using the satellite's attitude thrusters, it was eventually lifted into geosynchronous orbit above the equator south of South Korea. This use of fuel, however, shortened its service life by about half.

Additional South Korean communication satellites were launched in January 1996 and September 1999. The third will replace the first two beginning in January 2000, providing television and telephone communications to South Korea, using 11/14 GHz transponders.

August 29, 1995
N-Star A
Japan

The geosynchronous communications satellite N-Star A was the first launched by the Nippon Telegraph and Telephone Company of Japan, replacing the Sakura satellites (see **Sakura 3A–February 19, 1988**). It was positioned at the equator due south of Japan, providing Japanese telephone service. A second N-Star satellite was launched in January 1996 in order to increase the company's communications capacity. Both are still in operation as of December 1999.

August 31, 1995
Sich 1
Russia

Sich 1 was similar in design and purpose to the Soviet/Russian Okean satellites, providing detailed maps of the sea ice conditions in the Arctic and Antarctic Oceans. See **Cosmos 1500 (September 28, 1983)**.

Like many of the Russian missions in the mid-1990s, Sich 1 also carried a non-Russian, independently owned second satellite in order to pay for some of the launch's cost. In this case, the second satellite was the Chilean microsat FASat-Alfa, designed to monitor the ozone layer over Chile. Upon reaching orbit, however, FASat-Alfa could not be separated from Sich 1 and was terminated so as not to interfere with Sich 1's operation. Using the insurance money, Chile rebuilt the satellite and launched it as **FASat-Bravo (July 10, 1998)**.

As of December 1999, no additional Sich satellites have been launched by Russia.

September 3, 1995
Soyuz TM22
Russia / Europe

Crewed by Yuri Gidzenko, Sergei Avdeyev (**Soyuz TM15–July 27, 1992**; **Soyuz TM28–August 13, 1998**), and German Thomas Reiter, Soyuz TM22 docked with **Mir (February 20, 1986)** on September 5th. Its crew joined Nikolai Budarin and Anatoly Solovyov from **STS 71 (June 27, 1995)**, and began a six-day transfer of station operations. Budarin and Solovyov undocked **Soyuz TM21** (**March 14, 1995**) and returned to Earth on September 11, completing 76 days each in space.

Gidzenko, Avdeyev, and Reiter then continued the occupancy of Mir. Their mission, originally planned to last 4.5 months, ended up lasting six months, with Reiter's 179 days in orbit setting a new record for the longest space flight for any non-Russian. His record lasted until Shannon Lucid's flight on STS 76 (**March 22, 1976**). The six-week extension took place because a Russian shortage in funds delayed the construction of the rocket needed to launch their replacement crew to Mir.

Reiter's flight was part of the EuroMir project begun with **Soyuz TM20** (**October 3, 1994**). During his stay on Mir, he conducted research in several dozen experiments covering medicine, materials processing, astronomy, and technology. One medical experiment attempted to reduce the bone loss during weightlessness by mimicking the stress of walking, achieved by hitting the Reiter's heels 500 times in a 10-minute period each day. Much of his research work was restricted, however, because of Mir's limited electrical power resulting from the failure of its storage batteries (see **Soyuz TM20–October 3, 1994**).

Two space walks were conducted during this residency, placing and retrieving four packages on the outside of the **Spektr** (**May 20, 1995**) module. The packages tested the ability of several materials to withstand the rigors of space, as well as measured the flux of micrometeoroid impacts.

The crew was resupplied by two Progress freighters (see **Progress M1–August 23, 1989**), bringing the usual supplies of water, oxygen, and food. The crew also received a visit from the American space shuttle Atlantis, making its second docking visit to Mir. See **STS 74** (**November 12, 1995**) for further details.

September 7, 1995
STS 69, Endeavour Flight #9
Spartan 201 / Wake Shield 2
U.S.A.

The 10-day STS 69 mission was Endeavour's ninth flight and was crewed by David Walker (**STS 51A–November 8, 1984**; **STS 30–May 4, 1989**; **STS 53–December 2, 1992**), Kenneth Cockrell (**STS 56–April 9, 1993**; **STS 80–November 19, 1996**), James Voss (**STS 44–November 24, 1991**; **STS 53–December 2, 1992**), James Newman (**STS 51–September 12, 1993**; **STS 88–December 4, 1998**), and Michael Gernhardt (**STS 83–April 4, 1997**; **STS 94–July 1, 1997**). This science mission deployed and retrieved two satellites.

• *Spartan 201*   Released on September 8th for two days of independent observations of the Sun and its corona, this was Spartan 201's third mission, following **STS 64** (**September 9, 1994**). Its data were coordinated with the flight of Ulysses (see **Ulysses, First Solar Orbit–April 17, 1998**) as that spacecraft passed over the Sun's northern pole. See **STS 51G** (**June 17, 1985**) for Spartan program information.

The fourth flight of Spartan 201 was **STS 87** (**November 19, 1997**). A different Spartan spacecraft, Spartan 206, flew on **STS 72** (**January 11, 1996**).

Wake Shield in free orbit. *NASA*

• *Wake Shield 2*   Flown once before on **STS 60** (**February 3, 1994**), Wake Shield was designed by NASA to grow thin films of semiconductor alloys in the ultra-vacuum of the combined wake of the shuttle and Wake Shield. The spacecraft was released from the shuttle on September 11 and was retrieved on September 14th. Although a number of technical problems caused the satellite to grow only four of a planned seven films, one wafer of aluminum gallium arsenide was the purest ever produced.

The next flight of Wake Shield was **STS 80** (**November 19, 1996**).

Among the experiments carried in the cargo bay was a package of two ultraviolet telescopes, one covering the 200–1,700 angstrom range to monitor the Sun's ultraviolet energy output, and the second covering the 500–1,250 angstrom range to study unusual astronomical objects, from Jupiter's magnetic field to supernovae remnants.

Onboard experiments included one BRIC canister (see **STS 64–September 9, 1994**), studying how lack of gravity affected cell cultures of slime mold as they reproduced. Other experiments investigated water purification techniques, bone cell growth, and the growth of protein polymers for possible use as memory devices in computers. Protein crystal growth research (see **STS 26–September 29, 1988** and **STS 63–February 3, 1995**) used more than 400 samples of several different proteins to test five different protein crystal growth techniques, growing RNA crystals and crystals for breast cancer research.

Bone loss experiments studied why bone mineralization seemed to stop in space, continuing research begun on **STS 45** (**March 24, 1992**) and confirmed on **STS 56** (**April 8, 1993**). Osteoblast cells, which produce the tissue framework that eventually mineralizes into bones, somehow ceased their activity in space. Using refined equipment developed on **STS 59** (**April 9, 1994**), scientists tried to confirm this conclusion, and find out why it happened. During the flight, they recorded a drop in the production of the gene for making a protein called transforming-growth-factor-b (TGF-b), which is needed for communication between bone cells. After the flight, production quickly returned to normal. These tests were confirmed on two later shuttle flights in 1996 and 1997. See also STS 72 (**January 11, 1996**).

For the first time, the shuttle tested equipment for producing oxygen from water. Other technology experiments exposed materials to the space environment to test their durability and tested a new temperature control system using the capillary action. Heat was transferred by evaporating and condensing liquid ammonia in a sealed, pressurized loop.

Finally, after both Spartan 201 and Wake Shield were retrieved, astronauts Jim Voss and Michael Gernhardt performed an almost seven-hour space walk on September 16th to test a variety of construction techniques for use in building the International Space Station (see **Zarya–November 20, 1998** and **STS 88–December 4, 1998**). The space walk also tested new heated gloves and boots, redesigned based upon the experience of the astronauts from **STS 63 (February 3, 1995)**.

The next shuttle mission was STS 73 (**October 20, 1973**).

September 30, 1995
Pioneer 11, Shutdown
U.S.A.

After 22 years of operation and more than 4 billion miles of travel, Pioneer 11 was shut down on this date.

The spacecraft, launched in 1973, had been the first to fly past Jupiter and Saturn on its long journey out of the solar system (see **Pioneer 11–April 6, 1973**; **Pioneer 11, Jupiter Flyby–December 4, 1974**; **Pioneer 11, Saturn Flyby–September 1, 1979**). Following its Saturn flyby, Pioneer 11's flight direction was upstream of the Sun, heading towards the bow shock of the Sun's magnetic heliosphere. The spacecraft was also slowly rising above the solar system's ecliptic.

In the 1980s, data from the four spacecraft traveling outside the solar system—Pioneer 11, **Pioneer 10 (March 3, 1972)**, **Voyager 1 (March 5, 1979)**, and **Voyager 2 (July 9, 1979)**—further confirmed that the fast solar wind came from coronal holes, first indicated by **Helios 2 (January 15, 1976)** and later confirmed by **STS 56 (April 8, 1993)**. During the solar maximums in 1980–81 and 1990–91, coronal holes were scattered all across the surface of the Sun, and the wind as detected by these distant spacecraft was routinely fast. During the solar minimum in 1985–87, however, these holes migrated northward, and only Pioneer 11 traveling at higher latitudes could still detect the faster wind.

Because of Pioneer 11's higher latitude the spacecraft also confirmed the dipole nature of the Sun's magnetic field. The northern and southern hemispheres of the Sun's field were oppositely charged, with a wavy sheet of electrical current at the equator and rotating in unison with the Sun—almost like the extended skirt of a ballerina when she spins. As the pleats rolled past Pioneer 11, the spacecraft detected the change in the magnetic field's charge, alternating from negative to positive.

When Pioneer 11 was shut down, it had still not reached the heliosphere's bow shock, despite traveling more than 4 billion miles from the Sun. Though no longer operating, Pioneer 11 continues outward into interstellar space, its journey taking it towards the constellation of Aquila. See **Pioneer 10, Shutdown (March 31, 1997)** for further information.

October 20, 1995
STS 73, Columbia Flight #18
U.S.A.

The 16-day flight of STS 73, Columbia's eighteenth in space, was crewed by Kenneth Bowersox (**STS 61–December 2, 1993**; **STS 82–February 11, 1997**), Kent Rominger (**STS 80–November 19, 1996**; **STS 85–August 7, 1997**), Kathryn Thornton (**STS 33–November 22, 1989**; **STS 49–May 7, 1992**; **STS 61–December 2, 1993**), Catherine Coleman, Michael Lopez-Alegria, Fred Leslie, and Albert Sacco.

STS 73 was the second flight (see **STS 50–June 25, 1992**) of what NASA dubbed the U.S. Microgravity Laboratory (USML-2). Columbia's cargo bay was outfitted with a Spacelab module and configured to carry 16 different experiments in protein crystal growth, plant development, fluid behavior, the human body and materials research.

Astroculture continued plant greenhouse research, using the knowledge gained from the previous four flights (see **STS 50**, **STS 57–June 21, 1993**; **STS 60–February 3, 1994**; **STS 63–February 3, 1995**) to test a refined greenhouse, growing potato plants. The facility used LEDs for lighting to save on electricity and had sophisticated water and air circulation systems to make sure the plants received the proper amount of nutrients. Five potatoes were successfully grown. See **STS 77 (May 19, 1996)** for more plant research. Astroculture next flew on **STS 89 (January 22, 1998)**.

Protein crystal growth experiments (see STS 63) grew crystals of RNA, ribosome, plant and animal viruses, and a number of other bacteria and enzymes. The subsequent crystals were then used in a wide range of biological research, studying cancer treatment, the development of the artificial sweetener thaumatin (see **STS 78–June 20, 1996**), cell damage, and how chlorophyll turns light into energy.

The zeolite crystal growth experiment repeated the success of STS 50, producing larger and almost perfectly pure zeolite crystals, used by scientists to map the structure of zeolite and thereby understand how to use it as a filter and catalyst in the chemical industry.

In materials research, alloy crystals of cadmium zinc telluride, gallium arsenide, and mercury cadmium telluride were grown using two processes and four different techniques. The crystals grown were subsequently used in semiconductor and infrared sensor technology research.

The behavior of fluid droplets was also investigated, following up research from STS 50. Droplets of water were manipulated by sound waves, made to rotate, break apart, and even place a droplet of one liquid within the center of a second liquid's droplet.

The next shuttle flight was STS 74 (**November 12, 1995**).

November 4, 1995
Radarsat / Surfsat
Canada / U.S.A.

• *Radarsat* The most sophisticated radar imaging satellite yet launched, Radarsat was a joint project of Canada and the United States. Its synthetic aperture radar (see **Seasat–June**

27, 1978 for a definition) works in the 5.6-centimeter wavelength and can produce radar images in nine different resolution modes, ranging from 30 to 300 feet resolution and covering areas from 30 miles square to 300 miles square. Furthermore, in five of these modes, the radar look angle, which is what produces each image's pseudo shadows, can be adjusted to produce a different angle, from 10 to 59 degrees. Scientists can use this flexibility to squeeze the maximum amount of information from each image, depending on its location and terrain.

This flexibility was especially necessary because Canada's northern latitudes are difficult to map in visible and infrared light. The long arctic night combined with frequent bad weather makes monitoring the shipping lanes difficult at best. Radarsat, placed in a sun-synchronous orbit (see **Nimbus 1–April 28, 1964**), images the entire Earth's surface every 16 days, providing detailed maps of sea ice changes in shipping lanes. Scientists also use its radar photographs to study surface conditions during severe storms, tracking wind and water current changes as they happen.

Finally, Radarsat completed the radar mapping of the earth, mapping Antarctica for a month-long period in the fall of 1997. This radar map was the first such map of Antarctica and showed the large-scale features of glacier and ice flow across the continent.

• *Surfsat* This NASA test communications satellite Surfsat was designed and built by students in the Summer Undergraduate Research Fellowship (SURF) program at the California Institute of Technology. Using two different beacons, it did propagation experiments of the 32 GHz frequency band. This waveband allowed the use of more channels but was also more affected by weather and atmospheric changes. See **ATS 6 (May 30, 1974)**. NASA's Deep Space communications network, used to maintain contact with its planetary spacecraft, was being upgraded to use this frequency, and Surfsat provided NASA some necessary baseline data.

November 12, 1995
STS 74, Atlantis Flight #15
U.S.A. / Russia

The eight-day flight STS 74 was Atlantis's fifteenth mission and was its second docking with **Mir (February 20, 1986)**, following **STS 71 (June 27, 1995)**. The crew was Kenneth Cameron (STS 37–April 5, 1991; STS 56–April 8, 1993), James Halsell (STS 65–July 8, 1994; STS 83–April 4, 1997; STS 94–July 1, 1997), Chris Hadfield, Jerry Ross (STS 61B–November 26, 1985; STS 27–December 2, 1988; STS 37–April 5, 1991; STS 55–April 26, 1993; STS 88–December 4, 1998), and William McArthur (STS 58–October 18, 1993).

STS 74 docked on November 15th. For this docking, a new Russian-built docking module was carried in Atlantis's cargo bay for permanent addition to Mir. Without this module, the shuttle could only dock with the station if the Kristall module was repositioned to the main docking port on Mir's multidocking adapter, along the complex's main axis. See STS 71 for a diagram. This position placed Kristall's docking port clear of the station's solar panels. Because this main port was normally reserved for Soyuz spacecraft, if Kristall was placed there, the Soyuz craft then had to be placed at the rear docking port of Kvant 1, forcing the removal and premature abandonment of the Progress freighter (see **Progress M1–August 23, 1989**) that normally used the rear port.

With the installation of the docking module to Kristall, the module could remain at its normal radial port on Mir's multidocking adapter. The docking module extended the length of Kristall so that when the shuttle docked, it was clear of the station's solar panels. All future Shuttle-Mir dockings used this port.

Attached to the docking module were two solar panel arrays, one built jointly by the United States and Russia and the other entirely Russian-built. These panels, installed during space walks by the **Soyuz TM23 (February 21, 1996)** and Soyuz TM26 (see **Soyuz TM26–August 5, 1997** and **STS 86–September 25, 1997**) crews, extended the station's life and increased its total electrical capacity.

After docking, the five-man American crew joined the crew of **Soyuz TM22 (September 3, 1995)** on Mir—Yuri Gidzenko and Sergei Avdeyev of Russia and Thomas Reiter of Germany—for three days of joint operations. More than 2,000 pounds of supplies, including food, water, and hardware, were transferred into Mir, including 20 lithium hydroxide candle canisters used as a back-up system to scrub carbon dioxide from the station's atmosphere and replace it with oxygen. More than 800 pounds of equipment and scientific results were removed from Mir, including one of the station's failed batteries (see **Soyuz TM20–October 3, 1994**) and several hundred protein crystal samples first brought to Mir on STS 71.

During these operations, an IMAX camera in the shuttle's cargo bay photographed the docking operations with Mir. The footage was used in the film *Mission to Mir*.

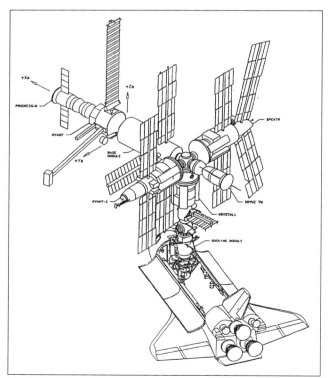

Mir Core with Kvant 1, Kvant 2, Kristall, Spektr, and Docking Module. Also shown is Progress M freighter, Soyuz TM spacecraft, and Space Shuttle. Compare with diagram on May 20, 1995. *NASA*

The next shuttle mission was **STS 72 (January 11, 1996)**. The next shuttle mission to Mir was **STS 76 (March 22, 1996)**.

Meanwhile, Gidzenko, Avdeyev, and Reiter remained on Mir, using the supplies brought to them by Atlantis to continue their residency. Because of Mir's lack of storage battery capacity due to failed batteries, they were forced to ration their use of power, and hence the research work continued to be limited.

Their occupancy lasted until the arrival of **Soyuz TM23 (February 21, 1996)**.

November 17, 1995
ISO
Europe

Built by the European Space Agency, the Infrared Space Observatory (ISO) continued the astronomical infrared observations first done by **IRAS (January 26, 1983)**. The spacecraft's telescope produced infrared images covering the wavelengths from 2 to 120 microns.

Like IRAS, ISO's infrared camera could observe the dust clouds between stars, regions of intense star formation. ISO, however, was much more sensitive. In one of its earliest images, ISO showed that in the Antennae, a weirdly shaped object of two colliding galaxies, star formation was intense in both the region of impact and a circle surrounding one galaxy's nucleus but was nonexistent in the second galaxy. This image proved that star formation varied from galaxy to galaxy, and that some galaxies were what astronomers dubbed starburst galaxies—places undergoing periods of heavy and intense starbirth.

ISO also detected water in many interstellar environments at greater amounts than predicted, including the Orion nebula, a hotbed of star formation. The data showed that water plays a significant part in the formation of stars, cooling the hot gaseous material of protostars so that they condense and then coalesce into solid bodies.

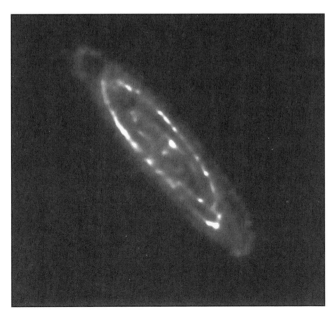

The Andromeda Galaxy in infrared. Compare with image in Introduction, as well as the infrared image taken by IRAS (January 26, 1983). *ESA*

In ISO's three years of operation, it also detected interstellar carbon soot, frozen carbon dioxide, and methane. It confirmed the existence of dust disks around Vega, Beta Pictoris, and several other stars seen by IRAS and the Hubble Space Telescope (see **STS 61–December 2, 1993**). It detected olivine crystals in very young stars. Since olivine is typically found in meteorites, its existence around young stars indicated that the most primitive objects in our solar system had much in common with newly born star systems.

In fact, ISO discovered hundreds of young stars previously unknown and hidden within nearby interstellar dust clouds. Many of these objects were merely dense shells of dust, prestellar cores of material that have not yet ignited into flame.

ISO also made observations of Comet Hale-Bopp, finding that though dominated by water, carbon dioxide was also an important constituent of the comet. The spacecraft also monitored the rising temperature of the comet's dust cloud as it approached the Sun, increasing from –184°F to –50°F as it dropped sunward from 435 to 260 million miles. ISO also found that the materials in this and other comets were the same in the dust clouds around both young and dying stars, containing water, carbon dioxide, and the minerals olivine and forsterite.

ISO also took images of Jupiter and Saturn, giving a clearer picture of the make-up of each planet's atmosphere. It also detected water vapor on Saturn's moon, Titan. In a sense, ISO functioned much like an Earth observation satellite like **Nimbus 7 (October 13, 1978)**, with its infrared images.

The next infrared telescope was **MSX (April 24, 1996)**.

December 2, 1995
Soho
Europe / U.S.A.

The Solar and Heliospheric Observatory (Soho) was the first spacecraft able to make observations of the Sun 24 hours a day year-round. It did this by being placed in a solar orbit at Lagrangian libration point L1, much as had been done with **ISEE 3 (August 12, 1978)**. In this position, 1 million miles closer to the Sun than the earth, the gravitational fields of the Sun and Earth balanced, and objects placed there tended to stay there. From this perch, Soho had a continuous unobstructed view of the Sun.

Built by the European Space Agency with support from NASA, Soho continued solar observations of **Yohkoh (August 30, 1991)**. It was also part of an international coordinated effort by the United States, Japan, the European Space Agency, and Russia to study the dynamics of the solar wind with the earth's magnetosphere. Earlier satellites in the program were **Geotail (July 24, 1992)**, **Wind (November 1, 1994)**, and **Interball 1 (August 2, 1995)**.

Soho carried 12 different sensors. Some imaged the Sun from the infrared to the ultraviolet and with both close-up and wide-angle camera lenses. Others monitored the particles and elements ejected from the Sun.

Through June 1998, the satellite made a host of discoveries. It detected rivers of plasma, or ionized gases, flowing about 12,000 miles below the surface of the Sun. The data also revealed the circulation pattern for the Sun's entire outer layer, to a depth of 15,000 miles, flowing from the equator to the poles at about 50 miles per hour.

December 2, 1995

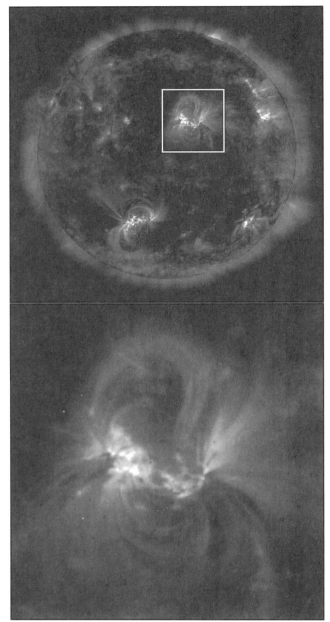

The full image clearly shows several active regions on the Sun's surface. The close-up shows how the hot plasma moves along loops of magnetic field lines. *NASA*

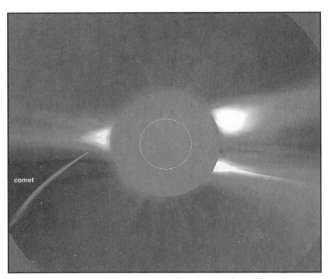

A comet is about to burn up as it falls into the Sun. Note the three "helmet" streamers flowing out from the Sun. *NASA*

the Sun's surface form, collide, and break up, usually in periods less than two days, they trigger huge bursts of energy outward that heat the corona.

Soho discovered more than 50 comets as they either grazed the Sun's surface or fell into it. Comet Hyakutake was photographed as it made its close approach in early May 1996, its tail day by day being redirected away from the Sun as the comet swung past.

Coronal mass ejections (CMEs), first discovered on **Skylab 2 (May 25, 1973)**, were observed with amazing detail. Soho

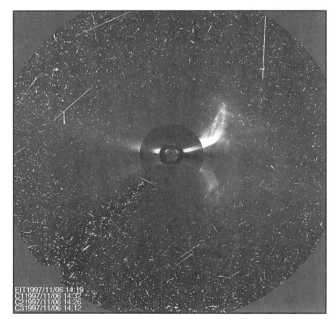

This composite image of four different Soho instruments allows one to see both the Sun's disk as well as nearby regions out to many solar radii. The white streak extending from the Sun's right side is a coronal mass ejection that took place on 6 November 1997. While the eruption would take several days to reach Earth's orbit, within an hour, high energy protons (more than 100 MeV) had arrived at Soho, causing the many tiny white streaks on the image. *NASA*

Its images indicated that the magnetic field lines close to the Sun directly affect the energy distribution in the Sun's corona. The images showed turbulent activity, even during the quiet part of the solar cycle. Plumes of charged gas could be seen streaming out of coronal holes, following the magnetic field lines, and gyrating wildly like gigantic tornadoes, producing the fast solar wind. Soho's observations also confirmed results from Spartan 201 (see **STS 56–April 8, 1993**) that the slow wind came from bright equatorial features called helmet streamers.

This magnetic turbulence was found to be the source of the high temperature of the Sun's corona, 100 times hotter than the Sun's surface. As the numerous magnetic spots on was able to shoot movies of these events, following the burst of matter as it rocketed outward off the Sun's surface. One burst was even spotted as it headed directly toward the earth.

By this time, solar scientists were beginning to realize that coronal mass ejections, *not* solar flares, were the cause of most of the electrical, radio, and ionospheric havoc on Earth. During solar maximum, these events can occur as often as several times per day. Scientists are still unsure what causes these ejections, nor do they yet completely understand their relationship to such solar features as magnetic clouds, solar flares, sunspots, filaments, and helmet streamers. About two-thirds of all coronal mass ejections seem associated with helmet streamers (see **OSO 7–September 29, 1971**), with the remaining third linked to solar flares and other surface prominences. When accompanied with flares, however, CMEs usually start first, making some scientists suspect that the CME sparks the flare. Soho also detected magnetic storms as they hit the earth. One such storm energized the earth's magnetic field 100-fold and possibly caused the failure of one of AT&T's **Telstar 401 (December 16, 1993)** geosynchronous satellites.

In June 1998, a ground controller error caused the spacecraft to lose power and cease operation. After five months of meticulous and dedicated work, however, Soho was reactivated, its instruments functioning with no obvious evidence of damage.

In December 1998, however, the spacecraft's last working gyroscope failed. By March 1999, new software made it possible for ground controllers to orient the satellite without gyros, using a minimum of fuel. As long as no other significant problems crop up, it is believed that Soho's operating life can continue for years.

The next satellite in this international program was **Polar (February 24, 1996)**. The next satellite to study the aurora was **Fast (August 21, 1996)**.

December 7, 1995
Galileo, Jupiter Orbit
U.S.A.

Launched in 1989 (see **STS 34–October 18, 1989**), Galileo and its probe finally arrived at Jupiter on this date, with the main spacecraft entering orbit around the giant planet while the probe dropped into its atmosphere.

Released from Galileo on July 13 to follow a separate course to Jupiter, the probe relayed data for just under an hour before burning up. Its data revealed an extremely dry atmosphere, much dryer than predicted by any researcher. It also detected significantly less helium, neon, and other heavy elements than those seen in the Sun, a wholly unexpected result. Most solar system formation theories postulated that the Sun and Jupiter would be nearly identical in make-up, having come from the same primordial cloud. Scientists theorized that these surprising results might be because the probe dropped into a clear, particularly dry area between clouds, hence giving them a false impression of the planet's overall make-up.

Wind speeds of more than 300 miles an hour were detected, much higher than expected. Lightning, however, was detected at about a 10th of the rate seen on Earth.

The main spacecraft, meanwhile, went into a long looping orbit around Jupiter. Over the next three years, it flew numer-

Close-up of Europa. The dark areas look like freshly frozen cracks where the ice had separated. *NASA*

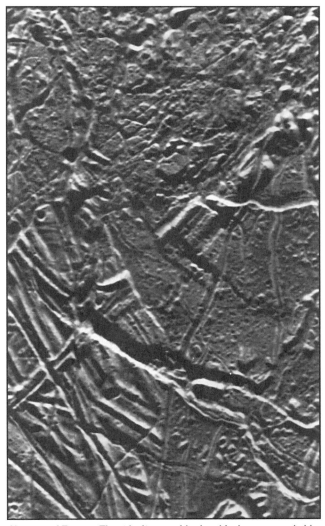
Close-up of Europa. The ridgelines and broken blocks are remarkably similar to the pressure ridges seen on the Arctic Ocean icecap. *NASA*

December 7, 1995

Callisto's dusty and eroding surface. *NASA*

ous times past the giant's four largest moons—Io, Europa, Ganymede, and Callisto—sending back hundreds of images.

Because Galileo's main radio antenna had failed to deploy, however, the download of data was seriously limited, with only a fraction of the hoped-for images relayed to Earth. Nonetheless, these images revealed a plethora of tantalizing facts, especially about the four moons.

Io, as seen during the **Voyager 1 (March 5, 1979)** and **Voyager 2 (July 9, 1979)** flybys, was home to numerous active volcanoes. Since the earlier missions, some volcanoes had become inactive, while others now spewed forth plumes on a regular basis. The plume of one volcano, Prometheus, had shifted about 46 miles west in the 17 years since it was first spotted by the Voyager spacecraft.

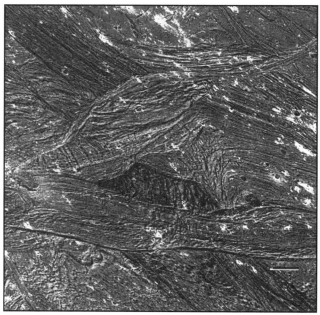
Ganymede's inexplicable grooves. *NASA*

Europa's surface appeared covered with an Arctic Ocean–like layer of ice. Some parts of that ice cover appeared both thin and young, suggesting that liquid water might still lurk below. The reddish lineaments seen by Voyager 2, some of which criss-crossed the moon's entire surface, appeared to be some form of upwelling. Close-up images of several of these lineaments revealed them to be triple ridgelines, with a higher centerline. One 500-mile-long lineament, located near Europa's south pole, was not unlike the San Andreas fault in California, where two continental plates were sliding horizontally past each other. On Europa, the plates were made of ice and appeared to have shifted about 50 miles along the entire length of the fault.

Callisto, following initial Voyager observations, was believed by scientists to be similar to Earth's Moon, with numerous craters and impact features. More detailed photographs, however, indicated a truly strange surface—all the features seemed eroded, as if some unknown process was causing them to corrode and dissolve. The existence of a faint and variable magnetic field, similar to Europa's, suggested that the satellite might also harbor a subsurface ocean.

Closer inspection of Ganymede revealed a strange topography, including patches of grooved terrain (not unlike the surface of a vinyl record) overlaying other patches of grooved terrain, the different patches oriented in random and totally unrelated directions. Moreover, the surface is overlain by bright and dark patches (the bright patches thought to be caused by water frost) that often had no apparent correspondence to topographical features. Planetary geologists could only scratch their heads in wonderment. The moon also has a magnetic field, and was found to harbor organic molecules.

Images and other readings about Jupiter documented the continuing evolution of its Red Spot and the nearby white spots *(see color section, Figure 12)*. In February 1998, photographs showed the merger of two of the planet's white spots, each almost as large as Earth itself.

Infrared data revealed the uppermost structure of the Red Spot, showing that the cloud structure near the spot's center was higher by about 6 miles than at its edges. Gas appeared to be flowing upward at the storm's center, then spraying outward at the top like water from a garden sprinkler.

The composition of the clouds in Jupiter's main cloud layer appeared to be frozen ammonia crystals, with an atmospheric pressure about half that of Earth's at sea level. The overall make-up of the atmosphere was 15 percent helium.

The gas giant's rings were found to be dust, not ice as seen in Saturn's rings. Jupiter's rings were formed when asteroids or small comets, accelerated to high speed by Jupiter's large gravitational field, impacted the planet's four innermost small moons. The ring material was merely ejecta from these impacts.

The possibility that liquid water might exist inside Europa, heated by the same tidal action that caused Io to be a hotbed of volcanoes, generated a great deal of excitement—the combination of liquid water and heat implying the possibility of life. As a result, much of Galileo's operation in 1998 was focused on observing this moon.

Furthermore, as of December 1999, NASA has been developing a spacecraft to orbit Europa, with a planned launch date in 2003. This spacecraft would determine whether a subsurface ocean exists and provide a better understanding of how the surface features formed.

December 28, 1995
Echostar 1
U.S.A. / China

The geosynchronous communications satellite Echostar 1, built by Lockheed Martin, was launched on a Chinese rocket. It was owned by EchoStar Communications and was positioned over the western United States to provide direct broadcast television programming on EchoStar's Dish network, which by December 1999 had more than 2.6 million subscribers.

Through September 1999, EchoStar launched five satellites in this constellation. The first two satellites covered the western United States, and the third the eastern. The fourth, launched in "September 1999, was intended as a replacement for Echostar 1, but because its solar panels failed to deploy properly it only operated at half capacity. The fifth replaced this failure.

The method by which EchoStar Communications launched these five satellites into space demonstrated how international the space industry was becoming in the 1990s. Its first satellite was launched by a Chinese Long March rocket, the second by the European Space Agency's Ariane rocket, the third and fifth by the American Atlas rocket, and the fourth by a Russian Proton rocket.

December 30, 1995
Rossi X-ray Timing Explorer (RXTE)
U.S.A.

Rossi (also called RXTE) was the first American space telescope since **Einstein (November 13, 1978)** dedicated solely to x-ray observations. Built by NASA and named after the American astronomer Bruno Rossi, its instruments monitored the spectrum in the energy range from 2 to 200 KeV, supplementing research of the Compton Gamma Ray Observatory (see **STS 37–April 5, 1991**). Two instruments made close-up observations, while a third acted as an all-sky monitor looking for sudden bursts of x-ray activity.

Rossi's instruments helped clarify the structure of x-ray stars. Within days of launch, Rossi was one of several space telescopes to discover a new type of object called a bursting pulsar. Previous x-ray telescopes had found two types of x-ray objects, pulsars that emit x-rays in regular precise beats (see **Uhuru–December 12, 1970**) and bursters that randomly erupt in powerful explosions of x-ray radiation (see **Granat–December 1, 1989**). Scientists had postulated that the two phenomena were mutually exclusive. With pulsars, the pulse is caused as a lighthouse beacon of radiation sweeps across our line of sight each time the neutron star rotates, produced by the steady stream of particles following the star's magnetic field lines down onto the star's surface near its poles.

With bursters, the eruption occurs when matter in an accretion disk surrounding the neutron star reaches a critical mass and collapses onto the star, causing a thermonuclear explosion. Such explosions take place randomly and with great fury, then die off until the mass builds up again.

Ignoring these theories, the new object, dubbed GRO J1744+28, both pulsed and burst. Furthermore, a careful analysis of its pulses revealed a quasi-periodic secondary pulse overlying the main pulse. Through the middle of 1996, the random bursts slowly died away, only to reappear the next year. As of December 1999, scientists are still unsure how a pulsing neutron star could also produce bursts.

Rossi also discovered a pulsar with the fastest known spin, more than 60 times a second. It also confirmed that objects called soft gamma-ray repeaters (see Compton), which periodically and randomly emit gamma radiation, are neutron stars with incredibly powerful magnetic fields. Astronomers named these objects magnetars.

Rossi made detailed observations of the microquasar GRS 1915+105, first discovered by **Granat (December 1, 1989)**. This very massive object, thought to be a black hole, periodically releases large blobs of material, moving at almost 90 percent the speed of light. Rossi's observations indicated that some regular but mysterious process is causing something to rip this material outward from the accretion disk surrounding the massive object at the center of GRS 1915+105.

Other observations indicated that as supermassive objects spin in space, they actually drag space with them, distorting our view of the light that passes through these regions.

As of December 1999, Rossi remains in operation.

The next x-ray telescope was on **BeppoSax (April 30, 1996)**.

1996

January 11, 1996
STS 72, Endeavour Flight #10
Space Flyer Unit / Spartan 206
U.S.A. / Japan

STS 72 was Endeavour's tenth space flight. It lasted nine days and was crewed by Brian Duffy (**STS 45–March 24, 1992**; **STS 57–June 21, 1993**), Brent Jett (**STS 81–January 12, 1997**), Leroy Chiao (**STS 65–July 8, 1994**), Daniel Barry, Winston Scott (**STS 87–November 19, 1997**), and from Japan, Koichi Wakata. This flight included the deployment and retrieval of one satellite and the retrieval of a second.

• *Space Flyer Unit*   Retrieved on January 13, the Space Flyer Unit (SFU) had been launched by Japan on **March 18, 1995**, to validate a number of engineering designs and perform research in orbit. See this date for results.

• *Spartan 206*   Deployed on January 14th and retrieved on January 16th, Spartan 206 was the third Spartan free-flyer, a simply designed NASA satellite temporarily deployed during space shuttle missions. See **STS 51G (June 17, 1985)** for program information. Spartan 206's experiments proved that global positioning system (GPS) sensors (see **Timation 1–May 31, 1967** and **Navstar B2 1–February 14, 1989**) could precisely locate a satellite's position in space. The next GPS test was on **STS 77** (**May 19, 1996**). The flyer also carried a technology test of a new kind of explosive bolt.

The next Spartan flight, on **STS 72 (January 11, 1996)**, used a different Spartan spacecraft, Spartan 207.

In addition, two space walks tested hardware being designed for the International Space Station (see **Zarya–November 20, 1998** and **STS 88–December 4, 1998**). Twice, Leroy Chiao entered the shuttle cargo bay, once with Daniel Barry and once with Winston Scott. The astronauts there tested a portable foot restraint, a toolbox design, handrail designs, and other equipment.

Onboard experiments included three different protein crystal growth packages (see **STS 63–February 3, 1995**). One sample grew crystals of human insulin (see **STS 60–February 3, 1994**), used by scientists to map the drug's structure. Other samples tested new refined methods for producing larger crystals more effectively.

Two medical experiments conducted preliminary neurological research leading to the first flight of Neurolab on **STS 90 (April 17, 1998)**. Six nursing female rats, each with 10 newborn rats, tested the operation of the rats' cage while also examining the rats' early development. Rat bone samples were also used to better understand why osteoblasts, the chemical used by the body to produce bone tissue, ceased production in weightlessness (see **STS 69–September 7, 1995**). The experiment also tested several candidate drugs for restarting osteoblast production.

Early in the mission, Commander Duffy was forced to make a small maneuver when ground controllers realized that the abandoned **MSTI 2 (May 9, 1994)** satellite would pass less than a mile from Endeavour on January 12. The thruster bursts widened the gap to more than 6 miles. This and other events (see **Cerise–July 7, 1995**) illustrated the rising problem of orbiting space debris and its potential threat to future space facilities.

The next shuttle flight was **STS 75 (February 22, 1995)**.

January 12, 1996
Measat 1
Malaysia / U.S.A. / Europe

Built by Hughes Space and Communications, launched on an Ariane rocket, and owned by a consortium of Malaysian and American companies, Measat 1 was Malaysia's first communications satellite. It was part of a two-satellite geosynchronous constellation. Both satellites provided telephone service in the 4/6 GHz frequency bands and direct television in the 11/14 GHz wavebands. Home television antennas could be as small as 20 inches across to receive the signal. This first satellite was positioned over the Indian Ocean, with the second, launched in December 1996, placed over New Guinea.

February 1, 1996
Palapa C-1
Indonesia / U.S.A.

The Indonesian third-generation geosynchronous communications satellite Palapa C-1, launched from the United States, was the first in a new satellite cluster to replace the Palapa B satellites (see **STS 7–June 18, 1983**). It carried increased television signal and telephone capacity, using 36 transponders in both the 4/6 GHz and 11/14 GHz wavebands. Through May 1996, Indonesia launched two Palapa C satellites, both of which are in use today, serving Indonesia and the Pacific and Asian markets, including Iran, parts of eastern Russian, Australia, and New Zealand. A fourth-generation satellite, **Telcom 1 (August 12, 1999),** later augmented the constellation.

February 17, 1996
NEAR
U.S.A.

Built by NASA, NEAR (Near Earth Asteroid Rendezvous) is intended to provide the first prolonged close-up look at an asteroid. Like a number of other spacecraft (see **Galileo–October 18, 1989**), the spacecraft used the earth's gravity well to slingshot it towards its final target, the asteroid Eros. Along the way, it flew past the asteroid Mathilde on June 27, 1997, taking close-up photographs (see **NEAR, Mathilde Flyby–June 27, 1997**).

NEAR's original flight plan had it entering orbit around Eros on January 10, 1999. An aborted engine burn on December 20, 1998, however, made this maneuver impossible. The spacecraft did a flyby of Eros on December 23, 1998 (see **NEAR, Eros Flyby–December 23, 1998**), with plans to enter orbit around the asteroid sometime around February 2000.

February 21, 1996
Soyuz TM23
Russia

Crewed by Yuri Onufrienko and Yuri Usachev (**Soyuz TM18–January 8, 1994**), Soyuz TM23 docked with **Mir (February 20, 1986)** on February 23rd, joining the crew of **Soyuz TM22 (September 3, 1995)**: Yuri Gidzenko, Sergei Avdeyev, and Thomas Reiter of Germany. For the next six days, the five men performed joint operations, and then, on February 29th, Soyuz TM22 undocked from Mir and returned to Earth, completing 179 days in space. Reiter's flight was the longest for any non-Russian to this date. Medical data from his flight showed that the rate of bone loss averaged between 0.5 to 1 percent per month.

Onufrienko and Usachev then began more than six months of residency on Mir. During their stay, they were visited by one shuttle docking (see **STS 76–March 22, 1996**), which brought refurbished storage batteries so that Mir could once again function at full power (see **Soyuz TM20–October 3, 1994**). STS 76 also brought several thousand pounds of supplies as well as a third crew member, Shannon Lucid.

Onufrienko and Usachev were also resupplied by two Progress freighters (see **Progress M1–August 23, 1989**), bringing supplies of water, oxygen, and food. The first freighter, arriving in May after Lucid had joined the Mir crew, also brought many of the experiments and equipment that she used during the remainder of her occupancy. The second, arriving in July, also brought her M&M candies. The launch of both supply ships was delayed because of cash shortages in Russia.

Also arriving at Mir during this crew's residency was **Priroda (April 23, 1996)**, Mir's last permanent module, completing the station's assembly. For later events in the Soyuz TM23 residency, see STS 76 and Priroda.

February 22, 1996
STS 75, Columbia Flight #19
TSS 1R
U.S.A. / Italy / Switzerland

STS 75 was Columbia's nineteenth flight, lasting just under 16 days and crewed by Andrew Allen (STS 46–July 31, 1992; STS 62–March 4, 1994), Scott Horowitz (STS 82–February 11, 1997), Jeffrey Hoffman (STS 51D–April 12, 1985; STS 35–December 2, 1990; STS 46–July 31, 1992; STS 61–December 2, 1993), Italian Maurizio Cheli, Swiss Claude Nicollier (STS 46–July 31, 1992; STS 61–December 2, 1993; STS 103–December 19, 1999), Franklin Chang-Diaz (STS 61C–January 12, 1986; STS 34–October 18, 1989; STS 46–July 31, 1992; STS 60–February 3, 1994; STS 91–June 2, 1998), and Italian Umberto Guidoni. STS 75's prime mission was the attempted reflight of the tethered satellite experiment from STS 46 (July 31, 1992).

• *TSS 1R*   Although the problems had been corrected that had caused the first tether test on STS 46 to jam, TSS (Tethered Satellite System) 1R was only marginally more successful. On February 25th, the satellite was slowly unreeled over a period of five hours to the planned distance of 13 miles. Just before it reached this distance, however, the tether broke near the shuttle and lost the satellite, which burned up in the atmosphere three weeks later.

Nonetheless, during this short period of tethered flight, sensors on the satellite measured electrical voltages as high as 3,500 volts from the build-up of an electrical charge as the tether moved through the earth's magnetosphere. By discharging the electricity back into space at the shuttle end of the tether, the scientists were also able to produce almost half an amp of current. These results proved the theory that the movement of an extended tether through the earth's magnetic field could be used to generate electricity.

The electrical current, combined with some frayed insulation, caused the tether failure. At the fray, an electrical arc developed, which in turn burned the tether apart.

The next tether experiment was on **STEX (October 3, 1998)**.

Cargo bay experiments included three furnaces to conduct materials research. One grew semiconductor crystals of lead-tin-telluride, an alloy used by infrared detectors and lasers. The second studied how a tin-bismuth mixture melted and mixed when solidified in weightlessness. The third studied how metals solidified, repeatedly melting and freezing a sample of succinonitrile, a transparent material that mimics the behavior of molten metals.

In-cabin experiments included three for studying the phenomenon of fire in space (see **STS 41–October 6, 1990**): how flames spread, what drove the ignition process, and how soot was transported. When paper burned, the fire did not spread in a circle but like the tentacles of an octopus. And like previous tests, flames were not pointed but spherical in shape. Further flame research occurred on **Priroda (April 23, 1996)**.

A package of protein crystal growth experiments (see **STS 63–February 3, 1995**) grew crystals used in developing drugs to treat cancer, hormone disorders, infections, and Chagas disease—a South and Central American disease normally spread by insects.

The next shuttle flight was **STS 76 (March 22, 1976)**. The next laboratory science shuttle flight was **STS 77 (May 19, 1996)**.

February 24, 1996
Polar
U.S.A.

Polar was part of an international effort by the United States, Japan, the European Space Agency, and Russia to study the dynamics of the solar wind with the earth's magnetosphere. Earlier satellites in the program were **Geotail (July 24, 1992)**, **Wind (November 1, 1994)**, and **Interball 1 (August 2, 1995)**.

Built by NASA and placed in an eccentric orbit (perigee: 3,451 miles; apogee: 31,331 miles) over the poles, Polar specifically observed the interaction of the earth's magnetosphere with the solar wind that produced the aurora, continuing research by **Freja (October 6, 1992)** and **Astrid 1 (January 24, 1995)**. Its instruments could photograph the aurora and upper atmosphere over the poles in visible (4,000–7,000 angstroms), ultraviolet (1,200–1,800 angstroms and 2,470–3,370 angstroms), and x-ray (1–100 KeV) wavelengths. It could also measure the flux of charged particles from the Sun impacting the upper atmosphere over the poles.

Polar's images of the aurora were the clearest and most detailed to date, confirming its halo shape from **Dynamics Explorer 1/Dynamics Explorer 2 (August 3, 1981)**. The pictures also confirmed data from **OGO 4 (July 28, 1967)**, showing that the day side part of the halo was sometimes brighter and more dynamic than the night side part.

Working in conjunction with the program's other spacecraft, Polar tracked the impact of solar magnetic storms upon

The Tethered Satellite System TSS 1R slowly being reeled from its base in the shuttle cargo bay. *NASA*

the magnetosphere, watching it get squeezed by each storm and then absorb each storm's energy.

Polar also made observations of Comet Hale-Bopp as it made its closest Earth approach in 1997 and discovered that the comet had a third tail made entirely of sodium atoms, in addition to its ion and dust tails.

Polar's images also continued the controversy over the existence of small comets impacting the top of the earth's atmosphere, appearing to confirm their existence (see **Dynamics Explorer 1/Dynamics Explorer 2 –August 3, 1981**). Nonetheless, many scientists still expressed skepticism, and the issue remains unresolved.

As of December 1999, Polar still operates, its auroral research augmented by **Fast (August 21, 1996)**, **Interball 2 (August 29, 1996)**, and **Astrid 2 (December 10, 1998)**.

March 22, 1996
STS 76, Atlantis Flight #16
U.S.A. / Russia

STS 76 was Atlantis's sixteenth flight, which lasted nine days. It was crewed by Kevin Chilton (STS 49–May 7, 1992; STS 59–April 9, 1994), Richard Searfoss (STS 58–October 18, 1993; STS 90–April 17, 1998), Linda Godwin (STS 37–April 5, 1991; STS 59–April 9, 1994), Ronald Sega (STS 60–February 3, 1994), Michael Clifford (STS 53–December 2, 1992; STS 59–April 9, 1994), and Shannon Lucid (STS 51G–June 17, 1985; STS 34–October 18, 1989; STS 43–August 2, 1991; STS 58–October 18, 1993).

This mission was the third American docking mission to **Mir (February 20, 1986)**, following STS 74–November 12, 1995. Its primary mission was to bring Shannon Lucid to the station to begin what was planned as two years of continuous American presence in space. Just as the operation of Mir was passed from Russian crew to crew, NASA would rotate six Americans in and out of Mir over the next six shuttle dockings, with the **Spektr (May 20, 1995)** module used as each astronaut's home and laboratory.

Atlantis docked on March 24th. In order to increase the amount of cargo that the shuttle could carry, its cargo bay was outfitted with the Spacehab module (see **STS 57–June 21, 1993**). In addition to food, water, and supplies, Atlantis brought a replacement gyroscope as well as three refurbished Russian batteries. These batteries finally solved the station's limited power problems (see **Soyuz TM20–October 3, 1994**). The cargo exchange also included replacing the samples of two protein crystal growth experiments, first brought to the station on **STS 74 (November 12, 1995)**.

During the five days of joint operation, Michael Clifford and Linda Godwin did a six-hour space walk to attach a science and technology package to the outside of the Mir docking module. This package, picked up during **STS 86 (September 25, 1997)**, collected micrometeoroid and orbital debris material so that scientists could further refine the estimated flux of material that might impact future large space stations. The packages also tested how samples of paint, fiber, and optical coatings managed in space.

The shuttle also carried a camera dubbed KidSat, controlled by students on the ground and taking digital images of the earth for direct download to the classroom. KidSat flew frequently on subsequent shuttle missions.

The next shuttle flight was STS 77 (**May 19, 1996**). The next shuttle docking mission to Mir was STS 79 (**September 16, 1996**). For more details about Lucid's Mir stay, see **Priroda (April 23, 1996)**.

April 3, 1996
Inmarsat 3 F1
International / U.S.A.

Launched from Cape Canaveral and owned by Inmarsat (see **Marisat 1–February 19, 1976**), Inmarsat 3 F1 supplemented the Inmarsat 2 satellite constellation (see **Inmarsat 2 F1–October 30, 1990**). Through December 1999, Inmarsat launched five Inmarsat 3 satellites, placing them over the Atlantic, Pacific, and Indian Oceans. In April 1999, Inmarsat incorporated, becoming a private company.

April 20, 1996
MSAT 1
Canada / Europe

Launched on an Ariane rocket, Mobile SATellite (MSAT) 1 was built by Telesat Mobile, Inc., a Canadian corporation formed by Telesat Canada (see **Anik A–November 10, 1972**). It provides mobile telephone communications in areas of Canada where cellular technology is unavailable, working in partnership with **AMSC 1 (April 7, 1995)**, which provides similar service for the United States, Hawaii, Alaska, and the Caribbean. Each satellite backs up the other.

April 23, 1996
Priroda
Russia / U.S.A.

The launch of the last module to **Mir (February 20, 1986)**, Priroda, had been planned several years earlier but was delayed because of lack of funds in Russia. The cash from the United States in the cooperative American-Russian Shuttle-Mir effort (see **STS 63–February 3, 1995**) made possible Priroda's launch to Mir.

Priroda ("nature" in Russian) added about 2,300 cubic feet to the Mir complex, bringing its total habitable volume to slightly less than 13,000 cubic feet, just exceeding the interior space of **Skylab (May 14, 1973)**. The module's weight of 20 tons raised Mir's overall mass to over 110 tons. When Soyuz TM and Progress M (see **Progress M1–August 23, 1989**) spacecraft were also docked to the station, the complex's total mass increased to more than 126 tons, making it the heaviest satellite to orbit the earth. And the addition of a docked space shuttle increased this tonnage to more than 225 tons.

Priroda initially docked with Mir's main port on April 26th. The next day, the robot arm on the multidocking adapter repositioned it to a side port, where it joined **Kvant 2 (November 26, 1989)**, **Kristall (May 31, 1990)**, and **Spektr (May 20, 1995)** to form a cross of modules attached perpendicular to the Mir and **Kvant 1 (March 31, 1987)** main axis.

Priroda conducted Earth resource observations, carrying three multispectral spectrometers, three multispectral radiometers, two multispectral scanners, and one synthetic aper-

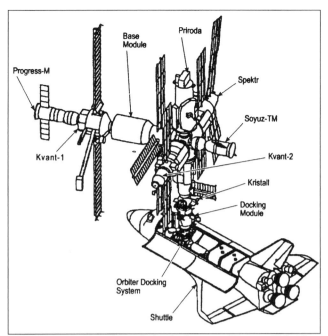

The complete Mir complex, with shuttle, Soyuz, and Progress M freighter also docked. Compare with illustration for November 12, 1995. *NASA*

ture radar. It also carried a French/Russian lidar sensor, similar to the lidar tested on **STS 64 (September 9, 1994)**, to study the atmosphere's vertical composition.

Priroda carried American equipment and experiments for use by American astronaut Shannon Lucid (see **STS 76–March 22, 1996**), including a spectrometer for analyzing medical specimens and a magnetic bottle for levitating samples during materials processing research.

The original plans had called for Priroda's attachment to Mir prior to Lucid's arrival, but delays in Russia due to funding shortages as well as a decision to bump the module's launch in order to give priority to a commercial satellite caused Lucid to arrive first.

Lucid, along with Yuri Onufrienko and Yuri Usachev of **Soyuz TM23 (February 21, 1996)**, remained on Mir for another four months. During this time period, Onufrienko and Usachev performed five space walks to install one of the two solar panels brought to Mir by **STS 74 (November 12, 1995)**, fitting it to the exterior of **Kvant 1 (March 31, 1987)** to replace an older worn-out panel. Other tasks included installing a multispectral Earth-scanning camera to the outside of Priroda, deploying the antenna for Priroda's synthetic aperture radar, and assembling and attaching a 20-foot boom to Kvant 1. This second Strela crane supplemented the first (see **Soyuz TM11–December 2, 1990**), moving objects to areas of Mir's exterior previously unreachable.

During their first space walk, the men also videotaped the first in-space television commercial, holding a 4-foot-high inflatable soft drink can.

Meanwhile, Lucid focused on scientific research. In one experiment, she burned 79 candles, studying how the flames burned differently in space (see **STS 41–October 6, 1990**). She noted that the flame consistently sat on the top of the wick like "a little blue igloo," rather than forming a point as seen on Earth. These flameballs were further studied on **STS 94 (July 1, 1997)**.

The arrival of **Soyuz TM24 (August 17, 1996)** ended Onufrienko's and Usachev's mission. Lucid's residency ended with the arrival of **STS 79 (September 16, 1996)**.

April 24, 1996
Midcourse Space Experiment (MSX)
U.S.A.

The U.S. military satellite MSX tested surveillance technology for identifying and tracking ballistic missiles after their engines had cut off and they were in their coast phase of flight. This period of time, lasting from 20 to 40 minutes, provided the best opportunity for pinpointing a missile's location and course.

To do this, MSX carried sensors for detecting missile targets against both the earth and sky, which were able to resolve objects less than one half inch in diameter. These instruments included ultraviolet, visible, and infrared imaging telescopes (0.11–0.9 microns and 2.5–28 microns), a scanning radiometer, and five spectrographs.

In February and August 1997, the satellite successfully tracked three different missiles launched from Wallops Island, Virginia, and Kauai, Hawaii. And beginning with **STS 79 (September 16, 1996)**, it routinely tested its tracking ability by locating each subsequent shuttle flight.

MSX's instruments were also used for astronomy. Continuing celestial infrared observations of **ISO (November 17, 1995)**, it obtained high-resolution infrared images of the galactic center of the Milky Way and the Small Magellanic Cloud. The wide angle nature of MSX's telescopes allowed it to capture each object entirely within its field of view.

The next ultraviolet telescope was Orfeus (see **STS 80–November 19, 1996**). The next infrared space telescope was the addition of the NICMOS infrared camera on Hubble (see **STS 82–February 11, 1997**). See also **SWAS (December 6, 1998)**.

April 30, 1996
BeppoSax
Italy / Netherlands / U.S.A.

BeppoSax was named in honor of Giuseppe (Beppo) Occhialini, who had been instrumental in the 1960s in establishing Italy's space science program (see **San Marco 1–December 15, 1964** and **San Marco 2–April 26, 1967**). BeppoSax was built by Italy, with help from The Netherlands, and launched by the United States. It was Italy's first x-ray telescope, its instruments covering the energy range from 0.1 to 300 KeV, with imaging capability in the 0.1 to 10 KeV range, augmenting the high-energy astronomy being done by **ASCA (February 20, 1993)**, **Compton** (see **STS 37–April 5, 1991**), and **Rossi (December 30, 1995)**.

In May 1998, BeppoSax made possible the first identification of a gamma-ray burst. These objects, first discovered by **Vela 5A/Vela 5B (May 23, 1969)** almost three decades earlier, had remained an enigma. Their make-up, distance, and cause were complete and total mysteries. BeppoSax, in conjunction with **Ulysses, First Solar Orbit (April 17, 1998)** and **NEAR, Mathilde Flyby (June 27, 1997)**, made it possible to quickly identify, with great precision, the location of burst GRB 970228. BeppoSax then made the first identification of a gamma-ray burst in wavelengths other than gamma rays, imaging the burst's x-ray afterglow. The Hubble Space Telescope (**STS 82–February 11, 1997**) followed with the first identification of a gamma-ray burst in optical wavelengths. Both Hubble and BeppoSax observed the optical and x-ray afterglow of the burst fade slowly away. By measuring the red shift of the light of another burst in May, scientists were then able to estimate the burst's distance, proving that it took place outside the Milky Way at many billions of light years distance. Hence, these gamma-ray bursts were some of the most powerful explosions ever seen. This discovery implied that all the other bursts spotted in the decades since Vela 5A/Vela 5B first identified them were as powerful.

Since then, BeppoSax has identified another dozen gamma-ray bursts, with the distance to one object so distant that cosmologists are at a loss to explain the power of its explosion. All previous theories for the cause of gamma-ray bursts, usually involving the merger of two neutron stars or the swallowing of a neutron star by a black hole, produced 100 times less energy than this explosion.

As of December 1999, BeppoSax remains in operation. The next x-ray telescope, the Chandra X-ray Observatory, was launched by the space shuttle. See **STS 93** (**July 23, 1999**).

May 16, 1996
Amos
Israel / Europe

Launched on an Ariane rocket and built by Israel Aircraft Industries, Amos was Israel's first geosynchronous communications satellite. It was positioned over the equator above the western coast of Africa, where it provided telephone, television, and Internet service using seven transponders operating in the 11/14 GHz frequencies. The satellite's operation includes the unusual policy that on Saturday, the Jewish day of Sabbath (as well as on other Jewish holidays), its thruster engines are not used in order to follow religious law.

May 17, 1996
MSTI 3
U.S.A.

Miniature Sensor Technology Integration (MSTI) 3 was the last of three U.S. Air Force research satellites (see **MSTI 1–November 21, 1992**) designed as low-cost tests of new advanced infrared imaging technology to track and image the launch of ballistic missiles.

During its one-year mission, MSTI 3 gathered extensive data on the background infrared radiation of the earth, so that infrared surveillance satellites could distinguish the heat of a rocket's plume against it.

Moreover, on October 17, 1997, the U.S. Air Force used this satellite as a target, firing a laser beam at it several times. The experiment was part of a U.S. military research program to see if lasers could be used to clear debris smaller than 4 inches in diameter from orbit by lowering their orbit and forcing them to re-enter the atmosphere sooner. This test proved the concept's feasibility.

May 19, 1996
STS 77, Endeavour Flight #11
Spartan 207 / PAMS
U.S.A. / Canada

Endeavour's eleventh flight, STS 77 lasted 10 days and was crewed by John Casper (**STS 36–February 28, 1990**; **STS 54–January 13, 1993**; **STS 62–March 4, 1994**), Curtis Brown (**STS 47–September 12, 1992**; **STS 66–November 3, 1994**; **STS 85–August 7, 1997**; **STS 95–October 29, 1998**; **STS 103–December 19, 1999**), Daniel Bursch (**STS 51–September 12, 1993**; **STS 68–September 30, 1994**), Mario Runco (**STS 44–November 24, 1991**; **STS 54–January 13, 1993**), Canadian Marc Garneau (**STS 41G–October 5, 1984**), and Andrew Thomas (**STS 89–January 22, 1997**). The mission deployed two research satellites and retrieved one.

• *Spartan 207*   Deployed on May 20th, Spartan 207 was one of three Spartan spacecraft simply designed by NASA and temporarily deployed from the shuttle. See **STS 51G** (**June 17, 1985**). Spartan 207 (also called the Spartan Service Module) had flown previously on **STS 63** (**February 3, 1995**), doing astronomical research under the name Spartan 204.

On this flight, Spartan 207 had been reconfigured, its sole payload an inflatable antenna to test technology for placing large-sized, lightweight, and inexpensive to build objects in space. The antenna inflated as designed, expanding in about five minutes to the size of a tennis court, 90 by 50 feet. After 90 minutes of data collection and picture taking, Spartan 207 ejected the antenna (which re-entered the atmosphere two days later). Spartan 207 was retrieved by Endeavour on May 21st.

Spartan 207's next flight is scheduled for sometime in 2000. The next Spartan flight was flown by Spartan 201, on STS 87 (**November 19, 1997**).

• *PAMS*   Deployed by the shuttle on May 22nd, PAMS demonstrated new technology for passively orienting a satellite's attitude in low-Earth orbit by using the spacecraft's center of gravity and its shape as it moves through the thin ionosphere. If incorporated into satellite design, the technique would save significant fuel and thereby extend spacecraft service life.

Endeavour's cargo bay was configured with the Spacehab (see **STS 57–June 21, 1993**) module, carrying more than a dozen different experiments. Eight protein crystal growth experiments (see **STS 63–February 3, 1995**) grew crystals for drug research relating to AIDS, diabetes, and a number of other diseases and chemicals. Influenza protein crystals grown were used to complete the mapping of the virus's structure; this knowledge was then used to develop new flu drugs. Other life sciences experiments tested drugs for treating cancer and malaria and studied bone and muscle deterioration using chicken embryos. A Canadian package of three experiments examined the effect of weightlessness on fertilized sea urchins and mussels.

A newly designed greenhouse, following up research by Chromex (see **STS 29–March 13, 1989** and **STS 63**) and Astroculture (see **STS 73–October 20, 1995**), grew spinach to see the effect of weightlessness on its production of starch, sugar, and fatty acid. Clover plants were also grown to see how they fixed nitrogen from the greenhouse's artificial soil. See **STS 79 (September 16, 1979)** for more greenhouse research.

Materials research grew gallium arsenide and gallium antinomide crystals for semiconductors, and oxide crystals for infrared and laser detectors.

Technology experiments included the first test of the global positioning system (GPS) to track the shuttle's orbital position (see **Timation 1–May 31, 1967**, **Navstar B2 1–February 14, 1989**, and **STS 72–January 11, 1996**). The mission also tested a new design for refueling tanks in space, a follow-up from **STS 57 (June 21, 1993)**.

A third technology experiment tested a new technique for cooling infrared sensors to near absolute zero. This method, using metal alloy powders called metal hydrides, was stable over many years, and if proven reliable could increase the operating life of orbiting infrared telescopes (such as **IRAS–January 26, 1983** and **ISO–November 17, 1995**) from at most three years to over a decade.

The next shuttle mission was **STS 78 (June 20, 1996)**.

June 20, 1996
STS 78, Columbia Flight #20
U.S.A. / Canada / France

STS 78 was Columbia's twentieth space flight; the 17-day flight set a new in-space shuttle record, exceeding the previous record set by **STS 67 (March 2, 1995)**. The crew was Terrence Henricks (**STS 44–November 24, 1991**; **STS 55–April 26, 1993**; **STS 70–July 13, 1995**), Kevin Kregel (**STS 70–July 13, 1995**; **STS 87–November 19, 1997**), Susan Helms (**STS 54–January 13, 1993**; **STS 64–September 9, 1994**), Richard Linnehan (**STS 90–April 17, 1998**), Charles Brady, Frenchman Jean-Jacques Favier, and Canadian Robert Thirsk.

STS 78's prime mission was the study of the human body in weightlessness, continued research from **STS 58 (October 18, 1993)**. The seven astronauts took regular samples of their blood, saliva, lung exhalations, urine, and fecal matter. Their muscle response and sleep cycles were monitored closely. Other tests studied how head movements affected their system of balance.

Of the non-human biological experiments, one used 12 rodents to study how hormones released by the adrenal glands might influence the loss of bone tissue in space. On Earth, an excess production of these hormones produced Cushing's Syndrome, whose symptoms included loss of bone minerals. Of the 12 rodents flown, six had had their adrenal glands removed to see if removing this hormone might aid bone production in weightlessness.

In two plant experiments, pine seedlings and daylilies were grown to see how they responded to microgravity. And by growing medaka fish embryos, scientists studied how weightlessness influenced their initial development after fertilization.

In materials research, a furnace was used to attempt to produce alloys of aluminum and indium, which normally did not mix in gravity. Experiments studied how bubbles behave in space, continuing research conducted first on **STS 5 (November 11, 1982)** and last on **STS 65 (July 8, 1994)**. See **STS 94 (July 1, 1997)** for later research.

Protein crystal growth research (see **STS 63–February 3, 1995**) tested three techniques for growing crystals. Crystals of the natural sugarless sweetener thaumatin, first attempted on **STS 73 (October 20, 1995)**, were 25 percent larger and more ordered than any grown on Earth.

The next shuttle flight was **STS 79 (September 16, 1996)**, the fourth docking with **Mir (February 20, 1986)**. The next science shuttle flight was **STS 80 (November 19, 1996)**. The next life sciences shuttle mission was **STS 90 (April 17, 1998)**.

July 2, 1996
TOMS
U.S.A.

Built by NASA, the Total Ozone Mapping Spectrometer (TOMS) carried ozone layer sensors identical to instruments first flown on **Nimbus 7 (October 13, 1978)**. It continued research from **STS 66 (November 3, 1994)**.

During the northern hemisphere's spring months, TOMS measured unusually low levels of ozone over the Arctic, 40 percent lower than earlier readings during the period from 1979 to 1982. The levels, however, were still two times greater than the lowest levels seen in the southern hemisphere when that hemisphere's ozone hole was at its greatest extent. Scientists theorized that the low levels in the north were caused not by CFCs but by an unusually cold and persistent winter. However, an increased link with CFC-caused chlorine seemed to strengthen the possibility that the decline was not solely caused by cold weather.

Through 1997, the ozone hole seen over Antarctica had yet to be found in the northern hemisphere, though an overall decrease since 1970 of about 5 to 7 percent had been tracked.

The decline, however, appeared to have bottomed out in 1995 and was focused mostly in Antarctica, with only slight changes in ozone levels in the northern hemisphere. These results continue to puzzle scientists—their theories said that the use of chlorofluorocarbons (CFCs) by northern hemisphere industrial nations would damage the ozone layer, yet the decline in ozone was mostly seen in the south. Since the two hemispheres were largely independent and it was difficulty at best for CFCs to migrate to the South Pole, the theories did not explain the facts very well.

This ozone layer monitoring was augmented by a second TOMS instrument on **ADEOS (August 17, 1996)**. TOMS itself functioned through December 1998, when its fuel supply was exhausted. A back-up orientation system, using magnets in conjunction with the earth's magnetic field, was then activated. As of December 1999, the satellite continued to function.

August 17, 1996, 1:53 GMT
ADEOS (Midori)
Japan / U.S.A. / France

Also called Midori ("green" in Japanese), the Advanced Earth Observing Satellite (ADEOS) monitored the earth's climate and environment. Placed in a sun-synchronous orbit (see **Nimbus 1–August 28, 1964**), it carried five Japanese, two NASA, and one French instrument for imaging the oceans, the ozone layer, and plant life, while also monitoring carbon dioxide, atmospheric content, and wind speeds above the oceans. The ocean wind measurements provided meteorologists with data from regions over the southern oceans that had never been monitored before.

August 17, 1996

The ozone layer instrument was identical to **TOMS** (**July 2, 1996**). Other instruments supplemented the earth monitoring of the NOAA satellites (see **NOAA 1–December 11, 1970**).

In September and October of 1996, both ADEOS and TOMS found that the annual growth of the ozone hole over Antarctica reached levels equal to the previous record, measured in 1993. Although the hole started to form slightly earlier than in previous years, it also started to decline earlier as well.

In May 1997, data from ADEOS confirmed measurements by **TOPEX–Poseidon** (**August 10, 1992**) that successfully predicted the arrival of El Niño in the winter of 1997–98.

Because of a failure in the spacecraft's solar panels, it shut down prematurely in July 1997 after only one year of operation.

Japan is presently building a follow-up ADEOS satellite, scheduled for launch in November 2000. The next ozone layer monitor was **SNOE** (**February 25, 1998**). See also **QuikScat** (**June 19, 1999**).

August 17, 1996, 13:18 GMT
Soyuz TM24
Russia / France / U.S.A.

Crewed by Valeri Korzun and Alexander Kaleri (**Soyuz TM14–March 17, 1992**) of Russia and Claudie Andre-Deshays of France, Soyuz TM24 docked with **Mir** (**February 20, 1986**) on August 19th. The crew joined Yuri Onufrienko and Yuri Usachev of **Soyuz TM23** (**February 21, 1996**) and Shannon Lucid of STS 76 (**March 22, 1996**).

Andre-Deshays was the first French female astronaut. Her flight completed a three-mission, joint French-Russian program, begun with **Soyuz TM15** (**July 27, 1992**) and continued with **Soyuz TM17** (**July 1, 1993**), in which France paid Russia a total of about $33 million to place three astronauts on Mir. Because of funding shortages in the Russian space program, however, Andre-Deshays's flight had been seriously delayed, and, as a result, many of the experiments left behind on Mir from Soyuz TM15 were never completed.

For two weeks, the four men and two women performed joint operations, with Onufrienko and Usachev transferring operation of Mir to Korzun and Kaleri while Lucid and Andre-Deshays conducted scientific experiments. The Frenchwoman noted later that she found a two-week stay on Mir insufficient to familiarize herself with the space laboratory and also complete her research.

On September 2nd, Onufrienko, Usachev, and Andre-Deshays returned to Earth in Soyuz TM23. The two Russians had completed a little more than six months in orbit.

Following Andre-Deshays's flight, the French and Russians immediately signed a new cooperative agreement, calling for two more French astronauts to occupy Mir through 1999, the first for a period of several weeks and the second for four months, with France paying Russia approximately $40 million. The first flight was delayed until **Soyuz TM27** (**January 29, 1998**) because of the many problems during **Soyuz TM25** (**February 10, 1997**). The second, **Soyuz TM29** (**February 20, 1999**), was part of the last mission to Mir as of December 1999 and would have French cosmonaut Jean-Pierre Haignere participate in preparing the station for deorbiting.

Meanwhile, on Mir, Korzun and Kaleri settled down to a routine six-month tour of duty, while Lucid began wrapping up her research in anticipation of the arrival of **STS 79** (**September 16, 1996**).

August 21, 1996
FAST
U.S.A.

Built by NASA, the Fast Auroral Snapshot Explorer (FAST) supplemented research by **Polar** (**February 24, 1996**), measuring both the earth's magnetic field and the particles directed and accelerated by that field into the atmosphere to cause the aurora. The spacecraft was placed in an eccentric polar orbit (perigee: 217 miles; apogee: 2,600 miles), flying through the region where the earth's magnetic field lines arched from a downward course towards the poles to an upward curve away from the earth. It was in this region that scientists believed that the aurora developed.

The next satellites to study the aurora were **Interball 2/Magion 5** (**August 29, 1996**).

August 29, 1996
Interball 2 / Magion 5 / Microsat (Musat)
Russia / Czech Republic / Argentina

• *Interball 2*   Also called the Interball Auroral Probe, this Russian research satellite Interball 2 was dedicated to studying the earth's aurora and its magnetosphere over the poles. It supplemented observations of **Polar** (**February 24, 1996**), **FAST** (**August 21, 1996**) and the Russian **Interball 1/Magion 4** (**August 2, 1995**). It carried 16 instruments for monitoring the radiation and particles of the magnetosphere and the solar wind, as well as their interaction.

• *Magion 5*   Like its predecessors (see Magion 4), this Czech Republic research satellite complemented the observations of its partner, Interball 2. It carried eight instruments for measuring the magnetic field and electron particle properties of the magnetosphere over the poles. Placed in nearly identical orbits, the two spacecraft provided scientists with data over a wider area.

The next satellite to study the aurora was **Astrid 2** (**December 10, 1998**). The next satellite to study the magnetosphere was **Equator S** (**December 2, 1997**).

• *Microsat (Musat)*   Built by the Instituto Universitario Aeronautico de Cordoba in Argentina, Microsat was a demonstration project to test new low-cost satellite technology designs. It was launched commercially by the Russians for $200,000.

September 8, 1996
GE 1
U.S.A.

Built and owned by GE American Communications (GE Americom), GE 1 was part of a private geosynchronous communications satellite cluster providing cable and broadcast television capacity as well as telephone and paging services. Through November 1999, GE Americom has placed five GE satellites in orbit over the United States. Furthermore, in 1997, GE Americom, in joint ownership with several European companies, placed two satellites over Europe to provide direct broadcast television, Internet services, and other broadcasting services.

September 16, 1996
STS 79, Atlantis Flight #17
U.S.A. / Russia

STS 79 was Atlantis's seventeenth space flight, lasting nine days. It was crewed by William Readdy (STS 42–January 22, 1992; STS 51–September 12, 1993), Terrence Wilcutt (STS 68–September 30, 1994; STS 89–January 22, 1998), Thomas Akers (STS 41–October 6, 1990; STS 49–May 7, 1992; STS 61–December 2, 1993), Jerome Apt (STS 37–April 5, 1991; STS 47–September 12, 1992; STS 59–April 9, 1994), Carl Walz (STS 51–September 12, 1993; STS 65–July 8, 1994), and John Blaha (STS 29–March 13, 1989; STS 33–November 22, 1989; STS 43–August 2, 1991; STS 58–October 18, 1993).

STS 79 was the fourth shuttle docking with **Mir (February 20, 1986)**, during which John Blaha took over for Shannon Lucid, who had been brought to Mir on **STS 76 (March 22, 1996)**. Lucid's flight of 188 days (extended because Atlantis's launch had been delayed by six weeks due to Hurricanes Bertha and Fran as well as NASA's decision to replace the shuttle's solid rocket boosters) set both an American and female space endurance record, exceeding the 169-day record of Elena Kondakova (see **Soyuz TM20–October 3, 1994**).

On September 19th, Atlantis docked with Mir, joining Lucid as well as Valeri Korzun and Alexander Kaleri of **Soyuz TM24 (August 17, 1996)**. The shuttle's cargo bay was outfitted with a double Spacehab module (see **STS 57–June 21, 1993**), adding 2,200 cubic feet of habitable space to the shuttle's 2,300 cubic feet and bringing the total habitable space of the combined Mir-Shuttle complex to more than 17,000 cubic feet. The total weight of the complex was a record 256 tons. An IMAX camera onboard Atlantis documented the shuttle's docking and undocking with Mir for use in the IMAX film *Mission to Mir*.

During the next four days, the nine astronauts from both crews transferred over 3,300 pounds of cargo to Mir and unloaded more than 2,100 pounds of cargo from the station. Besides water, food, and routine supplies, this cargo included a furnace and several experiments for Blaha during his four-month stay on Mir.

Atlantis undocked on September 23rd, leaving Valeri Korzun, Alexander Kaleri, and John Blaha behind. All three men remained on Mir until the arrival of **STS 81 (January 12, 1997)** and **Soyuz TM25 (February 10, 1997)**. In the interim, they were visited by one Progress freighter (see **Progress M1–August 23, 1989**), which brought two tons of water, oxygen, and food. This freighter also brought a replacement pump for emptying Mir's overflowing toilet tanks. A cash shortage in Russia had canceled the launch of several additional freighters, thereby limiting the supplies sent to Mir.

Korzun and Kaleri performed two EVAs (extra-vehicular activities) during their tour of duty, attaching a variety of electrical connections and adding a Kurs antenna to the docking module.

During Blaha's four months in space, he conducted several experiments, the most productive of which was the successful harvesting of 32 wheat stalks—one of the best harvests yet see in space. The results indicated that previous failures (see **Soyuz T5–May 13, 1982**) had occurred because of lack of light, and because weightlessness prevented the natural circulation of water through the artificial soil to the roots. By implanting moisture sensors near the roots, Blaha was able to tell when the plants needed water, and therefore he was able to get them to grow. See **STS 87 (November 19, 1997)** for more greenhouse research.

His other experiments included protein crystal growth studies and research into the behavior of high-temperature superconductor materials.

STS 81, the next manned mission to Mir, docked with the station on January 14, 1997.

November 19, 1996
STS 80, Columbia Flight #21
Orfeus / Wake Shield 3
U.S.A.

STS 80, Columbia's 21st space flight, set a new shuttle in-space endurance record of almost 18 days, exceeding the record set by **STS 78 (June 20, 1996)**. The crew was Kenneth Cockrell (STS 56–April 9, 1993; STS 69–September 7, 1995), Kent Rominger (STS 73–October 20, 1995; STS 85–August 7, 1997), Tamara Jernigan (STS 40–June 5, 1991; STS 52–October 22, 1992; STS 67–March 2, 1995), Thomas Jones (STS 59–April 9, 1994; STS 68–September 30, 1994), and Story Musgrave (STS 6–April 4, 1983; STS 51F–July 29, 1985; STS 33–November 22, 1989; STS 44–November 24, 1991; STS 61–December 2, 1993). Musgrave, at 61 years old, was at that time the oldest man to fly into space.

This shuttle science mission deployed and retrieved two research satellites.

• *Orfeus* Flown previously on **STS 51 (September 12, 1993)**, Orfeus was an ultraviolet telescope observing the celestial sky from 400 to 1,250 angstroms. It was deployed on November 20th and retrieved on December 4th. During its two weeks of flight, Orfeus made more than 100 observations of stars and galaxies, augmenting observations by **EUVE (June 7, 1992)**, and **MSX (April 24, 1996)**.

The next ultraviolet telescopes were flown on **STS 85 (August 7, 1997)** and **STS 95 (October 29, 1998)**. See also **Fuse (June 24, 1999)**.

• *Wake Shield 3* Deployed on November 22nd and retrieved on November 26th, this flight was Wake Shield's third test flight (see **STS 60–February 3, 1994** and **STS 69–September 7, 1995**), during which the technology for growing ultra-thin and pure membranes in space was further refined. On this flight, seven wafers were produced for use in transistors and electronic junctions.

One onboard experiment tested the effectiveness of a high-calcium diet for reducing the loss of bone mass as well as aiding the body in maintaining blood pressure during weightlessness. Seven rats were fed a high-calcium diet while seven received a low-calcium diet.

Another medical experiment made the third attempt to grow protein crystals of urokinase (see **STS 67–March 2, 1995**). Urokinase is an enzyme present during the spread of breast cancer. Scientists used the large-sized pure crystals grown in space to map its structure and develop drugs that could possibly inhibit cancer cell growth.

A BRIC canister (see **STS 64–September 9, 1994**) grew 200 genetically altered tomato and tobacco seedlings, testing how such alterations affected growth rates and plant mass.

Two planned space walks to practice space construction techniques were canceled when Columbia's airlock hatch refused to open—a loose screw had become embedded in the door's gearbox. These tasks were carried out on **STS 87 (November 19, 1997)**.

The next shuttle flight, **STS 81 (January 12, 1997)**, docked with **Mir (February 20, 1986)**. The next science shuttle mission was STS 83 (April 4, 1997).

December 24, 1996
Bion 11
Russia / International

Bion 11 was a biological research satellite, part of a program begun with **Bion 1 (October 31, 1973)**. Like **Bion 10 (December 29, 1992)**, Bion 11 carried a menagerie of biological life, including two monkeys, for its two weeks in space. Scientists from Russia, the United States, the Ukraine, Lithuania, and France participated.

Unfortunately, one of the two monkeys died one day after recovery during an operation to obtain tissue and bone samples. Because of the resulting protests in the United States, Congress voted to stop funding the Bion program, thereby forcing NASA to pull out. Although one more Bion flight had been planned, as of December 1999, it has not been launched.

While much of the results from these 11 missions suggested that weightlessness was harmful to life, scientists were still unsure if the overall negative symptoms were caused by the environment of space or the stress of launch and landing. The results from the rat dissections on **STS 58 (October 18, 1993)** suggested that many of the negative effects were caused by re-entry. More in-space autopsies would be required to settle the question.

1997

January 12, 1997
STS 81, Atlantis Flight #18
U.S.A. / Russia

The eighteenth flight of Atlantis, STS 81 made the fifth docking with **Mir (February 20, 1986)**. The flight lasted 10 days and was crewed by Michael Baker (**STS 43–August 2, 1991**; **STS 52–October 22, 1992**; **STS 68–September 30, 1994**), Brent Jett (**STS 72–January 11, 1996**), Peter Wisoff (**STS 57–June 21, 1993**; **STS 68–September 30, 1994**), John Grunsfeld (**STS 67–March 2, 1995**; **STS 103–December 19, 1999**), Marsha Ivins (**STS 32–January 9, 1990**; **STS 46–July 31, 1992**; **STS 62–March 4, 1994**), and Jerry Linenger (**STS 64–September 9, 1994**). Linenger relieved John Blaha on Mir, who had been stationed there with cosmonauts Valeri Korzun and Alexander Kaleri of **Soyuz TM24 (August 17, 1996)** since **STS 79 (September 16, 1996)**, a total of 128 days in space.

Atlantis remained docked with Mir for five days, transferring about 5,000 pounds of cargo onto Mir (including 1,400 pounds of water, a new gyroscope, and three batteries) while returning almost 2,500 pounds of cargo (including Blaha's wheat experiment for testing on Earth).

After undocking, Atlantis did a fly-around at a distance of about 550 feet, orbiting the complex twice to inspect and photograph its condition after more than a decade in space.

Once back on Earth, Blaha was "absolutely stunned" at how weak and wobbly he felt, even though he had maintained the Russian routine of two hours of exercise per day while in space. His recovery, however, was normal.

Linenger, Korzun, and Kaleri remained on Mir, though the two Russians were quickly replaced by the crew of **Soyuz TM25 (February 10, 1997)**.

The next shuttle flight to Mir was **STS 84 (May 15, 1997)**. The next shuttle flight was **STS 82 (February 11, 1997)**.

January 30, 1997
Nahuel 1A
Argentina / U.S.A. / Europe

Built by Aerospatiale, launched on an Ariane rocket, and owned by Nahuelsat, the geosynchronous communications satellite Nahuel 1A was Argentina's first commercial satellite. Nahuelsat was a private corporation backed by European and North and South American companies, including GE American Communications. The satellite initially provided television and telephone service to South America but later expanded its services to the United States.

February 10, 1997
Soyuz TM25
Russia / Germany / U.S.A.

Crewed by Vasily Tsibliyev (**Soyuz TM17–July 1, 1993**) and Alexander Lazutkin of Russia and Reinhold Ewald of Germany, Soyuz TM25 docked with **Mir (February 20, 1986)** on February 12 to join Valeri Korzun and Alexander Kaleri of **Soyuz TM24 (August 17, 1996)** and Jerry Linenger of **STS 81 (January 12, 1997)**. To make room for the docking of this second Soyuz spacecraft, the one Progress freighter (see **Progress M1–August 23, 1989**) that had resupplied the crew of Soyuz TM24 (see **STS 79–September 16, 1996**) was temporarily undocked from the station on February 4th with plans to redock it once Soyuz TM24 returned to Earth on March 2nd.

Tsibliyev and Lazutkin replaced Korzun and Kaleri, beginning what was to become one of the most eventful and difficult tours of duty for any Russian crew. As a harbinger of things to come, on February 24th a fire broke out while Lazutkin was activating one of Mir's lithium perchlorate candle canisters within the **Kvant 1** (**March 31, 1987**) module. These lithium candles were a back-up method for producing oxygen for the station's atmosphere. While Korzun fought the blaze, the other five men scrambled to obtain fire extinguishers from throughout the station. The fire's location in Kvant 1 blocked access to Soyuz TM25, which meant that escape was impossible for its crew of three. After using three extinguishers, Korzun contained the fire, limiting almost all of the damage to the canister.

During the next week, the two crews labored to bring the station back into full operation. Then, on March 2nd, Korzun, Kaleri, and Ewald undocked Soyuz TM24 from Mir and returned to Earth. Korzun and Kaleri had completed more than 6.5 months in space. Ewald's 20-day flight completed a German-Russian agreement whereby Germany paid Russia $60 million for the right to conduct research on Mir.

On Mir, Tsibliyev and Lazutkin, along with Linenger, continued their residency. On March 4th, their attempt to redock the Progress freighter failed. Unlike previous automatic dockings, which had used the Ukrainian-built Kurs docking system, this docking tested the newly designed Russian-built Toru docking system, controlled by Tsibliyev from within Mir. Because of the break-up of the Soviet Union, the Russians had developed the Toru system to avoid a dependence on the Ukrainian design.

During this failed test docking, the camera on the freighter malfunctioned, and Tsibliyev, who had been the pilot on **Soyuz TM17** (**July 1, 1993**) during Mir's only previous collision (see **Soyuz TM18–January 8, 1994**), prevented a crash by flying the freighter blind and missing the station by less than 600 feet. The freighter was then allowed to burn up in the atmosphere.

Soon thereafter, Mir's two electrolysis oxygen generating systems failed, with one leaking coolant into the station's atmosphere. To maintain their oxygen supply, the crew was forced to depend on lithium candles, their supply of which could only last two months. These failures were then followed by a shutdown for several days of the station's attitude control system because of a star sensor malfunction. Before they could install a replacement sensor, the cosmonauts had to power down the station and then use the rocket thrusters on their Soyuz TM25 spacecraft to reorient the station's solar panels to the Sun.

The next Progress freighter (**Progress M34–April 6, 1997**) carried a supply of lithium candles as well as repair equipment for patching the coolant leak in the oxygen generating system.

February 11, 1997
STS 82, Discovery Flight #22
Hubble Space Telescope, Refurbishment
U.S.A.

A 10-day mission, STS 82 was Discovery's 22nd flight. The crew was Kenneth Bowersox (STS 61–December 2, 1993; STS 73–October 20, 1995), Scott Horowitz (STS 75–February 22, 1996), Joseph Tanner (STS 66–November 3, 1994), Steven Hawley (STS 41D–August 30, 1984; STS 61C–January 12, 1986; STS 31–April 24, 1990), Gregory Harbaugh (STS 39–April 28, 1991; STS 54–January 13, 1993; STS

Saturn's aurora in ultraviolet. *NASA*

71–June 27, 1995), Mark Lee (STS 30–May 4, 1989; STS 47–September 12, 1992; STS 64–September 9, 1994), and Steven Smith (STS 68–September 30, 1994; STS 103–December 19, 1999). This flight was the second maintenance mission to the Hubble Space Telescope.

• *Hubble Space Telescope* First launched on STS 31 (**April 24, 1990**) and repaired on STS 61 (**December 2, 1993**), Hubble was retrieved on February 13. Steve Hawley used the robot arm to maneuver the space telescope into the shuttle's cargo bay. In teams of two, astronauts Mark Lee, Steven Smith, Greg Harbaugh, and Joe Tanner replaced two of Hubble's scientific instruments, installing a new spectrograph and a more sensitive infrared camera.

The astronauts also replaced one of the telescope's guidance sensors, installed a new computer and data recorder, and covered several areas of the spacecraft with new insulation. The insulation covering was an added repair, improvised when the astronauts spotted serious damage and rips in Hubble's original insulation.

The new infrared camera, dubbed NICMOS (for Near-Infrared and Multi-Object Spectrometer), extended Hubble's eyesight farther into the infrared (covering from 8,000 to 25,000 angstroms). Because the supercold cryogenic nitrogen dewars (needed to keep the infrared sensors at their necessary operating temperature of 58°K or –355°F) started losing nitrogen faster than expected, however, the instrument's life expectancy was reduced from five to less than two years. Hence, much of Hubble's research over the next few years centered on maximizing the results from NICMOS before it failed *(see color section, Figure 3)*. This in turn reduced the amount of observations performed by the Wide Field and Faint-Object cameras, both of which operated in the visible part of the spectrum.

Despite fewer optical observations, Hubble was almost immediately involved in the first optical identification of a gamma-ray burst, GRB 970228, spotted by **BeppoSax (April 30, 1996)**. Since then, Hubble has made extension observations of Comet Hale-Bopp, the Crab Nebula, and a host of other objects in space, from nearby planets to the most distant galaxies. Photographs of Saturn, Neptune, and Uranus continued mapping their weather and ring systems. A second deep field observation in the southern hemisphere was obtained, as well as infrared observations of the northern hemisphere deep field.

Hubble was designed to operate through 2010, with the next shuttle servicing mission originally planned for June 2000. Because of failing gyroscopes, however, an earlier shuttle repair mission, **STS 103 (December 9, 1999)**, took place.

February 12, 1997
Halca
Japan

The Highly Advanced Laboratory for Communications and Astronomy (Halca) was the first radio telescope placed in orbit since **Prognoz 9 (July 1, 1983)**. Its 26-foot-diameter antenna dish unfurled on February 28th. In addition to its own observations, Halca sometimes used interferometry to combine its data with 40 earthbound radio telescopes in order to increase the resolution of images.

Halca's telescope observes wavelengths at 1.3 cm (22 GHz), 6 cm (5 GHz), and 18 cm (1.6 GHz), snapping radio images of distant active galactic nuclei, blazars, radio galaxies, supermassive black holes, and quasars to find out why these various intergalactic objects emit so many different types of radiation at many different levels of energy. Present theory postulates that all these emissions might actually be caused by the same phenomenon. Because of each object's different orientation to our line of sight—some are angled with their disk edge-on, while others are turned so that they are face-on with their polar jets pointing directly at us—we see different radiation emissions.

As of December 1999, Halca is still in operation.

February 14, 1997
Gonets-D1 1 through Gonets-D1 3
Russia

The three satellites Gonets-D1 1, Gonets-D1 2, and Gonets-D1 3, a prototype of which was flown on **Informator 1 (January 29, 1991)**, were the first of a planned cluster of 12 commercial communications satellites. As of December 1999, however, no additional satellites have been launched.

March 1, 1997
Intelsat 801
International / Europe

The continuing increase in demand for international telephone and television service forced Intelsat (see **Early Bird–April 6, 1965**) to supplement its Intelsat 7 satellites (see **Intelsat 7 F1–October 22, 1993**) with a new generation of satellites, including Intelsat 801.

Intelsat 8 satellites, though similar to the Intelsat 7 satellites, had a slightly larger capacity, between 22,000 to 112,500 two-way voice circuits per satellite, depending on configuration. As of December 1999, Intelsat has launched six Intelsat 8 satellites, placing two over the Americas, with others over Africa, the Pacific, Atlantic, and Indian Oceans. Their maximum capacity of about 700,000 phone calls at any one time was a 3,000-fold increase in capacity from Intelsat's first satellite, **Early Bird (April 6, 1965)**. In those three decades, Intelsat itself had also grown, from a 23-nation consortium to more than 140 nations, with 27 satellites still in operation as of March 1999.

March 4, 1997
Zeya
Russia

Designed by students at the Mozhaisky military space engineering academy, Zeya was a geodetic and navigational research satellite performing military cartography. It was launched on the first flight of a new Russian rocket from a new Russian launch site at Svobodny, Russia. This launch site was intended as a back-up to the leased Baikonur launch facilities located in the now independent Kazakhstan.

March 8, 1997
Tempo 2
U.S.A.

Built by Loral Space Systems and launched in the United States for TCI Satellite Entertainment, this geosynchronous communications satellite was stationed over North America, providing direct broadcast television services to the United States and Canada for the Primestar satellite network. It was the first of a planned two-satellite cluster. By 1998, the network had more than 2 million subscribers, which was subsequently purchased by DirecTV, combining the two direct broadcast television satellite clusters (see **DBS 1–December 18, 1993**).

Tempo 2's high-power output, however, created electrical problems not seen previously. As the satellite traveled through the earth's magnetic field, the static electrical charge (see **Explorer 8–November 3, 1980**) that surrounded it interacted with the satellite's high current to burn holes in its insulation, causing short-circuits that eventually forced Tempo 2 to operate with 15 percent less power than expected. The solution was to keep solar cell voltage on future satellites below 60 volts.

March 31, 1997
Pioneer 10, Shutdown
U.S.A.

After 25 years of operation, routine telemetry and ground control with Pioneer 10 was terminated on March 31, 1997. Only one of 11 instruments still functioned, and its electrical power supply was almost gone. Further contacts with the spacecraft involved only training exercises or research into deep-space communications. Pioneer 10 at that moment was 6.7 billion miles from Earth, traveling at 28,000 miles per hour. In two million years, it will reach the red giant Aldeberan in the constellation of Taurus.

Pioneer 10, launched on March 3, 1972, was the first spacecraft to reach Jupiter, on December 3, 1973 (see **Pioneer 10–March 3, 1972; Pioneer 10, Jupiter Flyby–December 3, 1973**). It then moved outward, working in conjunction with **Pioneer 11 (April 6, 1973)**, **Voyager 1 (March 5, 1979)**, and **Voyager 2 (July 9, 1979)** to send back data of the Sun's solar wind and the outer reaches of the solar system. While Pioneer 11 headed out in the same direction the Sun travels through the galaxy, Pioneer 10 traveled in the opposite direction, moving into the Sun's wake. Meanwhile, Viking 1 and Viking 2 moved upstream, with Viking 2 moving off the ecliptic into the solar system's southern latitudes.

Pioneer 11 was shut down on September 30, 1995. At the time, it was 4 billion miles from Earth, heading for a flyby of the star Lambda Aquila in the constellation Aquila in a little less than four million years (see **Pioneer 11, Shutdown–September 30, 1995**).

Voyager 1 and Voyager 2, however, continue to operate, beaming back data. As of December 1999, they were about 7.2 billion and 5.6 billion miles from Earth, respectively. With power enough to last until about 2020, scientists hope that they will survive long enough to reach the heliopause, the boundary between the Sun's magnetosphere and interstellar space. If they do, they will send back the first direct data from regions beyond our solar system.

April 4, 1997
STS 83, Columbia Flight #22
U.S.A.

STS 83, the 22nd flight of Columbia, lasted only four days; it was the shortest shuttle flight in 12 years, since **STS 51C (January 24, 1985)**. Because of the failure of one of the spacecraft's three fuel cells, the flight had to be cut short for safety reasons.

The mission had been planned as a long microgravity research flight. Its crew was James Halsell (**STS 65–July 8, 1994; STS 74–November 12, 1995; STS 94–July 1, 1997**), Susan Still (**STS 94–July 1, 1997**), Janice Voss (**STS 57–June 21, 1993; STS 63–February 3, 1995; STS 94–July 1, 1997**), Michael Gernhardt (**STS 69–September 7, 1995; STS 94–July 1, 1997**), Donald Thomas (**STS 65–July 8, 1994; STS 70–July 13, 1995; STS 94–July 1, 1997**), Roger Crouch (**STS 94–July 1, 1997**), and Greg Lineris (**STS 94–July 1, 1997**).

After returning to Earth, the fuel cell was repaired, and the entire crew and spacecraft were rescheduled and reflown as STS 94 (**July 1, 1997**).

April 6, 1997
Progress M34
Russia / U.S.A.

Docking with **Mir (February 20, 1986)** on April 8th, the unmanned freighter Progress M34 (see **Progress M1–August 23, 1989**) carried several tons of supplies, including three fire extinguishers to replace those used to put out the February 24th fire (see **Soyuz TM25–February 10, 1997**), repair equipment to seal the leak in one of Mir's oxygen generators, about 50 oxygen-generating lithium candles, and a pair of new spacesuits for a forthcoming space walk.

During the next five weeks, leading up to the arrival of **STS 84 (May 15, 1997)**, Vasily Tsibliyev and Alexander Lazutkin of Soyuz TM25 and Jerry Linenger of **STS 81 (January 12, 1997)** worked to repair Mir's two oxygen systems. While they managed to get one operating, they were unable to find the coolant leaks in the second, preventing its use. To permanently solve the problem, a replacement unit was shipped to the United States and loaded into the cargo bay of Atlantis to be brought to Mir on STS 84.

On April 29th, Tsibliyev and Linenger performed a five-hour space walk to retrieve some experiments from the outside of Mir while also testing a new Russian spacesuit.

See STS 84 for later events during the Mir occupancy of Soyuz TM25.

April 16, 1997
BSAT 1A
Japan / U.S.A.

Built by Hughes Space and Communications for the Broadcast Satellite System Corporation of Japan, BSAT 1A was the first of two BSAT satellites providing direct broadcasting services to Japan from geosynchronous orbit. This satellite was positioned west of Japan, while the second was placed nearby to act as a back-up.

April 21, 1997
Minisat 1 / Celestis 1
Spain / U.S.A.

The two satellites Minisat 1 and Celestis 1 were launched by Orbital Sciences's Pegasus XL rocket (see **Pegsat–April 5, 1990**). Since the rocket took off from a runway on the Spanish island of Gran Canaria, these spacecraft were the first ever launched from European soil.

- *Minisat 1* A Spanish satellite, Minisat 1 was a technology test of a satellite design that Spain hoped to use on future commercial and governmental missions. Besides demonstrating the satellite's design, Minisat 1 carried an ultraviolet radiation sensor, a prototype gamma-ray telescope, and a materials processing experiment for monitoring fluid behavior in space.

- *Celestis 1* A private commercial capsule, Celestis 1 performed the first "burial" in space. The capsule carried into orbit the cremated remains of 24 individuals, including Gene Roddenberry (creator of Star Trek) and counterculture advocate Timothy Leary. For these "burials," Celestis, Inc., charged $4,800 per person. Its next launch was **Celestis 2 (February 10, 1998)**.

May 5, 1997
Iridium 4 through Iridium 8
U.S.A.

The five satellites Iridium 4, 5, 6, 7, and 8 were the first launch in Iridium's cluster of 66 low-Earth-orbiting satellites, offering worldwide communications from a handheld telephone. The satellites were built by Motorola, which owned one-quarter of Iridium, and used technology first tested as far back as 1965 (**LES 2–May 6, 1965**). See **Microsat 1 through Microsat 7 (July 17, 1991)** for more recent tests.

Through December 1999, Iridium completed 19 launches from facilities in the United States, Russia, and China, with each launch placing between two and seven satellites into orbit for a total of 83 satellites. Because eight satellites failed after launch, this higher total gave the company seven spare satellites should further failures take place.

The system formally began commercial operation in November 1998, charging from $2 to $8 per minute for phone calls from anywhere to anywhere in the world. Because of poor marketing and high prices, however, Iridium did not attract the number of customers it had expected, and on August 13, 1999, the company was forced to declare bankruptcy. Its future as of December 1999 remains unknown.

Despite this failure, Iridium ushered in a second wave of private commercial space development, following the initial wave of geosynchronous communications satellites (see **ASC 1 on STS 51I–August 27, 1985**). Additional low-Earth-orbit systems, such as Orbcomm's (see **Orbcomm FM2 through Orbcomm FM12–December 23, 1997**) and Globalstar's (see **Globalstar 1 through Globalstar 4–February 14, 1998**) quickly followed. Because these systems need to launch many satellites in a short time, they have also stimulated the development of lower cost and even reusable rockets, expected to start operating in the first years of the twenty-first century.

May 15, 1997
STS 84, Atlantis Flight #19
U.S.A. / Russia / France

STS 84 was Atlantis's nineteenth space flight. The nine-day mission performed the sixth docking to **Mir (February 20, 1986)** and was crewed by Charles Precourt (STS 55–April 26, 1993; STS 71–June 27, 1995; STS 91–June 2, 1998), Eileen Collins (STS 63–February 3, 1995, STS 103–December 11, 1999), Frenchman Jean-Francois Clervoy (STS 66–November 3, 1994; STS 103–December 19, 1999), Carlos Noriega, Edward Lu, Michael Foale (STS 45–**March 24, 1992**; STS 56–April 8, 1993; STS 63–February 3, 1995; STS 103–December 19, 1999), and Russian Elena Kondakova (**Soyuz TM20–October 3, 1994**). Foale replaced Jerry Linenger as the American on Mir. Linenger had been stationed there since **STS 81 (January 12, 1997)**, for a total of 132 days. With him on the station were Vasily Tsibliyev and Alexander Lazutkin of **Soyuz TM25 (February 10, 1997)**.

The Mir docking took place on May 17th. The joint mission lasted five days, during which the 10 crewmen transferred more than 5,000 pounds of supplies from Atlantis to the orbiting laboratory, including a 250-pound oxygen-generating electrolysis unit to replace one of Mir's two primary oxygen systems. The returned equipment included the remains of the disassembled lithium candle that had burned in the February 24th fire (see Soyuz TM25), as well as the failed oxygen generator.

The cargo transfer also included frozen blood and urine samples from experiments performed by Reinhold Ewald at the beginning of Soyuz TM25, as well as the replacement of several hundred protein crystal samples. STS 84 also conducted several onboard experiments during its nine-day flight, including a protein crystal growth experiment (see **STS 63–February 3, 1995**) that grew 15 different proteins for later study on Earth. Other experiments studied bone loss, the effect of weightlessness on yeast, sea urchin sperm, and lentil roots. The plant experiments attempted to find out why the presence of starch in plants affected their growth in weightlessness.

On May 24th, Atlantis returned to Earth. Linenger found that after 4.5 months in space, he had no trouble readapting to gravity and was able to walk off the shuttle. The next shuttle mission to Mir was **STS 86 (September 25, 1997)**. The next shuttle mission was **STS 94 (July 1, 1994)**.

On Mir, Tsibliyev and Lazutkin, along with Foale, continued their troubled residency. On June 24th, Tsibliyev attempted another test of the new Toru docking system (see Soyuz TM25) used by Progress freighters (see **Progress M1–August 23, 1989**). He undocked **Progress M34 (April 6, 1997)** and, as had happened in the first test in March, he had serious problems—this time leading to a collision. Ground control had learned that the malfunction of the television camera was because of interference from the radar system of the older Kurs docking system. To prevent this interference, mission control decided to turn off the radar, forcing Tsibliyev to judge the freighter's speed and distance by eye and with a hand-held laser rangefinder. The result was that Tsibliyev could not correctly judge the freighter's distance, and it collided with Mir. The freighter hit the solar panels on **Spektr (May 20, 1995)**, bounced off, and then impacted the module itself.

The collision caused a serious air leak in Spektr, forcing the three men to scramble to seal the module (which contained all of Foale's personal items and most of his experiments) in order to save the station. That entailed unhooking all the electrical wires that ran from Spektr's solar panels through the module's hatch to Mir's storage batteries, thereby cutting off half of Mir's power supply.

To conserve power, the crew shut down air conditioning and most of the scientific experiments onboard. Regaining this power required the construction on the ground of a special hatch for Spektr, with plugs on both sides that would allow the cosmonauts to reconnect the electrical cables to the rest of Mir's systems while still keeping the hatch closed. This new hatch was launched on **Progress M35 (July 5, 1997)**.

May 21, 1997
Thor 2
U.S.A.

Owned by Norway's state-owned telecommunications company Telenor, the geosynchronous communications satellite Thor 2 provided expanded television service to Scandinavia, supplementing a first satellite that had been purchased from British Sky Television (see **Marco Polo 1–August 27, 1989**). In June 1998, Telenor launched another Thor satellite, giving it three operating geosynchronous satellites with one stationed over the Atlantic and two over Europe.

June 10, 1997
Feng Yun 2B
China

Feng Yun 2B was China's third weather satellite (see **Feng Yun 1A–September 6, 1988**) and its first geosynchronous meteorological satellite. It was positioned over China, where it produced global multispectral weather images in both infrared and visible light. It operated until April 1998, when its communications antenna failed.

Feng Yun 3 (**May 10, 1999**) was China's next weather satellite.

June 27, 1997
NEAR, Mathilde Flyby
U.S.A.

NEAR (Near Earth Asteroid Rendezvous) was launched on February 17, 1996, with a mission to complete the first prolonged study of an asteroid, 433 Eros (see **NEAR–February 17, 1996**). Getting to Eros, however, required a round-about route out beyond Mars and then past Earth. It was a course that also took the spacecraft to within 750 miles of the asteroid 253 Mathilde.

First discovered on November 12, 1885, Earth-based studies of Mathilde had revealed an irregular-shaped object with an average diameter about 38 miles across. Changes in its albedo indicated that the asteroid had an extremely slow rotation rate, 17 days and 10 hours long.

Mathilde's reflective spectrum placed it in the C class of asteroids. C asteroids are very dark, blacker than coal, and they dominate the asteroid belt's outer half. Also called chondrites, they are very primitive objects and have remained unchanged since their molecules first clumped together out of the solar system's original nebula dust. While it was once thought C asteroids were made of carbon-based clays and organic compounds, this is not known for sure. Their sable color might come from other minerals, such as iron compounds like magnetite and ilmenite.

NEAR's flyby revealed an asteroid that was smaller than predicted, with an average diameter of only about 33 miles. More surprising, however, was the number of craters seen on this tiny object. Though scientists had expected to see a nonspherical irregular body, nothing prepared them for the number of pockmarked craters that dotted its surface. At least five craters larger than 12 miles across were seen.

The flyby yielded additional information. Mathilde is almost pitch black in color, twice as black as a lump of coal and reflecting less than 3 percent of the light that hits it. The imaged hemisphere is also quite uniform in color. Unlike the asteroids Gaspra and Ida (photographed by Galileo on October 29, 1991, and August 28, 1993, respectively), there are no discolorations along crater walls, rims, or floors (see **Galileo, Gaspra Flyby–October 29, 1991** and **Galileo, Ida Flyby–August 28, 1993**). The surface and interior of Mathilde are apparently made up of the same ebony-black material. And the solar wind seemingly has not altered the asteroid's surface as it blows against it.

By analyzing how Mathilde's tiny gravitational field changed NEAR's course through space, astronomers estimated the asteroid's mass to be about 110 trillion tons. This equaled a density of about 1.1–1.5 grams per cubic centimeter, or slightly denser than ice. Since most chondritic material should have a density twice this, the asteroid must either be very porous, with many internal cavities and spaces, or a gravel pile of primordial material only loosely held together.

The most astonishing discovery, however, was how older craters seemed undamaged by subsequent impacts. The expectation, based on what was seen on Gaspra and Ida (as well as in **Mariner 9, Mars Orbit–November 14, 1971** photos of the Martian moons Phobos and Deimos), was that on such a small surface, large impacts would shatter the surrounding terrain, causing cracks, fractures, and grooves. These, in turn, would act to erase earlier craters.

With Mathilde, this was not the case. Collisions had gouged deep craters in its surface, but nearby older craters were hardly affected. Somehow, the shock of impact was so localized that the loosely packed material could not shake apart adjacent geological features. Mathilde is not unlike a soft crunchy glob of partly mashed potatoes, holding its shape forever, easily molded when touched, but changing only in the places where it is poked.

Leaving Mathilde behind, NEAR flew past Earth on January 22, 1998, speeding within 335 miles at about 20,000 miles per hour and becoming the first interplanetary spacecraft visible to the naked eye. As the spacecraft whisked by, scientists calibrated the spacecraft's equipment by aiming its instruments homeward, taking global color pictures of the earth.

NEAR's Earth flyby modified its orbit, as planned, sending it toward its last destination, the asteroid 433 Eros. NEAR's original flight plan had it entering orbit around Eros on January 10, 1999. An aborted engine burn on December 20, 1998, however, made this maneuver impossible. The spacecraft instead did a flyby of Eros on December 23, 1998 (see **NEAR, Eros Flyby–December 23, 1998**), with plans to enter orbit around the asteroid around February 2000.

During its flight between the asteroids, NEAR's x-ray–gamma-ray spectrometer joined the deep-space network of satellites that triangulated the positions of gamma-ray bursts. Working in conjunction with gamma-ray burst detectors on **BeppoSax (April 30, 1996)** and **Ulysses, First Solar Orbit (April 17, 1998)**, astronomers were able to triangulate the

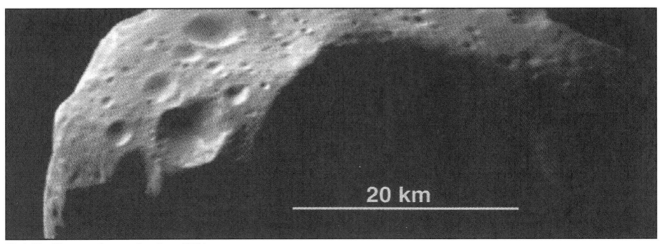

Mathilde. Close-up of cratered surface. *NASA*

July 1, 1997

location of GRB 970228. The **Hubble Space Telescope (February 11, 1997)** used these data to catch the optical afterglow of the explosion, the first such optical counterpart of a gamma-ray burst ever identified.

July 1, 1997
STS 94, Columbia Flight #23
U.S.A.

STS 94, the 23rd flight of Columbia, was the reflight of the aborted **STS 83 (April 4, 1997)** mission. The flight lasted almost 16 days and was crewed by James Halsell (**STS 65–July 8, 1994; STS 74–November 12, 1995; STS 83–April 4, 1997**), Susan Still (**STS 83–April 4, 1997**), Janice Voss (**STS 57–June 21, 1993; STS 63–February 3, 1995; STS 83–April 4, 1997**), Michael Gernhardt (**STS 69–September 7, 1995; STS 83–April 4, 1997**), Donald Thomas (**STS 65–July 8, 1994; STS 70–July 13, 1995; STS 83–April 4, 1997**), Roger Crouch (**STS 83–April 4, 1997**), and Greg Lineris (**STS 83–April 4, 1997**). This was the same crew that flew STS 83.

Columbia's cargo bay was outfitted with Spacelab, which was configured as a microgravity science laboratory for conducting 33 different experiments: three different protein crystal growth experiments (see **STS 63–February 3, 1995**) were conducted, growing more than 700 crystal samples. One experiment successfully grew crystals of a virus that causes pneumonia and upper respiratory infections in children. Scientists used these large crystals to map the virus's structure—information that is used to develop antibodies to fight the diseases.

Research into the nature of fire and flames in weightlessness continued (see **Priroda–April 23, 1996**), including the interaction of soot and flames and the structure of flame balls—the shape that flames take in microgravity. The tests included more than 200 different controlled fires.

The remaining experiments focused on materials research, using one furnace and two other test facilities to study how a variety of metals melted, mixed, and solidified in weightlessness. Alloys of lead-tin-telluride were examined. Other experiments continued work from **STS 78 (June 20, 1996)** into the behavior of bubbles and particles in fluids as they solidify. Electromagnetic levitation was used to control their motion and position.

The next shuttle mission was **STS 85 (August 7, 1997)**.

July 4, 1997
Mars Pathfinder
U.S.A.

Built by NASA and launched on December 4, 1996, Mars Pathfinder was the first spacecraft to land on Mars since **Viking 2 (August 7, 1976)**. It was also the first planetary probe to include a separate roving robot probe since the Soviet Union's **Luna 21 (January 8, 1973)**.

Pathfinder's landing technique was similar to that used by the early Soviet lunar landers (see **Luna 9–January 31, 1966**)—Pathfinder had no retro-rockets and was instead protected by giant airbags, on which it bounced numerous times during its landing. Once settled on the surface, at an outflow channel from the gigantic Valles Marineris called Ares Valles,

Sojourner takes spectroscopy of the rock dubbed "Yogi." *NASA*

Sojourner awaiting deployment. Note the two mountains on the horizon. *NASA*

which had been **Viking 1's (June 19, 1976)** original landing site, Pathfinder released a small robot rover, dubbed Sojourner. This rover was used to take photographs and to conduct a spectroscopic survey of a variety of rocks surrounding the landing site. Both rover and lander operated through September 1997.

Both probes found Ares Valles to be a place where gigantic floods had scoured the terrain between one and three billion years ago. Since then, little had changed. The data also suggested that Mars had been a wet planet between 3 and 4.5 billion years ago. This water had been the transport agent that had deposited many of the rocks to Pathfinder's landing site.

Sojourner's spectroscopic analysis of rocks found them to be similar to the silicon-rich volcanic Earth rock called andesite, formed in regions of tectonic plate activity. This composition was quite different from the Martian meteorites found on Earth. The andesite-like rock was unexpected because geologists did not think that there had been plate tectonic activity on Mars. As of December 1999, the rocks' composition remains unexplained.

Analysis of Pathfinder's images revealed atmospheric dust devils, plumes of whirling dust skittering past the landing site. On one day, five different dust devils were identified. These plumes were thought to be the initiators of the larger dust storms, such as those seen by the orbiting satellites **Mariner 9 (November 14, 1971)**, Viking 1, and Viking 2. The next American mission to arrive at Mars was **Mars Global Surveyor (September 12, 1997)**.

July 5, 1997
Progress M35
Russia / U.S.A.

Progress M35 was the second Progress freighter (see **Progress M1–August 23, 1989**) to resupply **Mir (February 20, 1986)** during the tour of duty of Vasily Tsibliyev and Alexander Lazutkin of **Soyuz TM25 (February 10, 1997)**.

Progress M35 brought two tons of supplies, as well as a specially constructed hatch for the **Spektr (May 20, 1995)** module. Because of the June 24th collision between Mir and **Progress M34 (April 6, 1997)**, Spektr had been depressurized and sealed. The electrical cables that ran from its solar panels to Mir's batteries were disconnected, thus depriving the station of about 50 percent of its electrical power. See **STS 84 (May 15, 1997)** for details about the collision.

The new hatch carried plugs on both sides. During a planned internal space walk, Tsibliyev and Lazutkin would seal off the station's multidocking adapter from the rest of Mir, depressurize it, replace Spektr's hatch, and connect cables on both sides. While they did this, Michael Foale of STS 84 would stand-by in Soyuz TM25. This work would restore about 30 percent of the station's power.

Soon after the arrival of Progress M35, however, Tsibliyev developed a heart murmur. Then, on July 16th, Lazutin accidentally unplugged Mir's guidance computer, causing the entire complex to drift in space, with its remaining solar panels no longer oriented toward the Sun. After shutting off almost all the station's power, the men replugged the lines and then used thrusters on their Soyuz TM25 spacecraft to maintain Mir's orientation until its batteries could recharge.

Because of these problems, mission control in Moscow decided to have the internal space walk performed by the crew of **Soyuz TM26 (August 5, 1997)**.

August 1, 1997
SeaWiFS (OrbView 2)
U.S.A.

Built and launched into a sun-synchronous orbit (see **Nimbus 1–August 28, 1964**) by Orbital Sciences, this NASA satellite carried what was dubbed the SEA-viewing WIde Field Sensor (SeaWiFS). The sensor was able to produce a global map every 48 hours of the earth's biological activity, both on land and sea. The sensor scanned in one optical and seven infrared wavebands, with a resolution as fine as 0.7 miles. Able to detect changing levels of chlorophyll, as well as subtle differences in ocean color, SeaWiFS was the first such sensor since the deactivation of the Coastal Zone Color Scanner on **Nimbus 7 (October 13, 1978)** in 1986.

Still in operation as of December 1999, data from SeaWiFS are used to monitor the effects of global weather patterns on plant and animal life. Commercial fishing operations are also able to buy the imagery to locate large fish schools and to identify toxic algae blooms known as red tides.

August 5, 1997
Soyuz TM26
Russia / U.S.A.

Crewed by Anatoly Solovyov (**Soyuz TM5–June 7, 1988; Soyuz TM9–February 11, 1990; Soyuz TM15–July 27, 1992; STS 71–June 27, 1995**) and Pavel Vinogradov, Soyuz TM26 docked with **Mir (February 20, 1986)** on August 7th. Its crew joined Vasily Tsibliyev and Alexander Lazutkin of **Soyuz TM25 (February 10, 1997)** and Michael Foale of **STS 84 (May 15, 1997)**.

After seven days of joint operations, Tsibliyev and Lazutkin returned to Earth on August 15th in Soyuz TM25. Both men faced harsh criticism for the many problems during their flight. They, in turn, laid much of the blame on faulty equipment, poor planning, and lack of support from ground control.

Meanwhile, on Mir, Solovyov and Vinogradov, with Foale as support, began repair operations. Over the next six months, the two Russian cosmonauts undertook seven internal and external space walks to repair the crippled space station. They began work on August 15th when all three men boarded Soyuz TM26 and flew around the station, videotaping it to assess its condition. On August 22, Solovyov and Vinogradov did an internal space walk, surveying the crippled Spektr's interior and retrieving some of Foale's equipment and belongings. On September 6th, Soloyvov and Foale did a six-hour EVA (extra-vehicular activity) to inspect the outside of Spektr, cutting away some insulation and manually realigning two solar panels so that they were better aimed toward the Sun.

During these operations, the station had a near collision with an abandoned satellite. On September 17, 1997, **MSTI 2 (May 9, 1994)** zipped within 1,500 feet of Mir.

With these preliminaries complete, the three men prepared for the arrival of **STS 86 (September 25, 1997)**.

August 7, 1997
STS 85, Discovery Flight #23
Crista-Spas 2
U.S.A. / Canada

STS 85, the shuttle Discovery's 23rd flight, lasted 12 days and was crewed by Curtis Brown (STS 47–September 12, 1992; STS 66–November 3, 1994; STS 77–May 19, 1996; STS 95–October 29, 1998; STS 103–December 19, 1999), Kent Rominger (STS 73–October 20, 1995; STS 80–November 19, 1996), Jan Davis (STS 47–September 12, 1992; STS 60–February 3, 1994), Stephen Robinson (STS 95–October 29, 1998), Robert Curbeam, and Canadian Bjarni Tryggvason. This space science mission deployed and retrieved one satellite.

• *Crista-Spas 2* Flown previously on STS 66 (**November 3, 1994**), Crista-Spas 2 studied Earth's middle atmosphere. It was deployed August 7th and recaptured August 16th, gathering data so that scientists could develop better models of atmospheric evolution. Among its results was the discovery of surprisingly high levels of water vapor in the upper atmosphere above the northern latitudes. These data supported the theory that small comets of water were constantly bombarding the earth, first indicated by results from **Dynamics Explorer 1/Dynamics Explorer 2 (August 3, 1981)** and further supported by **Polar (February 26, 1996)**.

Discovery's cargo bay also carried several other instruments. A new kind of infrared sensor was tested. Developed originally for military night-vision glasses, the Infrared Spectral Imaging Radiometer (ISIR) worked at room temperature, requiring no complicated supercold dewars to keep the sensor at temperatures below −350°F. The need for this refrigeration was the prime reason older infrared telescopes such as **IRAS (January 26, 1983)**, **ISO (November 17, 1995)**, and Nicmos on the Hubble Space Telescope (see **STS 82–February 11, 1997**) had such short operating life spans.

Another sensor measured the Sun's solar constant, calibrating its measurements with **Soho (December 2, 1995)**, while a third used laser altimetry to measure cloud heights and vegetation cover in a variety of remote regions.

Also onboard was a set of ultraviolet telescopes for observing astronomical targets, including globular clusters, supernovae remnants, and the magnetic torus surrounding Jupiter's moon Io.

The astronauts also conducted tests of a Japanese-built robot arm intended for use on the Japanese module of the International Space Station (see **Zarya–November 20, 1998**).

Onboard research continued the shuttle program's aggressive work in growing protein crystals for medical research (see **STS 63–February 3, 1995**), growing 63 protein crystal samples for later study on Earth.

The next shuttle mission, **STS 86 (September 25, 1997)**, docked with Mir. The next shuttle science mission was **STS 87 (November 19, 1997)**.

August 19, 1997
Agila 2
Philippines / China / U.S.A.

Built by Space Systems/Loral, launched on a Chinese rocket, and owned by the Mabuhay Philippines Satellite Corp., the geosynchronous communications satellite Agila 2 was the first launched to provide television and telephone service to the Philippines. Agila 1 had been a Russian-built Gorizont satellite (see **Gorizont 1–December 19, 1978**) that had been leased from Russia for two years for $9.5 million and moved to a position over the Philippines. Agila 2, also placed over the Philippines, supplemented and replaced this satellite. It carried 30 transponders in the 4/7 GHz wavebands and 24 in the 11/14 GHz wavebands.

Because of the increasingly crowded sky at geosynchronous orbit, Agila 2 had serious frequency conflicts with one of Japan's Superbird satellites (see **Superbird A–June 5, 1989**) and therefore has never been able to operate at full capacity.

August 25, 1997
ACE
U.S.A.

Built by NASA and positioned at the Lagrangian L1 point (see **ISEE 3–August 12, 1978**), the Advanced Composition Explorer (ACE) supplemented the solar wind monitoring of **Wind (November 1, 1994)**. ACE, however, was more sensitive, using 10 different instruments to detect solar ultraviolet, x-ray, and gamma-ray emissions, from 1 eV to 600 MeV.

The spacecraft monitored the speed, composition, and temperature of the solar wind and corona. It also observed solar activity on the Sun's surface, including solar flares and coronal mass ejections, and the effects of that activity on Earth's magnetosphere. These instruments provided scientists with advance warning of geomagnetic storms.

ACE also measured cosmic radiation hitting Earth's magnetosphere, discovering the first evidence of copper and zinc coming from galactic cosmic rays. ACE also detected rare isotopes of neon at higher levels than those found in the solar system, indicating that the interstellar medium was unlike that of Earth's, enriched by very hot and bright stars like Wolf-Rayet stars.

As of December 1999, ACE continues to operate, with a hoped-for service life of another four years.

August 28, 1997
Pan American Satellite 5
U.S.A.

The Pan American Satellite 5, part of Pan American Satellite's cluster of geosynchronous communications satellites (see **Pan American Satellite 1–June 15, 1988**), was unique in that it was the first private commercial spacecraft to use a xenon ion engine to maintain is attitude and position in space, replacing the heavier chemical thrusters used previously. This design, an advance from the ion engines used on **Sert 2 (February 4, 1970)** and AT&T's **Telstar 401 (December 16, 1993)**, was similar in design to the engine already in use on the Russian Gals and Express satellites (see **Gals 1–January 20, 1994** and **Express 1–October 13, 1994**). **Deep Space 1 (October 24, 1998)** later used this xenon engine to leave Earth orbit and rendezvous with an asteroid.

August 29, 1997
FORTE
U.S.A.

Launched on an Orbital Sciences Pegasus XL rocket, the Fast On-orbit Recording of Transient Events (FORTE) satellite conducted U.S. Air Force surveillance in conjunction with the U.S.-Russian START treaty, scanning for evidence of illegal nuclear tests.

September 12, 1997
Mars Global Surveyor
U.S.A.

Built by NASA and launched on November 7, 1996, the Mars Global Surveyor was the first probe launched to Mars by the United States since the unsuccessful **Mars Observer (September 25, 1992)**. (Although it arrived after **Mars Pathfinder–July 4, 1997**, it was launched first.) Arriving at Mars on September 12, 1997, it spent the next year and a half slowly adjusting its initial eccentric orbit into a sun-synchronous orbit (see **Nimbus 1–August 28, 1964**). This repositioning was done by dipping the satellite into the thin Martian atmosphere and using the satellite's solar panels like sails to brake its speed.

During this period, the satellite took about 2,000 images using its multispectral camera and a laser altimeter. A temperature spectrometer monitored variations in the planet's surface temperature, while a magnetometer searched for evidence of a Martian magnetic field.

Images recorded the entire birth and death of one of Mars's famous dust storms. This storm reached a height of 80 miles, covering an area larger than the Atlantic Ocean.

Global Surveyor's magnetometer also uncovered patches of a magnetic field over parts of the Martian surface. These small remnants indicated that, in its past, Mars had a global magnetic field and hence a molten core. The remnants also suggested the possibility that ancient Mars had plate tectonic activity similar to Earth's.

While the spacecraft confirmed evidence from **Mariner 9, Mars Orbit (November 14, 1971)**, **Viking 1 (June 19, 1976)**, and **Viking 2 (August 7, 1976)** that water had once been stable on or near the surface, Mars up close had a unique geology that was different from that of Earth. The laser altimeter found that vast areas of low elevation were incredibly flat, reminiscent of a dry ocean floor. Dried lakebeds were also identified. Remnant shorelines to ponds and oceans were seen, as well as erosional features resembling places on Earth where water seeps out from between cliff-face bedding planes.

Images also showed extensive evidence of ancient underground water drainage systems. Many valleys suggested ancient sustained liquid flow slowly cutting deep channels in the flat plains. And while earlier spacecraft had seen some evidence of a few thick layers or bedding plains below the Martian "topsoil," Global Surveyor saw many thin strata everywhere that resembled the sedimentary layers seen on Earth. Some scientists thought these layers might indicate that volcanism had been extensive during Mars's early history, which in turn might have produced the carbon dioxide necessary for making the planet warm enough for water to flow.

The photographs also revealed extensive wind-driven features, dominated by sand and sand dunes. Some craters were buried by sand; others were filled with it. And different sand dune complexes seemed composed of different materials, sorted by the wind.

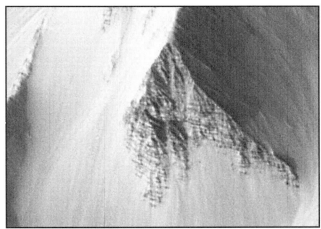

One of Valles Marineris's cliff faces. Note the many layers. *NASA*

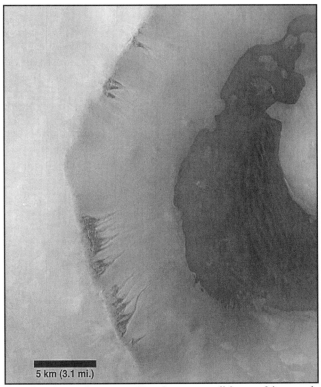

A southern hemisphere crater. Note the seepage off the top of the crater's rim. The crater floor, with sand dunes, also suggests the shoreline of a now-dried lake with islands. *NASA*

September 12, 1997

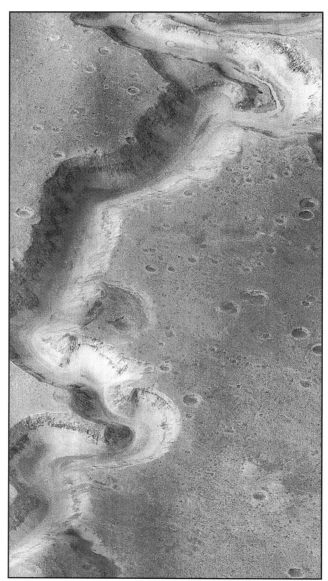

Nanedi Vallis. Features as small as 40 feet across can be seen. Near the top of the image note the central channel carved into the floor of the canyon. Also note what appears to be an abandoned meander in the middle of the image to the right of the canyon. The canyon wall also shows extensive layering. *NASA*

In March 1999, Global Surveyor finally reached its sun-synchronous orbit and began a full detailed survey of the entire surface of Mars, which continues as of December 1999. The three subsequent Martian probes launched through December 1999, **Nozomi (July 9, 1998), Mars Climate Orbiter (December 11, 1998),** and **Mars Polar Lander (January 3, 1998),** either failed or experienced problems. See those entries for details.

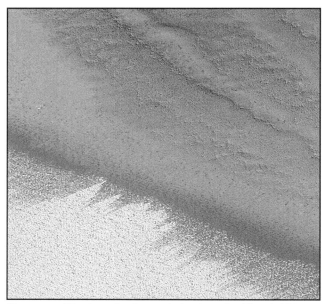

The retreat of Mars's north polar cap during the spring. The wind-blown bright material at the bottom of the page is the evaporating polar cap. *NASA*

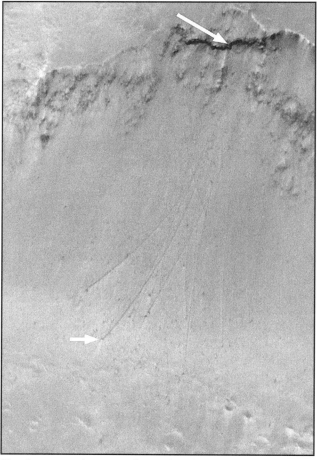

The short arrow shows one of several 60-foot-diameter boulders that apparently rolled down the slope after breaking free from the cliff face at the place indicated by the longer arrow. *NASA*

September 25, 1997
STS 86, Atlantis Flight #20
U.S.A. / Russia / France

STS 86 was Atlantis's twentieth space flight, and the seventh shuttle docking with **Mir (February 20, 1986)**. The flight lasted just under 11 days and was crewed by James Wetherbee (**STS 32–January 9, 1990; STS 52–October 22, 1992; STS 63– February 3, 1995**), Michael Bloomfield, Scott Parazynski (**STS 66–November 3, 1994; STS 95–October 29, 1998**), Wendy Lawrence (**STS 67–March 2, 1995; STS 91–June 2, 1998**), David Wolf (**STS 58–October 18, 1993**), Frenchman Jean-Loup Chretien (**Soyuz T6–June 24, 1982; Soyuz TM7–November 26, 1986**), and Russian Vladimir Titov (**Soyuz T8– April 20, 1983; Soyuz T10A–September 27, 1983; Soyuz TM4–September 21, 1987; STS 63–February 3, 1995**).

According to the original agreement between the Russians and Americans (see **STS 63–February 3, 1995**), this flight was intended to be the last shuttle mission to Mir. But because Russian cash shortages prevented them from fulfilling their part of the deal concerning the International Space Station (see **Zarya–November 20, 1998**), the United States agreed to add two more Mir-Shuttle flights (see **STS 89–January 22, 1998** and **STS 91–June 2, 1998**), and also to pay an additional $73 million to finance Mir's operation during this period. Hence, Michael Foale, who had been stationed there since STS 84 (**May 15, 1997**), was not the last American astronaut on Mir; instead, he was replaced by David Wolf.

Atlantis docked with Mir on September 27th, linking the seven-person shuttle crew with Foale from STS 84 and Anatoli Solovyov and Pavel Vinogradov of **Soyuz TM26 (August 5, 1997)**. The shuttle's cargo bay, outfitted with the 2,200-cubic-foot double Spacehab module, carried supplies of food, water, equipment, and scientific experiments weighing several thousand pounds.

A new guidance-control computer for operating the station's gyroscopes and keeping it oriented to the Sun was immediately unloaded and installed. The old computer had failed repeatedly in recent months, each time sending the station adrift and requiring the use of thruster fuel for realignment. One failure had even happened just prior to a docking.

The cargo also included more than 700 protein crystal growth samples (see STS 63). These samples remained on Mir for four months, returning in February when Soyuz TM26 returned to Earth (see **Soyuz TM27–January 29, 1998**). The shuttle also brought home a separate package of protein crystal growth samples.

On October 1st, while the two spacecraft were docked, Titov and Parazynski did a five-hour space walk, retrieving the science and technology package that had been attached to Mir's docking module by Linda Godwin and Michael Clifford during STS 76 (**March 22, 1996**). The two men also attached a specially built solar array cap to the outside of the station's docking module in anticipation of a later space walk during which Spektr's (**May 20, 1995**) damaged solar panel (see **STS 84–May 15, 1997**) would be removed and replaced with the cap, thereby sealing the module's air leak.

During the space walk, Parazynski and Titov also did more tests on the small emergency jetpack, SAFER, previously tested on **STS 64 (September 9, 1994)**.

On October 3rd, Atlantis undocked from Mir, leaving Anatoly Solovyov, Vinogradov, and Wolf behind. The shuttle then did a fly-around of Mir from a distance of about 240 feet. During this inspection tour, Solovyov pumped air into the Spektr module to help spot its leak. At one point, Titov on Atlantis and Vinogradov on Mir thought they saw particles oozing from the base of the damaged solar panels on Spektr, but this identification of the leak's location was considered inconclusive.

The shuttle then moved away, returning to Earth on October 6th after spending an extra day in orbit because cloud cover obscured the Florida landing site.

Meanwhile, on Mir, Solovyov and Vinogradov continued repair operations, with Wolf as support. In October, they did a second internal space walk lasting six hours, during which they installed the new hatch in Spektr and attached the power cables to it (see **Progress M35–July 5, 1997**). This last act restored about 30 percent of Mir's power supply. Then, in early November, they did two EVAs (extra-vehicular activities) to remove a solar array from Kvant and install the second of the two solar panels first brought to Mir by **STS 74 (November 12, 1995)**. During the first space walk, Vinogradov released **Sputnik 40 (November 3, 1997)** into orbit.

During the remainder of the Soyuz TM26 residency, the station was resupplied by two Progress freighters (see **Progress M1–August 23, 1989**), bringing food, oxygen, and water, as well as a second replacement guidance-control computer.

Although their repair effort had included plans for recovering Spektr by removing the damaged solar panel and capping the leak at its base, this operation did not take place, partly because its exact location could not be pinpointed and partly because other more urgent repairs took precedence (an airlock leak and a repair of the station's guidance-control computer).

See **STS 89 (January 22, 1998)** for later Mir events. The next shuttle mission, **STS 87 (November 19, 1997)**, was a science research flight.

The damage to Spektr. Note the twisted solar panel on the right. *NASA*

October 10, 1997
Mirka
Germany / Russia

Launched as part of a joint German-Russian Foton microgravity mission (see **Foton 6–April 11, 1990**), the German-made capsule Mirka tested experimental heat shield designs for returning samples from low-Earth orbit. After two weeks in space, Mirka was deorbited and recovered successfully in Kazakhstan.

October 15, 1997
Cassini/Huygens
U.S.A. / Europe

The double probe Cassini/Huygens, aimed at Saturn, was probably the most ambitious and complex unmanned planetary project ever attempted, costing more than $2.5 billion and involving 17 nations and hundreds of scientists from the U.S. and Europe. Like the Jupiter probe Galileo (see **STS 34–October 18, 1989**), Cassini/Huygens will use the gravitational fields of three planets to get to Saturn, looping twice past Venus, once past the Earth, and once past Jupiter on the eight-year journey outward. Over its planned lifetime, Cassini will orbit Saturn up to 60 times, sending back close-up photographs not only of Saturn's rings but of its 18 known moons.

Cassini was one of the heaviest planetary probes ever launched, weighing about 12,346 pounds fully fueled. Not only does it carry a sophisticated camera package for taking high-resolution photographs, it also includes 11 other instruments aimed at performing 19 different experiments. For example, one sensor will measure Saturn's magnetic field and its interactions with the solar wind, while another will gauge the surface and atmospheric composition of the planet, its satellites, and its rings. Other instruments will measure Saturn's gravitational field, the ice and cosmic dust in the Saturn system, and the circulation of plasma and ionized particles through the system.

Cassini also carries an entirely separate probe, Huygens. Its mission is to drop through the atmosphere of Titan, Saturn's largest satellite, and land on its surface. Huygens itself is almost as complicated a package as Cassini. Developed at a cost of approximately $500 million, it carries its own complement of six instruments. As it descends by parachute into Titan's nitrogen-methane-argon atmosphere, it will not only measure and sample its composition but will also take numerous pictures. If all goes well, scientists hope that Huygens will survive its landing on Titan's surface and send back data for as long as 90 minutes.

Because of the distance Cassini is traveling, the craft cannot depend on human help. Any command sent from Earth will take a minimum of 80 minutes to reach Saturn, much too long a response time for solving sudden emergencies. Instead, Cassini carries with it an extremely sophisticated computer system, actually able in some circumstances to react to problems on its own.

As of December 1999, the double probes have completed their first two swings past Venus, as well as an Earth flyby, and are on their way to a Jupiter flyby in 2000. Arrival at Saturn is scheduled for June 2004.

October 30, 1997
Ariane 5
Europe

The first successful launch of the European Space Agency's new Ariane 5 rocket (a first launch in June 1997 had exploded less than a minute after liftoff) placed in orbit two dummy payloads, providing ground controllers telemetry and launch data on the rocket's performance. Both payloads also carried several university experiments.

Because Ariane 5's main engine underperformed, shutting down early, one of the dummy payloads failed to reach geosynchronous orbit as planned. This failure required the European Space Agency to fly a third test of the rocket before making it available for commercial launches. See **Maqsat 3 (October 21, 1998)**.

November 3, 1997
Sputnik 40
Russia / France

Financed by donations from private and corporate sponsors and built by Russian and French ham radio operators and students, Sputnik 40 was a one-third-scale replica of **Sputnik (October 4, 1957)**. It had been delivered to **Mir (February 20, 1986)** by a Progress freighter on October 5th and was released into its own independent orbit on November 3rd by cosmonaut Pavel Vinogradov during a space walk.

Sputnik 40 operated for three months, sending out a radio beacon similar to that first broadcast 40 years earlier by the first Sputnik.

November 12, 1997, 17:00 GMT
Coupon
Russia

The geosynchronous communications satellite Coupon was owned by the Central Bank of the Russian Federation in Moscow. Built by the Russian company Lavochkin, which had in the past built Soviet planetary probes, Coupon was used for relaying financial data between Russian bank branches and its main headquarters.

November 12, 1997, 21:48 GMT
Cakrawarta 1
Indonesia / U.S.A. / Europe

Built by Orbital Sciences, launched on an Ariane rocket, and owned by Media Citra Indostar, Cakrawarta 1 was the first privately owned Indonesian geosynchronous communications satellite, following the government-sponsored Palapa satellites (see **Palapa 1–July 8, 1976**). Cakrawarta 1 provides television service to Indonesia, covering regions where service was previously unavailable. The satellite uses five transponders in the 2.6 GHz waveband, a frequency less affected by rainfall (a common occurrence in Indonesia).

November 19, 1997
STS 87, Columbia Flight #24
Spartan 201
U.S.A. / Japan / Ukraine

STS 87 was Columbia's 24th space flight. It lasted almost 16 days and was manned by Kevin Kregel (**STS 70–July 13,**

1995; STS 78–June 20, 1996), Steven Lindsey (STS 95–October 29, 1998), Kalpana Chawla, Winston Scott (STS 72–January 11, 1996), Japanese Takao Doi, and Ukrainian Leonid Kadenyuk. Kadenyuk was the first Ukrainian astronaut. The mission, mostly dedicated to science and the testing of new space technologies, released one free-flying satellite.

• *Spartan 201* Deployed on November 21st, Spartan 201 was one of three simply designed NASA spacecraft deployed for short periods from the space shuttle. See **STS 51G (June 17, 1985)**.

Spartan 201 was dedicated to solar research and had flown three previous times (see **STS 56–April 8, 1993**; **STS 64–September 9, 1994**; **STS 69–September 7, 1995**). On this flight, its data were to calibrate the data being received from **Soho (November 2, 1995)**. Because the astronauts failed to activate the spacecraft's automatic orientation system prior to deployment, however, Spartan 201 could not function. An attempt by Kalpana Chawla to recapture the satellite with the robot arm failed, sending Spartan 201 tumbling.

To save the satellite, Takao Doi and Winston Scott did a space walk on November 25th and recaptured it by hand, much as the astronauts had done on **STS 49 (May 7, 1992)**. Chawla then used the robot arm to place it in its berth for return to Earth.

This mission of Spartan 201 was reflown on **STS 95 (October 29, 1998)**.

Doi and Scott performed one other space walk on December 3rd, carrying out tests on an experimental crane that could not be done on **STS 80 (November 19, 1996)** because of a jammed airlock hatch.

On this second space walk, they also deployed and retrieved the AERCam Sprint remote-controlled camera, testing its ability to inspect and monitor the exterior of spacecraft, thereby making a human space walk unnecessary. The football-sized camera, weighing 35 pounds, flew free in the payload bay for an hour and 12 minutes, with Steve Lindsey inside Columbia using a laptop computer, television monitors, and a joystick to control it and take pictures. Inspector, a second camera design built by Daimler-Benz of Germany, was tested on December 17, 1997, from **Mir (February 20, 1986)**. See **Inspector (December 17, 1997)**.

Onboard experiments included three different furnaces to study the behavior of materials as they melted and solidified in weightlessness, including semiconductor alloys lead-tin-telluride and mercury-cadmium-telluride. Another materials experiment in a getaway special canister investigated how wet cement solidified in zero gravity.

In joint American-Ukrainian plant research, 11 different experiments studied the effect of weightlessness of plants and pollination, further confirming results from **STS 79 (September 16, 1996)**. In general, almost all the samples prospered, showing that plants could grow in space if the equipment in which they were housed maintained their water and air circulation. For instance, 65 flowers and almost 400 buds were successfully pollinated and bloomed. Of these, most survived the return to Earth and continued to grow once replanted.

For more greenhouse research, see the next shuttle mission **STS 89 (January 22, 1998)**, which docked with **Mir (February 20, 1986)**. The next shuttle science mission was STS 90 (April 17, 1998).

November 28, 1997
Kiku 7 (ETS 7) / TRMM
Japan / U.S.A.

• *Kiku 7* Also called Engineering Test Satellite (ETS) 7, Kiku 7 followed **Kiku 5 (August 27, 1987)** in testing new designs for Japanese satellites. Kiku 7 tested automatic docking techniques for eventual use in the International Space Station (see **Zarya–November 20, 1998**).

Kiku 7 was actually made of two sections, a target vehicle and a chase vehicle. In June, the two spacecraft separated by about eight inches and then redocked. In July, they separated by about seven feet and also redocked. In August, after separating by about 1,600 feet, computer problems prevented a redocking, thereby ending the tests.

Despite this failure, the two unmanned docking maneuvers made Japan only the second country, following the Soviet Union/Russia, to perform an automatic unmanned docking in space. See **Cosmos 186 (October 27, 1967)**.

• *TRMM* The Tropical Rainfall Measuring Mission (TRMM) was a NASA-built satellite designed to monitor global rainfall rates. Its sensors scanned across seven microwave frequencies from 10.7 to 85.5 GHz and infrared wavelengths from 0.3 to 50 microns.

TRMM confirmed a discovery of **OSO 2 (February 3, 1965)**—that 98 percent of all lightning occurs over land, caused by increased convection circulation over land surfaces. Its images also detailed the internal structure of large storms and hurricanes.

Other data provided scientists with global sea surface temperatures, spotting in 1998 the arrival of La Niña, a band of unusually cold temperatures over the Pacific equator region and the opposite global weather pattern to El Niño (see **TOPEX-Poseidon–August 10, 1992**).

The spacecraft still operates as of December 1999. Its data will eventually help scientists determine whether global climate change is actually taking place.

December 2, 1997
Equator S
Germany / Europe / U.S.A.

Built by the Max Planck Institute for Extraterrestrial Physics in Germany with NASA assistance, then launched on an Ariane rocket, Equator S studied the earth's magnetosphere along the equator at 40,000 miles altitude. It was part of an international coordinated effort by the United States, Japan, the European Space Agency, and Russia to study the dynamics of the solar wind with the earth's magnetosphere. Earlier satellites in the program were **Geotail (July 24, 1992)**, **Wind (November 1, 1994)**, **Interball 1 (August 2, 1995)**, and **Interball 2 (August 29, 1996)**.

The spacecraft collected data until May 1998, when its computer processor failed. **Astrid 2 (December 10, 1998)** continued auroral research.

December 17, 1997
Inspector
Germany / Russia

Brought to **Mir (February 20, 1986)** in a Progress freighter, Inspector was designed as a free-flying camera platform for inspecting the outside condition of the space station. It had been built by a German company, Daimler-Benz Aerospace, which hoped its design would be used in the construction of the International Space Station (see **Zarya–November 20, 1998**).

Unlike the American AERCam/Sprint camera tested on STS 87 (**November 19, 1997**), Inspector was much more sophisticated, four times heavier and six times more expensive to build. It was designed to operate at greater distances from a space station, offering more global views.

Immediately after deployment, one of the spacecraft's gyroscopes failed, however, preventing Inspector from orienting itself properly. Moreover, it did not respond to commands from the cosmonauts on Mir. Fearing a collision, the cosmonauts maneuvered the station away from the satellite.

Control of the spacecraft was later recovered, and during the next two months, both ground controllers and the cosmonauts on Mir tested its operation and cameras by taking pictures of Earth. Then its batteries failed, and Inspector ceased operation.

December 23, 1997
Orbcomm FM5 through Orbcomm FM12
U.S.A.

The eight satellites Orbcomm FM5–FM12 joined **Orbcomm FM1 through Orbcomm (April 3, 1995)** to begin the full implementation of Orbital Sciences corporation's 36-satellite communications cluster, providing paging and e-mail service worldwide. Orbital had previously tested the engineering of this system on **Microsat 1 through Microsat 7 (July 17, 1991)**.

This low-Earth-orbiting operational cluster followed the Iridium constellation of more than 66 satellites (see **Iridium 4 through Iridium 8–May 5, 1997**). Through December 1999, Orbital placed in low-Earth orbit 35 of the 36 satellites proposed for its cluster, essentially completing the constellation. Unlike Iridium, which used rockets from four different companies and four different countries, Orbital's cluster was launched entirely by its own Pegasus XL and Taurus rockets (see **Pegsat–April 5, 1990** and **Step 0/DarpaSat–March 13, 1994**), with one Taurus rocket placing two satellites in orbit and the Pegasus rockets launching the remainder.

Designed for message paging, tracking of packages, or e-mail messages, Orbital Sciences estimated that its constellation needed only 330,000 subscribers to turn a profit. Although the system was not yet complete, by December 1999, Orbcomm was in use in about 60 countries.

December 25, 1997
HGS 1
U.S.A. / Russia

Built by Hughes Space and Communications, the satellite HGS 1 was initially called Asiasat 3 and was owned by Asia Satellite Telecommunications (see **Asiasat 1–April 7, 1990**). When the last stage of its Russian Proton rocket malfunctioned, it failed to reach geosynchronous orbit, causing its insurance company to pay off Asia Satellite's claim and take possession.

At that point, Hughes made an unprecedented proposal, which the insurance companies accepted. At Hughes's expense, the spacecraft's attitude rockets were used to ease it into an eccentric orbit where it flew past the Moon twice so that the Moon's gravitational field was used to place the satellite in slightly inclined geosynchronous orbit. From this position, the satellite was usable to some communications customers. Hence, Hughes made some profit and the insurance company recouped some of its losses.

Thus, HGS 1 became the first satellite financed, built, owned, and launched by private commercial sources to travel to the environs of another world. For a satellite launched at the end of 1997 to have done this was especially fitting, because 1997 was the first year in the history of space exploration during which privately financed missions exceeded government-financed missions. Forty years after **Sputnik (October 4, 1957)**, private enterprise was beginning to dominate the exploration of space.

1998

January 7, 1998
Lunar Prospector
U.S.A.

Lunar Prospector was the first NASA mission to the Moon in 25 years, since **RAE 2 (June 10, 1973)**, and the first dedicated to lunar research since **Apollo 17 (December 7, 1972)**. It followed the U.S. Department of Defense mission of **Clementine (January 25, 1994)**.

The spacecraft was placed in lunar orbit to make a careful spectroscopic survey of the entire lunar surface, including its north and south poles. It soon confirmed what Clementine had found—that within some of the craters at the Moon's two poles were areas forever in shadow where about 6.6 trillion tons of probable water-ice existed, enough to sustain a human colony of several thousand people for several hundred years. In the north, the ice was scattered across a region from 3,600 to 18,000 square miles, while in the south, the area was about half that size. Because the spacecraft could only probe about two feet down, these numbers were considered conservative estimates by the scientists.

Other data indicated that the Moon had a small iron core, which in turn supported the theories that said it had been ripped free from the earth during a collision with another object. Magnetic field data confirmed that the Moon had no magnetic field, but these data also showed that remnant fields antipodal (the opposite hemisphere) to several mare basins proved that between 3.6 and 3.8 billion years ago the Moon did have a magnetic field. The remnant magnetic field opposite Mare Imbrium is so strong it actually deflects solar wind particles, forming its own small magnetosphere.

Lunar Prospector also carried a vial containing one ounce of the cremated remains of Eugene Shoemaker (see **Ranger 7–July 28, 1964**), one of the most important planetary scientists during the first three decades of planetary exploration, who had been killed in an automobile accident in July 1997.

In an effort to gather more proof that there was water at the lunar poles, Lunar Prospector was sent crashing into the lunar south pole on July 31, 1999. Scientists hoped that earthbound and orbiting instruments would detect water in the puff of debris caused by the impact. Analyses of all visual observations, however, revealed no evidence of water.

The next proposed lunar mission is Smart 1, intended to go into lunar orbit in late 2001.

January 22, 1998
STS 89, Endeavour Flight #12
U.S.A. / Russia / Kyrgyzstan

STS 89 was Endeavour's twelfth space flight. The mission lasted just under 10 days and was crewed by Terrence Wilcutt (STS 68–September 30, 1994; STS 79–September 16, 1996), Joe Edwards, James Reilly, Michael Anderson, Bonnie Dunbar (STS 61A–October 30, 1985; STS 32–January 9, 1990; STS 50–June 25, 1992; STS 71–June 27, 1995), Andrew Thomas (STS 77–May 19, 1996), and Salizhan Shakirovich Sharipov. Sharipov was from Kyrgyzstan, an independent state that was formerly part of the Soviet Union.

This flight was the eighth shuttle docking with **Mir (February 20, 1986)**, following **STS 86 (September 25, 1997)**, and the first not performed by the space shuttle Atlantis. Endeavour picked up David Wolf, who had been brought to the station on STS 86 and had completed 129 days in space, and dropped off Andrew Thomas, who joined cosmonauts Anatoly Solovyov and Pavel Vinogradov of **Soyuz TM26 (August 5, 1997)** for one short week until the arrival of the crew of **Soyuz TM27 (January 29, 1998)**. See that mission for events after Endeavour's return to Earth.

Endeavour docked with Mir on January 24th. During the next four days, more than two tons of food, water, and equipment were transferred to Mir, and almost a ton and a half of scientific equipment and experiment samples were transferred to Endeavour. In this cargo was a replacement of one of Mir's gyroscopes, as well as a package of frozen protein crystal growth samples to be grown during Andrew Thomas's four-month tour of duty. Also transferred to Mir was the Astroculture greenhouse (see **STS 50–June 25, 1992**), used to further test the ability of wheat plants to grow in space, as demonstrated by John Blaha on **STS 79 (September 16, 1996)**.

Experiments on Endeavour included an aquarium and a new design for maintaining the water and air circulation to plants when in weightlessness. Endeavour also carried a prototype design of an x-ray crystallography, used to grow and analyze protein crystals as they grew. This x-ray sensor was intended to be included on the International Space Station (see **Zarya–November 20, 1998**) and was being flown on STS 89 to evaluate its operation in space. During the Mir docking, it was transferred to the station for long-term testing in space, to be returned during the next shuttle docking mission.

Protein crystals growth experiments continued (see **STS 63–February 3, 1995**), carrying 162 samples, which were returned to Earth and used by scientists to better map their molecular structure for the development of pharmaceuticals. See Soyuz TM27 for more protein crystal research.

The behavior of granular materials, such as salt or sand, was studied. These studies were useful to building engineers, helping them to better understand how different density soils reacted during earthquakes.

The next (and last) shuttle flight to Mir was **STS 91 (June 2, 1998)**. The next shuttle flight, STS 90 (**April 17, 1998**), focused on neurological research.

January 29, 1998
Soyuz TM27
Russia / France / U.S.A.

Soyuz TM27 was crewed by Talgat Musabayev (**Soyuz TM19–July 1, 1994**), Nikolai Budarin (**STS 71–June 27, 1995**), and Leopold Eyharts. Eyharts was flown to Mir as part of a French-Russian agreement signed in 1996 (see **Soyuz TM24–August 17, 1996**). This first flight had originally been scheduled for **Soyuz TM26 (August 5, 1997)** but was delayed because of the many problems during **Soyuz TM25 (February 10, 1997)**.

Soyuz TM27 docked with Mir on January 31st, three days after the departure of **STS 89 (January 22, 1998)**, to join Anatoly Solovyov and Pavel Vinogradov of Soyuz TM26 and Andrew Thomas of STS 89 for three weeks of joint operations. Then, on February 19th, the crew of Soyuz TM26, along with Eyharts, returned to Earth. They carried with them more than 700 protein crystal samples that had been brought to Mir by **STS 86 (September 25, 1997)** and had been growing for four months under research sponsored by the Canadian Space Agency. By studying the structures of these crystals, scientists produced new drugs to treat diabetes, heart disease, Alzheimer's disease, and breast cancer.

Musabayev and Budarin then took over operation of Mir. During their tour of duty, they struggled to keep the station operating. An overheated fan started smoking in February. In early March, all space walks were canceled because the cosmonauts were unable to open the airlock hatch, breaking all three wrenches on hand in the process. Before they could continue their work, they had to wait until the first of two Progress freighters (see **Progress M1–August 23, 1989**) brought them a new wrench.

During the ensuing five space walks, they replaced a thruster engine used to orient the station. In June, they were visited by **STS 91 (June 2, 1998)**.

February 10, 1998
Geosat Follow-On / Celestis 2
U.S.A.

- *Geosat Follow-On* The U.S. Navy satellite Geosat Follow-On used radar altimetry to monitor precisely sea surface elevation changes, continuing the work of **Geosat (March 13, 1985)**.

- *Celestis 2* This satellite was the second "burial" flight of Celestis, Inc., following **Celestis 1 (April 21, 1997)**; it carried into orbit the cremated remains of 30 individuals.

February 14, 1998
Globalstar 1 through Globalstar 4
U.S.A.

The four satellites Globalstar 1, 2, 3, and 4 were the first in Globalstar's planned 44-satellite constellation of medium-Earth-orbit (about 900 miles altitude) communications satellites for providing voice and data links worldwide from both remote and home telephones. This system was planned as a direct competitor to Iridium's cluster (see **Iridium 4 through Iridium 8–May 5, 1997**). Despite the loss of 12 satellites in September 1998 due to the failure of a Russian Zenit rocket, Globalstar completed its constellation of 48 satellites in November 1999, using Boeing and Russian Soyuz rockets. The cluster is expected to go into full commercial operation by January 2000.

February 17, 1998
Cosmos 2349
Russia / U.S.A.

The Earth-imaging satellite Cosmos 2349 (also called Spin-2), a joint project between the Russian Space Agency and the American firm Aerial Images of Raleigh, North Carolina, was identical to many Russian military telephoto surveillance satellites (see **Cosmos 4–April 6, 1962**). In this case, however, the photographic images produced during the satellite's two-month operational life were purely for commercial sale. This mission photographed regions of the southeastern United States, Latin America, the Middle East, and many major cities throughout the world. Its images could resolve objects as small as 6 feet across.

On April 3, Cosmos 2349 was successfully recovered in Kazakhstan. Its film was processed in Russia then shipped to the United States where Eastman Kodak turned it into digital images for sale on the Internet. Images were sold at a price of $40 per square kilometer.

The satellite was the first of four planned flights, with later missions covering other parts of the United States and Mexico.

February 25, 1998
SNOE / Teledesic 1
U.S.A.

- *SNOE* The Student Nitric Oxide Explorer (SNOE) was built by students at the University of Colorado. It measured how the Sun's radiation affected the nitric oxide levels in the atmosphere's highest layers, the amount of which plays a direct role in the decline or growth of the ozone layer. As of December 1999, it continues to operate.

- *Teledesic 1* The prototype technology satellite Teledesic 1, owned and built by Teledesic Corp., was intended to test a variety of engineering designs related to Teledesic's proposed 312-satellite cluster, one of several planned private endeavors following the initial launches of the Iridium, Orbcomm, and Globalstar systems (see **Iridium 4 through Iridium 8–May 5, 1997, Orbcomm FM2 through FM12–December 23, 1997**, and **Globalstar 1 through Globalstar 4–February 14, 1998**, respectively).

This commercial space project, the most expensive ($9 billion) ever proposed, was backed by Bill Gates of Microsoft and cellular telephone pioneer Craig McCaw. Planned for completion by 2002, the cluster would be used to link computers worldwide with high-speed connections and to provide customers with video-conferencing capabilities.

Teledesic 1, however, was a failure. After two months of trying to solve a communications problem, the satellite was declared a loss, and the company filed an insurance claim to recover $14 million in expenses.

March 23, 1998
Spot 4
France / Europe

Launched on an Ariane rocket, Spot 4 is a second-generation Earth resource satellite, following **Spot 1 (February 22, 1986)**. It carries two identical camera systems imaging the earth in four spectral bands, two visible and two near-infrared. The addition of a second infrared band in the 1.58–1.75 micron wavelengths allows Spot 4 to image moisture content and cloud layers, and to see through thin clouds.

The spacecraft also carries a sensor for large-scale monitoring of the earth's vegetation. This instrument, covering two visible wavelengths (0.43–0.47 microns and 0.61–0.68 microns), one near-infrared (0.78–0.89 microns), and one mid-infrared (1.58–1.75 microns), scans the ground in 1,400-mile swathes, with a resolution of about two-thirds of a mile. These wavebands record different plant types, their quantities, and their health.

While Spot 4 records and transmits by radio its images to ground stations, it also carries an experimental high-speed laser transmission system for beaming its data directly to France.

The next earth resources satellite was **TMSat (July 10, 1998)**. See also **Landsat 7 (April 15, 1999)**.

April 1, 1998
Trace
U.S.A.

Built by NASA, the Transition Region and Coronal Explorer (Trace) augmented the solar observations of **Wind (November 1, 1994)**, **Soho (December 2, 1995)**, and **ACE (August 25, 1997)**, studying the transitional region just above the Sun's surface where the three-dimensional magnetic structures that emerge from the Sun's surface—such as coronal mass ejections, solar flares, and other magnetic loop events—raised the temperature of the Sun's atmosphere or corona from about 6,000°F to over 16,000,000°F.

Trace used a CCD telescope (see **KH-11 1–December 19, 1976**) that observed across a number of ultraviolet wavelengths from 171 to 1,600 angstroms and was able to take images every half-minute. Its images showed the Sun's magnetic field as it rearranged itself within the corona, heating it to high temperatures. Loops of magnetic field lines revealed complex structures and many threads of hot material. These events appeared closely linked to the generation of solar flares and coronal mass ejections, which developed much faster than scientists had expected.

Trace still operates, and remains the last solar research satellite launched as of December 1999.

April 17, 1998
STS 90, Columbia Flight #25
U.S.A.

The 16-day mission of STS 90 was Columbia's 25th space flight, crewed by Richard Searfoss (**STS 58–October 18, 1993**; **STS 76–March 22, 1996**), Scott Altman, Kathryn Hire, Richard Linnehan (**STS 78–June 20, 1996**), Dafydd Rhys Williams, Jay Buckey, and James Pawelczyk.

STS 90, continuing research from **STS 78 (June 20, 1996)**, was dedicated to studying the effects of weightlessness on the human neurological system, with the astronauts serving as both researchers and experimental subjects. Columbia's cargo bay was outfitted with Spacelab, which carried a menagerie of rats, mice embryos, toadfish, fresh water snails, swordtail fish, and crickets.

Four experiments used the astronauts to study how the cardiovascular and nervous systems interacted to adapt to weightlessness, and then readapted to gravity upon return to Earth. Several experiments examined changes in perception, including the confusion experienced by some astronauts in determining up and down, as well as distance.

Five experiments, using astronauts, laboratory rats, and aquatic life, studied the body's system of balance. Each attempted to better understand the ability of the nervous cells in the inner ear to quickly grow and shrink depending on the level of gravity, a phenomenon first discovered on **STS 40 (June 4, 1991)**. In two experiments, the astronauts sat blindfolded in a rotating chair that spun them at 45 revolutions a minute to mimic 1g. How their balance and perceptions changed during and after each session was measured. The toadfish and snail experiments tried to see if these animals' otolith organs, which they used to sense gravity and body position, also grew and shrank depending on gravity. The cricket experiment tried to determine if the insects' cerci organ, which it uses to sense gravity, developed normally in weightlessness. One rat experiment had the rodents running along a track called the Escher staircase, which returned to its starting point after only three 90-degree turns (an impossibility on Earth but doable in zero gravity because of the ability to run on walls and ceilings).

Two experiments studied the human sleep cycle. Another attempted to find out if gravity affected the production of thyroid hormones and hence contributed to the deterioration of muscles. Four experiments examined the development of the nervous and muscular systems in space, using pregnant mice and young rats. Some of the young rats were dissected while still in orbit.

The next shuttle flight, **STS 91 (June 2, 1998)**, was the last docking mission to **Mir (February 20, 1986)**. The next shuttle science mission was STS 95 (October 29, 1998).

April 17, 1998
Ulysses, First Solar Orbit
U.S.A. / Europe

Launched by the space shuttle (see **STS 41–October 6, 1990**), Ulysses had swung past Jupiter on February 8, 1992 (see **Ulysses, Jupiter Flyby–February 8, 1992**), in order to place it in a solar orbit above the Sun's poles, a region never before traveled. Flying over the Sun's south pole in 1994 and its north pole in 1997, Ulysses completed its first polar orbit in the spring of 1998. This orbit had circled the Sun during the solar cycle's quiet period.

The spacecraft found that although the Sun's fast wind came from coronal holes (regions confined to both poles during this quiet period), it quickly fanned out to fill two-thirds of the heliosphere. It dominated the north and south hemispheres at latitudes as low as 30 degrees, with the slow wind restricted to the equatorial regions. Moreover, the boundary between the equatorial slow wind and the two polar fast winds was very sharp, with the two polar fast winds remarkably alike in speed and flux.

At Ulysses' distance, the Sun's magnetic field strength surprised scientists by being very uniform, both at the ecliptic and above the poles. Theory had predicted that its strength above the poles would be twice that at the equator. Events close to the Sun helped to evenly distribute this flux.

Nonetheless, strong waves of magnetic energy periodically swept past as the spacecraft moved from 40 to 70 degrees latitude. Ulysses also detected low-energy electrons at great distances and above the poles. Since these particles were produced by solar activity near the Sun and at the equator, their presence at the poles indicated a link between equatorial events and the solar environment at the poles.

Unlike the earth's magnetic field, which allows a greater entrance of outside particles at the poles as ionized particles follow magnetic field lines down to the surface (see **OGO 4–July 28, 1967**), Ulysses found no increase in intergalactic cosmic rays above the Sun's poles. The fast wind and the magnetic energy waves acted to block the entrance of these particles.

Ulysses' gamma-ray burst detector also worked with **BeppoSax (April 30, 1996)** and the Hubble Space Telescope (see **STS 82–February 11, 1997**) to make the first optical identification of a gamma-ray burst.

The spacecraft is now in its second solar orbit, to be completed by the end of 2001. During this orbit, Ulysses would be observing the Sun during the solar maximum.

April 28, 1998
Nilesat 1
Egypt / France / Europe

Built by Matra Marconi Space of France and launched on an Ariane rocket, Nilesat 1 was the first satellite dedicated to providing communications to Africa. It was owned by Nilesat Co. of Egypt and delivered direct broadcasting television to Egypt with a capacity to broadcast 100 digital television channels.

June 2, 1998
STS 91, Discovery Flight #24
U.S.A. / Russia

STS 91, Discovery's 24th space flight, lasted 10 days and was the last shuttle docking to **Mir (February 20, 1986)**. The crew was Charles Precourt (**STS 55–April 26, 1993; STS 71–June 27, 1995; STS 84–May 15, 1997**), Dominic Gorie, Wendy Lawrence (**STS 67–March 2, 1995; STS 86–September 25, 1997**), Franklin Chang-Diaz (**STS 61C–January 12, 1986; STS 34–October 18, 1989; STS 46–July 31, 1992; STS 60–February 3, 1994; STS 75–February 22, 1996**), Janet

Kavandi, and Russian Valery Ryumin (**Soyuz 25–October 9, 1977**; **Soyuz 32–February 25, 1979**; **Soyuz 35–April 9, 1980**). The flight also brought Andrew Thomas back to Earth after 142 days in space. Thomas had been brought to Mir on **STS 89 (January 22, 1998)**.

Discovery docked with Mir on June 4, 1998, for four days of joint operations, joining Thomas and cosmonauts Talgat Musabayev and Nikolai Budarin of **Soyuz TM27 (January 29, 1998)**. Discovery's cargo bay was outfitted with a single Spacehab module, adding 1,110 cubic feet for Mir cargo. This cargo included water, food, clothing, three storage batteries, a gyroscope, and an air pressurization unit. Returning to Earth were the wheat plant growth and protein crystal growth experiments that Andrew Thomas had been tending during his four months on Mir.

After undocking on June 8th, Precourt piloted Discovery in a fly-around of Mir, while Musabayev and Budarin pumped a special tracer gas into the depressurized and damaged Spektr module (see **STS 84–May 15, 1997**). Although the astronauts tried to spot the precise location of the module's air leak, no one saw any evidence of a gas leak, preventing further repair work. The remainder of Musabayev's and Budarin's Mir residency focused on routine maintenance, their tour of duty ending with the arrival of **Soyuz TM28 (August 13, 1998)**.

On Discovery, materials processing research continued, using one furnace. The shuttle also tested a variety of new equipment designs intended for installation on the International Space Station (see **Zarya–November 20, 1998**), the most significant of which was the Alpha Magnetic Spectrometer, designed to detect the highest energy cosmic ray particles, including anti-matter, left over from the "big bang" and percolating into the solar system from intergalactic space. This mission checked out the spectrometer's operation, which is scheduled to be installed on the space station in 2002, where it will operate for three to five years.

Discovery also grew a large number of protein crystal growth samples (see **STS 63–February 3, 1995**), used to develop new antibacterial drugs as well as treat diseases like streptococcus pneumonia, influenza, and bronchitis. Three crystal samples were potential drugs for the treatment of Chagas disease, continuing research begun on **STS 75 (February 22, 1996)**.

The next shuttle mission, **STS 95 (October 29, 1998)**, was the last purely scientific mission in the shuttle program, as scheduled in December 1999.

July 3, 1998
Nozomi
Japan

Aimed at Mars, Nozomi ("hope" in Japanese) is the first planetary mission by a country other than the United States or the Soviet Union/Russia. Like **International Cometary Explorer (September 11, 1985)**, Nozomi did not depend solely on rocket engines to get out of Earth orbit. Instead, the spacecraft was first placed in an extremely elliptical Earth orbit, 211 miles by 250,000 miles, taking it past the Moon on September 24, 1998, and December 18, 1998. Each lunar flyby increased the spacecraft's velocity. Then, two days after the second lunar flyby, Nozomi streaked past Earth, using Earth's gravity—as well as an engine burn—to send it toward Mars.

Unfortunately, a stuck engine valve limited the engine burn, requiring a second burn to get Nozomi out of Earth orbit. This use of extra fuel meant that the spacecraft did not have the fuel to enter Mars's orbit as planned in October 1999. Only by flying past Earth two more times and delaying its arrival at Mars until December 2003 could its limited fuel supply place it in Mars's orbit. As of December 1999, this is Nozomi's flight plan.

Unlike previous Mars orbital missions (see **Mariner 9, Mars Orbit–November 14, 1971**; **Viking 1–June 19, 1976**; **Viking 2–August 2, 1976**; **Mars Global Surveyor–September 12, 1997**), Nozomi is not designed to study the Martian surface. Instead, it will try to understand how the planet's faint atmosphere interacts with the Sun's solar wind.

Scientists believe that this interaction is causing Mars a net atmospheric loss, but they are not sure how. In 1988, **Phobos 2 (July 12, 1988)** had detected a flow of ionized particles leaving the planet's atmosphere in a stream flying away from Mars's night side at a rate of escape large enough to have contributed significantly to the depletion of the planet's atmosphere over time. It is not clear, however, what causes this stream. Unlike Earth, Mars has no global magnetic field to direct the flow. Some scientists suspect that the solar wind itself is directing the particles, but this is uncertain. Whether the solar wind is flinging particles from the Martian atmosphere, or whether those particles are redirecting the solar wind around the planet, is the question Nozomi hopes to answer.

July 7, 1998
Tubsat N / Tubsat N1
Germany / Russia

Tubsat N and Tubsat N1 were two microsatellites built by the Technical University of Berlin (TUB), the third launch by this university (see **Tubsat 1–July 17, 1994** and **Tubsat 2–January 25, 1994**). The two satellites were also the first commercial launch by the Russian Navy, launched from a submerged Russian submarine in the Barents Sea north of Norway. Tubsat N/Tubsat N1 tested communications technologies. They were also used to track wildlife, much like Tubsat 1.

As of December 1999, Tubsat N and Tubsat N1 both remain operational and have been augmented by **DLR-Tubsat (May 26, 1999)**.

July 10, 1998
FASat-Bravo / TMSat / Gurwin Techsat 1B / Westpac / Safir 2
Chile / U.K. / Thailand / Israel / Australia / Germany / Russia

• *FASat-Bravo*   Placed in a sun-synchronous orbit (see **Nimbus 1–August 28, 1964**) by a Russian rocket, FASat-Bravo carried an ultraviolet and an infrared sensor to monitor ozone levels over Chile. Built for the Chilean Air Force by Surrey Satellite Technology Ltd. of Guildford, England, it has a planned service life of from 5 to 10 years. Chilean scientists expect to use it to measure the yearly ozone hole variations over their country, along with the resulting changes in ultraviolet radiation reaching the surface.

- *TMSat* Built by Surrey Satellite Technology for Thailand, the remote sensing satellite TMSat used multispectral infrared cameras for making earth resource observations of Thailand. It followed **Spot 4 (March 23, 1998)** and was followed by **Landsat 7 (April 15, 1999)**.

- *Gurwin Techsat 1B* Built by students at Israel's Technion Institute of Technology in Haifa, Gurwin Techsat 1B demonstrated a variety of Israeli-built space technologies. It carried an ultraviolet ozone detector, a television camera, e-mail data relay equipment, and an x-ray sensor for astronomical observations.

- *Westpac* Built by the Australian government, Westpac was a geodetic satellite for more accurately pinpointing the location of the Australian landmass. It was similar in design to the German GFZ 1 geodetic laser reflecting satellite (see **Soyuz TM21–March 14, 1995**).

- *Safir 2* Owned by OHB-System of Germany, Safir 2 provided a data message relay service for the German Air Force and scientific organizations.

July 18, 1998
Sinosat 1
China / Germany / France

Built by Alcatel Space of France and owned by EurasSpace, a Chinese-German joint venture, Sinosat 1 was launched by a Chinese Long March 3B rocket. This geosynchronous communications satellite provides television, telephone, and data transmission service to China, Vietnam, and the Philippines, using 24 4/6 GHz transponders and 14 11/14 GHz transponders. It has an expected service life of 15 years.

August 13, 1998
Soyuz TM28
Russia

Soyuz TM28 was crewed by Sergei Avdeyev (**Soyuz TM15–July 27, 1992, Soyuz TM22–September 3, 1995**), Gennady Padalka, and Yuri Baturin. Baturin was the former security aide to Russian President Boris Yeltsin.

On August 15th, Soyuz TM28 docked with **Mir (February 20, 1986)**. Its crew then joined the crew of **Soyuz TM27 (January 28, 1998)**, Talgat Musabayev and Nikolai Budarin, for 10 days of joint operations, after which Musabayev, Budarin, and Baturin returned to Earth.

During the Soyuz TM28 residency, one Progress freighter (see **Progress M1–August 23, 1989**) visited Mir, bringing several tons of supplies. It also brought an 82-foot-wide mirror, dubbed Znamya ("banner" in Russian). The mirror was larger but similar in concept to the mirror experiment from **Soyuz TM16 (January 24, 1993)**. After two tries, however, the mirror failed to unfurl as planned, and the experiment was abandoned on February 4, 1999, with both mirror and freighter burning up in the atmosphere.

Avdeyev and Padalka did one space walk, placing a micro-meteorite package to measure the impact rate during the Leonid meteorite shower. Although the shower was intense, it was not as powerful as expected. During this space walk, they also released a second mini-Sputnik amateur radio satellite, similar to **Sputnik 40 (November 3, 1997)**. The two men also did an internal space walk in the depressurized **Spektr (May 20, 1995)** module to tighten cable connections installed during **Soyuz TM26 (August 5, 1997)**. See also STS 86 (**September 25, 1997**).

Soyuz TM28 returned to Earth on February 28, 1999, soon after the arrival of Soyuz TM29. See **Soyuz TM29 (February 20, 1999)**.

August 25, 1998
ST 1
Singapore / Taiwan / France / Europe

Manufactured by Matra Marconi of France and launched on an Ariane rocket, the geosynchronous communications satellite ST 1 was a joint project of companies from Singapore and Taiwan. The satellite carried 14 transponders in the 4/6 GHz frequencies and 16 in the 11/14 GHz frequencies and was positioned over the Indian Ocean to provide service to China and the Far East.

October 3, 1998
STEX
U.S.A.

Built by Lockheed Martin Astronautics for the U.S. National Reconnaissance Office (NRO), the Space Technology EXperiment (STEX) satellite tested 29 new spacecraft designs, including an almost four-mile-long tether, advanced solar panels, and an ion engine test.

The tether experiment was similar to **Seds (March 30, 1993)**, studying how tethers behaved in orbit. Unlike previous more ambitious tests that had failed (see **STS 46–July 31, 1992** and **STS 75–February 22, 1996**), this experiment made no attempt to measure the electric current produced along the tether's length. Instead, laser reflectors at the end of the tether allowed ground controllers to monitor its behavior during maneuvers.

The solar panels used mirrors to concentrate the Sun's light, thereby increasing the spacecraft's electrical output by 20 percent.

The xenon ion engine, originally developed in the United States in the 1960s, continued research by Soviet/Russian technicians and used on its Gals and Express spacecraft (see **Gals 1–January 20, 1994**, and **Express 1–October 13, 1994**). It was also similar to the engine used commercially by Pan American Satellites, Inc., on **Pan American Satellite 5 (August 28, 1997)** as well as by NASA on **Deep Space 1 (October 24, 1998)**.

October 21, 1998
ARD / Maqsat 3
Europe

The launch of ARD/Maqsat 3 was the third test flight of the Ariane 5 and the first flight that was completely successful. The first had exploded less than a minute after launch, while the second (see **Ariane 5–October 30, 1997**) had failed to reach its planned orbit when its main engine shut down early. This test carried two payloads.

- *ARD* Developed by the European Space Agency, this reentry test capsule was part of Europe's first program to build a manned-rated spacecraft. Remarkably similar in shape to the American Apollo capsule (see **Apollo 4–November 9, 1967**), the ARD (Atmospheric Re-entry Demonstrator) was about 20 percent smaller.

  Though unmanned, it completed one orbit before splashing down in the Pacific Ocean only three miles from its target, where the French Navy successfully recovered it. Inside, it carried more than 200 sensors to monitor the temperatures and stresses experienced by the craft during its flight. The capsule also used a GPS (global positioning system) sensor to track its course.

- *Maqsat 3* This satellite was a dummy payload that transmitted telemetry and launch data as the Ariane 5 rocket lifted it into geosynchronous orbit.

October 24, 1998
Deep Space 1 / Sedsat 1
U.S.A.

- *Deep Space 1* The NASA technology test spacecraft Deep Space 1 evaluated a dozen advanced spacecraft engineering designs, from mirror-enhanced solar panels to the first use of an ion engine to leave Earth orbit and rendezvous with the asteroid Braille.

  The xenon engine used by Deep Space 1 resembled engines used on several communications satellites (see **Gals 1–January 20, 1994** and **Pan American Satellite 1–June 15, 1988**) as well as the engine tested on **STEX (October 3, 1998)**. Unlike these other engines, which merely maintained the spacecraft's orbital position, Deep Space 1's engine was used to propel the spacecraft from Earth. Through July 1999, all systems operated as designed, the ion engine having pushed the satellite more than 115 million miles from Earth to its rendezvous with Braille. The spacecraft's velocity through space had been increased by more than 300 miles per hour, and at one point, the ion engine had operated for nearly two weeks without stopping, exceeding its planned longest burn time.

  See **Deep Space 1, Braille Flyby (July 28, 1999)** for further information.

- *Sedsat 1* Developed and built by Students for the Exploration and Development of Space (Seds) at the University of Alabama at Huntsville, Sedsat 1 carried a multispectral camera for taking Earth images, which was able to resolve objects as small as 600 feet across. It also included a store-and-forward communications package for use by amateur radio enthusiasts.

October 28, 1998
AfriStar
Africa / France / Europe

Owned by WorldSpace Corporation and launched on an Ariane rocket, the geosynchronous communications satellite AfriStar was built by Alcatel Espace of Paris. The first of a planned four-satellite cluster, it provides radio service to Africa and the Middle East. Customers use specially built receivers, available for about $200, to receive between 50 and 100 broadcast channels. Later satellites are intended to expand service worldwide.

October 29, 1998
STS 95, Discovery Flight #25
Spartan 201 / Pansat
U.S.A. / Japan

STS 95 was Discovery's 25th space flight. The mission lasted nine days and was crewed by Curtis Brown (**STS 47–September 12, 1992**; **STS 66–November 3, 1994**; **STS 77–May 19, 1996**; **STS 85–August 7, 1997**; **STS 103–December 19, 1999**), Steven Lindsey (**STS 87–November 19, 1997**), Scott Parazynski (**STS 66–November 3, 1994**; **STS 86–September 25, 1997**), Stephen Robinson (**STS 85–August 7, 1997**), Pedro Duque, Japanese Chiaki Mukai (**STS 65–July 8, 1994**), and John Glenn (**Friendship 7–February 20, 1962**).

Except for a radar mission planned for January 2000, STS 95 was the last shuttle mission dedicated solely to scientific research, at least as planned in December 1999.

Following this flight, the bulk of space shuttle activity was ferry and construction flights to the International Space Station (see **Zarya–November 20, 1998** and **STS 88–December 4, 1998**).

STS 95 was also significant in that it heralded the return to space of John Glenn, the fifth man to fly into space and the first American to orbit the earth. After a retirement from the space program stretching over three-and-a-half decades, Glenn volunteered for one last mission, partly because of his own desire to fly again and partly to aid NASA in its research into the effects of weightlessness on the human body. At 77 years old, Glenn was by far the oldest man to fly in space, exceeding the previous record of 61 years set by Story Musgrave on **STS 80** (**November 19, 1996**).

The mission's work was divided into three research areas: astronomy, life sciences, and materials. The shuttle released and retrieved one satellite and deployed a second.

- *Spartan 201* One of three simply designed NASA spacecraft deployed for short periods from the space shuttle (see **STS 51G–June 17, 1985**), Spartan 201 was dedicated to solar research and had flown four previous times (see **STS 56–April 8, 1993**; **STS 64–September 9, 1994**; **STS 69–September 7, 1995**; **STS 87–November 19, 1987**). Because of problems during its last flight, however, Spartan 201 failed to operate, and so it was reflown on STS 95. Deployed on November 1st for two days of free-flying, Spartan 201's data were used by scientists to calibrate the information being received from the recently reactivated **Soho (November 2, 1995)**.

- *Pansat* Deployed on October 30th, the Petite Amateur Naval Satellite (Pansat) was developed by the Naval Postgraduate School in California to test new technology for detecting very weak radio signals, useful in rescuing downed pilots.

Discovery also carried six additional astronomical instruments in its cargo bay, as well as a variety of materials and equipment intended for use in the next servicing mission to the Hubble Space Telescope (see **STS 103–December 19, 1999**). The astronomical instruments included two extreme ultraviolet telescopes, one for observing the celestial sky from 850 to 1,250 angstroms, and the second for studying the Sun from 250 to 1,700 angstroms. Another solar instrument, Solcon, monitored the Sun's solar constant in order to further calibrate other satellite observations.

Among the technology being tested for Hubble was a new cooling system for the Nicmos infrared telescope (see STS 82), which NASA hoped would repair its problems and extend its life expectancy to at least five years.

A single module of Spacehab was installed in Discovery's cargo bay for the onboard experiments. Protein crystal research (see **STS 63–February 3, 1995**) included almost 400 different samples for later study on Earth. For instance, the pike parvalbumin protein is found in muscles, but scientists remain unsure of its purpose. Although crystals had been grown on two previous shuttle flights, researchers grew additional crystals on this flight to better understand the protein's structure and purpose.

Protein crystals of urokinase, a key enzyme in breast cancer, were also grown (see **STS 80–November 19, 1996**), as were crystals of human insulin (see **STS 60–February 3, 1994** and **STS 72–January 11, 1996**). Also grown was a crystal of a neutralizing antibody to respiratory syncytial virus, an infection that attacks the human respiratory system.

Three microencapsulation experiments were performed, continuing research from **STS 53 (December 2, 1992)** and **STS 70 (July 13, 1995)**. If successfully produced, these tiny capsules could be used to time-release insulin, or to precisely release cancer chemotherapy treatments in much smaller dosages to tumors at precisely timed intervals, thereby reducing secondary symptoms to the patient.

Two plant experiments were carried in the Astroculture greenhouse, first flown on **STS 50 (June 25, 1992)** and last used for four months on **Mir (February 20, 1986)**. See **STS 89 (June 22, 1998)**. One tested how the flavor and fragrance of flowers were affected by weightlessness. The second tested whether a gene useful for medical applications could be transplanted into soybean seedlings during weightlessness.

During the flight, blood and urine samples from John Glenn were collected daily for later study. Other sensors measured his changing bone and muscle density, immune system, metabolism, perception, and system of balance.

The next shuttle mission, **STS 88 (December 4, 1998)**, was the first assembly mission of the International Space Station (see **Zarya–November 20, 1998**).

November 20, 1998
Zarya (International Space Station Module)
U.S.A. / Russia

Launched on a Russian rocket, Zarya ("sunrise" in Russian) was built by Russia and paid for by the United States. It was the first component in the American-led International Space Station project. If completed as planned, this station will weigh more than 500 tons (more than twice as heavy as the combined Mir-Shuttle complex) and include 26 separate modules (plus its solar panel structure), with more than 43,000 cubic feet of habitable space—almost four times more than **Mir (February 20, 1986)**. Six humans will occupy it continuously, doing scientific and industrial research and development.

Zarya, weighing 21 tons, contained power, fuel storage, propulsion, and multiple docking ports. Until the arrival of the third module, being built by Russia, Zarya provided the communications, electrical power, and attitude control for the station. Its docking ports will be used by Soyuz and Progress freighters (see **Progress M1–August 23, 1989**), both of which can refuel its 16 fuel tanks.

Zarya was mated to the station's second module during STS 88 (December 4, 1998).

November 22, 1998
Bonum 1
Russia / U.S.A.

Built by Hughes Space and Communications and launched on a Boeing rocket, the geosynchronous communications satellite Bonum 1 was owned by Media-Most, a private Russian company. Until this launch, the company had operated a five-channel commercial television cable network, called NTV plus, using the Russian government's Gals satellites (see **Gals 1– January 20, 1994**).

Bonum 1 was significant because it was the first foreign-built communications satellite used by a private Russian company to broadcast to Russia. Prior to this launch, all Russian communications satellites had been built in Russia and had been controlled in some way by the Russian (and before that, the Soviet) government.

With Bonum 1, however, the Russian government had no control over the satellite's operation. Media-Most positioned the satellite over western Russia, where it offered western Russian citizens up to 30 cable channels of both Russian and non-Russian content.

A second satellite is presently under construction for Media-Most by a Russian firm to expand the service to eastern Russia.

December 4, 1998
STS 88, Endeavour Flight #13
Unity (International Space Station Module) / SAC-A / MightySat 1
U.S.A. / Russia / Argentina

STS 88 was Endeavour's thirteenth space flight. It lasted 12 days and was crewed by Robert Cabana (STS 41–October 6, 1990; STS 53–December 2, 1992; STS 65–July 8, 1994), Fred Sturckow, Nancy Sherlock Currie (STS 57–June 21, 1993; STS 70–July 13, 1995), Jerry Ross (STS 61B–November 26, 1985; STS 27–December 2, 1988; STS 37–April 5, 1991; STS 55–April 26, 1993; STS 74–November 12, 1995), James Newman (STS 51–September 12, 1993; STS 69–September 7, 1995), and Russian Sergei Krikalev (Soyuz TM7–November 26, 1988; Soyuz TM12–May 18, 1991; STS 60–February 3, 1994).

STS 88 was the first assembly mission of the International Space Station. It deployed the station's second module.

- *Unity (International Space Station Module)* The American-built module Unity provided six docking ports, with the

December 4, 1998

Zarya *(left)* and Unity *(right)*, linked in orbit. NASA

The International Space Station, as planned in December 1999. Unity is now hidden behind the long central truss structure that supports the 16 large solar panels. Zarya is visible near the center, just to the right and above that central truss. NASA

fuel, atmospheric, and electrical lines designed to interconnect between every docked module.

Unity was linked to the station's first module, **Zarya (November 20, 1998)**, using Endeavour's robot arm, several docking maneuvers, and three space walks. An IMAX camera photographed these events for use in the fifth IMAX documentary filmed in space. As of December 1999, this film is still in production.

Then, on December 11th, the astronauts activated both modules, and, after confirming they were functioning as planned, Bob Cabana and Sergei Krikalev entered the station together, with the rest of the shuttle crew soon to follow. While Ross and Newman activated the station's communications system and other astronauts unstowed gear, Krikalev and Currie uninstalled one of Zarya's failed batteries and replaced it with a new battery that had been added to Endeavour's cargo.

The two modules were then left in orbit unmanned, awaiting the launch of additional modules and shuttle missions.

Discovery also deployed two small satellites.

• *SAC-A* Built for Argentina, SAC-A tested satellite equipment and designs for use in future satellite construction.

• *MightySat 1* An Air Force satellite, MightySat 1 was designed to conduct a one-year mission to test new solar cell designs and the ability of new composite materials to tolerate the environment of space.

The next shuttle mission, STS 96 (**May 27, 1999**), continued the construction of the International Space Station.

December 6, 1998, 0:43 GMT
Satmex 5
Mexico / U.S.A. / Europe

Unlike the previous Mexican geosynchronous communications satellites (see **Solidaridad 1–November 20, 1993**), Satmex 5 was privately owned. Built by Hughes Communications and launched on an Ariane rocket, it provides television and telephone service to Spanish-speaking customers from the northwest United States to southern Chile.

December 6, 1998, 0:57 GMT
SWAS
U.S.A.

Built by NASA, the Submillimeter Wave Astronomy Satellite (SWAS) studies interstellar molecular clouds within 3,500 light years of the earth, such as the star-forming regions in Orion, Taurus, Ophiuchi, and Perseus. The satellite operates in the far-infrared (100–750 microns), looking for evidence of water, oxygen, and carbon within these clouds. It continues research by **ISO (November 17, 1995)** and **MSX (April 24, 1996)**. From these data, astronomers hope to better understand the process by which these clouds collapse to form stars.

SWAS's research was augmented by **Fuse (June 24, 1999)**.

December 10, 1998
Astrid 2
Sweden / Russia

Launched piggyback on a Russian rocket, Astrid 2 continued the auroral research of **Astrid 1 (January 24, 1995)** and **Equator S (December 2, 1997)**. Later magnetospheric-auroral research continued on **Orsted (February 23, 1999)**.

December 11, 1998
Mars Climate Orbiter
U.S.A.

Built by NASA, this Mars probe was similar to **Mars Global Surveyor (September 12, 1997)**. Climate Orbiter's intended focus, however, was to study the Martian weather in its lower atmosphere, measuring daily temperature, water, and dust content. To obtain these data, the spacecraft carried two instruments, a color imaging system, and an infrared radiometer.

The Climate Orbiter was lost as it approached Mars on September 23, 1999. A lack of communication between two different spacecraft teams, one of which was using English units of measurement while the other was using metric units, caused Climate Orbiter's path through space to be calculated incorrectly. The error placed the spacecraft only 35 miles above the surface of Mars as it flew past, close enough for the planet's atmosphere to destroy the probe.

December 23, 1998
NEAR, Eros Flyby
U.S.A.

Launched on February 17, 1996, NEAR is designed to make the first long-term observations of an asteroid, 433 Eros. On its way to Eros, it also flew past a second asteroid, Mathilde, on June 27, 1997. The original flight plan had called for an engine burn on December 10th, placing the spacecraft into orbit around Eros on January 10, 1999. NEAR's computers aborted the burn, however, forcing a change in plans (see **NEAR–February 17, 1996; NEAR, Mathilde Flyby–June 27, 1997**).

On December 23rd, the spacecraft flew within 2,400 miles of Eros, taking 222 photographs of about two-thirds of the asteroid's surface. Eros was smaller than predicted, 21 by 8 by 8 miles. It rotates once every 5 hours and 16 minutes, and its surface has at least two medium-sized craters as well as a prominent ridgeline running at least 12 miles. Its density is about the same as the earth's crust, similar to that of Ida (see **Galileo–August 28, 1993**) and about twice that of Mathilde's. Since both Eros and Ida were S type (or stony) asteroids and Mathilde was a C type asteroid, this result was expected.

Overall, the data indicated that Eros was not a rubble pile like Mathilde, but a solid body.

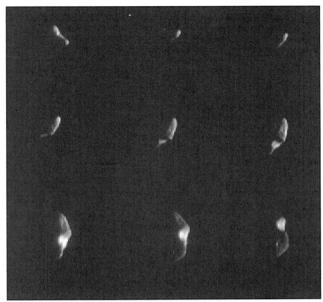

Eros rotating as NEAR flew by at a distance of 3,300 miles. *NASA*

After correcting the software problem that had aborted the December 10th burn, NEAR performed a second engine burn on January 3, 1999. This course change will bring the spacecraft back to Eros in February 2000, when it will go into orbit.

Once there, NEAR's close look will hopefully solve one of the biggest mysteries of asteroid and meteorite study: to what asteroid class do meteorites belong? Although the S and C classifications of asteroids make up almost 75 percent of all know asteroids, the spectrum of neither type corresponds very closely to the spectrum of most meteorites found on Earth. In fact, only a handful of asteroids fit precisely the spectral class of 90 percent of all known meteorites.

Even more puzzling, the closer one gets to the earth, the worse the spectral match. C asteroids, orbiting from the middle of the asteroid belt outward, match poorly; S asteroids, orbiting from the middle of the asteroid belt inward, do not match at all. While almost all the near-Earth asteroids belong to the S class, only 10 percent of the meteorites discovered on Earth resemble these asteroids even vaguely.

In many ways, NEAR's journey to Eros is very comparable to the arrival of **Mariner 9 (November 14, 1971)** at Mars. Before that first Martian orbital mission, our knowledge of that planet was mostly based on its albedo and reflective spectra. Our only good views had been limited to three quick flybys, **Mariner 4, Mars Flyby (July 15, 1965), Mariner 6, Mars Flyby (July 31, 1969)**, and **Mariner 7, Mars Flyby (August 5, 1969)**. Overall, we knew little and understood less.

Similarly, other than the tantalizing glimpses that NEAR and Galileo have given us (see **Galileo, Gaspra Flyby–October 29, 1991; Galileo, Ida Flyby–August 28, 1993**), no asteroids have yet been imaged up close and in detail. NEAR's observations of Eros will hopefully change this.

1999

January 3, 1999
Mars Polar Lander
U.S.A.

Built by NASA, Mars Polar Lander was designed to soft-land about 600 miles from the Martian south pole soon after the winter's frost layer of carbon dioxide had evaporated from the ground. In addition, two small javelin-like probes (dubbed Amundsen and Scott, in honor of the two terrestrial polar explorers) were designed to impale themselves three feet into the Martian soil somewhere within 120 miles of Polar Lander's landing site and radio back additional data. Scientists hoped that these three probes would give them information about the past history of Mars's climate.

On December 3, 1999, something went wrong during landing, and communication with the lander and two probes was lost. Mars Polar Lander's failure ended a series of 11 successful American planetary soft-landings in a row, a winning streak that had lasted more than 30 years, beginning when **Surveyor 5 (September 8, 1967)** gently touched down in the Moon's Sea of Tranquility.

As of December 1999, only **Nozomi (July 3, 1998)** is on course toward Mars. NASA plans a series of additional Martian probes, however, to be launched regularly through the first decade of 2000. Some missions will be designed to return samples to Earth, and all will test technologies necessary for an eventual manned Mars mission.

January 26, 1999
Rocsat
Taiwan / U.S.A.

Built by TRW Space and Electronics in California and launched by Lockheed Martin's new Athena rocket, Rocsat (Republic of China SATellite) is Taiwan's first earth resource satellite, using six spectral bands to monitor ocean color as well as track plankton growth for Taiwan's fishing industry.

February 7, 1999
Stardust
U.S.A.

The NASA satellite Stardust will rendezvous with Comet Wild-2 (pronounced "Vilt-2") in January 2004, taking the detailed pictures of a comet's nucleus as well as picking up dust and particle samples from its tail. On a flyby of Earth in January 2006, it will jettison the re-entry capsule holding the samples, which will be recovered on Earth for analysis.

February 20, 1999
Soyuz TM29
Russia / France / Slovakia

The eight-day joint mission of Soyuz TM29 was crewed by Russian Victor Afanasyev (**Soyuz TM11–December 2, 1990; Soyuz TM18–January 8, 1994**), Frenchmen Jean-Pierre Haignere (**Soyuz TM17–July 1, 1993**), and Slovak Ivan Bella. It docked with **Mir (February 20, 1986)** on February 22, 1999, joining Padalka and Avdeyev of **Soyuz TM28 (August 13, 1998)**. Haignere's flight completed a two-mission Russian-French cooperative effort (see **Soyuz TM24–August 17, 1996**).

After six days of joint operations, Soyuz TM28 returned to Earth with Padalka and Bella, leaving Avdeyev, Afanasyev, and Haignere behind to begin preparations for the deorbiting of Mir. According to the plans announced by Russian space officials in July 1999, this de-orbiting was scheduled for sometime in February 2000. These plans call for the crew of Soyuz TM29 to return to Earth in August 1999 and the station to remain unoccupied until a last crew goes up to complete preparations just prior to de-orbiting. However, because of a significant Russian reluctance to abandon Mir, it is very uncertain whether these de-orbiting plans will take place as announced.

During the following six months, this three-man crew was visited by two Progress freighters, the second of which brought computer equipment necessary for Mir to be operated unmanned by ground controllers. The crew also performed three space walks, to retrieve experiments on the outside of the station and to attempt to locate a small air leak in the hull of the **Kvant 2 (November 26, 1989)** module. Unfortunately, they were unsuccessful in locating the leak.

On August 28, 1999, Soyuz TM29 returned to Earth, ending an occupancy of Mir that had lasted just eight days short of 10 years. Avdeyev's mission, beginning with Soyuz TM28, had lasted one year, two weeks, and two days, making him the second human to spend more than a year continuously in space. His record is second only to Valery Polyakov's (see **Soyuz TM18–January 8, 1994, Soyuz TM20–October 3, 1994, and Soyuz TM21–March 14, 1995**). Avdeyev's total cumulative time in space, 739 days, is a record, however, making him the world's most experienced space traveler.

February 23, 1999
Argos / Orsted / Sunsat
U.S.A. / Denmark / South Africa

• *Argos* Built by the U.S. Air Force, Argos tests ion engine designs and high temperature superconductivity research. It also gathers data on the upper atmosphere, as well as measure the speed, mass, and trajectory of micrometeorite and man-made debris in low-Earth polar orbit. Argos also carries an x-ray telescope for celestial astronomy, covering wavelengths from 1 to 10 angstroms.

• *Orsted* Denmark's first satellite, Orsted was placed in a sun-synchronous orbit (see **Nimbus 1–August 28, 1964**) to map the earth's magnetic field as well as do magnetospheric and auroral research. It augmented research of **Astrid 2 (December 10, 1998)**.

• *Sunsat* A communications satellite, Sunsat was built by and serves amateur radio enthusiasts in South Africa.

March 4, 1999
WIRE
U.S.A.

The infrared telescope WIRE's four-month mission was intended to study the formation of starburst galaxies, continuing earlier research by **ISO (November 17, 1995)**, **MSX (April 24, 1996)**, and **SWAS (December 6, 1998)**. Unfortunately, the mission failed when the telescope's cover opened prematurely, allowing the Sun to heat and evaporate its hydrogen coolant within days of launch. The satellite was subsequently used for controller training and for advanced ground-control operations.

One scientist, however, was able to use WIRE's two-inch attitude telescope (normally used to orient the spacecraft) to do astronomy, detecting starquakes on the surface of the nearby star Dubhe.

March 27, 1999
DemoSat
U.S.A. / Ukraine / Russia

This demonstration dummy satellite was launched on a Russian/Ukrainian Zenit rocket from a privately owned floating launch platform located in the Pacific Ocean south of Hawaii. Built and owned by Sea Launch, a joint venture led by the Boeing Company, DemoSat was placed in orbit to prove the usability of the rocket platform. This launch was also the first from a floating launch platform since **San Marco D/L (March 25, 1988)**.

Sea Launch began full operations with its first commercial launch, **DirecTV 1R (October 10, 1999)**.

April 15, 1999
Landsat 7
U.S.A.

Built by NASA to replace Landsat 5 (see **Landsat 4–July 16, 1982**), Landsat 7 also followed **Spot 4 (March 23, 1998)** and **TMSat (July 10, 1998)**. It was placed in a sun-synchronous orbit (see **Nimbus 1–August 28, 1964**) and is able to photograph the entire globe every 16 days. Its sensors cover eight spectral bands in the far infrared and microwave wavelengths, from 0.45 to 2.35 mm, with images covering an area 105 miles by 114 miles. Black and white optical resolution at launch was under 50 feet, with multispectral color images having a resolution of under 100 feet. Sensors will provide global information on annual vegetation growth, agriculture, and land use; snow accumulation and melt, and the associated freshwater reservoir replenishment; and changes to urban areas.

Designed with a 5- to 10-year life expectancy, NASA intends Landsat 7 to be its last Earth resources satellite. The space agency hopes that in the future, such American satellites will be built and operated for profit by the private sector.

The next Earth resources satellite was **UoSat 12 (April 21, 1999)**.

April 21, 1999
UoSat 12
United Kingdom / Russia

Built for only $10 million by Surrey Satellite Technology in England (which had developed out of the satellite program at the University of Surrey, England—see **UoSat 1–October 6, 1981**) and launched by Russia, UoSat 12 was Surrey's largest research satellite yet. Primarily an Earth resources satellite, its black-and white camera produced images 6.2 miles square with a 33-foot resolution. Color images have a resolution of 107 feet.

UoSat 12 also tested a variety of radical new technologies, including new attitude control thruster engines and the use of GPS receivers to maintain the satellite's position and orientation.

This launch was also the first for the Russian Dnepr rocket, which had been a SS-18 nuclear missile. Scheduled for destruction under U.S.-Russian arms-control treaties, the missile was refurbished with its warhead removed so that it could be used to launch commercial payloads. Thus, Russia hoped to make a profit from its stockpile of SS-18 missiles.

The next Earth resources satellite was IRS-P4 Oceansat (**May 26, 1999**).

May 10, 1999
Feng Yun 3 / Shijian 5
China

• *Feng Yun 3*   China's third weather satellite, Feng Yun 3 replaced **Feng Yun 2 (June 10, 1997)**, which had failed in April 1998.

• *Shijian 5*   This satellite followed up scientific research of Shijian 4 (**February 8, 1994**).

May 21, 1999
Nimiq 1
Canada / Russia

Built by Telesat Canada and launched by Russia, Nimiq 1 is a geosynchronous communications satellite, positioned over North America and providing direct broadcast services to Canada. With a life expectancy of 12 years, Nimiq 1 supplements the Anik E satellites (see **Anik E2–April 4, 1991**).

May 26, 1999
IRS-P4 Oceansat / Uribyol 3 (Kitsat 3) / DLR-Tubsat
India / South Korea / Germany

All three vehicles were launched by India on its Polar rocket from its launch facility in Sriharikota, India, with Uribyol 3 and DLR-Tubsat being India's first commercial payloads.

• *IRS-P4 Oceansat*   A follow-up of the Indian Remote Sensing (IRS) satellites (see **IRS 1A–March 17, 1988**), IRS-P4 Oceansat was built by the Indian Space Research Organization (ISRO) as the first in a series of ocean-monitoring Earth resource satellites. Its imaging sensors have a ground resolution of 18 feet for black and white images, the sharpest of any remote sensing Earth resources satellite. Other multispectral sensors obtain data on ocean color, as well as the temperature, wind, and water vapor changes at the ocean's surface.

• *Uribyol 3 (Kitsat 3)*   Also called Kitsat 3, Uribyol 3 was a follow-up of **Uribyol 2 (September 26, 1993)**. It was built entirely in South Korea, testing the operation of a variety of Korean-built satellite technologies. Placed in a sun-synchronous orbit (see **Nimbus 1–August 28, 1964**), Uribyol 3 carries a multispectral camera covering three wavebands from 5,200 to 9,000 angstroms, with a swath 31 miles wide and a resolution of under 50 feet.

- *DLR-Tubsat* The fourth satellite built by the Technical University of Berlin (see **Tubsat N/Tubsat N1–July 7, 1998**), DLR-Tubsat was a one-foot-wide box weighing just under 100 pounds. In this box was a camera package of three lens, capable of taking Earth photographs with resolutions of 1,214, 394, and 20 feet respectively. Images are downloadable with a standard satellite dish having a diameter of at least 10 feet.

May 27, 1999
STS 96, Discovery Flight #26
Starshine
U.S.A. / Russia / Canada / International

Discovery's 26th flight, STS 96 was an 11-day mission. It was crewed by Kent V. Rominger (STS 73–October 20, 1995; STS 80–November 19, 1996; STS 85–August 7, 1997), Rick Husband, Ellen Ochoa (STS 56–April 9, 1993; STS 66–November 3, 1994), Tamara Jernigan (STS 40–June 4, 1991; STS 52–October 22, 1992; STS 67–March 2, 1995; STS 80–November 19, 1996), Daniel Barry (STS 72–January 11, 1996), Canadian Julie Payette, and Russian Valery Tokarev.

STS 96's primary goal was to bring two tons of supplies and equipment to the International Space Station (see **STS 88–December 4, 1998**). As with previous resupply missions to **Mir (February 20, 1986)** (see **STS 91–June 2, 1998**), the shuttle's cargo bay was outfitted with a double module of Spacehab, thereby increasing Discovery's habitable space by around 2,200 cubic feet.

During the mission, Daniel Barry and Tamara Jernigan performed a space walk lasting almost eight hours to install two exterior cranes to the station, one American-built and one Russian-built. They also installed a variety of tools and other equipment for use by future astronauts.

During the 3.5 days in which Discovery was docked with the station, the crew found that if they stayed inside the station for too long a period, they experienced headaches, nausea, and eye irritation. They alleviated these symptoms by returning to Discovery for several hours. Based on the crew's description, NASA engineers concluded that the problem stemmed from stagnant air inside the station and a build-up of too much carbon dioxide. Because the astronauts did not reveal these problems until after the flight, however, nothing could be done to pinpoint the cause of the problem. The next flight to the station, scheduled for no earlier than March 2000, will carry sensors and fans to both study the problem and try and prevent its reoccurrence.

STS 96 also deployed one satellite, Starshine, on June 4th.

- *Starshine* The first satellite since **Atmosphere 1 (September 3, 1990)** dedicated to studying atmospheric density changes, Starshine was designed and built under the supervision of the Naval Research Laboratory in Washington for use by students through the world. This beachball-sized satellite is covered entirely with 878 aluminum mirrors one inch in diameter—machined by students in Utah and polished by students in Argentina, Austria, Australia, Belgium, Canada, China, Denmark, England, Finland, Japan, Mexico, New Zealand, Pakistan, South Africa, Spain, Turkey, the United States, and Zimbabwe. Student observers will look for the twinkle of Starshine's reflection in the twilight sky, tracking its motion and the slow change of the satellite's orbit as the atmosphere causes it to decay.

The next shuttle mission, STS 93 (**July 23, 1999**), launched the Chandra X-ray Observatory.

June 19, 1999
QuikScat
U.S.A.

Built by NASA as an ocean wind monitoring satellite, QuikScat was also intended to fill the void in Earth resource observations caused by the premature failure of **ADEOS (August 17, 1996)**. Built in only 11 months for $39 million, QuikSat was placed in a sun-synchronous orbit (see **Nimbus 1–August 28, 1964**), where it could obtain readings of wind speeds and wind direction over 90 percent of the earth's oceans each day.

June 24, 1999
Fuse
U.S.A.

Fuse (Far Ultraviolet Spectroscopic Explorer) is an ultraviolet space telescope, supplementing **EUVE (June 5, 1992)**, Alexis (**April 25, 1993**), and SWAS (**December 6, 1998**), as well as following up research by Orfeus (see **STS 80–November 19, 1996**). Fuse's instruments observed the celestial sky from 905 to 1,185 angstroms, a part of the electromagnetic spectrum that only one previous permanent space telescope, **Copernicus (August 21, 1972)**, had studied before. Fuse, however, is 10,000 times more sensitive.

In these wavelengths, Fuse will map the chemical elements of the Milky Way and other nearby galaxies, studying the molecular clouds of hot and cold gas found in the spiral arms where starbirth occurs. Fuse will also be available to astronomers to study many other specific objects, from supernova remnants to active galactic nuclei, to the other planets in the solar system.

July 23, 1999
STS 93, Columbia Flight #26
Chandra X-ray Observatory
U.S.A. / France

Columbia's 26th flight, this five-day mission was crewed by Eileen Collins (STS 63–February 3, 1995; STS 84–May 15, 1997), Jeffrey Ashby, Steven Hawley (STS 41D–August 30, 1984; STS 61C–January 12, 1986; STS 31–April 24, 1990; STS 82–February 11, 1997), Catherine Coleman (STS 73–October 20, 1995), and Frenchman Michel Tognini (Soyuz TM15–July 27, 1992). Collins was the first woman to command a space mission. STS 93's main goal was the deployment of one science satellite on July 24th—the Chandra X-ray Observatory.

- *Chandra X-ray Observatory* The third in the United States Great Observatory program, Chandra followed the **Hubble Space Telescope (STS 31–April 24, 1990)** and the **Compton Gamma Ray Observatory (STS 37–April 5, 1991)**. Chandra's telescope, using a mirror assembly over 47 inches across, will observe the x-ray sky in the energy range from

0.09 to 10 KeV. Its ability to produce x-ray images is expected to have 50 times the resolution of previous x-ray telescopes.

Because of an unexpected degradation of some of Chandra's CCD chips, the telescope's capabilities were somewhat hampered, requiring more time to obtain data than originally planned. Nonetheless, early images of a number of supernova remnants, such as the Crab Nebula and Cassiopeia A, revealed previously unknown details about their inner regions, including vast rings of ionized gas expanding outward from the central neutron stars.

The next x-ray telescope was **XMM (December 10, 1999)**.

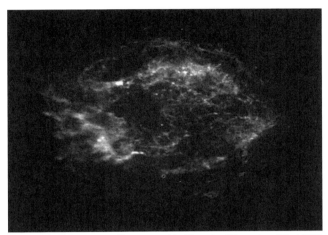

Cassiopeia A in x-rays. Compare with Einstein (November 13, 1978) image. *NASA*

Onboard experiments included one BRIC canister (see **STS 64–September 9, 1994**) performing plant research. During the flight, over 100,000 arabidopsis seeds were germinated in darkness and in a sterile environment in an attempt to understand how weightlessness affects plant development. Other plant experiments tested equipment that had been designed to maintain water and air supplies to mature plants while in weightlessness.

Several experiments tested a variety of in-space engineering techniques. Two experiments tested new protein crystal growth techniques. Another experimented with methods for producing the drug taxol in microgravity. Another attempted new water purification technology.

Of the two insect experiments, one tested the reaction of fruit fly larvae to weightlessness. The second studied the development of wings in space of the painted lady butterfly.

July 28, 1999
Deep Space 1, Braille Flyby
U.S.A.

Launched as an engineering test satellite experimenting with prototype spacecraft designs (see **Deep Space 1–October 24, 1998**), Deep Space 1 flew within 16 miles of the asteroid Braille on July 28, 1999, the closest rendezvous yet attempted with an asteroid.

Data from the flyby, including two black-and-white photographs, indicated that Braille, 0.6 mile by 1.3 miles in size, was very similar in makeup to the 300-mile-wide asteroid Vesta, a rare type of asteroid. Scientists wondered whether Braille and Vesta had formed together and later split apart.

As of December 1999, Deep Space 1 is on course to rendezvous with Comet Wilson-Harrington in January 2001, to be followed by a rendezvous with Comet Borrelly in September 2001. Comet Wilson-Harrington, whose coma and tail disappeared in 1949, is thought to be a dying comet that is now changing into an inactive asteroid. Comet Borrelly, meanwhile, is one of the most active comets to frequent the inner solar system.

August 12, 1999
Telcom 1 (Palapa D-1)
Indonesia / U.S.A. / Europe

This first satellite in Indonesia's fourth-generation cluster of geosynchronous communications satellites was built by Lockheed Martin and launched on an Ariane rocket. Placed over Indonesia, it augments the Palapa C satellites (see **Palapa C-1–February 1, 1996**), using 36 transponders in the 4/6 GHz waveband.

September 6, 1999
Yamal 101 / Yamal 102
Russia

These two geosynchronous communications satellites were designed to provide telecommunications for Gazprom, the Russian gas industry monopoly. After launch, however, Yamal 101, intended for placement over western Russian, failed to obey any commands and was considered a loss. Yamal 102, meanwhile, was placed over eastern Russian, where—as of December 1999—it was operating as expected.

September 24, 1999
Ikonos
U.S.A.

Built by Lockheed Martin, Ikonos was designed to compete with commercial Earth resource satellites like **IRS 1A (March 17, 1988)** and **Spot 4 (March 23, 1998)**. See also **Spot 1 (February 22, 1986)**. It provides the highest-resolution commercial pictures of the earth's surface yet available commercially, with black-and-white images resolving objects as small as 3 feet across and color images resolving objects 12 feet across. Its first image showing the Mall in Washington, D.C., clearly showed individual cars and even the lines of people waiting to visit the Washington Monument.

September 26, 1999
LMI 1
U.S.A. / Russia

Built by Lockheed Martin and launched on a Russian rocket, this geosynchronous communications satellite followed **Bonum 1 (November 22, 1998)** as the second foreign-built communications satellite serving Russia and the first to provide long-distance telephone service. Placed over the Indian Ocean, it carries 44 transponders in the 4/6 GHz and 11/14 GHz wavebands. As of December 1999, however, the Russian long-

October 10, 1999
DirecTV 1R
U.S.A. / Ukraine / Russia

distance telephone market was only about one-third of what the satellite's owners had expected.

October 10, 1999
DirecTV 1R
U.S.A. / Ukraine / Russia

As the first commercial launch from the Sea Launch platform built by a Boeing-led consortium (see **DemoSat–March 27, 1999**), DirecTV 1R was placed in geosynchronous orbit over the United States to supplement DirecTV's direct broadcasting satellite cluster. See **DBS 1 (December 18, 1993)** and **Tempo 2 (March 8, 1997)**.

October 14, 1999
CBERS 1 / SACI 1
Brazil / China

- **CBERS 1** The China-Brazil Earth Resources Satellite (CBERS) 1 is an Earth resource satellite operating in optical and infrared wavebands and able to resolve objects as small as 65 feet across with an image swathe of 75 miles. It was placed in a sun-synchronous polar orbit where every 26 days it completes a sweep of the earth. The spacecraft's high-resolution cameras are also capable of taking stereoscopic images.

- **SACI 1** The first scientific research satellite built entirely in Brazil, SACI 1 (the Portuguese acronym for Satellite of Scientific Application) was designed to conduct research on Earth's magnetic field. After reaching orbit, however, the spacecraft failed to respond to commands, and as of December 1999, it was considered a loss.

November 19, 1999
Shenzhou
China

Shenzhou (translated as either "beautiful land" or "divine craft") was the first unmanned test of China's manned capsule. Similar in design to the Soviet/Russian Soyuz capsule, Shenzhou had an orbital module, a descent module, and an equipment module with two solar panels attached (though on this test flight, the solar panels were not deployed). The spacecraft completed 14 orbits before its descent capsule returned to Earth, landing in China about 260 miles east of its launchpad, where it was successfully recovered.

December 10, 1999
XMM
Europe

The first commercial use of the heavy-lift Ariane 5 booster (see **Ariane 5–August 30, 1997**) put in orbit XMM (for X-ray Multiple Mirrors), the European Space Agency's second x-ray telescope (following **Exosat–May 26, 1983**). XMM carries three main instruments. An x-ray CCD camera will produce x-ray images covering the electromagnetic spectrum from 0.5 to 10 KeV. An optical/ultraviolet telescope (1,700–5,000 angstroms) with a 12-inch mirror produces images designed to match those produced by the x-ray CCD camera. A third instrument will allow scientists to obtain x-ray spectra, making XMM the first space telescope with this capability.

December 18, 1999
Terra
U.S.A.

Terra is the first in a planned 10-satellite project aimed at studying Earth and its environment. Placed in a sun-synchronous orbit, its five instruments measure Earth's radiation budget (the total amount of energy received from the Sun, minus what is dissipated back into space) while also taking detailed images in 14 multispectral wavelengths ranging from the optical (0.3 microns) to the far infrared (14.4 microns). Image resolution can be as good as 50 feet, depending on the instrument and circumstances.

Over its planned six-year life span, Terra will provide data on cloud temperatures, atmospheric content, pollution, vegetation growth, forest fires, variations in ice and snow cover, ocean currents, and Earth's surface temperature.

December 19, 1999
STS 103, Discovery Flight #27
Hubble Space Telescope, Repair
U.S.A. / Switzerland / France

The third maintenance mission to the Hubble Space Telescope (see **STS 61–December 2, 1993**; **STS 82–February 11, 1997**) since its deployment from **STS 31 (April 24, 1990)**, STS 103 had originally been scheduled for June 2000. Its launch was moved up because of the deterioration to Hubble's gyroscopes, the last of which failed in November 1999. Because these gyroscopes were not functioning, the telescope was placed in safe mode, unable to make observations.

The mission was crewed by Curtis Brown (STS 47–September 12, 1992; STS 66–November 3, 1994; STS 77–May 19, 1996; STS 85–August 7, 1997; STS 95–October 29, 1998), Scott Kelly, Steven Smith (STS 68–September 30, 1994; STS 82–February 11, 1997), Michael Foale (STS 45–March 24, 1992; STS 56–April 8, 1993; STS 63–February 3, 1995; STS 84–May 15, 1997), John Grunsfeld (STS 67–March 2, 1995; STS 81–January 12, 1997), Swiss Claude Nicollier (STS 46–July 31, 1992; STS 61–December 2, 1993; STS 75–February 22, 1996), and Frenchman Jean-Francois Clervoy (STS 66–November 3, 1994; STS 84–May 15, 1997).

Lasting eight days, the mission included three space walks to install the six new gyroscopes, as well as a new guidance sensor, a new computer, a voltage/temperature kit for the spacecraft's batteries, a new transmitter, and a new solid state recorder. In addition, thermal insulation blankets were installed to further protect the telescope from the space environment.

December 21, 1999, 7:13 GMT
Acrimsat / Kompsat / Celestis 3
U.S.A. / South Korea

- *Acrimsat* Designed to measure the total solar constant in order to find out how much the variations in the Sun's output contribute to climate change, Acrimsat (Active Cavity Radiometer Irradiance Monitor Satellite) continues work begun with **Solar Max (February 14, 1980)**. It is designed to gather data for five years.

- *Kompsat* Built by the Korea Aerospace Research Institute of South Korea, Kompsat carries three instruments and was placed in a sun-synchronous orbit. Its primary instrument is an optical camera whose images cover a swath 10 miles wide with a resolution of less than 22 feet. Its data will be used to produce detailed digital topographic maps of Korea.

  The second instrument covers a multispectral range from the optical (4,000 angstroms) to the infrared (9,000 angstroms) and will be used to monitor the ocean colors on Earth, which in turn indicate variations in biological life.

  The third instrument tests the effect of high energy particles in low-earth orbit on modern microelectronic equipment.

- *Celestis 3* Celestis 3, the third "burial" flight of the private company Celestis, carried the cremated remains of 36 individuals. See **Celestis 1 (April 21, 1997)** for more information.

December 21, 1999, 0:50 GMT (December 22, 1999)
Galaxy 11
U.S.A.

As of December 1999, Galaxy 11 is the largest private communications satellite ever launched. Built by Hughes Communications for Pan American Satellites, it was positioned over Central America where it provides video, telephone, and data services to North America and Brazil, joining the Pan American Satellite cluster of satellites (see **Pan American Satellite 1–June 15, 1988**).

Fully deployed, Galaxy 11 is over 100 feet in length and almost 30 feet wide. In order to produce over 10 kilowatts of power, its solar array also carries along its perimeter angled-mirrored reflectors to concentrate the Sun's light on the panels, using designs first tested on **Deep Space 1 (October 24, 1998)**. The satellite also uses xenon ion engines to maintain its orbital position (see **Sert 2–February 4, 1970** and **Gals 1–January 20, 1994**).

Appendix 1
Satellites and Missions Listed Alphabetically

38C–September 28, 1963
83C–December 13, 1964
ACE–August 25, 1997
Acrimsat–December 21, 1999
ACTS–September 12, 1993
ADEOS (Midori)–August 17, 1996
Aeros 1–December 16, 1972
Aeros 2–July 16, 1974
AfriStar–October 28, 1998
Agila 2–August 19, 1997
Ajisai (EGP)–August 12, 1986
Akebono (EXOS D)–February 21, 1989
Alexis–April 25, 1993
Almaz 1–March 31, 1991
Alouette 1–September 29, 1962
Alouette 2–November 29, 1965
Altair 1 (Cosmos 1700)–October 25, 1985
Amos–May 16, 1996
AMSC 1–April 7, 1995
Anik A1 (Telesat 1)–November 10, 1972
Anik B (Telesat 4)–December 15, 1978
Anik C1–April 12, 1985
Anik C2 (Telesat 7)–June 18, 1983
Anik C3 (Telesat 6)–November 11, 1982
Anik D1 (Telesat 5)–August 26, 1982
Anik D2–November 8, 1984
Anik E2–April 4, 1991
Anna 1B–October 31, 1962
ANS 1–August 30, 1974
APEX–August 3, 1994
Apollo 1–January 27, 1967
Apollo 4–November 9, 1967
Apollo 5–January 22, 1968
Apollo 6–April 4, 1968
Apollo 7–October 11, 1968
Apollo 8–December 21, 1968
Apollo 9–March 3, 1969
Apollo 10–May 18, 1969
Apollo 11–July 16, 1969
Apollo 12–November 14, 1969
Apollo 13–April 11, 1970
Apollo 14–January 31, 1971
Apollo 15–July 26, 1971
Apollo 16–April 16, 1972
Apollo 17–December 7, 1972
Apollo 18 (ASTP)–July 15, 1975
Apple–June 19, 1981
Apstar 1–July 21, 1994
Arabsat 1A–February 8, 1985
Arabsat 1B–June 17, 1985

ARD–October 21, 1998
Argos–February 23, 1999
Ariane 1–June 19, 1981
Ariane 5–October 30, 1997
Ariel 1–April 26, 1962
Ariel 2–March 27, 1964
Ariel 3–May 5, 1967
Ariel 4–December 11, 1971
Ariel 5–October 5, 1974
Ariel 6 (UK 6)–June 2, 1979
Arsene–May 12, 1993
Aryabhata–April 19, 1975
ASC 1–August 27, 1985
ASCA (Astro-D)–February 20, 1993
Asiasat 1–April 7, 1990
Asterix–November 26, 1965
ASTEX, see STP 1
ASTP, see Apollo 18, Soyuz 19
Astra 1A–December 11, 1988
Astrid 1–January 24, 1995
Astrid 2–December 10, 1998
Astro-A, see Hinotori
Astro-B, see Tenma
Astro-C, see Ginga
Astro-D, see ASCA
Astron–March 23, 1983
Astron 2, see Granat
Atlantis, see STS 51J (STS 21), STS 61B
    (STS 23), STS 27, STS 30, STS 34, STS
    36, STS 38, STS 37, STS 43, STS 44,
    STS 45, STS 46, STS 66, STS 71, STS
    74, STS 76, STS 79, STS 81, STS 84,
    STS 86
Atmosphere 1–September 3, 1990
Atmosphere Explorer, see Explorer 17
Atmosphere Explorer B, see Explorer 32
Atmosphere Explorer C, see Explorer 51
Atmosphere Explorer D, see Explorer 54
Atmosphere Explorer E, see Explorer 55
ATS 1–December 7, 1966
ATS 3–November 5, 1967
ATS 5–August 12, 1969
ATS 6–May 30, 1974
Aura–September 27, 1975
Aureol 1–December 27, 1971
Aurora 1 (October 28, 1982), see RCA
    Satcom 5
Aurora 1–June 29, 1967
Aurora 2–May 29, 1991
Aurora 7 (Mercury 7)–May 24, 1962

Aurorae–October 3, 1968
Aussat A1–August 27, 1985
Aussat A2–November 26, 1985
Aussat B1 (Optus B1)–August 14, 1992
Azur–November 8, 1969
Badr A–July 16, 1990
Beacon Explorer B, see Explorer 22
Beacon Explorer C, see Explorer 27
BeppoSax–April 30, 1996
Bhaskara 1–June 7, 1979
Big Joe–September 9, 1959
Bion 1 (Cosmos 605)–October 31, 1973
Bion 2 (Cosmos 690)–October 22, 1974
Bion 3 (Cosmos 782)–November 25, 1975
Bion 4 (Cosmos 936)–August 3, 1977
Bion 5 (Cosmos 1129)–September 25, 1979
Bion 6 (Cosmos 1514)–December 14, 1983
Bion 7 (Cosmos 1667)–July 10, 1985
Bion 8 (Cosmos 1887)–September 29, 1987
Bion 9 (Cosmos 2044)–September 15, 1989
Bion 10 (Cosmos 2229)–December 29, 1992
Bion 11–December 24, 1996
Biosatellite 2–September 7, 1967
Biosatellite 3–June 29, 1969
Bonum 1–November 22, 1998
BOR-4, see Cosmos 1374
Boreas–October 1, 1969
Brasilsat A1–February 8, 1985
Brasilsat B1–August 10, 1994
BremSat 1–February 3, 1994
BSAT 1A–April 16, 1997
Buran–November 15, 1988
Cakrawarta 1–November 12, 1997
Calsphere–December 13, 1962
CAMEO–October 13, 1978
Cannonball 1 (OV1-16)–July 11, 1968
Cannonball 2–August 7, 1971
Canyon 1–August 6, 1968
Cassini–October 15, 1997
Castor–May 17, 1975
CAT 1–December 24, 1979
CAT 3–June 19, 1981
CAT 4–December 20, 1981
CBERS 1–October 14, 1999
CCE–August 16, 1984
Celestis–April 21, 1997
Celestis 2–February 10, 1998
Celestis 3–December 21, 1999
Cerise–July 7, 1995
Chalet 1–June 10, 1978

Appendix 1. Satellites and Missions Listed Alphabetically

Challenger, see STS 6, STS 7, STS 8, STS 41B (STS 10), STS 41C (STS 11), STS 41G (STS 13), STS 51B (STS 17), STS 51F (STS 19), STS 51G (STS 18), STS 61A (STS 22), STS 51L (STS 25)
Chandra–July 23, 1999
China 9/China 10/China 11–September 19, 1981
China 20–August 5, 1987
China 23–August 5, 1988
China 35–October 6, 1992
Clementine–January 25, 1994
COBE–November 18, 1989
Columbia, see STS 1, STS 2, STS 3, STS 4, STS 5, STS 9, STS 61C (STS 24), STS 28, STS 2, STS 35, STS 40, STS 50, STS 52, STS 55, STS 58, STS 62, STS 65, STS 73, STS 75, STS 78, STS 80, STS 83, STS 94, STS 87, STS 90, STS 93
Compton Gamma Ray Observatory–April 5, 1991
Comstar 1–May 13, 1976
Conestoga–September 9, 1982
Copernicus (OAO 3)–August 21, 1972
Coronas I, see Intercosmos 26
CORSA B, see Hakucho
COS B–August 9, 1975
Cosmos 1–March 16, 1962
Cosmos 2 (Sputnik 12)–April 6, 1962
Cosmos 3–April 24, 1962
Cosmos 4–April 26, 1962
Cosmos 5 (Sputnik 15)–May 28, 1962
Cosmos 7–July 28, 1962
Cosmos 17–May 22, 1963
Cosmos 26–March 18, 1964
Cosmos 41–August 22, 1964
Cosmos 44–August 28, 1964
Cosmos 49–October 24, 1964
Cosmos 51–December 10, 1964
Cosmos 110–February 22, 1966
Cosmos 122–June 25, 1966
Cosmos 133–November 28, 1966
Cosmos 140–February 7, 1967
Cosmos 144–February 28, 1967
Cosmos 148–March 16, 1967
Cosmos 156–April 27, 1967
Cosmos 166–June 16, 1967
Cosmos 186–October 27, 1967
Cosmos 188–October 30, 1967
Cosmos 198–December 27, 1967
Cosmos 248–October 19, 1968
Cosmos 249–October 20, 1968
Cosmos 252–November 1, 1968
Cosmos 343, see Strela 1
Cosmos 373–October 20, 1970
Cosmos 375–October 30, 1970
Cosmos 398–February 26, 1971
Cosmos 434–August 12, 1971
Cosmos 605, see Bion 1
Cosmos 613–November 30, 1973
Cosmos 637–March 26, 1974
Cosmos 638–April 3, 1974
Cosmos 690, see Bion 2
Cosmos 699–December 24, 1974
Cosmos 782, see Bion 3
Cosmos 881/Cosmos 882–December 15, 1976
Cosmos 929–July 17, 1977
Cosmos 936, see Bion 4
Cosmos 954–September 18, 1977
Cosmos 1001–April 4, 1978

Cosmos 1076 (Okean-E)–February 12, 1979
Cosmos 1129, see Bion 5
Cosmos 1267–April 25, 1981
Cosmos 1374 (BOR-4)–June 4, 1982
Cosmos 1383 (COSPAS 1)–June 30, 1982
Cosmos 1402–August 27, 1982
Cosmos 1413/Cosmos 1414/Cosmos 1415–October 12, 1982
Cosmos 1443–March 2, 1983
Cosmos 1500 (Okean-OE)–September 28, 1983
Cosmos 1514, see Bion 6
Cosmos 1645, see Foton 1
Cosmos 1667, see Bion 7
Cosmos 1686–September 27, 1985
Cosmos 1700, see Altair 1
Cosmos 1809, see Ionosonde
Cosmos 1870–July 25, 1987
Cosmos 1887, see Bion 8
Cosmos 1989–January 10, 1989
Cosmos 2024–May 31, 1989
Cosmos 2044, see Bion 9
Cosmos 2176–January 24, 1992
Cosmos 2229, see Bion 10
Cosmos 2268 through Cosmos 2273–February 12, 1994
Cosmos 2349–February 17, 1998
COSPAS 1, see Cosmos 1383
Coupon–November 12, 1997
Courier 1B–October 4, 1960
Crista-Spas 1–November 3, 1994
Crista-Spas 2–August 7, 1997
CRRES–July 25, 1990
CTA–October 22, 1992
CTS 1, see Hermes
DarpaSat–March 13, 1994
Dash 1–May 9, 1963
Dash 2–October 5, 1966
DATS–July 1, 1967
DBS 1–December 18, 1993
DEBUT, see Orizuru
Deep Space 1–October 24, 1998
Deep Space 1, Braille Flyby–July 28, 1999
DemoSat–March 27, 1999
DFH (Dong Fang Hong) 3A–November 29, 1994
Diadème 1–February 8, 1967
Diadème 2–February 15, 1967
DIAL-MIKA–March 10, 1970
DIAL-WIKA–March 10, 1970
DirecTV 1R–October 10, 1999
Discoverer 1–February 28, 1959
Discoverer 13–August 10, 1960
Discoverer 14–August 18, 1960
Discoverer 17–November 12, 1960
Discoverer 18–December 8, 1960
Discoverer 36–December 12, 1961
Discovery, see STS 41D (STS 12), STS 51A (STS 14), STS 51C (STS 15), STS 51D (STS 16), STS 51G (STS 18), STS 51I (STS 20), STS 26, STS 29, STS 33, STS 31, STS 41, STS 39, STS 48, STS 42, STS 53, STS 56, STS 51, STS 60, STS 64, STS 63, STS 70, STS 82, STS 86, STS 91, STS 95, STS 96, STS 103
DLR-Tubsat–May 26, 1999
DMSP Block-4A F1–January 19, 1965
Dodecapole 1–March 9, 1965
Dodge 1–July 1, 1967
Dong Fang Hong 3A, see DFH 3A
DSCS II 1/DSCS II 2–November 3, 1971
DSCS III–October 30, 1982

DSP 16–November 24, 1991
DSP-647 1, see IMEWS 1
Dynamics Explorer 1/Dynamics Explorer 2–August 3, 1981
Early Bird (Intelsat 1)–April 6, 1965
Echo 1–August 12, 1960
Echo 2–January 25, 1964
Echostar 1–December 28, 1995
ECS 1, see Eutelsat 1
EGP, see Ajisai
Einstein (HEAO 2)–November 13, 1978
Ekran 1–October 26, 1976
Elektro–October 31, 1994
Elektron 1/Elektron 2–January 30, 1964
Elektron 3/Elektron 4–July 11, 1964
Endeavour, see STS 49, STS 47, STS 54, STS 57, STS 61, STS 59, STS 68, STS 67, STS 69, STS 72, STS 77, STS 89, STS 88
Enterprise, Flight 4–August 12, 1977
Eole–August 16, 1971
Equator S–December 2, 1997
ERBS–October 5, 1984
ERS 1–July 17, 1991
ERS 2–April 21, 1995
ERS 16 (ORS 2)–June 9, 1966
ERS 18–April 28, 1967
ESA-GEOS 1–April 20, 1977
ESA-GEOS 2–July 14, 1978
ESRO 4–November 22, 1972
ESSA 1–February 3, 1966
ETS 2, see Kiku 2
ETS 3, see Kiku 4
ETS 4, see Kiku 3
ETS 5, see Kiku 5
ETS 7, see Kiku 7
EURECA–July 31, 1992; June 21, 1993
Eutelsat 1 (ECS 1)–June 16, 1983
Eutelsat 2 F1–August 30, 1990
EUVE–June 7, 1992
EXOS A, see Kyokko
EXOS B, see Jikiken
EXOS C, see Ohzora
EXOS D, see Akebono
Exosat–May 26, 1983
Explorer 1–February 1, 1958
Explorer 3–March 26, 1958
Explorer 4–July 6, 1958
Explorer 6–August 7, 1959
Explorer 7–October 13, 1959
Explorer 8–November 3, 1960
Explorer 9–February 16, 1961
Explorer 10–March 25, 1961
Explorer 11–April 27, 1961
Explorer 12–August 16, 1961
Explorer 14–October 2, 1962
Explorer 15–October 27, 1962
Explorer 16–December 16, 1962
Explorer 17 (Atmosphere Explorer)–April 3, 1963
Explorer 18, see Imp A
Explorer 19–December 19, 1963
Explorer 20–August 25, 1964
Explorer 22 (Beacon Explorer B)–October 10, 1964
Explorer 23–November 6, 1964
Explorer 24–November 21, 1964
Explorer 25 (Injun 4)–November 21, 1964
Explorer 26–December 21, 1964
Explorer 27 (Beacon Explorer C)–April 29, 1965
Explorer 29 (GEOS 1)–November 6, 1965

Appendix 1. Satellites and Missions Listed Alphabetically

Explorer 30 (Solrad 8)–November 19, 1965
Explorer 31–November 29, 1965
Explorer 32 (Atmosphere Explorer B)–May 25, 1966
Explorer 33, *see* Imp D
Explorer 35, *see* Imp E
Explorer 39–August 8, 1968
Explorer 42, *see* Uhuru
Explorer 45–November 15, 1971
Explorer 46–August 13, 1972
Explorer 48 (SAS 2)–November 15, 1972
Explorer 51 (Atmosphere Explorer C)–December 16, 1973
Explorer 52, *see* Hawkeye
Explorer 53 (SAS 3)–May 7, 1975
Explorer 54 (Atmosphere Explorer D)–October 6, 1975
Explorer 55 (Atmosphere Explorer E)–November 20, 1975
Express 1–October 13, 1994
Express RV–January 15, 1995
Eyesat 1–September 26, 1993
FAISAT–January 24, 1995
Faith 7 (Mercury 9)–May 15, 1963
Fanhui Shu Weixing 12–October 5, 1990
FASat-Bravo–July 10, 1998
FAST–August 21, 1996
Feng Yun 1A–September 6, 1988
Feng Yun 2B–June 10, 1997
Feng Yun 3–May 10, 1999
Ferret 1–June 2, 1962
FLTSATCOM 1–February 9, 1978
FORTE–August 29, 1997
Foton 1 (Cosmos 1645)–April 16, 1985
Foton 5–April 26, 1989
Foton 6–April 11, 1990
France 1–December 6, 1965
Freedom 7 (Mercury 3)–May 5, 1961
Freja–October 6, 1992
Friendship 7 (Mercury 6)–February 20, 1962
Fuji 1 (Oscar 12)–August 12, 1986
Fuse–June 24, 1999
Galaxy 1–June 28, 1983
Galaxy 7–October 28, 1992
Galaxy 11–December 21, 1999
Galileo–October 18, 1989
Galileo, Gaspra Flyby–October 29, 1991
Galileo, Ida Flyby–August 28, 1993
Galileo, Jupiter Orbit–December 7, 1995
Gals 1–January 20, 1994
Gamma–July 11, 1990
GE 1–September 8, 1996
GE Satcom–January 12, 1986
Gemini 2–January 19, 1965; November 3, 1966
Gemini 3–March 23, 1965
Gemini 4–June 3, 1965
Gemini 5–August 21, 1965
Gemini 6–December 15, 1965
Gemini 7–December 4, 1965
Gemini 8–March 16, 1966
Gemini 8 Agena Target–March 16, 1966
Gemini 9–June 3, 1966
Gemini 9 ATDA–June 1, 1966
Gemini 10–July 18, 1966
Gemini 10 Agena Target–July 18, 1966
Gemini 11–September 12, 1966
Gemini 11 Agena Target–September 12, 1966
Gemini 12–November 11, 1966
Gemini 12 Agena Target–November 11, 1966
Geo-IK–November 29, 1994

GEOS 1, *see* Explorer 29
Geos 2–January 11, 1968
Geos 3–April 9, 1975
Geosat–March 13, 1985
Geosat Follow-On–February 10, 1998
Geotail–July 24, 1992
GGSE 1–January 11, 1964
GGSE 2/GGSE 3–March 9, 1965
GGTS 1–June 16, 1966
Ginga (Astro-C)–February 5, 1987
Giotto, Halley's Comet Flyby–March 13, 1986
Globalstar 1 through Globalstar 4–February 14, 1998
GLOMR–April 29, 1985; October 30, 1985
GOES 1–October 16, 1975
Gonets-D1 1 through Gonets-D1 3–February 14, 1997
Gorizont 1–December 19, 1978
Granat (Astron 2)–December 1, 1989
Greb, *see* Solrad 3
Gridsphere 1/Gridsphere 2–August 7, 1971
GRS–June 28, 1963
GStar 1–May 7, 1985
Gurwin Techsat 1B–July 10, 1998
Hagoromo–January 24, 1990
Hakucho (CORSA B)–February 21, 1979
Halca–February 12, 1997
Hawkeye (Explorer 52)–June 3, 1974
HCMM–April 26, 1978
Healthsat 2–September 26, 1993
HEAO 1–August 12, 1977
HEAO 2, *see* Einstein
HEAO 3–September 20, 1979
Helios 1–December 10, 1974
Helios 1A–July 7, 1995
Helios 2–January 15, 1976
HEOS 1–December 5, 1968
HEOS 2–January 31, 1972
Hermes (CTS 1)–January 17, 1976
HGS 1–December 25, 1997
HiLat (P83-1)–June 27, 1983
Himawari 1–July 14, 1977
Hinotori (Astro-A)–February 21, 1981
Hipparcos–August 8, 1989
Hispasat 1A–September 10, 1992
Hitch Hiker 1–June 27, 1963
Hiten (Muses A)–January 24, 1990
Hot Bird 1–March 28, 1995
Hubble Space Telescope–April 24, 1990; December 2, 1993; February 11, 1997; December 19, 1999
Huygens–October 15, 1997
ICE, *see* International Cometary Explorer
IDCSP 1 through IDCSP 7–June 16, 1966
Ikonos–September 24, 1999
IMEWS 1 (DSP-647 1)–May 5, 1971
IMEWS 6–June 26, 1976
Imp A (Explorer 18)–November 27, 1963
Imp D (Explorer 33)–July 1, 1966
Imp E (Explorer 35)–July 19, 1967
Informator 1–January 29, 1991
Injun 1–June 29, 1961
Injun 3–December 13, 1962
Injun 4, *see* Explorer 25
Injun 5–August 8, 1968
Inmarsat 2 F1–October 30, 1990
Inmarsat 3 F1–April 3, 1996
Insat 1A–April 10, 1982
Insat 1B–August 30, 1983
Insat 2A–July 9, 1992
Inspector–December 17, 1997

Intasat 1–November 15, 1974
Intelsat 1, *see* Early Bird
Intelsat 2A–October 26, 1966
Intelsat 2B–January 11, 1967
Intelsat 2C–March 23, 1967
Intelsat 2D–September 28, 1967
Intelsat 3B–December 19, 1968
Intelsat 4 F2–January 26, 1971
Intelsat 4A F1–September 26, 1975
Intelsat 5 F2–December 6, 1980
Intelsat 5A F10–March 22, 1985
Intelsat 6 F3–May 7, 1992
Intelsat 6A F2–October 27, 1989
Intelsat 7 F1–October 22, 1993
Intelsat 801–March 1, 1997
Intelsat K–June 10, 1992
Interball 1–August 2, 1995
Interball 2–August 29, 1996
Intercosmos 1–October 14, 1969
Intercosmos 6–April 7, 1972
Intercosmos 15–June 19, 1976
Intercosmos 18–October 24, 1978
Intercosmos 20–November 1, 1979
Intercosmos 21–February 6, 1981
Intercosmos 23, *see* Prognoz 10
Intercosmos 24–September 28, 1989
Intercosmos 25–December 18, 1991
Intercosmos 26 (Coronas I)–March 2, 1994
International Cometary Explorer (ICE)–September 11, 1985
International Space Station, *see* Unity, Zarya
Ionosonde (Cosmos 1809)–December 18, 1986
Ion Release Module–August 16, 1984
IRAS–January 26, 1983
Iridium 4 through Iridium 8–May 5, 1997
IRIS–May 17, 1968
IRS 1A–March 17, 1988
IRS-P4 Oceansat–May 26, 1999
ISAS–January 25, 1994
ISEE 1/ISEE 2–October 22, 1977
ISEE 3–August 12, 1978
ISIS 1–January 30, 1969
Iskra 2–May 17, 1982
Iskra 3–November 18, 1982
ISO–November 17, 1995
ISS 1–February 29, 1976
ISS 2, *see* Ume 2
Italsat 1–January 15, 1991
Itamsat–September 26, 1993
ITOS 1–January 23, 1970
IUE–January 26, 1978
JCSat 1–March 6, 1989
JERS 1–February 11, 1992
JETS 1, *see* Kiku 1
Jikiken (EXOS B)–September 16, 1978
Jindai (MABES)–August 12, 1986
Jumpseat 1–March 21, 1971
KH-4 9032–April 18, 1962
KH-7 1–July 12, 1963
KH-7-10–August 14, 1964
KH-9 1–June 15, 1971
KH-11 1–December 19, 1976
KH-12 1–February 28, 1990
Kiku 1 (JETS 1)–September 9, 1975
Kiku 2 (ETS 2)–February 23, 1977
Kiku 3 (ETS 4)–February 11, 1981
Kiku 4 (ETS 3)–September 3, 1982
Kiku 5 (ETS 5)–August 27, 1987
Kiku 7 (ETS 7)–November 28, 1997
Kompsat–December 21, 1999
Kopernikus 1–June 5, 1989

347

Appendix 1. Satellites and Missions Listed Alphabetically

Koreasat 1, see Mugunghwa 1
Kristall–May 31, 1990
Kvant 1–March 31, 1987
Kvant 2–November 26, 1989
Kyokko (EXOS A)–February 4, 1978
LACE–February 14, 1990
Lacrosse 1–December 2, 1988
Lageos 1–May 4, 1976
Lageos 2–October 22, 1992
Landsat 1–July 23, 1972
Landsat 4–July 16, 1982
Landsat 7–April 15, 1999
LCS 4–August 7, 1971
LDEF–April 6, 1984; January 9, 1990
LEASAT 1–November 8, 1984
LEASAT 2–August 30, 1984
LEASAT 3–April 12, 1985
LEASAT 4–August 27, 1985
LEASAT 5–January 9, 1990
LES 1–February 11, 1965
LES 2–May 6, 1965
LES 3/LES 4–December 21, 1965
LES 5–July 1, 1967
LES 6–September 26, 1968
LES 8/LES 9–March 15, 1976
Liberty Bell 7 (Mercury 4)–July 21, 1961
Little Joe 3–December 4, 1959
LMI 1–September 26, 1991
Lofti 2A–June 15, 1963
Losat X–July 4, 1991
Luch 1–December 16, 1994
Luna 1–January 2, 1959
Luna 2–September 12, 1959
Luna 3–October 4, 1959
Luna 9–January 31, 1966
Luna 10–March 31, 1966
Luna 11–August 24, 1966
Luna 12–October 22, 1966
Luna 13–December 21, 1966
Luna 14–April 7, 1968
Luna 15–July 13, 1969
Luna 16–September 12, 1970
Luna 17–November 10, 1970
Luna 18–September 2, 1971
Luna 19–September 28, 1971
Luna 20–February 14, 1972
Luna 21–January 8, 1973
Luna 22–May 29, 1974
Luna 23–October 28, 1974
Luna 24–August 9, 1976
Lunar Orbiter 1–August 10, 1966
Lunar Orbiter 2–November 6, 1966
Lunar Orbiter 3–February 5, 1967
Lunar Orbiter 4–May 4, 1967
Lunar Orbiter 5–August 1, 1967
Lunar Prospector–January 7, 1998
MABES, see Jindai
Macsat 1–May 9, 1990
Macsat 2–May 9, 1990
Magellan–May 4, 1989
Magellan, Venus Orbit–August 10, 1990
Magion 1–October 24, 1978
Magion 2–September 28, 1989
Magion 3–December 18, 1991
Magion 4–August 2, 1995
Magion 5–August 29, 1996
Magnum 1–January 24, 1985
Magnum 2–November 22, 1989
Magnum 3–November 15, 1990
Magsat–October 30, 1979
Mao 1–April 24, 1970
Maqsat 3–October 21, 1998

Marco Polo 1–August 27, 1989
MARECS A–December 20, 1981
Mariner 2–August 27, 1962
Mariner 2, Venus Flyby–December 14, 1962
Mariner 4–November 28, 1964
Mariner 4, Mars Flyby–July 15, 1965
Mariner 5, Venus Flyby–October 19, 1967
Mariner 6, Mars Flyby–July 31, 1969
Mariner 7, Mars Flyby–August 5, 1969
Mariner 9, Mars Orbit–November 14, 1971
Mariner 10–November 3, 1973
Mariner 10, Mercury Flyby #3–March 16, 1975
Mariner 10, Venus Flyby–February 5, 1974
Marisat 1–February 19, 1976
Mars 1–November 1, 1962
Mars 2–November 27, 1971
Mars 3–December 2, 1971
Mars 4, Mars Flyby–February 10, 1974
Mars 5, Mars Orbit–February 12, 1974
Mars 6, Mars Landing–March 12, 1974
Mars 7, Mars Flyby–March 9, 1974
Mars Climate Orbiter–December 11, 1998
Mars Global Surveyor–September 12, 1997
Mars Observer–September 25, 1992
Mars Pathfinder–July 4, 1997
Mars Polar Lander–January 3, 1999
Measat 1–January 12, 1996
Mercury 1A–December 19, 1960
Mercury 2–January 31, 1961
Mercury 2A–March 24, 1961
Mercury 3, see Freedom 7
Mercury 4, see Liberty Bell 7
Mercury 5–November 29, 1961
Mercury 6, see Friendship 7
Mercury 7, see Aurora 7
Mercury 8, see Sigma 7
Mercury 9, see Faith 7
Meteor 1-01–March 26, 1969
Meteor 1-30–June 18, 1980
Meteor 2-01–July 11, 1975
Meteor 3-01–October 24, 1985
Meteor 3-05–August 15, 1991
Meteosat 1–November 23, 1977
Meteosat 3–June 15, 1988
MHV–September 13, 1985
Microlab 1–April 3, 1995
Microsat–August 29, 1996
Microsat 1 through Microsat 7–July 17, 1991
Midas 2–May 24, 1960
Midas 4–October 21, 1961
Midori, see ADEOS
Midcourse Space Experiment (MSX)–April 24, 1996
MightySat 1–December 4, 1998
Milstar 1–February 7, 1994
Minisat 1–April 21, 1997
Mir–February 20, 1986
Miranda–March 9, 1974
Mirka–October 10, 1997
Molniya 1-01–April 23, 1965
Molniya 1-02–October 14, 1965
Molniya 1-03–April 25, 1966
Molniya 1S–July 30, 1974
Molniya 2-01–November 24, 1971
Molniya 3-01–November 21, 1974
Momo 1 (MOS 1)–February 19, 1987
Morelos 1–June 17, 1985
Morelos 2–November 26, 1985
MOS 1, see Momo 1
MSAT 1–April 20, 1996
MSTI 1–November 21, 1992

MSTI 2–May 9, 1994
MSTI 3–May 17, 1996
MSX, see Midcourse Space Experiment
Mugunghwa 1 (Koreasat 1)–August 5, 1995
Muses A, see Hiten
Musketball 1–August 7, 1971
Mylar Balloon–August 7, 1971
Myojo–February 3, 1994
Nadezhda 1–July 4, 1989
Nahuel 1A–January 30, 1997
NATO 3A–April 22, 1976
NATO 4A–January 8, 1991
Navstar 1–February 22, 1978
Navstar 5–February 9, 1980
Navstar B2 1–February 14, 1989
NEAR–February 17, 1996
NEAR, Eros Flyby–December 23, 1998
NEAR, Mathilde Flyby–June 27, 1997
Nilesat 1–April 28, 1998
Nimbus 1–August 28, 1964
Nimbus 7–October 13, 1978
Nimiq 1–May 21, 1999
NOAA 1–December 11, 1970
Northern Utah Satellite, see NUSAT
NOSS 1 (Whitecloud 1)–April 30, 1976
Nozomi–July 3, 1998
N-Star A–August 29, 1995
NTS 1 (Timation 3)–July 14, 1974
NTS 2–June 23, 1977
NUSAT (Northern Utah Satellite)–April 29, 1985
OAO 2–December 7, 1968
OAO 3, see Copernicus
ODERACS 1A-1F–February 3, 1994
ODERACS 2A-2F–February 3, 1995
OEX, see Orbiter Experiment
Ofeq 1–September 19, 1988
Ofeq 2–April 3, 1990
Ofeq 3–April 5, 1995
OFO–November 9, 1970
OGO 1–September 5, 1964
OGO 2–October 14, 1965
OGO 3–June 7, 1966
OGO 4–July 28, 1967
OGO 5–March 4, 1968
OGO 6–June 5, 1969
Ohsumi–February 11, 1970
Ohzora (EXOS C)–February 14, 1984
Okean-E, see Cosmos 1076
Okean-OE, see Cosmos 1500
Olympus 1–July 12, 1989
Optus B1, see Aussat B1
Orbcomm FM1/Orbcomm FM2–April 3, 1995
Orbcomm FM5 through Orbcomm FM12–December 23, 1997
Orbcomm X–July 17, 1991
Orbiscal 2, see OV1-17A
Orbiter Experiment (OEX)–November 26, 1985
OrbView 2, see SeaWiFS
Orfeus–November 19, 1996
Orfeus-Spas–September 12, 1993
Orion 1–November 29, 1994
Orizuru (DEBUT)–February 7, 1990
ORS 2–June 9, 1966
ORS 3–July 20, 1965
Orsted–February 23, 1999
Oscar 1–December 12, 1961
Oscar 2–June 2, 1962
Oscar 3–March 9, 1965
Oscar 4–December 21, 1965

Oscar 5–January 23, 1970
Oscar 6–October 15, 1972
Oscar 7–November 15, 1974
Oscar 8–March 5, 1978
Oscar 9, see UoSat 1
Oscar 10–June 16, 1983
Oscar 11, see UoSat 2
Oscar 12, see Fuji 1
Oscar 13–June 15, 1988
Oscar 14/Oscar 15, see UoSat 3/UoSat 4
Oscar 16/Oscar 17/Oscar 19–January 22, 1990
Oscar 18, see Webersat
OSO 1–March 7, 1962
OSO 2–February 3, 1965
OSO 3–March 8, 1967
OSO 4–October 15, 1967
OSO 5–January 22, 1969
OSO 6–August 9, 1969
OSO 7–September 29, 1971
OSO 8–June 21, 1975
OTS 2–May 11, 1978
OV1-2–October 5, 1965
OV1-4–March 30, 1966
OV1-5–March 30, 1966
OV1-8–July 14, 1966
OV1-13–April 6, 1968
OV1-15, see Spades
OV1-16, see Cannonball 1
OV1-17–March 18, 1969
OV1-17A (Orbiscal 2)–March 18, 1969
OV1-18–March 18, 1969
OV1-19–March 18, 1969
OV1-20–August 7, 1971
OV1-21–August 7, 1971
OV2-5/OV5-2–September 26, 1968
OV3-1–April 22, 1966
OV3-2–October 28, 1966
OV3-3–August 4, 1966
OV3-4–June 10, 1966
OV3-6–December 5, 1967
OV4-1T/OV4-1R–November 3, 1966
OV5-1/OV5-3–April 28, 1967
OV5-5/OV5-6/OV5-9–May 23, 1969
P76-5–May 22, 1976
P78-1, see Solwind
P83-1, see HiLat
PAC 1–August 9, 1969
Pageos 1–June 24, 1966
Palapa 1–July 8, 1976
Palapa B1–June 18, 1983
Palapa B2–February 3, 1984
Palapa B2R–April 13, 1990
Palapa C-1–February 1, 1996
Palapa D-1, see Telcom 1
PAMS–May 19, 1996
Pan American Satellite 1–June 15, 1988
Pan American Satellite 5–August 28, 1997
Pansat–October 29, 1998
Pegasus 1–February 16, 1965
Pegasus 2–May 25, 1965
Pegasus 3–July 30, 1965
Pegsat–April 5, 1990
Peole–December 12, 1970
Phobos 1–July 7, 1988
Phobos 2–July 12, 1988
Pion 1/Pion 2–May 25, 1989
Pioneer 1–October 11, 1958
Pioneer 3–December 6, 1958
Pioneer 4–March 3, 1959
Pioneer 5–March 11, 1960
Pioneer 6–December 16, 1965

Pioneer 7–August 17, 1966
Pioneer 8–December 13, 1967
Pioneer 9–November 8, 1968
Pioneer 10–March 3, 1972
Pioneer 10, Jupiter Flyby–December 3, 1973
Pioneer 10, Shutdown–March 31, 1997
Pioneer 11–April 6, 1973
Pioneer 11, Jupiter Flyby–December 4, 1974
Pioneer 11, Saturn Flyby–September 1, 1979
Pioneer 11, Shutdown–September 30, 1995
Pioneer Venus Orbiter–December 4, 1978
Pioneer Venus Probe–December 9, 1978
PIX 1–March 5, 1978
POGS–April 11, 1990
Polar Bear–November 14, 1986
Polar–February 24, 1996
Pollux–May 17, 1975
Polyot 1–November 1, 1963
Polyot 2–April 12, 1964
Posat 1–September 26, 1993
Priroda–April 23, 1996
Prognoz 1–April 14, 1972
Prognoz 9–July 1, 1983
Prognoz 10 (Intercosmos 23)–April 26, 1985
Progress 1–January 20, 1978
Progress M1–August 23, 1989
Progress M34–April 6, 1997
Progress M35–July 5, 1997
Prospero–October 28, 1971
Proton 1–July 16, 1965
Proton 2–November 2, 1965
Proton 3–July 6, 1966
Proton 4–November 16, 1968
QuikScat–June 19, 1999
Radarsat–November 4, 1995
Radcat–October 2, 1972
Radio-Rosto–December 26, 1994
Radio Sputnik 1/Radio Sputnik 2–October 26, 1978
Radose–June 15, 1963
Radsat–October 2, 1972
Raduga 1–December 22, 1975
Raduga 28–December 19, 1991
RAE 1–July 4, 1968
RAE 2–June 10, 1973
Ranger 3–January 26, 1962
Ranger 4–April 23, 1962
Ranger 6–January 30, 1964
Ranger 7–July 28, 1964
Ranger 8–February 17, 1965
Ranger 9–March 21, 1965
RCA Satcom 1–December 13, 1975
RCA Satcom 5 (Aurora 1)–October 28, 1982
RCA Satcom Ku2–November 26, 1985
Relay 1–December 13, 1962
Relay 2–January 21, 1964
Resurs F1–May 25, 1989
Resurs-500–November 16, 1992
Rex 1–June 29, 1991
Rhyolite 1–March 6, 1973
Rigidsphere–August 7, 1971
RM 1–November 9, 1970
RME–February 14, 1990
Rocsat–January 26, 1999
Rohini 1B–July 18, 1980
Rosat–June 1, 1990
Rossi X-ray Timing Explorer (RXTE)–December 30, 1995
RXTE, see Rossi X-ray Timing Explorer
Ryusei–February 3, 1994
S80-T–August 10, 1992
SAC-A–December 4, 1998

SACI 1–October 14, 1999
Safir 2–July 10, 1998
SAGE–February 18, 1979
Sakigake–January 7, 1985
Sakigake, Halley's Comet Flyby–March 11, 1986
Sakura 1–December 15, 1977
Sakura 2A–February 4, 1983
Sakura 3A–February 19, 1988
Salyut 1–April 19, 1971
Salyut 3–June 25, 1974
Salyut 4–December 26, 1974
Salyut 5–June 22, 1976
Salyut 6–September 29, 1977
Salyut 7–April 19, 1982
Samos 2–January 31, 1961
Sampex–July 3, 1992
San Marco 1–December 15, 1964
San Marco 2–April 26, 1967
San Marco 3–April 24, 1971
San Marco 4–February 18, 1974
San Marco D/L–March 25, 1988
SARA–July 17, 1991
SAS 1, see Uhuru
SAS 2, see Explorer 48
SAS 3, see Explorer 53
Satcom C1–November 20, 1990
Satmex 5–December 6, 1998
SBS 1–November 15, 1980
SBS 3–November 11, 1982
SBS 4–August 30, 1984
SCATHA–January 30, 1979
SCD 1–February 9, 1993
SCE–April 11, 1990
SCORE–December 18, 1958
SDS 1–June 2, 1976
Seasat–June 27, 1978
SeaWiFS (OrbView 2)–August 1, 1997
Secor 1–January 11, 1964
SECS–April 5, 1990
Seds–March 30, 1993
Sedsat 1–October 24, 1998
SERT 2–February 4, 1970
Shenzhou–November 19, 1999
Shijian 4–February 8, 1994
Shijian 5–May 10, 1999
Shinsei–September 28, 1971
Shiyan Tongxin Weixing 1, see STW 1
Sich 1–August 31, 1995
Sigma 7 (Mercury 8)–October 3, 1962
SIGNE 3–June 17, 1977
Sinosat 1–July 18, 1998
Sirio 1–August 25, 1977
Skylab–May 14, 1973
Skylab 2–May 25, 1973
Skylab 3–July 28, 1973
Skylab 4–November 16, 1973
Skynet 1–November 22, 1969
Skynet 2–November 23, 1974
Skynet 4B–December 11, 1988
SME–October 6, 1981
SMS 1–May 17, 1974
SNOE–February 25, 1998
Soho–December 2, 1995
SOICAL Cone/SOICAL Cylinder–September 30, 1969
Solar-A, see Yohkoh
Solar Max–April 6, 1984; February 14, 1980
Solidaridad 1–November 20, 1993
Solrad 1–June 22, 1960
Solrad 3 (Greb)–June 29, 1961
Solrad 6–June 15, 1963

Appendix 1. Satellites and Missions Listed Alphabetically

Solrad 7A–January 11, 1964
Solrad 8, see Explorer 30
Solrad 9–March 5, 1968
Solrad 10–July 8, 1971
Solrad 11A/Solrad 11B–March 15, 1976
Solwind (P78-1)–February 24, 1979
Soyuz 1–April 23, 1967
Soyuz 2–October 25, 1968
Soyuz 3–October 26, 1968
Soyuz 4–January 14, 1969
Soyuz 5–January 15, 1969
Soyuz 6–October 11, 1969
Soyuz 7–October 12, 1969
Soyuz 8–October 13, 1969
Soyuz 9–June 1, 1970
Soyuz 10–April 23, 1971
Soyuz 11–June 6, 1971
Soyuz 12–September 27, 1973
Soyuz 13–December 18, 1973
Soyuz 14–July 3, 1974
Soyuz 15–August 26, 1974
Soyuz 16–December 2, 1974
Soyuz 17–January 11, 1975
Soyuz 18A–April 5, 1975
Soyuz 18B–May 24, 1975
Soyuz 19 (ASTP)–July 15, 1975
Soyuz 20–November 17, 1975
Soyuz 21–July 6, 1976
Soyuz 22–September 15, 1976
Soyuz 23–October 14, 1976
Soyuz 24–February 7, 1977
Soyuz 25–October 9, 1977
Soyuz 26–December 10, 1977
Soyuz 27–January 10, 1978
Soyuz 28–March 2, 1978
Soyuz 29–June 15, 1978
Soyuz 30–June 27, 1978
Soyuz 31–August 26, 1978
Soyuz 32–February 25, 1979
Soyuz 33–April 10, 1979
Soyuz 34–June 6, 1979
Soyuz 35–April 9, 1980
Soyuz 36–May 26, 1980
Soyuz 37–July 23, 1980
Soyuz 38–September 18, 1980
Soyuz 39–March 22, 1981
Soyuz 40–May 14, 1981
Soyuz T1–December 16, 1979
Soyuz T2–June 5, 1980
Soyuz T3–November 27, 1980
Soyuz T4–March 12, 1981
Soyuz T5–May 13, 1982
Soyuz T6–June 24, 1982
Soyuz T7–August 19, 1982
Soyuz T8–April 20, 1983
Soyuz T9–June 27, 1983
Soyuz T10A–September 27, 1983
Soyuz T10B–February 8, 1984
Soyuz T11–April 3, 1984
Soyuz T12–July 17, 1984
Soyuz T13–June 6, 1985
Soyuz T14–September 17, 1985
Soyuz T15–March 13, 1986
Soyuz TM1–May 21, 1986
Soyuz TM2–February 5, 1987
Soyuz TM3–July 22, 1987
Soyuz TM4–December 21, 1987
Soyuz TM5–June 7, 1988
Soyuz TM6–August 29, 1988
Soyuz TM7–November 26, 1988
Soyuz TM8–September 5, 1989
Soyuz TM9–February 11, 1990

Soyuz TM10–August 1, 1990
Soyuz TM11–December 2, 1990
Soyuz TM12–May 18, 1991
Soyuz TM13–October 2, 1991
Soyuz TM14–March 17, 1992
Soyuz TM15–July 27, 1992
Soyuz TM16–January 24, 1993
Soyuz TM17–July 1, 1993
Soyuz TM18–January 8, 1994
Soyuz TM19–July 1, 1994
Soyuz TM20–October 3, 1994
Soyuz TM21–March 14, 1995
Soyuz TM22–September 3, 1995
Soyuz TM23–February 21, 1996
Soyuz TM24–August 17, 1996
Soyuz TM25–February 10, 1997
Soyuz TM26–August 5, 1997
Soyuz TM27–January 29, 1998
Soyuz TM28–August 13, 1998
Soyuz TM29–February 20, 1999
Space Flyer Unit–January 11, 1996; March 18, 1995
Spacenet 1–May 23, 1984
Spades (OV1-15)–July 11, 1968
Spartan 1–June 17, 1985
Spartan 201–April 8, 1993; September 9, 1994; September 7, 1995; November 19, 1997; October 29, 1998
Spartan 204–February 3, 1995
Spartan 206–January 11, 1996
Spartan 207–May 19, 1996
SPAS (Shuttle PAllet Satellite) 1–June 18, 1983
Spektr–May 20, 1995
Spot 1–February 22, 1986
Spot 4–March 23, 1998
Sputnik–October 4, 1957
Sputnik 2–November 3, 1957
Sputnik 3–May 15, 1958
Sputnik 4–May 15, 1960
Sputnik 5–August 19, 1960
Sputnik 6–December 1, 1960
Sputnik 9–March 9, 1961
Sputnik 10–March 25, 1961
Sputnik 12, see Cosmos 2
Sputnik 15, see Cosmos 5
Sputnik 40–November 3, 1997
SRATS, see Taiyo
ST 1–August 25, 1998
Starad–October 26, 1962
Stardust–February 7, 1999
Starfish Test–July 9, 1962
Starlette–February 6, 1975
Starshine–May 27, 1999
Stella–September 26, 1993
STEP 0 (TAOS)–March 13, 1994
STEX–October 3, 1998
STP 1 (ASTEX)–October 17, 1971
Strela 1 (Cosmos 343)–April 25, 1970
STRV-1A/STRV-1B–June 17, 1994
STS 1–April 12, 1981
STS 2–November 12, 1981
STS 3–March 22, 1982
STS 4–June 27, 1982
STS 5–November 11, 1982
STS 6–April 4, 1983
STS 7–June 18, 1983
STS 8–August 30, 1983
STS 9–November 28, 1983
STS 10, see STS 41B
STS 11, see STS 41C
STS 12, see STS 41D

STS 13, see STS 41G
STS 14, see STS 51A
STS 15, see STS 51C
STS 16, see STS 51D
STS 17, see STS 51B
STS 18, see STS 51G
STS 19, see STS 51F
STS 20, see STS 51I
STS 21, see STS 51J
STS 22, see STS 61A
STS 23, see STS 61B
STS 24, see STS 61C
STS 25, see STS 51L
STS 26–September 29, 1988
STS 27–December 2, 1988
STS 28–August 8, 1989
STS 29–March 13, 1989
STS 30–May 4, 1989
STS 31–April 24, 1990
STS 32–January 9, 1990
STS 33–November 22, 1989
STS 34–October 18, 1989
STS 35–December 2, 1990
STS 36–February 28, 1990
STS 37–April 5, 1991
STS 38–November 15, 1990
STS 39–April 28, 1991
STS 40–June 5, 1991
STS 41–October 6, 1990
STS 41B (STS 10)–February 3, 1984
STS 41C (STS 11)–April 6, 1984
STS 41D (STS 12)–August 30, 1984
STS 41G (STS 13)–October 5, 1984
STS 42–January 22, 1992
STS 43–August 2, 1991
STS 44–November 24, 1991
STS 45–March 24, 1992
STS 46–July 31, 1992
STS 47–September 12, 1992
STS 48–September 12, 1991
STS 49–May 7, 1992
STS 50–June 25, 1992
STS 51–September 12, 1993
STS 51A (STS 14)–November 8, 1984
STS 51B (STS 17)–April 29, 1985
STS 51C (STS 15)–January 24, 1985
STS 51D (STS 16)–April 12, 1985
STS 51F (STS 19)–July 29, 1985
STS 51G (STS 18)–June 17, 1985
STS 51I (STS 20)–August 27, 1985
STS 51J (STS 21)–October 3, 1985
STS 51L (STS 25)–January 28, 1986
STS 52–October 22, 1992
STS 53–December 2, 1992
STS 54–January 13, 1993
STS 55–April 26, 1993
STS 56–April 8, 1993
STS 57–June 21, 1993
STS 58–October 18, 1993
STS 59–April 9, 1994
STS 60–February 3, 1994
STS 61–December 2, 1993
STS 61A (STS 22)–October 30, 1985
STS 61B (STS 23)–November 26, 1985
STS 61C (STS 24)–January 12, 1986
STS 62–March 4, 1994
STS 63–February 3, 1995
STS 64–September 9, 1994
STS 65–July 8, 1994
STS 66–November 3, 1994
STS 67–March 2, 1995
STS 68–September 30, 1994

Appendix 1. Satellites and Missions Listed Alphabetically

STS 69–September 7, 1995
STS 70–July 13, 1995
STS 71–June 27, 1995
STS 72–January 11, 1996
STS 73–October 20, 1995
STS 74–November 12, 1995
STS 75–February 22, 1996
STS 76–March 22, 1996
STS 77–May 19, 1996
STS 78–June 20, 1996
STS 79–September 16, 1996
STS 80–November 19, 1996
STS 81–January 12, 1997
STS 82–February 11, 1997
STS 83–April 4, 1997
STS 84–May 15, 1997
STS 85–August 7, 1997
STS 86–September 25, 1997
STS 87–November 19, 1997
STS 88–December 4, 1998
STS 89–January 22, 1998
STS 90–April 17, 1998
STS 91–June 2, 1998
STS 93–July 23, 1999
STS 94–July 1, 1997
STS 95–October 29, 1998
STS 96–May 27, 1999
STS 103–December 19, 1999
STW (Shiyan Tongxin Weixing) 1–April 8, 1984
Suisei, Halley's Comet Flyby–March 8, 1986
Sunsat–February 23, 1999
Superbird A–June 5, 1989
Surcal 1A–December 13, 1962
Surfsat–November 4, 1995
Surveyor 1–May 30, 1966
Surveyor 3–April 17, 1967
Surveyor 5–September 8, 1967
Surveyor 6–November 7, 1967
Surveyor 7–January 7, 1968
SWAS–December 6, 1998
Symphonie 1–December 18, 1974
Syncom 1–June 26, 1963
Syncom 2–July 26, 1963
Syncom 3–August 19, 1964
Tacsat 1–February 9, 1969
Taiyo (SRATS)–February 24, 1975
Tansei 2–February 16, 1974
TAOS, see STEP 0
TDF 1–October 28, 1988
TDRS 1–April 4, 1983
TDRS 3–September 29, 1988
TDRS 4–March 13, 1989
TDRS 5–August 2, 1991
TDRS 6–January 13, 1993
TDRS 7–July 13, 1995
Telcom 1 (Palapa D-1)–August 12, 1999
Telecom 1A–August 4, 1984
Telecom 2A–December 16, 1991
Teledesic 1–February 25, 1998
Telesat 1, see Anik A1
Telesat 4, see Anik B
Telesat 5, see Anik D1
Telesat 6, see Anik C3
Telesat 7, see Anik C2
Telstar 1–July 10, 1962
Telstar 2–May 7, 1963
Telstar 3A–July 28, 1983
Telstar 3B–August 30, 1984
Telstar 3C–June 17, 1985
Telstar 401–December 16, 1993
Temisat–August 31, 1993

Tempo 2–March 8, 1997
Tenma (Astro-B)–February 20, 1983
Terra–December 18, 1999
Test and Training Satellite 1–December 13, 1967
Tethered Satellite System 1, see TSS 1
Tethered Satellite System 1R, see TSS 1R
TEX–April 11, 1990
Thaicom 1A–December 18, 1993
Thor 2–May 21, 1997
Thor Delta 1A–March 12, 1972
Timation 1–May 31, 1967
Timation 2–September 30, 1969
Timation 3, see NTS 1
Tiros 1–April 1, 1960
Tiros 2–November 23, 1960
Tiros 3–July 12, 1961
Tiros 4–February 8, 1962
Tiros 5–June 19, 1962
Tiros 6–September 18, 1962
Tiros 7–June 19, 1963
Tiros 8–December 21, 1963
Tiros 9–January 22, 1965
Tiros N–October 13, 1978
TMSat–July 10, 1998
TOMS–July 2, 1996
TOPEX-Poseidon–August 10, 1992
TRAAC–November 15, 1961
Trace–April 1, 1998
Transit 1B–April 13, 1960
Transit 2A–June 22, 1960
Transit 4A–June 29, 1961
Triad 1–September 2, 1972
TRMM–November 28, 1997
TRS 4–July 18, 1963
Trumpet 1–May 3, 1994
TSS 1–July 31, 1992
TSS 1R–February 22, 1996
Tubsat 1–July 17, 1991
Tubsat 2–January 25, 1994
Tubsat N/Tubsat N1–July 7, 1998
Turksat 1B–August 10, 1994
UARS–September 12, 1991
UHF Follow-On 1–September 3, 1993
Uhuru (Explorer 42 or SAS 1)–December 12, 1970
UK 6, see Ariel 6
UKS–August 16, 1984
Ulysses–October 6, 1990
Ulysses, First Solar Orbit–April 17, 1998
Ulysses, Jupiter Flyby–February 8, 1992
Ume 2 (ISS 2)–February 16, 1978
Unity–December 4, 1998
UoSat 1 (Oscar 9)–October 6, 1981
UoSat 2 (Oscar 11)–March 1, 1984
UoSat 3/UoSat4 (Oscar 14/Oscar 15)–January 22, 1990
UoSat 5–July 17, 1991
UoSat 12–April 21, 1999
UPM-LBSAT 1–July 7, 1995
Uribyol–August 10, 1992
Uribyol 2–September 26, 1993
Uribyol 3–May 26, 1999
USA 40–August 8, 1989
USA 41–August 8, 1989
USA 59–June 8, 1990
USA 89–December 2, 1992
Vanguard 1–March 17, 1958
Vanguard 2–February 17, 1959
Vanguard 3–September 18, 1959
Vega 1, Halley's Comet Flyby–March 6, 1986

Vega 1, Venus Flyby–June 10, 1985
Vega 2, Halley's Comet Flyby–March 9, 1986
Vega 2, Venus Flyby–June 14, 1985
Vela 1A/Vela 1B–October 17, 1963
Vela 2A/Vela 2B–July 17, 1964
Vela 5A/Vela 5B–May 23, 1969
Venera 1–February 12, 1961
Venera 2–November 12, 1965
Venera 3–November 16, 1965
Venera 4, Venus Landing–October 18, 1967
Venera 5, Venus Landing–May 16, 1969
Venera 6, Venus Landing–May 17, 1969
Venera 7, Venus Landing–December 15, 1970
Venera 8, Venus Landing–July 22, 1972
Venera 9, Venus Landing–October 22, 1975
Venera 10, Venus Landing–October 25, 1975
Venera 11, Venus Landing–December 25, 1978
Venera 12, Venus Landing–December 21, 1978
Venera 13, Venus Landing–March 1, 1982
Venera 14, Venus Landing–March 4, 1982
Venera 15, Venus Orbit–October 10, 1983
Venera 16, Venus Orbit–October 14, 1983
Viking–February 22, 1986
Viking 1, Mars Orbit/Landing–June 19, 1976
Viking 2–August 7, 1976
Voskhod 1–October 12, 1964
Voskhod 2–March 18, 1965
Vostok 1–April 12, 1961
Vostok 2–August 6, 1961
Vostok 3–August 11, 1962
Vostok 4–August 12, 1962
Vostok 5–June 14, 1963
Vostok 6–June 16, 1963
Voyager 1, Jupiter Flyby–March 5, 1979
Voyager 1, Saturn Flyby–December 12, 1980
Voyager 2, Jupiter Flyby–July 9, 1979
Voyager 2, Neptune Flyby–August 25, 1989
Voyager 2, Saturn Flyby–August 26, 1981
Voyager 2, Uranus Flyby–January 24, 1986
Wake Shield 1–February 3, 1994
Wake Shield 2–September 7, 1995
Wake Shield 3–November 19, 1996
Webersat (Oscar 18)–January 22, 1990
West Ford 1–October 21, 1961
West Ford 2–May 9, 1963
Westar 1–April 13, 1974
Westar 4–February 26, 1982
Westar 6–February 3, 1984
Westpac–July 10, 1998
Whitecloud 1, see NOSS 1
Wind–November 1, 1994
WIRE–March 4, 1999
Wresat 1–November 29, 1967
Yamal 101/Yamal 102–September 6, 1999
Yohkoh (Solar-A)–August 30, 1991
Yuri 1–April 7, 1978
Zarya–November 20, 1998
Zeya–March 4, 1997
Zond 1–April 2, 1964
Zond 2–November 30, 1964
Zond 3–July 18, 1965
Zond 4–March 2, 1968
Zond 5–September 15, 1968
Zond 6–November 10, 1968
Zond 7–August 8, 1969
Zond 8–October 20, 1970

Appendix 2
Satellites and Missions Listed by Subject

Subjects

Amateur Radio Communications 352
Astronomical Research 352
Atmosphere/Environment/Earth Resources Research 353
Atmospheric Density Research 354
Biological Research 354
Burial of Human Remains 355

Communication Propagation and Design Research 355
Communications Satellites 355
Geodetic/Navigational Research 357
Ionosphere/Magnetosphere Research 358
Manned Missions 359
Manned Space Laboratories/Modules 361

Materials Research 361
Micrometeoroid Research 362
Planetary Missions 362
Solar Research 363
Spacecraft Design Research 364
Surveillance 366
Weather Satellites 366

Amateur Radio Communications

Alphabetical Listing
Arsene–May 12, 1993
Badr A–July 16, 1990
Fuji 1 (Oscar 12)–August 12, 1986
Iskra 2–May 17, 1982
Iskra 3–November 18, 1982
Itamsat–September 26, 1993
Oscar 1–December 12, 1961
Oscar 2–June 2, 1962
Oscar 3–March 9, 1965
Oscar 4–December 21, 1965
Oscar 5–January 23, 1970
Oscar 6–October 15, 1972
Oscar 7–November 15, 1974
Oscar 8–March 5, 1978
Oscar 9, see UoSat 1
Oscar 10–June 16, 1983
Oscar 11, see UoSat 2
Oscar 12, see Fuji 1
Oscar 13–June 15, 1988
Oscar 14/Oscar 15, see UoSat 3/UoSat 4
Oscar 16/Oscar 17/Oscar 19–January 22, 1990
Oscar 18, see Webersat
Radio-Rosto–December 26, 1994
Radio Sputnik 1/Radio Sputnik 2–October 26, 1978
Sedsat 1–October 24, 1998
Sputnik 40–November 3, 1997
Tubsat 1–July 17, 1991
UoSat 1 (Oscar 9)–October 6, 1981
UoSat 2 (Oscar 11)–March 1, 1984
UoSat 3/UoSat 4 (Oscar 14/Oscar 15)–January 22, 1990

UoSat 5–July 17, 1991
Webersat (Oscar 18)–January 22, 1990

Chronological Listing
Oscar 1–December 12, 1961
Oscar 2–June 2, 1962
Oscar 3–March 9, 1965
Oscar 4–December 21, 1965
Oscar 5–January 23, 1970
Oscar 6–October 15, 1972
Oscar 7–November 15, 1974
Oscar 8–March 5, 1978
Radio Sputnik 1/Radio Sputnik 2–October 26, 1978
UoSat 1 (Oscar 9)–October 6, 1981
Iskra 2–May 17, 1982
Iskra 3–November 18, 1982
Oscar 10–June 16, 1983
UoSat 2 (Oscar 11)–March 1, 1984
Fuji 1 (Oscar 12)–August 12, 1986
Oscar 13–June 15, 1988
Oscar 16/Oscar 17/Oscar 19–January 22, 1990
UoSat 3/UoSat 4 (Oscar 14/Oscar 15)–January 22, 1990
Webersat (Oscar 18)–January 22, 1990
Badr A–July 16, 1990
Tubsat 1–July 17, 1991
UoSat 5–July 17, 1991
Arsene–May 12, 1993
Itamsat–September 26, 1993
Radio-Rosto–December 26, 1994
Sputnik 40–November 3, 1997
Sedsat 1–October 24, 1998

Astronomical Research

Alphabetical Listing
83C–December 13, 1964
ACE–August 25, 1997
Alexis–April 25, 1993
ANS 1–August 30, 1974
Argos–February 23, 1999
Ariel 2–March 27, 1964
Ariel 5–October 5, 1974
Ariel 6 (UK 6)–June 2, 1979
Aryabhata–April 19, 1975
ASCA (Astro-D)–February 20, 1993
Astro-B, see Tenma
Astro-C, see Ginga
Astro-D, see ASCA
Astron–March 23, 1983
BeppoSax–April 30, 1996
Chandra–July 23, 1999
COBE–November 18, 1989
Compton Gamma Ray Observatory–April 5, 1991
Copernicus (OAO 3)–August 21, 1972
CORSA B, see Hakucho
COS B–August 9, 1975
Cosmos 51–December 10, 1964
Einstein (HEAO 2)–November 13, 1978
EUVE–June 7, 1992
Exosat–May 26, 1983
Explorer 11–April 27, 1961
Explorer 42, see Uhuru
Explorer 48 (SAS 2)–November 15, 1972
Explorer 53 (SAS 3)–May 7, 1975
Fuse–June 24, 1999
Gamma–July 11, 1990
Ginga (Astro-C)–February 5, 1987
Granat (Astron 2)–December 1, 1989

Hakucho (CORSA B)–February 21, 1979
Halca–February 12, 1997
HEAO 1–August 12, 1977
HEAO 2, see Einstein
HEAO 3–September 20, 1979
Hipparcos–August 8, 1989
Hubble Space Telescope–April 24, 1990; December 2, 1993; February 11, 1997; December 19, 1999
Intercosmos 6–April 7, 1972
IRAS–January 26, 1983
ISEE 3–August 12, 1978
ISO–November 17, 1995
IUE–January 26, 1978
Midcourse Space Experiment (MSX)–April 24, 1996
MSX, see Midcourse Space Experiment
NEAR, Mathilde Flyby–June 27, 1997
OAO 2–December 7, 1968
OAO 3, see Copernicus
Orfeus–November 19, 1996
Orfeus-Spas–September 12, 1993
OSO 1–March 7, 1962
OSO 2–February 3, 1965
OSO 3–March 8, 1967
OSO 7–September 29, 1971
OSO 8–June 21, 1975
Prognoz 9–July 1, 1983
RAE 1–July 4, 1968
RAE 2–June 10, 1973
Ranger 3–January 26, 1962
Rosat–June 1, 1990
Rossi X-ray Timing Explorer (RXTE)–December 30, 1995
SARA–July 17, 1991
SAS 1, see Uhuru
SAS 2, see Explorer 48
SAS 3, see Explorer 53
SIGNE 3–June 17, 1977
Solrad 11A/Solrad 11B–March 15, 1976
Space Flyer Unit–January 11, 1996; March 18, 1995
Spartan 1–June 17, 1985
Spartan 204–February 3, 1995
SWAS–December 6, 1998
Tenma (Astro-B)–February 20, 1983
Thor Delta 1A–March 12, 1972
Uhuru (Explorer 42 or SAS 1)–December 12, 1970
UK 6, see Ariel 6
Vela 5A/Vela 5B–May 23, 1969

Chronological Listing
Explorer 11–April 27, 1961
Ranger 3–January 26, 1962
OSO 1–March 7, 1962
Ariel 2–March 27, 1964
Cosmos 51–December 10, 1964
83C–December 13, 1964
OSO 2–February 3, 1965
OSO 3–March 8, 1967
RAE 1–July 4, 1968
OAO 2–December 7, 1968
Vela 5A/Vela 5B–May 23, 1969
Uhuru (Explorer 42 or SAS 1)–December 12, 1970
OSO 7–September 29, 1971
Thor Delta 1A–March 12, 1972
Intercosmos 6–April 7, 1972
Copernicus (OAO 3)–August 21, 1972
Explorer 48 (SAS 2)–November 15, 1972
RAE 2–June 10, 1973

ANS 1–August 30, 1974
Ariel 5–October 5, 1974
Aryabhata–April 19, 1975
Explorer 53 (SAS 3)–May 7, 1975
OSO 8–June 21, 1975
COS B–August 9, 1975
Solrad 11A/Solrad 11B–March 15, 1976
SIGNE 3–June 17, 1977
HEAO 1–August 12, 1977
IUE–January 26, 1978
ISEE 3–August 12, 1978
Einstein (HEAO 2)–November 13, 1978
Hakucho (CORSA B)–February 21, 1979
Ariel 6 (UK 6)–June 2, 1979
HEAO 3–September 20, 1979
IRAS–January 26, 1983
Tenma (Astro-B)–February 20, 1983
Astron–March 23, 1983
Exosat–May 26, 1983
Prognoz 9–July 1, 1983
Spartan 1–June 17, 1985
Ginga (Astro-C)–February 5, 1987
Hipparcos–August 8, 1989
COBE–November 18, 1989
Granat (Astron 2)–December 1, 1989
Hubble Space Telescope–April 24, 1990
Rosat–June 1, 1990
Gamma–July 11, 1990
Compton Gamma Ray Observatory–April 5, 1991
SARA–July 17, 1991
EUVE–June 7, 1992
ASCA (Astro-D)–February 20, 1993
Alexis–April 25, 1993
Orfeus-Spas–September 12, 1993
Hubble Space Telescope–December 2, 1993
Spartan 204–February 3, 1995
Space Flyer Unit–March 18, 1995
ISO–November 17, 1995
Rossi X-ray Timing Explorer (RXTE)–December 30, 1995
Space Flyer Unit–January 11, 1996
Midcourse Space Experiment (MSX)–April 24, 1996
BeppoSax–April 30, 1996
Orfeus–November 19, 1996
Hubble Space Telescope–February 11, 1997
Halca–February 12, 1997
NEAR, Mathilde Flyby–June 27, 1997
ACE–August 25, 1997
SWAS–December 6, 1998
Argos–February 23, 1999
Fuse–June 24, 1999
Chandra–July 23, 1999
Hubble Space Telescope–December 19, 1999

Atmosphere/Environment/ Earth Resources Research

Alphabetical Listing
ADEOS (Midori)–August 17, 1996
Almaz 1–March 31, 1991
Argos–February 23, 1999
Ariel 2–March 27, 1964
Atmosphere Explorer E, see Explorer 55
Bhaskara 1–June 7, 1979
CBERS 1–October 14, 1999
Cosmos 1076 (Okean-E)–February 12, 1979
Cosmos 1500 (Okean-OE)–September 28, 1983

Cosmos 1870–July 25, 1987
Cosmos 2349–February 17, 1998
Crista-Spas 1–November 3, 1994
Crista-Spas 2–August 7, 1997
DLR-Tubsat–May 26, 1999
ERBS–October 5, 1984
ERS 1–July 17, 1991
ERS 2–April 21, 1995
EURECA–July 31, 1992
Explorer 55 (Atmosphere Explorer E)–November 20, 1975
FASat-Bravo–July 10, 1998
Geosat–March 13, 1985
Geosat Follow-On–February 10, 1998
HCMM–April 26, 1978
Ikonos–September 24, 1999
Intercosmos 21–February 6, 1981
IRS 1A–March 17, 1988
IRS-P4 Oceansat–May 26, 1999
JERS 1–February 11, 1992
Kompsat–December 21, 1999
Landsat 1–July 23, 1972
Landsat 4–July 16, 1982
Landsat 7–April 15, 1999
Meteor 1-30–June 18, 1980
Meteor 3-05–August 15, 1991
Microlab–April 3, 1995
Midori, see ADEOS
Momo 1 (MOS 1)–February 19, 1987
MOS 1, see Momo 1
Nimbus 7–October 13, 1978
NOAA 1–December 11, 1970
Okean-E, see Cosmos 1076
Okean-OE, see Cosmos 1500
OrbView 2, see SeaWiFS
OSO 2–February 3, 1965
QuikScat–June 19, 1999
Radarsat–November 4, 1995
Resurs F1–May 25, 1989
Resurs-500–November 16, 1992
Rocsat–January 26, 1999
SAGE–February 18, 1979
Sampex–July 3, 1992
San Marco D/L–March 25, 1988
SCD 1–February 9, 1993
Seasat–June 27, 1978
SeaWiFS (OrbView 2)–August 1, 1997
Sedsat 1–October 24, 1998
Sich 1–August 31, 1995
SME–October 6, 1981
SNOE–February 25, 1998
Solar Max–April 6, 1984; February 14, 1980
Spot 1–February 22, 1986
Spot 4–March 23, 1998
Terra–December 18, 1999
Tiros N–October 13, 1978
TMSat–July 10, 1998
TOMS–July 2, 1996
TOPEX-Poseidon–August 10, 1992
TRMM–November 28, 1997
UARS–September 12, 1991
UoSat 12–April 21, 1999
Uribyol 3–May 26, 1999

Chronological Listing
Ariel 2–March 27, 1964
OSO 2–February 3, 1965
NOAA 1–December 11, 1970
Landsat 1–July 23, 1972
Explorer 55 (Atmosphere Explorer E)–November 20, 1975
HCMM–April 26, 1978

Appendix 2. Satellites and Missions Listed by Subject

Seasat–June 27, 1978
Nimbus 7–October 13, 1978
Tiros N–October 13, 1978
Cosmos 1076 (Okean-E)–February 12, 1979
SAGE–February 18, 1979
Bhaskara 1–June 7, 1979
Solar Max–February 14, 1980
Meteor 1-30–June 18, 1980
Intercosmos 21–February 6, 1981
SME–October 6, 1981
Landsat 4–July 16, 1982
Cosmos 1500 (Okean-OE)–September 28, 1983
Solar Max–April 6, 1984
ERBS–October 5, 1984
Geosat–March 13, 1985
Spot 1–February 22, 1986
Momo 1 (MOS 1)–February 19, 1987
Cosmos 1870–July 25, 1987
IRS 1A–March 17, 1988
San Marco D/L–March 25, 1988
Resurs F1–May 25, 1989
Almaz 1–March 31, 1991
ERS 1–July 17, 1991
Meteor 3-05–August 15, 1991
UARS–September 12, 1991
JERS 1–February 11, 1992
Sampex–July 3, 1992
EURECA–July 31, 1992
TOPEX-Poseidon–August 10, 1992
Resurs-500–November 16, 1992
SCD 1–February 9, 1993
Crista-Spas 1–November 3, 1994
Microlab 1–April 3, 1995
ERS 2–April 21, 1995
Sich 1–August 31, 1995
Radarsat–November 4, 1995
TOMS–July 2, 1996
ADEOS (Midori)–August 17, 1996
SeaWiFS (OrbView 2)–August 1, 1997
Crista-Spas 2–August 7, 1997
TRMM–November 28, 1997
Geosat Follow-On–February 10, 1998
Cosmos 2349–February 17, 1998
SNOE–February 25, 1998
Spot 4–March 23, 1998
FASat-Bravo–July 10, 1998
TMSat–July 10, 1998
Sedsat 1–October 24, 1998
Rocsat–January 26, 1999
Argos–February 23, 1999
Landsat 7–April 15, 1999
UoSat 12–April 21, 1999
DLR-Tubsat–May 26, 1999
IRS-P4 Oceansat–May 26, 1999
Uribyol 3–May 26, 1999
QuikScat–June 19, 1999
Ikonos–September 24, 1999
CBERS 1–October 14, 1999
Terra–December 18, 1999
Kompsat–December 21, 1999

Atmospheric Density Research

Alphabetical Listing
Atmosphere 1–September 3, 1990
Atmosphere Explorer, *see* Explorer 17
Atmosphere Explorer B, *see* Explorer 32
Atmosphere Explorer C, *see* Explorer 51
Atmosphere Explorer E, *see* Explorer 55

Cannonball 1 (OV1-16)–July 11, 1968
Cannonball 2–August 7, 1971
Castor–May 17, 1975
Dash 1–May 9, 1963
Dash 2–October 5, 1966
Echo 1–August 12, 1960
Echo 2–January 25, 1964
Explorer 9–February 16, 1961
Explorer 17 (Atmosphere Explorer)–April 3, 1963
Explorer 19–December 19, 1963
Explorer 24–November 21, 1964
Explorer 29 (GEOS 1)–November 6, 1965
Explorer 32 (Atmosphere Explorer B)–May 25, 1966
Explorer 39–August 8, 1968
Explorer 51 (Atmosphere Explorer C)–December 16, 1973
Explorer 55 (Atmosphere Explorer E)–November 20, 1975
GEOS 1, *see* Explorer 29
Musketball 1–August 7, 1971
Mylar Balloon–August 7, 1971
OV1-15, *see* Spades
OV1-16, *see* Cannonball 1
Pageos 1–June 24, 1966
Pion 1/Pion 2–May 25, 1989
Rigidsphere–August 7, 1971
San Marco 1–December 15, 1964
San Marco 2–April 26, 1967
San Marco 3–April 24, 1971
San Marco 4–February 18, 1974
San Marco D/L–March 25, 1988
Spades (OV1-15)–July 11, 1968
Starshine–May 27, 1999
Vanguard 1–March 17, 1958

Chronological Listing
Vanguard 1–March 17, 1958
Echo 1–August 12, 1960
Explorer 9–February 16, 1961
Explorer 17 (Atmosphere Explorer)–April 3, 1963
Dash 1–May 9, 1963
Explorer 19–December 19, 1963
Echo 2–January 25, 1964
Explorer 24–November 21, 1964
San Marco 1–December 15, 1964
Explorer 29 (GEOS 1)–November 6, 1965
Explorer 32 (Atmosphere Explorer B)–May 25, 1966
Pageos 1–June 24, 1966
Dash 2–October 5, 1966
San Marco 2–April 26, 1967
Cannonball 1 (OV1-16)–July 11, 1968
Spades (OV1-15)–July 11, 1968
Explorer 39–August 8, 1968
San Marco 3–April 24, 1971
Cannonball 2–August 7, 1971
Musketball 1–August 7, 1971
Mylar Balloon–August 7, 1971
Rigidsphere–August 7, 1971
Explorer 51 (Atmosphere Explorer C)–December 16, 1973
San Marco 4–February 18, 1974
Castor–May 17, 1975
Explorer 55 (Atmosphere Explorer E)–November 20, 1975
San Marco D/L–March 25, 1988
Pion 1/Pion 2–May 25, 1989
Atmosphere 1–September 3, 1990
Starshine–May 27, 1999

Biological Research

Alphabetical Listing
Bion 1 (Cosmos 605)–October 31, 1973
Bion 2 (Cosmos 690)–October 22, 1974
Bion 3 (Cosmos 782)–November 25, 1975
Bion 4 (Cosmos 936)–August 3, 1977
Bion 5 (Cosmos 1129)–September 25, 1979
Bion 6 (Cosmos 1514)–December 14, 1983
Bion 7 (Cosmos 1667)–July 10, 1985
Bion 8 (Cosmos 1887)–September 29, 1987
Bion 9 (Cosmos 2044)–September 15, 1989
Bion 10 (Cosmos 2229)–December 29, 1992
Bion 11–December 24, 1996
Biosatellite 2–September 7, 1967
Biosatellite 3–June 29, 1969
China 35–October 6, 1992
Cosmos 110–February 22, 1966
Cosmos 605, *see* Bion 1
Cosmos 690, *see* Bion 2
Cosmos 782, *see* Bion 3
Cosmos 936, *see* Bion 4
Cosmos 1129, *see* Bion 5
Cosmos 1514, *see* Bion 6
Cosmos 1667, *see* Bion 7
Cosmos 1887, *see* Bion 8
Cosmos 2044, *see* Bion 9
Cosmos 2229, *see* Bion 10
Fanhui Shu Weixing 12–October 5, 1990
Little Joe 3–December 4, 1959
Mercury 2–January 31, 1961
Mercury 5–November 29, 1961
OFO–November 9, 1970
OV1-4–March 30, 1966
OV3-4–June 10, 1966
Sputnik 2–November 3, 1957
Sputnik 5–August 19, 1960
Sputnik 6–December 1, 1960
Sputnik 9–March 9, 1961
Sputnik 10–March 25, 1961

Chronological Listing
Sputnik 2–November 3, 1957
Little Joe 3–December 4, 1959
Sputnik 5–August 19, 1960
Sputnik 6–December 1, 1960
Mercury 2–January 31, 1961
Sputnik 9–March 9, 1961
Sputnik 10–March 25, 1961
Mercury 5–November 29, 1961
Cosmos 110–February 22, 1966
OV1-4–March 30, 1966
OV3-4–June 10, 1966
Biosatellite 2–September 7, 1967
Biosatellite 3–June 29, 1969
OFO–November 9, 1970
Bion 1 (Cosmos 605)–October 31, 1973
Bion 2 (Cosmos 690)–October 22, 1974
Bion 3 (Cosmos 782)–November 25, 1975
Bion 4 (Cosmos 936)–August 3, 1977
Bion 5 (Cosmos 1129)–September 25, 1979
Bion 6 (Cosmos 1514)–December 14, 1983
Bion 7 (Cosmos 1667)–July 10, 1985
Bion 8 (Cosmos 1887)–September 29, 1987
Bion 9 (Cosmos 2044)–September 15, 1989
Fanhui Shu Weixing 12–October 5, 1990
China 35–October 6, 1992
Bion 10 (Cosmos 2229)–December 29, 1992
Bion 11–December 24, 1996

Appendix 2. Satellites and Missions Listed by Subject

Burial of Human Remains

Alphabetical Listing
Celestis–April 21, 1997
Celestis 2–February 10, 1998
Celestis 3–December 21, 1999
Lunar Prospector–January 7, 1998

Chronological Listing
Celestis–April 21, 1997
Lunar Prospector–January 7, 1998
Celestis 2–February 10, 1998
Celestis 3–December 21, 1999

Communication Propagation and Design Research

Alphabetical Listing
ACTS–September 12, 1993
Alexis–April 25, 1993
Apple–June 19, 1981
ATS 1–December 7, 1966
ATS 3–November 5, 1967
ATS 5–August 12, 1969
ATS 6–May 30, 1974
Cosmos 41–August 22, 1964
Cosmos 637–March 26, 1974
Courier 1B–October 4, 1960
CTS 1, see Hermes
DATS–July 1, 1967
Echo 1–August 12, 1960
Echo 2–January 25, 1964
ETS 2, see Kiku 2
ETS 4, see Kiku 3
ETS 5, see Kiku 5
FAISAT–January 24, 1995
GLOMR–April 29, 1985; October 30, 1985
Gridsphere 1/Gridsphere 2–August 7, 1971
Gurwin Techsat 1B–July 10, 1998
Hermes (CTS 1)–January 17, 1976
HiLat (P83-1)–June 27, 1983
Informator 1–January 29, 1991
Intasat 1–November 15, 1974
Intercosmos 24–September 28, 1989
Kiku 2 (ETS 2)–February 23, 1977
Kiku 3 (ETS 4)–February 11, 1981
Kiku 5 (ETS 5)–August 27, 1987
LES 1–February 11, 1965
LES 2–May 6, 1965
LES 3/LES 4–December 21, 1965
LES 5–July 1, 1967
LES 6–September 26, 1968
LES 8/LES 9–March 15, 1976
Lofti 2A–June 15, 1963
Macsat 1/Macsat 2–May 9, 1990
Magion 2–September 28, 1989
Microsat 1 through Microsat 7–July 17, 1991
Northern Utah SATellite, see NUSAT
NUSAT (Northern Utah SATellite)–April 29, 1985
Orbcomm FM1/Orbcomm FM2–April 3, 1995
Orbcomm X–July 17, 1991
Orbiscal 2, see OV1-17A
Oscar 11, see UoSat 2
OTS 2–May 11, 1978
OV1-8–July 14, 1966
OV1-17A (Orbiscal 2)–March 18, 1969
OV1-18–March 18, 1969
OV4-1T/OV4-1R, November 3, 1966
P76-5–May 22, 1976
P83-1, see HiLat
Pansat–October 29, 1998
Polar Bear–November 14, 1986
Posat 1–September 26, 1993
Relay 1–December 13, 1962
Relay 2–January 21, 1964
Rex 1–June 29, 1991
S80-T–August 10, 1992
SCE–April 11, 1990
SCORE–December 18, 1958
SDS 1–June 2, 1976
SECS–April 5, 1990
Sirio 1–August 25, 1977
Spartan 207–May 19, 1996
STP 1 (ASTEX)–October 17, 1971
SURFSAT–November 4, 1995
Syncom 1–June 26, 1963
Syncom 2–July 26, 1963
Syncom 3–August 19, 1964
Teledesic 1–February 25, 1998
Telstar 1–July 10, 1962
Telstar 2–May 7, 1963
Test and Training Satellite 1–December 13, 1967
TEX–April 11, 1990
Tubsat 1–July 17, 1991
Tubsat N/Tubsat N1–July 7, 1998
UoSat 2 (Oscar 11)–March 1, 1984
UPM-LBSAT 1–July 7, 1995
Uribyol–August 10, 1992
Uribyol 2–September 26, 1993
USA 89–December 2, 1992
West Ford 1–October 21, 1961
West Ford 2–May 9, 1963

Chronological Listing
SCORE–December 18, 1958
Echo 1–August 12, 1960
Courier 1B–October 4, 1960
West Ford 1–October 21, 1961
Telstar 1–July 10, 1962
Relay 1–December 13, 1962
Telstar 2–May 7, 1963
West Ford 2–May 9, 1963
Lofti 2A–June 15, 1963
Syncom 1–June 26, 1963
Syncom 2–July 26, 1963
Relay 2–January 21, 1964
Echo 2–January 25, 1964
Syncom 3–August 19, 1964
Cosmos 41–August 22, 1964
LES 1–February 11, 1965
LES 2–May 6, 1965
LES 3/LES 4–December 21, 1965
OV1-8–July 14, 1966
OV4-1T/OV4-1R, November 3, 1966
ATS 1–December 7, 1966
DATS–July 1, 1967
LES 5–July 1, 1967
ATS 3–November 5, 1967
Test and Training Satellite 1–December 13, 1967
LES 6–September 26, 1968
OV1-17A (Orbiscal 2)–March 18, 1969
OV1-18–March 18, 1969
ATS 5–August 12, 1969
Gridsphere 1/Gridsphere 2–August 7, 1971
STP 1–October 17, 1971
Cosmos 637–March 26, 1974
ATS 6–May 30, 1974
Intasat 1–November 15, 1974
Hermes (CTS 1)–January 17, 1976
LES 8/LES 9–March 15, 1976
P76-5–May 22, 1976
SDS 1–June 2, 1976
Kiku 2 (ETS 2)–February 23, 1977
Sirio 1–August 25, 1977
OTS 2–May 11, 1978
Kiku 3 (ETS 4)–February 11, 1981
Apple–June 19, 1981
HiLat (P83-1)–June 27, 1983
UoSat 2 (Oscar 11)–March 1, 1984
GLOMR–April 29, 1985
Northern Utah SATellite (NUSAT)–April 29, 1985
GLOMR–October 30, 1985
Polar Bear–November 14, 1986
Kiku 5 (ETS 5)–August 27, 1987
Intercosmos 24–September 28, 1989
Magion 2–September 28, 1989
SECS–April 5, 1990
SCE–April 11, 1990
TEX–April 11, 1990
Macsat 1/Macsat 2–May 9, 1990
Informator 1–January 29, 1991
Rex 1–June 29, 1991
Orbcomm X–July 17, 1991, 1:46 GMT
Tubsat 1–July 17, 1991, 1:46 GMT
Microsat 1 through Microsat 7–July 17, 1991, 17:33 GMT
S80-T–August 10, 1992
Uribyol–August 10, 1992
USA 89–December 2, 1992
Alexis–April 25, 1993
ACTS–September 12, 1993
Posat 1–September 26, 1993
Uribyol 2–September 26, 1993
FAISAT–January 24, 1995
Orbcomm FM1/Orbcomm FM2–April 3, 1995
UPM-LBSAT 1–July 7, 1995
SURFSAT–November 4, 1995
Spartan 207–May 19, 1996
Teledesic 1–February 25, 1998
Tubsat N/Tubsat N1–July 7, 1998
Gurwin Techsat 1B–July 10, 1998
Pansat–October 29, 1998

Communications Satellites

Alphabetical Listing
AfriStar–October 28, 1998
Agila 2–August 19, 1997
Altair 1 (Cosmos 1700)–October 25, 1985
Amos–May 16, 1996
AMSC 1–April 7, 1995
Anik A1 (Telesat 1)–November 10, 1972
Anik B (Telesat 4)–December 15, 1978
Anik C1–April 12, 1985
Anik C2 (Telesat 7)–June 18, 1983
Anik C3 (Telesat 6)–November 11, 1982
Anik D1 (Telesat 5)–August 26, 1982
Anik D2–November 8, 1984
Anik E2–April 4, 1991
Apstar 1–July 21, 1994
Arabsat 1A–February 8, 1985
Arabsat 1B–June 17, 1985
ASC 1–August 27, 1985
Asiasat 1–April 7, 1990
Astra 1A–December 11, 1988
ATS 1–December 7, 1966
ATS 3–November 5, 1967
ATS 6–May 30, 1974

355

Appendix 2. Satellites and Missions Listed by Subject

Aurora 1–October 28, 1982
Aurora 2–May 29, 1991
Aussat A1–August 27, 1985
Aussat A2–November 26, 1985
Aussat B1 (Optus B1)–August 14, 1992
Bonum 1–November 22, 1998
Brasilsat A1–February 8, 1985
Brasilsat B1–August 10, 1994
BSAT 1A–April 16, 1997
Cakrawarta 1–November 12, 1997
Comstar 1–May 13, 1976
Cosmos 41–August 22, 1964
Cosmos 1700, see Altair 1
Cosmos 2268 through Cosmos 2273–February 12, 1994
Coupon–November 12, 1997
Courier 1B–October 4, 1960
DBS 1–December 18, 1993
DFH (Dong Fang Hong) 3A–November 29, 1994
DirectTV 1R–October 10, 1999
Dong Fang Hong 3A, see DFH 3A
DSCS II 1/DSCS II 2–November 3, 1971
DSCS III–October 30, 1982
Early Bird (Intelsat 1)–April 6, 1965
Echo 1–August 12, 1960
Echo 2–January 25, 1964
Echostar 1–December 28, 1995
ECS 1, see Eutelsat 1
Ekran 1–October 26, 1976
Eutelsat 1 (ECS 1)–June 16, 1983
Eutelsat 2 F1–August 30, 1990
Express 1–October 13, 1994
FLTSATCOM 1–February 9, 1978
Galaxy 1–June 28, 1983
Galaxy 7–October 28, 1992
Galaxy 11–December 21, 1999
Gals 1–January 20, 1994
GE 1–September 8, 1996
GE Satcom–January 12, 1986
Globalstar 1 through Globalstar 4–February 14, 1998
Gonets-D1 1 through Gonets-D1 3–February 14, 1997
Gorizont 1–December 19, 1978
GStar 1–May 7, 1985
Healthsat 2–September 26, 1993
HGS 1–December 25, 1997
Hispasat 1A–September 10, 1992
Hot Bird 1–March 28, 1995
IDCSP 1 through IDCSP 7–June 16, 1966
Inmarsat 2 F1–October 30, 1990
Inmarsat 3 F1–April 3, 1996
Insat 1A–April 10, 1982
Insat 1B–August 30, 1983
Insat 2A–July 9, 1992
Intelsat 1, see Early Bird
Intelsat 2A–October 26, 1966
Intelsat 2B–January 11, 1967
Intelsat 2C–March 23, 1967
Intelsat 2D–September 28, 1967
Intelsat 3B–December 19, 1968
Intelsat 4 F2–January 26, 1971
Intelsat 4A F1–September 26, 1975
Intelsat 5 F2–December 6, 1980
Intelsat 5A F10–March 22, 1985
Intelsat 6 F3–May 7, 1992
Intelsat 6A F2–October 27, 1989
Intelsat 7 F1–October 22, 1993
Intelsat 801–March 1, 1997
Intelsat K–June 10, 1992
Iridium 4 through Iridium 8–May 5, 1997

Italsat 1–January 15, 1991
JCSat 1–March 6, 1989
Kopernikus 1–June 5, 1989
Koreasat 1, see Mugunghwa 1
LEASAT 1–November 8, 1984
LEASAT 2–August 30, 1984
LEASAT 3–April 12, 1985
LEASAT 4–August 27, 1985
LEASAT 5–January 9, 1990
LMI–September 26, 1999
Luch 1–December 16, 1994
Marco Polo 1–August 27, 1989
MARECS A–December 20, 1981
Marisat 1–February 19, 1976
Measat 1–January 12, 1996
Milstar 1–February 7, 1994
Molniya 1-01–April 23, 1965
Molniya 1-02–October 14, 1965
Molniya 1-03–April 25, 1966
Molniya 1S–July 30, 1974
Molniya 2-01–November 24, 1971
Molniya 3-01–November 21, 1974
Morelos 1–June 17, 1985
Morelos 2–November 26, 1985
MSAT 1–April 20, 1996
Mugunghwa 1 (Koreasat 1)–August 5, 1995
Nahuel 1A–January 30, 1997
NATO 3A–April 22, 1976
NATO 4A–January 8, 1991
Nilesat 1–April 28, 1998
Nimiq 1–May 21, 1999
N-Star A–August 29, 1995
Olympus 1–July 12, 1989
Orbcomm FM1/Orbcomm FM2–April 3, 1995
Orbcomm FM5 through Orbcomm FM12–December 23, 1997
Orion 1–November 29, 1994
OTS 2–May 11, 1978
Palapa 1–July 8, 1976
Palapa B1–June 18, 1983
Palapa B2R–April 13, 1990
Palapa C-1–February 1, 1996
Pan American Satellite 1–June 15, 1988
Pan American Satellite 5–August 28, 1997
Raduga 1–December 22, 1975
Raduga 28–December 19, 1991
RCA Satcom 1–December 13, 1975
RCA Satcom 5 (Aurora 1)–October 28, 1982
RCA Satcom Ku2–November 26, 1985
Relay 1–December 13, 1962
Relay 2–January 21, 1964
Safir 2–July 10, 1998
Sakura 1–December 15, 1977
Sakura 2A–February 4, 1983
Sakura 3A–February 19, 1988
Satcom C1–November 20, 1990
Satmex 5–December 6, 1998
SBS 1–November 15, 1980
SBS 3–November 11, 1982
SBS 4–August 30, 1984
SCORE–December 18, 1958
SDS 1–June 2, 1976
Shiyan Tongxin Weixing 1, see STW 1
Sinosat 1–July 18, 1998
Skynet 1–November 22, 1969
Skynet 2–November 23, 1974
Skynet 4B–December 11, 1988
Solidaridad 1–November 20, 1993
Spacenet 1–May 23, 1984
ST 1–August 25, 1998
Strela 1 (Cosmos 343)–April 25, 1970

STW (Shiyan Tongxin Weixing) 1–April 8, 1984
Sunsat–February 23, 1999
Superbird A–June 5, 1989
Symphonie 1–December 18, 1974
Syncom 1–June 26, 1963
Syncom 2–July 26, 1963
Syncom 3–August 19, 1964
Tacsat 1–February 9, 1969
TDF 1–October 28, 1988
TDRS 1–April 4, 1983
TDRS 3–September 29, 1988
TDRS 4–March 13, 1989
TDRS 5–August 2, 1991
TDRS 6–January 13, 1993
TDRS 7–July 13, 1995
Telcom (Palapa D-1)–August 12, 1999
Telecom 1A–August 4, 1984
Telecom 2A–December 16, 1991
Telesat 1, see Anik A1
Telesat 4, see Anik B
Telesat 5, see Anik D1
Telesat 6, see Anik C3
Telesat 7, see Anik C2
Telstar 1–July 10, 1962
Telstar 2–May 7, 1963
Telstar 3A–July 28, 1983
Telstar 3B–August 30, 1984
Telstar 3C–June 17, 1985
Telstar 401–December 16, 1993
Tempo 2–March 8, 1997
Thaicom 1A–December 18, 1993
Thor 2–May 21, 1997
Turksat 1B–August 10, 1994
UHF Follow-On 1–September 3, 1993
USA 89–December 2, 1992
Westar 1–April 13, 1974
Westar 4–February 26, 1982
Yamal 101/Yamal 102–September 6, 1999
Yuri 1–April 7, 1978

Chronological Listing
SCORE–December 18, 1958
Echo 1–August 12, 1960
Courier 1B–October 4, 1960
Telstar 1–July 10, 1962
Relay 1–December 13, 1962
Telstar 2–May 7, 1963
Syncom 1–June 26, 1963
Syncom 2–July 26, 1963
Relay 2–January 21, 1964
Echo 2–January 25, 1964
Syncom 3–August 19, 1964
Cosmos 41–August 22, 1964
Early Bird (Intelsat 1)–April 6, 1965
Molniya 1-01–April 23, 1965
Molniya 1-02–October 14, 1965
Molniya 1-03–April 25, 1966
IDCSP 1 through IDCSP 7–June 16, 1966
Intelsat 2A–October 26, 1966
ATS 1–December 7, 1966
Intelsat 2B–January 11, 1967
Intelsat 2C–March 23, 1967
Intelsat 2D–September 28, 1967
ATS 3–November 5, 1967
Intelsat 3B–December 19, 1968
Tacsat 1–February 9, 1969
Skynet 1–November 22, 1969
Strela 1 (Cosmos 343)–April 25, 1970
Intelsat 4 F2–January 26, 1971
DSCS II 1/DSCS II 2–November 3, 1971
Molniya 2-01–November 24, 1971

356

Anik A1 (Telesat 1)–November 10, 1972
Westar 1–April 13, 1974
ATS 6–May 30, 1974
Molniya 1S–July 30, 1974
Molniya 3-01–November 21, 1974
Skynet 2–November 23, 1974
Symphonie 1–December 18, 1974
Intelsat 4A F1–September 26, 1975
RCA Satcom 1–December 13, 1975
Raduga 1–December 22, 1975
Marisat 1–February 19, 1976
NATO 3A–April 22, 1976
Comstar 1–May 13, 1976
SDS 1–June 2, 1976
Palapa 1–July 8, 1976
Ekran 1–October 26, 1976
Sakura 1–December 15, 1977
FLTSATCOM 1–February 9, 1978
Yuri 1–April 7, 1978
OTS 2–May 11, 1978
Anik B (Telesat 4)–December 15, 1978
Gorizont 1–December 19, 1978
SBS 1–November 15, 1980
Intelsat 5 F2–December 6, 1980
MARECS A–December 20, 1981
Westar 4–February 26, 1982
Insat 1A–April 10, 1982
Anik D1 (Telesat 5)–August 26, 1982
RCA Satcom 5 (Aurora 1)–October 28, 1982
DSCS III–October 30, 1982
Anik C3 (Telesat 6)–November 11, 1982
SBS 3–November 11, 1982
Sakura 2A–February 4, 1983
TDRS 1–April 4, 1983
Eutelsat 1 (ECS 1)–June 16, 1983
Anik C2 (Telesat 7)–June 18, 1983
Palapa B1–June 18, 1983
Galaxy 1–June 28, 1983
Telstar 3A–July 28, 1983
Insat 1B–August 30, 1983
STW (Shiyan Tongxin Weixing) 1–April 8, 1984
Spacenet 1–May 23, 1984
Telecom 1A–August 4, 1984
LEASAT 2–August 30, 1984
SBS 4–August 30, 1984
Telstar 3B–August 30, 1984
Anik D2–November 8, 1984
LEASAT 1–November 8, 1984
Arabsat 1A–February 8, 1985
Brasilsat A1–February 8, 1985
Intelsat 5A F10–March 22, 1985
Anik C1–April 12, 1985
LEASAT 3–April 12, 1985
GStar 1–May 7, 1985
Arabsat 1B–June 17, 1985
Morelos 1–June 17, 1985
Telstar 3C–June 17, 1985
ASC 1–August 27, 1985
Aussat A1–August 27, 1985
LEASAT 4–August 27, 1985
Altair 1 (Cosmos 1700)–October 25, 1985
Aussat A2–November 26, 1985
Morelos 2–November 26, 1985
RCA Satcom Ku2–November 26, 1985
GE Satcom–January 12, 1986
Sakura 3A–February 19, 1988
Pan American Satellite 1–June 15, 1988
TDRS 3–September 29, 1988
TDF 1–October 28, 1988
Astra 1A–December 11, 1988
Skynet 4B–December 11, 1988

JCSat 1–March 6, 1989
TDRS 4–March 13, 1989
Kopernikus 1–June 5, 1989
Superbird A–June 5, 1989
Olympus 1–July 12, 1989
Marco Polo 1–August 27, 1989
Intelsat 6A F2–October 27, 1989
LEASAT 5–January 9, 1990
Asiasat 1–April 7, 1990
Palapa B2R–April 13, 1990
Eutelsat 2 F1–August 30, 1990
Inmarsat 2 F1–October 30, 1990
Satcom C1–November 20, 1990
NATO 4A–January 8, 1991
Italsat 1–January 15, 1991
Anik E2–April 4, 1991
Aurora 2–May 29, 1991
TDRS 5–August 2, 1991
Telecom 2A–December 16, 1991
Raduga 28–December 19, 1991
Intelsat 6 F3–May 7, 1992
Intelsat K–June 10, 1992
Insat 2A–July 9, 1992
Aussat B1 (Optus B1)–August 14, 1992
Hispasat 1A–September 10, 1992
Galaxy 7–October 28, 1992
USA 89–December 2, 1992
TDRS 6–January 13, 1993
UHF Follow-On 1–September 3, 1993
Healthsat 2–September 26, 1993
Intelsat 7 F1–October 22, 1993
Solidaridad 1–November 20, 1993
Telstar 401–December 16, 1993
DBS 1–December 18, 1993
Thaicom 1A–December 18, 1993
Gals 1–January 20, 1994
Milstar 1–February 7, 1994
Cosmos 2268 through Cosmos 2273–February 12, 1994
Apstar 1–July 21, 1994
Brasilsat B1–August 10, 1994
Turksat 1B–August 10, 1994
Express 1–October 13, 1994
Orion 1–November 29, 1994, 10:21 GMT
DFH (Dong Fang Hong) 3A–November 29, 1994, 17:02 GMT
Luch 1–December 16, 1994
Hot Bird 1–March 28, 1995
Orbcomm FM1/Orbcomm FM2–April 3, 1995
AMSC 1–April 7, 1995
TDRS 7–July 13, 1995
Mugunghwa 1 (Koreasat 1)–August 5, 1995
N-Star A–August 29, 1995
Echostar 1–December 28, 1995
Measat 1–January 12, 1996
Palapa C-1–February 1, 1996
Inmarsat 3 F1–April 3, 1996
MSAT 1–April 20, 1996
Amos–May 16, 1996
GE 1–September 8, 1996
Nahuel 1A–January 30, 1997
Gonets-D1 1 through Gonets D1 3–February 14, 1997
Intelsat 801–March 1, 1997
Tempo 2–March 8, 1997
BSAT 1A–April 16, 1997
Iridium 4 through Iridium 8–May 5, 1997
Thor 2–May 21, 1997
Agila 2–August 19, 1997
Pan American Satellite 5–August 28, 1997
Coupon–November 12, 1997, 17:00 GMT

Cakrawarta 1–November 12, 1997, 21:48 GMT
Orbcomm FM5 through Orbcomm FM12–December 23, 1997
HGS 1–December 25, 1997
Globalstar 1 through Globalstar 4–February 14, 1998
Nilesat 1–April 28, 1998
Safir 2–July 10, 1998
Sinosat 1–July 18, 1998
ST 1–August 25, 1998
AfriStar–October 28, 1998
Bonum 1–November 22, 1998
Satmex 5–December 6, 1998
Sunsat–February 23, 1999
Nimiq 1–May 21, 1999
Telcom (Palapa D-1)–August 12, 1999
Yamal 101/Yamal 102–September 6, 1999
LMI–September 26, 1999
DirecTV 1R–October 10, 1999
Galaxy 11–December 21, 1999

Geodetic/Navigational Research

Alphabetical Listing
Ajisai (EGP)–August 12, 1986
Anna 1B–October 31, 1962
Beacon Explorer B, see Explorer 22
Beacon Explorer C, see Explorer 27
Cosmos 1383 (COSPAS 1)–June 30, 1982
Cosmos 1413/Cosmos 1414/Cosmos 1415–October 12, 1982
Cosmos 1989–January 10, 1989
Cosmos 2024–May 31, 1989
COSPAS 1, see Cosmos 1383
Diademe 1–February 8, 1967
Diademe 2–February 15, 1967
Echo 1–August 12, 1960
Echo 2–January 25, 1964
EGP, see Ajisai
Explorer 22 (Beacon Explorer B)–October 10, 1964
Explorer 27 (Beacon Explorer C)–April 29, 1965
Explorer 29 (GEOS 1)–November 6, 1965
Geo-IK–November 29, 1994
GEOS 1, see Explorer 29
Geos 2–January 11, 1968
Geos 3–April 9, 1975
Lageos 1–May 4, 1976
Lageos 2–October 22, 1992
Nadezhda 1–July 4, 1989
Navstar 1–February 22, 1978
Navstar 5–February 9, 1980
Navstar B2 1–February 14, 1989
NTS 1 (Timation 3)–July 14, 1974
NTS 2–June 23, 1977
Pageos 1–June 24, 1966
Secor 1–January 11, 1964
Starlette–February 6, 1975
Stella–September 26, 1993
Timation 1–May 31, 1967
Timation 2–September 30, 1969
Timation 3, see NTS 1
Transit 1B–April 13, 1960
Transit 2A–June 22, 1960
Transit 4A–June 29, 1961
Westpac–July 10, 1998
Zeya–March 4, 1997

Appendix 2. Satellites and Missions Listed by Subject

Chronological Listing
Transit 1B–April 13, 1960
Transit 2A–June 22, 1960
Echo 1–August 12, 1960
Transit 4A–June 29, 1961
Anna 1B–October 31, 1962
Secor 1–January 11, 1964
Echo 2–January 25, 1964
Explorer 22 (Beacon Explorer B)–October 10, 1964
Explorer 27 (Beacon Explorer C)–April 29, 1965
Explorer 29 (GEOS 1)–November 6, 1965
Pageos 1–June 24, 1966
Diademe 1–February 8, 1967
Diademe 2–February 15, 1967
Timation 1–May 31, 1967
GEOS 2–January 11, 1968
Timation 2–September 30, 1969
NTS 1 (Timation 3)–July 14, 1974
Starlette–February 6, 1975
Geos 3–April 9, 1975
Lageos 1–May 4, 1976
NTS 2–June 23, 1977
Navstar 1–February 22, 1978
Navstar 5–February 9, 1980
Cosmos 1383 (COSPAS 1)–June 30, 1982
Cosmos 1413/Cosmos 1414/Cosmos 1415–October 12, 1982
Ajisai (EGP)–August 12, 1986
Cosmos 1989–January 10, 1989
Navstar B2 1–February 14, 1989
Cosmos 2024–May 31, 1989
Nadezhda 1–July 4, 1989
Lageos 2–October 22, 1992
Stella–September 26, 1993
Geo-IK–November 29, 1994
Zeya–March 4, 1997
Westpac–July 10, 1998

Ionosphere/Magnetosphere Research

Alphabetical Listing
38C–September 28, 1963
83C–December 13, 1964
Aeros 1–December 16, 1972
Aeros 2–July 16, 1974
Akebono (EXOS D)–February 21, 1989
Alouette 1–September 29, 1962
Alouette 2–November 29, 1965
Ariel 1–April 26, 1962
Ariel 3–May 5, 1967
Ariel 4–December 11, 1971
Astrid 1–January 24, 1995
Astrid 2–December 10, 1998
Atmosphere Explorer D, see Explorer 54
Aureol 1–December 27, 1971
Aurorae–October 3, 1968
Azur–November 8, 1969
Beacon Explorer B, see Explorer 22
Beacon Explorer C, see Explorer 27
Boreas–October 1, 1969
BremSat 1–February 3, 1994
CAMEO–October 13, 1978
CCE–August 16, 1984
Coronas I, see Intercosmos 26
Cosmos 1–March 16, 1962
Cosmos 2 (Sputnik 12)–April 6, 1962
Cosmos 3–April 24, 1962
Cosmos 5 (Sputnik 15)–May 28, 1962
Cosmos 17–May 22, 1963
Cosmos 26–March 18, 1964
Cosmos 49–October 24, 1964
Cosmos 1809, see Ionosonde
CRRES–July 25, 1990
DIAL-WIKA–March 10, 1970
Dynamics Explorer 1/Dynamics Explorer 2–August 3, 1981
Elektron 1/Elektron 2–January 30, 1964
Elektron 3/Elektron 4–July 11, 1964
Equator S–December 2, 1997
ESA-GEOS 1–April 20, 1977
ESA-GEOS 2–July 14, 1978
ESRO 4–November 22, 1972
EXOS A, see Kyokko
EXOS B, see Jikiken
EXOS C, see Ohzora
EXOS D, see Akebono
Explorer 1–February 1, 1958
Explorer 3–March 26, 1958
Explorer 4–July 6, 1958
Explorer 6–August 7, 1959
Explorer 7–October 13, 1959
Explorer 8–November 3, 1960
Explorer 10–March 25, 1961
Explorer 12–August 16, 1961
Explorer 14–October 2, 1962
Explorer 15–October 27, 1962
Explorer 18, see Imp A
Explorer 20–August 25, 1964
Explorer 22 (Beacon Explorer B)–October 10, 1964
Explorer 25 (Injun 4)–November 21, 1964
Explorer 26–December 21, 1964
Explorer 27 (Beacon Explorer C)–April 29, 1965
Explorer 31–November 29, 1965
Explorer 33, see Imp D
Explorer 45–November 15, 1971
Explorer 52, see Hawkeye
Explorer 54 (Atmosphere Explorer D)–October 6, 1975
FAST–August 21, 1996
France 1–December 6, 1965
Freja–October 6, 1992
Geotail–July 24, 1992
GRS–June 28, 1963
Hawkeye (Explorer 52)–June 3, 1974
HEOS 1–December 5, 1968
HEOS 2–January 31, 1972
Hitch Hiker 1–June 27, 1963
Imp A (Explorer 18)–November 27, 1963
Imp D (Explorer 33)–July 1, 1966
Injun 1–June 29, 1961
Injun 3–December 13, 1962
Injun 4, see Explorer 25
Injun 5–August 8, 1968
Intasat 1–November 15, 1974
Interball 1–August 2, 1995
Interball 2–August 29, 1996
Intercosmos 1–October 14, 1969
Intercosmos 18–October 24, 1978
Intercosmos 24–September 28, 1989
Intercosmos 25–December 18, 1991
Intercosmos 26 (Coronas I)–March 2, 1994
Ionosonde (Cosmos 1809)–December 18, 1986
Ion Release Module–August 16, 1984
ISAS–January 25, 1994
ISEE 1/ISEE 2–October 22, 1977
ISEE 3–August 12, 1978
ISIS 1–January 30, 1969
ISS 1–February 29, 1976
ISS 2, see Ume 2
Jikiken (EXOS B)–September 16, 1978
Kyokko (EXOS A)–February 4, 1978
Luna 1–January 2, 1959
Luna 2–September 12, 1959
Luna 10–March 31, 1966
Magion 1–October 24, 1978
Magion 2–September 28, 1989
Magion 3–December 18, 1991
Magion 4–August 2, 1995
Magion 5–August 29, 1996
Magsat–October 30, 1979
OGO 1–September 5, 1964
OGO 2–October 14, 1965
OGO 3–June 7, 1966
OGO 4–July 28, 1967
OGO 5–March 4, 1968
OGO 6–June 5, 1969
Ohzora (EXOS C)–February 14, 1984
ORS 3–July 20, 1965
Orsted–February 23, 1999
OV2-5/OV5-2–September 26, 1968
Pegsat–April 5, 1990
Pioneer 1–October 11, 1958
Pioneer 3–December 6, 1958
Pioneer 4–March 3, 1959
Pioneer 5–March 11, 1960
POGS–April 11, 1990
Polar–February 24, 1996
Prognoz 1–April 14, 1972
SACI 1–October 14, 1999
Sampex–July 3, 1992
Sputnik 2–November 3, 1957
Sputnik 3–May 15, 1958
SRATS, see Taiyo
Starad–October 26, 1962
Starfish Test–July 9, 1962
Taiyo (SRATS)–February 24, 1975
UKS–August 16, 1984
Ume 2 (ISS 2)–February 16, 1978
Vanguard 3–September 18, 1959
Vela 1A/Vela 1B–October 17, 1963
Vela 2A/Vela 2B–July 17, 1964
Vela 5A/Vela 5B–May 23, 1969
Viking–February 22, 1986
Wind–November 1, 1994

Chronological Listing
Sputnik 2–November 3, 1957
Explorer 1–February 1, 1958
Explorer 3–March 26, 1958
Sputnik 3–May 15, 1958
Explorer 4–July 6, 1958
Pioneer 1–October 11, 1958
Pioneer 3–December 6, 1958
Luna 1–January 2, 1959
Pioneer 4–March 3, 1959
Explorer 6–August 7, 1959
Luna 2–September 12, 1959
Vanguard 3–September 18, 1959
Explorer 7–October 13, 1959
Pioneer 5–March 11, 1960
Explorer 8–November 3, 1960
Explorer 10–March 25, 1961
Injun 1–June 29, 1961
Explorer 12–August 16, 1961
Cosmos 1–March 16, 1962
Cosmos 2 (Sputnik 12)–April 6, 1962
Cosmos 3–April 24, 1962
Ariel 1–April 26, 1962

358

Appendix 2. Satellites and Missions Listed by Subject

Cosmos 5 (Sputnik 15)–May 28, 1962
Starfish Test–July 9, 1962
Alouette 1–September 29, 1962
Explorer 14–October 2, 1962
Starad–October 26, 1962
Explorer 15–October 27, 1962
Injun 3–December 13, 1962
Cosmos 17–May 22, 1963
Hitch Hiker 1–June 27, 1963
GRS–June 28, 1963
38C–September 28, 1963
Vela 1A/Vela 1B–October 17, 1963
Imp A (Explorer 18)–November 27, 1963
Elektron 1/Elektron 2–January 30, 1964
Cosmos 26–March 18, 1964
Elektron 3/Elektron 4–July 11, 1964
Vela 2A/Vela 2B–July 17, 1964
Explorer 20–August 25, 1964
OGO 1–September 5, 1964
Explorer 22 (Beacon Explorer B)–October 10, 1964
Cosmos 49–October 24, 1964
Explorer 25 (Injun 4)–November 21, 1964
83C–December 13, 1964
Explorer 26–December 21, 1964
Explorer 27 (Beacon Explorer C)–April 29, 1965
ORS 3–July 20, 1965
OGO 2–October 14, 1965
Alouette 2–November 29, 1965
Explorer 31–November 29, 1965
France 1–December 6, 1965
Luna 10–March 31, 1966
OGO 3–June 7, 1966
Imp D (Explorer 33)–July 1, 1966
Ariel 3–May 5, 1967
OGO 4–July 28, 1967
OGO 5–March 4, 1968
Injun 5–August 8, 1968
OV2-5/OV5-2–September 26, 1968
Aurorae–October 3, 1968
HEOS 1–December 5, 1968
ISIS 1–January 30, 1969
Vela 5A/Vela 5B–May 23, 1969
OGO 6–June 5, 1969
Boreas–October 1, 1969
Intercosmos 1–October 14, 1969
Azur–November 8, 1969
DIAL-WIKA–March 10, 1970
Explorer 45–November 15, 1971
Ariel 4–December 11, 1971
Aureol 1–December 27, 1971
HEOS 2–January 31, 1972
Prognoz 1–April 14, 1972
ESRO 4–November 22, 1972
Aeros 1–December 16, 1972
Hawkeye (Explorer 52)–June 3, 1974
Aeros 2–July 16, 1974
Intasat 1–November 15, 1974
Taiyo (SRATS)–February 24, 1975
Explorer 54 (Atmosphere Explorer D)–October 6, 1975
ISS 1–February 29, 1976
ESA-GEOS 1–April 20, 1977
ISEE 1/ISEE 2–October 22, 1977
Kyokko (EXOS A)–February 4, 1978
Ume 2 (ISS 2)–February 16, 1978
ESA-GEOS 2–July 14, 1978
ISEE 3–August 12, 1978
Jikiken (EXOS B)–September 16, 1978
CAMEO–October 13, 1978
Intercosmos 18–October 24, 1978

Magion 1–October 24, 1978
Magsat–October 30, 1979
Dynamics Explorer 1/Dynamics Explorer 2–August 3, 1981
Ohzora (EXOS C)–February 14, 1984
CCE–August 16, 1984
UKS–August 16, 1984
Ion Release Module–August 16, 1984
Viking–February 22, 1986
Ionosonde (Cosmos 1809)–December 18, 1986
Akebono (EXOS D)–February 21, 1989
Intercosmos 24–September 28, 1989
Magion 2–September 28, 1989
Pegsat–April 5, 1990
POGS–April 11, 1990
CRRES–July 25, 1990
Intercosmos 25–December 18, 1991
Magion 3–December 18, 1991
Sampex–July 3, 1992
Geotail–July 24, 1992
Freja–October 6, 1992
ISAS–January 25, 1994
BremSat 1–February 3, 1994
Intercosmos 26 (Coronas I)–March 2, 1994
Wind–November 1, 1994
Astrid 1–January 24, 1995
Interball 1–August 2, 1995
Magion 4–August 2, 1995
Polar–February 24, 1996
FAST–August 21, 1996
Interball 2–August 29, 1996
Magion 5–August 29, 1996
Equator S–December 2, 1997
Astrid 2–December 10, 1998
Orsted–February 23, 1999
SACI 1–October 14, 1999

Manned Missions
See also **Manned Space Laboratories/Modules**

Alphabetical Listing
Apollo 1–January 27, 1967
Apollo 7–October 11, 1968
Apollo 8–December 21, 1968
Apollo 9–March 3, 1969
Apollo 10–May 18, 1969
Apollo 11–July 16, 1969
Apollo 12–November 14, 1969
Apollo 13–April 11, 1970
Apollo 14–January 31, 1971
Apollo 15–July 26, 1971
Apollo 16–April 16, 1972
Apollo 17–December 7, 1972
Apollo 18 (ASTP)–July 15, 1975
ASTP, *see* Apollo 18, Soyuz 19
Augmented Target Docking Adapter (ATDA)–June 1, 1966
Aurora 7 (Mercury 7)–May 24, 1962
Enterprise, Flight 4–August 12, 1977
Faith 7 (Mercury 9)–May 15, 1963
Freedom 7 (Mercury 3)–May 5, 1961
Friendship 7 (Mercury 6)–February 20, 1962
Gemini 3–March 23, 1965
Gemini 4–June 3, 1965
Gemini 5–August 21, 1965
Gemini 6–December 15, 1965
Gemini 7–December 4, 1965
Gemini 8–March 16, 1966
Gemini 8 Agena Target–March 16, 1966

Gemini 9–June 3, 1966
Gemini 10–July 18, 1966
Gemini 10 Agena Target–July 18, 1966
Gemini 11–September 12, 1966
Gemini 11 Agena Target–September 12, 1966
Gemini 12–November 11, 1966
Gemini 12 Agena Target–November 11, 1966
Liberty Bell 7 (Mercury 4)–July 21, 1961
Mercury 3, *see* Freedom 7
Mercury 6, *see* Friendship 7
Mercury 7, *see* Aurora 7
Mercury 8, *see* Sigma 7
Mercury 9, *see* Faith 7
Progress 1–January 20, 1978
Progress M1–August 23, 1989
Progress M34–April 6, 1997
Progress M35–July 5, 1997
Sigma 7 (Mercury 8)–October 3, 1962
Skylab 2–May 25, 1973
Skylab 3–July 28, 1973
Skylab 4–November 16, 1973
Soyuz 1–April 23, 1967
Soyuz 3–October 26, 1968
Soyuz 4–January 14, 1969
Soyuz 5–January 15, 1969
Soyuz 6–October 11, 1969
Soyuz 7–October 12, 1969
Soyuz 8–October 13, 1969
Soyuz 9–June 1, 1970
Soyuz 10–April 23, 1971
Soyuz 11–June 6, 1971
Soyuz 12–September 27, 1973
Soyuz 13–December 18, 1973
Soyuz 14–July 3, 1974
Soyuz 15–August 26, 1974
Soyuz 16–December 2, 1974
Soyuz 17–January 11, 1975
Soyuz 18A–April 5, 1975
Soyuz 18B–May 24, 1975
Soyuz 19 (ASTP)–July 15, 1975
Soyuz 20–November 17, 1975
Soyuz 21–July 6, 1976
Soyuz 22–September 15, 1976
Soyuz 23–October 14, 1976
Soyuz 24–February 7, 1977
Soyuz 25–October 9, 1997
Soyuz 26–December 10, 1977
Soyuz 27–January 10, 1978
Soyuz 28–March 2, 1978
Soyuz 29–June 15, 1978
Soyuz 30–June 27, 1978
Soyuz 31–August 26, 1978
Soyuz 32–February 25, 1979
Soyuz 33–April 10, 1979
Soyuz 34–June 6, 1979
Soyuz 35–April 9, 1980
Soyuz 36–May 26, 1980
Soyuz 37–July 23, 1980
Soyuz 38–September 18, 1980
Soyuz 39–March 22, 1981
Soyuz 40–May 14, 1981
Soyuz T2–June 5, 1980
Soyuz T3–November 27, 1980
Soyuz T4–March 12, 1981
Soyuz T5–May 13, 1982
Soyuz T6–June 24, 1982
Soyuz T7–August 19, 1982
Soyuz T8–April 20, 1983
Soyuz T9–June 27, 1983
Soyuz T10A–September 27, 1983
Soyuz T10B–February 8, 1984

Appendix 2. Satellites and Missions Listed by Subject

Soyuz T11–April 3, 1984
Soyuz T12–July 17, 1984
Soyuz T13–June 6, 1985
Soyuz T14–September 17, 1985
Soyuz T15–March 13, 1986
Soyuz TM1–May 21, 1986
Soyuz TM2–February 5, 1987
Soyuz TM3–July 22, 1987
Soyuz TM4–December 21, 1987
Soyuz TM5–June 7, 1988
Soyuz TM6–August 29, 1988
Soyuz TM7–November 26, 1988
Soyuz TM8–September 5, 1989
Soyuz TM9–February 11, 1990
Soyuz TM10–August 1, 1990
Soyuz TM11–December 2, 1990
Soyuz TM12–May 18, 1991
Soyuz TM13–October 2, 1991
Soyuz TM14–March 17, 1992
Soyuz TM15–July 27, 1992
Soyuz TM16–January 24, 1993
Soyuz TM17–July 1, 1993
Soyuz TM18–January 8, 1994
Soyuz TM19–July 1, 1994
Soyuz TM20–October 3, 1994
Soyuz TM21–March 14, 1995
Soyuz TM22–September 3, 1995
Soyuz TM23–February 21, 1996
Soyuz TM24–August 17, 1996
Soyuz TM25–February 10, 1997
Soyuz TM26–August 5, 1997
Soyuz TM27–January 29, 1998
Soyuz TM28–August 13, 1998
Soyuz TM29–February 20, 1999
STS 1–April 12, 1981
STS 2–November 12, 1981
STS 3–March 22, 1982
STS 4–June 27, 1982
STS 5–November 11, 1982
STS 6–April 4, 1983
STS 7–June 18, 1983
STS 8–August 30, 1983
STS 9–November 28, 1983
STS 26–September 29, 1988
STS 27–December 2, 1988
STS 28–August 8, 1989
STS 29–March 13, 1989
STS 30–May 4, 1989
STS 31–April 24, 1990
STS 32–January 9, 1990
STS 33–November 22, 1989
STS 34–October 18, 1989
STS 35–December 2, 1990
STS 36–February 28, 1990
STS 37–April 5, 1991
STS 38–November 15, 1990
STS 39–April 28, 1991
STS 40–June 5, 1991
STS 41–October 6, 1990
STS 41B (STS 10)–February 3, 1984
STS 41C (STS 11)–April 6, 1984
STS 41D (STS 12)–August 30, 1984
STS 41G (STS 13)–October 5, 1984
STS 42–January 22, 1992
STS 43–August 2, 1991
STS 44–November 24, 1991
STS 45–March 24, 1992
STS 46–July 31, 1992
STS 47–September 12, 1992
STS 48–September 12, 1991
STS 49–May 7, 1992
STS 50–June 25, 1992

STS 51–September 12, 1993
STS 51A (STS 14)–November 8, 1984
STS 51B (STS 17)–April 29, 1985
STS 51C (STS 15)–January 24, 1985
STS 51D (STS 16)–April 12, 1985
STS 51F (STS 19)–July 29, 1985
STS 51G (STS 18)–June 17, 1985
STS 51I (STS 20)–August 27, 1985
STS 51J (STS 21)–October 3, 1985
STS 51L (STS 25)–January 28, 1986
STS 52–October 22, 1992
STS 53–December 2, 1992
STS 54–January 13, 1993
STS 55–April 26, 1993
STS 56–April 8, 1993
STS 57–June 21, 1993
STS 58–October 18, 1993
STS 59–April 9, 1994
STS 60–February 3, 1994
STS 61–December 2, 1993
STS 61A (STS 22)–October 30, 1985
STS 61B (STS 23)–November 26, 1985
STS 61C (STS 24)–January 12, 1986
STS 62–March 4, 1994
STS 63–February 3, 1995
STS 64–September 9, 1994
STS 65–July 8, 1994
STS 66–November 3, 1994
STS 67–March 2, 1995
STS 68–September 30, 1994
STS 69–September 7, 1995
STS 70–July 13, 1995
STS 71–June 27, 1995
STS 72–January 11, 1996
STS 73–October 20, 1995
STS 74–November 12, 1995
STS 75–February 22, 1996
STS 76–March 22, 1996
STS 77–May 19, 1996
STS 78–June 20, 1996
STS 79–September 16, 1996
STS 80–November 19, 1996
STS 81–January 12, 1997
STS 82–February 11, 1997
STS 83–April 4, 1997
STS 84–May 15, 1997
STS 85–August 7, 1997
STS 86–September 25, 1997
STS 87–November 19, 1997
STS 88–December 4, 1998
STS 89–January 22, 1998
STS 90–April 17, 1998
STS 91–June 2, 1998
STS 93–July 23, 1999
STS 94–July 1, 1997
STS 95–October 29, 1998
STS 96–May 27, 1999
STS 103–December 19, 1999
Voskhod 1–October 12, 1964
Voskhod 2–March 18, 1965
Vostok 1–April 12, 1961
Vostok 2–August 6, 1961
Vostok 3–August 11, 1962
Vostok 4–August 12, 1962
Vostok 5–June 14, 1963
Vostok 6–June 16, 1963

Chronological Listing
Vostok 1–April 12, 1961
Freedom 7 (Mercury 3)–May 5, 1961
Liberty Bell 7 (Mercury 4)–July 21, 1961
Vostok 2–August 6, 1961

Friendship 7 (Mercury 6)–February 20, 1962
Aurora 7 (Mercury 7)–May 24, 1962
Vostok 3–August 11, 1962
Vostok 4–August 12, 1962
Sigma 7 (Mercury 8)–October 3, 1962
Faith 7 (Mercury 9)–May 15, 1963
Vostok 5–June 14, 1963
Vostok 6–June 16, 1963
Voskhod 1–October 12, 1964
Voskhod 2–March 18, 1965
Gemini 3–March 23, 1965
Gemini 4–June 3, 1965
Gemini 5–August 21, 1965
Gemini 7–December 4, 1965
Gemini 6–December 15, 1965
Gemini 8 Agena Target–March 16, 1966, 15:07 GMT
Gemini 8–March 16, 1966, 16:48 GMT
Augmented Target Docking Adapter (ATDA)–June 1, 1966
Gemini 9–June 3, 1966
Gemini 10 Agena Target–July 18, 1966, 20:38 GMT
Gemini 10–July 18, 1966, 22:22 GMT
Gemini 11 Agena Target–September 12, 1966, 13:12 GMT
Gemini 11–September 12, 1966, 14:38 GMT
Gemini 12 Agena Target–November 11, 1966, 19:12 GMT
Gemini 12–November 11, 1966, 20:53 GMT
Apollo 1–January 27, 1967
Soyuz 1–April 23, 1967
Apollo 7–October 11, 1968
Soyuz 3–October 26, 1968
Apollo 8–December 21, 1968
Soyuz 4–January 14, 1969
Soyuz 5–January 15, 1969
Apollo 9–March 3, 1969
Apollo 10–May 18, 1969
Apollo 11–July 16, 1969
Soyuz 6–October 11, 1969
Soyuz 7–October 12, 1969
Soyuz 8–October 13, 1969
Apollo 12–November 14, 1969
Apollo 13–April 11, 1970
Soyuz 9–June 1, 1970
Apollo 14–January 31, 1971
Soyuz 10–April 23, 1971
Soyuz 11–June 6, 1971
Apollo 15–July 26, 1971
Apollo 16–April 16, 1972
Apollo 17–December 7, 1972
Skylab 2–May 25, 1973
Skylab 3–July 28, 1973
Soyuz 12–September 27, 1973
Skylab 4–November 16, 1973
Soyuz 13–December 18, 1973
Soyuz 14–July 3, 1974
Soyuz 15–August 26, 1974
Soyuz 16–December 2, 1974
Soyuz 17–January 11, 1975
Soyuz 18A–April 5, 1975
Soyuz 18B–May 24, 1975
Soyuz 19 (ASTP)–July 15, 1975, 10:30 GMT
Apollo 18 (ASTP)–July 15, 1975, 18:00 GMT
Soyuz 21–July 6, 1976
Soyuz 22–September 15, 1976
Soyuz 23–October 14, 1976
Soyuz 24–February 7, 1977
Soyuz 26–December 10, 1977
Soyuz 27–January 10, 1978

360

Appendix 2. Satellites and Missions Listed by Subject

Progress 1–January 20, 1978
Soyuz 28–March 2, 1978
Soyuz 29–June 15, 1978
Soyuz 30–June 27, 1978
Soyuz 31–August 26, 1978
Soyuz 32–February 25, 1979
Soyuz 33–April 10, 1979
Soyuz 34–June 6, 1979
Soyuz 35–April 9, 1980
Soyuz 36–May 26, 1980
Soyuz T2–June 5, 1980
Soyuz 37–July 23, 1980
Soyuz 38–September 18, 1980
Soyuz T3–November 27, 1980
Soyuz T4–March 12, 1981
Soyuz 39–March 22, 1981
STS 1–April 12, 1981
Soyuz 40–May 14, 1981
STS 2–November 12, 1981
STS 3–March 22, 1982
Soyuz T5–May 13, 1982
Soyuz T6–June 24, 1982
STS 4–June 27, 1982
Soyuz T7–August 19, 1982
STS 5–November 11, 1982
STS 6–April 4, 1983
Soyuz T8–April 20, 1983
STS 7–June 18, 1983
Soyuz T9–June 27, 1983
STS 8–August 30, 1983
Soyuz T10A–September 27, 1983
STS 9–November 28, 1983
STS 41B (STS 10)–February 3, 1984
Soyuz T10B–February 8, 1984
Soyuz T11–April 3, 1984
STS 41C (STS 11)–April 6, 1984
Soyuz T12–July 17, 1984
STS 41D (STS 12)–August 30, 1984
STS 41G (STS 13)–October 5, 1984
STS 51A (STS 14)–November 8, 1984
STS 51C (STS 15)–January 24, 1985
STS 51D (STS 16)–April 12, 1985
STS 51B (STS 17)–April 29, 1985
Soyuz T13–June 6, 1985
STS 51G (STS 18)–June 17, 1985
STS 51F (STS 19)–July 29, 1985
STS 51I (STS 20)–August 27, 1985
Soyuz T14–September 17, 1985
STS 51J (STS 21)–October 3, 1985
STS 61A (STS 22)–October 30, 1985
STS 61B (STS 23)–November 26, 1985
STS 61C (STS 24)–January 12, 1986
STS 51L (STS 25)–January 28, 1986
Soyuz T15–March 13, 1986
Soyuz TM1–May 21, 1986
Soyuz TM2–February 5, 1987
Soyuz TM3–July 22, 1987
Soyuz TM4–December 21, 1987
Soyuz TM5–June 7, 1988
Soyuz TM6–August 29, 1988
STS 26–September 29, 1988
Soyuz TM7–November 26, 1988
STS 27–December 2, 1988
STS 29–March 13, 1989
STS 30–May 4, 1989
STS 28–August 8, 1989
Progress M1–August 23, 1989
Soyuz TM8–September 5, 1989
STS 34–October 18, 1989
STS 33–November 22, 1989
STS 32–January 9, 1990
Soyuz TM9–February 11, 1990

STS 36–February 28, 1990
STS 31–April 24, 1990
Soyuz TM10–August 1, 1990
STS 41–October 6, 1990
STS 38–November 15, 1990
STS 35–December 2, 1990, 6:49 GMT
Soyuz TM11–December 2, 1990, 8:13 GMT
STS 37–April 5, 1991
STS 39–April 28, 1991
Soyuz TM12–May 18, 1991
STS 40–June 5, 1991
STS 43–August 2, 1991
STS 48–September 12, 1991
Soyuz TM13–October 2, 1991
STS 44–November 24, 1991
STS 42–January 22, 1992
Soyuz TM14–March 17, 1992
STS 45–March 24, 1992
STS 49–May 7, 1992
STS 50–June 25, 1992
Soyuz TM15–July 27, 1992
STS 46–July 31, 1992
STS 47–September 12, 1992
STS 52–October 22, 1992
STS 53–December 2, 1992
STS 54–January 13, 1993
Soyuz TM16–January 24, 1993
STS 56–April 8, 1993
STS 55–April 26, 1993
STS 57–June 21, 1993
Soyuz TM17–July 1, 1993
STS 51–September 12, 1993
STS 58–October 18, 1993
STS 61–December 2, 1993
Soyuz TM18–January 8, 1994
STS 60–February 3, 1994
STS 62–March 4, 1994
STS 59–April 9, 1994
Soyuz TM19–July 1, 1994
STS 65–July 8, 1994
STS 64–September 9, 1994
STS 68–September 30, 1994
Soyuz TM20–October 3, 1994
STS 66–November 3, 1994
STS 63–February 3, 1995
STS 67–March 2, 1995
Soyuz TM21–March 14, 1995
STS 71–June 27, 1995
STS 70–July 13, 1995
Soyuz TM22–September 3, 1995
STS 69–September 7, 1995
STS 73–October 20, 1995
STS 74–November 12, 1995
STS 72–January 11, 1996
Soyuz TM23–February 21, 1996
STS 75–February 22, 1996
STS 76–March 22, 1996
STS 77–May 19, 1996
STS 78–June 20, 1996
Soyuz TM24–August 17, 1996
STS 79–September 16, 1996
STS 80–November 19, 1996
STS 81–January 12, 1997
Soyuz TM25–February 10, 1997
STS 82–February 11, 1997
STS 83–April 4, 1997
Progress M34–April 6, 1997
STS 84–May 15, 1997
STS 94–July 1, 1997
Progress M35–July 5, 1997
Soyuz TM26–August 5, 1997
STS 85–August 7, 1997

STS 86–September 25, 1997
STS 87–November 19, 1997
STS 89–January 22, 1998
Soyuz TM27–January 29, 1998
STS 90–April 17, 1998
STS 91–June 2, 1998
Soyuz TM28–August 13, 1998
STS 95–October 29, 1998
STS 88–December 4, 1998
Soyuz TM29–February 20, 1999
STS 96–May 27, 1999
STS 93–July 23, 1999
STS 103–December 19, 1999

Manned Space Laboratories/Modules

Alphabetical Listing
Cosmos 1267–April 25, 1981
Cosmos 1443–March 2, 1983
Cosmos 1686–September 27, 1985
International Space Station, see Unity, Zarya
Kristall–May 31, 1990
Kvant 1–March 31, 1987
Kvant 2–November 26, 1989
Mir–February 20, 1986
Priroda–April 23, 1996
Salyut 1–April 19, 1971
Salyut 3–June 25, 1974
Salyut 4–December 26, 1974
Salyut 5–June 22, 1976
Salyut 6–September 29, 1977
Salyut 7–April 19, 1982
Skylab–May 14, 1973
Spektr–May 20, 1995
Unity–December 4, 1998
Zarya–November 20, 1998

Chronological Listing
Salyut 1–April 19, 1971
Skylab–May 14, 1973
Salyut 3–June 25, 1974
Salyut 4–December 26, 1974
Salyut 5–June 22, 1976
Salyut 6–September 29, 1977
Cosmos 1267–April 25, 1981
Salyut 7–April 19, 1982
Cosmos 1443–March 2, 1983
Cosmos 1686–September 27, 1985
Mir–February 20, 1986
Kvant 1–March 31, 1987
Kvant 2–November 26, 1989
Kristall–May 31, 1990
Spektr–May 20, 1995
Priroda–April 23, 1996
Zarya–November 20, 1998
Unity–December 4, 1998

Materials Research

Alphabetical Listing
China 20–August 5, 1987
China 23–August 5, 1988
Cosmos 1645, see Foton 1
ERS 16 (ORS 2)–June 9, 1966
EURECA–July 31, 1992
Foton 1 (Cosmos 1645)–April 16, 1985
Foton 5–April 26, 1989
Foton 6–April 11, 1990
LDEF–January 9, 1990

361

Appendix 2. Satellites and Missions Listed by Subject

Long-Duration Exposure Facility (LDEF)–
    April 6, 1984
OV1-13–April 6, 1968
Space Flyer Unit–January 11, 1996; March
    18, 1995
Wake Shield 1–February 3, 1994
Wake Shield 2–September 7, 1995
Wake Shield 3–November 19, 1996

Chronological Listing
ERS 16 (ORS 2)–June 9, 1966
OV1-13–April 6, 1968
Long-Duration Exposure Facility (LDEF)–
    April 6, 1984
Foton 1 (Cosmos 1645)–April 16, 1985
China 20–August 5, 1987
China 23–August 5, 1988
Foton 5–April 26, 1989
LDEF–January 9, 1990
Foton 6–April 11, 1990
EURECA–July 31, 1992
Wake Shield 1–February 3, 1994
Space Flyer Unit–March 18, 1995
Wake Shield 2–September 7, 1995
Wake Shield 3–November 19, 1996
Space Flyer Unit–January 11, 1996

Micrometeoroid Research

Alphabetical Listing
Argos–February 23, 1999
Ariel 2–March 27, 1964
Explorer 16–December 16, 1962
Explorer 23–November 6, 1964
Explorer 46–August 13, 1972
Pegasus 1–February 16, 1965
Pegasus 2–May 25, 1965
Pegasus 3–July 30, 1965
RM 1–November 9, 1970

Chronological Listing
Explorer 16–December 16, 1962
Ariel 2–March 27, 1964
Explorer 23–November 6, 1964
Pegasus 1–February 16, 1965
Pegasus 2–May 25, 1965
Pegasus 3–July 30, 1965
RM 1–November 9, 1970
Explorer 46–August 13, 1972
Argos–February 23, 1999

Planetary Missions

Alphabetical Listing
Cassini–October 15, 1997
Clementine–January 25, 1994
Deep Space 1–October 24, 1998
Deep Space 1, Braille Flyby–July 28, 1999
Explorer 35, *see* Imp E
Galileo–October 18, 1989
Galileo, Gaspra Flyby–October 29, 1991
Galileo, Ida Flyby–August 28, 1993
Galileo, Jupiter Orbit–December 7, 1995
Giotto, Halley's Comet Flyby–March 13,
    1986
Hagoromo–January 24, 1990
Hiten (Muses A)–January 24, 1990
Huygens–October 15, 1997
Imp E (Explorer 35)–July 19, 1967
Luna 1–January 2, 1959
Luna 2–September 12, 1959

Luna 3–October 4, 1959
Luna 9–January 31, 1966
Luna 10–March 31, 1966
Luna 11–August 24, 1966
Luna 12–October 22, 1966
Luna 13–December 21, 1966
Luna 14–April 7, 1968
Luna 15–July 13, 1969
Luna 16–September 12, 1970
Luna 17–November 10, 1970
Luna 18–September 2, 1971
Luna 19–September 28, 1971
Luna 20–February 14, 1972
Luna 21–January 8, 1973
Luna 22–May 29, 1974
Luna 23–October 28, 1974
Luna 24–August 9, 1976
Lunar Orbiter 1–August 10, 1966
Lunar Orbiter 2–November 6, 1966
Lunar Orbiter 3–February 5, 1967
Lunar Orbiter 4–May 4, 1967
Lunar Orbiter 5–August 1, 1967
Lunar Prospector–January 7, 1998
Magellan–May 4, 1989
Magellan, Venus Orbit–August 10, 1990
Mariner 2–August 27, 1962
Mariner 2, Venus Flyby–December 14, 1962
Mariner 4–November 28, 1964
Mariner 4, Mars Flyby–July 15, 1965
Mariner 5, Venus Flyby–October 19, 1967
Mariner 6, Mars Flyby–July 31, 1969
Mariner 7, Mars Flyby–August 5, 1969
Mariner 9, Mars Orbit–November 14, 1971
Mariner 10–November 3, 1973
Mariner 10, Mercury Flyby #3–March 16,
    1975
Mariner 10, Venus Flyby–February 5, 1974
Mars 1–November 1, 1962
Mars 2–November 27, 1971
Mars 3–December 2, 1971
Mars 4, Mars Flyby–February 10, 1974
Mars 5, Mars Orbit–February 12, 1974
Mars 6, Mars Landing–March 12, 1974
Mars 7, Mars Flyby–March 9, 1974
Mars Climate Orbiter–December 11, 1998
Mars Global Surveyor–September 12, 1997
Mars Observer–September 25, 1992
Mars Pathfinder–July 4, 1997
Mars Polar Lander–January 3, 1999
Muses A, *see* Hiten
NEAR–February 17, 1996
NEAR, Eros Flyby–December 23, 1998
NEAR, Mathilde Flyby–June 27, 1997
Nozomi–July 3, 1998
Phobos 1–July 7, 1988
Phobos 2–July 12, 1988
Pioneer 10–March 3, 1972
Pioneer 10, Jupiter Flyby–December 3, 1973
Pioneer 10, Shutdown–March 31, 1997
Pioneer 11–April 6, 1973
Pioneer 11, Jupiter Flyby–December 4, 1974
Pioneer 11, Saturn Flyby–September 1, 1979
Pioneer 11, Shutdown–September 30, 1995
Pioneer Venus Orbiter–December 4, 1978
Pioneer Venus Probe–December 9, 1978
Ranger 3–January 26, 1962
Ranger 4–April 23, 1962
Ranger 6–January 30, 1964
Ranger 7–July 28, 1964
Ranger 8–February 17, 1965
Ranger 9–March 21, 1965
Sakigake–January 7, 1985

Sakigake, Halley's Comet Flyby–March 11,
    1986
Stardust–February 7, 1999
Suisei, Halley's Comet Flyby–March 8, 1986
Surveyor 1–May 30, 1966
Surveyor 3–April 17, 1967
Surveyor 5–September 8, 1967
Surveyor 6–November 7, 1967
Surveyor 7–January 7, 1968
Ulysses–October 6, 1990
Ulysses, First Solar Orbit–April 17, 1998
Ulysses, Jupiter Flyby–February 8, 1992
Vega 1, Halley's Comet Flyby–March 6,
    1986
Vega 1, Venus Flyby–June 10, 1985
Vega 2, Halley's Comet Flyby–March 9,
    1986
Vega 2, Venus Flyby–June 14, 1985
Venera 1–February 12, 1961
Venera 2–November 12, 1965
Venera 3–November 16, 1965
Venera 4, Venus Landing–October 18, 1967
Venera 5, Venus Landing–May 16, 1969
Venera 6, Venus Landing–May 17, 1969
Venera 7, Venus Landing–December 15,
    1970
Venera 8, Venus Landing–July 22, 1972
Venera 9, Venus Landing–October 22, 1975
Venera 10, Venus Landing–October 25, 1975
Venera 11, Venus Landing–December 25,
    1978
Venera 12, Venus Landing–December 21,
    1978
Venera 13, Venus Landing–March 1, 1982
Venera 14, Venus Landing–March 4, 1982
Venera 15, Venus Orbit–October 10, 1983
Venera 16, Venus Orbit–October 14, 1983
Viking 1, Mars Orbit/Landing–June 19, 1976
Viking 2–August 7, 1976
Voyager 1, Jupiter Flyby–March 5, 1979
Voyager 1, Saturn Flyby–December 12, 1980
Voyager 2, Jupiter Flyby–July 9, 1979
Voyager 2, Neptune Flyby–August 25, 1989
Voyager 2, Saturn Flyby–August 26, 1981
Voyager 2, Uranus Flyby–January 24, 1986
Zond 1–April 2, 1964
Zond 2–November 30, 1964
Zond 3–July 18, 1965
Zond 4–March 2, 1968
Zond 5–September 15, 1968
Zond 6–November 10, 1968
Zond 7–August 8, 1969
Zond 8–October 20, 1970

Chronological Listing
Luna 1–January 2, 1959
Luna 2–September 12, 1959
Luna 3–October 4, 1959
Venera 1–February 12, 1961
Ranger 3–January 26, 1962
Ranger 4–April 23, 1962
Mariner 2–August 27, 1962
Mars 1–November 1, 1962
Mariner 2, Venus Flyby–December 14, 1962
Ranger 6–January 30, 1964
Zond 1–April 2, 1964
Ranger 7–July 28, 1964
Mariner 4–November 28, 1964
Zond 2–November 30, 1964
Ranger 8–February 17, 1965
Ranger 9–March 21, 1965
Mariner 4, Mars Flyby–July 15, 1965

362

Zond 3–July 18, 1965
Venera 2–November 12, 1965
Venera 3–November 16, 1965
Luna 9–January 31, 1966
Luna 10–March 31, 1966
Surveyor 1–May 30, 1966
Lunar Orbiter 1–August 10, 1966
Luna 11–August 24, 1966
Luna 12–October 22, 1966
Lunar Orbiter 2–November 6, 1966
Luna 13–December 21, 1966
Lunar Orbiter 3–February 5, 1967
Surveyor 3–April 17, 1967
Lunar Orbiter 4–May 4, 1967
Imp E (Explorer 35)–July 19, 1967
Lunar Orbiter 5–August 1, 1967
Surveyor 5–September 8, 1967
Venera 4, Venus Landing–October 18, 1967
Mariner 5, Venus Flyby–October 19, 1967
Surveyor 6–November 7, 1967
Surveyor 7–January 7, 1968
Zond 4–March 2, 1968
Luna 14–April 7, 1968
Zond 5–September 15, 1968
Zond 6–November 10, 1968
Venera 5, Venus Landing–May 16, 1969
Venera 6, Venus Landing–May 17, 1969
Luna 15–July 13, 1969
Mariner 6, Mars Flyby–July 31, 1969
Mariner 7, Mars Flyby–August 5, 1969
Zond 7–August 8, 1969
Luna 16–September 12, 1970
Zond 8–October 20, 1970
Luna 17–November 10, 1970
Venera 7, Venus Landing–December 15, 1970
Luna 18–September 2, 1971
Luna 19–September 28, 1971
Mariner 9, Mars Orbit–November 14, 1971
Mars 2–November 27, 1971
Mars 3–December 2, 1971
Luna 20–February 14, 1972
Pioneer 10–March 3, 1972
Venera 8, Venus Landing–July 22, 1972
Luna 21–January 8, 1973
Pioneer 11–April 6, 1973
Mariner 10–November 3, 1973
Pioneer 10, Jupiter Flyby–December 3, 1973
Mariner 10, Venus Flyby–February 5, 1974
Mars 4, Mars Flyby–February 10, 1974
Mars 5, Mars Orbit–February 12, 1974
Mars 7, Mars Flyby–March 9, 1974
Mars 6, Mars Landing–March 12, 1974
Luna 22–May 29, 1974
Luna 23–October 28, 1974
Pioneer 11, Jupiter Flyby–December 4, 1974
Mariner 10, Mercury Flyby #3–March 16, 1975
Venera 9, Venus Landing–October 22, 1975
Venera 10, Venus Landing–October 25, 1975
Viking 1, Mars Orbit/Landing–June 19, 1976
Viking 2–August 7, 1976
Luna 24–August 9, 1976
Pioneer Venus Orbiter–December 4, 1978
Pioneer Venus Probe–December 9, 1978
Venera 12, Venus Landing–December 21, 1978
Venera 11, Venus Landing–December 25, 1978
Voyager 1, Jupiter Flyby–March 5, 1979
Voyager 2, Jupiter Flyby–July 9, 1979
Pioneer 11, Saturn Flyby–September 1, 1979

Voyager 1, Saturn Flyby–December 12, 1980
Voyager 2, Saturn Flyby–August 26, 1981
Venera 13, Venus Landing–March 1, 1982
Venera 14, Venus Landing–March 4, 1982
Venera 15, Venus Orbit–October 10, 1983
Venera 16, Venus Orbit–October 14, 1983
Vega 1, Venus Flyby–June 10, 1985
Vega 2, Venus Flyby–June 15, 1985
International Cometary Explorer (ICE)–September 11, 1985
Voyager 2, Uranus Flyby–January 24, 1986
Vega 1, Halley's Comet Flyby–March 6, 1986
Suisei, Halley's Comet Flyby–March 8, 1986
Vega 2, Halley's Comet Flyby–March 9, 1986
Sakigake, Halley's Comet Flyby–March 11, 1986
Giotto, Halley's Comet Flyby–March 13, 1986
Phobos 1–July 7, 1988
Phobos 2–July 12, 1988
Magellan–May 4, 1989
Voyager 2, Neptune Flyby–August 25, 1989
Galileo–October 18, 1989
Hagoromo–January 24, 1990
Hiten (Muses A)–January 24, 1990
Magellan, Venus Orbit–August 10, 1990
Ulysses–October 6, 1990
Galileo, Gaspra Flyby–October 29, 1991
Ulysses, Jupiter Flyby–February 8, 1992
Mars Observer–September 25, 1992
Galileo, Ida Flyby–August 28, 1993
Clementine–January 25, 1994
Pioneer 11, Shutdown–September 30, 1995
Galileo, Jupiter Orbit–December 7, 1995
NEAR–February 17, 1996
Pioneer 10, Shutdown–March 31, 1997
NEAR, Mathilde Flyby–June 27, 1997
Mars Pathfinder–July 4, 1997
Mars Global Surveyor–September 12, 1997
Cassini–October 15, 1997
Huygens–October 15, 1997
HGS 1–December 25, 1997
Lunar Prospector–January 7, 1998
Nozomi–July 3, 1998
Deep Space 1–October 24, 1998
Mars Climate Orbiter–December 11, 1998
NEAR, Eros Flyby–December 23, 1998
Mars Polar Lander–January 3, 1999
Stardust–February 7, 1999
Deep Space 1, Braille Flyby–July 28, 1999

Solar Research

Alphabetical Listing
ACE–August 25, 1997
Acrimsat–December 21, 1999
Aeros 2–July 16, 1974
Astro-A, *see* Hinotori
Atmosphere Explorer E, *see* Explorer 55
Aura–September 27, 1975
Coronas I, *see* Intercosmos 26
Cosmos 166–June 16, 1967
Discoverer 17–November 12, 1960
ERBS–October 5, 1984
EURECA–July 31, 1992
Explorer 30 (Solrad 8)–November 19, 1965
Explorer 55 (Atmosphere Explorer E)–November 20, 1975
Greb, *see* Solrad 3
Helios 1–December 10, 1974

Helios 2–January 15, 1976
HEOS 1–December 5, 1968
Hinotori (Astro-A)–February 21, 1981
Intercosmos 23, *see* Prognoz 10
Intercosmos 26 (Coronas I)–March 2, 1994
IRIS–May 17, 1968
ISEE 1/ISEE 2–October 22, 1977
ISEE 3–August 12, 1978
Mariner 2–August 27, 1962
Mariner 4–November 28, 1964
OSO 1–March 7, 1962
OSO 2–February 3, 1965
OSO 3–March 8, 1967
OSO 4–October 15, 1967
OSO 5–January 22, 1969
OSO 6–August 9, 1969
OSO 7–September 29, 1971
OSO 8–June 21, 1975
OV5-5/OV5-6/OV5-9–May 23, 1969
P78-1, *see* Solwind
Pioneer 5–March 11, 1960
Pioneer 6–December 16, 1965
Pioneer 7–August 17, 1966
Pioneer 8–December 13, 1967
Pioneer 9–November 8, 1968
Pioneer 10–March 3, 1972
Prognoz 1–April 14, 1972
Prognoz 9–July 1, 1983
Prognoz 10 (Intercosmos 23)–April 26, 1985
San Marco D/L–March 25, 1988
Shinsei–September 28, 1971
Soho–December 2, 1995
Solar Max–April 6, 1984; February 14, 1980
Solar-A, *see* Yohkoh
Solrad 1–June 22, 1960
Solrad 3 (Greb)–June 29, 1961
Solrad 6–June 15, 1963
Solrad 7A–January 11, 1964
Solrad 8, *see* Explorer 30
Solrad 9–March 5, 1968
Solrad 10–July 8, 1971
Solrad 11A/Solrad 11B–March 15, 1976Solwind (P78-1)–February 24, 1979
Spartan 201–April 8, 1993
November 19, 1997; October 29, 1998; September 7, 1995; September 9, 1994
SRATS, *see* Taiyo
Taiyo (SRATS)–February 24, 1975
Trace–April 1, 1998
UARS–September 12, 1991
Ulysses, First Solar Orbit–April 17, 1998
Venera 2–November 12, 1965
Venera 3–November 16, 1965
Wresat 1–November 29, 1967
Yohkoh (Solar-A)–August 30, 1991
Zond 2–November 30, 1964

Chronological Listing
Pioneer 5–March 11, 1960
Solrad 1–June 22, 1960
Discoverer 17–November 12, 1960
Solrad 3 (Greb)–June 29, 1961
OSO 1–March 7, 1962
Mariner 2–August 27, 1962
Solrad 6–June 15, 1963
Solrad 7A–January 11, 1964
Mariner 4–November 28, 1964
Zond 2–November 30, 1964
OSO 2–February 3, 1965
Venera 2–November 12, 1965
Venera 3–November 16, 1965

Appendix 2. Satellites and Missions Listed by Subject

Explorer 30 (Solrad 8)–November 19, 1965
Pioneer 6–December 16, 1965
Pioneer 7–August 17, 1966
OSO 3–March 8, 1967
Cosmos 166–June 16, 1967
OSO 4–October 15, 1967
Wresat 1–November 29, 1967
Pioneer 8–December 13, 1967
Solrad 9–March 5, 1968
IRIS–May 17, 1968
Pioneer 9–November 8, 1968
HEOS 1–December 5, 1968
OSO 5–January 22, 1969
OV5-5/OV5-6/OV5-9–May 23, 1969
OSO 6–August 9, 1969
Solrad 10–July 8, 1971
Shinsei–September 28, 1971
OSO 7–September 29, 1971
Pioneer 10–March 3, 1972
Prognoz 1–April 14, 1972
Aeros 2–July 16, 1974
Helios 1–December 10, 1974
Taiyo (SRATS)–February 24, 1975
OSO 8–June 21, 1975
Aura–September 27, 1975
Explorer 55 (Atmosphere Explorer E)–November 20, 1975
Helios 2–January 15, 1976
Solrad 11A/Solrad 11B–March 15, 1976
ISEE 1/ISEE 2–October 22, 1977
ISEE 3–August 12, 1978
Solwind (P78-1)–February 24, 1979
Solar Max–February 14, 1980
Hinotori (Astro-A)–February 21, 1981
Prognoz 9–July 1, 1983
Solar Max–April 6, 1984
ERBS–October 5, 1984
Prognoz 10 (Intercosmos 23)–April 26, 1985
San Marco D/L–March 25, 1988
Yohkoh (Solar-A)–August 30, 1991
UARS–September 12, 1991
EURECA–July 31, 1992
Spartan 201–April 8, 1993
Intercosmos 26 (Coronas I)–March 2, 1994
Spartan 201–September 9, 1994
Spartan 201–September 7, 1995
Soho–December 2, 1995
ACE–August 25, 1997
Spartan 201–November 19, 1997
Trace–April 1, 1998
Ulysses, First Solar Orbit–April 17, 1998
Spartan 201–October 29, 1998
Acrimsat–December 21, 1999

Spacecraft Design Research

Alphabetical Listing
83C–December 13, 1964
Alexis–April 25, 1993
APEX–August 3, 1994
Apollo 4–November 9, 1967
Apollo 5–January 22, 1968
Apollo 6–April 4, 1968
ARD–October 21, 1998
Argos–February 23, 1999
Ariane 1–June 19, 1981
Ariane 5–October 30, 1997
Asterix–November 26, 1965
Aurora 1–June 29, 1967
Big Joe–September 9, 1959
Buran–November 15, 1988
Calsphere–December 13, 1962

Cannonball 2–August 7, 1971
CAT 1–December 24, 1979
CAT 3–June 19, 1981
CAT 4–December 20, 1981
Conestoga–September 9, 1982
Cosmos 133–November 28, 1966
Cosmos 140–February 7, 1967
Cosmos 186–October 27, 1967
Cosmos 188–October 30, 1967
Cosmos 248–October 19, 1968
Cosmos 249–October 20, 1968
Cosmos 252–November 1, 1968
Cosmos 373–October 20, 1970
Cosmos 375–October 30, 1970
Cosmos 398–February 26, 1971
Cosmos 434–August 12, 1971
Cosmos 613–November 30, 1973
Cosmos 638–April 3, 1974
Cosmos 881/Cosmos 882–December 15, 1976
Cosmos 929–July 17, 1977
Cosmos 1001–April 4, 1978
Cosmos 1374 (BOR-4)–June 4, 1982
CTA–October 22, 1992
DarpaSat–March 13, 1994
Dash 1–May 9, 1963
DEBUT, see Orizuru
Deep Space 1–October 24, 1998
DemoSat–March 27, 1999
Diademe 1–February 8, 1967
Diademe 2–February 15, 1967
DIAL-MIKA–March 10, 1970
Discoverer 13–August 10, 1960
DLR-Tubsat–May 26, 1999
Dodecapole 1–March 9, 1965
Dodge 1–July 1, 1967
Enterprise, Flight 4–August 12, 1977
ERS 18–April 28, 1967
ETS 3, see Kiku 4
ETS 7, see Kiku 7
EURECA–July 31, 1992; June 21, 1993
Explorer 1–February 1, 1958
Express 1–January 15, 1995
Gemini 2–January 19, 1965; November 3, 1966
GGSE 1–January 11, 1964
GGSE 2/GGSE 3–March 9, 1965
GGTS 1–June 16, 1966
Gurwin Techsat 1B–July 10, 1998
Inspector–December 17, 1997
Intercosmos 15–June 19, 1976
Intercosmos 21–February 6, 1981
ISAS–January 25, 1994
JETS 1, see Kiku 1
Jindai (MABES)–August 12, 1986
Kiku 1 (JETS 1)–September 9, 1975
Kiku 4 (ETS 3)–September 3, 1982
Kiku 7 (ETS 7)–November 28, 1997
LACE–February 14, 1990
LCS 4–August 7, 1971
Little Joe 3–December 4, 1959
Losat X–July 4, 1991
MABES, see Jindai
Mao 1–April 24, 1970
Maqsat 3–October 21, 1998
Mercury 1A–December 19, 1960
Mercury 2–January 31, 1961
Mercury 2A–March 24, 1961
Mercury 5–November 29, 1961
MHV–September 13, 1985
Microsat–August 29, 1996
Midcourse Space Experiment (MSX)–April 24, 1996

MightySat 1–December 4, 1998
Minisat 1–April 21, 1997
Miranda–March 9, 1974
Mirka–October 10, 1997
MSTI 1–November 21, 1992
MSTI 2–May 9, 1994
MSTI 3–May 17, 1996
Musketball 1–August 7, 1971
Mylar Balloon–August 7, 1971
Myojo–February 3, 1994
Nimiq 1–May 21, 1999
ODERACS 1A-1F–February 3, 1994
ODERACS 2A-2F–February 3, 1995
OEX, see Orbiter Experiment
Ofeq 1–September 19, 1988
Ofeq 2–April 3, 1990
Ohsumi–February 11, 1970
Orbiter Experiment (OEX)–November 26, 1985
Orizuru (DEBUT)–February 7, 1990
ORS 3–July 20, 1965
OV1-2–October 5, 1965
OV1-4–March 30, 1966
OV1-5–March 30, 1966
OV1-13–April 6, 1968
OV1-17–March 18, 1969
OV1-19–March 18, 1969
OV1-20–August 7, 1971
OV3-1–April 22, 1966
OV3-2–October 28, 1966
OV3-3–August 4, 1966
OV3-6–December 5, 1967
OV5-1/OV5-3–April 28, 1967
P78-1, see Solwind
PAC 1–August 9, 1969
Palapa B2–February 3, 1984
PAMS–May 19, 1996
Pegsat–April 5, 1990
Pioneer 1–October 11, 1958
Pioneer 4–March 3, 1959
PIX 1–March 5, 1978
Pollux–May 17, 1975
Polyot 1–November 1, 1963
Polyot 2–April 12, 1964
Prospero–October 28, 1971
Proton 1–July 16, 1965
Proton 2–November 2, 1965
Proton 3–July 6, 1966
Proton 4–November 16, 1968
Radcat–October 2, 1972
Radose–June 15, 1963
Radsat–October 2, 1972
Rigidsphere–August 7, 1971
RM 1–November 9, 1970
RME–February 14, 1990
Rohini 1B–July 18, 1980
Ryusei–February 3, 1994
SAC-A–December 4, 1998
Sakigake–January 7, 1985
SCATHA–January 30, 1979
Seds–March 30, 1993
SERT 2–February 4, 1970
Shenzhou–November 19, 1999
Shijian 4–February 8, 1994
Shijian 5–May 10, 1999
SOICAL Cone/SOICAL Cylinder–September 30, 1969
Solwind (P78-1)–February 24, 1979
Soyuz 2–October 25, 1968
Soyuz 20–November 17, 1975
Soyuz T1–December 16, 1979
Soyuz TM1–May 21, 1986

Appendix 2. Satellites and Missions Listed by Subject

Space Flyer Unit–January 11, 1996; March 18, 1995
Spartan 1–June 17, 1985
Spartan 206–January 11, 1996
Spartan 207–May 19, 1996
SPAS (Shuttle PAllet Satellite) 1–June 18, 1983
Sputnik–October 4, 1957
Sputnik 2–November 3, 1957
Sputnik 4–May 15, 1960
Sputnik 5–August 19, 1960
Sputnik 6–December 1, 1960
Sputnik 9–March 9, 1961
Sputnik 10–March 25, 1961
STEP 0 (TAOS)–March 13, 1994
STEX–October 3, 1998
STRV-1A/STRV-1B–June 17, 1994
Surcal 1A–December 13, 1962
Tansei 2–February 16, 1974
Tethered Satellite System (TSS) 1–July 31, 1992
Tethered Satellite System (TSS) 1R–February 22, 1996
TRAAC–November 15, 1961
Triad 1–September 2, 1972
TRS 4–July 18, 1963
Tubsat 2–January 25, 1994
UoSat 5–July 17, 1991
UoSat 12–April 21, 1999
Vanguard 1–March 17, 1958
Vanguard 2–February 17, 1959
Vanguard 3–September 18, 1959
Westar 6–February 3, 1984
WIRE–March 4, 1999

Chronological Listing
Sputnik–October 4, 1957
Sputnik 2–November 3, 1957
Explorer 1–February 1, 1958
Vanguard 1–March 17, 1958
Pioneer 1–October 11, 1958
Vanguard 2–February 17, 1959
Pioneer 4–March 3, 1959
Big Joe–September 9, 1959
Vanguard 3–September 18, 1959
Little Joe 3–December 4, 1959
Sputnik 4–May 15, 1960
Discoverer 13–August 10, 1960
Sputnik 5–August 19, 1960
Sputnik 6–December 1, 1960
Mercury 1A–December 19, 1960
Mercury 2–January 31, 1961
Sputnik 9–March 9, 1961
Mercury 2A–March 24, 1961
Sputnik 10–March 25, 1961
TRAAC–November 15, 1961
Mercury 5–November 29, 1961
Calsphere–December 13, 1962
Surcal 1A–December 13, 1962
Dash 1–May 9, 1963
Radose–June 15, 1963
TRS 4–July 18, 1963
Polyot 1–November 1, 1963
GGSE 1–January 11, 1964
Polyot 2–April 12, 1964
83C–December 13, 1964
Gemini 2–January 19, 1965
Dodecapole 1–March 9, 1965
GGSE 2/GGSE 3–March 9, 1965
Proton 1–July 16, 1965
ORS 3–July 20, 1965
OV1-2–October 5, 1965

Proton 2–November 2, 1965
Asterix–November 26, 1965
OV1-4–March 30, 1966
OV1-5–March 30, 1966
OV3-1–April 22, 1966
GGTS 1–June 16, 1966
Proton 3–July 6, 1966
OV3-3–August 4, 1966
OV3-2–October 28, 1966
Gemini 2–November 3, 1966
Cosmos 133–November 28, 1966
Cosmos 140–February 7, 1967
Diademe 1–February 8, 1967
Diademe 2–February 15, 1967
ERS 18–April 28, 1967
OV5-1/OV5-3–April 28, 1967
Aurora 1–June 29, 1967
Dodge 1–July 1, 1967
Cosmos 186–October 27, 1967
Cosmos 188–October 30, 1967
Apollo 4–November 9, 1967
OV3-6–December 5, 1967
Apollo 5–January 22, 1968
Apollo 6–April 4, 1968
OV1-13–April 6, 1968
Cosmos 248–October 19, 1968
Cosmos 249–October 20, 1968
Soyuz 2–October 25, 1968
Cosmos 252–November 1, 1968
Proton 4–November 16, 1968
OV1-17–March 18, 1969
OV1-19–March 18, 1969
PAC 1–August 9, 1969
SOICAL Cone/SOICAL Cylinder–September 30, 1969
SERT 2–February 4, 1970
Ohsumi–February 11, 1970
DIAL-MIKA–March 10, 1970
Mao 1–April 24, 1970
Cosmos 373–October 20, 1970
Cosmos 375–October 30, 1970
RM 1–November 9, 1970
Cosmos 398–February 26, 1971
Cannonball 2–August 7, 1971
LCS 4–August 7, 1971
Musketball 1–August 7, 1971
Mylar Balloon–August 7, 1971
OV1-20–August 7, 1971
Rigidsphere–August 7, 1971
Cosmos 434–August 12, 1971
Prospero–October 28, 1971
Triad 1–September 2, 1972
Radcat–October 2, 1972
Radsat–October 2, 1972
Cosmos 613–November 30, 1973
Tansei 2–February 16, 1974
Miranda–March 9, 1974
Cosmos 638–April 3, 1974
Pollux–May 17, 1975
Kiku 1 (JETS 1)–September 9, 1975
Soyuz 20–November 17, 1975
Intercosmos 15–June 19, 1976
Cosmos 881/Cosmos 882–December 15, 1976
Cosmos 929–July 17, 1977
Enterprise, Flight 4–August 12, 1977
PIX 1–March 5, 1978
Cosmos 1001–April 4, 1978
SCATHA–January 30, 1979
Solwind (P78-1)–February 24, 1979
Soyuz T1–December 16, 1979
CAT 1–December 24, 1979

Rohini 1B–July 18, 1980
Intercosmos 21–February 6, 1981
Ariane 1–June 19, 1981
CAT 3–June 19, 1981
CAT 4–December 20, 1981
Cosmos 1374 (BOR-4)–June 4, 1982
Kiku 4 (ETS 3)–September 3, 1982
Conestoga–September 9, 1982
SPAS (Shuttle PAllet Satellite) 1–June 18, 1983
Palapa B2–February 3, 1984
Westar 6–February 3, 1984
Sakigake–January 7, 1985
Spartan 1–June 17, 1985
MHV–September 13, 1985
Orbiter Experiment (OEX)–November 26, 1985
Soyuz TM1–May 21, 1986
Jindai (MABES)–August 12, 1986
Ofeq 1–September 19, 1988
Buran–November 15, 1988
Orizuru (DEBUT)–February 7, 1990
LACE–February 14, 1990
RME–February 14, 1990
Ofeq 2–April 3, 1990
Pegsat–April 5, 1990
Losat X–July 4, 1991
UoSat 5–July 17, 1991
EURECA–July 31, 1992
Tethered Satellite System (TSS) 1–July 31, 1992
CTA–October 22, 1992
MSTI 1–November 21, 1992
Seds–March 30, 1993
Alexis–April 25, 1993
EURECA–June 21, 1993
Tubsat 2–January 25, 1994, 0:25 GMT
ISAS–January 25, 1994, 16:34 GMT
Myojo–February 3, 1994, 22:20 GMT
Ryusei–February 3, 1994, 22:20 GMT
ODERACS 1A-1F–February 3, 1994, 12:10 GMT
Shijian 4–February 8, 1994
DarpaSat–March 13, 1994
STEP 0 (TAOS)–March 13, 1994
MSTI 2–May 9, 1994
STRV-1A/STRV-1B–June 17, 1994
APEX–August 3, 1994
Express 1–January 15, 1995
ODERACS 2A-2F–February 3, 1995
Space Flyer Unit–March 18, 1995
Space Flyer Unit–January 11, 1996
Spartan 206–January 11, 1996
Tethered Satellite System (TSS) 1R–February 22, 1996
Midcourse Space Experiment (MSX)–April 24, 1996
MSTI 3–May 17, 1996
PAMS–May 19, 1996
Spartan 207–May 19, 1996
Microsat–August 29, 1996
Minisat 1–April 21, 1997
Mirka–October 10, 1997
Ariane 5–October 30, 1997
Kiku 7 (ETS 7)–November 28, 1997
Inspector–December 17, 1997
Gurwin Techsat 1B–July 10, 1998
STEX–October 3, 1998
ARD–October 21, 1998
Maqsat 3–October 21, 1998
Deep Space 1–October 24, 1998
MightySat 1–December 4, 1998

365

Appendix 2. Satellites and Missions Listed by Subject

SAC-A–December 4, 1998
Argos–February 23, 1999
WIRE–March 4, 1999
DemoSat–March 27, 1999
UoSat 12–April 21, 1999
Shijian 5–May 10, 1999
Nimiq 1–May 21, 1999
DLR-Tubsat–May 26, 1999
Shenzhou–November 19, 1999

Surveillance

Alphabetical Listing

38C–September 28, 1963
Canyon 1–August 6, 1968
Cerise–July 7, 1995
Chalet 1–June 10, 1978
China 9/China 10/China 11–September 19, 1981
Cosmos 4–April 26, 1962
Cosmos 7–July 28, 1962
Cosmos 148–March 16, 1967
Cosmos 198–December 27, 1967
Cosmos 699–December 24, 1974
Cosmos 954–September 18, 1977
Cosmos 1402–August 27, 1982
Cosmos 2176–January 24, 1992
Discoverer 1–February 28, 1959
Discoverer 13–August 10, 1960
Discoverer 14–August 18, 1960
Discoverer 17–November 12, 1960
Discoverer 18–December 8, 1960
Discoverer 36–December 12, 1961
DSP 16–November 24, 1991
DSP-647 1, *see* IMEWS 1
Ferret 1–June 2, 1962
FORTE–August 29, 1997
Geosat–March 13, 1985
Geosat Follow-On–February 10, 1998
Helios 1A–July 7, 1995
IMEWS 1 (DSP-647 1)–May 5, 1971
IMEWS 6–June 26, 1976
Jumpseat 1–March 21, 1971
KH-4 9032–April 18, 1962
KH-7 1–July 12, 1963
KH-7-10–August 14, 1964
KH-9 1–June 15, 1971
KH-11 1–December 19, 1976
KH-12 1–February 28, 1990
Lacrosse 1–December 2, 1988
Magnum 1–January 24, 1985
Magnum 2–November 22, 1989
Magnum 3–November 15, 1990
Midas 2–May 24, 1960
Midas 4–October 21, 1961
NOSS 1 (Whitecloud 1)–April 30, 1976
Ofeq 3–April 5, 1995
Rhyolite 1–March 6, 1973
Samos 2–January 31, 1961
Trumpet 1–May 3, 1994
USA 40–August 8, 1989
USA 41–August 8, 1989
USA 59–June 8, 1990
Vela 1A/Vela 1B–October 17, 1963
Vela 2A/Vela 2B–July 17, 1964
Vela 5A/Vela 5B–May 23, 1969
Whitecloud 1, *see* NOSS 1

Chronological Listing

Discoverer 1–February 28, 1959
Midas 2–May 24, 1960
Discoverer 13–August 10, 1960
Discoverer 14–August 18, 1960
Discoverer 17–November 12, 1960
Discoverer 18–December 8, 1960
Samos 2–January 31, 1961
Midas 4–October 21, 1961
Discoverer 36–December 12, 1961
KH-4 9032–April 18, 1962
Cosmos 4–April 26, 1962
Ferret 1–June 2, 1962
Cosmos 7–July 28, 1962
KH-7 1–July 12, 1963
38C–September 28, 1963
Vela 1A/Vela 1B–October 17, 1963
Vela 2A/Vela 2B–July 17, 1964
KH-7-10–August 14, 1964
Cosmos 148–March 16, 1967
Cosmos 198–December 27, 1967
Canyon 1–August 6, 1968
Vela 5A/Vela 5B–May 23, 1969
Jumpseat 1–March 21, 1971
IMEWS 1 (DSP-647 1)–May 5, 1971
KH-9 1–June 15, 1971
Rhyolite 1–March 6, 1973
Cosmos 699–December 24, 1974
NOSS 1 (Whitecloud 1)–April 30, 1976
IMEWS 6–June 26, 1976
KH-11 1–December 19, 1976
Cosmos 954–September 18, 1977
Chalet 1–June 10, 1978
China 9/China 10/China 11–September 19, 1981
Cosmos 1402–August 27, 1982
Magnum 1–January 24, 1985
Geosat–March 13, 1985
Lacrosse 1–December 2, 1988
USA 40–August 8, 1989
USA 41–August 8, 1989
Magnum 2–November 22, 1989
KH-12 1–February 28, 1990
USA 59–June 8, 1990
Magnum 3–November 15, 1990
DSP 16–November 24, 1991
Cosmos 2176–January 24, 1992
Trumpet 1–May 3, 1994
Ofeq 3–April 5, 1995
Cerise–July 7, 1995
Helios 1A–July 7, 1995
FORTE–August 29, 1997
Geosat Follow-On–February 10, 1998

Weather Satellites

Alphabetical Listing

Cosmos 44–August 28, 1964
Cosmos 122–June 25, 1966
Cosmos 144–February 28, 1967
Cosmos 156–April 27, 1967
DMSP Block-4A F1–January 19, 1965
Elektro–October 31, 1994
Eole–August 16, 1971
ESSA 1–February 3, 1966
Eyesat 1–September 26, 1993
Feng Yun 1A–September 6, 1988
Feng Yun 2B–June 10, 1997
Fen Yun 3–May 10, 1999
GOES 1–October 16, 1975
Himawari 1–July 14, 1977
Insat 1A–April 10, 1982
Insat 1B–August 30, 1983
Intercosmos 20–November 1, 1979
ITOS 1–January 23, 1970
Meteor 1-01–March 26, 1969
Meteor 2-01–July 11, 1975
Meteor 3-01–October 24, 1985
Meteor 3-05–August 15, 1991
Meteosat 1–November 23, 1977
Meteosat 3–June 15, 1988
Nimbus 1–August 28, 1964
NOAA 1–December 11, 1970
Peole–December 12, 1970
Posat 1–September 26, 1993
SCD 1–February 9, 1993
SMS 1–May 17, 1974
Temisat–August 31, 1993
Tiros 1–April 1, 1960
Tiros 2–November 23, 1960
Tiros 3–July 12, 1961
Tiros 4–February 8, 1962
Tiros 5–June 19, 1962
Tiros 6–September 18, 1962
Tiros 7–June 19, 1963
Tiros 8–December 21, 1963
Tiros 9–January 22, 1965
Tubsat 2–January 25, 1994
Uribyol 2–September 26, 1993
Uribyol 3–May 26, 1999
USA 59–June 8, 1990

Chronological Listing

Tiros 1–April 1, 1960
Tiros 2–November 23, 1960
Tiros 3–July 12, 1961
Tiros 4–February 8, 1962
Tiros 5–June 19, 1962
Tiros 6–September 18, 1962
Tiros 7–June 19, 1963
Tiros 8–December 21, 1963
Nimbus 1–August 28, 1964, 8:52 GMT
Cosmos 44–August 28, 1964, 16:19 GMT
DMSP Block-4A F1–January 19, 1965
Tiros 9–January 22, 1965
ESSA 1–February 3, 1966
Cosmos 122–June 25, 1966
Cosmos 144–February 28, 1967
Cosmos 156–April 27, 1967
Meteor 1-01–March 26, 1969
ITOS 1–January 23, 1970
NOAA 1–December 11, 1970
Peole–December 12, 1970
Eole–August 16, 1971
SMS 1–May 17, 1974
Meteor 2-01–July 11, 1975
GOES 1–October 16, 1975
Himawari 1–July 14, 1977
Meteosat 1–November 23, 1977
Intercosmos 20–November 1, 1979
Insat 1A–April 10, 1982
Insat 1B–August 30, 1983
Meteor 3-01–October 24, 1985
Meteosat 3–June 15, 1988
Feng Yun 1A–September 6, 1988
USA 59–June 8, 1990
Meteor 3-05–August 15, 1991
SCD 1–February 9, 1993
Posat 1–September 26, 1993
Temisat–August 31, 1993
Uribyol 2–September 26, 1993
Eyesat 1–September 26, 1993
Tubsat 2–January 25, 1994
Elektro–October 31, 1994
Feng Yun 2B–June 10, 1997
Fen Yun 3–May 10, 1999
Uribyol 3–May 26, 1999

Appendix 3
Satellites and Missions Listed by Nation or Group of Nations

Afghanistan
Soyuz TM6–August 29, 1988

Africa
See also **South Africa**
AfriStar–October 28, 1998

Arab States
Arabsat 1A–February 8, 1985
Arabsat 1B–June 17, 1985

Argentina
Microsat–August 29, 1996
Nahuel 1A–January 30, 1997
SAC-A–December 4, 1998

Australia
Wresat 1–November 29, 1967
Oscar 5–January 23, 1970
Aussat A1–August 27, 1985
Aussat A2–November 26, 1985
Aussat B1 (Optus B1)–August 14, 1992
Westpac–July 10, 1998

Austria
Soyuz TM13–October 2, 1991

Brazil
Brasilsat A1–February 8, 1985
SCD 1–February 9, 1993
Brasilsat B1–August 10, 1994
CBERS 1–October 14, 1999
SACI 1–October 14, 1999

Bulgaria
Soyuz 33–April 10, 1979
Soyuz TM5–June 7, 1988

Canada
Alouette 1–September 29, 1962
Alouette 2–November 29, 1965
Anik A1 (Telesat 1)–November 10, 1972
Hermes (CTS 1)–January 17, 1976
Anik B (Telesat 4)–December 15, 1978
Anik D1 (Telesat 5)–August 26, 1982
Anik C3 (Telesat 6)–November 11, 1982
Anik C2 (Telesat 7)–June 18, 1983
STS 41G (STS 13)–October 5, 1984
Anik D2–November 8, 1984
Anik C1–April 12, 1985
Anik E2–April 4, 1991
CTA–October 22, 1992
Brasilsat B1–August 10, 1994
Radarsat–November 4, 1995
MSAT 1–April 20, 1996
STS 77–May 19, 1996
STS 78–June 20, 1996
STS 85–August 7, 1997
Nimiq 1–May 21, 1999
STS 96–May 27, 1999

Chile
FASat-Bravo–July 10, 1998

China
Mao 1–April 24, 1970
China 9/China 10/China 11–September 19, 1981
STW (Shiyan Tongxin Weixing) 1–April 8, 1984
China 20–August 5, 1987
China 23–August 5, 1988
Feng Yun 1A–September 6, 1988
Asiasat 1–April 7, 1990
Badr A–July 16, 1990
Atmosphere 1–September 3, 1990
Fanhui Shu Weixing 12–October 5, 1990
Aussat B1 (Optus B1)–August 14, 1992
China 35–October 6, 1992
Freja–October 6, 1992
Shijian 4–February 8, 1994
Apstar 1–July 21, 1994
DFH (Dong Fang Hong) 3A–November 29, 1994
Echostar 1–December 28, 1995
Feng Yun 2B–June 10, 1997
Agila 2–August 19, 1997
Sinosat 1–July 18, 1998
Feng Yun 3–May 10, 1999
Shijian 5–May 10, 1999
CBERS 1–October 14, 1999
SACI 1–October 14, 1999
Shenzhou–November 19, 1999

Cuba
Soyuz 38–September 18, 1980

Czechoslovakia
See also **Czech Republic**
Soyuz 28–March 2, 1978
Magion 1–October 24, 1978
Prognoz 9–July 1, 1983
Prognoz 10 (Intercosmos 23)–April 26, 1985
Magion 2–September 28, 1989
Magion 3–December 18, 1991

Czech Republic
See also **Czechoslovakia**
Magion 4–August 2, 1995
Magion 5–August 29, 1996

Denmark
Orsted–February 23, 1999

East Europe
See also individual nations
Intercosmos 1–October 14, 1969
Intercosmos 15–June 19, 1976
Intercosmos 20–November 1, 1979
Intercosmos 21–February 6, 1981
Intercosmos 24–September 28, 1989
Intercosmos 25–December 18, 1991

East Germany
See also **Germany**
Soyuz 31–August 26, 1978

Egypt
Nilesat 1–April 28, 1998

Europe
See also individual nations
IRIS–May 17, 1968
Aurorae–October 3, 1968
HEOS 1–December 5, 1968
Boreas–October 1, 1969
HEOS 2–January 31, 1972
Thor Delta 1A–March 12, 1972
ESRO 4–November 22, 1972
COS B–August 9, 1975
Hermes (CTS 1)–January 17, 1976
ESA-GEOS 1–April 20, 1977
ISEE 1/ISEE 2–October 22, 1977
Meteosat 1–November 23, 1977

Appendix 3. Satellites and Missions Listed by Nation or Group of Nations

IUE–January 26, 1978
OTS 2–May 11, 1978
ESA-GEOS 2–July 14, 1978
ISEE 3–August 12, 1978
CAT 1–December 24, 1979
Apple–June 19, 1981
Ariane 1–June 19, 1981
CAT 3–June 19, 1981
CAT 4–December 20, 1981
MARECS A–December 20, 1981
Exosat–May 26, 1983
Eutelsat 1 (ECS 1)–June 16, 1983
Oscar 10–June 16, 1983
STS 9–November 28, 1983
Brasilsat A1–February 8, 1985
Telecom 1A–August 4, 1984
Arabsat 1A–February 8, 1985
STS 51B (STS 17)–April 29, 1985
STS 51F (STS 19)–July 29, 1985
Spot 1–February 22, 1986
Viking–February 22, 1986
Giotto, Halley's Comet Flyby–March 13, 1986
Meteosat 3–June 15, 1988
Phobos 1–July 7, 1988
Phobos 2–July 12, 1988
Astra 1A–December 11, 1988
Skynet 4B–December 11, 1988
Kopernikus 1–June 5, 1989
Superbird A–June 5, 1989
Olympus 1–July 12, 1989
Hipparcos–August 8, 1989
Intelsat 6A F2–October 27, 1989
UoSat 3/Uosat 4 (Oscar 14/Oscar 15)–January 22, 1990
Oscar 16/Oscar 17/Oscar 19–January 22, 1990
Webersat (Oscar 18)–January 22, 1990
Eutelsat 2 F1–August 30, 1990
Ulysses–October 6, 1990
Italsat 1–January 15, 1991
Anik E2–April 4, 1991
ERS 1–July 17, 1991
Telecom 2A–December 16, 1991
Insat 2A–July 9, 1992
EURECA–July 31, 1992
S80-T–August 10, 1992
TOPEX-Poseidon–August 10, 1992
Uribyol–August 10, 1992
Arsene–May 12, 1993
EURECA–June 21, 1993
Eyesat 1–September 26, 1993
Healthsat 2–September 26, 1993
Itamsat–September 26, 1993
Posat 1–September 26, 1993
Stella–September 26, 1993
Uribyol 2–September 26, 1993
Intelsat 7 F1–October 22, 1993
Solidaridad 1–November 20, 1993
DBS 1–December 18, 1993
Thaicom 1A–December 18, 1993
STRV-1A/B–June 17, 1994
Brasilsat B1–August 10, 1994
Turksat 1B–August 10, 1994
Soyuz TM20–October 3, 1994
Hot Bird 1–March 28, 1995
ERS 2–April 21, 1995
Cerise–July 7, 1995
Helios 1A–July 7, 1995
UPM-LBSAT 1–July 7, 1995
Soyuz TM22–September 3, 1995
ISO–November 17, 1995

Soho–December 2, 1995
Measat 1–January 12, 1996
MSAT 1–April 20, 1996
AMOS–May 16, 1996
Nahuel 1A–January 30, 1997
Intelsat 801–March 1, 1997
Cassini–October 15, 1997
Huygens–October 15, 1997
Ariane 5–October 30, 1997
Cakrawarta 1–November 12, 1997
Equator S–December 2, 1997
Spot 4–March 23, 1998
Ulysses, First Solar Orbit–April 17, 1998
Nilesat 1–April 28, 1998
ST 1–August 25, 1998
ARD–October 21, 1998
Maqsat 3–October 21, 1998
AfriStar–October 28, 1998
Satmex 5–December 6, 1998
Telcom 1 (Palapa D-1)–August 12, 1999

France
Asterix–November 26, 1965
France 1–December 6, 1965
Diademe 1–February 8, 1967
Diademe 2–February 15, 1967
DIAL-MIKA–March 10, 1970
Peole–December 12, 1970
Eole–August 16, 1971
Aureol 1–December 27, 1971
Symphonie 1–December 18, 1974
Starlette–February 6, 1975
Castor–May 17, 1975
Pollux–May 17, 1975
Aura–September 27, 1975
SIGNE 3–June 17, 1977
Soyuz T6–June 24, 1982
Astron–March 23, 1983
Prognoz 9–July 1, 1983
Telecom 1A–August 4, 1984
STS 51G (STS 18)–June 17, 1985
Spot 1–February 22, 1986
China 20–August 5, 1987
TDF 1–October 28, 1988
Soyuz TM7–November 26, 1988
Foton 5–April 26, 1989
Granat–December 1, 1989
Foton 6–April 11, 1990
Gamma–July 11, 1990
SARA–July 17, 1991
Telecom 2A–December 16, 1991
Soyuz TM15–July 27, 1992
S80-T–August 10, 1992
TOPEX-Poseidon–August 10, 1992
Arsene–May 12, 1993
Soyuz TM17–July 1, 1993
Stella–September 26, 1993
Turksat 1B–August 10, 1994
STS 66–November 3, 1994
Hot Bird 1–March 28, 1995
Cerise–July 7, 1995
Helios 1A–July 7, 1995
STS 78–June 20, 1996
ADEOS (Midori)–August 17, 1996
Soyuz TM24–August 17, 1996
STS 84–May 15, 1997
STS 86–September 25, 1997
Sputnik 40–November 3, 1997
Soyuz TM27–January 29, 1998
Spot 4–March 23, 1998
Nilesat 1–April 28, 1998
Sinosat 1–July 18, 1998

ST 1–August 25, 1998
AfriStar–October 28, 1998
Soyuz TM29–February 20, 1999
STS 93–July 23, 1999
STS 103–December 19, 1999

Germany
See also **East Germany** *and* **West Germany**
Rosat–June 1, 1990
Tubsat 1–July 17, 1991
Soyuz TM14–March 17, 1992
STS 55–April 26, 1993
Temisat–August 31, 1993
Orfeus-Spas–September 12, 1993
Tubsat 2–January 25, 1994
BremSat 1–February 3, 1994
STS 59–April 9, 1994
Crista-Spas 1–November 3, 1994
Express RV–January 15, 1995
Orfeus–November 19, 1996
Soyuz TM25–February 10, 1997
Crista-Spas 2–August 7, 1997
Mirka–October 10, 1997
Equator S–December 2, 1997
Inspector–December 17, 1997
Tubsat N/Tubsat N1–July 7, 1998
Safir 2–July 10, 1998
Sinosat 1–July 18, 1998
DLR-Tubsat–May 26, 1999

Hong Kong
Asiasat 1–April 7, 1990
Apstar 1–July 21, 1994

Hungary
Soyuz 36–May 26, 1980

India
Aryabhata–April 19, 1975
Bhaskara 1–June 7, 1979
Rohini 1B–July 18, 1980
Apple–June 19, 1981
Insat 1A–April 10, 1982
Insat 1B–August 30, 1983
Soyuz T11–April 3, 1984
IRS 1A–March 17, 1988
Insat 2A–July 9, 1992
DLR-Tubsat–May 26, 1999
IRS-P4 Oceansat–May 26, 1999
Uribyol 3–May 26, 1999

Indonesia
Palapa 1–July 8, 1976
Palapa B1–June 18, 1983
Palapa B2–February 3, 1984
Palapa B2R–April 13, 1990
Palapa C-1–February 1, 1996
Cakrawarta 1–November 12, 1997
Telcom 1 (Palapa D-1)–August 12, 1999

International
See also individual nations
Early Bird (Intelsat 1)–April 6, 1965
Intelsat 2A–October 26, 1966
Intelsat 2B–January 11, 1967
Intelsat 2C–March 23, 1967
Intelsat 2D–September 28, 1967
Intelsat 3B–December 19, 1968

Appendix 3. Satellites and Missions Listed by Nation or Group of Nations

Intelsat 4 F2–January 26, 1971
Oscar 6–October 15, 1972
Bion 2 (Cosmos 690)–October 22, 1974
Oscar 7–November 15, 1974
Intelsat 4A F1–September 26, 1975
Bion 4 (Cosmos 936)–August 3, 1977
Oscar 8–March 5, 1978
Bion 5 (Cosmos 1129)–September 25, 1979
Intelsat 5 F2–December 6, 1980
Eutelsat 1 (ECS 1)–June 16, 1983
Oscar 10–June 16, 1983
STS 9–November 28, 1983
Bion 6 (Cosmos 1514)–December 14, 1983
Intelsat 5A F10–March 22, 1985
Vega 1, Venus Flyby–June 10, 1985
Vega 2, Venus Flyby–June 14, 1985
Vega 1, Halley's Comet Flyby–March 6, 1986
Vega 2, Halley's Comet Flyby–March 9, 1986
Bion 8 (Cosmos 1887)–September 29, 1987
Oscar 13–June 15, 1988
Bion 9 (Cosmos 2044)–September 15, 1989
Intelsat 6A F2–October 27, 1989
Oscar 16/Oscar 17/Oscar 19–January 22, 1990
Inmarsat 2 F1–October 30, 1990
STS 42–January 22, 1992
STS 45–March 24, 1992
Intelsat 6 F3–May 7, 1992
Intelsat K–June 10, 1992
Bion 10 (Cosmos 2229)–December 29, 1992
Intelsat 7 F1–October 22, 1993
Intercosmos 26 (Coronas I)–March 2, 1994
STS 65–July 8, 1994
Wind–November 1, 1994
Inmarsat 3 F1–April 3, 1996
Bion 11–December 24, 1996
Intelsat 801–March 1, 1997
Starshine–May 27, 1999

Israel
Ofeq 1–September 19, 1988
Ofeq 2–April 3, 1990
Ofeq 3–April 5, 1995
AMOS–May 16, 1996
Gurwin Techsat 1B–July 10, 1998

Italy
San Marco 1–December 15, 1964
San Marco 2–April 26, 1967
Uhuru (Explorer 42 or SAS 1)–December 12, 1970
San Marco 3–April 24, 1971
Explorer 45–November 15, 1971
Explorer 48 (SAS 2)–November 15, 1972
San Marco 4–February 18, 1974
Explorer 53 (SAS 3)–May 7, 1975
Sirio 1–August 25, 1977
San Marco D/L–March 25, 1988
Italsat 1–January 15, 1991
Tethered Satellite System (TSS) 1–July 31, 1992
Lageos 2–October 22, 1992
Temisat–August 31, 1993
Itamsat–September 26, 1993
STS 59–April 9, 1994
Helios 1A–July 7, 1995
UPM-LBSAT 1–July 7, 1995
Tethered Satellite System (TSS) 1R–February 22, 1996
BeppoSax–April 30, 1996

Japan
Ohsumi–February 11, 1970
Shinsei–September 28, 1971
Tansei 2–February 16, 1974
Taiyo (SRATS)–February 24, 1975
Kiku 1 (JETS 1)–September 9, 1975
ISS 1–February 29, 1976
Kiku 2 (ETS 2)–February 23, 1977
Himawari 1–July 14, 1977
Sakura 1–December 15, 1977
Kyokko (EXOS A)–February 4, 1978
Ume 2 (ISS 2)–February 16, 1978
Yuri 1–April 7, 1978
Jikiken (EXOS B)–September 16, 1978
Hakucho (CORSA B)–February 21, 1979
Kiku 3 (ETS 4)–February 11, 1981
Hinotori (Astro-A)–February 21, 1981
Kiku 4 (ETS 3)–September 3, 1982
Sakura 2A–February 4, 1983
Tenma (Astro-B)–February 20, 1983
Ohzora (EXOS C)–February 14, 1984
Sakigake–January 7, 1985
Suisei, Halley's Comet Flyby–March 8, 1986
Sakigake, Halley's Comet Flyby–March 11, 1986
Ajisai (EGP)–August 12, 1986
Fuji 1 (Oscar 12)–August 12, 1986
Jindai (MABES)–August 12, 1986
Ginga (Astro-C)–February 5, 1987
Momo 1 (MOS 1)–February 19, 1987
Kiku 5 (ETS 5)–August 27, 1987
Sakura 3A–February 19, 1988
Akebono (EXOS D)–February 21, 1989
JCSat 1–March 6, 1989
Superbird A–June 5, 1989
Hagoromo–January 24, 1990
Hiten (Muses A)–January 24, 1990
Orizuru (DEBUT)–February 7, 1990
Yohkoh (Solar-A)–August 30, 1991
JERS 1–February 11, 1992
Geotail–July 24, 1992
STS 47–September 12, 1992
ASCA (Astro-D)–February 20, 1993
Myojo–February 3, 1994
Ryusei–February 3, 1994
Express RV–January 15, 1995
Space Flyer Unit–March 18, 1995
N-Star A–August 29, 1995
Space Flyer Unit–January 11, 1996
ADEOS (Midori)–August 17, 1996
Halca–February 12, 1997
BSAT 1A–April 16, 1997
STS 87–November 19, 1997
Kiku 7 (ETS 7)–November 28, 1997
TRMM–November 28, 1997
Nozomi–July 3, 1998
STS 95–October 29, 1998

Kazakhstan
See also **Soviet Union (U.S.S.R.)**
Soyuz TM13–October 2, 1991
Soyuz TM19–July 1, 1994

Kyrgyzstan
See also **Soviet Union (U.S.S.R.)**
STS 89–January 22, 1998

Luxembourg
Astra 1A–December 11, 1988

Malaysia
Measat 1–January 12, 1996

Mexico
Morelos 1–June 17, 1985
Morelos 2–November 26, 1985
Solidaridad 1–November 20, 1993
Satmex 5–December 6, 1998

Mongolia
Soyuz 39–March 22, 1981

NATO
NATO 3A–April 22, 1976
NATO 4A–January 8, 1991

Netherlands
ANS 1–August 30, 1974
IRAS–January 26, 1983
STS 61A (STS 22)–October 30, 1985
BeppoSax–April 30, 1996

Pakistan
Badr A–July 16, 1990

Philippines
Agila 2–August 19, 1997

Poland
Intercosmos 6–April 7, 1972
Soyuz 30–June 27, 1978
Ionosonde (Cosmos 1809)–December 18, 1986

Portugal
Posat 1–September 26, 1993

Romania
Soyuz 40–May 14, 1981

Russia
See also **Soviet Union (U.S.S.R.)**
Cosmos 2176–January 24, 1992
Soyuz TM14–March 17, 1992
Soyuz TM15–July 27, 1992
Resurs-500–November 16, 1992
Bion 10 (Cosmos 2229)–December 29, 1992
Soyuz TM16–January 24, 1993
Soyuz TM17–July 1, 1993
Temisat–August 31, 1993
Soyuz TM18–January 8, 1994
Gals 1–January 20, 1994
Tubsat 2–January 25, 1994
STS 60–February 3, 1994
Cosmos 2268 through Cosmos 2273–February 12, 1994
Intercosmos 26 (Coronas I)–March 2, 1994
Soyuz TM19–July 1, 1994
Soyuz TM20–October 3, 1994
Express 1–October 13, 1994
Elektro–October 31, 1994
Geo-IK–November 29, 1994
Luch 1–December 16, 1994
Radio-Rosto–December 26, 1994
Express RV–January 15, 1995
Astrid 1–January 24, 1995

Appendix 3. Satellites and Missions Listed by Nation or Group of Nations

FAISAT–January 24, 1995
STS 63–February 3, 1995
Soyuz TM21–March 14, 1995
Spektr–May 20, 1995
STS 71–June 27, 1995
Interball 1–August 2, 1995
Magion 4–August 2, 1995
Sich 1–August 31, 1995
Soyuz TM22–September 3, 1995
STS 74–November 12, 1995
Soyuz TM23–February 21, 1996
STS 76–March 22, 1996
Priroda–April 23, 1996
Soyuz TM24–August 17, 1996
Interball 2–August 29, 1996
Magion 5–August 29, 1996
Microsat–August 29, 1996
STS 79–September 16, 1996
Bion 11–December 24, 1996
STS 81–January 12, 1997
Soyuz TM25–February 10, 1997
Gonets-D1 1/Gonets-D1 2/Gonets-D1 3–February 14, 1997
Zeya–March 4, 1997
Progress M34–April 6, 1997
STS 84–May 15, 1997
Progress M35–July 5, 1997
Soyuz TM26–August 5, 1997
STS 86–September 25, 1997
Mirka–October 10, 1997
Sputnik 40–November 3, 1997
Coupon–November 12, 1997
Inspector–December 17, 1997
HGS 1–December 25, 1997
STS 89–January 22, 1998
Soyuz TM27–January 29, 1998
Cosmos 2349–February 17, 1998
STS 91–June 2, 1998
Tubsat N/Tubsat N1–July 7, 1998
FASat-Bravo–July 10, 1998
Gurwin Techsat 1B–July 10, 1998
Safir 2–July 10, 1998
TMSat–July 10, 1998
Westpac–July 10, 1998
Soyuz TM28–August 13, 1998
Zarya–November 20, 1998
Bonum 1–November 22, 1998
STS 88–December 4, 1998
Astrid 2–December 10, 1998
Soyuz TM29–February 20, 1999
DemoSat–March 27, 1999
UoSat 12–April 21, 1999
Nimiq 1–May 21, 1999
STS 96–May 27, 1999
Yamal 101/Yamal 102–September 6, 1999
LMI 1–September 26, 1999
DirecTV 1R–October 10, 1999

Scandinavia
See also individual nations
TDF 1–October 28, 1988

Singapore
ST 1–August 25, 1998

Slovakia
Soyuz TM29–February 20, 1999

South Africa
See also **Africa**
Sunsat–February 23, 1999

South Korea
Uribyol–August 10, 1992
Uribyol 2–September 26, 1993
Mugunghwa 1 (Koreasat 1)–August 5, 1995
Uribyol 3–May 26, 1999
Kompsat–December 21, 1999

Soviet Union (U.S.S.R)
See also **Kazakhstan; Kyrgyzstan; Russia; Ukraine**
Sputnik–October 4, 1957
Sputnik 2–November 3, 1957
Sputnik 3–May 15, 1958
Luna 1–January 2, 1959
Luna 2–September 12, 1959
Luna 3–October 4, 1959
Sputnik 4–May 15, 1960
Sputnik 5–August 19, 1960
Sputnik 6–December 1, 1960
Venera 1–February 12, 1961
Sputnik 9–March 9, 1961
Sputnik 10–March 25, 1961
Vostok 1–April 12, 1961
Vostok 2–August 6, 1961
Cosmos 1–March 16, 1962
Cosmos 2 (Sputnik 12)–April 6, 1962
Cosmos 3–April 24, 1962
Cosmos 4–April 26, 1962
Cosmos 5 (Sputnik 15)–May 28, 1962
Cosmos 7–July 28, 1962
Vostok 3–August 11, 1962
Vostok 4–August 12, 1962
Mars 1–November 1, 1962
Cosmos 17–May 22, 1963
Vostok 5–June 14, 1963
Vostok 6–June 16, 1963
Polyot 1–November 1, 1963
Elektron 1/Elektron 2–January 30, 1964
Cosmos 26–March 18, 1964
Zond 1–April 2, 1964
Polyot 2–April 12, 1964
Elektron 3/Elektron 4–July 11, 1964
Cosmos 41–August 22, 1964
Cosmos 44–August 28, 1964
Voskhod 1–October 12, 1964
Cosmos 49–October 24, 1964
Zond 2–November 30, 1964
Cosmos 51–December 10, 1964
Voskhod 2–March 18, 1965
Molniya 1-01–April 23, 1965
Proton 1–July 16, 1965
Zond 3–July 18, 1965
Molniya 1-02–October 14, 1965
Proton 2–November 2, 1965
Venera 2–November 12, 1965
Venera 3–November 16, 1965
Luna 9–January 31, 1966
Cosmos 110–February 22, 1966
Luna 10–March 31, 1966
Molniya 1-03–April 25, 1966
Cosmos 122–June 25, 1966
Proton 3–July 6, 1966
Luna 11–August 24, 1966
Luna 12–October 22, 1966
Cosmos 133–November 28, 1966
Luna 13–December 21, 1966
Cosmos 140–February 7, 1967
Cosmos 144–February 28, 1967
Cosmos 148–March 16, 1967
Soyuz 1–April 23, 1967
Cosmos 156–April 27, 1967

Cosmos 166–June 16, 1967
Venera 4, Venus Landing–October 18, 1967
Cosmos 186–October 27, 1967
Cosmos 188–October 30, 1967
Cosmos 198–December 27, 1967
Zond 4–March 2, 1968
Luna 14–April 7, 1968
Zond 5–September 15, 1968
Cosmos 248–October 19, 1968
Cosmos 249–October 20, 1968
Soyuz 2–October 25, 1968
Soyuz 3–October 26, 1968
Cosmos 252–November 1, 1968
Zond 6–November 10, 1968
Proton 4–November 16, 1968
Soyuz 4–January 14, 1969
Soyuz 5–January 15, 1969
Meteor 1-01–March 26, 1969
Venera 5, Venus Landing–May 16, 1969
Venera 6, Venus Landing–May 17, 1969
Luna 15–July 13, 1969
Zond 7–August 8, 1969
Soyuz 6–October 11, 1969
Soyuz 7–October 12, 1969
Soyuz 8–October 13, 1969
Intercosmos 1, October 14, 1969
Strela 1 (Cosmos 343)–April 25, 1970
Soyuz 9–June 1, 1970
Luna 16–September 12, 1970
Cosmos 373–October 20, 1970
Zond 8–October 20, 1970
Cosmos 375–October 30, 1970
Luna 17–November 10, 1970
Venera 7, Venus Landing–December 15, 1970
Cosmos 398–February 26, 1971
Salyut 1–April 19, 1971
Soyuz 10–April 23, 1971
Soyuz 11–June 6, 1971
Cosmos 434–August 12, 1971
Luna 18–September 2, 1971
Luna 19–September 28, 1971
Molniya 2-01–November 24, 1971
Mars 2–November 27, 1971
Mars 3–December 2, 1971
Aureol 1–December 27, 1971
Luna 20–February 14, 1972
Intercosmos 6–April 7, 1972
Prognoz 1–April 14, 1972
Venera 8, Venus Landing–July 22, 1972
Luna 21–January 8, 1973
Soyuz 12–September 27, 1973
Bion 1 (Cosmos 605)–October 31, 1973
Cosmos 613–November 30, 1973
Soyuz 13–December 18, 1973
Mars 4, Mars Flyby–February 10, 1974
Mars 5, Mars Orbit–February 12, 1974
Mars 7, Mars Flyby–March 9, 1974
Mars 6, Mars Landing–March 12, 1974
Cosmos 637–March 26, 1974
Cosmos 638–April 3, 1974
Luna 22–May 29, 1974
Salyut 3–June 25, 1974
Soyuz 14–July 3, 1974
Molniya 1S–July 30, 1974
Soyuz 15–August 26, 1974
Bion 2 (Cosmos 690)–October 22, 1974
Luna 23–October 28, 1974
Molniya 3-01–November 21, 1974
Soyuz 16–December 2, 1974
Cosmos 699–December 24, 1974
Salyut 4–December 26, 1974

Appendix 3. Satellites and Missions Listed by Nation or Group of Nations

Soyuz 17–January 11, 1975
Soyuz 18A–April 5, 1975
Aryabhata–April 19, 1975
Soyuz 18B–May 24, 1975
Meteor 2-01–July 11, 1975
Soyuz 19 (ASTP)–July 15, 1975
Venera 9, Venus Landing–October 22, 1975
Venera 10, Venus Landing–October 25, 1975
Soyuz 20–November 17, 1975
Bion 3 (Cosmos 782)–November 25, 1975
Raduga 1–December 22, 1975
Intercosmos 15–June 19, 1976
Salyut 5–June 22, 1976
Soyuz 21–July 6, 1976
Luna 24–August 9, 1976
Soyuz 22–September 15, 1976
Soyuz 23–October 14, 1976
Ekran 1–October 26, 1976
Cosmos 881/Cosmos 882–December 15, 1976
Soyuz 24–February 7, 1977
SIGNE 3–June 17, 1977
Cosmos 929–July 17, 1977
Bion 4 (Cosmos 936)–August 3, 1977
Cosmos 954–September 18, 1977
Salyut 6–September 29, 1977
Soyuz 25–October 9, 1977
Soyuz 26–December 10, 1977
Soyuz 27–January 10, 1978
Progress 1–January 20, 1978
Soyuz 28–March 2, 1978
Cosmos 1001–April 4, 1978
Soyuz 29–June 15, 1978
Soyuz 30–June 27, 1978
Soyuz 31–August 26, 1978
Intercosmos 18–October 24, 1978
Magion 1–October 24, 1978
Radio Sputnik 1/Radio Sputnik 2–October 26, 1978
Gorizont 1–December 19, 1978
Venera 12, Venus Landing–December 21, 1978
Venera 11, Venus Landing–December 25, 1978
Cosmos 1076 (Okean-E)–February 12, 1979
Soyuz 32–February 25, 1979
Soyuz 33–April 10, 1979
Soyuz 34–June 6, 1979
Bhaskara 1–June 7, 1979
Bion 5 (Cosmos 1129)–September 25, 1979
Intercosmos 20–November 1, 1979
Soyuz T1–December 16, 1979
Soyuz 35–April 9, 1980
Soyuz 36–May 26, 1980
Soyuz T2–June 5, 1980
Meteor 1-30–June 18, 1980
Soyuz 37–July 23, 1980
Soyuz 38–September 18, 1980
Soyuz T3–November 27, 1980
Intercosmos 21,February 6, 1981
Soyuz T4–March 12, 1981
Soyuz 39–March 22, 1981
Cosmos 1267–April 25, 1981
Soyuz 40–May 14, 1981
Venera 13, Venus Landing–March 1, 1982
Venera 14, Venus Landing–March 4, 1982
Salyut 7–April 19, 1982
Soyuz T5–May 13, 1982
Iskra 2–May 17, 1982
Cosmos 1374 (BOR-4)–June 4, 1982
Soyuz T6–June 24, 1982
Cosmos 1383 (COSPAS 1)–June 30, 1982

Soyuz T7–August 19, 1982
Cosmos 1402–August 27, 1982
Cosmos 1413/Cosmos 1414/Cosmos 1415– October 12, 1982
Iskra 3–November 18, 1982
Cosmos 1443–March 2, 1983
Astron–March 23, 1983
Soyuz T8–April 20, 1983
Soyuz T9–June 27, 1983
Prognoz 9–July 1, 1983
Soyuz T10A–September 27, 1983
Cosmos 1500 (Okean-OE)–September 28, 1983
Venera 15, Venus Orbit–October 10, 1983
Venera 16, Venus Orbit–October 14, 1983
Bion 6 (Cosmos 1514)–December 14, 1983
Soyuz T10B–February 8, 1984
Soyuz T11–April 3, 1984
Soyuz T12–July 17, 1984
Foton 1 (Cosmos 1645)–April 16, 1985
Prognoz 10 (Intercosmos 23)–April 26, 1985
Soyuz T13–June 6, 1985
Vega 1, Venus Flyby–June 10, 1985
Vega 2, Venus Flyby–June 14, 1985
Bion 7 (Cosmos 1667)–July 10, 1985
Soyuz T14–September 17, 1985
Cosmos 1686–September 27, 1985
Meteor 3-01–October 24, 1985
Altair 1 (Cosmos 1700)–October 25, 1985
Mir–February 20, 1986
Vega 1, Halley's Comet Flyby–March 6, 1986
Vega 2, Halley's Comet Flyby–March 9, 1986
Soyuz T15–March 13, 1986
Soyuz TM1–May 21, 1986
Ionosonde (Cosmos 1809)–December 18, 1986
Soyuz TM2–February 5, 1987
Kvant 1–March 31, 1987
Soyuz TM3–July 22, 1987
Cosmos 1870–July 25, 1987
Bion 8 (Cosmos 1887)–September 29, 1987
Soyuz TM4–December 21, 1987
IRS 1A–March 17, 1988
Soyuz TM5–June 7, 1988
Phobos 1–July 7, 1988
Phobos 2–July 12, 1988
Soyuz TM6–August 29, 1988
Buran–November 15, 1988
Soyuz TM7–November 26, 1988
Cosmos 1989–January 10, 1989
Foton 5–April 26, 1989
Pion 1/Pion 2–May 25, 1989
Resurs F1–May 25, 1989
Cosmos 2024–May 31, 1989
Nadezhda 1–July 4, 1989
Progress M1–August 23, 1989
Soyuz TM8–September 5, 1989
Bion 9 (Cosmos 2044)–September 15, 1989
Intercosmos 24–September 28, 1989
Magion 2–September 28, 1989
Kvant 2–November 26, 1989
Granat (Astron 2)–December 1, 1989
Soyuz TM9–February 11, 1990
Foton 6–April 11, 1990
Kristall–May 31, 1990
Gamma–July 11, 1990
Soyuz TM10–August 1, 1990
Soyuz TM11–December 2, 1990
Informator 1–January 29, 1991
Almaz 1–March 31, 1991

Soyuz TM12–May 18, 1991
Meteor 3-05–August 15, 1991
Soyuz TM13–October 2, 1991
Intercosmos 25–December 18, 1991
Magion 3–December 18, 1991
Raduga 28–December 19, 1991

Spain
Intasat 1–November 15, 1974
Hispasat 1A–September 10, 1992
Helios 1A–July 7, 1995
Minisat 1–April 21, 1997

Sweden
Viking–February 22, 1986
Freja–October 6, 1992
Astrid 1–January 24, 1995
Astrid 2–December 10, 1998

Switzerland
STS 46–July 31, 1992
STS 61–December 2, 1993
STS 75–February 22, 1996
STS 103–December 19, 1999

Syria
Soyuz TM3–July 22, 1987

Taiwan
ST 1–August 25, 1998
Rocsat–January 26, 1999

Thailand
Thaicom 1A–December 18, 1993
TMSat–July 10, 1998

Turkey
Turksat 1B–August 10, 1994

Ukraine
See also **Soviet Union (U.S.S.R.)**
STS 87–November 19, 1997
DemoSat–March 27, 1999
DircTV 1R–October 10, 1999

United Kingdom (U.K.)
Ariel 1–April 26, 1962
Ariel 2–March 27, 1964
Ariel 3–May 5, 1967
Skynet 1–November 22, 1969
Prospero–October 28, 1971
Ariel 4–December 11, 1971
Miranda–March 9, 1974
Ariel 5–October 5, 1974
Skynet 2–November 23, 1974
IUE–January 26, 1978
Ariel 6 (UK 6)–June 2, 1979
UoSat 1 (Oscar 9)–October 6, 1981
IRAS–January 26, 1983
UoSat 2 (Oscar 11)–March 1, 1984
UKS–August 16, 1984
Skynet 4B–December 11, 1988
Marco Polo 1–August 27, 1989
UoSat 3/UoSat 4 (Oscar 14/Oscar 15)– January 22, 1990
Rosat–June 1, 1990
Soyuz TM12–May 18, 1991

Appendix 3. Satellites and Missions Listed by Nation or Group of Nations

UoSat 5–July 17, 1991
Yohkoh (Solar-A)–August 30, 1991
Uribyol–August 10, 1992
Healthsat 2–September 26, 1993
STRV-1A/B–June 17, 1994
FASat-Bravo–July 10, 1998
TMSat–July 10, 1998
UoSat 12–April 21, 1999

Union of Soviet Socialist Republics
See **Soviet Union (U.S.S.R); Kazakhstan; Kyrgyzstan; Russia; Ukraine**

United States of America (U.S.A.)

Explorer 1–February 1, 1958
Vanguard 1–March 17, 1958
Explorer 3–March 26, 1958
Explorer 4–July 6, 1958
Pioneer 1–October 11, 1958
Pioneer 3–December 6, 1958
SCORE–December 18, 1958
Vanguard 2–February 17, 1959
Discoverer 1–February 28, 1959
Pioneer 4–March 3, 1959
Explorer 6–August 7, 1959
Big Joe–September 9, 1959
Vanguard 3–September 18, 1959
Explorer 7–October 13, 1959
Little Joe 2–December 4, 1959
Pioneer 5–March 11, 1960
Tiros 1–April 1, 1960
Transit 1B–April 13, 1960
Midas 2–May 24, 1960
Solrad 1–June 22, 1960
Transit 2A–June 22, 1960
Discoverer 13–August 10, 1960
Echo 1–August 12, 1960
Discoverer 14–August 18, 1960
Courier 1B–October 4, 1960
Explorer 8–November 3, 1960
Discoverer 17–November 12, 1960
Tiros 2–November 23, 1960
Discoverer 18–December 8, 1960
Mercury 1A–December 19, 1960
Mercury 2–January 31, 1961
Samos 2–January 31, 1961
Explorer 9–February 16, 1961
Mercury 2A–March 24, 1961
Explorer 10–March 25, 1961
Explorer 11–April 27, 1961
Freedom 7 (Mercury 3)–May 5, 1961
Injun 1–June 29, 1961
Solrad 3 (Greb)–June 29, 1961
Transit 4A–June 29, 1961
Tiros 3–July 12, 1961
Liberty Bell 7 (Mercury 4)–July 21, 1961
Explorer 12–August 16, 1961
Midas 4–October 21, 1961
West Ford 1–October 21, 1961
TRAAC–November 15, 1961
Mercury 5–November 29, 1961
Discoverer 36–December 12, 1961
Oscar 1–December 12, 1961
Ranger 3–January 26, 1962
Tiros 4–February 8, 1962
Friendship 7 (Mercury 6)–February 20, 1962
OSO 1–March 7, 1962
KH-4 9032–April 18, 1962

Ranger 4–April 23, 1962
Ariel 1–April 26, 1962
Aurora 7 (Mercury 7)–May 24, 1962
Ferret 1–June 2, 1962
Oscar 2–June 2, 1962
Tiros 5–June 19, 1962
Starfish Test–July 9, 1962
Telstar 1–July 10, 1962
Mariner 2–August 27, 1962
Tiros 6–September 18, 1962
Alouette 1–September 29, 1962
Explorer 14–October 2, 1962
Sigma 7 (Mercury 8)–October 3, 1962
Starad–October 26, 1962
Explorer 15–October 27, 1962
Anna 1B–October 31, 1962
Calsphere–December 13, 1962, 4:04 GMT
Injun 3–December 13, 1962, 4:04 GMT
Surcal 1A–December 13, 1962, 4:04 GMT
Relay 1–December 13, 1962, 23:30 GMT
Mariner 2, Venus Flyby–December 14, 1962
Explorer 16–December 16, 1962
Explorer 17 (Atmosphere Explorer–April 3, 1963
Telstar 2–May 7, 1963
Dash 1–May 9, 1963
West Ford 2–May 9, 1963
Faith 7 (Mercury 9)–May 15, 1963
Lofti 2A–June 15, 1963
Radose–June 15, 1963
Solrad 6–June 15, 1963
Tiros 7–June 19, 1963
Syncom 1–June 26, 1963
Hitch Hiker 1–June 27, 1963
GRS–June 28, 1963
KH-7 1–July 12, 1963
TRS 4–July 18, 1963
Syncom 2–July 26, 1963
38C–September 28, 1963
Vela 1A/Vela 1B–October 17, 1963
Imp A (Explorer 18)–November 27, 1963
Explorer 19–December 19, 1963
Tiros 8–December 21, 1963
GGSE 1–January 11, 1964
Secor 1–January 11, 1964
Solrad 7A–January 11, 1964
Relay 2–January 21, 1964
Echo 2–January 25, 1964
Ranger 6–January 30, 1964
Ariel 2–March 27, 1964
Vela 2A/Vela 2B–July 17, 1964
Ranger 7–July 28, 1964
KH-7-10–August 14, 1964
Syncom 3–August 19, 1964
Explorer 20–August 25, 1964
Nimbus 1–August 28, 1964
OGO 1–September 5, 1964
Explorer 22 (Beacon Explorer B)–October 10, 1964
Explorer 23–November 6, 1964
Explorer 24–November 21, 1964
Explorer 25 (Injun 4)–November 21, 1964
Mariner 4–November 28, 1964
83C–December 13, 1964
San Marco 1–December 15, 1964
Explorer 26–December 21, 1964
DMSP Block-4A F1–January 19, 1965
Gemini 2–January 19, 1965
Tiros 9–January 22, 1965
OSO 2–February 3, 1965
LES 1–February 11, 1965
Pegasus 1–February 16, 1965

Ranger 8–February 17, 1965
Dodecapole 1–March 9, 1965
GGSE 2/GGSE 3–March 9, 1965
Oscar 3–March 9, 1965
Ranger 9–March 21, 1965
Gemini 3–March 23, 1965
Early Bird (Intelsat 1)–April 6, 1965
Explorer 27 (Beacon Explorer C)–April 29, 1965
LES 2–May 6, 1965
Pegasus 2–May 25, 1965
Gemini 4–June 3, 1965
Mariner 4, Mars Flyby–July 15, 1965
ORS 3–July 20, 1965
Pegasus 3–July 30, 1965
Gemini 5–August 21, 1965
OV1-2–October 5, 1965
OGO 2–October 14, 1965
Explorer 29 (GEOS 1)–November 6, 1965
Explorer 30 (Solrad 8)–November 19, 1965
Alouette 2–November 29, 1965
Explorer 31–November 29, 1965
Gemini 7–December 4, 1965
France 1–December 6, 1965
Gemini 6–December 15, 1965
Pioneer 6–December 16, 1965
LES 3/LES 4–December 21, 1965
Oscar 4–December 21, 1965
ESSA 1–February 3, 1966
Gemini 8 Agena Target–March 16, 1966, 15:07 GMT
Gemini 8–March 16, 1966, 16:48 GMT
OV1-4–March 30, 1966
OV1-5–March 30, 1966
OV3-1–April 22, 1966
Explorer 32 (Atmosphere Explorer B)–May 25, 1966
Surveyor 1–May 30, 1966
Gemini 9 ATDA–June 1, 1966
Gemini 9–June 3, 1966
OGO 3–June 7, 1966
ERS 16 (ORS 2)–June 9, 1966
OV3-4–June 10, 1966
IDCSP 1 through IDCSP 7–June 16, 1966
GGTS 1–June 16, 1966
Pageos 1–June 24, 1966
Imp D (Explorer 33)–July 1, 1966
OV1-8–July 14, 1966
Gemini 10 Agena Target–July 18, 1966, 20:38 GMT
Gemini 10–July 18, 1966, 22:22 GMT
OV3-3–August 4, 1966
Lunar Orbiter 1–August 10, 1966
Pioneer 7–August 17, 1966
Gemini 11 Agena Target–September 12, 1966, 13:12 GMT
Gemini 11–September 12, 1966, 14:38 GMT
Dash 2–October 5, 1966
Intelsat 2A–October 26, 1966
OV3-2–October 28, 1966
OV4-1T/OV4-1R–November 3, 1966
Lunar Orbiter 2–November 6, 1966
Gemini 2–November 3, 1966
Gemini 12 Agena Target–November 11, 1966, 19:12 GMT
Gemini 12–November 11, 1966, 20:53 GMT
ATS 1–December 7, 1966
Intelsat 2B–January 11, 1967
Apollo 1–January 27, 1967
Lunar Orbiter 3–February 5, 1967
OSO 3–March 8, 1967
Intelsat 2C–March 23, 1967

Appendix 3. Satellites and Missions Listed by Nation or Group of Nations

Surveyor 3–April 17, 1967
San Marco 2–April 26, 1967
ERS 18–April 28, 1967
OV5-1/OV5-3–April 28, 1967
Lunar Orbiter 4–May 4, 1967
Ariel 3–May 5, 1967
Timation 1–May 31, 1967
Aurora 1–June 29, 1967
DATS–July 1, 1967
Dodge 1–July 1, 1967
LES 5–July 1, 1967
Imp E (Explorer 35)–July 19, 1967
OGO 4–July 28, 1967
Lunar Orbiter 5–August 1, 1967
Biosatellite 2–September 7, 1967
Surveyor 5–September 8, 1967
Intelsat 2D–September 28, 1967
OSO 4–October 15, 1967
Mariner 5, Venus Flyby–October 19, 1967
ATS 3–November 5, 1967
Surveyor 6–November 7, 1967
Apollo 4–November 9, 1967
Wresat 1–November 29, 1967
OV3-6–December 5, 1967
Pioneer 8–December 13, 1967
Test and Training Satellite 1–December 13, 1967
Surveyor 7–January 7, 1968
Geos 2–January 11, 1968
Apollo 5–January 22, 1968
OGO 5–March 4, 1968
Solrad 9–March 5, 1968
Apollo 6–April 4, 1968
OV1-13–April 6, 1968
IRIS–May 17, 1968
RAE 1–July 4, 1968
Cannonball 1 (OV1-16)–July 11, 1968
Spades (OV1-15)–July 11, 1968
Canyon 1–August 6, 1968
Explorer 39–August 8, 1968
Injun 5–August 8, 1968
LES 6–September 26, 1968
OV2-5/OV5-2–September 26, 1968
Aurorae–October 3, 1968
Apollo 7–October 11, 1968
Pioneer 9–November 8, 1968
HEOS 1–December 5, 1968
OAO 2–December 7, 1968
Intelsat 3B–December 19, 1968
Apollo 8–December 21, 1968
OSO 5–January 22, 1969
ISIS 1–January 30, 1969
Tacsat 1–February 9, 1969
Apollo 9–March 3, 1969
OV1-17–March 18, 1969
OV1-18–March 18, 1969
OV1-19–March 18, 1969
OV1-17A (Orbiscal 2)–March 18, 1969
Apollo 10–May 18, 1969
OV5-5/OV5-6/OV5-9–May 23, 1969
Vela 5A/Vela 5B–May 23, 1969
OGO 6–June 5, 1969
Biosatellite 3–June 29, 1969
Apollo 11–July 16, 1969
Mariner 6, Mars Flyby–July 31, 1969
Mariner 7, Mars Flyby–August 5, 1969
OSO 6–August 9, 1969
PAC 1–August 9, 1969
ATS 5–August 12, 1969
SOICAL Cone/SOICAL Cylinder–September 30, 1969
Timation 2–September 30, 1969

Boreas–October 1, 1969
Azur–November 8, 1969
Apollo 12–November 14, 1969
Skynet 1–November 22, 1969
ITOS 1–January 23, 1970
Oscar 5–January 23, 1970
SERT 2–February 4, 1970
Apollo 13–April 11, 1970
OFO–November 9, 1970
RM 1–November 9, 1970
NOAA 1–December 11, 1970
Uhuru (Explorer 42 or SAS 1)–December 12, 1970, 10:53 GMT
Peole–December 12, 1970, 12:57 GMT
Intelsat 4 F2–January 26, 1971
Apollo 14–January 31, 1971
Jumpseat 1–March 21, 1971
San Marco 3–April 24, 1971
IMEWS 1 (DSP-647 1)–May 5, 1971
KH-9 1–June 15, 1971
Solrad 10–July 8, 1971
Apollo 15–July 26, 1971
Cannonball 2–August 7, 1971
Gridsphere 1/Gridsphere 2–August 7, 1971
LCS 4–August 7, 1971
Musketball 1–August 7, 1971
Mylar Balloon–August 7, 1971
OV1-20–August 7, 1971
OV1-21–August 7, 1971
Rigidsphere–August 7, 1971
Eole–August 16, 1971
OSO 7–September 29, 1971
STP 1 (ASTEX)–October 17, 1971
DSCS II 1/DSCS II 2–November 3, 1971
Mariner 9, Mars Orbit–November 14, 1971
Explorer 45–November 15, 1971
Ariel 4–December 11, 1971
HEOS 2–January 31, 1972
Pioneer 10–March 3, 1972
Thor Delta 1A–March 12, 1972
Apollo 16–April 16, 1972
Landsat 1–July 23, 1972
Explorer 46–August 13, 1972
Copernicus (OAO 3)–August 21, 1972
Triad 1–September 2, 1972
Radsat–October 2, 1972
Radcat–October 2, 1972
Oscar 6–October 15, 1972
Anik A1 (Telesat 1)–November 10, 1972
Explorer 48 (SAS 2)–November 15, 1972
ESRO 4–November 22, 1972
Apollo 17–December 7, 1972
Aeros 1–December 16, 1972
Rhyolite 1–March 6, 1973
Pioneer 11–April 6, 1973
Skylab–May 14, 1973
Skylab 2–May 25, 1973
RAE 2–June 10, 1973
Skylab 3–July 28, 1973
Mariner 10–November 3, 1973
Skylab 4–November 16, 1973
Pioneer 10, Jupiter Flyby–December 3, 1973
Explorer 51 (Atmosphere Explorer C)–December 16, 1973
Mariner 10, Venus Flyby–February 5, 1974
San Marco 4–February 18, 1974
Miranda–March 9, 1974
Westar 1–April 13, 1974
SMS 1–May 17, 1974
ATS 6–May 30, 1974
Hawkeye (Explorer 52)–June 3, 1974
NTS 1 (Timation 3)–July 14, 1974

Aeros 2–July 16, 1974
ANS 1–August 30, 1974
Ariel 5–October 5, 1974
Intasat 1–November 15, 1974
Oscar 7–November 15, 1974
Skynet 2–November 23, 1974
Pioneer 11, Jupiter Flyby–December 4, 1974
Helios 1–December 10, 1974
Symphonie 1–December 18, 1974
Mariner 10, Mercury Flyby #3–March 16, 1975
Geos 3–April 9, 1975
Explorer 53 (SAS 3)–May 7, 1975
OSO 8–June 21, 1975
Apollo 18 (ASTP)–July 15, 1975
COS B–August 9, 1975
Intelsat 4A F1–September 26, 1975
Explorer 54 (Atmosphere Explorer D)–October 6, 1975
GOES 1–October 16, 1975
Explorer 55 (Atmosphere Explorer E)–November 20, 1975
Bion 3 (Cosmos 782)–November 25, 1975
RCA Satcom 1–December 13, 1975
Helios 2–January 15, 1976
Hermes (CTS 1)–January 17, 1976
Marisat 1–February 19, 1976
LES 8/LES 9–March 15, 1976
Solrad 11A/Solrad 11B–March 15, 1976
NATO 3A–April 22, 1976
NOSS 1 (Whitecloud 1)–April 30, 1976
Lageos 1–May 4, 1976
Comstar 1–May 13, 1976
P76-5–May 22, 1976
SDS 1–June 2, 1976
Viking 1, Mars Orbit/Landing–June 19, 1976
IMEWS 6–June 26, 1976
Palapa 1–July 8, 1976
Viking 2–August 7, 1976
KH-11 1, U.S.A–December 19, 1976
ESA-GEOS 1–April 20, 1977
NTS 2–June 23, 1977
HEAO 1–August 12, 1977, 6:39 GMT
Enterprise, Flight 4–August 12, 1977, 15:00 GMT
Sirio 1–August 25, 1977
ISEE 1/ISEE 2–October 22, 1977
Meteosat 1–November 23, 1977
Sakura 1–December 15, 1977
IUE–January 26, 1978
FLTSATCOM 1–February 9, 1978
Navstar 1–February 22, 1978
Oscar 8–March 5, 1978
PIX 1–March 5, 1978
Yuri 1–April 7, 1978
HCMM–April 26, 1978
OTS 2–May 11, 1978
Chalet 1–June 10, 1978
Seasat–June 27, 1978
ESA-GEOS 2–July 14, 1978
ISEE 3–August 12, 1978
CAMEO–October 13, 1978
Nimbus 7–October 13, 1978
Tiros N–October 13, 1978
Einstein (HEAO 2)–November 13, 1978
Pioneer Venus Orbiter–December 4, 1978
Pioneer Venus Probe–December 9, 1978
Anik B (Telesat 4)–December 15, 1978
SCATHA–January 30, 1979
SAGE–February 18, 1979
Solwind (P78-1)–February 24, 1979
Voyager 1, Jupiter Flyby–March 5, 1979

373

Appendix 3. Satellites and Missions Listed by Nation or Group of Nations

Ariel 6 (UK 6)–June 2, 1979
Voyager 2, Jupiter Flyby–July 9, 1979
Pioneer 11, Saturn Flyby–September 1, 1979
HEAO 3–September 20, 1979
Magsat–October 30, 1979
Navstar 5–February 9, 1980
Solar Max–February 14, 1980
SBS 1–November 15, 1980
Intelsat 5 F2–December 6, 1980
Voyager 1, Saturn Flyby–December 12, 1980
STS 1–April 12, 1981
Dynamics Explorer 1/Dynamics Explorer 2–August 3, 1981
Voyager 2, Saturn Flyby–August 26, 1981
SME–October 6, 1981
UoSat 1 (Oscar 9)–October 6, 1981
STS 2–November 12, 1981
Westar 4–February 26, 1982
STS 3–March 22, 1982
Insat 1A–April 10, 1982
STS 4–June 27, 1982
Landsat 4–July 16, 1982
Anik D1 (Telesat 5)–August 26, 1982
Conestoga–September 9, 1982
RCA Satcom 5 (Aurora 1)–October 28, 1982
DSCS III–October 30, 1982
Anik C3 (Telesat 6)–November 11, 1982
SBS 3–November 11, 1982
STS 5–November 11, 1982
IRAS–January 26, 1983
STS 6–April 4, 1983
TDRS 1–April 4, 1983
Exosat–May 26, 1983
Anik C2 (Telesat 7)–June 18, 1983
Palapa B1–June 18, 1983
SPAS (Shuttle PAllet Satellite) 1–June 18, 1983
STS 7–June 18, 1983
HiLat (P83-1)–June 27, 1983
Galaxy 1–June 28, 1983
Telstar 3A–July 28, 1983
Insat 1B–August 30, 1983
STS 8–August 30, 1983
STS 9–November 28, 1983
Palapa B2–February 3, 1984
STS 41B (STS 10)–February 3, 1984
Westar 6–February 3, 1984
UoSat 2 (Oscar 11)–March 1, 1984
STS 41C (STS 11)–April 6, 1984
LDEF–April 6, 1984
Solar Max–April 6, 1984
Spacenet 1–May 23, 1984
CCE–August 16, 1984
Ion Release Module–August 16, 1984
UKS–August 16, 1984
LEASAT 2–August 30, 1984
SBS 4–August 30, 1984
STS 41D (STS 12)–August 30, 1984
Telstar 3B–August 30, 1984
ERBS–October 5, 1984
STS 41G (STS 13)–October 5, 1984
Anik D2–November 8, 1984
LEASAT 1–November 8, 1984
STS 51A (STS 14)–November 8, 1984
Magnum 1–January 24, 1985
STS 51C (STS 15)–January 24, 1985
Geosat–March 13, 1985
Intelsat 5A F10–March 22, 1985
Anik C1–April 12, 1985
LEASAT 3–April 12, 1985
STS 51D (STS 16)–April 12, 1985
GLOMR–April 29, 1985

NUSAT–April 29, 1985
STS 51B (STS 17)–April 29, 1985
GStar 1–May 7, 1985
Arabsat 1B–June 17, 1985
Morelos 1–June 17, 1985
Spartan 1–June 17, 1985
STS 51G (STS 18)–June 17, 1985
Telstar 3C–June 17, 1985
STS 51F (STS 19)–July 29, 1985
ASC 1–August 27, 1985
Aussat A1–August 27, 1985
LEASAT 4–August 27, 1985
STS 51I (STS 20)–August 27, 1985
International Cometary Explorer (ICE)–September 11, 1985
MHV–September 13, 1985
STS 51J (STS 21)–October 3, 1985
GLOMR–October 30, 1985
STS 61A (STS 22)–October 30, 1985
Aussat A2–November 26, 1985
Morelos 2–November 26, 1985
Orbiter Experiment (OEX)–November 26, 1985
RCA Satcom Ku2–November 26, 1985
STS 61B (STS 23)–November 26, 1985
GE Satcom–January 12, 1986
STS 61C (STS 24)–January 12, 1986
Voyager 2, Uranus Flyby–January 24, 1986
STS 51L (STS 25)–January 28, 1986
Polar Bear–November 14, 1986
San Marco D/L–March 25, 1988
Meteosat 3–June 15, 1988
Oscar 13–June 15, 1988
Pan American Satellite 1–June 15, 1988
STS 26–September 29, 1988
TDRS 3–September 29, 1988
Lacrosse 1–December 2, 1988
STS 27–December 2, 1988
Navstar B2 1–February 14, 1989
STS 29–March 13, 1989
TDRS 4–March 13, 1989
Magellan–May 4, 1989
STS 30–May 4, 1989
STS 28–August 8, 1989
USA 40–August 8, 1989
USA 41–August 8, 1989
Voyager 2, Neptune Flyby–August 25, 1989
Marco Polo 1–August 27, 1989
Intercosmos 24–September 28, 1989
Galileo–October 18, 1989
STS 34–October 18, 1989
COBE–November 18, 1989
LDEF–January 9, 1990
LEASAT 5–January 9, 1990
STS 32–January 9, 1990
Magnum 2, U.S.A.–November 22, 1989
STS 33–November 22, 1989
Webersat (Oscar 18)–January 22, 1990
LACE–February 14, 1990
RME–February 14, 1990
KH-12 1–February 28, 1990
STS 36–February 28, 1990
Pegsat–April 5, 1990
SECS–April 5, 1990
POGS–April 11, 1990
SCE–April 11, 1990
TEX–April 11, 1990
Palapa B2R–April 13, 1990
Hubble Space Telescope–April 24, 1990
STS 31–April 24, 1990
Macsat 1/Macsat 2–May 9, 1990
Rosat–June 1, 1990

USA 59–June 8, 1990
CRRES–July 25, 1990
Magellan, Venus Orbit–August 10, 1990
STS 41–October 6, 1990
Ulysses–October 6, 1990
Inmarsat 2 F1–October 30, 1990
Magnum 3–November 15, 1990
STS 38–November 15, 1990
Satcom C1–November 20, 1990
STS 35–December 2, 1990
NATO 4A–January 8, 1991
Compton Gamma Ray Observatory–April 5, 1991
STS 37–April 5, 1991
STS 39–April 28, 1991
Aurora 2–May 29, 1991
STS 40–June 5, 1991
Rex 1–June 29, 1991
Losat X–July 4, 1991
ERS 1–July 17, 1991, 1:46 GMT
Orbcomm X–July 17, 1991, 1:46 GMT
SARA–July 17, 1991, 1:46 GMT
Tubsat 1–July 17, 1991, 1:46 GMT
UoSat 5–July 17, 1991, 1:46 GMT
Microsat 1 through Microsat 7–July 17, 1991, 17:33 GMT
STS 43–August 2, 1991
TDRS 5–August 2, 1991
Meteor 3-05–August 15, 1991
Yohkoh (Solar-A)–August 30, 1991
STS 48–September 12, 1991
UARS–September 12, 1991
Galileo, Gaspra Flyby–October 29, 1991
DSP 16–November 24, 1991
STS 44–November 24, 1991
STS 42–January 22, 1992
Ulysses, Jupiter Flyby–February 8, 1992
STS 45–March 24, 1992
Intelsat 6 F3–May 7, 1992
STS 49–May 7, 1992
EUVE–June 7, 1992
Intelsat K–June 10, 1992
STS 50–June 25, 1992
Sampex–July 3, 1992
Geotail–July 24, 1992
EURECA–July 31, 1992
STS 46–July 31, 1992
Tethered Satellite System (TSS) 1–July 31, 1992
TOPEX-Poseidon–August 10, 1992
Aussat B1 (Optus B1)–August 14, 1992
Hispasat 1A–September 10, 1992
STS 47–September 12, 1992
Mars Observer–September 25, 1992
CTA–October 22, 1992
Lageos 2–October 22, 1992
STS 52–October 22, 1992
Galaxy 7–October 28, 1992
MSTI 1–November 21, 1992
STS 53–December 2, 1992
USA 89–December 2, 1992
STS 54–January 13, 1993
TDRS 6–January 13, 1993
SCD 1–February 9, 1993
Seds–March 30, 1993
Spartan 201–April 8, 1993
STS 56–April 8, 1993
Alexis–April 25, 1993
STS 55–April 26, 1993
EURECA–June 21, 1993
STS 57–June 21, 1993
Galileo, Ida Flyby–August 28, 1993

Appendix 3. Satellites and Missions Listed by Nation or Group of Nations

UHF Follow-On 1–September 3, 1993
ACTS–September 12, 1993
Orfeus-Spas–September 12, 1993
STS 51–September 12, 1993
Eyesat 1–September 26, 1993
Healthsat 2–September 26, 1993
STS 58–October 18, 1993
Hubble Space Telescope–December 2, 1993
STS 61–December 2, 1993
Telstar 401–December 16, 1993
DBS 1–December 18, 1993
Clementine–January 25, 1994
ISAS–January 25, 1994
BremSat 1–February 3, 1994
ODERACS 1A-1F–February 3, 1994
STS 60–February 3, 1994
Wake Shield 1–February 3, 1994
Milstar 1–February 7, 1994
STS 62–March 4, 1994
DarpaSat–March 13, 1994
STEP 0 (TAOS)–March 13, 1994
STS 59–April 9, 1994
Trumpet 1–May 3, 1994
MSTI 2–May 9, 1994
STS 65–July 8, 1994
Apstar 1–July 21, 1994
APEX–August 3, 1994
Spartan 201–September 9, 1994
STS 64–September 9, 1994
STS 68–September 30, 1994
Wind–November 1, 1994
Crista-Spas 1–November 3, 1994
STS 66–November 3, 1994
Orion 1–November 29, 1994
FAISAT–January 24, 1995
ODERACS 2A-2F–February 3, 1995
Spartan 204–February 3, 1995
STS 63–February 3, 1995
STS 67–March 2, 1995
Soyuz TM21–March 14, 1995
Space Flyer Unit–March 18, 1995
Microlab 1–April 3, 1995
Orbcomm FM1-FM2–April 3, 1995
AMSC 1–April 7, 1995
Spektr–May 20, 1995
STS 71–June 27, 1995
STS 70–July 13, 1995
TDRS 7–July 13, 1995
Mugunghwa 1 (Koreasat 1)–August 5, 1995
Pioneer 11, Shutdown–September 30, 1995
Spartan 201–September 7, 1995
STS 69–September 7, 1995
Wake Shield 2–September 7, 1995
STS 73–October 20, 1995
Radarsat–November 4, 1995
SURFSAT–November 4, 1995
STS 74–November 12, 1995
Soho–December 2, 1995
Galileo, Jupiter Orbit–December 7, 1995
Echostar 1–December 28, 1995
Rossi X-ray Timing Explorer (RXTE)–December 30, 1995
Space Flyer Unit–January 11, 1996
Spartan 206–January 11, 1996
STS 72–January 11, 1996
Measat 1–January 12, 1996
Palapa C-1–February 1, 1996
NEAR–February 17, 1996
STS 75–February 22, 1996
Tethered Satellite System (TSS) 1R–February 22, 1996

Polar–February 24, 1996
STS 76–March 22, 1996
Inmarsat 3 F1–April 3, 1996
Priroda–April 23, 1996
Midcourse Space Experiment (MSX)–April 24, 1996
BeppoSax–April 30, 1996
MSTI 3–May 17, 1996
PAMS–May 19, 1996
Spartan 207–May 19, 1996
STS 77–May 19, 1996
STS 78–June 20, 1996
TOMS–July 2, 1996
ADEOS (Midori)–August 17, 1996
Soyuz TM24–August 17, 1996
FAST–August 21, 1996
GE 1–September 8, 1996
STS 79–September 16, 1996
Orfeus–November 19, 1996
STS 80–November 19, 1996
Wake Shield 3–November 19, 1996
STS 81–January 12, 1997
Nahuel 1A–January 30, 1997
Soyuz TM25–February 10, 1997
Hubble Space Telescope–February 11, 1997
STS 82–February 11, 1997
Tempo 2–March 8, 1997
Pioneer 10, Shutdown–March 31, 1997
STS 83–April 4, 1997
Progress M34–April 6, 1997
BSAT 1A–April 16, 1997
Celestis–April 21, 1997
Minisat–April 21, 1997
Iridium 4 through Iridium 8–May 5, 1997
STS 84–May 15, 1997
Thor 2–May 21, 1997
NEAR, Mathilde Flyby–June 27, 1997
STS 94–July 1, 1997
Mars Pathfinder–July 4, 1997
Progress M35–July 5, 1997
SeaWiFS (OrbView 2)–August 1, 1997
Crista-Spas 2–August 7, 1997
STS 85–August 7, 1997
Agila 2–August 19, 1997
ACE–August 25, 1997
Pan American Satellite 5–August 28, 1997
FORTE–August 29, 1997
Mars Global Surveyor–September 12, 1997
STS 86–September 25, 1997
Cassini–October 15, 1997
Huygens–October 15, 1997
Cakrawarta 1–November 12, 1997
Spartan 201–November 19, 1997
STS 87–November 19, 1997
TRMM–November 28, 1997
Equator S–December 2, 1997
Orbcomm FM5-FM12–December 23, 1997
HGS 1–December 25, 1997
Lunar Prospector–January 7, 1998
STS 89–January 22, 1998
Soyuz TM27–January 29, 1998
Celestis 2, Feburary 10, 1998
Geosat Follow-On–February 10, 1998
Globalstar 1 through Globalstar 4–February 14, 1998
SNOE–February 25, 1998
Teledesic 1–February 25, 1998
Trace–April 1, 1998
STS 90–April 17, 1998

Ulysses, First Solar Orbit–April 17, 1998
STS 91–June 2, 1998
STEX–October 3, 1998
Deep Space 1–October 24, 1998
Sedsat 1–October 24, 1998
Pansat–October 29, 1998
Spartan 201–October 29, 1998
STS 95–October 29, 1998
Zarya–November 20, 1998
Bonum 1–November 22, 1998
MightySat 1–December 4, 1998
SAC-A–December 4, 1998
STS 88–December 4, 1998
Unity–December 4, 1998
Satmex 5–December 6, 1998, 0:43 GMT
SWAS–December 6, 1998, 0:57 GMT
Mars Climate Orbiter–December 11, 1998
NEAR, Eros Flyby–December 23, 1998
Mars Polar Lander–January 3, 1999
Rocsat–January 26, 1999
Stardust–February 7, 1999
Argos–February 23, 1999
Orsted–February 23, 1999
Sunsat–February 23, 1999
WIRE–March 4, 1999
DemoSat–March 27, 1999
Landsat 7–April 15, 1999
Starshine–May 27, 1999
STS 96–May 27, 1999
QuikScat–June 19, 1999
Fuse–June 24, 1999
Chandra X-ray Observatory–July 23, 1999
STS 93–July 23, 1999
Deep Space 1, Braille Flyby–July 28, 1999
Telcom 1 (Palapa D-1)–August 12, 1999
Ikonos–September 24, 1999
LMI 1–September 26, 1999
DirecTV 1R–October 10, 1999
Terra–December 18, 1999
STS 103–December 19, 1999
Acrimsast–December 21, 1999
Celestis 3–December 21, 1999
Galaxy 11–December 21, 1999

U.S.A.
See **United States of America (U.S.A.)**

U.S.S.R.
See **Soviet Union (U.S.S.R.)**

West Germany
See also **Germany**
Azur–November 8, 1969
DIAL-WIKA–March 10, 1970
Aeros 1–December 16, 1972
Aeros 2–July 16, 1974
Helios 1–December 10, 1974
Symphonie 1–December 18, 1974
Helios 2–January 15, 1976
SPAS (Shuttle PAllet Satellite) 1–June 18, 1983
Ion Release Module–August 16, 1984
STS 61A (STS 22)–October 30, 1985
San Marco D/L–March 25, 1988
China 23–August 5, 1988
TDF 1–October 28, 1988
Kopernikus 1–June 5, 1989

Glossary

"all-up" test flight NASA jargon for a test flight that uses a spacecraft's or a rocket's entire complement of components.

ablative material Used in the heat shields of non-reusable capsules in order to safely dissipate the extreme heat caused by the satellite's re-entry into Earth's atmosphere. The heat shield's ablative outside layers break off, carrying away the heat of re-entry so that the spacecraft interior remains protected.

albedo The optical light reflected off a planet's surface in all directions.

androgynous docking The standard spacecraft docking systems employed by the U.S. and the U.S.S.R./Russia use the same concept: a male probe on the active spacecraft searches out and locks with a female drogue on the passive craft. An androgynous docking design, however, is neither male nor female, and allows either spacecraft to be the active port, so that any craft with this unit could dock with any other craft. See **Apollo 18 (July 15, 1975)**.

anti-satellite (ASAT) system A military weapon system using either satellites, rockets, or ground-based lasers for seeking out and destroying orbiting satellites.

apogee The point in a satellite's orbit at which it is farthest from Earth. The term is also sometimes used loosely for satellites orbiting other planets.

asteroid Sometimes called a minor planet or planetoid, an asteroid is any object orbiting the Sun that is smaller than a planet but larger than a meteoroid. The specific dividing lines between planets, asteroids, and meteoroids have not yet been clearly established.

astronaut A traveler above Earth's atmosphere; the term is generally applied to individuals flying on U.S. spacecraft.

astronomical unit (AU) A unit of length used by astronomers, based on the mean distance of Earth from the Sun, about 93 million miles.

attitude The orientation of a satellite or spacecraft relative to Earth, the Sun, or any other object of interest.

aurora A luminous phenomenon consisting of curtains or streamers of light that are seen periodically in the night sky at Earth's higher latitudes.

ballistic re-entry Unlike a controlled re-entry from orbit, a spacecraft making a ballistic re-entry does nothing to guide its trajectory, enabling ambient forces such as gravity and atmospheric pressure to control its flight.

blazar Belonging to the class of galaxies also known as BL Lac objects, blazars vary in brightness across a wide range of wavelengths, sometimes fluctuating radically in time spans as short as a single day. The data imply that a single object, possibly a super-massive black hole, is the radiation source.

bow shock The shock wave caused by the interaction of the supersonic solar wind with the magnetic field of a planet.

capsule A small, self-contained spacecraft capable of supporting life in the vacuum of space for short periods.

carbonaceous chondrite A meteorite composed of a large amount of carbon, giving it a very dark color. The meteorite also contains chondrules, small spherical bodies formed from the minerals pyroxene or olivine.

"chibis" suit A Soviet-built garment designed to be worn by cosmonauts while on long-term space missions. The suit was designed to force blood and fluid to circulate to the lower limbs, attempting to mimic the effect of gravity on Earth.

circumlunar An orbit around Earth that causes a spacecraft to swing closely past the Moon. The route usually passes the Moon by cutting in front, swinging around behind, and cutting around the front again, like a figure 8. However, other routes are also possible.

co-orbital Circumstances where two satellites have similar or identical orbits.

coronal mass ejection Vast eruptions of matter flung outward from the Sun at great speed and emitting energy in ultraviolet, visible, and x-ray radiation. See **Skylab 2 (May 25, 1973)**.

cosmic dust Term used to describe the smallest solid particles found in interplanetary space and thought to originate from beyond the solar system.

Glossary

cosmic rays Very energetic electrons or atomic nuclei (mostly hydrogen), traveling at nearly the speed of light and arriving from all parts of the sky as they enter the solar system.

cosmonaut A traveler above Earth's atmosphere; the term is generally applied to individuals flying on Soviet or Russian spacecraft.

decay (as in *orbit decay*) Although space is usually considered a vacuum above 100 miles altitude, the atmosphere is still thick enough to exert a drag on any satellite whose orbit crosses below approximately 500 miles altitude. Over time, that drag will cause the satellite's orbit to shrink or "decay," and eventually the satellite will re-enter the atmosphere.

double-skip re-entry maneuver A technique using Earth's atmosphere to slow the speed of a spacecraft returning from lunar distances during re-entry. The spacecraft's angle of re-entry is made shallow enough that it cuts through the atmosphere like a stone skipping across a pond, bouncing out once before finally dropping back down for a final descent. See **Apollo 4 (November 9, 1967)**.

dummy satellite During test launches of new rocket designs, engineers test the rocket's behavior by loading a dead weight equivalent to an average satellite payload. Often this dead weight or dummy satellite carries sensors for recording the rocket's performance and telemetry.

eccentric orbit Although most satellites follow generally circular orbits, sometimes (usually for research purposes) a satellite will be placed in an oval, elliptical, or "eccentric" orbit, with its apogee greatly exceeding its perigee.

ecliptic The plane in which all the planets of the solar system orbit the Sun, describing a flat, disk-like region.

ejecta The material thrown out when an asteroid or comet impacts the surface of a planet or moon and forms a crater.

eleven-year solar cycle *See* solar cycle.

escape velocity The minimum speed needed to escape the gravitational field of a planet or moon. For Earth, escape velocity equals 7 miles per second, or 25,200 miles per hour.

EVA (extra-vehicular activity) A human activity taking place in space but outside of a spacecraft. Also called a space walk when in space, and a Moon-walk when on the surface of the Moon.

exosphere The outermost region of Earth's atmosphere, between 300 to 600 miles elevation.

faculae Bright patches on the Sun's surface in the vicinity of sunspots.

film magazine A reloadable cartridge for holding film used in still or motion photography.

flux The quantity of energy, radiation, or particles flowing past a given area, measured over a period of time.

g A unit of measure equal to the standard force of gravity at Earth's surface, an acceleration equaling approximately 32 feet per second per second.

galactic plane The ecliptic region in which most of the stars in the Milky Way orbit its center, forming a flat, disk-like spiral.

gamma-ray telescope A telescope designed to study the celestial sky in the gamma rays, with energies between 1 MeV and 1 GeV.

geodesy The study of Earth's precise shape and size and its gravitational field. Geodesy also involves the exact placement of surface features on a global coordinate system.

geodetic Pertaining to the science of geodesy.

geomagnetic field Earth's magnetic field.

geomagnetic flux The strength of Earth's magnetic field, as measured at any particular point.

geosynchronous orbit A circular orbit about 22,300 miles high with 0 degrees inclination (matching the Earth's equator) and moving from west to east. A satellite's speed in such an orbit matches Earth's rotational speed from west to east; therefore, the satellite's orbit lasts exactly 24 hours and remains fixed above the same location on Earth's surface.

getaway special A self-contained canister launched in the cargo bay of the space shuttle, which provides a low-cost method for placing scientific experiments in space. See **STS 4 (June 27, 1982)**.

globular clusters A group of many thousands of stars generally forming a spherical shape and traveling together as a unit in orbit around the Milky Way's center.

gravity well The gravitational field of any planet or moon. The deeper any object drops into the gravity well of a planet, the more energy it must expend to climb out.

hard docking When any two satellites attempt to link-up in space, they initially perform a soft docking, whereby a probe on one spacecraft merely makes contact with a target on the second spacecraft. A hard docking follows when the two spacecraft join together, connecting electrical and mechanical systems, permitting joint operation of the two crafts.

heliopause The dividing region between the Sun's heliosphere and interstellar space.

heliosphere The region of space where the Sun's solar wind controls the environment of interplanetary space.

hot spot A long-lasting point on a planet's crust where volcanic activity occurs. Such hot spots appear to be independent of any tectonic plate movement and the volcanic activity caused by that process. On Earth, the hot spot that created the Hawaiian island chain is the best-known example.

hydrogen Lyman-alpha radiation Specific ultraviolet radiation (wavelength: 1,215 angstroms) produced by hydrogen, which can be detected by spectroscopy.

inclination The angle that a satellite's orbit is tilted from the Earth's equator.

interplanetary cosmic ray flux The amount of cosmic rays detected within the solar system.

interplanetary dust The very fine dust particles found drifting between the planets and thought to originate in the solar system.

interplanetary medium The environment of space between the planets and within the solar system.

interplanetary space The region of space between the planets and within the solar system.

interstellar medium The environment of space outside the solar system but within the Milky Way galaxy.

ion engine A spacecraft engine that uses small amounts of electricity to generate charged particles, which are then beamed away from the spacecraft to produce a thrust. Although the acceleration is small, an ion engine requires very little fuel and is continuous over long periods. The result is that tremendous speeds can eventually be achieved. See **SERT 2 (February 4, 1970)**.

ionosphere The part of Earth's upper atmosphere, beginning at about 40 to 50 miles elevation and extending to infinity, where solar ultraviolet radiation causes enough ionization of atmospheric molecules to affect the propagation of radio waves.

Jovian Pertaining to the planet Jupiter.

Large Magellanic Cloud Visible to the naked eye—and first identified by the sixteenth-century Portuguese navigator Ferdinand Magellan—the Large Magellanic Cloud is an irregular dwarf galaxy of stars about 30,000 light years across and orbiting the Milky Way galaxy at a distance of between 155,000 and 170,000 light years.

limb The outer edge of the disk of any celestial body.

magnetars A pulsar or neutron star with a super-strong magnetic field (about one thousand trillion times stronger than Earth's). The field is so strong that it heats the neutron star's surface to about 18 million degrees Fahrenheit and periodically causes starquakes in the star's crust. These quakes, in turn, are thought to cause periodic but random bursts of soft or weak gamma-ray energy. See **ASCA (February 20, 1993)**.

magnetometer An instrument for measuring the strength and direction of a magnetic field.

magnetopause The boundary marking the transitional region between Earth's magnetosphere and the interplanetary medium beyond.

magnetosphere The region of space where Earth's magnetic field dominates the environment.

mare The large dark regions on the lunar surface visible to the unaided eye. The term comes from the Latin word for "sea."

mascon A high density concentration of matter under a mare region of the Moon, which distorts the Moon's gravitational field.

mass spectrometer A device to identify the components in all types of gas, liquid, and solid samples and measure their various masses.

metallic hydrogen When subjected to enough pressure, such as in the central core of giant planets like Jupiter and Saturn, hydrogen will take on metallic properties.

meteor The flash of light, or "shooting star," that is seen as a meteorite flies through Earth's atmosphere.

meteorite Any natural object that reaches Earth's surface and is recoverable.

meteoroid Any object in interplanetary space smaller than an asteroid but larger than a micrometeoroid. The specific dividing line between asteroids and meteoroids has not yet been established.

micrometeoroid Any object in interplanetary space less than a millimeter in diameter.

Milky Way's halo The spherical region that surrounds the spiral of the Milky Way and contains that galaxy's population of globular clusters.

module A spacecraft unit designed to link up with others to form a larger structure in space.

multispectral cameras By using special films and filters, cameras can record images in wavelengths other than visible light, extending into both the infrared and the ultraviolet regions of the electromagnetic spectrum.

NASA The National Aeronautics and Space Administration (NASA) was formed on July 19, 1958, to plan and operate the civilian space programs of the United States government.

neutron star The dying corpse of a star following a supernova explosion. Very small (usually less than 10 miles across), very massive (about 1.3 solar masses), and composed mostly of a super-dense mass of neutrons, the neutron star has used up all its fuel, so its nuclear fire no longer burns.

nuclear batteries These batteries provide long-term electrical power for satellites that either operate too far away from the Sun to use solar power, or whose power requirements are greater than solar power's generating capacity. As radioactive material decays, it releases heat, which is then used to produce electricity. Such batteries can provide sufficient operating power for several decades.

orbit The path an object takes in circling another celestial body.

outer space Pertaining to the regions of space above Earth's atmosphere. Although the term was used extensively in the early years of space exploration, its use has since declined, generally replaced by the simpler term, "space."

parking orbit An orbit in which a satellite is placed for temporary storage, out of the way of other spacecraft but unsuitable for its intended work.

payload The useful operating cargo package that is launched into orbit. The term can refer to a single satellite, or to several different units within a single satellite.

"penguin" suit Soviet-built garment to be worn by cosmonauts while on long-term space missions. The suit was designed to make wearers use their muscles—otherwise, the suit forced the wearer's body into a fetal position.

Glossary

perigree The point in a satellite's orbit at which it is closest to Earth. The term is also sometimes used loosely for satellites orbiting other planets.

perihelion The point in a satellite's solar orbit in which it is closest to the Sun.

plasma An ionized gas in which the motion of the particles of electrons, protons, neutral atoms, and molecules is dominated by electromagnetic interactions.

polar cusps The complex and turbulent region where Earth's magnetic field lines transition from lying parallel to Earth's surface to pointing in toward the poles. These regions are closely linked to the formation of the auroral halos above Earth's poles.

prominences Streamers or other features seen to emanate outward from the surface of the Sun along magnetic field lines.

pulsar A fast-rotating neutron star that emits pulses of short, intense radio waves. The radio emissions occur with each rotation of the pulsar, sweeping past our line of sight like a lighthouse beacon.

radiation pressure The pressure exerted on a surface by any electromagnetic emission, including ordinary light.

radio galaxy A galaxy radiating strongly in radio frequencies.

re-entry The act of returning to the Earth's atmosphere after being in orbit.

regolith The top layer of rocks, dust, or any unconsolidated material that overlies the bedrock of any planet or moon.

retro-rocket The engines used by spacecraft to slow their orbital speed in order to re-enter Earth's atmosphere.

rille A narrow gorge, usually sinuous, on the Moon's surface, thought to be caused by lava flows.

robot mission An unmanned space project.

robot ship An unmanned space probe performing humanlike actions.

sample return mission A space mission that goes to another planet or moon and returns to Earth with material from the visited body.

Seyfert galaxy A galaxy that has a compact, bright nucleus from which significant amounts of ultraviolet, x-ray, and other emissions radiate.

soft docking When any two spacecraft attempt to link in space, they initially perform a soft docking, whereby a probe on one spacecraft merely makes contact with a target on the second spacecraft. A hard docking follows when the two spacecraft join together, connecting electrical and mechanical systems, permitting joint operation of the two crafts.

soft landing With unmanned probes, a soft landing is any landing on the surface of a planet or moon in which the spacecraft still operates afterward. With manned spacecraft, it is any landing on the surface or a planet or moon in which the passengers land safely.

solar cells A device for converting the radiant light of the Sun into electrical energy.

solar cycle The 11-year sunspot cycle of the Sun, during which sunspot activity goes from minimum to maximum and the Sun's magnetic field reverses polarity.

solar flares A sudden and intense eruption on the Sun's surface of ultraviolet and x-ray emissions, tracing magnetic field lines and occurring in connection with sunspot activity. Flares can last from two to four hours. Flares release significantly less matter than coronal mass ejections.

solar mass A unit of weight used by astronomers, based on the estimated mass of the Sun.

solar orbit The path an object takes in circling the Sun.

solar radii A unit of measure used by astronomers based on the Sun's radius, equal to about 432,000 miles.

solar telescope A telescope designed to study the Sun.

solar wind The continuous outflow from the Sun of ionized hydrogen and helium, moving at speeds from 650,000 to 2 million miles per hour.

solid rocket fuel Rocket fuel in solid form, usually prepared with both the fuel and the oxidizer combined or mixed together. Although solid rocket fuel is much simpler to store than liquid fuels, once the fuel is ignited the burn rate cannot be adjusted, nor can it be stopped until all the fuel is burned up.

space sickness Nausea and vomiting often experienced by astronauts during their first few days in space.

space walk An human activity taking place in space but outside of a spacecraft. Also called an EVA (extra-vehicular activity).

space Generally used to describe all regions above Earth's atmosphere.

spectrograph An instrument for photographing the light spectrum of a celestial object.

spicules Relatively short (compared with other features) jets or spikes that shoot up from the Sun's surface. Spicules usually occur in bunches.

spin stabilization A method for stabilizing the orientation of a satellite while in orbit. If the satellite is shaped like a drum and the drum is made to spin about its axis of symmetry, the spacecraft will act like a gyroscope and maintain its orientation.

stage In order to reach orbit in the most efficient manner, rockets are built with multiple stages, generally either three or four. As each stage uses up its fuel, it is cast off, reducing the rocket's total weight so that the fuel in later stages has to lift the least amount of necessary mass.

starquake The equivalent of an earthquake on the surface of a neutron star.

suborbital A rocket launch that takes a spacecraft into space but does not travel fast enough to remain in orbit.

subsolar point The position on a planet or moon that is closest to the Sun at any moment.

sunspot flares *See* solar flares.

telemetry   The transmission of flight data from a rocket or satellite, detailing, for example, the spacecraft's operation and location.

tesserae   High plateau regions on Venus that look bright in radar images and are covered with many complex, crisscrossing ridges and fractures. They are thought to be among the oldest surface features on Venus, though their origin is not yet clearly understood.

thermosphere   The upper part of Earth's atmosphere, in which there is a steady increase in temperature with altitude, beginning at about 40 to 50 miles elevation. This layer includes both the exosphere and ionosphere.

torus   A donut-shaped plasma field that surrounds some planets or moons, such as Saturn and Jupiter's moon Io, formed by that body's magnetic field.

ultraviolet telescope   A telescope designed to study the celestial sky in the ultraviolet wavelengths, between 100 and 4,000 angstroms.

winter pole   Either the North or South Pole during the wintertime.

x-ray bursters   These objects release x-rays randomly in short staccato bursts, and are thought to be binary star systems with one star a neutron star or pulsar. The bursts are somehow caused when the neutron star pulls material from its binary companion.

x-ray pulsars   A pulsar that emits pulses both in radio frequencies and x-rays.

x-ray telescope   A telescope designed to study the celestial sky in the x-ray wavelengths, between 0.01 and 100 angstroms.

x-ray transients   An astronomical object whose x-ray emissions vary over time.

zodiacal light   A diffuse band of light that can sometimes be seen in the sky along the ecliptic or zodiac. It is thought to be caused by the diffraction and reflection of sunlight off of dust particles.

Bibliography

Two resources will be of particular value to many readers and researchers: *Mark Wade's Encyclopedia Astronautica* and the *NSSDC Master Catalog*. Readers are also advised to refer to articles in the periodicals *Advances in Space Research*, *Astronomy and Astrophysics*, *Astrophysical Journal*, *Space News*, and *Space Research*.

Aerospace Corporation, Its Work: 1960–1980. El Segundo, CA: Aerospace Corporation, 1980.

Ball, A. W. "Orbiting Solar Observatory." *Spaceflight* 12 (6), June 1970: 244–47.

Barker, John, ed. *Landsat 4, Science Investigations Summary, Including December 1983 Workshop Results*, volumes 1 and 2. Washington, DC: NASA, 1984.

Barth, C. A., et al. "Solar Mesosphere Explorer: Scientific Objectives and Results." *Geophysical Research Letters* 10 (4), April 1983: 237–40.

Blagonravov, A. A. *U.S.S.R. Achievements in Space Research (First Decade in Space, 1957–1967)*. Springfield, VA: Joint Publications Research Service, 1969.

Bougher, S. W., D. M. Hunten, and R. J. Phillips, eds. *Venus II, Geology, Geophysics, Atmosphere, and Solar Wind Environment*. Tucson: University of Arizona Press, 1997.

Bradt, Hale, Takaya Ohashi, and Kenneth Pounds. "X-Ray Astronomy Missions." *Annual Review of Astronomy and Astrophysics* 30, 1992: 391–427.

Brandt, John C. "OSO 3: Preliminary Scientific Results." *Solar Physics* 6 (2), 1969: 171–75.

Burgess, Eric, and Douglass Torr, *Into the Thermosphere: The Atmospheric Explorers*. NASA SP-490. Washington, DC: NASA, 1987.

Burrows, William E. *Deep Black: Space Espionage and National Security*. New York: Random House, 1986.

Campell, James B. *Introduction to Remote Sensing*, 2nd edition. New York: Guilford Press, 1996.

Caramanolis, Stratis. *Oscar, Amateur Radio Satellites*. Munich, West Germany: Carmanolis Verlag, 1976.

Chaikin, Andrew. *A Man on the Moon, the Voyages of the Apollo Astronauts*, New York: Penguin Books, 1998.

Chetty, P. R. K. *Satellite Technology and Its Applications*, 2nd edition. Blue Ridge Summit, PA: TAB Books, 1991.

Clark, Phillip. *The Soviet Manned Space Program: An Illustrated History of the Men, the Missions, and the Spacecraft*. New York: Orion Books, 1988.

———. "From Whence Almaz?" *Space* 8 (6), December 1992–January 1993: 31–33.

———, ed. *Jane's Space Directory, 1997–1998*. Surrey, UK: Jane's Information Group, 1997.

Code, Arthur D. *The Scientific Results from the Orbiting Astronomical Observatory (OAO-2)*. NASA SP-310. Washington, DC: NASA, 1972.

Collins, Stewart A. *The Mariner 6 and 7 Pictures of Mars*. NASA SP-263. Washington, DC: NASA, 1971.

Dunne, James A., and Eric Burgess. *The Voyage of Mariner 10, Mission to Venus and Mercury*. NASA SP-424. Washington, DC: NASA, 1978.

Eddy, John A. *A New Sun: Solar Results from Skylab*. NASA SP-402. Washington, DC: NASA, 1979.

Elder, Donald C. *Out from Behind the Eight-Ball: A History of Project Echo*. San Diego: American Astronautical Society, 1995.

Fairfield, D. H. "Whistler Waves Observed Upstream from Collisionless Shocks." *Journal of Geophysical Research* 79, 1974: 1368–78.

Fimmel, Richard O., James Van Allen, and Eric Burgess. *Pioneer, First to Jupiter, Saturn, and Beyond*. NASA SP-446. Washington, DC: NASA, 1980.

Ford, J. P., ed. "Advances in Shuttle Imaging Radar-B Research." *International Journal of Remote Sensing* 9 (5), May 1988: 835–1049.

Fruitkin, Arnold. *International Cooperation in Space*. Englewood Cliffs, NJ: 1965.

Gagarin, Yuri, and Vladimir Lebedev. *Survival in Space*. New York: Frederick A. Praeger, 1969.

Green, Constance, and Milton Lomask. *Vanguard: A History*. Washington, DC: Smithsonian Institution Press, 1971.

Grewing, Michael, Francoise Praderie, and Rudeger Reinhard, eds. *Exploration of Halley's Comet*. Heidelberg: Springer-Verlag, 1988.

Hall, R. Cargill. *Project Ranger: A Chronology*. Pasadena, CA: Jet Propulsion Laboratories, 1971.

———. *Lunar Impact, a History of Project Ranger*. Washington, DC: NASA, 1977.

Hess, Wilmot. *The Radiation Belt and Magnetosphere*. Waltham, MA: Blaisdell Publishing Co., 1968.

Hudson, Heather. *Communication Satellites, Their Development and Impact*. New York: The Free Press, 1990.

Hughes, J. Kendrick, and Gerald W. Brewer. *Lunar Orbiter 1: Preliminary Results*. NASA SP-197. Washington, DC: NASA, 1969.

Bibliography

Hunten, D. M., L. Colin, T. M. Donahue, and V. I. Moroz. *Venus*. Tucson: University of Arizona Press, 1983.

International Satellite Directory, 1998, Sonoma, CA: Design Publishers, 1998.

Jacchia, L. G., and J. W. Slowey. "A Supplemental Catalog of Atmospheric Densities from Satellite-Drag Analysis." *Smithsonian Astrophysical Observatory, Special Report 348*. Cambridge, MA: Smithsonian Institution Astrophysical Observatory, December 1972.

Jackson, John E., and James I. Vette. *OGO Program Summary*. NASA SP-7601. Washington, DC: NASA, 1975.

Jackson, John E. *OGO Program Summary, Supplement 1*. NASA SP-7601. Washington, DC: NASA, 1978.

Jastrow, R., and S. I. Rasool, eds. *The Venus Atmosphere*. New York: Gordon and Breach, 1969.

Johnson, Nicholas L. *The Handbook of Soviet Lunar and Planetary Exploration*. San Diego: American Astronomical Society, 1980.

———. *The Handbook of Soviet Manned Space Flight*. San Diego: American Astronomical Society, 1980.

———. *Soviet Space Programs, 1980–1985*. San Diego: American Astronomical Society, 1987.

Johnston, Richard S., and Lawrence F. Dietlein. *Biomedical Results from Skylab*. NASA SP-377. Washington, DC: NASA, 1977.

Kelly, P. T., and W. A. Rense. "Solar Flares in the EUV Observed from OSO 5." *Solar Physics* 26, 1972: 431–40.

Klass, Philip J. *Secret Sentries in Space*. New York: Random House, 1971.

Kopal, Zedenek. *Telescopes in Space*. New York: Hart Publishing Co., 1968.

Kramer, H. J. *Observation of the Earth and Its Environment, Survey of Missions and Sensors*, 3rd enlarged edition. Heidelberg: Springer-Verlag, 1996.

Kreplin, R. W. "NRL Solar Radiation Monitoring Satellite: Description of Instrumentation and Preliminary Results." *Space Research V, Proceedings of the Fifth International Space Science Symposium, Florence, May 12–16, 1964*, 951–65. Amsterdam: North Holland Publishing Co., 1965.

Kreplin, R. W., T. A Chubb, and H. Friedman. "X-Ray and Lyman-Alpha Emission from the Sun As Measured from the NRL SR-1 Satellite." *Journal of Geophysical Research* 67 (6), June 1962: 2231–53.

Kreplin, R. W., and R. G. Taylor. "Localization of the Source of Flare X-ray Emission during the Eclipse of 7 March, 1970." *Solar Physics* 21, 1971: 452–59.

Link, Mae Mills. *Space Medicine in Project Mercury*. Washington, DC: NASA, 1965.

Logsdon, Tom. *Understanding the Navstar GPS, GIS, and IVHS*. New York: Van Nostrand Reinhold, 1995.

Mark Wade's Encyclopedia Astronautica <http://solar.rtd.utk.edu/~mwade/spaceflt.htm>.

Martin, Donald. *Communication Satellites, 1958–1992*. El Segundo, CA: Aerospace Corporation, 1991.

Mason, John W. *Comet Halley: Investigations, Results, Interpretations*, volumes 1 and 2. New York: Ellis Horwood Limited, 1990.

McCormac, Billy, ed. *Radiation Trapped in the Earth's Magnetic Field: Proceedings of the Advanced Study Institute Held at the Chr. Michelsen Institute, Bergen, Norway, August 16–September 3, 1965*. Dordrecht, Netherlands: D. Reidel Publishing Co., 1966.

Morgenthaler, George W., and Gerald E. Simonson, eds. *Skylab Science Experiments*. Tarzana, CA: AAS Publications Office, 1975.

Morrison, David, and Jane Samz. *Voyage to Jupiter*. NASA SP-439. Washington, DC: NASA, 1980.

NASA. *Analysis of Surveyor 3 Material and Photographs Returned by Apollo 12*. NASA SP-284. Washington, DC: NASA, 1972.

NASA. *Mariner-Venus 1962: Final Project Report*. Washington, DC: NASA, 1965.

———. *SP-121: Gemini Midprogram Conference, Including Experiment Results, February 23–25, 1966*. Washington, DC: NASA, 1966.

———. *SP-126: Surveyor 1, A Preliminary Report*. Washington, DC: NASA, 1966.

———. *Significant Achievements in Satellite Meteorology, 1958–1964*. Washington, DC: NASA, 1966.

———. *SP-138: Gemini Summary Conference, February 1–2, 1967*. Washington, DC: NASA, 1967.

———. *SP-146: Surveyor 3, A Preliminary Report*. Washington, DC: NASA, 1967.

———. *SP-163: Surveyor 5, A Preliminary Report*. Washington, DC: NASA, 1967.

———. *Mariner-Mars 1964: Final Project Report*. Washington, DC: NASA, 1967.

———. *SP-166: Surveyor 6, A Preliminary Report*. Washington, DC: NASA, 1968.

———. *SP-173: Surveyor 7, A Preliminary Report*. Washington, DC: NASA, 1968.

———. *JSC-09423: Apollo Program Summary Report*. Houston, TX: NASA, 1975.

———. *The Earth's Trapped Radiation Belts*. NASA SP-8116, March 1975.

———. *Apollo-Soyuz Test Project Preliminary Science Report*. NASA TM X-58173. Washington, DC: NASA, 1976.

NSSDC (National Space Science Data Center) Master Catalog, NASA Goddard Space Flight Center <http://nssdc.gsfc.nasa.gov/nmc/sc-query.html>.

National Weather Satellite Center of the U.S. Weather Bureau. *Reduction and Use of Data Obtained by Tiros Meteorological Satellites*. Geneva, Switzerland: World Meteorological Organization, 1963.

Naumann, Robert J., and Harvey W. Herring. *Material Processing in Space: Early Experiments*. NASA SP-443. Washington, DC: NASA, 1980.

Global Positioning System: Papers Published in Navigation, Volume 1. Washington, DC: Institute of Navigation, 1980.

Newell, Homer. *High Altitude Rocket Research*. New York: Academic Press, 1953.

Newkirk, Roland W., and Ivan D. Ertel. *Skylab, A Chronology*. NASA SP-4011. Washington, DC: NASA, 1977.

Newlan, Irl. *First to Venus: The Story of Mariner 2*. New York: McGraw-Hill, 1963.

Pitts, John A. *The Human Factor*. NASA SP-4213. Washington, DC: NASA, 1985.

Portee, David. *Mir Hardware Heritage*. NASA RP-1357. Washington, DC: NASA, 1995

Rao, P. Krishna, Susan J. Holmes, Ralph K. Anderson, Jay S. Winston, and Paul E. Lehr. *Weather Satellites: Systems, Data, and Environmental Applications*. Boston: American Meteorological Society, 1990.

Reeves, E. M., and W. H. Parkinson. "An Atlas of Extreme Ultra-violet Spectroheliograms from OSO 4." *The Astrophysical Journal Supplement* 21 (181), August 1970: 1–30.

Russian Space History, Sale 6516, December 11, 1993, Auction Results. New York: Sotheby's, 1993.

Sabins, Floyd F. *Remote Sensing, Principles and Interpretation*, 3rd edition. New York: W. H. Freeman and Co., 1997.

Saunders, Joseph F. *The Experiments of Biosatellite 2*. NASA SP-204. Washington, DC: NASA, 1971.

Schnapf, A. "Tiros: The Television and Infra-red Observation Satellite." *Journal of the British Interplanetary Society* 19, May–June 1964: 386–409.

Schneider, William C., and Thomas E. Hanes. *The Skylab Results, Volume 31 of the Advances in Astronautical Sciences*. Tarzana, CA: American Astronautical Society, 1975.

Sheehan, William. *The Planet Mars: A History of Observation and Discovery.* Tucson: University of Arizona Press, 1996.

Short, Nicholas M., Paul D. Lowman Jr., Stanley C. Freden, and William A. Finch Jr. *Mission to Earth: Landsat Views the World.* NASA SP-360. Washington, DC: NASA, 1976.

Smith, James R. *Introduction to Geodesy: The History and Concepts of Modern Geodesy.* New York: John Wiley and Sons, Inc., 1997.

Soviet Space Programs, 1971–1975, Overview, Facilities, and Hardware, Manned and Unmanned Flight Programs, Bioastronautics, Civil and Military Applications, Projection of Future Plans. Staff Report prepared for the use of the Committee on Aeronautical and Space Sciences, United States Senate. Washington, DC: U.S. Government Printing Office, 1976.

Spitzer, Cary R., ed. *Viking Orbiter Views of Mars.* NASA SP-441. Washington, DC: NASA, 1980.

Thrane, E., ed. *Electron Density Distribution in Ionosphere and Exosphere, Proceedings of the NATO Advanced Study Institute, Skeikampen, Norway, April 17–26, 1963.* Amsterdam: North-Holland Publishing Co., 1964.

Tousey, R. "The Solar Corona." *Space Research XIII, Proceedings of Open Meetings of Working Groups on Physical Sciences of the Fifteenth Plenary Meeting of COSPAR,* 713–30. Berlin: Akademie-Verlag, 1973.

Van Allen, James A., Bruce Randall, and Stamatios Krimigis. "Energetic Carbon, Nitrogen, and Oxygen Nuclei in the Earth's Outer Radiation Zone." *Journal of Geophysical Research* 75 (31), November 1, 1970: 6085.

Vinogradov, A. P., Yu. A. Surkov, and F. F. Kirnov. "The Content of Uranium, Thorium, and Potassium in the Rocks of Venus As Measured by Venera 8." *Icarus* 20, 1973: 253–59.

Wade, Mark. See *Mark Wade's Encyclopedia Astronautica.*

Widger, William K. *Meteorological Satellites,* New York: Holt, Rinehart, and Winston, Inc, 1966.

Zimmerman, Robert. *Genesis: The Story of Apollo 8.* New York: Four Walls Eight Windows, 1998.

Zmuda, A. J., et al. "The Scalar Magnetic Intensity at 1,100 Kilometers in Middle and Low Latitudes." *Journal of Geophysical Research* 73, April 1968: 2495–2503.

Index

by Virgil Diodato

Bold page references in this index refer to major entries in the chronology section. *Italicized* page references refer to illustrations. When an index entry refers to "appendix listing(s)," the reader will find detailed satellite and mission lists in Appendix 2 and Appendix 3 (for research subjects and nations, respectively).

38C (satellite), **29**
83C (satellite), **36**

Acceleration sled, 215
Accelerations and vibrations, monitoring of, 241, 277, 284
ACCESS construction method, 215
ACE, 290, **322**
Acoustic research, 199, 216, 265, 269, 299. *See also* Communication propagation and design research
Acrimsat, **342**
ACSN (Appalachian Community Services Network), 123, 137
Active repeater satellites. *See* Communications satellites
Acton, Loren, 211
ACTS, **276**
Acupuncture, 215
Adamson, James, 235, 258
ADEOS, **311–12,** 340
Adrastea, 168
ADSF, 211, 216, 230
Advanced Communication Technology Satellite. *See* ACTS
Advanced Composition Explorer. *See* ACE
Advanced Earth Observing Satellite. *See* ADEOS
Advanced Photovoltaic and Electronic Experiments. *See* APEX
Advanced Very High Resolution Radiometer, 158
Advertising, 248, 309
AERCam/Sprint camera, 327, 328
Aerial Images, 330
Aeros 1, **112**
Aeros 2, **124–25**
Aerosols, 162, 273
Aerospatiale, 206, 287, 294, 314
Aetna, 173
Afanasyev, Victor, 248, 252, 256, 281, 286, 338
Afghanistan
  appendix listing for, 367
  Intersputnik consortium, 161
  Soyuz crew member from, 230

Africa, appendix listing for, 367. *See also individual nations*
AfriStar, **334**
Age. *See* Humans, oldest in space
Agila 2, **322**
Air traffic control, 208
Airbags, 320
Airglow, 18–19, *19,* 54, 77, 84, 108
Ajisai, **224**
Akebono, **233**
Akers, Thomas, 250, 263, 264, *264,* 278, 279, 313
Akiyama, Toyohiro, 248, 252
Aksenov, Vladimir, 145, 172
Al Sa'ud, Salman Abdul Aziz, 210, 211
Alcatel Space/Alcatel Espace, 333, 334
Aldrin, Buzz, 56, *80,* 80–81, 202
Alenia, 275
Alexandrov, Alexander, 194, 224, 225, 226, 227
Alexandrov, Alexander P., 228
Alexis, 272, **274**
Algae, 12, 49, 154, 157, 187, 226
Algeria, 161
Allen, Andrew, 267, 284, 307
Allen, Joe, 190, 205
Alloy research. *See* Materials research
Almaz 1, **253,** 263
Alouette 1, **23,** 33, 46
Alouette 2, **45**
Alpha Magnetic Spectrometer, 332
Altair 1, 138, **214,** 219, 223, 291
Altman, Scott, 331
Amalthea, 119, 163
Amaranth seeds, 215
Amateur Club de l'Espace, 274
Amateur radio communications, 212, 247, 274, 277, 291, 334, 338
  appendix listings for, 352
  Iskra satellites, 186, 187, 191
  mini-Sputnik satellites, 326, 333
  OSCAR satellites, 18, 21, 39, 47, 88, 108, 126, 154, 159, 181, 193, 200, 224, 229, 241–42, 257–58
American Mobile Satellite Corporation 1. *See* AMSC 1

American Satellite Corporation, 202
Amos, **310**
Amphibians, 63, 90, 128, 151, 211, 237, 252, 269, 274, 286, 294
AMSAT (Radio Amateur Satellite Corporation), 108, 126, 154, 193, 229. *See also* Amateur radio communications
AMSC 1, **294,** 308
Anders, Bill, 75
Anderson, Michael, 329
Andre-Deshays, Claudie, 312
Andromeda Galaxy, *xvii,* 159, 191, 279, *279, 301*
Angry alligator, 50
Anik A, 189, 190
Anik A1, **108,** 122, 145, 161
Anik B, **161,** 189, 190
Anik C, 62, 189, 253
Anik C1, **207**
Anik C2, **194**
Anik C3, **190–91,** 194
Anik D1, **189,** 190, 253
Anik D2, **205**
Anik E2, 190, **253**
Anita (spider), 117
Anna 1B, 9, **24,** 29, 35, 44
ANS 1, **125,** 148
Antennas, 62, 70, 123, 163, 310. *See also* Communication propagation and design research
Antibiotics, 187
Anti-matter, 169, 332
Anti-satellite (ASAT) systems, 29, 32, 72, 73, 90, 213. *See also* Missile detection systems
Ants, 194, 237
APEX, **287**
Apollo 1, **57,** 67, 72
Apollo 4, **67,** 67, 70
Apollo 5, **69,** 94
Apollo 6, **70,** 75
Apollo 7, **72,** 74
Apollo 8, 30, 74, **75,** *75*
Apollo 9, **77,** *77,* 106
Apollo 10, **78**

385

Apollo 11, 30, *80*, **80–81**, *81*, 86, 141
Apollo 12, 58, 59, **85–87**, *85–87*, 93
Apollo 13, **88–89**
Apollo 14, **93**, *93*, 103, 105, 188
Apollo 15, 81, *96*, **96–98**, *97*, 103
Apollo 16, **104–06**, *105*, *106*
Apollo 17, 98, *109*, **109–12**, *110*, *111*, 112
Apollo 18 (ASTP), 84, 126, *133*, **133–34**, 188, 215, 273
Apollo program, 40, 67, 112
Apollo-Soyuz Test Project (ASTP), *133*, **133–34**
Appalachian Community Services Network (ACSN), 123, 137
Apple, **178**, 185
Applications Technology Satellite. *See* ATS
Apstar 1, **287**
Apt, Jerome, 253, 254, 268, 285, 313
Arab Satellite Communications Organizations, 206
Arab States, appendix listing for, 367
Arabella (spider), 117
Arabidopsis plants, 172, 176, 186, 272, 277, 288, 341. *See also* Cress plants
Arabsat 1A, **206**
Arabsat 1B, **211**
ARD, **334**
Argentina, 99, 186, 205, 214, 223, 336, 340
   appendix listing for, 367
Argos, **338**
Ariane 1 rocket, 170, **178**
Ariane 5 rocket, **326**, 333, 342
Ariane Passenger PayLoad Experiment. *See* Apple
Ariel 1, **20**, 21, 23, 33, 46, 61
Ariel 2, **31**, 61
Ariel 3, **61**, 102
Ariel 4, **102–03**
Ariel 5, **125–26**, 148, 153
Ariel 6, **166**
Ariel (moon), *217*, 218
Armstrong, Neil, xix, 48, *80*, 80–81
Array of Low-Energy X-ray Imaging Sensors. *See* Alexis
Arsene, **274**
Artificial eclipse, 133
Artificial gravity, 89, 148, 286–87
Artificial radiation belt, 20, 21, 22, 23–24, 31, 33, 36
Artsebarski, Anatoly, 256, 260
Artyukhin, Yuri, 124
Aryabhata, **131**
ASAT (anti-satellite) systems, 29, 32, 72, 73, 90, 213
ASC 1, **212**
ASCA, **272–73**, 277
Ashby, Jeffrey, 340
Asia Satellite Telecommunications Company, 244, 328
Asiasat 1, 198, **244**
Asiasat 3, 328
Asterix, **45**
Asteroids
   belt of, 36, 103, 104, 112, 130, 280
   classes of, 337
   comets and, 192, 222
   flybys, 260, *260*, 276, *276*, 306, 318–20, *321*, 334, 337, 341
   moon of, 276
ASTEX, **99**

ASTP (Apollo-Soyuz Test Project), *133*, **133–34**
Astra 1A, **232**, 237
Astrid 1, **292**
Astrid 2, **337**
Astro-A, **176**
Astro-B, **192**
Astro-C, **224**
Astro-D, **272–73**
Astroculture, 265, 275, 283, 292, 299, 310, 329, 335
Astron, **192**
Astron 2, **240**
Astronomical Netherlands Satellite. *See* ANS
Astronomical research, appendix lists for, 352–53. *See also individual research areas, such as* Solar research
AT&T, 21, 25, 140, 173, 195, 202, 204, 210, 280, 281, 303
AT&T Wireless, 294
ATCOS 2, 67
Athena rocket, 338
Atkov, Oleg, 199, 200
Atlantis, flights of, 214, 215, 232, 234, 238, 243, 251, 253–55, 258, 260–61, 263, 267, 290–91, 295–96, 308, 313, 314, 318, 325
Atmosphere, of spacecraft, 57, 127, 133, 144, 146, 177, 223, 340
Atmosphere 1, **250**
Atmosphere Explorer, **25**
Atmosphere Explorer B, **49**
Atmosphere Explorer C, **120**
Atmosphere Explorer D, **134–35**
Atmosphere Explorer E, **137**
Atmospheric Composition Satellite. *See* ATCOS
Atmospheric density research, 4, 10, 13, 25, 30, 31, 36, 37, 45, 49, 51, 54, 60, 61, 71, 79, 94, 98, 112, 120, 121–22, 131, 132, 137, 228, 234, 250, 340
   appendix listings for, 354
Atmospheric Re-entry Demonstrator. *See* ARD
Atmospheric research
   appendix listings for, 353–54
   Earth. *See* Earth, atmosphere of
   Halley's Comet, 223
   Io, 119
   Jupiter, xiv–xv, 119, 127, 164, 166–68, 238, 301, 303, 304
   Mars, xiii–xiv, 42–43, 74, 82, 101, 102, 121, 141–44, 229, 321, 332
   Mercury, xiii
   Moon, 85, 109, 290
   Neptune, 236
   Saturn, xv, 153, 169, 175, 179, 301
   Titan, xv, 174, 301
   Triton, 236
   Uranus, xv–xvi, 216, 217
   Venus, xiii, 25, 64–65, 78, 92, 106, 120–21, 136–37, 160–61, 183–84, 210, 248, 249–50
ATS 1, **57**
ATS 2, 66
ATS 3, 57, **66**
ATS 5, **83**
ATS 6, **123–24**, 152

Attitude control systems, 48, 56, 58, 59–60, 83, 108, 113, 139
   three-axis design, 268, 270, 282
Aubakirov, Toktar, 260
Auctions, 89, 91, 147, 192, 248
Aura, **134**
Aureol 1, 84, **103**, 109
Aurora 1, **61**, 190
Aurora 2, **256**
Aurora 7, **20–21**, 23
Aurora (phenomenon)
   artificial, 255
   Earth. *See* Earth, aurora of
   Jupiter, 160, 164, 166, 280
   Saturn, 176
   Uranus, 153
Aurorae, **72**, 79, 109
Aussat A1, **212**
Aussat A2, **215**
Aussat B1, **268**
Aussat Proprietary Limited, 212
Australia
   appendix listing for, 367
   ham (amateur) radio in, 126, 154
   images of, 117, 118, 238
   launch from, 100
   satellite service to, 306
   telescope from, 262
Austria
   appendix listing for, 367
   experiments and equipment from, 210, 340
   images of, 285
   satellite service to, 232
   Soyuz crew member from, 260
Automated Directional Solidification Furnace. *See* ADSF
Autopilot software, 215
Avdeyev, Sergei, 266, 272, 297, 298, 300, 301, 306, 338
Azur, 84, **85**

Back pain, 116, 200, 262, 270
Background radiation, cosmic, 238–39
Bacteria, 12, 63, 86, 105, 118, 120, 127, 128, 187, 269, 272, 275, 293
Badr A, **247**
Bagian, James, 233, 256
Baker, Ellen, 238, 265, 286, 295
Baker, Michael, 258, 269–70, 288, 314
Balance/inner ear research
   aquatic life, 331
   humans, 17, 44, 90, 116, 198, 200, 211, 256, 278, 289, 311, 331, 335
   rats, 290, 291, 331
Balandin, Alexander, 242
Bangladesh, 205
Barium clouds, 158, 243
Barley seeds, 7
Barry, Daniel, 305, 306, 340
Bartoe, John-David, 211
BATSE, 253
Baturin, Yuri, 333
Baudry, Patrick, 210, 211
Beacon Explorer B, **35**
Beacon Explorer C, **41**
Bean, Alan, 85–87, 116–17
Bean plant seedlings, 185
Bees, 185, 202
Beetles, 7, 63, 118, 132

Belgium
    Spektr components, 295
    Spot satellite, 221
    Starshine satellite, 340
    STS crew member from, 263
Belka (dog), 11
Bell Labs, 10
Bella, Ivan, 338
Belyaev, Pavel, 39
BeppoSax, **309–10**, 316, 319, 331
Beregovoi, Georgi, 72–73
Berezovoi, Anatoly, 186
Bhaskara 1, **166**
Big bang theory, xvii, 15, 19, 107, 195, 238–39, 332
Big Bird, 20, 96
Big Joe, **6**, 7
Billboards, 248. *See also* Advertising
Biofeedback, 208
Biological activity scanners, 157, 321. *See also* Wildlife tracking
Biological materials research, 234. *See also* Crystals research (proteins)
    Apollo missions, 93, 134
    Soyuz-Mir missions, 226, 228, 232, 296
    STS missions, 187–88, 193, 194, 196, 199, 204, 207, 215, 230–31, 233, 241, 245, 255, 259, 262, 264, 265, 269, 270, 272, 273, 275, 283, 284, 286, 288, 290, 293, 297
Biological research
    appendix lists for, 354
    existence of life, xiii, 43, 82, 141–42
    humans. *See* Human biology and medicine
    non-human organisms. *See individual organisms*
    plants. *See* Plant research; *and individual plants*
Biological Research in Canisters. *See* BRIC
Bion 1, **118**, 208
Bion 2, **126**
Bion 3, **137**
Bion 4, **148**
Bion 5, **170**
Bion 6, **198**, 208
Bion 7, **211**
Bion 8, **227**
Bion 9, **237**
Bion 10, **271**
Bion 11, **314**
Biopolymers, 186
Biosatellite 2, **63**
Biosatellite 3, **79**
Birds, 163, 233, 240, 242, 285, 293, 294
Black holes, 91, 99, 107, 109, 125, 128, 132, 148, 153, 159, 170, 240, 246, 254, 260, 279, 305, 316
Blacks, first in space, 195, 268
Blaha, John, 233, 239, 258, 277, 313, 314, 329
Blood biology. *See* Cardiovascular research
Bloomfield, Michael, 325
Blue Planet, The (film), 234, 238, 241, 245
Bluford, Guion, 195, 214, 255, 271
Bobko, Karol, 192, 207, 214

Body fluids
    blood. *See* Cardiovascular research
    redistribution, 79, 95, 114, 120, 157, 163, 196, 199, 274, 278, 289
    saliva, 256, 311
    urine, 18, 42, 256, 311, 318, 335
Body mass measuring device, 114
Body temperature, 18, 22, 27, 46
Boeing Company, 330, 335, 339, 342
Bolden, Charles, 216, 244, 263, 283
Bondar, Roberta, 261
Bone research, 298–99, 318
    chickens, 285, 293
    dogs, 48
    fractures and disease, 233, 250
    frogs, 252
    humans, Gemini missions, 42, 46
    humans, Skylab missions, 115, 117, 118
    humans, Soyuz-Mir missions, 289, 298, 306
    humans, Soyuz-Salyut missions, 128, 163, 173, 200
    humans, STS missions, 256, 263, 270, 335
    monkey, 79
    rats, 118, 126, 227, 233, 250, 256, 274, 275, 285, 290, 293, 306, 314
    rodents, 311
Bonnie (pigtail monkey), 79
Bonum 1, **335**
Booms
    for antennas, 70
    for stabilization, 30, 39, 51, 62, 70, 242
Booster explosion, 218
BOR-4, **187**
Boreas, **83**
Borman, Frank, 46, 75, 77
Botanical research. *See* Plant research
Bowersox, Kenneth, 278, 299, 315
Brady, Charles, 311
Brahe, Tycho, 125
Braille (asteroid), 334, 341
Brain research
    dogs, 48
    humans, 27, 46, 120, 124, 128, 144, 173, 187
    monkey, 79
Braking, solar panels for, 323
Brand, Vance, 133, 190, 198, 251
Brandenstein, Daniel, 195, 210, 241, 263, 264
Brasilsat A1, **206**
Brasilsat B1, **287**
Brazil
    appendix listing for, 367
    communications for, 343
    radar scans of, 285
Brazil Space Agency, 287
Bread mold, 241
Breathing. *See* Respiratory research
BremSat 1, **283**
BRIC canisters, 288, 297, 298, 314, 341
Bridges, Roy, 211
Britain. *See* United Kingdom
British Satellite Broadcasting, 237
British Sky Television, 237
Broadcast Satellite System Corporation, 317

Brown, Curtis, 268, 290, 310, 322, 334, 342
Brown, Kenneth, 265
Brown, Mark, 235, 259
BSAT 1A, **317**
Bubble research, 191, 216, 286, 320
Buchli, James, 206, 214, 233, 259, 260
Buckey, Jay, 331
Budarin, Nikolai, 296, 297, 329, 332, 333
Bulgaria, 84, 148
    appendix listing for, 367
    Intercosmos satellites, 238, 261, 284
    Intersputnik consortium, 161
    Soyuz crew members from, 165–66, 228
    Vega program, 210
Bullfrogs, 90
Buran shuttle, 187, 227, **231**, *231*, 246, 296
Burial of human remains, 317, 328, 329, 343, 355
Burma, 112, 145, 200
Bursch, Daniel, 276, 288, 310
Burst and Transient Source Experiment. *See* BATSE
Bursts. *See* Gamma radiation; Ultraviolet radiation; X-ray radiation
Butterfly, 341
Bykovsky, Valeri, 26–27, 145, 157

Cabana, Robert, 250, 271, 286, 335, 336
Cacti, 137
Cakrawarta 1, **326**
California Institute of Technology, 300
Callisto, xv, 119–20, 127, 163, 165, *166*, 168, *168*, 304, *304*
Calsphere, **24**, 108
CAMEO, **158**
Cameras
    IMAX. *See* IMAX cameras
    for spacecraft inspection, 327, 328
Cameron, Kenneth, 253, 273, 300
Canada. *See also* Telesat Canada
    appendix listing for, 367
    Cosmos 954 scattered over, 149
    equipment and experiments from, 206, 286, 288, 340
    ham (amateur) radio in, 126, 154
    remote manipulator arm built in, 181, 184
    search-and-rescue agreement, 188
    STS crew members from, 204, 270, 310, 311, 322, 340
Canadian Space Agency, 329
Canadian Target Assembly. *See* CTA
Canary Islands, 268
Candles, lithium, 315, 318
Cannonball 1, **71**
Cannonball 2, **98**
Canyon 1, 21, **71**, 96, 112, 155
Cardiovascular research
    dogs, 11, 13, 48
    humans, Apollo missions, 93, 98
    humans, Gemini missions, 40, 42, 44, 46, 54
    humans, Mercury missions, 18, 23, 26
    humans, Skylab missions, 114, 117
    humans, Soyuz-Mir missions, 224, 226, 289, 318
    humans, Soyuz mission, 120
    humans, Soyuz-Salyut missions, 95, 124, 128, 132, 144, 151, 155, 156, 172, 178, 187, 200

Index

Cardiovascular research *(continued)*
    humans, STS missions, 206, 207, 211, 216, 230, 256, 257, 274, 278, 311, 331, 335
    humans, Voskhod mission, 35
    humans, Vostok missions, 22, 27
    monkeys, 7, 79, 198, 211
    rats, 126, 211, 227, 250, 314
Carl Zeiss camera factory, 157
Carpenter, Scott, 19, 20–21, 23
Carpenter ants, 194
Carr, Jerry, 118
Carter, Manley, 239
Casper, John, 243, 271, 284, 310
Cassen, Patrick, 164
Cassini/Huygens, **326**
Cassini's Division, xv, 169
Cassiopeia A, 341, *341*
Castor, **131**
CAT 1, **170**
CAT 3, **178**
CAT 4, **182**
CBERS 1, **342**
CCDs (charge-coupled devices), 146, 242, 257–58, 262, 273
CCE, 200, **203**
Celestis, Inc., 317
Celestis 1, **317**
Celestis 2, **329**
Celestis 3, **342–43**
Cenker, Robert, 216
Centaurus A Galaxy, 125, 159
Central Bank of the Russian Federation, 326
Cepheid variable stars, 235–36
Cerise, **296**
Cernan, Eugene, 50, 52, 78, 109–11, 117
CFCs (chlorofluorocarbons), 137, 158, 258, 259, 273, 311
CFES, 187–88, 193, 194, 196, 204, 207, 215
Chad, 285
Chaffee, Roger, 57, 72
Chalet 1, **155**, 206, 251
Challenger
    destruction of, 218–19
    flights of, 192–94, 195–96, 198–99, 201–02, 204–05, 208–09, 211–12, 213–14
Chandra X-ray Observatory, **340–41**
Chang-Diaz, Franklin, 216, 238, 267, 283, 307, 331
Charge Composition Explorer. *See* CCE
Charge-coupled devices. *See* CCDs
Charon, 290
Chawla, Kalpana, 327
Cheli, Maurizio, 307
Chemically Active Material Ejected into Orbit. *See* CAMEO
Chiao, Leroy, 286, 305, 306
Chickens, 233, 285, 293
Chile, 189, 197, 336
    appendix listing for, 367
Chilton, Kevin, 263, 285, 308
Chimpanzees, 12, 17–18
China
    appendix listing for, 367
    experiments from, 149, 288
    Iridium launches, 317
    monitoring of, 112, 155
    satellite service to, 277
    Starshine satellite, 317
    unmanned test of manned capsule, 342
China 9 through China 11, **180**
China 20, **226**
China 23, **230**
China 35, **269**
China-Brazil Earth Resources Satellite. *See* CBERS
Chlorine monoxide, 259, *259*
Chlorofluorocarbons. *See* CFCs
Chlorophyll, 157–58
Chretien, Jean-Loup, 187, 231, 232, 266, 325
Chromex facility, 233, 250–51, 272, 277, 288, 292, 310
Chromosome and Plant Cell Experiment. *See* Chromex facility
Circinus X-1, 159
Clarke, Arthur C., viii
Cleave, Mary, 215, 234
Clementine, **282**, 328
Clervoy, Jean-Francois, 290, 318, 342
Clifford, Michael, 271, 285, 308, 325
Climate. *See* Atmospheric research; Earth resources research; Weather satellites
Clover plants, 310–11
CMEs. *See* Solar research, corona mass ejections
Co-orbital satellites, 72
Coastal Zone Color Scanner, 157
Coats, Michael, 203, 233, 255
COBE, **238–39**
Coca-Cola company, 292
Cockrell, Kenneth, 273, 298, 313
Cold War, xix
Colds caught by astronauts, 72
Coleman, Catherine, 299, 340
Collins, Eileen, 292, 318, 340
Collins, Mike, 52, 80, *80*
Collisions, 281, 318
    near misses, 306, 315, 321
    with small object, 194
Columbia, flights of, 177, 181–82, 184–85, 187–88, 190–91, 197–98, 216, 235, 241, 251–52, 256–57, 265, 269–70, 274, 277–78, 284, 286–87, 299, 307, 311, 313–14, 317, 320, 326–27, 331, 340–41
Columbia Communications, 193
Columbus, Christopher, 271
Combined Release and Radiation Effects Satellite. *See* CRRES
Comets, 74, 118, 120, 153, 162, 192, 201
    Borrelly, 341
    Bradfield, 153
    Giacobini-Zinner, 157, 209, 213
    Hale-Bopp, 290, 301, 308, 316
    Halley's, 153, 206, 209, 210, 218, 221–23, 286, 300, 317, 320
    Hyakutake, 302
    impacting Earth's atmosphere, 308
    impacting Sun's surface, 302, *302*
    Kohoutek, 118, 120
    Sergeant, 153
    Wild-2, 338
    Wilson-Harrington, 341
Command module/service module design, 67, 77
Commercial investment. *See* Private investment in space; *and individual companies*
Communication propagation and design research, 120, 123, 140, 147, 149, 153, 154–55, 161, 195, 224, 244, 268, 274, 276, 300. *See also* Antennas
    appendix lists for, 355
Communications satellites, viii, xviii. *See also* Amateur radio communications
    active repeater design, 5
    appendix listings for, 355–57
    competition among owners, 291
    first air-to-air and air-to-ground relay, 57
    first direct broadcast technology, 154
    first from Intelsat, 40
    first geosynchronous orbit, 29
    first geosynchronous "stationary" orbit, 33
    first global communications service from one organization, 75
    first in orbit, 5
    first nuclear power source, 139
    first officially announced from Soviet Union, 40–41
    first privately built, 21–22
    geosynchronous versus low-Earth orbit, 258
    gravity assist from Moon for, 328
    low-Earth orbit, 21, 258, 317–18, 328
    Open Skies policy, 122
    passive reflector design, 10–11
    political issues, 26, 122
    solid state technology, 62
    for space shuttle communications. *See* TDRS
    store-and-forward memory, 200
    three-axis design, 268, 270
Communications Technology Satellite. *See* CTS
Compton, Arthur Holly, 253
Compton Gamma Ray Observatory, **253–54**, *254*
Compton Telescope. *See* COMTEL
Computers
    failures of, 296, 321
    laptop, 234
Comsat Corporation, 123, 173
Comsat General Corporation, 140
Comstar 1, **140**, 195
COMTEL, 253, 354
Conestoga rocket, **189**, 243
Conrad, Pete, 43–44, 54, 85–87, 114–15
Construction work. *See* Work in space
Contel American Satellite Company, 212
Continuous Flow Electrophoresis System. *See* CFES
Convection currents, 191
Cooper, Gordon, 26, 43–44
Copernicus, **107–08**, 125, 148, 152, 264
Corona satellite program, 18
Coronae
    of other stars, 125, 273
    of the Sun. *See* Solar research
Coronas I, **284**
CORSA B, **162**
COS B, **134**
Cosmic B, **134**
Cosmic Background Explorer. *See* COBE
Cosmic dust, 6, 23, 251–52. *See also* Micrometeorite/micrometeoroid research
Cosmic particles, 104, 105

388

Cosmic radiation, 3, 36, 42, 43, 45, 47, 60, 70, 73, 74, 86, 94, 104, 105, 109, 123, 131, 166, 169, 211, 265, 273, 322, 331, 332
Cosmos 1, **19**
Cosmos 2, **19**
Cosmos 3, **20**, 21, 26
Cosmos 4, **20**, 22
Cosmos 5, 20, **21**, 26
Cosmos 7, **22**
Cosmos 17, 20, **26**
Cosmos 26, **31**, 35, 36, 79
Cosmos 41, **33**, 40
Cosmos 44, **34**
Cosmos 49, **35**, 36, 44, 79
Cosmos 51, **36**
Cosmos 110, **48**
Cosmos 122, **52**
Cosmos 133, **56**
Cosmos 140, **58**
Cosmos 144, **58**, 60
Cosmos 148, **58**
Cosmos 156, 58, **60**
Cosmos 166, **61**
Cosmos 186, **66**
Cosmos 188, **66**
Cosmos 198, **67–68**
Cosmos 248, **72**, 73
Cosmos 249, **72**, 73
Cosmos 252, **73**
Cosmos 343, **89**
Cosmos 373, **90**
Cosmos 375, **90**
Cosmos 398, **94**, 99
Cosmos 434, 94, **99**
Cosmos 605, **118**
Cosmos 613, **119**
Cosmos 637, **122**, 125, 138
Cosmos 638, **122**
Cosmos 690, **126**
Cosmos 699, 88, **127–28**
Cosmos 782, **137**
Cosmos 881, **146**
Cosmos 882, **146**
Cosmos 929, **147–48**
Cosmos 936, **148**
Cosmos 954, **149**
Cosmos 1001, **154**
Cosmos 1076, **162**
Cosmos 1129, **170**
Cosmos 1267, 150, *177*, **177–78**, 186
Cosmos 1374, **187**
Cosmos 1383, **188**
Cosmos 1402, **189**
Cosmos 1413, **189–90**
Cosmos 1414, **189–90**
Cosmos 1415, **189–90**
Cosmos 1443, 186, **192**, 194
Cosmos 1500, **162**, 196
Cosmos 1514, **198**
Cosmos 1645, **208**
Cosmos 1667, **211**
Cosmos 1686, 186, *213*, **213–14**
Cosmos 1700, **214**
Cosmos 1809, **224**
Cosmos 1870, **226**
Cosmos 1887, **227**
Cosmos 1989, **233**
Cosmos 2024, **234**
Cosmos 2044, **237**
Cosmos 2176, **262**
Cosmos 2229, **271**

Cosmos 2268 through Cosmos 2273, **284**
Cosmos 2349, **330**
COSPAS 1, **188**
Cotton plants, 213
Coupon, **326**
Courier 1B, **11**
Covey, Richard, 212, 230, 251, 278
Crab Nebula, 74, 91, 99, 104, 108, 128, 159, 169, 187, 280, 316, 341
Craters, vii
    Callisto, 165, 168
    Dione, 175
    Earth, 130, 285
    Ganymede, 165, 168
    Gaspra, 260, 319
    Halley's Comet, 222, *223*
    Ida, 319
    Mars, *42*, 43, *81*, 82, *82*, *83*, 100–01, 130, 142, 323
    Mathilde, 319
    Mercury, 129–30
    Mimas, 175
    Miranda, 218
    Moon. See Moon, craters of
    Rhea, 175
    Tethys, 175, 179
    Triton, 236
    Venus, 160, 197, 249
Creighton, John, 210, 243, 259
Cremated remains in space, 317, 328, 329
Cress plants, 287, 288. See also Arabidopsis plants
Crickets, 331
Crimson Satellite, 265
Crippen, Robert, 177, 193, 201, 204
Crista-Spas 1, **290**
Crista-Spas 2, **322**
Crocker, James, 279
Crouch, Roger, 317, 320
CRRES, **247**
Crystals research. See also Materials research
    Skylab missions, 115, 118
    Soyuz-Mir missions, 224, 225, 228, 232, 237, 242, 248, 260, 267, 275, 296
    Soyuz-Salyut missions, 144, 151, 155, 157, 172, 173, 177, 178, 186
    STS missions, 194, 197–98, 205, 207, 208, 212, 215, 231, 233, 234, 238, 241, 245, 255, 259, 262, 263, 264, 265, 270, 272, 273, 275, 283, 284, 285, 286, 288, 290, 293, 294, 297, 298, 299, 307
    unmanned mission, 244
Crystals research (proteins)
    Soyuz-Mir missions, 228
    STS-Mir missions, 313, 318, 325, 329, 332
    STS missions, 298, 299, 306, 307, 310, 311, 314, 320, 322, 325, 329, 332, 335, 341
CTA, **270**
CTS 1, **138–39**
Cuba
    appendix listing for, 367
    Intercosmos satellite, 84
    Interputnik consortium, 161
    Soyuz cosmonaut from, 173
Cucumbers, 171
Culbertson, Frank, 251, 276

Cultural knowledge, xviii–xix, 75
Cunningham, Walt, 72
Curbeam, Robert, 322
Currie, Nancy Sherlock, 296, 335, 336
Cygnus X-1, 91, 107, 125, 132, 159, 169, 260
Cygnus X-3, 107
Czech Republic
    appendix listing for, 367
    Intercosmos satellite, 284
Czechoslovakia, 104, 126, 148, 210
    appendix listing for, 367
    Intercosmos satellites, 84, 140, 170, 176, 238, 261
    Intersputnik consortium, 161
    Soyuz cosmonaut from, 154

Dactyl, 276, *276*
Daimant B rocket test, 88
Daimler-Benz, 327, 328
DarpaSat, **285**
Dash 1, **26**
Dash 2, **54**
Data Collection Satellite. See SCD
DATS, **62**
Davis, Jan, 268, 283, 322
Day-night cycle, 30, 32, 154, 189, 241, 273
Daylilies, 233, 297, 311
DBS 1, **281**
Deaths. See Fatalities
Debris, 114, 189, 234, 271, 283, 292, 308, 310, 338
    collisions or near collisions with, 296, 306, 321
    radioactive, 149, 189
DEBUT, **242**
Deep Space 1, 88, **334**
Deep Space 1, Braille Flyby, **341**
Defecation and fecal matter, 42, 208, 311. See also Waste management
Defense Satellite Communications System. See DSCS
Defense Support Program. See DSP
Deimos, 100
DeLucas, Lawrence, 265
DemoSat, **339**
Demin, Lev, 125
Denmark, 232, 288, 340
    appendix listing for, 367
Density and Scale Height. See Dash
Deployable Boom and Umbrella Test. See DEBUT
Despun Antenna Test Satellite. See DATS
Destiny in Space (film), 245, 262, 267, 277, 279
Dezhurov, Vladimir, 293, 294, 295, 296
DFH 3A, **291**
Diademe 1, **58**
Diademe 2, **58**
DIAL-MIKA, **88**
DIAL-WIKA, **88**
Diet. See Food; Metabolism
Diffusive Mixing of Organic Solutions. See DMOS
Digestive system research, 207. See also Eating; Food
Dill plants, 171
Dione, 175, *175*, 179
Direct Broadcast Satellite 1. See DBS 1
Direct broadcast technology, 154, 232, 281, 282, 316, 342

Index

DirecTV 1R, **342**
Discoverer 1, **5**, 9
Discoverer 3, 11
Discoverer 13, 5, **9**, 11
Discoverer 14, **11**
Discoverer 17, **11**–12
Discoverer 18, **12**
Discoverer 36, **18**
Discoverer launches, release of information about, 13
Discovery, flights of, 203–04, 205, 206, 207–08, 210–11, 212, 230–31, 233–34, 239, 244–45, 250–51, 255–56, 259–60, 261–62, 271, 273–74, 276–77, 283, 287–88, 292–93, 296–97, 322, 331–32, 334–35, 340, 342
Dissections, first in space, 278
DLR-Tubsat, **340**. *See also* Tubsat
DMOS, 205, 215
DMSP Block-4A F1, 34, **37**
Dnepr rocket, 339
Dobrovolsky, Georgi, 95, 95–96, 117, 124, 170
Docking. *See* Renvezvous and docking
Doctors. *See* Physicians in space
Dodecapole 1, **39**
Dodge 1, **62**
Dog ticks, 216
Dogs, 3, 11, 12, 13, 48, 196
Doi, Takao, 327
Dong Fang Hong 3A. *See* DFH 3A
Doppler shift, 8–9, 61, 83, 155–56, 257
Double skip re-entry maneuver, 67, 69, 70, 73
Dream Is Alive, The (film), 202, 204, 205, 215
Drug research. *See* Pharmaceutical materials research
DSCS II 1, 62, 76, 87, **100**, 126, 139
DSCS II 2, 62, 76, 87, **100**, 126, 139
DSCS III, **190**, 214, 252, 284
DSP 16, **261**
DSP-647 1, **95**
Dual missions, 22, 26–27, 81–82. *See also* Three-spacecraft mission
Duckweed, 49
Duffy, Brian, 263, 275, 305, 306
Duke, Charles, 105, *105*
Dunbar, Bonnie, 214, 241, 265, 286, 295, 329
Duque, Pedro, 334
Durrance, Samuel, 251, 293
Dynamics Explorer 1, **178**–**79**, 221, 307
Dynamics Explorer 2, **178**–**79**, 221, 307
Dzhanibekov, Vladimir, 152, 176, 187, 200, 202, 209, 213

Eagle Nebula, *xvii*
Early Bird, 33, **40**, 55, 57, 58, 64
Early-warning satellite, first, 9. *See also* Surveillance
Earth. *See also* Earth resources research; Geodetic/navigational research; Geological research
  1957 knowledge about, xviii
  craters, 130, 285
  gravitational field, 22, 108, 129, 157, 301
  lightning, 38, 188, 234, 238, 274, 294, 327
  polar regions, 62, *69*, 85, 103, 134–35, 158, *158*, 162, *178*, 221, 224, 233

radiation budget, 158, 204
seismic research, 140
space dragged by, 270
surface. *See* Earth resources research
volcanic activity, 140, 158, 162, 189, 207, 257, 260, 261, 270, 273, 287, 288
weather. *See* Earth resources research; Weather satellites
winds, 120, 156, 259, 294, 311, 340
Earth, atmosphere of, 7, 12, 178–79, 294
  composition, 31, 61, 62, 67, 71, 79, 84, 91, 94, 98, 120, 122, 127, 128, 132, 146, 158, 263, 273, 290, 311, 322
  density. *See* Atmospheric density research
  dirtiness/clarity of, 176, 260, 261
  drag of, vii, 24, 26, 27, 36, 37, 45, 54, 60, 71, 94, 98
  exosphere, 79
  ionosphere. *See* Earth, ionosphere of
  magnetosphere. *See* Earth, magnetosphere of
  mesosphere, 181
  nitric oxide, 290, 330
  nitrogen oxide, 62, 132, 146
  oxygen, 61, 98, 132
  ozone, 31, 61, 91, 128, 132, 137, 146, 158, 162, 181, 201, 258, 259, 266, 273, 290, 294–95, 311–12, 330, 332, 333
  solar activity and, 4, 79, 112, 122, 124, 129, 134, 137, 158, 259, 261, 273. *See also* Solar research
  stratosphere, 158, 162
  thermosphere, 79
Earth, aurora of, xi, xviii, 4, 7, 16, 24, 32, 34, 37, 61, 62, 70, 72, 79, 83, 84, 85, 103, 109, 124, 132, 153, 178, 208–09, 221, 233, 255, 269, 292, 307, 312, 337, 338
Earth, images of
  as almost-full disk, 62
  earthrise, 75
  first from space, xix, 6
  first of Earth above lunar horizon, 53
  first of Earth as globe, 66
  for sale, 330
  from vicinity of the Moon, 53, *53*, *73*, 74, 75, *75*
Earth, ionosphere of, xi, xviii, 3, 4, 9, 11, 19, 20, 23, 25, 28, 33, 35, 37, 45, 46, 61, 76, 77, 83, 84, 94, 99, 103, 109, 137, 139, 153, 154, 158, 178, 184–85, 187, 200, 211–12, 224, 228, 238, 242, 243, 255, 267, 277, 283, 292. *See also* Earth, magnetosphere of; Earth, Van Allen belts of; Ionosphere/magnetosphere research
Earth, magnetic field of, xviii, 4, 5, 6, 7, 8, 12, 14, 16, 17, 21, 24, 29–30, 31, 32, *34*, 34–35, 36, 44, 49, 51, 52, 54, 62, 70, 77, 79, 83, 88, 134–35, 238, 244, 297, 307, 312, 338
  gas clouds released into, 243, 247
  northern versus southern halves, 157
  polarity reversal, 170
  solar activity and, 73, 135, 157, 261. *See also* Solar research

Earth, magnetosphere of, 5, 6, 8, 11, 21, 23, 29–30, 31, 34–35, 37, 43, 44, 51, 54, 60, 62, 64, *69*, 71, 84, 88, 99, 156, 158, 238, 255, 263, 284, 292, 312, 337, 338. *See also* Earth, ionosphere of; Ionosphere/magnetosphere research
  bow shock, 103
  magnetopause, 14, 17, 32, 69, 290
  polar cusp, 103
  solar activity and, 74, 79, 91, 102, 103, 104, 124, 134, 135, 150, 157, 203, 208, 265–66, 269, 290, 307–08, 322, 327. *See also* Solar research
  structural detail, 178
Earth, Van Allen belts of, 3, 4, 5, 6, 7, 16, 17, 20, 21–22, 23–24, 25–26, 27, 28, 29, 31, 32, 33, 34–35, 36, 37, 42, 49, 51, 55, 69, 71, 72, 77, 85, 131, 132, 157, 247, 266, 282, 284, 287
Earth Limb Radiance test, 231
Earth Radiation Budget Satellite. *See* ERBS
Earth Radiation Budget sensor, 158
Earth Remote-sensing Satellite. *See* ERS
Earth resources research, 166, 170, 172, 226, 262–63, 327, 333, 334, 338, 340, 341, 342. *See also* Geodetic/navigational research; Geological research; Weather satellites
  appendix listings for, 353–54
  biological activity scanners, 157, 321
  Gemini missions, 42, 44
  Geos satellites, 44–45, 68–69, 131
  GOES satellites, 135
  IRS satellites, 228
  Landsat satellites, 106–07, 188–89, 339
  Mercury missions, 18, 26
  Nimbus satellite, 157–58
  oceans. *See* Oceanic research
  Skylab missions, 113, 115, 117, 118
  Soyuz-Mir missions, 225, 227, 228, 230, 232, 242, 256, 260, 267, 275, 308–09
  Soyuz missions, 73, 89
  Soyuz-Salyut missions, 117, 120, 124, 132, 144, 145, 146, 155, 163, 171, 172, 173, 177, 194, 200, 209
  Spot satellites, 221, 330
  STS missions, 181–82, 198, 204–05, 208–09, 234, 238, 257, 261, 275, 285, 288, *288*, 290, 322
Earthrise photo, 75. *See also* Earth, images of
EASE construction method, 215
East Europe, appendix listing for, 367. *See also individual nations*
East Germany, 145, 148, 210. *See also* Germany; West Germany
  appendix listing for, 367
  Intercosmos satellites, 84, 140, 170, 176, 238, 261
  Intersputnik consortium, 161
  Soyuz cosmonaut from, 157
Eastman Kodak, 330
Eating, 40, 42, 46, 118. *See also* Digestive system research; Food; Metabolism

Eavesdropping. *See* Surveillance
Ebola virus, 277
Echo 1, 4, 5, *10*, **10–11**, 11, 25, 30, 51
Echo 2, **31**, 51
Echocardiograph Experiment, 211
Echostar 1, **305**
EchoStar Communications, 305
Eclipses
    artificial, 133
    solar, 55, 56, 59, 64, 118, 133
ECS 1, **193**
Edwards, Joe, 329
Efficiency research, 144–45
EGP, **224**
EGRET, 253, 254
Egypt, 331
Einstein, Albert, 270
Einstein (spacecraft), **159**, 162
Eisenhower, Dwight, 10
Eisle, Don, 72
Ekran 1, **146**, 161, 282
El Niño, 311, 327
Elected officials in space, 207, 216
Electrical build-up on spacecraft, 11, 42, 52, 154, 162, 185, 307, 316. *See also* Wake, from shuttles
Electronic Ocean Reconnaissance Satellites. *See* EORSATs
Electrophoresis
    Apollo missions, 93, 134
    Soyuz-Mir missions, 226, 228, 232
    Soyuz-Salyut missions, 195
    STS missions, 187–88, 193, 194, 196, 204, 207, 215, 230–31, 269, 286
Elektro, **289**
Elektro 1, 135
Elektron 1, **31**
Elektron 2, **31**, 51, 61
Elektron 3, **32**
Elektron 4, **32**, 51
Embratel, 206, 287
Emergency jetpack, 287–88, 325. *See also* Maneuvering units
Enceladus, 175, 179, *180*
Endeavour, flights of, 263–64, 268–69, 271–72, 274, 278–80, 285, 288, 293, 298–99, 305–06, 310–11, 329, 335–36
Energetic Gamma-Ray Experiment Telescope. *See* EGRET
Engineering Test Satellite. *See* ETS
England. *See* United Kingdom
England, Anthony, 211
Engle, Joe, 181, 212
Enos (chimpanzee), 17–18
Enterprise, Flight 4, **149**
Environmental research, appendix listings for, 353–54. *See also* Atmospheric research; Earth resources research
Environmental Research Satellite. *See* ERS
Environmental Science and Services Administration. *See* ESSA
Eole, 91, **99**
EORSATs, 68, 88, 127–28
Equator S, **327**
ERBS, 204
Eros, 306, 319, 337, *337*
ERS 1, **257**, 263, 295
ERS 2, **294–95**
ERS 16, **51**, 84
ERS 18, **60**

ESA (European Space Agency), 70, 100, 134, 138, 147, 150, 152, 154
    Ariane 1 rocket, 170
    Ariane 5 rocket, 326
    Challenger accident and, 219
    EuroMir project, 289
    international agreement on weather satellites, 135
    NASA and, 156
    Spacelab, 197
ESA-GEOS 1, **147**, 156
ESA-GEOS 2, 153, **156**, 157
ESRO 4, **109**
ESRO (European Space Research Organization), 70, 72, 74, 83, 103, 104, 109
ESSA 1, 34, **47–48**, 59, 91
Ethiopia, 215
ETS 2, **147**
ETS 3, **189**
ETS 4, **176**
ETS 5, **226**
ETS 7, **327**
EurasSpace, 333
EURECA, 235, **267**, 275
EuroMir, 289, 298
Europa, xv, 119–20, 127, 163, *164*, *165*, *166*, *167*, 168, *303*, 304–05
Europe, appendix listing for, 367–68. *See also* ESA; ESRO; *and individual nations*
European Communications Satellite. *See* ECS
European Remote-sensing Satellite. *See* ERS
European Retrievable Carrier. *See* EURECA
European Space Agency. *See* ESA
European Space Research Organization. *See* ESRO
Eutelsat 1, **193**, 203
Eutelsat 2 F1, **250**
Eutelsat consortium, 155, 193, 250, 294
EUVE, **264–65**
EVA (extra-vehicular activity)
    on Moon, 80–81, 85–86, 93, 96–98, 105, 109–11
    in space. *See* Space walks
Evans, Ron, 109, 111, *111*
Ewald, Reinhold, 314, 315, 318
Exercise
    Gemini missions, 42
    Skylab missions, 114, 116, 117, 118
    Soyuz-Mir missions, 226, 227, 228, 232, 242, 260
    Soyuz missions, 89, 95
    Soyuz-Salyut missions, 128, 132, 144, 151, 155, 156, 163, 187, 199, 256
    STS-Mir missions, 314
    STS missions, 190, 256, 278
EXOS A, **153**
EXOS B, **157**
EXOS C, **200**
EXOS D, **233**
Exosat, **193**
Exosphere, 79
Experimental Geophysical Payload. *See* EGP
Explorer 1, **4**, 5, 6
Explorer 2, 4
Explorer 3, **4**, 5, 6

Explorer 4, **4**, 5, 6
Explorer 6, **6**, 7
Explorer 7, **7**, 8
Explorer 8, **11**, 33, 42
Explorer 9, **13**, 30, 36
Explorer 10, **14**, 17, 22
Explorer 11, **15**, 19
Explorer 12, **17**, 23, 32
Explorer 13, 23
Explorer 14, 17, **23**
Explorer 15, **24**
Explorer 16, 6, **25**, 35
Explorer 17, **25**, 49
Explorer 18, **29–30**
Explorer 19, **30**, 35
Explorer 20, **33**
Explorer 22, **35**
Explorer 23, **35**, 38
Explorer 24, **36**
Explorer 25, **36**
Explorer 26, **37**
Explorer 27, 35, **41**, 45
Explorer 29, **44–45**
Explorer 30, **45**
Explorer 31, **45**
Explorer 32, **49**
Explorer 33, **52**
Explorer 35, **62**
Explorer 36, **68–69**
Explorer 39, **71**, 94
Explorer 42, **91–92**
Explorer 45, **102**
Explorer 46, **107**
Explorer 48, **108–09**
Explorer 51, **120**, 122
Explorer 52, **124**
Explorer 53, **131**
Explorer 54, **134–35**
Explorer 55, **137**
Express 1, **289**
Express RV, **291**
Extrasensory perception, 93
Extra-vehicular activity. *See* EVA; Space walks
Extreme Ultra-Violet Explorer. *See* EUVE
Eye cells, 11. *See also* Vision research
Eyesat 1, **277**
Eyharts, Leopold, 329

Fabian, John, 194, 210
Faculae, 267
FAISAT, **291**
Faith 7, **26**, 42
Fanhui Shu Weixing 12, **250**
Far Ultraviolet Spectroscopic Explorer. *See* Fuse
Faris, Mohammed, 225, 226
Farkas, Bertalan, 172
FASat-Alfa, 297
FASat-Bravo, 297, **332**
FAST, **312**
Fast Auroral Snapshot Explorer. *See* FAST
Fast On-orbit Recording of Transient Events. *See* FORTE
Fatalities
    dogs, 3, 12
    frogs, 90
    humans, 57, 60, 95–96, 97, 218–19
    monkeys, 79, 314
Favier, Jean-Jacques, 311
Fax machine, 234
Fecal matter and defecation, 311

Index

Females. *See* Humans, female
Feng Yun 1A, **230**
Feng Yun 2B, **318**
Feng Yun 3, **339**
Feoktistov, Konstantin, 35
Ferret 1, **21**, 58, 71, 94
Fettman, Martin, 277, 278
Filipchenko, Anatoli, 84, 126
Final Analysis, Inc., 291
Finland, 340
Fire research, 251, 265, 269, 272, 307, 309, 320
"Fireflies," 19, 21, 35
Fires
 on launchpads, 57, 196
 prevention of, 57, 133
 in space, 314–15
Fish, 137, 144, 198, 227, 237, 257, 274, 286, 287, 311, 331
Fisher, Anna, 205
Fisher, William, 212
Flade, Klaus-Dietrich, 263, 282
Flame research. *See* Fire research
Fleet Satellite Communication. *See* FLTSATCOM
Flies, 11, 63, 118, 128, 132, 137, 185, 211, 237
Floating platform launches, 60, 91, 94, 102, 108, 121, 131, 228, 339, 342
Flour beetles, 63
Flowers, 171, 172, 176, 335. *See also* Plant research
FLTSATCOM 1, 7, 139, **153**, 190, 203, 276
Fluids, body. *See* Body fluids
Flying, by insects, 185
Flywheels, 83, 224
Foale, Michael, 263, 273, 292, 318, 321, 325, 342
Food, 40, 42, 46, 120, 132, 156, 292, 314. *See also* Digestive system research; Eating; Metabolism
Ford Aerospace and Communications, 185, 206
FORTE, **323**
Foton 1, **208**
Foton 3, 244
Foton 5, **234**
Foton 6, **244**
France
 appendix listing for, 368
 Bion satellites, 148, 314
 Eutelsat consortium, 155
 Intercosmos satellite, 284
 international agreements, 188, 312
 Prognoz satellite, 104
 satellite service to, 232
 Soyuz crew members from, 187, 231, 266, 275, 312, 329, 338
 STS crew members from, 210, 211, 290, 311, 318, 325, 340, 342
 STS research, 263, 286
 Venera program, 161
France 1, **46**
Freedom 7, **15**, *15*
Freja, **269**
French Guiana
 launches from, 58, 88, 156, 170
 satellite service to, 203
Friction tests, 60, 70. *See also* Materials research
Friendship 7, **18–19**, 23

Frimout, Dirk, 263
Frogs, 63, 90, 128, 252, 269. *See also* Tadpoles
Fruit flies, 63, 118, 128, 132, 137, 237
FSW-14, 269
Fuel cells, 44, 181
Fuel pump designs, 144
Fuji 1, **224**
Fullerton, Gordon, 149, 184, 211
Fungi, 11, 127
Furnaces for materials research
 Soyuz-Mir missions, 219, 224, 225, 246
 Soyuz-Salyut missions, 151, 154, 155, 163, 172, 187
 Space Flyer Unit mission, 294
 STS missions, 197, 199, 211, 216, 230, 241, 262, 269, 307, 311, 320, 327
Furrer, Reinhard, 214
Fuse, **340**

Gaffney, Francis, 256
Gagarin, Yuri, 9, 13, *14*, 14–15, 17
Galaxies, *xvii, 279, 301. See also* Milky Way Galaxy
 1957 knowledge about, xvii–xviii
 clusters and superclusters, 91–92
 distance of, 235–36
 gamma ray research, 69, 108–09, 254. *See also* Gamma radiation
 Hubble research, 279, 280
 infrared research, 191–92, 301. *See also* Infrared radiation
 radio astronomy, 195, 316
 radio noise, 31, 61, 91
 Seyfert, 92, 134, 153, 159
 ultraviolet research, 52, 153, 192, 251, 265, 277, 340. *See also* Ultraviolet radiation
 x-ray research, 91–92, 125–26, 159, 211, 224, 253, 254, 273. *See also* X-ray radiation
Galaxy 1, **195**
Galaxy 7, **270**
Galaxy 11, **343**
Galilean satellites. *See* Callisto; Europa; Ganymede; Io
Galileo, **238**
Galileo, Gaspra Flyby, **260**, *260*, 337
Galileo, Ida Flyby, **276**, *276*, 337
Galileo, Jupiter Orbit, *303*, **303–05**, *304*
Galileo Galilei
 gravity experiment, 97
 telescopic discoveries, xv
Gals 1, **282**, 335
Gamma radiation, 44, 94, 163, 186, 284, 317
 beyond solar system, 15, 18, 19, 69, 79, 108–09, 134, 138, 139, 147, 149, 157, 169, 201, 224, 240, 247, 331
 bursts, 69, 79, 108–09, 138, 139, 149, 157, 224, 240, 253, 254, *255*, 277, 290, 309–10, 316, 319–20, 331
 Compton Observatory, 253–54
 soft gamma-ray repeaters, 273, 277, 305
 solar system, Earth, 29, 37, 43, 58, 60, 63, 96, 108
 solar system, Moon, 54–55, 98, 123

 solar system, Sun. *See* Solar research, on solar radiation
 solar system, Venus, 106, 136, 161
Gamma (spacecraft), **247**
Ganymede, xv, 119–20, 127, 163, *164, 165, 167*, 168, 304, *304*
Garbage, 152. *See also* Debris; Waste management
Gardner, Dale, 195, 205
Gardner, Guy, 232, 251
Garlic plants, 171, 172
Garn, Jake, 207
Garneau, Marc, 204, 310
Garriott, Owen, 116–17, 197
Gaspra, 260, *260*, 319
Gates, Bill, 330
Gazprom, 341
GE 1, **312–13**
GE American Communications, 312–13, 314
GE Satcom, **216**
Gemar, Charles, 251, 259, 284
Gemini 2, viii, **37**, 55
Gemini 3, **40**, 120
Gemini 4, vii, 39, 40, *41*, **41–42**
Gemini 5, **43–44**, 95
Gemini 6, **46**
Gemini 7, **46**, *46*, 89
Gemini 8, **48**, *48*, **48–49**
Gemini 8 Agena Target, **48**, *48*, 52
Gemini 9, **50–51**, 52
Gemini 9 ATDA, **50**, *50*
Gemini 10, 48, 52
Gemini 10 Agena Target, 52
Gemini 11, 40, **54**
Gemini 11 Agena Target, **54**
Gemini 12, **56**, 74, 202
Gemini 12 Agena Target, **56**
Genentech, Inc., 250
General Electric, 154, 190, 212, 215, 216, 251, 256, 265
General Telephone and Electronics (GTE), 202, 209, 212
Genesis Rock, 97, 111
Genetics, 13, 132, 195, 273–74. *See also* Chromex facility
Geo-IK, **291**
Geodetic Earth Orbiting Satellite. *See* Geos
Geodetic/navigational research, 4, 9, 10, 20, 24, 30–31, 35, 39, 41, 44–45, 51, 58, 69, 131, 140, 206–07, 224, 233, 234, 241, 270, 277, 291, 293, 316, 333. *See also* Earth resources research; Navigation
 appendix listings for, 357–58
Geodetic Satellite. *See* Geosat
Geodynamics Experimental Ocean Satellite. *See* Geos
Geological research, xviii, 28, 54, 79, 106–07, 170, 182, 189. *See also* Earth resources research; *and other planets and moons*
GEOS 1, **44–45**
Geos 2, **68–69**
Geos 3, 123, **131**
Geosat, **206–07**, *207*
Geosat Follow-On, **329**
Geostationary Operational Environmental Satellite. *See* GOES
Geosynchronous orbit
 first communications satellite in, 29

392

first communications stationary satellite in, 33
first non-communications satellite in, 72
versus low-Earth orbit, 258
Geotail, **266**, 290, 297
German Space Agency, 286
Germany. *See also* East Germany; West Germany
  appendix listing for, 368
  equipment and experiments from, 194, 269, 286, 293
  international agreement, 315
  Soyuz crew members from, 263, 282, 289, 297, 314
  STS crew members from, 197, 214, 261, 274
Gernhardt, Michael, 298, 299, 317, 320
Getaway specials, 187, 191, 193, 194, 196, 199, 208, 211, 215, 216, 256, 262, 263, 269, 275, 285, 288, 327
GFZ 1, 293
GGSE 1, **30**, 108
GGSE 2, **39**
GGSE 3, 30, **39**
GGTS 1, **51**, 62
Ghana, 291
Gibson, Ed, 118
Gibson, Robert, 198, 216, 232, 268, 295
Gidzenko, Yuri, 297, 298, 300, 301, 306
Ginga, **224**, 240
Giotto, Halley's Comet Flyby, **222–23**
Glenn, John, 17, 18–19, 20, 23, 35, 334, 335
Global Low Orbiting Message Relay. *See* GLOMR
Global positioning system. *See* GPS
Global warming, 295
Globalstar, 318
Globalstar 1 through Globalstar 4, **330**
Globular clusters, 322
GLOMR, **208**, **214**
Glonass, 171, 189–90. *See also* GPS
Glow, on shuttle surfaces, 184–85, 255, 292
Godwin, Linda, 253, 285, 308, 325
GOES 1, **135**, *135*, 147, 150, 185, 188
GOES satellites, 57, 66, 123
Goldfish, 286
Golf, on the Moon, 93
Gonets, 153
Gonets-D1 1 through Gonets-D1 3, **316**
Gorbatko, Viktor, 84, 172
Gordon, Richard, 54, 85, 87
Gorie, Dominic, 331
Gorizont 1, 146, **161**, 282, 289, 322
GOSAMR, 262
GPS (global positioning system), xviii, 9, 61, 69, 83, 124, 147, 153, 171, 233, 285, 305, 311, 334, 339. *See also* Glonass
Grabe, Ron, 214, 234, 261, 275
Granat, **240**, 246
Gravity
  artificial, 89, 148, 286–87
  Earth, 22, 108, 129, 157, 301
  Jupiter, 103, 104, 119, 304
  Lagrangian libration points, 157, 301, 322
  Mathilde, 319
  Moon, 4, 49, 53, 63, 70, 80, 97, 98, 99, 109, 145, 242
  Sun, 4, 157, 301
  Venus, 248
Gravity assists, 103, 118, 120, 129, 210, 238, 262, 306, 319, 326, 328, 332
Gravity Gradient Stabilization Experiment. *See* GGSE
Gravity Gradient Test Satellite. *See* GGTS
Gravity readaptation. *See* Readaptation to gravity
Great Britain. *See* United Kingdom
Great Nebula in Orion, *xviii*
Greb, 16
Grechko, Georgi, 128, 151, 152, 209, 213
Greenhouse effect, 65, 184
Greenhouses, 265, 275, 283, 288, 292, 299, 310–11, 327, 329, 335. *See also* Plant research
Greenland, 189
Gregory, Fred, 208, 239, 260, 261
Gregory, William, 293
Gridsphere 1, **98**
Gridsphere 2, **98**
Griggs, David, 207
Grissom, Gus, 16, 18, 40, 57, 72
GRS, **28**
Grunsfeld, John, 293, 314, 342
GStar 1, 202, **209**
GTE Sprint, 202, 209, 212
Gubarev, Alexei, 128, 154
Guidoni, Umberto, 307
Guppies, 137, 144, 198, 286
Gurragcha, Jugderdemidiyn, 177
Gurwin Techsat 1B, **333**
Gutierrez, Sidney, 256, 285
Gypsy moths, 216, 288
Gyroscopes, 83, 113, 122, 152, 225, 279, 342

H1 rocket, 224, 226, 228
H2 rocket, 283, 284
Hadfield, Chris, 300
Hagoromo, **242**
Haignere, Jean-Pierre, 275, 312, 338
Haise, Fred, 89, 149
Hakucho, **162**
Halca, **316**
Halley's Comet, 153, 206, 209, 210, 218, 221–23, *223*
Halsell, James, 286, 300, 317, 320
Ham (chimpanzee), 12
Ham radio. *See* Amateur radio communications
Hammond, Blaine, 255, 287
Harbaugh, Gregory, 255, 271, 272, 295, 315
Harris, Bernard, 274, 292
Hart, Terry, 201
Hartsfield, Henry, 187, 203, 214
Hatches, 16, 57, 242, 248, 252, 318, 329
Hauck, Fred, 193, 205, 230
Hawkeye, **124**
Hawley, Steven, 203, 216, 244, 315, 340
HBO (Home Box Office), 123, 137
HCMM, **154**
Healthsat 2, **277**
HEAO 1, *148*, **148–49**, 153, 157, 159, 162
HEAO 2, **159**
HEAO 3, **169–70**
Heart research. *See* Cardiovascular research

Heat Capacity Mapping Mission. *See* HCMM
Heat detection systems, 9. *See also* Missile detection systems
Heat flow systems, 97, 105, 109, 185, 191, 255, 269, 299
Heat shields, 6, 9, 12, 18, 55, 58, 76, 194, 326
Heavy Cosmos modules, 147, 150, 177, 186, 192, 213
Height, changes in, 200, 262, 270, 287. *See also* Posture
Heliopause, 176, 317
Helios 1, **127**
Helios 1A, **296**
Helios 2, **138**, 139, 157, 273
Heliosphere, 120, 169, 299, 331
Helms, Susan, 271, 272, 287, 311
Henize, Karl, 211
Hennen, Tom, 261
Henricks, Terrence, 260, 274, 296, 311
HEOS 1, **74**, 103, 150
HEOS 2, 84, **103**
Hercules X-1, 99, 132, 159
Hermaszewski, Miroslav, 156
Hermes, 123, **138–39**, 155, 173, 190
HGS 1, **328**
Hieb, Richard, 255, 263, 264, *264*, 286, 287
High Energy Astronomical Observatory. *See* HEAO
Highly Advanced Laboratory for Communications and Astronomy. *See* Halca
Highly Eccentric Orbit Satellite. *See* HEOS
HiLat, **195**
Hilmers, David, 214, 230, 243, 261
Himawari 1, **147**, 150, 188
Hinotori, **176**
Hipparcos, **235–36**
Hire, Kathryn, 331
Hispasat 1A, **268**
Hispasat SA, 268
Hitch Hiker 1, **28**
Hiten, **242**
Hoffman, Jeffrey, 207, 251, 267, 278, *278*, 279, 307
Holland. *See* Netherlands
Holograms, 177
Home Box Office (HBO), 123, 137
Honeybees, 185
Hong Kong
  appendix listing for, 368
  satellite service to, 244, 287
Hormone production, 173, 196, 228, 233, 331. *See also* Biological materials research
Hornets, 269
Horowitz, Scott, 307, 315
Hot Bird 1, **294**
Houseflies, 185
HST. *See* Hubble Space Telescope
Hubble, Edwin, xvii
Hubble Space Telescope, 146, **244–45**, **279–80**, **315–16**, 342
  image of, *278*
  images from, *245*, *279*, *315*
  repairs and upgrades, 245, 277, 279, 315, 342
  research, 301, 310, 315–16, 320, 331

Index

Hughes Aircraft, 108, 189, 190, 287
Hughes Communications Inc., 145, 173, 182, 195, 203, 205, 207, 210, 212, 237, 241, 268, 270, 281, 294, 336, 343
Hughes-Fulford, Millie, 256
Hughes Space and Communications, 244, 281, 291, 306, 317, 328, 335
Human biology and medicine. *See also* Biological materials research
   acceleration effects, 215
   artificially grown cells, 11, 12
   atmosphere of spacecraft, 57, 127, 133, 144, 146, 177, 223, 340
   back pain, 116, 200, 262, 270
   balance/inner ear research, 17, 44, 90, 116, 198, 200, 211, 256, 278, 289, 311, 331, 335
   body fluid redistribution, 95, 114, 120, 157, 163, 196, 199, 274, 278, 289
   bone research, 42, 46, 115, 117, 118, 128, 163, 173, 200, 256, 263, 270, 289, 298, 306, 335
   brain research, 27, 46, 120, 124, 128, 144, 173, 187
   cardiovascular research. *See* Cardiovascular research
   defecation, 42, 311
   eating, 40, 42, 46, 118
   exercise. *See* Exercise
   fatalities, 57, 60, 95–96, 97, 218–19
   height changes, 200, 262, 270, 287
   illnesses in space, 72, 213, 224, 226, 321, 340. *See also* Space sickness
   immune research, 173, 335
   mental difficulties, 151
   metabolism, 173, 256, 278, 335
   mineral balance, 173
   muscle research. *See* Muscle research
   neurological research, 257, 331
   readaptation. *See* Readaptation to gravity
   respiratory research, 18, 22, 27, 46, 95, 124, 128, 132, 144, 154, 257, 311
   sleep research, 17, 26, 46, 54, 187, 311, 331
   space sickness. *See* Space sickness
   urination, 18, 42, 256, 311, 335
   vision research, 11, 26, 44, 187
   water balance, 173
   weight changes, 128, 132, 151, 155, 189, 232
   weight measurements, 114, 144
Humans
   changing self-concept of, xix
   female, first in space, 27, 189, 194
   female, first to command a space mission, 340
   female, first to walk in space, 202
   first born of two cosmonauts, 27
   first husband-wife team on same spacecraft, 268
   first to land on another world, 80–81
   first to reach and orbit another world, 75
   first to sleep in space, 17
   first to witness earthrise, 75
   male, first in space, 14–15, 16, 18–19
   male, first to walk in space, 39, *39*
   oldest in space, 195, 313, 334

Hungary, 104, 148, 210
   appendix listing for, 368
   Intercosmos satellites, 84, 140, 170, 176, 238, 261
   Intersputnik consortium, 161
   Soyuz cosmonaut from, 172
Hurricanes. *See also* Weather satellites
   Earth, 16, 28, 117
   Venus, 120
Husband, Rick, 340
Huygens, 326
Hydrazine thrusters, 131
Hyperion, 179, *180*

Iapetus, xv, 179, *180*
IBM, 173
Ice
   Earth's seas, 162, 196, 225, 226, 253, 257, 294, 297, 300
   Jupiter's moons, 120, 304
   Moon, 282, 328
   near shuttle's wing, 204
ICE (spacecraft), 157, **213**
Ida, 276, *276*, 319
IDCSP 1 through 7, **51,** 62, 76, 87, 89, 100
Ikonos, **341**
Illness in space. *See also* Fatalities
   colds, 72
   eye irritation, 340
   headaches, 340
   heart problems, 224, 226, 321
   inflamation and fever, 213
   nausea. *See* Space sickness
IMAX cameras, 202, 204, 205, 215, 234, 238, 241, 245, 262, 267, 277, 279, 292, 296, 300, 313, 336
IMEWS 1, **95**
IMEWS 6, **144**
IML-1, 261
IML-2, 286
Immune system research
   humans, 173, 335
   rats, 227, 250, 283, 290
Imp 1. *See* Imp A
Imp 8, 30
Imp A, 8, 17, 22, **29–30,** 32, 36, 46
Imp D, **52**
Imp E, 30, **62**
Imp satellites, 84, 99, 102, 103, 104, 109, 124, 150
India
   appendix listing for, 368
   satellite service to, 123, 135, 188
   Soyuz crew member from, 200
India Remote Sensing satellites. *See* IRS
Indian Space Research Organization, 178, 185, 266, 339
Indonesia
   appendix listing for, 368
   images of, 182, 238
Inert gases, on Venus versus Earth, 184
Inflationary big bang theory, 239. *See also* Big bang theory
Informator 1, **253,** 316
InfraRed Astronomical Satellite. *See* IRAS
Infrared radiation
   beyond solar system, 12, 191–92, 238–39, *239*, 294, 301, *301*, 315, 336–37, 338–39
   new technology, 322
   in vicinity of Earth, 9, 18, 21, 34, 37, 49, 309, 310, 330, 332, 342

Infrared Space Observatory. *See* ISO
Infrared Spectral Imaging Radiometer. *See* ISIR
Infrared Telescope in Space. *See* IRTS
Initial Defense Communications Satellite Program. *See* IDCSP
Injun 1, **16,** 21, 22, 24, 25
Injun 2, 24
Injun 3, **24,** 28, 32
Injun 4, **36**
Injun 5, **71,** 124
Injuries, healing of, 237. *See also* Wounds, healing of
Inmarsat, 182, 251, 308
Inmarsat 2 F1, **251**
Inmarsat 3 F1, **308**
Inmarsat Organization, 139
Inner ear. *See* Balance/inner ear research
Insat 1A, 135, **185**
Insat 1B, 185, **196**
Insat 1C, 185
Insat 1D, 185
Insat 2, 188
Insat 2A, **266**
Insects, 7, 11, 63, 105, 128, 132, 185, 194, 216, 237, 269, 286, 288, 297, 331, 341
   Bion satellites, 118, 137, 211, 227, 237, 271
Inspector, **328**
Instituto Universitario Aeronautico de Cordoba, 312
Intasat 1, **126**
Integrated Missile Early Warning Satellite. *See* IMEWS
Intelsat 1, **40**
Intelsat 2A, **55,** 57
Intelsat 2B, viii, **57,** 92
Intelsat 2C, **58,** 64
Intelsat 2D, **64**
Intelsat 3B, 64, **75,** 92
Intelsat 4 F2, **92,** 173
Intelsat 4A F1, **134,** 173
Intelsat 5 F2, **173–74**
Intelsat 5A F10, **207**
Intelsat 6 F3, **264,** *264*
Intelsat 6A F2, **238**
Intelsat 7 F1, **278**
Intelsat 801, **316**
Intelsat consortium, 40, 55, 173, 193, 207, 229, 238, 251, 264, 265, 278, 291, 316
Intelsat K, **265**
Interball 1, **297,** 312
Interball 2, **312**
Interceptor missions, 72, 73, 90. *See also* Anti-satellite systems
Intercosmos 1, **84,** 99, 102, 103, 109, 150
Intercosmos 6, **104**
Intercosmos 15, **140**
Intercosmos 18, **158**
Intercosmos 20, **170**
Intercosmos 21, **176**
Intercosmos 23, **208**
Intercosmos 24, **238**
Intercosmos 25, **261**
Intercosmos 26, **284,** 297
International Cometary Explorer. *See* ICE
International Maritime Satellite Organization (Inmarsat), 139
International Microgravity Laboratory, 261, 286. *See also* Spacelab

International Radiation Investigation Satellite. *See* IRIS
International Satellite for Ionospheric Studies. *See* ISIS
International satellites and missions, appendix listing for, 368–69. *See also individual nations*
International Space Station, vii, xix, 40
  modules, 322, 335, 336, *336*
  supplies, 340
International Space Station, tests for construction techniques, 306
  docking techniques, 327
  equipment, 299, 328, 329, 332
International Sun-Earth Explorers. *See* ISEE
International Telecommunications Satellite Consortium, 40. *See also* Intelsat
International Ultraviolet Explorer. *See* IUE
Internet, images for sale on, 330
Interplanetary Monitoring Platform. *See* Imp
Interplanetary space
  1957 knowledge about, xvii
  energy fluxes, 5, 6, 7, 13, 14, 23, 36, 43, 47. *See also* Solar research, on solar radiation; Solar research, on solar wind
  particles and dust, 8, 23, 24, 25, 36, 262. *See also* Cosmic dust; Cosmic particles
Intersputnik consortium, 161
Interstage Adapter Satellite. *See* ISAS
Interstellar cosmic dust, particles, and rays. *See* Cosmic dust; Cosmic particles; Cosmic radiation
Io, xv, 127, *163*, *164*, 165
  ionosphere, 119
  magnetic field, 252, 322
  magnetosphere, 166–67
  mass, 119–20
  volcanic activity, 163–64, 168, 262, 304
Ion engines, 88, 127, 280, 282, 289, 322–23, 333, 334, 338, 343
Ion Release Module. *See* IRM
Ionized particles on spacecraft. *See* Electrical build-up on spacecraft
Ionosonde, **224**
Ionosphere/magnetosphere research. *See also* Magnetic fields
  appendix listings for, 358–59
  Earth. *See* Earth, ionosphere of; Earth, magnetosphere of
  Gaspra, 260
  Io, 119, 166–67
  Jupiter, 116, 119, 127, 164–65, 168, 262
  Mercury, 129
  Saturn, 116, 168, 176
  Sun, 8, 112, 176. *See also* Solar research
  Venus, 65
Ionospheric Sounding Satellite. *See* ISS
Iran, 294, 306
Iraq, 161, 294
IRAS, *191*, **191–92**, 246, 301, 322
Iridium, 317–18, 328, 330
Iridium 4 through 8, **317–18**
IRIS, **70**, 72
IRM, **203**
IRS 1A, **228**

IRS-P4 Oceansat, **339**
IRTS, 294
Irwin, Jim, 96–98, *97*
ISAS, **282**
ISEE 1, **150**, 153, 156, 157, 171
ISEE 2, **150**, 153, 156, 157, 171
ISEE 3, **157**, 213
ISIR, 322
ISIS 1, **76**
ISIS 2, **76**
Iskra 2, 186, **187**
Iskra 3, 186, **191**
Iskra satellites, 159
ISO, **301**, *301*
Isoelectric Focusing Experiment, 199, 230–31
Israel, appendix listing for, 369
Israel Aircraft Industries, 310
ISS 1, **139**
ISS 2, **153**
Italsat 1, **252–53**
Italy
  appendix listing for, 369
  Eutelsat consortium, 155
  satellite service to, 232
  satellites launched for U.S., 91–92, 102, 108–09, 131
  STS crew members from, 267, 307
Itamsat, **277**
ITOS 1, **88**, 91
IUE, 122, **152–53**, 222
Ivanchenkov, Alexander, 155, 187
Ivanov, Georgi, 165–66
Ivins, Marsha, 241, 267, 284, 314

Jahn, Sigmund, 157
Japan
  appendix listing for, 369
  images of, 245
  international agreements, 135
  Soyuz crew member-journalist from, 248, 252
  Soyuz experiments from, 240
  STS crew members from, 268, 286, 305, 327, 334
  STS experiments and equipment from, 263, 286, 288, 340
Japan Communications Satellite, 233
Japanese American Satellite. *See* JAS
Japanese Earth Resource Satellite. *See* JERS
Japanese Engineering Test Satellites. *See* JETS
Jarvis, Greg, 218, *218*
JAS 1, **224**
JCSat 1, **233**
Jellyfish, 257, 286, 287
Jemison, Mae, 268
Jernigan, Tamara, 256, 270, 293, 313, 340
JERS 1, **262–63**
Jetpacks. *See* Maneuvering units
JETS 1, **134**
Jett, Brent, 305, 314
Jikiken, 153, **157**
Jindai, **224**
Johnson, Lyndon B., 21
Joint tissue, 11–12. *See also* Bone research; Muscle research
Jones, Thomas, 285, 288, 313
Jordan, 189
Journalist in space, 248, 252
Jumping, by frogs, 252

Jumpseat 1, 21, **94**
Jupiter, *xiv*, *163*
  1957 knowledge about, xiv–xv
  atmosphere, xiv–xv, 119, 127, 164, 166–68, 238, 301, 303, 304
  aurora, 160, 164, 166, 280
  bands belts, 119
  composition, 119
  gravitational field, 304
  gravity well, 103, 104, 119
  lightning, 164, 303
  magnetic field, 119, 168, 252, 262, 298
  magnetosphere, 116, 119, 127, 164–65, 168, 262
  moons of, xv, 119–20, 127, 163, *163*, *164*, 165, *165*, *166*, *167*, 168, *168*, 262, 304–05
  particles to Earth from, 266
  polar regions, 164, 166
  Red Spot, xiv–xv, 119, 127, 164, 166–67, 304
  rings, 163, 164, 168, 262, 304
  ultraviolet radiation, 74, 166, 167, 298, 322
  winds, 164, 166, 303
Jupiter, spacecraft to
  flybys, 103–04, 112, 119–20, 127, 163–65, 166–68, 262
  orbiting, 238, 303–05

Kadenyuk, Leonid, 327
Kaleri, Alexander, 263, 266, 312, 313, 314, 315
Kappel, Fred, 21
Kavandi, Janet, 331–32
Kazakhstan. *See also* Soviet Union
  appendix listing for, 369
  landing in, 326
  launches from, 260, 316
  Soyuz crew members from, 260, 286
Kelly, Scott, 342
Kennedy, John F., xix, 13, 80
Kenya
  radar images of, 205
  satellites launched off coast of, 60, 91, 94, 102, 108, 121, 131, 228
Kerwin, Joe, 114–15
Keyhole satellites. *See* KH
KH satellites, 18
KH-4, 28
KH-4 9032, **19–20**, 106
KH-4A, 20, 28, 106
KH-4B, 20, 28, 106
KH-5, 19, 20, 28, 106
KH-6, 20
KH-7, 20
KH-7 1, **28**, 146
KH-7-10, **33**
KH-8, 20, 28
KH-9, 20, 106
KH-9 1, **96**, 146
KH-11, 28, 140, 243
KH-11 1, **146**
KH-12 1, **243**
Khrunov, Yevgeni, 76
Khrushchev, Nikita, xix, 35, 118
Kidney cells and functions, 134, 196, 256
KidSat, 308
Kiku 1, **134**
Kiku 2, 123, **147**
Kiku 3, **176**

Kiku 4, **189**
Kiku 5, **226**
Kiku 7, **327**
Kitsat 1, **268**
Kitsat 2, **277**
Kitsat 3, **339**
Kizim, Leonid, 173, 199–200, 202, 223
Klebesadel, Ray, 79
Klimuk, Pyotr, 120, 131, 156
Knowledge, scientific. *See* Scientific knowledge
Kolodin, Pyotr, 95
Komarov, Vladimir, 35, 59–60, *60*, 66, 72
Kompsat, **342–43**
Kondakova, Elena, 286, 289, 292, 293, 313, 318
Kopernikus 1, **234–35**
Korea. *See* North Korea; South Korea
Koreasat 1, **297**
Korolev, Sergie, *27*
Korzun, Valeri, 312, 313, 314, 315
Kovalyonok, Vladimir, 150, 155, 176
Kregel, Kevin, 296, 311, 326
Krikalev, Sergie, 231, 232, 256, 260, 263, 283, 335, 336
Kristall, 242, **246**, *246*, 248, 252, 267, 272, 281, 295, 296, 300, 308
Kubasov, Valery, 84, 95, 133, 151, 172
Kuiper, Gerard, 33
Kuiper belt, 280. *See also* Asteroids
Kurs docking system, 286, 315. *See also* Rendezvous and docking
Kvant 1, **225**, *225*, 232, *246*, 252, 295, 308, 315
Kvant 2, 237, **239–40**, *240*, 242, *246*, 248, 252, 256, 272, 275, 295, 308, 338
Kyokko, **153**, 156, 157
Kyrgyzstan. *See also* Soviet Union
  appendix listing for, 369
  STS crew member from, 329

La Niña, 327. *See also* El Niño
Laboratories. *See* International Space Station; Mir; Salyut; Skylab; Spacehab; Spacelab
  shuttles as, 185. *See also* STS
LACE, **243**
Lacrosse 1, **232**
Lageos 1, **140**, 270
Lageos 2, **270**
Lagrangian libration points, 157, 301, 322
Laika (dog), 3, 17
Landsat 1, **106–07**, *107*, 115, 135, 151, 166, 221
Landsat 4, 115, 135, **188–89**, 262
Landsat 5, 188, 189
Landsat 6, 189
Landsat 7, **339**
Laos, 161
Laptop computers, 234
Large Format Camera, 204
Large Magellanic Cloud, 125, 201, 224, 225, 232, 247, 262
Laser Geodetic Satellite. *See* Lageos
Laser Geodynamic Satellite. *See* Lageos
Lasers, 41, 46, 111, 128, 132, 140, 211, 224, 233, 234, 243, 270, 271, 277, 285, 287, 293, 310, 322
  Moon-Earth, 81, 93, 97, 112
Latex reactor experiment, 185, 199
Latitude/Longitude Locator, 241

Launch aborts, 131, 196
Launch detection. *See* Missile detection systems
Launchpad fires, 57, 196
Lava. *See* Volcanic activity
Lavochkin, 326
Lawrence, Wendy, 293, 325, 331
Lazarev, Vasili, 117, 131
Lazutkin, Alexander, 314, 315, 317, 318, 321
LCS 4, **99**
LDEF, **202**, 241
Learning Channel, The, 123, 137
Leary, Timothy, 317
LEASAT 1, **205**
LEASAT 2, **203**
LEASAT 3, **207**
LEASAT 4, **212**
LEASAT 5, **241**
LEASAT satellites, 153, 190, 238, 276
Lebedev, Valentin, 120, 171, 186
Lee, Mark, 234, 268, 287, 315
Leetsma, David, 204, 205, 235, 263
Lem. *See* Lunar module
Lenoir, William, 190
Lentil plants, 215, 261, 318
Leonov, Alexei, viii, 39, *39*, 41, 76, 95, 133
LES 1, **38**, 41
LES 2, **41**, 47, 76
LES 3, 38, **47**
LES 4, **47**, 76
LES 5, **62**, 72, 76
LES 6, **72**, 76
LES 8, **139**
LES 9, **139**
Levchenko, Anatoly, 227
Leveykin, Alexander, 224, 225, 226, 227
Liberty Bell 7, **16**, 40
Lichtenberg, Byron, 197, 263
Lidar, 287, 309
Lidar In-Space Technology Experiment. *See* LITE
Life, research on existence of
  Mars, 43, 82, 141–42
  Venus, xiii
Lifting body concept, 149, 177, 187
Lightning
  Earth, 38, 85, 188, 234, 238, 274, 294, 327
  Jupiter, 164, 303
  Saturn, 179
  Venus, 161–62, 184
Lightning strikes to spacecraft, 37, 85
Lincoln Calibration Sphere. *See* LCS
Lincoln Experimental Satellite. *See* LES
Lincoln Laboratory, MIT, 72. *See also* LES
Lind, Don, 208
Lindsey, Steven, 327, 334
Linenger, Jerry, 287, 314, 317, 318
Lineris, Greg, 317, 320
Linneham, Richard, 311, 331
Liquids research. *See also* Body fluids, redistribution; Materials research; Pumps; Refueling
  Apollo missions, 93
  Soyuz-Mir missions, 266
  STS missions, 191, 194, 198, 199, 211, 215, 216, 255, 259, 261, 265, 269, 270, 274, 284, 285, 299
  unmanned missions, 317

LITE, **287**
Lithium candles, 315, 318
Lithuania, 314
Little Joe, 7
Little Joe 3, 7
Lloyd's of London, 244
LM. *See* Lunar module
LMI 1, **341–42**
Local Bubble, 254, 264
Local Space Systems, 316
Lockheed Martin, 291, 297, 305, 338, 341
Lockheed Martin Astronautics, 333
Lofti 2A, **27**
Long Duration Exposure Facility. *See* LDEF
Long March rockets, 89, 247
Lopez-Alegria, Michael, 299
Loral Skynet, 281
Losat X, **257**
Lounge, Mike, 212, 230, 251
Lousma, Jack, 116–17, 184
Lovell, Jim, 46, 56, 75, 88–89
Low, David, 241, 258, 275
Low Altitude Satellite. *See* Losat
Low-Earth orbit, for communications satellites, 21, 258, 317–18, 328
Low-power Atmospheric Compensation Experiment. *See* LACE
Lower back pain. *See* Back pain
Lu, Edward, 318
Luch 1, 138, 214, 219, **291**
Lucid, Shannon, 210, 211, 238, 258, 277, 278, 298, 306, 308, 309, 312, 313
Luna 1, **5**, 6, 8, 45
Luna 2, **6**, 7, 13, 20, 22
Luna 3, **6–7**, 43
Luna 5, 47
Luna 7, 47
Luna 8, 47
Luna 9, 33, **47**, 49, 57
Luna 10, 49, 53, 62
Luna 11, **54**
Luna 12, **54–55**
Luna 13, **57**
Luna 14, **70**
Luna 15, **80**
Luna 16, **89**, 103
Luna 17, *90*, **90–91**
Luna 18, **99**
Luna 19, **99**
Luna 20, **103**
Luna 21, **112**
Luna 22, **123**
Luna 23, **126**
Luna 24, **145**
Lunar module (LM), 69, 77, 78, 80, 85–87, 88, 93, 94, 96–98, 105–06
Lunar Orbiter 1, **53**, *53*, 58, 61
Lunar Orbiter 2, **55**, 58, 61, 66
Lunar Orbiter 3, **58**, 61
Lunar Orbiter 4, **61**
Lunar Orbiter 5, 61, **62–63**, 66, 68
Lunar Prospector, **328–29**
Lunar research. *See* Moon
Lungs. *See* Respiratory research
Lunokhod 1, **90–91**
Luxembourg, 232
Lyakhov, Vladimir, 163, 166, 194, 230

MABES, **224**
Mabuhay Philippines Satellite Corp., 322

Index

MacLean, Steven, 270
Macsat 1, **245**
Macsat 2, **245**
Magellan, **234**
Magellan, Venus Orbit, 205, **248–50**
Magion 1, 84, **158**
Magion 2, **238**
Magion 3, **261**
Magion 4, **297**, 312
Magion 5, **312**
Magnetars, 273
Magnetic bearing flywheel experimental system. *See* MABES
Magnetic fields. *See also* Ionosphere/magnetosphere research; Magnetospheres
    asteroids, 260, 276
    Callisto, 304
    Earth. *See* Earth, magnetic field of
    Europa, 304
    gamma-ray repeaters, 273
    Ganymede, 304
    Io, 252, 322
    Jupiter, 119, 168, 252, 262, 298
    Mars, 43, 82, 229, 323
    Mercury, 129
    Moon, 5, 6, 29, 49, 62, 93, 97, 98, 106, 328
    Neptune, 237
    Saturn, 168, 176
    stars beyond solar system, 305
    Sun. *See* Solar research, on magnetic field
    Uranus, 216–17
    Venus, 25, 65, 121
Magnetopause, 14, 17, 32, 69, 290
Magnetospheres. *See also* Ionosphere/magnetosphere research; Magnetic fields
    Earth. *See* Earth, magnetosphere of
    Gaspra, 260
    Io, 166–67
    Jupiter, 116, 119, 127, 164–65, 168, 262
    Mercury, 129
    Saturn, 116, 168, 176
    Sun, 8, 112, 176
Magnum 1, **206**
Magnum 2, **239**
Magnum 3, **251**
Magsat, **170**
Maize plants, 137
Mak 2, 267
Makarov, Oleg, 117, 130, 152, 173
Malaysia
    appendix listing for, 369
    satellite service to, 145
Malenchenko, Yuri, 286, 289
Malerba, Franco, 267
Males. *See* Humans, male
Malyshev, Yuri, 172, 200
Mammals, 170. *See also individual mammals*
Manakov, Gennady, 248, 272, 275
Manarov, Musa, 227, *227*, 228, 230, 231, 232, 248, 252, 256, 281
Maneuvering units, 41–42, 50, 117, 237
    emergency, 287–88, 325
    MMU, 198, 201, 205, 240
Manned maneuvering unit (MMU), 198, 201, 205, 240
Manned missions. *See also* Apollo; Gemini; Mercury; Mir; Skylab; Soyuz; STS; Voskhod; Vostok
    appendix listings for, 359–61
    biology and medicine. *See* Human biology and medicine
    fatalities, 57, 60, 95–96, 97, 218–19
    to Moon. *See* Moon, spacecraft to
    permanent human presence in space, goal of, 219
    rendezvous and docking. *See* Rendezvous and docking
    repairs. *See* Repairs
    research. *See individual research areas*
    space walks. *See* Space walks
    suborbital flights, 15, 16, 131
    work. *See* Work in space
Manned Orbital Laboratory, 51, 55
Manned space laboratories/modules. *See also* International Space Station; Mir; Salyut; Skylab; Spacehab; Spacelab
    appendix listings for, 361
    shuttles as, 185. *See also* STS
Manned spacecraft, problems with
    air leaks, 96, 318
    attitude control system failures, 48, 59–60, 315
    automatic control system failures, 18, 26
    booster explosion/spacecraft destruction, 218
    capsule separation failures, 27, 76
    collisions, 194, 281, 318
    computer failures, 296, 321
    coolant system leak, 289
    docking failures, 72–73, 84, 94, 125, 146, 150, 166, 193, 289, 318
    engine failures, 166, 211
    environmental control system odor, 145
    fires, 57, 196, 315
    fuel cell failures, 44, 181, 317
    fuel leaks, 116, 163, 186, 194, 199–200
    hatch door balkiness, 242, 248, 252, 329
    insulation damage, 242
    landing bag sensor light malfunction, 19
    launch aborts, 131, 196
    launchpad engine shut down, 46
    launchpad fires, 57, 196
    lightning strike, 85
    oxygen generation failures, 315
    oxygen tank explosion, 88–89
    parachute failure, 60
    power failures, 289, 318, 321
    radio control failure, 186
    retro-rocket failures, 39, 166, 230
    safety time limit exceeded, 166
    sinking of capsule, 16
    solar panel damage, 113–14, 318
Mao 1, **89**
Maqsat 3, **334**
Marco Polo 1, **237**
Mare regions of the Moon, xiii. *See also* Moon, surface of
MARECS A, 139, **182**, 251
Marine Observation Satellite. *See* MOS
Mariner 1, 22
Mariner 2, 13, **22–23**, 24, 32, 36, 54, 73
Mariner 2, Venus Flyby, **25, 65**
Mariner 3, 36
Mariner 4, 24, **36**, 54, 69
Mariner 4, Mars Flyby, **42–43**, 74, 101
Mariner 5, Venus Flyby, **65**, 78, 92, 121
Mariner 6, Mars Flyby, **81**, *81*, 82, *82*, 100, 101
Mariner 7, Mars Flyby, vii, 81, **82**, *82*, *83*, 100, 101
Mariner 8, 100, 269
Mariner 9, Mars Orbit, 98, *100*, **100–02**, *101*, *102*, 141
Mariner 10, vii, **118**
Mariner 10, Mercury Flyby #3, 101, *129*, **129–30**, *130*
Mariner 10, Venus Flyby, **120–21**, *121*, 160
Marisat 1, 72, 76, **139**
Maritime European Communications Satellite. *See* MARECS
Mars 1, **24**, 36
Mars 2, 100, **102**
Mars 3, 100, **102**
Mars 4, Mars Flyby, **121**
Mars 5, Mars Orbit, **121**
Mars 6, Mars Landing, **122**
Mars 7, Mars Flyby, **122**
Mars Climate Orbiter, **337**
Mars Global Surveyor, *323*, **323–24**, *324*
Mars Observer, **269**
Mars Pathfinder, *320*, **320–21**
Mars (planet), *xiv, 320, 323, 324*
    1957 knowledge about, xiii–xiv
    atmosphere, xiii–xiv, 42–43, 74, 82, 101, 102, 121, 141–44, 229, 321, 332
    canyons, 100–01, 142
    composition of, 101, 141–44, 321
    craters, *42*, 43, *81*, 82, *82*, *83*, 100–01, 130, 142, 323
    dust storms, 100, 101, 102, 142, 144, 321, 323
    life, research on existence of, 43, 82, 141–42
    magnetic field, 43, 82, 229, 323
    moons of, xiv, 100, 229
    polar regions, 142–44
    surface, xiv, *42*, 43, *81–83*, 82, 100–01, 121, 141–44, 145, 320–21, 323–24
    ultraviolet radiation, 74
    volcanic activity, 100, 101, 130, 142, 323
    water, evidence of, 101, 142, 143, 321, 323
    winds, 142, 323
Mars (planet), spacecraft to
    on course, 332
    failures, 24, 100, 121, 122, 229, 269, 337, 338
    flybys, 36, *42*, 42–43, 81, *81*, 121
    hard impacts, 102, 122
    landings, 102, 140–44, 145, 320–21
    orbiting, 100–02, 121, 140–44, 145, 229, 323–24
Mars Polar Lander, **338**
Mascons, 55, 70, 99
Materials research
    Apollo missions, 93
    appendix listings for, 361–62
    biological. *See* Biological materials research
    crystals. *See* Crystals research
    Foton missions, 208, 234, 244
    liquids. *See* Liquids research

Index

Materials research *(continued)*
  ORS missions, 51
  OV missions, 60, 70
  Skylab missions, 115, 118
  Soyuz-Mir missions, 225–26, 227, 228, 240, 248, 260, 298
  Soyuz missions, 84
  Soyuz-Salyut missions, 144, 151, 154, 155, 156, 157, 163, 172, 173, 178, 186, 187, 202, 289
  STS missions, 185, 194, 196, 197–98, 199, 208, 211, 215, 216, 230, 231, 238, 241, 245, 250, 251, 255, 256, 258, 262, 265, 267, 269, 270, 271, 274, 275, 277, 284, 285, 286, 287, 299, 307, 311, 320, 327, 329, 332
Mathilde, 306, 318–20, *319*
Matra, 226, 234
Matra Marconi Space, 331, 333
Mattingly, Ken, 105, 187, 206
Max Planck Institute for Extraterrestrial Physics, 327
McArthur, William, 277, 300
McAuliffe, Christa, 218, *218*
McBride, Jon, 204
McCandless, Bruce, 198, *199*, 237, 244
McCaw, Craig, 330
McCulley, Michael, 238
McDivitt, Jim, 41–42, 77
McDonnell Douglas, 188, 193, 194, 196, 204, 207, 215, 231, 234, 269, 275
MCI Communications, 173, 190, 212
McMonagle, Donald, 255, 271, 290
McNair, Ronald, 198, 218, *218*, 288
Meade, Carl, 265, 287
Measat 1, **306**
Medaka fish, 286, 311
Media Citra Indostar, 326
Media-Most, 335
Medical research. *See* Biological research; Human biology and medicine
Medicines, research on. *See* Pharmaceutical materials research
Melbourne University, 88
Melnick, Bruce, 250, 263, 264
Membranes, polymer, 245, 250, 258, 275, 277. *See also* Thin-film research
Men. *See* Humans, male
Mendez, Arnaldo, 173
Merbold, Ulf, 197, 261, 289
Merck & Co., 270
Mercury 1A, **12**
Mercury 2, **12–13**
Mercury 2A, **13**
Mercury 3, **15**, *15*
Mercury 4, **16**
Mercury 5, **17–18**
Mercury 6, **18–19**
Mercury 7, **20–21**
Mercury 8, **23**
Mercury 9, **26**
Mercury capsule tests, 6, 7, 12–13, 17–18
Mercury (planet), xi, *129*, *130*
  1957 knowledge about, xi, xiii
  atmosphere, xiii
  composition of, 129
  craters, 129–30
  magnetic field, 129
  magnetosphere, 129
  seismic research, 130

  spacecraft flyby, 118, 129–30
  surface, 129–30
Merkur capsule, 146
Messerschmid, Ernst, 214
Metabolism, 173, 256, 278, 335. *See also* Digestive system research; Eating; Food
Metallurgy. *See* Materials research
Meteor 1-01, **77–78**, 89, 124, 132
Meteor 1-30, **172**
Meteor 2-01, **132**
Meteor 3-01, **214**
Meteor 3-05, **258**, 259
Meteorites, 275. *See also* Micrometeorite/ micrometeoroid research
  asteroids, 337
  comets, 223
  Earth, 68
  Moon, 53, 55, 58, 61, 64, 66, 68, 75, 86, 98, 109, 123
  origins of, 68
Meteorology. *See* Atmospheric research; Earth resources research; Weather satellites
Meteosat 1, **150–51**
Meteosat 3, **229**
Metis, 168
Mexico
  appendix listing for, 369
  images of, 118
  Starshine satellite, 340
  STS crew member from, 215
MHV, **213**
Mice, 11, 331. *See also* Rats
Microencapsulated products, 271, 297, 335
Microgravity, 241. *See also* Weightlessness
Microlab 1, **294**
Micrometeorite/micrometeoroid research. *See also* Cosmic dust; Meteorites
  1957 knowledge about, xvii
  appendix listings for, 362
  comets, 209
  Earth orbit, 4, 5, 11, 25, 31, 35, 38, 41, 43, 52, 56, 90, 100, 107, 119
  Earth orbit, Soyuz-Mir missions, 333
  Earth orbit, Soyuz-Salyut missions, 94, 154, 185, 209, 297, 308
  Earth orbit, STS missions, 185
  interplanetary space, 24
  lunar orbit, 55, 58, 61, 98, 123
  lunar surface, 66, 75, 86, 109
Microsat, **312**
Microsat 1 through Microsat 7, **258**
Microsoft, 330
Microwave radiation, background, 239
Midas 1, 9
Midas 2, **9**, 44, 106
Midas 4, **17**
Midas satellites, 13, 17, 71, 95, 112
Midcourse Space Experiment, **309**
Midori, **311–12**
MightySat 1, **336**
Miles, Judith, 117
Military Man in Space, 261
Military reconnaissance. *See* Anti-satellite systems; Surveillance
Milkweed bug, 288
Milky Way Galaxy, xvii, 52, 69, 107, 153, 169

  gamma ray research, 108–09, 134, 169, 240, 254. *See also* Gamma radiation
  infrared research, 191, 211, 239, 309. *See also* Infrared radiation
  radio astronomy, 71
  ultraviolet research, 107–08, 153, 159, 246, 277, 340. *See also* Ultraviolet radiation
  x-ray research, 91, 104, 107, 131, 148, 159, 211, 240, 246, *247*, 253. *See also* X-ray radiation
Milstar 1, **284**
Mimas, 175, *175*, 179
Mineral balance, 173
Miniature Homing Vehicle, 213
Miniature Sensor Technology Integration. *See* MSTI
Minisat 1, **317**
Mir, 40, 122, *219*, **219–21**, *225*, *240*, *295*, *300*, *309*, *325*
  collisions with, 281, 318
  collisions with, near misses, 306, 315, 321
  communications satellites for, 214, 219
  compared to Skylab, 112
  modules added to, 225, *225*, 239–40, *240*, 246, *246*, 295, *295*, 308–09, *309*
  Soyuz missions to, 223, 224–26, 227, 228, 230, 231–32, 237, 242, 248, 252, 256, 260, 263, 266–67, 272, 281–82, 286, 289, 293–94, 297–98, 306, 312, 321, 329, 333, 338
  STS missions to, 292, 295–96, 300–301, 308, 313, 314, 318, 325, 329, 331–32
Miranda, **122**, *217*, 218
Mirka, **326**
Mirror experiments, 243, 267, 272, 333, 340
Missile Detection and Surveillance. *See* Midas
Missile detection systems, 9, 21, 71, 94, 112, 243, 257, 271, 285, 309, 310. *See also* Anti-satellite systems; Nuclear explosion detection systems
Mission to Mir (film), 292, 296, 300, 313
MIT Lincoln Laboratory, 72. *See also* LES
Mitchell, Edgar, 93
MLR, 185, 199
MMU (manned maneuvering unit), 198, 201, 205, 240
Mobile Satellite 1. *See* MSAT 1
Mohmand, Abdul Ahad, 230
Mohri, Mamoru, 268
Molds, 7, 241, 285, 286, 298
Molecular clouds, 191, 336–37, 340
Molniya 1-01, 33, **40–41**, 94, 161
Molniya 1-02, **44**
Molniya 1-03, **49**
Molniya 1S, **125**, 138
Molniya 2, 41
Molniya 2-01, **102**, 146, 161
Molniya 3-01, 41, **126**, 146, 161
Momo 1, **225**
Mongolia, 51, 161, 244
  appendix listing for, 369
  Soyuz crew member from, 177

398

Monkeys, 7, 79, 208
    Bion satellites, 198, 211, 227, 237, 271, 314
Moon, 282
    1957 knowledge about, xiii
    age of, 81, 86, 97, 98, 109
    atmosphere, 85, 109, 290
    colors, 59, 68, 78, 81, 109
    composition of, 54, 57, 59, 62–63, 64, 66, 68, 86, 89, 90–91, 97–98, 103, 105, 106, 109–11, 123, 145, 328
    cosmic radiation, 86, 109, 123
    density, 57, 59
    distance to, 81, 97
    far side, atlas of, 90
    far side, first photographs of, 6–7, 7
    Galileo's experiment, 97
    gamma rays, 54–55, 98, 123
    Genesis Rock, 97
    geophones, 93
    gravitational field, 4, 49, 53, 63, 70, 80, 97, 98, 99, 109, 145, 242
    heat flow experiments, 97, 105, 109
    ice, 282
    lasers, 81, 93, 97, 111, 112
    magnetic field, 5, 6, 29, 49, 62, 93, 97, 98, 106, 328
    magnetic material on, 64, 66, 68
    mascons, 55, 70, 99, 145
    meteorites, 53, 55, 58, 61, 64, 66, 68, 75, 86, 98, 109, 123
    orbital motion, 81
    rovers, manned, 96–97, 105, 109–11
    rovers, unmanned, 90–91, 112
    samples returned from, manned missions, 80, 86, 93, 96–98, 104–05, 109–11
    samples returned from, unmanned missions, 89, 103, 145
    seismic research, 81, 85, 86, 87, 93, 97, 98, 105, 106
    shape, 53, 55
    solar wind, 81, 85, 93, 97, 99, 105
    ultraviolet radiation, 43, 112
    volcanic activity, xiii, 33, 38, 39, 53, 55, 64, 66, 68, 86, 97, 98, 105, 109, 111, 130
    water, evidence of, 282, 328, 329
    x-rays, 91, 98, 112
Moon, craters of, xiii, 282, 328
    end to debate on formation of, 105
    studied by manned missions, 80–81, 86, 93, 97, 105, 105, 109, 111
    studied by unmanned missions, 6, 33, 38, 39, 50, 53, 57, 61, 62–63, 63, 64, 66, 68, 75, 75, 112
Moon, spacecraft to
    flybys, 43, 332
    hard impacts, 6, 20, 31, 33, 38, 39–40, 47, 53, 80, 87, 93, 99, 329
    landings, manned, 80–81, 85–87, 93, 96–98, 104–06, 109–12
    landings, unmanned, 47, 49–50, 55, 59, 63–64, 66, 68, 89, 90–91, 103, 112, 126, 145
    misses, 18, 20
    orbiting, manned, 75, 78
    orbiting, unmanned, 49, 53, 54–55, 58, 62–63, 70, 73–74, 80, 83, 90, 99, 116, 123, 242, 282, 328–29

"swing around," manned, 88–89
"swing around," unmanned, 6–7
Moon, surface of, 328–29
    dust layer, 33, 39, 47, 50, 59, 80–81, 86, 97
    erosion, pace of, 68, 86
    strength, 33, 38, 39, 47, 49, 59
    studied by manned missions, 80–81, 85–87, 93, 96–98, 105, 109–11
    studied by unmanned missions, 6–7, 33, 38, 39, 47, 49, 50, 53, 54, 57, 59, 61, 62–63, 63, 64, 66, 68, 75, 90–91, 99, 112, 122, 145
Moon-walks, 80–81, 85–86, 93, 96–98, 105, 109–11
Morelos 1, **210–11**, 278
Morelos 2, **215**, 278
Moroz, V. I., 183
MOS 1, **225**
Moths, 185, 216, 288
Motion sickness. See Space sickness
Mozhaisky academy, 316
MSAT 1, **308**
MSTI 1, **271**
MSTI 2, **285–86**, 306, 321
MSTI 3, **310**
MSX, **309**
Mugunghwa 1, **297**
Mukai, Chiaki, 286, 289, 334
Mullane, Richard, 203, 232, 243
Multiple Access Communications Satellites. See Mascat
Multispectral imaging, 91, 113, 115, 117, 118. See also Earth resources research
Musabayev, Talgat, 286, 289, 329, 332, 333
Musat, **312**
Muscle research, 237, 285, 293. See also Exercise
    dogs, 48
    humans, Skylab missions, 116, 117
    humans, Soyuz-Mir missions, 232, 289
    humans, Soyuz-Salyut missions, 95, 128, 132, 151, 155, 163, 173, 200
    humans, STS missions, 256, 311, 331, 335
    mice, 331
    rats, 118, 148, 196, 208, 227, 275, 278, 290, 331
    rodents, 259
    worms, 297
Muses A, **242**
Musgrave, Story, 192, 211, 239, 261, 278, 278, 279, 313, 334
Mushroom bed, 118
Musketball 1, **98**
Mussels, 310
Myanmar, 112
Mylar Balloon, **98**
Mylar umbrella, 113–14
Myojo, **284**

N-Star A, **297**
N2 rocket, 176, 192
Nadezhda 1, 188, **235**
Nagel, Steven, 210, 214, 253, 274
Nahuel 1A, **314**
Nahuelsat, 314
NASA (National Aeronautics and Space Administration)

commercial launches, no longer involved in, 216, 218–19
    first satellite launched by, 4–5
    manned missions. See Apollo; Gemini; Mercury; Skylab; STS
NASDA (National Space Development Agency), 151, 154, 192, 228
National Aeronautics and Space Administration. See NASA
National Oceanic and Atmospheric Administration. See NOAA
National Reconnaissance Office, 333
National Space Development Agency. See NASDA
NATO 1, 87
NATO 2, 87, 139
NATO 3A, 87, **139**, 152
NATO 4A, **252**
NATO (North Atlantic Treaty Organization), 87, 203
    appendix listing for, 369
Nausea. See Space sickness
Naval Postgraduate School, 334
Naval Remote Ocean Sensing System. See NROSS
Naval Research Laboratory, 340
Navigation. See also Geodetic/navigational research; Gravity assists
    Doppler shift technique, 8–9, 61, 83
    Glonass technique, 171, 189–90
    GPS technique, 9, 61, 69, 83, 124. See also GPS
    gravity-gradient attitude control, 108
Navigation Technology Satellite. See NTS
Navstar 1, 147, **153–54**, 171
Navstar 5, **171**, 189
Navstar B2 1, **233**
Navy Ocean Sensing System. See NOSS
Navy Ocean Surveillance Satellite. See NOSS
Navy Remote Ocean Sensing System. See NROSS
NEAR, **306**
NEAR, Eros Flyby, **337**, 337
NEAR, Mathilde Flyby, 309, **318–20**, 319
Near Earth Asteroid Rendezvous. See NEAR
Near-Infrared and Multi-Object Spectrometer. See NICMOS
Nebulae, xvii, xvii, xviii, 20, 74, 91, 99, 104, 108, 120, 153, 192, 279–80, 301. See also Crab Nebula
Nelson, Bill, 216
Nelson, George, 201, 201, 216, 230
Nematode worm, 286
Neptune
    1957 knowledge about, xvi
    atmosphere, 236
    bands, 236
    Great Dark Spot, 236
    magnetic field, 237
    moons of, xvi, 236
    rings, 236
    spacecraft to, 236–37
Nereid, xvi
Netherlands
    appendix listing for, 369
    satellite service to, 232
    Spacelab research by, 263
    STS crew member from, 214
Neurolab, 306

Index

Neurological research
  human, 257, 331
  mice, 331
  rats, 290, 331
Neutrinos, 252
Neutron particles, 29
Neutron stars, 109, 153, 159, 170, 240, 246, 254, 273, 305, 341
New Guinea, 287, 306
New Zealand, 306, 340
Newman, James, 276, 277, 298, 335, 336
Newts, 211, 237, 286, 294
Next Generation Space Telescope, 146
Nicaragua, 161
NICMOS, 315, 322, 335
Nicollier, Claude, 267, 279, 307, 342
Nikolayev, Andrian, 22, 27, *27*, 89
Nilesat 1, **331**
Nilesat Co., 331
Nimbus 1, 30, **34**, 37, 106
Nimbus 7, **157,** *158,* 162, 171, 204, 258, 311
Nimiq 1, **339**
Nippon Telegraph and Telephone Company, 297
Nixon, Richard, 81
NOAA 1, **91,** 151, 170, 188, 204
NOAA (National Oceanic and Atmospheric Administration), 91, 135
Noble gases, on Venus versus Earth, 184
Noriega, Carlos, 318
North Atlantic Treaty Organization. *See* NATO
North Korea, 161. *See also* South Korea
Northern Utah Satellite. *See* NUSAT
Norway, 237, 318
NOSS 1, **140,** 247
Nova, 74. *See also* Supernovae
Nozomi, **332,** 338
NROSS, 140, 247
NTS 1, **124**
NTS 2, **147**
Nuclear explosion detection systems, 4, 21, 23, 24, 29, 32, 78–79, 95. *See also* Missile detection systems
Nuclear power sources, 16, 68, 139, 149
NUSAT, **208**

OAO 1, **74**
OAO 2, **74**
OAO 3, **107–08**
Oat seedlings, 185
Oberon, 218
Occhialini, Giuseppe, 309
Ocean platform launches. *See* Floating platform launches
Oceanic research, xviii, 69, 131, 155–56, 157–58, 162, 176, 179, 182, 205, 206–07, 209, 247, 268, 275, 311, 321, 329, 339, 340, 343. *See also* Earth resources research
  1957 knowledge about, xviii
  Jupiter's moons, 304
  sea ice conditions, 162, 196, 225, 226, 253, 257, 294, 297, 300
Ochoa, Ellen, 273, 290, 340
Ockels, Wubbo, 214
O'Connor, Bryan, 215, 256
ODERACS 1A though ODERACS 1F, **283**
ODERACS 2A though ODERACS 2F, **292**
OEX, 215
Ofeq 1, **230**
Ofeq 2, 243

Ofeq 3, **294**
OFO, **90**
OGO 1, 32, **34–35,** 51, 69, 84, 150
OGO 2, **44,** 62, 79
OGO 3, **51,** 69
OGO 4, **62,** 70, 72, 79, 307
OGO 5, 62, **69**
OGO 6, 62, 69, **79,** 170
OHB-System, 333
Ohsumi, **88**
Ohzora, **200**
Okean-E, **162**
Okean-OE, **196**
Okean satellites, 176
Oldest humans in space, 195, 313, 334
Olson, Ray, 79
Olympus 1, **235**
Oman, 189
Onion plants, 132, 171, 172, 226
Onizuka, Ellison, 206, 218, *218*
Onufrienko, Yuri, 306, 309, 312
Open Skies policy, 122
Optus A1, **212**
Optus A2, **215**
Optus B1, **268**
Orbcomm, 318
Orbcomm FM1, **294**
Orbcomm FM2, **294**
Orbcomm FM5 through Orbcomm FM12, **328**
Orbcomm X, 257
Orbiscal 2, **77**
Orbital Debris Radar Calibration Spheres. *See* ODERACS
Orbital Re-entry Experiment Vehicle. *See* OREX
Orbital Refueling System, 205
Orbital Sciences Corporation, 243, 257, 284, 294, 317, 321, 323, 326, 328
Orbital Test Satellite. *See* OTS
Orbiter Experiment. *See* OEX
Orbiting Astronomical Observatory. *See* OAO
Orbiting Frog Otolith. *See* OFO
Orbiting Geophysical Observatory. *See* OGO
Orbiting Research Satellite. *See* ORS
Orbiting Solar Observatory. *See* OSO
Orbiting Vehicle. *See* OV
OrbView 2, **321**
Orchids, 172, 176
OREX, 283
Orfeus, **313**
Orfeus-Spas, **276–77**
Oriented Scintillation Spectrometer Experiment. *See* OSSE
Orion 1, **291**
Orion nebula, 279–80, 301
Orion Satellite Corporation, 291
Orizuru, **242**
ORS 2, **51**
ORS 3, **43**
Orsted, **338**
Ortho Pharmaceuticals Corporation, 188
Oscar 1, **18,** 21
Oscar 2, **21**
Oscar 3, **39**
Oscar 4, **47**
Oscar 5, **88**
Oscar 6, **108**
Oscar 7, **126,** 154
Oscar 8, **154**

Oscar 9, **181**
Oscar 10, **193**
Oscar 11, **200**
Oscar 12, **224**
Oscar 13, **229**
Oscar 14, **241–42**
Oscar 15, **241–42**
Oscar 16, **242**
Oscar 17, **242**
Oscar 18, **242**
Oscar 19, **242**
Oscar 22, 257–58
OSO 1, **19**
OSO 2, **38,** 327
OSO 3, **58**
OSO 4, **64**
OSO 5, **76**
OSO 6, **83**
OSO 7, **99**
OSO 8, **132,** 148
OSSE, 253
Oswald, Stephen, 261, 273, 293
Otolith sensor cells, 90
OTS 1, **154**
OTS 2, **154–55,** 193
OV1-2, **44**
OV1-4, **49**
OV1-5, **49**
OV1-8, **52**
OV1-13, **70**
OV1-15, **71**
OV1-16, **71**
OV1-17, **77**
OV1-17A, **77**
OV1-18, **77**
OV1-19, **77**
OV1-20, **98**
OV1-21, **98**
OV2-5, **72**
OV3-1, **49**
OV3-2, **55**
OV3-3, **52**
OV3-4, **51**
OV3-6, **67**
OV4-1R, **55**
OV4-1T, **55**
OV5-1, **60**
OV5-2, **72**
OV5-3, **60**
OV5-5, **78**
OV5-6, **78**
OV5-9, **78**
Overmyer, Robert, 190, 208
Oxygen production
  from electrolysis, 315, 318
  from lithium candles, 315, 318
  from water, 299
Ozone, 31, 61, 91, 128, 132, 137, 158, 162, 181, 201, 258, 259, 266, 273, 290, 294–95, 311–12, 330, 332, 333

P76-5, **140**
P78-1, **162**
P83-1, **195**
PAC 1, **83,** 113
Package Attitude Control. *See* PAC
Padalka, Gennady, 333, 338
Pageos 1, **51–52,** 54
Pailes, William, 214
Pakistan
  appendix listing for, 369
  Starshine satellite, 340

400

Pakistan Amateur Radio Society, 247
Palapa 1, **145**
Palapa B1, **194**
Palapa B2, **198**, 205
Palapa B2R, 198, **244**
Palapa C-1, **306**
Palapa D-1, **341**
PAMS, **310**
Pan American Satellite, 193, 229
Pan American Satellite 1, 88, **229**
Pan American Satellite 5, **322–23**
Pansat, **334–35**
Parachutes, 60, 74, 196
Paraguay, 117
Parallax, 235–36
Paramecia, 152
Parasol, 113–14, 116, 117
Parazynski, Scott, 290, 325, 334
Parise, Ronald, 251
Parker, Robert, 197, 251
Parsley plants, 171
Particles and Fields Subsatellite. *See* PFS 1
Passive reflector satellites. *See* Communications satellites
Patsayev, Viktor, *95*, 95–96, 117, 124, 170
Pawelczyk, James, 331
Payette, Julie, 340
Payload Flight Test Article. *See* PFTA
Payload Systems, Inc., 237
Payton, Gary, 206
PDP, 211–12
Pea plants, 128, 132, 171
Peale, Stanton, 164
Pegasus 1, **38**
Pegasus 2, **41**
Pegasus 3, **43**
Pegasus launch system, 243
Pegsat, **243**
Peole, 91, **92**
People's Republic of China. *See* China
Perch, 274
Peru, 117
Peterson, Don, 192
Petite Amateur Navel Satellite. *See* Pansat
PFS 1, 98
PFS 2, 105–06
PFTA, 196
Pham Tuan, 172
Pharmaceutical materials research, 134, 187–88, 193, 194, 204, 207, 215, 226, 230, 265, 270, 283, 286, 293, 297, 307, 310, 332, 341. *See also* Biological materials research
Phase Partitioning Experiment, 231
Philippines
    appendix listing for, 369
    radar scans of, 285
    satellite service to, 145, 333
Phobos 1, **229**
Phobos 2, **229**
Phobos (moon), 100, 229, *229*
Phoebe, xv, 179
Photosurveillance. *See* Surveillance
Physicians in space, 196, 199, 230, 270, 274, 278, 281
Pigtail money, 79
Pine seedlings, 185, 311
Pion 1, **234**
Pion 2, **234**
Pioneer 1, **4–5**, 6
Pioneer 3, 4, **5**, 6, 8

Pioneer 4, **5–6**, 8
Pioneer 5, **8**, 23, 29
Pioneer 6, 45, **46–47**
Pioneer 7, 53, **54**
Pioneer 8, **67**, 86
Pioneer 9, **73**, 86, 102
Pioneer 10, 73, 102, **103–04**, 116
Pioneer 10, Jupiter Flyby, **119–20**, 127, 163, 262, 299
Pioneer 10, Shutdown, 120, **316–17**
Pioneer 11, 104, **112**
Pioneer 11, Jupiter Flyby, **127**
Pioneer 11, Saturn Flyby, **168–69**
Pioneer 11, Shutdown, **299**, 316, 317
Pioneer Division, 169
Pioneer Venus Orbiter, **159–61**, *160*, 205
Pioneer Venus Probe, **161**
PIX 1, **154**
Planetary missions, appendix listings for, 362–63. *See also individual planets and moons*
Planetary nebulae, xvii, 120, 153, 280
Plant research, vii, 7, 11, 12, 49, 63, 105, 127, 128, 132, 154, 250, 269. *See also* Earth resources research
    biological activity scanners, 157, 321
    Bion satellites, 118, 137, 237, 271
    importance of water and air circulation, 327
    Landsat satellites, 107, 188, 339
    Soyuz-Mir missions, 226
    Soyuz-Salyut missions, 171–72, 176, 186, 195, 213
    STS-Mir missions, 313, 318, 329
    STS missions, 185, 187, 193, 194, 215, 233, 238, 241, 250–51, 256, 261, 265, 272, 275, 277, 283, 287, 288, 292, 297, 299, 310–11, 314, 327, 335, 341
Plaques, 80, 97
Plasma Diagnostics Package. *See* PDP
Plasma Interaction Experiment. *See* PIX
Plate tectonics
    Earth, 140, 207, 270, 277
    Europa, 304
    Mars, 142, 321, 323
    Venus, 249
Pluto, *xvi*, 280
    1957 knowledge about, xvi–xvii
    moon of, 280
POGS, **244**
Pogue, Bill, 118
Poland
    appendix listing for, 369
    Bion satellite, 148
    Intercosmos satellites, 84, 104, 140, 238, 261, 284
    Intersputnik consortium, 161
    Ionosonde satellite, 224
    Soyuz cosmonaut from, 156
    Vega mission, 210
Polar, 178, **307–08**, 312
Polar Bear, **224**
Polar Orbiting Geomagnetic Survey. *See* POGS
Polar rocket, 339
Polarity reversal, 170
Poleshchuk, Alexander, 272, 275
Politics. *See also* Space race
    Arab States, 206
    Russia, 263

    Soviet Union, xix, 35, 59, 118, 133, 219, 226, 260
    United States, 26, 122, 221
Pollux, **131**
Polyakov, Valery, 230, 231, 232, 281, *281*, 282, 286, 289, 292, 293, 338
Polymer membranes, 245, 250, 258, 275, 277, 292. *See also* Thin-film research
Polyot 1, **29**
Polyot 2, **32**
Popov, Leonid, 171, 178, 189
Popovich, Pavel, 22, *27*, 124
Portugal, 277
Portuguese Satellite. *See* Posat
Posat 1, **277**
Poseidon. *See* TOPEX-Poseidon
Posture
    frog, 252
    human, 211. *See also* Height, changes in
Potatoes, 299
Powers, Gary, xix
Precourt, Charles, 274, 295, 318, 331, 332
Preparation for Eole. *See* Peole
Primates. *See* Chimpanzees; Humans; Monkeys
Primestar satellite network, 316
Priroda, 306, **308–09**, *309*
Private investment in space, 212. *See also individual companies*
    Challenger accident and, 218–19
    first wave of, 212
    second wave of, 318
Prognoz 1, 84, **104**, 109, 124, 150
Prognoz 9, **195**
Prognoz 10, **208**, 297
Progress 1, 150, 151, **152**, *152*
Progress M1, **236**, *236*
Progress M34, **317**
Progress M35, **321**
Project Argus, 4, 21
Propagation research. *See* Communication propagation and design research
Proplyd stars, 279–80
Prospero, **100**
Protein crystals. *See* Crystals research
Proton 1, **43**
Proton 2, **44**
Proton 3, **52**
Proton 4, **74**
Protostars, 191
Protozoa cysts, 105
Prunariu, Dumitru, 178
Psychology
    extrasensory perception, 93
    feeling of exposure, 115
    long-term missions, 187
Pulmonary functions. *See* Respiratory research
Pulsars, 91, 109, 148, 159, 224, 254, 273, 305
Pumps, 144, 205, 211, 216, 270. *See also* Liquids research; Refueling

Quail eggs, 163, 240, 242, 294
Quasars, 91, 134, 153, 159, 246, 254, 305, 316
QuikScat, **340**

Radar
  detection of, 67–68, 127–28, 149, 189
  imaging/mapping, 155, 159
  imaging/mapping, of Earth, 182, 205, 206, 226, 232, 243, 253, 257, 262–63, 268, 285, 288, 294–95, 299–300. *See also* Earth resources research
  imaging/mapping, of Venus, 137, 159–60, 197, 205, 234, 248–49
  rendezvous and, 44, 72
  side-looking, 196, 232, 248
  synthetic aperture, 155–56, 196, 226, 257, 262–63, 285, 294, 299–300
  tracking and identification, 83, 90, 98, 99, 127, 283, 292
  wide-beam, 294
Radar Ocean Reconnaissance Satellites. *See* RORSATs
Radarsat, **299–300**
Radcat, **108**
Radiation. *See* Cosmic radiation; Gamma radiation; Infrared radiation; Solar research, on solar radiation; Ultraviolet radiation; X-ray radiation
Radiation belts
  artificial, 20, 21, 22, 23–24, 31, 33, 36
  Van Allen. *See* Earth, Van Allen belts of
Radiation budget, 158, 204
Radiation Experiment. *See* Rex
Radiation/Meteoroid. *See* RM
Radiator, 255
Radio Amateur Satellite Corporation. *See* AMSAT
Radio astronomy, 70–71, 116, 195, 316
Radio Astronomy Explorer. *See* RAE
Radio Corporation of America (RCA), 25, 137, 190, 215
Radio-M, 291
Radio noise
  galactic, 31, 61, 71
  terrestrial, 61
Radio-Rosto, **291**
Radio Sputnik 1, **159**
Radio Sputnik 2, **159**
Radioactive debris, 149, 189
Radish plants, 171, 194, 195, 199
Radose, 27
Radsat, **108**
Raduga 1, **138**, 161, 214, 291
Raduga 28, **261**
RAE 1, **70–71**
RAE 2, **116**, 163
Rail transportation systems, 255
Rainfall rates, 327
Ranger 1, 18
Ranger 2, 18
Ranger 3, **18**
Ranger 4, **20**
Ranger 5, 20
Ranger 6, 20, **31**
Ranger 7, vii, 20, **33**, *33*, 38
Ranger 8, **38**, 55
Ranger 9, **39–40**
Rats. *See also* Rodents
  Bion satellites, 7, 11, 118, 126, 137, 148, 170, 198, 211, 227, 237
  dissections of, 278, 314
  STS missions, 196, 208, 233, 250, 256, 274, 275, 278, 285, 286, 290–91, 293, 297, 306, 314, 331

RCA Alascom, 190
RCA Astro Electronics, 229
RCA (Radio Corporation of America), 25, 137, 190, 215
RCA Satcom 1, 123, **137–38**
RCA Satcom 5, **190**, 251
RCA Satcom Ku2, **215**
Re-entry
  double skip maneuver, 67, 69, 70, 73
  heat shields, 6, 9, 12, 18, 55, 58, 76, 194, 326
  problems, 12, 19, 39, 58, 60, 71, 76, 96, 166
  stress of, 278
  tests, by China, 342
  tests, by Europe, 334
  tests, by Japan, 283–84
  tests, by Soviet Union, 11, 12, 13, 35, 56, 58, 69, 71, 117
  tests, by United States, 6, 7, 12, 37, 55, 67
Readaptation to gravity
  dogs, 48
  humans, Apollo missions, 98
  humans, Gemini missions, 42, 44, 46
  humans, Mercury missions, 23, 26
  humans, Skylab missions, 116, 117, 118
  humans, Soyuz-Mir missions, 293
  humans, Soyuz missions, 89
  humans, Soyuz-Salyut missions, 124, 128, 132, 151, 155, 163, 172, 187, 189, 200
  humans, STS-Mir missions, 226, 232, 242, 314, 318
  humans, STS missions, 256, 331
  humans, Vostok missions, 17, 22
  rats, 274, 291
Readdy, William, 261, 276, 313
Reconnaissance. *See* Surveillance
Recovery. *See* Re-entry; Readaptation to gravity
Redstone rocket, 13
Refueling, 151, 152, 154, 194, 198–99, 205, 223, 271, 275, 311. *See also* Liquids research; Pumps
Reightler, Kenneth, 259, 283
Reilly, James, 329
Reiter, Thomas, 297, 298, 300, 301, 306
Relay 1, 16, **25**, 28, 29, 33
Relay 2, **31**
Relay Mirror Experiment. *See* RME
Remek, Vladimir, 154
Remote manipulator arm tests, 181, 184, 194, 196, 198. *See also* Strela cranes
Rendezvous and docking
  anti-satellite system, 72
  Apollo-LM, 77, 78
  Apollo-Skylab, 114
  Apollo-Soyuz, 126–27, 133
  Gemini and target vehicles, 40, 42, 44, 46, 48, 50, 52, 54, 56
  Kiku target and chase vehicles, 327
  Kristall-Mir, 246
  Kvant-Mir, 225, 240
  Priroda-Mir, 308
  Progress-Mir, 236, 252, 263, 286, 315, 318
  Progress-Salyut, 152
  Soyuz and target vehicles, 66, 72–73, 76, 84
  Soyuz-Mir, 281, 289

Soyuz-Salyut, 94, 95, 125, 146, 150, 166, 193, 194, 209
Spektr-Mir, 295
STS-Mir, 292, 296, 300, 313, 314, 318, 325, 329, 331–32
Repairs. *See also* Work in space
  Skylab missions, 114, 115, 116
  Soyuz-Mir missions, 225, 227, 242, 248, 252, 256, 263, 266, 272, 309, 321, 325, 329, 333
  Soyuz-Salyut missions, 163, 173, 176, 186, 199–200, 209
  STS missions, 198, 201, 207, 211, 245, 253, 263, 277, 279, 315, 342
Reptiles, 118, 137, 271
Republic of China. *See* Taiwan
Rescue maneuvering unit, 287–88, 325
Rescue missions. *See* Repairs; Search-and-rescue systems
Resnik, Judith, 203, 204, 218, *218*
Resonance, vibrational, 70, 152, 223
Respiratory research
  dogs, 11, 13, 48
  humans, 18, 22, 27, 46, 95, 124, 128, 132, 144, 154, 257, 311
  monkey, 7
  rats, 118, 126
Resurs-500, **271**
Resurs F1, **234**
Reused spacecraft
  orbital flights, 55, 181–82, 193–94, 205, 211, 215, 244, 267, 268–69, 273, 283, 290, 298, 307, 310, 322
  suborbital tests, 149, 231
Rex 1, **257**
Reynolds, Ray, 164
Rhea, 175, *175*, 179
Rhesus monkey, 7
Rhyolite 1, **112**
Richards, Richard, 235, 250, 265, 287
Ride, Sally, 193–94, 204
Rigidsphere, **98**
Rings
  Jupiter, 163, 164, 168, 262, 304
  Neptune, 236
  Saturn, xv, *xv*, 163, 168, *168*, 169, *169*, 174, *174*, *176*, 179
  Uranus, *216*, 217–18
RM 1, **90**
RME, **243**
Robinson, Stephen, 322, 334
Robot arm tests, 181, 184, 194, 196, 198. *See also* Strela cranes
Robot exploration. *See individual planets and moons*
Robot-Operated Materials Processing System. *See* Romps
Rock samples. *See* Sample returns
Rocket booster explosion, 218
Rockets, detection of. *See* Missile detection systems
Rocsat, **338**
Roddenberry, Gene, 317
Rodents, 11, 259, 311, 331. *See also* Rats
Roentgen Satellite. *See* Rosat
Rohini 1B, **172**
Romanenko, Yuri, 151, 152, 173, 224, 225, 226, 227, 232
Romania
  appendix listing for, 369

Bion satellites, 126, 148
Intercosmos satellites, 84, 170, 176, 238, 261
Intersputnik consortium, 161
Soyuz cosmonaut from, 178
Rominger, Kent, 299, 313, 322, 340
Romps, 287
Roosa, Stuart, 93, 105
RORSATs, 67–68, 127–28, 149, 189
Rosat, **246,** *247,* 272
Ross, Jerry, 215, 232, 253, 254, *254,* 255, 274, 300, 335, 336
Rossi, Bruno, 305
Rossi X-ray Timing Explorer, **305**
Rovers
    manned, Moon, 96–97, 105, 109–11
    unmanned, Mars, 320–21
    unmanned, Moon, 90–91, 112
Rowell, Galen, 75
Rozdestvensky, Valery, 146
Rukavishnikov, Nikolai, 94, 126, 165
Runco, Mario, 260, 261, 271, 272, 310
Russia. *See also* Soviet Union
    appendix listing for, 369–70
    economic issues, 276
    international agreements and treaties, 292, 312, 315, 323, 325
    political issues, 263
    STS crew members from, 283, 292, 318, 325, 335, 340
Russian Space Agency, 330
RXTE, **305**
Ryumin, Valery, 150, 163, 166, 171, 172, 332
Ryusei, **283–84**

S80-T, **268**
SAC-A, **336**
Sacco, Albert, 299
SACI 1, **342**
Safer maneuvering unit, 287–88, 325
Safir 2, **333**
SAGE, 162
SAGE 2, 204
Sakigake, **206**
Sakigake, Halley's Comet Flyby, **222**
Sakura 1, **151–52,** 192
Sakura 2A, **192,** 228
Sakura 3A, **228,** 297
Salamanders, 286
Saliva, 256, 311
SALT treaty, 73
Salyut 1, **94,** *94,* 95–96, 108, 112, 124, 128
Salyut 2, 124
Salyut 3, 96, 119, **124,** 125, 128, 144, 177–78, 186
Salyut 4, 84, **128,** 130, 131–32, 133
Salyut 5, **144,** 146–47, 178, 186
Salyut 6, 147, *149,* **149–50,** 151, 152, 154, 155, 156, 157, 163, 166, 171, 173, 176, 178, 185, 186, 223, 272
Salyut 7, 147, 159, **185–86,** 186–87, 192, 193, 194, 199–200, 202, *209,* 209–10, 213, 223, 272
Salyut program, 40
Sam (rhesus monkey), 7
Samos 2, **13**
Samos satellites, 13, 18, 19
Sampex, **265–66**
Sample returns
    from Moon, manned, 80, 86, 93, 96–98, 104–05, 109–11

from Moon, unmanned, 89, 103, 145
San Marco 1, **37**
San Marco 2, **60**
San Marco 3, **94**
San Marco 4, 120, **121–22**
San Marco D/L, **228**
SARA, **257**
Sarafanov, Gennady, 125
SARSAT-COSPAS, 91, 135, 147, 188, 235, 266
SAS 1, **91–92**
SAS 2, **108–09**
SAS 3, **131**
Satcom C1, **251**
Satellite Business Systems, 173, 190, 195, 203. *See also* SBS
Satellite Data Systems. *See* SDS
Satellite Italiano Ricerca Industriale Orientata. *See* Sirio
Satellite of Scientific Application. *See* SACI
Satellites, artificial. *See individual satellites*
Satellites, natural. *See* Moon; *and names of other planets' moons*
Satmex, **336**
Saturn, *xv, 168, 169, 174, 176*
    1957 knowledge about, xv
    atmosphere, xv, 153, 169, 175, 179, 301
    aurora, 176
    bands, 175
    composition of, 169
    lightning, 179
    magnetic field, 168, 176
    magnetosphere, 116, 168, 176
    moons of, xv, *174,* 174–75, *175,* 179, *179, 180,* 301
    rings, xv, *xv,* 163, *168,* 169, *169, 174, 176,* 179
    winds, 175, 179
Saturn, spacecraft to
    on course, 326
    flybys, 112, 168–69, 174–76, 179–80
Saturn 5 rocket tests, 67, 70
Saudi Arabia, STS crew member from, 210
Savinykh, Viktor, 176, 209, 213, 228
Savitskaya, Svetlana, 189, 202, 265
SBS 1, **173,** 238
SBS 3, **190**
SBS 4, **203**
SBUV, 158. *See also* SSBUV
Scale, for weight measurements, 114, 144
Scandinavia. *See also* Denmark; Finland; Norway; Sweden
    appendix listing for, 370
    satellite service to, 318
SCATHA, **162**
SCD 1, **272**
SCE, **244**
Schiaparelli, Giovanni, xiii
Schirra, Wally, 23, 46, 72
Schlegel, Hans, 274
Schmidt, Harrison, *109,* 109–11, *111*
Schweickart, Rusty, 77
Scientific knowledge
    1957 baseline for, xi–xix
    growth of, vii, xi
Scobee, Francis, 201, 218, *218*
SCORE, **5,** 10
Scorpius-Centaurus Association, 149
Scorpius X-1, 91, 132, 148, 159

Scorpius X-2, 211
Scott, Dave, 48, 77, 96–98, *97*
Scott, Winston, 305, 306, 327
Scully-Power, Paul, 204, 205, 241
SDS 1, **140**
SDS 2. *See* USA 40
Sea ice conditions, 162, 196, 225, 226, 253, 257, 294, 297, 300. *See also* Oceanic research
Sea Launch, 339, 342
Sea launch systems. *See* Floating platform launches
Sea urchins, 310, 318
Sea-viewing Wide Field Sensor. *See* SeaWiFS
Search-and-rescue systems, 91, 135, 147, 188, 235, 266, 334
Searfoss, Richard, 277, 308, 331
Seasat, **155–56,** *156*
SeaWiFS, **321**
Secor 1, **30–31**
SECS, **243**
Seddon, Rhea, 207, 256, 277, 278
Seds, 273
Sedsat 1, **334**
Sega, Ronald, 283, 308
Seismic research
    Earth, 140
    Mercury, 130
    Moon, 81, 85, 86, 87, 93, 97, 98, 105, 106
    stars, 273, 339
    Venus, 184
Selective Communications Experiment. *See* SCE
Serebrov, Alexander, 189, 193, 237, 275, 281
SERT 2, **88,** 128
Service module/command module design, 67, 77
Sevastyanov, Vitali, 89, 131
Seyfert galaxies, 92, 134, 153, 159
Sharipov, Salizhan Shakirovich, 329
Sharma, Rakesh, 200
Sharman, Helen, 256
Shatalov, Vladimir, 76, 84, 94
Shaw, Brewster, 197, 215, 235
Shenzhou, **342**
Shepard, Alan, 12, 13, 15, *15,* 16, 18, 93, *93*
Shepherd, William, 232, 250, 270
Shepherding moons, 174. *See also* Rings
Sherlock, Nancy, 275
Shijian 4, **284**
Shijian 5, **339**
Shinsei, **99**
Shiyan Tongxin Weixing. *See* STW
Shoemaker, Eugene, 33, 328–29
Shonin, Georgi, 84
Shrimp, 105, 187, 216, 255, 262
Shriver, Loren, 206, 244, 267
Shuttle Imaging Radar. *See* SIR
Shuttle Pallet Satellite. *See* SPAS
Shuttle Pointed Autonomous Research Tool for Astronomy. *See* Spartan
Shuttle Solar Backscatter Ultraviolet sensor. *See* SSBUV
Shuttles
    glow on, 184–85, 255, 292
    Soviet Union. *See* Buran shuttle
    United States. *See* Atlantis; Challenger; Columbia; Discovery; Endeavour;

Index

Shuttles *(continued)*
    Enterprise; STS
    wake from, 184, 187, 211, 298, 313.
        See also Electrical build-up on
        spacecraft
Sich 1, **297**
Sigma 7, **23**, 42
Signal Communications by Orbiting Relay
    Equipment. *See* SCORE
Signal propagation research. *See* Communication propagation and design research
SIGNE 3, **147**
Sill, Godfrey, 121
Simplified Aid for EVA Rescue. *See* Safer maneuvering unit
Singapore
    appendix listing for, 370
    satellite service to, 145
Singapore Telecommunications, 294
Sinosat 1, **333**
SIR-A, 182
SIR-B, 205
SIR C, 285, 288
Sirio 1, **149**
Sirius, 125
Sky Television, 237
Skylab, 40, 71, 96, **112–14**, *113*
Skylab 2, **114–16**, *116*
Skylab 3, **116–17**, *117*
Skylab 4, 116, **118–19**
Skynet, 291
Skynet 1, 51, **87**, 100, 126
Skynet 2, **126**, 232
Skynet 4B, **232**
Slayton, Deke, 133
Sleep research, 17, 26, 46, 54, 187, 311, 331
Slime mold, 285, 286, 298
Slingshot effect. *See* Gravity assists
Slinkys, 208. *See also* Toys
Slovak Republic, 284
Slovakia, Soyuz crew member from, 338
Small Astronomical Satellite. *See* SAS
Small Expendable-tether Deployer System.
    See Seds
Small Experimental Communications
    Satellite. *See* SECS
Small Magellanic Cloud, 309
SME, **181**, 201
Smelting, 151
Smith, Michael, 218, *218*
Smith, Steven, 288, 315, 342
SMS 1, **123**
SN1987a, 153, 201, 224, 225, 232, 247, *279*, 280. *See also* Supernovae
Snails, 331
SNOE, **330**
Snowflakes, 193, 196
Societe Europeenne des Satellites, 232
Soderblom, Larry, 168
Soft Gamma-Ray Repeaters, 273
Soho, 64, **301–03**, *302*, 322, 327
SOICAL Cone, **83**
SOICAL Cylinder, **83**
Sojourner, *320*, 321
Solar-A, **258–59**
Solar and Heliospheric Observatory. *See* Soho
Solar Anomalous and Magnetosphere
    Particle Explorer. *See* Sampex

Solar Backscatter Ultraviolet sensor. *See* SBUV; SSBUV
Solar cells, 17, 24, 26, 29, 36, 77, 108, 178, 228, 242, 258, 336
Solar Interplanetary Gamma-Neutron
    Experiment. *See* SIGNE
Solar Max, **171**, 176, 204, 250, 273
    repair of, 198, 199, **201**, *201*
Solar Mesosphere Explorer. *See* SME
Solar panels
    design, 70, 99, 139, 154, 185, 204, 266, 287, 333
    ESA-GEOS 2, 156
    Hubble Space Telescope, 279
    Mars Global Surveyor, 323
    Mir, 219, 225, 227, 252, 295, 300, 309, 318, 325
    Progress, 236
    Salyut, 124, 152, 186, 194, 200, 209, 213
    Skylab, 113, 114, 115
    Soyuz, 56, 60, 117, 120, 126
    STS, 204, 251
Solar Radiation and Thermospheric
    Satellite. *See* SRATS
Solar research. *See also* Galaxies; Stars
    1957 knowledge about, xi
    appendix listings for, 363–64
    circulation pattern, 301
    comets, 162, 201, 206, 213, 222, 223
    corona, 64, 99, 105, 115, 133, 162, 258–59, 273, 284, 287, 298, 301
    corona mass ejections (CMEs), 116, 117, 118, 201, 259, 290, *302*, 302–03, 322, 330
    coronal holes, 138, *138*, 299, 302, 331
    eclipses, 55, 56, 59, 64, 118, 133
    faculae, 128, 132
    gravitational field, 4, 157, 301
    heliopause, 176, 317
    heliosphere, 120, 169, 299, 331
    helmet streamers, 99, 273, 302, *302*
    magnetosphere, 8, 112, 176
    solar constant, xi, 158, 273, 322, 335
    solar cycle, 4, 10, 17, 19, 23, 31, 32, 36, 37, 45, 69, 73, 76, 79, 137, 139, 201, 204, 267, 273
    solar system history, 130
    sunspots, xi, 9, 73, 118, 128, 132, 267
    variable star, Sun as, 158, 204, 267
Solar research, on magnetic field, xi, 8, 22, 47, 51, 62, 65, 74, 118, 132, 171, 211, 262, 299, *302*, 331
    comets and, 213, 222, 223
    corona and, 201, 258, 302, 330
    dipole nature of, 299
    structure of, 138, *138*
Solar research, on solar flares, xi, 11–12, 17, 28, 53, 60, 61, 69, 79, 83, 99, 102, 112, 115, 122, 135, 139, 147, 162, 176, 258, 303, 322, 330
    appendix listings on solar research, 363–64
    beyond solar system, 265
    Explorer satellites, 7, 17, 45, 102
    Intercosmos satellites, 84, 104
    ISEE satellite, 150
    magnetic field and, 171
    Mariner mission, 36
    OSO satellites, 19, 38, 58, 76, 83, 132
    Pioneer missions, 8, 104, 112
    Prognoz satellites, 104, 195

Soho mission, 303
Solar Max satellite, 171, 201
Solrad satellites, 9, 31, 45, 70, 139
Solwind satellite, 162
Solar research, on solar radiation
    appendix listings on solar research, 363–64
    gamma, 38, 76, 99, 104, 131, 176, 322
    gamma, OSO satellites, 76, 99
    gamma, Prognoz satellite, 104
    gamma, Solar Max satellite, 171
    infrared, 301
    ultraviolet, 67, 112, 124, 134, 141, 147, 158, 174, 181, 228, 229, 238, 273, 298, 322, 330, 335
    ultraviolet, Explorer satellites, 30, 45, 49, 137
    ultraviolet, Intercosmos satellite, 84
    ultraviolet, OSO satellites, 19, 38, 58, 64, 76, 83, 99, 132
    ultraviolet, Skylab missions, 113, 115
    ultraviolet, Soho mission, 301
    ultraviolet, Solar Max satellite, 171
    ultraviolet, Solrad satellites, 9, 27, 31, 45, 96, 139
    ultraviolet, STS missions, 185, 198, 211
    x-ray, 11, 12, 60, 61, 67, 112, 129, 135, 258–59, *259*, 322
    x-ray, Explorer satellite, 45
    x-ray, Intercosmos satellites, 84, 104
    x-ray, OSO satellites, 19, 38, 58, 64, 76, 83, 99, 132
    x-ray, Prognoz satellite, 195
    x-ray, Skylab missions, 113, 115
    x-ray, Solar Max satellite, 171
    x-ray, Solrad satellites, 9, 16, 27, 31, 45, 70, 96, 139
    x-ray, STS missions, 185
Solar research, on solar wind, xi, 6, 7, 13, 16, 24, 25, 34, 47, 51, 62, 65, 69, 70, 71, 72, 74, 79, 81, 85, 88, 91, 93, 97, 99, 103, 105, 109, 119, 121, 127, 162, 168, 184, 203, 206, 213, 222, 223, 260, 262, 266, 269, 276, 290, 307–08, 312, 322, 327, 332
    appendix listings on solar research, 363–64
    Explorer satellites, 14, 17, 29, 124, 134
    fast and slow, 73, 138, 273, 299, 301, 331
    first detection of, 22–23
    Helios satellite, 138
    Imp satellite, 29, 30
    Intercosmos satellite, 84
    ISEE satellite, 150, 157
    Mariner missions, 22–23, 36
    Pioneer missions, 8, 46, 54, 67, 73, 104, 112
    Prognoz satellite, 104, 195
    Soho mission, 301–02
    Solwind satellite, 162
    Spartan spacecraft, 273, 287
    Vela satellites, 29, 32
    Venera mission, 45
    Wind satellite, 290
Solcon, 335
Soldering, 194, 198, 256
Solid-fuel rocket, first satellite orbited with, 13

404

Solid state technology in communications satellites, 62
Solidaridad 1, 211, **278**
Solovyov, Anatoly, 228, 242, 266, 272, 296, 297, 321, 325, 329
Solovyov, Vladimir, 199–200, 202, 223
Solrad 1, **9**, 58
Solrad 3, **16**
Solrad 6, **27**
Solrad 7A, **31**
Solrad 8, **45**
Solrad 9, **70**, 76, 96
Solrad 10, **96**
Solrad 11A, **139**, 157, 162, 171
Solrad 11B, **139**, 157, 162, 171
Solrad satellites, 84, 103, 109, 124
Solwind, **162**, 213
Somalia, 215
Sound wave research. *See* Acoustic research; Communication propagation and design research
South Africa
  appendix listing for, 370
  Starshine satellite, 340
South Atlantic anomaly, 20, 26, 36, 44, 90
South Korea. *See also* North Korea
  appendix listing for, 370
  mapping of, 343
  satellite service to, 244
Southern Pacific Communications Company (SPCC), 202
Soviet Union (U.S.S.R.). *See also* Kazakhstan; Kyrgyzstan; Russia; Ukraine
  appendix listing for, 370–71
  break-up of, 219, 237, 256
  economic issues, 228, 232, 234, 237, 242, 248, 256
  international agreements and treaties, 73, 135, 188
  joint programs with communist allies, 84, 104, 126, 140, 148, 170, 176, 210, 238, 261
  political issues, xix, 35, 59, 118, 133, 219, 226, 260
  space race, xix, 12–13, 73
Soybeans, 292, 335
Soyuz, test flights of, 56, 58, 66, 69, 71, 83, 117, 119, 122, 137
Soyuz 1, 32, **59–60**, 72
Soyuz 2, **72**
Soyuz 3, **72–73**
Soyuz 4, **76**
Soyuz 5, **76**
Soyuz 6, **84**, 151, 187
Soyuz 7, **84**, 187
Soyuz 8, **84**, 187
Soyuz 9, 46, **89**
Soyuz 10, **94**, 95
Soyuz 11, 94, *95*, **95–96**, 114, 124, 170
Soyuz 12, 96, **117–18**
Soyuz 13, **120**
Soyuz 14, **124**
Soyuz 15, **125**
Soyuz 16, **126–27**, 145
Soyuz 17, **128**, 132, 151
Soyuz 18A, **130–31**
Soyuz 18B, 84, **131–32**
Soyuz 19 (ASTP), 84, **133**
Soyuz 20, **137**
Soyuz 21, **144–45**, 146
Soyuz 22, **145–46**

Soyuz 23, **146**
Soyuz 24, **146–47**
Soyuz 25, **150**
Soyuz 26, 118, **151**, 152, 154, 155
Soyuz 27, 151, **152**
Soyuz 28, 151, **154**
Soyuz 29, **155**, 156, 157, 163
Soyuz 30, 155, **156**
Soyuz 31, 155, **157**
Soyuz 32, **163**, 166, 171, 172
Soyuz 33, 163, **165–66**
Soyuz 34, 163, **166**
Soyuz 35, vii, 163, **171–72**, 173, 176, 272
Soyuz 36, 171, **172**
Soyuz 37, 171, 172, **172**
Soyuz 38, 171, **173**
Soyuz 39, **176–77**
Soyuz 40, 176, **178**
Soyuz T, 154
Soyuz T1, 163, **170**, *170*, 172, 178
Soyuz T2, 96, 171, **172**
Soyuz T3, **173**
Soyuz T4, **176**, 178, 272
Soyuz T5, **186–87**, 189, 191
Soyuz T6, 186, **187**, 207
Soyuz T7, 186, **189**
Soyuz T8, 192, **193**
Soyuz T9, 192, **194–95**, 200, 226, 272
Soyuz T10A, 195, **196**
Soyuz T10B, **199–200**, 201, 202
Soyuz T11, 199, **200**, 201
Soyuz T12, 199, 200, **202**, *203*, 286
Soyuz T13, *209*, **209–10**, 213, 223
Soyuz T14, 209, *209*, **213**, 223
Soyuz T15, 213, 219, **223**, 286
Soyuz TM1, **223–24**
Soyuz TM2, 219, **224–25**, 226, 228, 232
Soyuz TM3, 219, 224, **225–26**, 228
Soyuz TM4, 219, 226, **227**, *227*, 228
Soyuz TM5, 219, 227, **228**, 230
Soyuz TM6, 219, 227, **230**, 232
Soyuz TM7, 219, 227, **231–32**, 237
Soyuz TM8, 220, **237**, 242
Soyuz TM9, 163, 220, **242–43**, 248
Soyuz TM10, 220, 243, **248**, 252
Soyuz TM11, 220, 248, **252**, 256
Soyuz TM12, 220, **256**, 260, 263
Soyuz TM13, 220, 256, **260**, 263
Soyuz TM14, 220, 256, **263**
Soyuz TM15, 220, 266–67, 272, 275, 312
Soyuz TM16, 220, **272**, 275
Soyuz TM17, 220, **275–76**, 281, 312
Soyuz TM18, 220, **281–82**, 286
Soyuz TM19, 220, **286**, 289
Soyuz TM20, 220, 286, **289**, 293
Soyuz TM21, 220, 286, 289, **293–94**, 296, 297
Soyuz TM22, 220, **297–98**, 306
Soyuz TM23, 220, **306**, 312
Soyuz TM24, 220, 275, 309, **312**, 313, 314, 315
Soyuz TM25, 220, 312, 313, **314–15**, 318, 321
Soyuz TM26, 220, **321**, 329
Soyuz TM27, 221, 312, **329**, 332, 333
Soyuz TM28, 221, **333**, 338
Soyuz TM29, 219, 221, 312, **338**
Space Communications Corporation, 235
Space Electric Rocket Test. *See* SERT
Space Flyer Unit, **294**, 305
Space plane, 187

Space race, xix, 12–13, 73
Space Radar Laboratory, 285, 288
Space Services, Inc., 189
Space shuttles. *See* Shuttles
Space sickness, 7, 16–17, 18, 21, 22, 75, 77, 89
  effective treatment for, 191
  Skylab missions, 114, 116, 118
  Soyuz-Mir missions, 289
  Soyuz-Salyut missions, 177
  STS missions, 196, 198, 207, 208, 215
Space stations. *See* International Space Station; Mir; Salyut; Skylab; Spacehab; Spacelab
Space Systems/Loral, 322
Space Technology and Research Vehicle. *See* STRV
Space Technology Experiment. *See* STEX
Space Technology Project. *See* STP
Space Test Experiment Program. *See* STEP 0
Space Transportation System. *See* STS
Space walks. *See also* Moon-walks
  Apollo missions, 77, 98, 105, 111, *111*
  first, viii, 35, 39, *39*
  first untethered, 198
  first woman, 202
  Gemini missions, 41–42, 50, 52, 54, 56
  Mir-STS missions, 308, 325
  Skylab missions, 115–16, 117, 118–19
  Soyuz-Mir missions, 223, 225, 227, 232, 237, 240, 242, 248, 252, 256, 263, 266–67, 272, 275, 281, 286, 294, 295, 298, 309, 313, 321, 325, 333, 338
  Soyuz missions, 76, 151, 155, 163
  Soyuz-Salyut missions, 186–87, 194, 199–200, 202, 209, 213, 317, 329
  STS missions, 192, 198–99, 201, 205, 207, 212, 215, 253, 254–55, 264, 271, 277, 279, 287–88, 292, 299, 306, 327, 340, 342
Spacecraft Charging at High Altitudes. *See* SCATHA
Spacecraft design research, appendix listings for, 364–66. *See also* Manned missions; Manned spacecraft, problems with
Spacehab, 275, 283, 292, 308, 310, 313, 325, 332, 335, 340
Spacelab, 197, 208, 211, 214–15, 256, 261, 262, 263, 265, 268–69, 274, 277, 299, 320, 331
Spacenet 1, **202**
Spacenet 4, 212
Spacesuits
  Apollo missions, 77, 80
  Gemini missions, 41–42, 44, 50, 52, 54
  Skylab missions, 114
  Soyuz-Mir missions, 317
  Soyuz-Salyut missions, 95, 124, 128, 132, 155, 177
  STS missions, 187, 192, 274, 299
Spades, 71
Spain
  appendix listing for, 371
  getaway specials from, 288
  images of, 189
  Starshine satellite, 340

Index

Spar Aerospace, 189, 206, 287
Spartan 1, **211**
Spartan 201, 211, **273**, **287**, 292, **298**, 302, **327**, **334**
Spartan 204, 211, **292**, 310
Spartan 206, 211, **305**
Spartan 207, **310**
Spartan Service Module. *See* Spartan 204
SPAS 1, **194**, 198
SPAS 2, 255
SPCC (Southern Pacific Communications Company), 202
Speaking, difficulty in, 163. *See also* Body fluids, redistribution
Spectroheliograms, 64, 76, 99
Spektr, 294, **295**, *295*, 308, 318, 321, 325, *325*, 332, 333
Spheres, latex, 185, 199
Spider web, 117
Spin-2, 330
Spin stabilization, 62, 89. *See also* Geosynchronous orbit
Spinach plants, 310
Spinal column, 262, 270, 287. *See also* Back pain
Spine length, 200. *See also* Height, changes in
Spot 1, 115, **221**
Spot 4, **330**
Spring, Sherwood, 215
Springer, Robert, 233, 251
Sputnik, **3**, *3*
  politics and, xix
  replicas, 326, 328
  revolution begun by, vii
Sputnik 2, **3**, 6, 17, 55
Sputnik 3, **4**, 6, 16
Sputnik 4, **9**
Sputnik 5, **11**, 31
Sputnik 6, **12**
Sputnik 9, **13**
Sputnik 10, **13**
Sputnik 12, **19**
Sputnik 15, **21**
Sputnik 40, **326**, 333
Spy satellites. *See* Surveillance
Squirrel monkeys, 208
SRATS, **129**
SS433, 169–70
SSBUV, 238. *See also* SBUV
ST 1, **333**
Stabilization booms, 30, 39, 51, 62, 70, 242
Stafford, Tom, 46, 50, 78, 133
Star Trek, 317
Starad, **23–24**
Stardust, **338**
Starfish Test, 16, 17, 20, **21**, 22, 23–24, 28, 31, 33, 36
Starlette, **128–29**, 140, 277
Starquakes, 273, 339
Stars. *See also* Galaxies; Solar research
  1957 knowledge about, xvii
  evolutionary history of, 153
  formation of, 191, 279–80, 301, 335–36, 340
  parallax, 235–36
  radiation, 36, 54, 56, 74, 125, 153, 159, 191–92, *192*, 301. *See also* Gamma radiation; Infrared radiation; Ultraviolet radiation; X-ray radiation

Starshine, **340**
START treaty, 323
Static electricity. *See* Electrical build-up
Steady state theory, xvii, 15, 19
Stella, **277**
STEP 0, **285**
Stewart, Robert, 198, 214, 237
STEX, **333**
Still, Susan, 317, 320
Store-and-forward technology, 200, 208, 241, 243, 245, 253, 257, 258, 277, 291
STP 1, 70, **99**
Stratospheric Aerosol and Gas Experiment. *See* SAGE
Strekalov, Gennady, 173, 193, 196, 200, 248, 293, 294, 295, 296
Strela 1, **89**, 253
Strela cranes, 252, 256, 272, 309. *See also* Robot arm tests
Strelka (dog), 11
Strong, Ian, 79
STRV-1A, **286**
STRV-1B, **286**
STS 1, Columbia Flight #1, **177**, *177*
STS 2, Columbia Flight #2, *181*, **181–82**, 263
STS 3, Columbia Flight #3, **184–85**, *185*
STS 4, Columbia Flight #4, **187–88**
STS 5, Columbia Flight #5, **190–91**
STS 6, Challenger Flight #1, **192–93**
STS 7, Challenger Flight #2, **193–94**
STS 8, Challenger Flight #3, **195–96**
STS 9, Columbia Flight #6, **197–98**
STS 10. *See* STS 41B
STS 11. *See* STS 41C
STS 12. *See* STS 41D
STS 13. *See* STS 41G
STS 14. *See* STS 51A
STS 15. *See* STS 51C
STS 16. *See* STS 51D
STS 17. *See* STS 51B
STS 18. *See* STS 51G
STS 19. *See* STS 51F
STS 20. *See* STS 51I
STS 21. *See* STS 51J
STS 22. *See* STS 61A
STS 23. *See* STS 61B
STS 24. *See* STS 61C
STS 25. *See* STS 51L
STS 26, Discovery Flight #7, **230–31**
STS 27, Atlantis Flight #3, **232**
STS 28, Columbia Flight #8, **235**
STS 29, Discovery Flight #8, **233–34**, 272
STS 30, Atlantis Flight #4, **234**
STS 31, Discovery Flight #10, **244–45**
STS 32, Columbia Flight #9, **241**
STS 33, Discovery Flight #9, **239**
STS 34, Atlantis Flight #5, **238**
STS 35, Columbia Flight #10, **251–52**
STS 36, Atlantis Flight #6, **243**
STS 37, Atlantis Flight #8, **253–55**, *254*
STS 38, Atlantis Flight #7, **251**
STS 39, Discovery Flight #12, **255–56**
STS 40, Columbia Flight #11, **256–57**
STS 41, Discovery Flight #11, **250–51**, 272
STS 41B (STS 10), Challenger Flight #4, **198–99**, 205, 240, 244
STS 41C (STS 11), Challenger Flight #5, 84, *201*, **201–02**, 241

STS 41D (STS 12), Discovery Flight #1, **203–04**
STS 41G (STS 13), Challenger Flight #6, 199, **204–05**
STS 42, Discovery Flight #14, **261–62**
STS 43, Atlantis Flight #9, **258**
STS 44, Atlantis Flight #10, **260–61**
STS 45, Atlantis Flight #11, **263**
STS 46, Atlantis Flight #12, **267**
STS 47, Endeavour Flight #2, **268–69**
STS 48, Discovery Flight #13, **259–60**
STS 49, Endeavour Flight #1, **263–64**, *264*
STS 50, Columbia Flight #12, **265**
STS 51, Discovery Flight #17, **276–77**
STS 51A (STS 14), Discovery Flight #2, **205**
STS 51B (STS 17), Challenger Flight #7, **208–09**
STS 51C (STS 15), Discovery Flight #3, **206**
STS 51D (STS 16), Discovery Flight #4, **207–08**, 231
STS 51F (STS 19), Challenger Flight #8, **211–12**
STS 51G (STS 18), Discovery Flight #5, **210–11**
STS 51I (STS 20), Discovery Flight #6, **212**
STS 51J (STS 21), Atlantis Flight #1, **214**
STS 51L (STS 25), Challenger Flight #10, 211, *218*, **218–19**, 288
STS 52, Columbia Flight #13, **269–70**
STS 53, Discovery Flight #15, **271**
STS 54, Endeavour Flight #3, **271–72**
STS 55, Columbia Flight #14, **274**
STS 56, Discovery Flight #16, 211, **273–74**
STS 57, Endeavour Flight #4, **275**
STS 58, Columbia Flight #15, **277–78**
STS 59, Endeavour Flight #6, **285**
STS 60, Discovery Flight #18, **283**
STS 61, Endeavour Flight #5, 245, 277, *278*, **278–80**, *279*
STS 61A (STS 22), Challenger Flight #9, **214–15**
STS 61B (STS 23), Atlantis Flight #2, **215**
STS 61C (STS 24), Columbia Flight #7, **216**
STS 62, Columbia Flight #16, **284**
STS 63, Discovery Flight #20, 211, **292–93**
STS 64, Discovery Flight #19, **287–88**
STS 65, Columbia Flight #17, **286–87**
STS 66, Atlantis Flight #13, **290–91**
STS 67, Endeavour Flight #8, **293**
STS 68, Endeavour Flight #7, **288**
STS 69, Endeavour Flight #9, **298–99**
STS 70, Discovery Flight #21, **296–97**
STS 71, Atlantis Flight #14, 220, **295–96**
STS 72, Endeavour Flight #10, 211, **305–06**
STS 73, Columbia Flight #18, **299**
STS 74, Atlantis Flight #15, 220, **300–301**, *301*
STS 75, Columbia Flight #19, **307**, *307*
STS 76, Atlantis Flight #16, 220, **308**, 312
STS 77, Endeavour Flight #11, 211, **310–11**
STS 78, Columbia Flight #20, **311**

406

STS 79, Atlantis Flight #17, 220, **313**
STS 80, Columbia Flight #21, **313–14**
STS 81, Atlantis Flight #18, 220, 313, **314**
STS 82, Discovery Flight #22, **315–16**
STS 83, Columbia Flight #22, **317**
STS 84, Atlantis Flight #19, 220, **318**
STS 85, Discovery Flight #23, **322**
STS 86, Atlantis Flight #20, 220–21, **325**
STS 87, Columbia Flight #24, **326–27**
STS 88, Endeavour Flight #13, **335–36**
STS 89, Endeavour Flight #12, 221, **329**
STS 90, Columbia Flight #25, **331**
STS 91, Discovery Flight #24, 221, **331–32**
STS 93, Columbia Flight #26, **340–41**
STS 94, Columbia Flight #23, **320**
STS 95, Discovery Flight #25, **334–35**
STS 96, Discovery Flight #26, **340**
STS 103, Discovery Flight #27, **342**
Student getaway special experiments. *See* Getaway specials
Student Nitric Oxide Explorer. *See* SNOE
Students for the Exploration and Development of Space, 334
Sturckow, Fred, 335
STW 1, **202**, 291
Submarine, launch from, 332
Submillimeter Wave Astronomy Satellite. *See* SWAS
Suborbital manned flights, 15, 16, 131
Suction-cup footwear, 187
Suisei, Halley's Comet Flyby, **222**
Sullivan, Kathryn, 204, 205, 244, 263
Sun. *See* Solar research
Sunsat, **338**
Superbird A, **235**, 322
Supernovae, 125, 148–49, 153, 159, 201, 224, 225, 232, 247, 251, 273, *279*, 280, 298, 322, 341. *See also* Nova
Surcal 1A, **24**
Surface (spacecraft) treatments, 77, 108, 116, 241, 277
Surfsat, **300**
Surrey Satellite Technology Ltd., 332, 333, 339
Surveillance, xix. *See also* Anti-satellite systems
    appendix listings for, 366
    detection of missiles, 9, 21, 71, 94, 112, 243, 257, 271, 285, 309, 310
    detection of nuclear explosions, 4, 21, 23, 24, 29, 32, 78–79, 95
    detection of radar, 67–68, 127–28, 149, 189
Surveyor 1, *49*, **49–50**, 57, 58
Surveyor 2, 59
Surveyor 3, 30, 58, **59**, *59*, 68, 81, 85–86, *85–86*
Surveyor 4, 63
Surveyor 5, **63–64**
Surveyor 6, 55, **66**
Surveyor 7, **68**, 75, 81
SWAS, **336–37**
Sweden
    appendix listing for, 371
    Prognoz satellite and, 104
Swigert, Jack, 89
Swimming, 144, 151, 269, 286
Switzerland
    appendix listing for, 371

    international studies of Earth, 263
    satellite service to, 232
    STS crew member from, 267, 279, 307, 342
Swordtail fish, 331
Symphonie 1, **127**, 203
Synchronous Meteorological Satellite. *See* SMS
Syncom 1, **28**
Syncom 2, **29**, 33
Syncom 3, **33**, 40
Syria
    appendix listing for, 371
    images of, 294
    Soyuz crew member from, 225

Tacsat 1, 72, **76**, 92
Tadpoles, 151, 269, 274. *See also* Frogs
Tail, magnetosphere, 266, 297. *See also* Magnetospheres
Taiwan, appendix listing for, 371
Taiyo, **129**
Tanner, Joseph, 290, 315
Tansei 2, **121**
TAOS, 285
Taste, of food, 156, 292. *See also* Eating; Food
Taurus rocket, 284
TCI Satellite Entertainment, 316
TDF 1, **231**
TDRS 1, **192–93**
TDRS 3, **230**
TDRS 4, **233**
TDRS 5, **258**
TDRS 6, **271**
TDRS 7, **297**
TDRS satellites, 188, 205, 218, 267
Teacher-in-Space program, 218
Technical University of Berlin, 257, 282, 332, 340
Technion Institute of Technology, 333
Tectonic plate activity. *See* Plate tectonics
Telcom 1, **341**
Telecom 1A, **203**
Telecom 2A, **261**
Teledesic 1, **330**
Teledesic Corp., 330
Telediffusion de France. *See* TDF
Telenor, 237, 318
Telesat 1, **108**
Telesat 4, **161**
Telesat 5, **189**
Telesat 6, **190–91**
Telesat 7, **194**
Telesat Canada, 108, 161, 189, 190, 205, 207, 215, 253, 308, 339
Telescopes, space-based. *See also* Astronomical research
    Chandra X-ray Observatory, 340–41
    Compton Gamma Ray Observatory, 253, 254
    gamma ray, 15, 43, 44, 52, 94, 95–96, 109, 163, 186, 240, 247, 253–54, 317. *See also* Gamma radiation
    Hubble. *See* Hubble Space Telescope
    infrared, 191–92, 238–39, 294, 301, 309, 311, 338–39. *See also* Infrared radiation
    IRIS, 70, 72
    IUE, 152–53
    on Moon, 91

    Next Generation, 146
    radio, 70–71, 116, 195, 316
    Spartan, 287
    ultraviolet, 74, 107, 120, 125, 131, 152–53, 192, 198, 246, 251, 262, 264–65, 293, 298, 313, 322, 335, 340, 342. *See also* Ultraviolet radiation
    x-ray, 91, 107, 113, 125, 128, 131, 132, 148, 159, 162, 186, 192, 193, 211, 224, 225, 240, 246, 251, 253, 272–73, 305, 309–10, 338, 340–41, 342. *See also* X-ray radiation
Television and Infra-Red Observation Satellite. *See* Tiros
Telstar 1, 11, **21–22**, *22*, 25, 280–81
Telstar 2, **25–26**
Telstar 3A, **195**
Telstar 3B, **204**
Telstar 3C, **211**
Telstar 401, 88, **280–81**, 303
Temisat, **276**
Temperature, body. *See* Body temperature
Tempo 2, **316**
Tenma, **192**
Tereshkova, Valentina, 27, *27*, 170, 189
Terra, **342**
Terra Scout, 261
Test and Training Satellite 1, **67**
Tethered Satellite System. *See* TSS
Tethered satellites, 20–21, 54, 56, 108, 140, 267, 273, 307, 333
Tethys, 175, *175*, 179, *179*
TEX, **244**
Thagard, Norman, 118, 194, 208, 234, 261, 289, 293, 294, 296
Thaicom 1A, **281**
Thailand
    appendix listing for, 371
    satellite service to, 145, 244
Thale cress plants, 261
Thematic mapper, 188, 189
Thermal (spacecraft) treatments, 77, 108, 116, 241, 277
Thermosphere, 79
Thin-film research, 230, 277, 283, 298, 313. *See also* Polymer membranes
Thirsk, Robert, 311
Thomas, Andrew, 310, 329, 331
Thomas, Donald, 286, 296, 317, 320
Thor 1, 237
Thor 2, **318**
Thor Delta 1A, **104**, 107
Thornton, Kathryn, 239, 263, 264, 278, 279, 299
Thornton, William, 195, 208
Three-axis design, 268, 270, 282
Three-spacecraft mission, 84. *See also* Dual missions
3M Company, 205, 212, 215, 230, 238, 262
Thuot, Pierre, 243, 263, 264, *264*, 284
Tides, 129
Tiles, shuttle, 177. *See also* Surface (spacecraft) treatments
Timation 1, **61**, 189
Timation 2, **83**
Timation 3, **124**
Time, efficient use of, 144–45
Tiros 1, **8**, *8*, 106
Tiros 2, **12**

Tiros 3, **16**, *16*, 18
Tiros 4, **18**, 20, 52
Tiros 5, **21**
Tiros 6, **23**, 28
Tiros 7, 28
Tiros 8, **30**, 34
Tiros 9, 34, **37**, *38*
Tiros N, 91, **158**
Tiros Operational Vertical Sounder, 158
Titan, xv, 174–75, 179, 326
  atmosphere, xv, 174, 301
Titania, *217*, 218
Titov, Gherman, 16–17, 18, 21, 22, 77, 116
Titov, Vladimir, 193, 196, 227, *227*, 228, 230, 231, 232, 281, 292, 325
TMSat, **333**
Toadfish, 331
Tobacco hornworm, 297
Tobacco seedlings, 314
Tognini, Michel, 266, 267, 275, 340
Tokarev, Valery, 340
Tokyo Broadcasting System, 252
Tomatoe plants, 195, 314
TOMS, 158, 258, 259, **311**, 312
Tools. *See also* Work in space
  Skylab missions, 115, 116
  Soyuz-Mir missions, 223
  Soyuz-Salyut missions, 163, 202
  STS missions, 192, 207, 277, 340
TOPEX-Poseidon, **268**, 312
Tortoises, 118, 137
Toru docking system, 315. *See also* Rendezvous and docking
Total Ozone Mapping Sensor. *See* TOMS
Total Ozone Mapping Spectrometer. *See* TOMS
Toys, 208, 272
TRAAC, **17**, 21, 24
Trace, **330**
Tracking and Data Relay Satellites. *See* TDRS
Transceiver Experiment. *See* TEX
Transistors, 17, 24, 25, 29, 36
Transit 1A, 8
Transit 1B, **8–9**, 61
Transit 2A, **9**
Transit 4A, **16**
Transit Research and Attitude Control. *See* TRAAC
Transition Region and Coronal Explorer. *See* Trace
Tree frogs, 252
Triad 1, **108**
Trinh, Eugene, 265
Triton, xvi, *236*, 236–37
TRMM, 38, **327**
Tropical Rainfall Measuring Mission. *See* TRMM
TRS 4, **29**
Truly, Richard, 181, 195
Trumpet 1, **285**
TRW Space and Electronics, 338
Tryggvason, Bjarni, 322
Tsibliyev, Vasily, 275, 281, 314, 315, 317, 318, 321
Tsiolkovsky, Konstantin, 7
TSS 1, **267**
TSS 1R, **307**, *307*
Tubsat 1, **257**
Tubsat 2, **282**
Tubsat-DLR. *See* DLR-Tubsat

Tubsat N, **332**
Tubsat N1, **332**
Turkey
  appendix listing for, 371
  Starshine satellite, 340
Turksat 1B, **287**
Turtles. *See* Tortoises

UARS, **259**
Ugolek (dog), 48
UHF Follow-On, 190
UHF Follow-On 1, **276**
Uhuru, **91–92**, *92*, 99, 107, 125, 148, 153
UK 6, **166**
Ukraine
  appendix listing for, 371
  Bion satellite, 314
  Intercosmos satellite, 284
  Kurs docking system, 315
  STS crew member from, 327
UKS, **203**
Ultraviolet radiation, 251, 317, 342. *See also* Telescopes, space-based
  beyond solar system, 36, 52, 54, 56, 74, 104, 107–08, 120, 125, 152–53, 192, 198, 216, 246, 262, 263, 264–65, 274, 276, 293, 298, 313, 322, 335, 340
  bursts, 264–65, 274
  solar system, comets, 118, 222
  solar system, Earth, 30, 37, 108, 131, 153, 158, 284, 292, 307, 332, 333
  solar system, Jupiter, 74, 166, 167, 298, 322
  solar system, Mars, 74
  solar system, Moon, 43, 112
  solar system, Sun. *See* Solar research, on solar radiation
  solar system, Venus, 121
Ulysses, **250**, 287, 298
Ulysses, First Solar Orbit, 309, **331**
Ulysses, Jupiter Flyby, **262**
Umbrellas
  Orizuru, 242
  Skylab, 113–14, 116, 117
Umbriel, *217*, 218
Ume 2, **153**
Union of Soviet Socialist Republics (U.S.S.R.). *See* Soviet Union
United Kingdom
  appendix listing for, 371–72
  Eutelsat consortium, 155
  Hong Kong and, 287
  Intercosmos satellite, 284
  Soyuz crew member from, 256
  Spacelab research, 263
  Starshine satellite, 340
United Kingdom Satellite. *See* UKS
United States Microgravity Laboratory. *See* USML
United States of America (U.S.A.)
  appendix listing for, 372–75
  Challenger accident, 218–19
  economic issues, 291
  elected officials in space, 207, 216
  international agreements and treaties, 135, 188, 292, 323, 325
  Mir crew members from, 293, 308, 312, 313, 314, 318, 325, 329, 332

  political issues, 26, 122, 221
  space race, xix, 12–13, 73
Unity (International Space Station Module), **335–36**, *336*. *See also* International Space Station
Universe, nature and structure of, xvii–xviii, 15, 107, 235–36, 238–39. *See also* Big bang theory; Steady state theory
Universidad Politecnia de Madrid, 296
University of Alabama, 334
University of Bremen, 283
University of Colorado, 330
University of Surrey, 181, 200, 241, 242, 257, 268, 277, 339
Unmanned planetary exploration. *See individual planets and moons*
UoSat 1, **181**
UoSat 2, **200**
UoSat 3, **241–42**
UoSat 4, **241–42**
UoSat 5, **257–58**
UoSat 12, **339**
UPM-LBSAT 1, **296**
Upper Atmosphere Research Satellite. *See* UARS
Uranus, *xvi*, *216*
  1957 knowledge about, xv–xvi
  atmosphere, xv–xvi, 216, 217
  aurora, 153
  composition of, 216
  magnetic field, 216–17
  moons of, xvi, *217*, 218
  rings, *216*, 217–18
  spacecraft flyby, 216–18
Uribyol, **268**
Uribyol 2, **277**
Uribyol 3, **339**
Urine and urination, 18, 42, 256, 311, 318, 335. *See also* Waste management
U.S.A. *See* United States of America
USA 40, **235**
USA 41, **235**
USA 59, **247**
USA 89, **271**
Usachev, Yuri, 281, 286, 306, 309, 312
USML-1, 265
USML-2, 299
U.S.S.R. *See* Soviet Union

Van Allen, James, 4
Van Allen radiation belts. *See* Earth, Van Allen belts of
van den Berg, Lodwijk, 208
van Hoften, James, 201, *201*, 212
Vanguard 1, vii, **4**, 9, 10
Vanguard 2, **5**
Vanguard 3, **6**, 7
Vasyutin, Vladimir, 209–10, 213
Veach, Charles, 255, 270
Vega 1, Halley's Comet Flyby, **221–22**
Vega 1, Venus Flyby, **210**
Vega 2, Halley's Comet Flyby, 221, **222**
Vega 2, Venus Flyby, **210**
Vegetables, 171–72. *See also* Plant research
Vegetation. *See* Earth resources research; Plant research
Vehicle Evaluation Payload. *See* VEP
Vela 1A, 22, **29**
Vela 1B, 22, **29**

Vela 2A, 29, **32**, 35, 36
Vela 2B, 29, **32**, 35, 36
Vela 5A, **78–79**, 109
Vela 5B, **78–79**, 109
Vela X-1, 128
Vela, Rodolfo Neri, 215
Venera 1, **13**
Venera 2, **45**, 47
Venera 3, **45**, 47
Venera 4, Venus Landing, **64–65**, 78, 121
Venera 5, Venus Landing, **78**, 92
Venera 6, Venus Landing, **78**, 92
Venera 7, Venus Landing, **92**
Venera 8, Venus Landing, **106**, 121
Venera 9, Venus Landing, **136**, *136*, 160, 182
Venera 10, Venus Landing, *136*, **136–37**, 160, 182
Venera 11, Venus Landing, 161, **162**
Venera 12, Venus Landing, **161–62**, 184
Venera 13, Venus Landing, **182**, *183*
Venera 14, Venus Landing, **182–84**, *183*
Venera 15, Venus Orbit, 196, **197**, 205
Venera 16, Venus Orbit, 196, **197**, 205
Venus, *xiii, 183, 197, 248, 249*
    1957 knowledge about, xiii
    atmosphere, xiii, 25, 64–65, 78, 92, 106, 120–21, 136–37, 160–61, 183–84, 210, 248, 249–50
    composition of, 106, 121, 136, 183, 210, 249
    craters, 160, 197, 249
    gravitational field, 248
    life on, speculation, xiii
    lightning, 161–62, 184
    magnetic field, 25, 65, 121
    propagation measurements, 120
    radar mapping, 137, 159–60, 197, 205, 234, 248–49
    seismic research, 184
    solar wind, 121, 184
    surface, 65, 78, 92, 106, 120, 136, 159–60, 161, 182–83, 197, 210, 248–49
    volcanic activity, 136, 160, 183–84, 210, 248
    water, evidence of, 65, 121, 184, 210
    winds, 120, 137, 210
Venus, spacecraft to
    flybys, 23, 25, 32, 45, 65, 120–21, *121*, 210
    landings, 45, 64–65, 78, 92, 106, 136–37, 161–62, 182–84, 210
    orbiting, 136, 159–61, 197, 234, 248–49
    radio contact lost, 13
VEP, 284
Vesta, 341
Veterok (dog), 48
Vibrational resonance, 70, 152, 223
Vibrations and accelerations, monitoring of, 241, 277, 284
Viehbock, Franz, 260
Vietnam
    Intersputnik consortium, 161
    satellite service to, 333
    Soyuz cosmonaut from, 172
Viking, **221**
Viking 1, 316
Viking 1, Mars Orbit/Landing, **140–44**, *141, 142, 143*
Viking 2, 140, 141, **145**, *145*, 316

Viktorenko, Alexander, 225, 226, 237, 263, 266, 289, 292, 293
Vinogradov, Pavel, 321, 325, 326, 329
Viruses, 233, 277, 293, 320
Vision research, 11, 26, 44, 187
Volcanic activity
    Earth, 140, 158, 162, 189, 207, 257, 260, 261, 270, 273, 287, 288
    Io, 163–64, 168, 262, 304
    Mars, 100, 101, 130, 142, 323
    Moon, xiii, 33, 38, 39, 53, 55, 64, 66, 68, 86, 97, 98, 105, 109, 111, 130
    Triton, 236
    Venus, 136, 160, 183–84, 210, 248
Volk, Igor, 202
Volkov, Alexander, 209–10, 214, 231, 232, 260, 263
Volkov, Vladislav, 84, *95*, 95–96, 117, 124, 170
Volynov, Boris, 76, 144
von Braun, Wernher, 13
Voskhod 1, **35**, 39
Voskhod 2, viii, 35, **39**, 41, 145
Voss, James, 260, 271, 298, 299
Voss, Janice, 275, 292, 317, 320
Vostok 1, **14–15**
Vostok 2, **16–17**
Vostok 3, **22**, 39, 42
Vostok 4, 22, **22**, 39, 42
Vostok 5, **26–27**, 42
Vostok 6, 26, **27–28**, 42
Vostok capsule tests, 9, 11, 12, 13
Voyager 1, Jupiter Flyby, *163*, **163–65**, *164, 165*, 252, 299, 304, 317
Voyager 1, Saturn Flyby, *174*, **174–76**, *175, 176*
Voyager 2, Jupiter Flyby, *166*, **166–68**, *167, 168*, 252, 304
Voyager 2, Neptune Flyby, *236*, **236–37**
Voyager 2, Saturn Flyby, *179*, **179–80**, *180*
Voyager 2, Uranus Flyby, *216*, **216–18**, *217*, 299, 317

Wakata, Koichi, 305
Wake
    magnetospheric, 54. *See also* Magnetospheres
    from shuttles, 184, 187, 211, 298, 313. *See also* Electrical build-up on spacecraft
Wake Shield 1, **283**
Wake Shield 2, **298**, *298*
Wake Shield 3, **313**
Walker, Charles, 203, 204, 207, 215
Walker, David, 205, 234, 271, 298
Walter, Ulrich, 274
Walz, Carl, 276, 277, 286, 313
Wang, Taylor, 208
Waste management, 18, 27, 42, 208
Water, evidence of
    comets, 153
    Earth, 117, 128
    Europa, 304
    interstellar environments, 301
    Mars, 101, 142, 143, 321, 323
    Moon, 282, 328, 329
    Titan, 301
    Venus, 65, 121, 184, 210
Water infall, to Earth, 178–79
Weapons Research Establishment, 67

Weather satellites, xviii. *See also* Atmospheric research; Earth resources research
    appendix listings for, 366
    geosynchronous, 91, 123
    international agreement on, 135
Weber, Mary Ellen, 296
Weber State College, 242
Webersat, **242**
Webs, spider, 117
Weight, changes in
    dogs, 48
    humans, Soyuz-Mir missions, 232
    humans, Soyuz-Salyut missions, 128, 132, 151, 155, 189
    humans, STS missions, 191
    monkeys, 227
    rats, 198
Weight measurements in space, 114, 144
Weightlessness
    biology, medicine, and. *See* Biological research; Human biology and medicine; Plant research; *and names of individual organisms*
    compared to re-entry stress, 278
    materials research. *See* Biological materials research; Crystals research; Materials research
    no longer a concern for short-term missions, 191
    recovery from. *See* Readaptation to gravity
    as therapy, 206
Weitz, Paul, 114–15, 192
Welding, 51, 84, 115, 202, 223
West Ford 1, **17**
West Ford 2, **26**, 52
West Germany. *See also* East Germany; Germany
    appendix listing for, 375
    Eutelsat consortium, 155
    ham (amateur) radio in, 126, 154, 193, 229
    satellite service to, 205, 232
    STS crew members from, 214
    Vega mission, 210
Westar 1, 26, **122**, 145
Westar 3, 182
Westar 4, **182**
Westar 5, 182
Westar 6, 182, **198**, 205, 244
Western Union, 122, 137, 182, 193, 195, 198, 244
Westpac, **333**
Wetherbee, James, 241, 269, 292, 325
Wheat plants, 63, 215, 277, 313, 332
Whispering Gallery, 55
White, Ed, 41–42, 57, 72
Whitecloud 1, **140**
Wilcutt, Terrence, 288, 313, 329
Wildlife tracking, 332. *See also* Biological activity scanners
Williams, Dafydd Rhys, 331
Williams, Don, 207, 238
Wind (spacecraft), **290**
Winds, and star types, 153
Winds, planetary
    Earth, 120, 156, 259, 294, 311, 340
    Jupiter, 164, 166, 303
    Mars, 142, 323
    Neptune, 236
    Saturn, 175, 179

Winds, planetary *(continued)*
    Uranus, 217
    Venus, 120, 137, 210
Winds, solar. *See* Solar research, on solar wind
WIRE, **338–39**
Wisoff, Peter, 275, 288, 314
Wolf, David, 277, 278, 325, 329
Women. *See* Humans, female
Worden, Al, 96–98, 109
Work in space. *See also* Repairs
    Gemini missions, 56
    Skylab missions, 114, 115, 117
    Soyuz-Mir missions, 223, 232, 256, 275, 294, 295, 313
    Soyuz-Salyut missions, 187, 194, 202
    STS missions, 204, 215, 254–55, 264, 271, 284, 292, 299, 336, 340
WorldSpace Corporation, 334
Worms, 237, 286, 297
Wounds, healing of, 211. *See also* Injuries, healing of
Wresat 1, **67**

X-ray crystallography, 329
X-ray radiation, 113, 186, 284, 333, 342. *See also* Telescopes, space-based
    beyond solar system, 91–92, *92*, 99, 104, 107, 125–26, 128, 131, 132, 134, 148–49, 153, 159, 162, 166, 192, 193, 195, 198, 211, 224, 225, 232, 240, 246, 251, 253–54, 260, 272, 273, 274, 305, 309–10, 319, 338, 340–41
    bursters, 125–26, 131, 159, 195, 198, 224, 240, 246, 254
    soft, 9, 31, 246, 272, 274
    solar system, comets, 222
    solar system, Earth, 29, 37, 43, 44, 60, 78, 91, 131, 307
    solar system, Moon, 91, 98, 112
    solar system, Sun. *See* Solar research, on solar radiation
Xenon ion engines. *See* Ion engines
XMM, **342**

Yamal 101, **341**
Yamal 102, **341**
Yeast, 318
Yegorov, Boris, 35
Yeliseyev, Alexei, 76, 84, 94
Yeltsin, Boris, 333
Yemen, 161
Yo-yos, 208. *See also* Toys
Yoga, 200

Yohkoh, **258–59**
Young, John, 40, 52, 78, 104–05, *105*, 177, 197
Yuri 1, **154**

Zaire, 189
Zarya (International Space Station Module), **335,** 336, *336*. *See also* International Space Station
Zenit rocket, 339
Zero gravity. *See* Microgravity; Weightlessness
Zeya, **316**
Zholobov, Vitaly, 144
Zimbabwe, 340
Zodiacal light, 38, 74
Zond 1, **32**
Zond 2, **36**
Zond 3, **43,** 45, 53
Zond 4, **69,** 70
Zond 5, **71**
Zond 6, **73,** 73–74
Zond 7, **83**
Zond 8, **90**
Zudov, Vyacheslav, 146
Zvezdochka (dog), 13

Robert Zimmerman is also the author of *Genesis: The Story of Apollo 8*, an acclaimed history of the 1960s space race. With a bachelor's degree from Brooklyn College and a master's degree from New York University, he has taught at New York University, the New School for Social Research in New York, and the Stevens Institute of Technology in Hoboken, New Jersey. In 1987, he was chair of the New York chapter of the National Space Society and co-chair of the New York public school system's Challenger Space Fair. Zimmerman's essays on astronomy, space exploration, and science history have been published in *The Sciences, The Wall Street Journal, Fortune, Astronomy, Invention & Technology, Sky & Telescope*, and *Stardate*.